Fundamentals of Human Sexuality

Fundamentals of Human Sexuality

4th Edition

Herant A. Katchadourian, M.D.
Stanford University

HOLT, RINEHART AND WINSTON
New York   Chicago   San Francisco   Philadelphia
Montreal   Toronto   London   Sydney
Tokyo   Mexico City   Rio de Janeiro   Madrid

For Stina, Nina, and Kai

Library of Congress Cataloging in Publication Data
Katchadourian, Herant A.
  Fundamentals of human sexuality.

  Bibliography: p.
  Includes indexes.
  1. Sex.   I. Title.   [DNLM: 1. Sex Behavior.
HQ 21 K185f]
HQ31.K36 1985   612'.6   84-25219
ISBN 0-03-060429-X

Credits for previously published material appear with
illustrations and extracts and on page 612.

Address correspondence to:
383 Madison Avenue
New York, N.Y. 10017
All rights reserved
Printed in the United States of America
Published simultaneously in Canada
5 6 7 8 016 9 8 7 6 5 4 3 2 1

CBS COLLEGE PUBLISHING
Holt, Rinehart and Winston
The Dryden Press
Saunders College Publishing

Foreword

It is a curious and poignant fact that some of the subjects that affect human lives most powerfully receive little attention in scientific research or in higher education. Sex is such a subject. This situation is beginning to change for the better, and this book is a remarkable contribution to the improvement. I know of no other book on human sexuality that matches it in clarity, cogency, and dependability of information.

The scientific literature on sex is in an early stage of development. Although popular books abound, full of flashy claims and disputable speculations, there are few adequate textbooks on human sexuality for college-level instruction. This is in pronounced contrast to the large number of excellent textbooks in such subjects as biology and psychology.

So far as I know, this book was the first successful attempt to produce a textbook on human sexuality that, in scope, reliability, and level of sophistication, compares favorably with the better textbooks in the biological and social sciences. What began as a very good book has been further refined in successive revisions so that the present edition represents the best effort to date of this highly successful venture.

The primary emphasis in this book is on information, not advice. The information has been selected with care and the reader is clearly told when empirical data are being presented, what the limitations of the evidence are, and when one must rely upon hypothetical constructs. Whenever pertinent, alternative explanations are offered for the same behavior or phenomenon. The material is presented in a fashion that should make it readily comprehensible for beginning college students, yet the information is sufficiently sophisticated to make the book valuable for advanced students as well. Since people of all ages have generally lacked scientifically based education about sex, this book can be useful to a broad range of readers.

There are few topics where the need for an integrated view is as great as in the area of sex. Although much remains to be learned about the biological, psychological, and cultural aspects of sex, we know enough to realize that these elements cannot be separated from each other, nor can one be ignored in favor of the other. One of the major strengths of this book is that it approaches human sexuality as a comprehensive and integrated topic by thoughtfully distributing emphasis across different areas, and by viewing sexual behavior in an evolutionary, historical, and cross-cultural perspective.

The author has ranged widely across biological sciences, social sciences, the humanities, and clinical medicine in arriving at his assessment of the current state of knowledge about sex. He is not inhibited by the narrow confines of a single discipline, nor preoccupied with any doctrinaire ideological position. Open-minded, broadly informed, and freely inquisitive, his assessment of evidence is consistently judicious and useful and his appraisals are expressed with wit and grace. The book is fascinating to read, a genuine contribution to our understanding of human sexuality.

David A. Hamburg, M.D.
President
Carnegie Corporation of New York
1984

Preface to the Fourth Edition

The first edition of this text appeared over a decade ago when courses in human sexuality were beginning to be offered in colleges and universities. Since then, it has been used extensively through three editions and translated into French, Spanish, and Portuguese.

The fourth edition retains the basic strengths of previous editions. But it also breaks fresh ground with the addition of several new chapters and extensive revisions of previous ones. In many respects, this is a new book.

Organization of the Text

Chapter 1 is an introduction to the subject of human sexuality. It describes the disciplinary contexts and the historical perspectives for the study of the subject. The following 19 chapters are divided about equally into three sections representing the *biological, behavioral,* and *cultural* perspectives on sexuality. This disciplinary approach taps into the scientific and scholarly literature which is organized along similar lines. It allows the three major perspectives to be presented in a discrete and systematic way, but it is not intended to reinforce the artificial separation of the biological from the psychosocial variables in sexual behavior and function.

Chapter 8 in the third edition, Sexual Behavior Through the Life Span, has been divided into Sexual Behavior (chapter 8) and Sexual Development (chapter 9), thus allowing for more focused coverage. Chapter 12 in the fourth edition is a new effort to deal more extensively with the interpersonal and institutional aspects of sexual relationships. Unlike previous editions, homosexuality is dealt with in a separate chapter (chapter 13) while the paraphilias and sexual aggression are discussed as a unit (chapter 14).

Human sexuality texts generally pay inadequate attention to the historical and cultural aspects of sexuality. This text attempts to overcome that weakness through new chapters on sexuality in the ancient world (chapter 17) and in Western culture (chapter 18). The use of fine art in these chapters and in the rest of the book further enhances the cultural dimension.

The text attempts to be comprehensive without being exhaustive in coverage. Many important aspects of sex are dealt with but the main emphasis is on what is likely to be of concern to the majority of readers. The level of discourse is comparable to those of college texts in the biological and behavioral sciences. The language used is academic English, its tone objective. Matters are stated and illustrated frankly and explicitly, but there is no intention to shock the reader or to give offense. There is much material for rejoicing in these pages, but also a great deal to be saddened or outraged by. That is because sex is portrayed as it is, not as it ought to be according to the author's or someone else's preferences. There is an attempt made to state all reasonable sides of an issue, without taking a stand. This does not mean, however, that every argument that is presented has equal merit nor that the author has no personal convictions about what is desirable, sensible, and edifying.

The effectiveness of a text depends on its content first, and on its manner of presentation next. If there is no substance to a book, no amount of pedagogical gimmickry will make it worthwhile. On the other hand, if the material presented does not engage the reader's attention, it will do little good. To that end, each topic is first defined, then described, and then discussed. With respect to biological functions, key concepts are clarified before being applied to the subject at hand. When dealing with a particular behavior, cross-cultural and cross-species comparisons are provided to place it in proper perspective.

The material is presented at several levels of detail. The primary information is in the main text under three orders of headings. Key terms and names appear in bold type. Extensive footnotes provide bibliographical sources and a variety of interesting tidbits. Major sources of information are listed, but there is no attempt to document every fact and statement. The style of presentation should also make it clear when research findings are being reported and when reflective comments or speculative statements are being made.

Boxes are used to highlight special topics and to present literary excerpts. A large number of illustrations likewise supplement the text, ranging from graphs to reproductions of fine art. The purpose of these illustrations is to add substance to the text, not to decorate it. Instead of a glossary, the use of bold type in the index shows where a term is defined or first discussed in the text.

The Author

I am a psychiatrist by training. But my research work in psychiatry has been mainly in cross-cultural epidemiology and therefore, quite close to the perspective of the behavioral sciences. Over the past 15 years, I have been primarily engaged in undergraduate education at Stanford University, both as a member of the faculty of the Program in Human Biology and as Dean and Vice Provost for Undergraduate Education. I have been heavily involved in studying the process by which students make academic decisions.

My primary interest in the field of human sexuality is with respect to teaching. I initiated the course in human sexuality at Stanford in 1968 and have since taught over 10,000 students. I have also taught medical students and thousands of behavioral scientists and other professionals at the postgraduate level. I consider my authorship of this text as an extension of my efforts as a teacher.

To my mind, teachers fulfill three primary roles. First, a teacher acts as a *condenser* by gathering and funneling to the student the information in a field. Second, a teacher acts as a *filter,* sorting sense from nonsense. Third, and perhaps most important, a teacher raises questions and challenges the student to examine his or her attitudes, assumptions, and convictions. Education involves the learning of critical thinking and the acquiring of self-knowledge as much as the acquisition of facts.

If I am successful as a teacher, I cannot fail but to touch my students' lives, but it is not appropriate that I try to inculcate in them my own sexual values. Even though I consider sex to be a wonderful gift in life, it is not my job as a teacher to sell it to my students. So far as it is feasible, I have therefore tried to keep my own personal sexual preferences, values, and prejudices out of this book; the reader will have to judge how well I have succeeded.

Acknowledgments

Through successive editions of this text numerous colleagues and reviewers have contributed greatly to its success. Space limitations do not permit the reiteration of acknowledgments from previous editions but that does not diminish my appreciation for their help.

Donald T. Lunde was co-author of the three previous editions of this text with primary responsibility for the chapters on sex hormones; reproduction; contraception; sexual disorders; homosexuality and other sexual behaviors; sex and society; sex and the law; and sex and morality. The present edition reiterates the basic information in these areas and retains a substantial number of the illustrations from previous editions. Nonetheless, all of these chapters have been extensively revised and virtually rewritten in the fourth edition. Though circumstances made it necessary that we dissolve our partnership, I continue to value our earlier collaboration.

The fourth edition has benefited from the contributions of many. Lynda Anderson (University of North Carolina); John B. Black (Augusta College); Nancy Cozzens (De Anza College); and Susan Fleischer (Queens College), all of whom teach courses in human sexuality, read early and revised drafts and made numerous helpful suggestions with respect to organization and substance. I received valuable advice and assistance with respect to more specialized topics from the following Stanford colleagues: George and Phyllis Brown (literature and medieval studies); Carl Djerassi (contraception); Julian Davidson (physiology and endocrinology); Mark Edwards (classics); William J. Goode (sociology); John Kaplan (law); Iris Litt (contraception, adolescent medicine); and Arthur Wolf (anthropology).

For their help in research and preparation of materials, I would like to thank: Sue Abkowitz; Roger Broadman; Alan Fausel; Carol King; and Ruth-Andrea Levinson. The typing and retyping of a manuscript is no mean task. Margaret Gelatt worked on early drafts; Andrea Garwood and Nancy Bovee on various segments, but the major effort was exerted by Laurie Burmeister. Above and beyond her competent and tireless work, her patience and cheerfulness sustained me during many difficult periods.

My appreciation goes also to the professionals at Holt, Rinehart and Winston who provided essential help—especially my editor, Earl McPeek; his assistant, Lucy MacMillan Stitzer; Jeanette Ninas Johnson, whose editorial hand guided the project from manuscript through to bound books; Production Manager Annette Mayeski; and Design Supervisor Robert Kopelman.

Finally, the forbearance of Provost Albert H. Hastorf made it possible for me to complete this work in the midst of my vice-provostial duties. My wife and our two children, to all of whom this book is dedicated, made my labors worthwhile. To all of them, and to other friends and colleagues, too numerous to mention, I express my gratitude.

H.A.K.

Brief Contents

Contents

Part II BEHAVIOR

Part III CULTURE

Boxed Features

Chapter 1

Learning About Human Sexuality

As long as sex is dealt with in the current confusion of ignorance and sophistication, denial and indulgence, suppression and stimulation, punishment and exploitation, secrecy and display, it will be associated with a duplicity and indecency that lead neither to intellectual honesty nor human dignity.

Alan Gregg[1]

Sex is the source of human life and an integral part of life from birth to death. Whether or not we actively engage in it, sexuality is part of our everyday thoughts and feelings; it is rooted in our dreams and longings, fears and frustrations.

While the ubiquity of sex in human interactions can hardly be denied, this is not to say that sex is a dominant influence in each and every life at all times. Like parenthood, the quest for power, or altruism, sex defines people in variable degrees. For some, it is the guiding principle of their lives, taking up much of their time and energy. For others sex plays a less significant role, either by choice or by virtue of circumstance. For most of us sexual attractions wax and wane, depending on a host of internal needs and the dictates of external circumstances. At each phase of life, sexuality unfolds in distinctive ways. For the child, sex is a form of play; for the adolescent, a source of intense yearning; for the adult, a bond of love and a means of procreation. For everyone it is a potential source of pleasure or sometimes of suffering. Sex is a part of every culture, and depictions of sexual activity can be found from all periods of history, ranging from prim-

itive rock carvings of prehistoric times to the elaborate erotic statuary of Indian temples (figure 1.1).

Sex (Latin: *sexus*) is difficult to define because it means so many things at so many levels. It designates being male or female, and by extension, differentiates between gender characteristics of masculinity and femininity. Sex refers to the genital organs, the sexual and reproductive functions. Its behavioral meanings are legion. In evolutionary terms, the primary purpose of sex is reproduction but it also serves a host of other biological, psychological, and social purposes. To find out what sex is, is one of the key issues we pursue throughout this text.

If sex is a natural function and so much a part of our life, why do we need to learn about it? There is no simple answer to that question; people in fact disagree over which aspects of sex, if any, should be taught, by whom, to whom, where, under what circumstances, and for what purposes.

---

[1]In Kinsey et al. (1948), p. vii. Gregg was Director for Medical Sciences of the Rockefeller Foundation, the major financial supporter of the Kinsey study.

FIGURE 1.1 Temple reliefs, sixteenth century A.D., Srirangam Temple, Trichinopoly, India.

Fundamental Questions About Human Sexuality

However one chooses to proceed with respect to the specifics of learning about sexuality, three fundamental questions need to be addressed: Why do we behave sexually? How do we behave sexually? How should we behave sexually? We need to elaborate on these questions briefly before we seek to answer them in subsequent chapters.

Why We Behave Sexually

There are certain physiological functions, like eating, whose primary purpose is obvious. Even though food preferences vary widely among individuals and groups, everyone must eat something: fish or fowl, raw or cooked, with or without fork and knife. Similarly, waste products must be eliminated from the body regardless of the social customs attached to the process. No society can interfere with the physiological functions of eating and elimination without jeopardizing life.

What about sex? Can it be compared to either eating or elimination? Both comparisons have been tried, but neither is quite adequate. We can, of course, refer metaphorically to "sexual hunger." In fact, at certain times people and animals may prefer copulating to eating. Nevertheless, sex is not necessary for sustaining life, except in the broad sense of preserving the species. In fact, there is no evidence that abstinence from sexual activity is necessarily detrimental to health.

The comparison to elimination also has some superficial validity, since most men and women experience a buildup of sexual tension; so we can speak of periodic neurophysiological "discharge" through orgasm, with the understanding that sexual activity is not literally the "discharging" of anything that would otherwise be "dammed up" like the waste products of the body.

Sexual behavior is ultimately the interactional outcome of three types of forces—biological, psychological, and social. This should not imply a hierarchy among these factors as to which one is more important: Let us assume that these forces are complementary and integrated rather than mutually exclusive.

Biological Origins of Sexuality   Biological explanations of sexual behavior have often been based on the concept of *instinct*, an innate force that compels people to act a certain way. Yet it has been difficult to define, let alone demonstrate, the existence of such motivating forces in humans.

Sex and reproduction are so intimately linked in most living beings that one explanation of sexual behavior is simply the need to reproduce. Sex, in this sense, is part of a deep-rooted biological incentive for animals to mate and perpetuate their species. But since lower animals cannot know that mating results in reproduction, what mysterious force propels them to mate? Despite the reproductive consequences of copulation, sexual behavior cannot be scientifically explained in teleological terms, as behavior in which animals engage for the purpose of reproduction.

People of course do engage in sex at an individual level in order to reproduce. Yet a great deal of human sexual activity serves no reproductive function.

A simpler explanation is that humans and animals engage in sex for physical pleasure. The incentive is in the act itself, rather than in its possible consequences. Sexual behavior in this sense arises from a psychological drive, associated with sensory pleasure, and its reproductive consequences are a by-product. We are only beginning to understand the neurophysiological basis of pleasure. It has been demonstrated, for example, that there are "pleasure centers" in the brain that, when electrically stimulated, cause animals to experience intense pleasure. Such pleasure centers may also exist among humans.

What about *sex hormones*? They are a fascinating and problematic subject of study. We know, for instance, that they begin to exert their influence before birth and are vital in sexual development. Although these hormones may also be intimately linked to sexual behavior, the link is not a simple one, and we have yet to discover a substance that might represent a true "sex fuel."

Thus, the biological basis of our sexual behavior involves certain physical givens, including sex organs, hormones, intricate networks of nerves, and brain centers. How these components are constructed and how they work occupies us in part I of this book.

Psychological Determinants of Sexual Behavior   In one sense, psychological or social forces are merely reflections and manifestations of underlying biological processes. For example, Freud argued that the *libido*, or sex drive, is the psychological representation of a biological sex instinct.

In another sense psychological factors are independent, even though they must be mediated through the neurophysiological mechanisms of the brain. But these mechanisms are considered only as the intermediaries through which thought and emotions operate, rather than as their primary determinants. Let us again use hunger as an example. When hunger motivates a person to eat, the behavior of the individual is relatively independent of psychological and social factors. Although such factors may influence the individual's eating behavior, they are not the main determinants of why a person eats. Yet if a person dislikes pork or is expected to abstain from eating it, he acts from personal preference or religious conviction. His motivation is still mediated through the brain, but it originates in learned patterns of behavior rather than in biological factors. The same may be true for sex, although a biological imperative is less clear in this case.

Theories of the psychological motivation for sexual behavior are therefore fundamentally of two types. In the first, psychological factors are considered to be representations or extensions of biological forces. In the second, it is assumed that patterns of sexual behavior are largely acquired through a variety of psychological and social mechanisms.

If we accept that the primary goals of sexuality are *reproduction* and *erotic pleasure*, we can also discern a number of secondary, though by no means unimportant, goals that are "nonsexual." Paramount among these is the use of sex as a vehicle of expressing and obtaining **love.** This may include deep and genuine affection or its shallow and stereotyped parodies. Adolescent girls may indulge in joyless sexual encounters to maintain popularity and acceptance by their peers, and as a defense against loneliness. Teenage boys may be driven to sexual exploits to defend against self-doubts about masculinity and power. The contribution of sex to *self-esteem* is important. Each of us needs a deep and firmly rooted conviction of personal worth and a reaffirmation of such worth from significant others. An important component of a person's self-esteem is his or her sexual standing. A man who is impotent is likely to lose confidence in himself and may feel uncertain and incompetent even in areas in which sexual virility has no direct bearing. Others may be driven to compensate for their sexual weaknesses by acquiring power, wealth, or through antisocial behavior.

For women, the association between sex

and self-esteem has seemed less direct than for men. Women have tended to be more concerned with being attractive, desirable, and lovable, and deficiencies in these attributes are more apt to damage their self-esteem than sexual prowess or responsiveness as such. But these attributes and sexual responsiveness are often mutually enhancing: a woman who feels admired and desired is more likely to respond sexually, and a responsive woman in turn is more ardently desired. Currently, for increasing numbers of women, the importance of being a sexually alive and responsive person is becoming more central to their lives.

Sexuality is clearly an important component of an individual's self-concept or *sense of identity.* Awareness of sexual differentiation precedes that of all other social attributes in the child: The child knows the self as a boy or girl long before it learns to associate the self with national, ethnic, religious, and other cultural groupings.

Although developing an awareness of one's biological sex is a relatively simple matter, the acquisition of a sense of sexual identity is a more complex, culturally relative process. Traditionally we have assumed that biological sex (maleness and femaleness) and its psychological attributes (masculinity and femininity) are two sides of the same coin. In traditional and stable societies this assumption may have been (and may still be) valid. But in the technologically advanced and rapidly changing societies such direct correspondence between biological sex and psychological attributes is being challenged more and more vigorously as occupational and social roles for men and women become progressively blurred.

For some people sexual activity is a form of *self-expression* in a creative, or esthetic, sense. What matters most is not simple physical pleasure, but the broadening of sensual horizons with each experience, and the opportunity to express and share these experiences in a very special and intimate way with another person. Such feelings, though very real, are difficult to describe.

Sex also functions as a form of interpersonal *communication.* The very act of giving oneself to another fully, willingly, and joyfully can be an exquisite means of expressing affection. But sex can also be used as a weapon to convey *dominance* and *anger.* When, after coming home from a party, a wife refuses to sleep with her husband because he has been flirting with other women, she may be using sex to communicate a message ("I am angry") and a lesson ("Next time behave yourself"). In other situations a husband may impose sex on his wife or even discreetly entice her into it primarily to show his mastery over her. Similarly, indiscriminate sexual activity may communicate messages like "I am lonely," "I am not impotent," "I dare to misbehave," and so on.

The association of sex and *aggression* is very broad and encompasses biological, psychological, and social considerations. We need only point out here that aggressive impulses may be expressed through sexual behavior and vice versa, and that there is an intimate relation between these two powerful drives in all kinds of sexual liaisons.

Social Factors in Sexual Behavior   Just as psychological functions are intimately linked with biological forces, so also they are tied to social factors. In fact, distinctions between what is primarily psychological and what is social often tend to be arbitrary. As a rule, in referring to social or cultural factors, we emphasize the interpersonal over the intrapsychic and group processes over personal ones.

Sexuality is often considered a *cohesive* force that binds the family together. In this sense it serves a social goal, which is why society facilitates sexual aims by providing opportunities for contacts with sexual partners. Sexuality can also have a *divisive* influence, and this potential may be one reason for the ambivalence with which sex has been viewed in many societies, especially our own.

Sex symbolizes *social status.* Like dominant male primates in a troop, men with most power have often had first choice of the more desirable females. Beauty is naturally pleasing to the eye, but the company of a beautiful woman is also a testimony and a tribute to a man's social standing. A woman's status is more often enhanced by her man's social importance than by his looks. As women gain true social parity with men, these attitudes are likely to undergo profound changes.

Finally, sex figures prominently in an individual's *moral* standing. In Western cultures, sex is used as a moral yardstick more consistently than any other form of behavior, at both the personal and public levels. Many of us feel greater guilt and are often punished more severely for sexual transgressions than for other offenses. As a result sexual behavior or its restraint becomes the means to social acceptability.

In various places and at various times sex has been used for the loftiest, as well as for the basest, ends. Although sexuality has been to various degrees suspect in the major Western religious traditions, other forms of religious worship have had distinctly sexual components, as attested by the erotic statuary adorning Indian temples (figure 1.1) and phallic monuments from classical times (figure 17.6).

On balance, sex has been more crassly and mercilessly *exploited* than any other human need. The female body in particular has been used as a commodity since remotest antiquity. Prostitution is the most flagrant example of the use of sex for practical gain, but it is by no means the only one. Sexual favors are exchanged for other services between spouses and friends. The overt and covert use of sex in advertising is blaringly obvious.

Other uses of sex are legion. The ancient Romans wore phallic amulets as a protection against evil. Some people use sex to cure headaches, to calm their nerves, or to end insomnia. Sex can thus be used for nonsexual as well as sexual ends. Conversely, erotic gratification can be achieved through sex as well as by the displacement and sublimation of the sexual drive into countless ordinary and extraordinary activities.

How We Behave Sexually

Since human motivation is difficult to fathom, it is understandable that we lack satisfactory answers to our first basic question about sexuality. We might hope, however, that answers to our second question would be relatively easy. But, in fact, we are frequently in the dark on questions of how we behave sexually as well, and this ignorance may be an important cause of our ignorance of motivation. After all, if we do not know enough about *how* people behave, how can we investigate *why* they behave as they do?

A comprehensive picture of how we behave sexually would require knowledge of current patterns of sexual behavior in our own society, the historical background of such behavior, its cross-cultural variations, and its evolutionary background. Given the universality of sex, we would expect a great deal to be known about all this. But this is not the case, partly because all societies regulate sexual activity. This social control restricts both the observation of sexual behavior and the access to information about it. We can therefore neither generalize from personal experience nor investigate freely.

Although it is customary (for good reason) to decry our ignorance in these matters, we actually do have a great deal of information (and a great deal of misinformation) on human sexual behavior. Most of this information can be found in art, history, and literature; clinical reports; ethnographic accounts; and statistical surveys of sexual behavior. We discuss these approaches individually later on, and the information obtained from them will be presented in parts II and III.

How We Should Behave Sexually

Most difficult, yet also most critical, are the answers to our third question. In addition to more strictly moral judgments, the answer must consider the effects on health, the factors that enhance sexual satisfaction, and the social customs and conventions that carry no moral weight but define courtesy, decency, and the like. At worst these codes may be hollow rituals that needlessly complicate life, offer arrogant badges of status, and serve as tools of intimidation. At best they make social intercourse more comfortable and gracious, and define us as truly human.

The application of *value judgments* (that is, making distinctions between good and bad) to sexual behavior in heterogeneous and pluralistic societies has resulted in much confusion. The meaning of the statistical norm (how common is a given behavior?) has been distorted, partly through ignorance and partly deliberately. Medical judgments (is it healthy?) have often lacked scientific support. Morality has been confused with tradition and the idiosyncrasies of those in power. Statutes and ordinances have frozen into law many dubious factual claims and moral conclusions.

Sexual morality requires judgments on a wide range of specific activities. Some activities are exceedingly rare. Others are so destructive or disruptive that no society, no matter how tolerant, can possibly condone them. We certainly cannot allow "free rape," and so far no society has managed "free love" (in which anyone can sleep with anyone else whenever they choose). For practical purposes, moral issues become matters of general concern when disapproved behaviors are attractive to many people and not barred by overwhelming practical considerations.

There are several well-known approaches to the discussion of sexual ethics. First is the straitlaced attitude, which at its worst stands for dullness, hypocritical double standards, tedious lists of don'ts with a few do's (that are hardly worth doing), and an outlook that generally inhibits all that is spontaneous, imaginative, and exciting in sex. Its opposite is a cavalier insistence on discarding all sexual restraints irrespective of the consequences.

More typical is the intermediary stance that is becoming increasingly prevalent, whereby permissiveness is tempered with consideration for others. There is little concern with abstract principle in this stance; the attitude is rather that anything is fine as long as it gives pleasure to all concerned and hurts no one.

In the search for a sexual ethic it is tempting to bypass basic moral issues by dismissing them as philosophical abstractions of no practical immediacy and instead to try to go directly to the specifics of sexual behavior. This shortcut usually fails, for before we can sensibly discuss whether premarital sex, for example, is right or wrong, we must agree on what "right" and "wrong" mean. Nor can we be satisfied with the mere enunciation of moral generalities, for the specifics of sexual acts must also be recognized: agreeing that sexual exploitation is wrong is insufficient to determine whether, for instance, premarital sex is wrong. Premarital sex may or may not be exploitative, and besides, exploitation need not be the only basis of moral judgment.

In contemporary Western society, neither the principles underlying moral judgment nor their application is generally agreed on. There are marked differences of opinion in these matters among groups divided along generational, educational, and socioeconomic lines. Even within the clergy of the same denomination, opinions are likely to be in conflict.

In the light of such diversity, there is increasing awareness of the need to distinguish between *private* and *public* sexual behavior, especially when the former involves acts between consenting adults conducted in private. The trend now is toward viewing such behavior as beyond the right of society to regulate it. But because a behavior is legal does not necessarily mean it is healthy or moral.

Sexual behavior becomes public when it is carried out before others or when others' individual rights are violated. Although standards obviously vary, all societies attempt to regulate public sexual behavior in one way or another. There is no record of a community totally free in this respect. Judgments, moral as well as legal, are important in both private and public life, but the two spheres should not be confused. These are the sorts of issues that occupy us in part III.

The Study of Human Sexuality

Sexuality is such a pervasive element in human life that there is virtually no field of study that could not be related to it in one manner or another. This very scope of the topic has resulted in its diffusion: by being everybody's business, the study of sex has ended up being nobody's business.

If one defines the field of human sexuality in terms of the store of sexual information and misinformation gathered over the centuries, then the field is vast and its origins ancient.[2] But this sexual information which has been mainly produced as an incidental by-product of other fields of study lacks the coherence of a primary field of inquiry. Geneticists studying sex chromosomes, anthropologists charting lineages, and art historians analyzing erotic symbolism have very little to do with each other's work, even though each of them may be said to be dealing with the field of sexuality.

The study of sexuality as a specialized field is by comparison very modest in scope and has a much shorter history. It is like a tiny statue standing on the massive pedestal formed by the broader approaches. But what it lacks in size, it makes up in a clearer identity.

The information in this text comes from both sources. Parts I and III mainly draw from the general contributions of various fields; part II largely reflects the findings of sex research. We first briefly review all the disciplinary approaches that have contributed to this area. Then we dwell in greater detail on the field of sex research itself.

Biological Perspectives

The more obvious contributions from the *biomedical sciences* to the study of sexuality have been with respect to the structure, functions, and diseases of the sexual organs. But there are also important linkages to the study of sexual behavior through medical specialties like psychiatry, and through the comparative studies of animal sexual behavior by biologists and primatologists.

Medical Approaches to Sexuality Medicine is as old as civilization but we trace the origins of our basic medical concepts to the Greeks. *Aristotle* (384–322 B.C.) was the great codifier of ancient science who laid the foundations of biological study; words like *species* or *genus* are Latin translations of terms Aristotle employed. *Hippocrates* (c. 460–377 B.C.) (the "Father of Medicine") established medicine as an empirical science. The links between behavior and bodily humors that he elaborated on are the precursors of our modern concepts of the influence of hormones on sexual behavior.[3] The discovery of Greek medicine during the Renaissance through retranslations of Arabic texts laid the foundation of modern medical sciences. The works of great anatomists like *Andreas Vesalius* (1514–1564) and artists like *Leonardo da Vinci* (1452–1518) transformed the representation of the body, including the genital organs, into an accurate science (figure 1.2).

With the rise of medical specialities during the 18th and 19th centuries, a number of fields took on a more active interest in sexual functions. Currently, sexuality remains part of the concerns of every aspect of medical practice, but some specialities have a more direct involvement. *Anatomists* study genital structures; *embryologists*, intrauterine development; *physiologists*, sexual functions; *geneticists*, the hereditary mechanisms underlying sexual development and behavior. These fields represent the **basic sciences** in medicine. Investigators working in these areas are often not physicians but specialists in these particular sciences.

More strictly medical are the applied or **clinical fields.** *Endocrinology* is the study of hormones and their disturbances; *urology*

---

[2]The catalog of the Library of Congress has over 500 subject headings listed under "sex."

[3]For a short history of anatomy and physiology from the Greeks to the 17th century, see Singer (1957).

FIGURE 1.2   Leonard da Vinci: Figures in coitus and anatomical sketches.

is the study of diseases of the reproductive tract; *obstetrics-gynecology*, of female reproductive functions and disorders; *dermatology*, of sexually transmitted diseases (or VD) as part of broader concerns with skin diseases. (This association is due to the fact that sexually transmitted diseases often manifest themselves through skin lesions.)

Specialists in *epidemiology* and *public health* are concerned with the prevention of ill-ness. They have a particular interest in the sexually transmitted diseases because of the large numbers of individuals afflicted by them and because of the availability of effective means of preventing them.

The medical study of sexual behavior and the treatment of disturbed sexual functions and behavior are the domain of *psychiatry*. Particularly influential have been the contributions of *psychoanalysis* because of the centrality that sexuality plays in its theories. In its European origins, psychoanalysis was not an exclusively medical field, but in the United States, most psychoanalytic practitioners are required to be M.D.s, and the great majority have had psychiatric training as well.

Biological Approaches to Sexuality   Biologists have been primarily concerned with the study of plants and animals, not human beings. Yet the study of animals is of enormous significance to our understanding of sexual functions and behavior.[4] First, the study of animals leads to the development of methods that can be adapted to humans; most medical experiments and treatment trials are first worked out with animals. Since animal patterns of behavior are simpler, they can be more readily described and analyzed; the insights gained can then be used for the study of human social behavior.

Second, animals can be used in experiments where ethical considerations exclude the use of humans. For instance, the effects of separation of monkey infants from their mothers have led to important findings about sexual development. Similarly, the administration of sex hormones to pregnant animals have led to the elucidation of how the reproductive system develops and the effects of hormones on gender identity; such experiments could not conceivably be carried out with human mothers and infants.

Third, the study of animal behavior can generate principles and rules of behavior whose applicability to humans can be tested. For example, the discovery that the "contact comfort" infant monkeys obtain by clinging to their mother has led to a better appreciation of the importance of early nurturance for human babies. The study of social associations between male and female primates has likewise suggested ways of understanding certain aspects of human sexual relationships.[5]

There are, however, important limitations in applicability of animal studies to human sexuality or any other form of behavior. Even though we are genetically very similar to our close primate cousins, like the chimpanzees, there is a chasm that separates humans from animals: our language and culture set us apart from the rest of the living world. This distinction is further enhanced by the religious belief that human beings have a soul that inhabits their animal-like bodies.

There is nothing we learn from the study of animals that can be automatically extended to people: what applies to humans must be demonstrated among humans. Furthermore, animals are so diverse in their behaviors that evidence can be found to bolster the "naturalness" of any sexual pattern, such as monogamy or promiscuity. Superficial behavioral comparisons and use of animal behavior as a metaphor are likely to lead to false conclusions. The seemingly same sexual behavior among various species many serve different purposes; likewise, the same basic sexual purpose may be attained by different behaviors.

The basic organizing principle in biology is the *theory of evolution* elaborated by Charles Robert Darwin (1809–1882), one of the most revolutionary concepts in the history of human thought. The importance of evolution is not only to the understanding of the emergence of physical forms but of behavior as well. The primary importance of biological study is hence to elucidate the nature of sexual evolution from the simplest organisms to the awesome complexity of human behavior.

Over the past several decades new fields have opened up within biology that focus more directly on behavioral study. *Ethology* studies the social behavior of animals. *Sociobiology* is an integrative approach to the study of human behavior based on evolutionary principles.[6] With respect to sexuality, the sources from which it draws include studies of the social behavior of

---

[4]The following discussion is based on Hinde (1974).
[5]Symons (1979).

[6]Wilson (1975).

animals, especially *primates;*[7] the study of the fossils of early humans (*hominids*); cross-cultural data from the ethnographic literature; and current studies of human behavior. The purpose of this approach is to achieve a synthesis of the biological and social aspects of human interactions and their integration with evolutionary antecedents.[8]

The study of animal sexuality is not the exclusive domain of biologists. Experimental psychologists also do much of their research with animals. Similarly, many primatologists were trained in anthropology. And some sociologists now follow a biosocial orientation. The biological perspective on sexuality is therefore not a narrowly defined enterprise but the product of a multidisciplinary approach to the study of sexuality across the entire spectrum of living organisms.

Psychosocial Perspectives

Various disciplinary approaches to the study of sexuality must be seen as complementary rather than competitive. However, because of historical reasons, conceptual differences, methodological diversity, and disciplinary self-interest, this ideal has yet to be achieved. Thus, even though there is wide recognition of the interaction and interdependence of the biological and psychosocial determinants of sexual behavior, there has been much tension with respect to the relative importance of such variables (historically known as the "nature versus nurture" controversy).

At its worst, the biological approach portrays individuals as soulless machines, whereas the psychosexual approach depicts them as disembodied souls. Yet, for most behavioral scientists, the role of biology in sexual behavior is similar to its role

in the acquisition of language. We have a system of vocal cords that allow the articulation of human sounds, and our brain has the capacity to learn languages. But whether or not as individuals we learn to speak a given language is a function of whether or not we are taught that language. Whether or not we learn English or Swahili has nothing to do with biological factors but simply reflects the culture into which we are born. By the same reasoning, one would say that we are born with sex organs and the capacity to behave sexually, but our sexual behaviors and orientation are largely acquired through the social circumstances of our lives.

Psychological Approaches to Sexuality
Sexual behavior includes the sexual interactions between people as well as the more private world of erotic fantasies and dreams. *Psychologists* are concerned with the scientific study of all observable sexual behavior. *Developmental psychologists* focus on the emergence of sexuality as a child grows up. *Social psychologists* concern themselves with issues such as sexual attitudes. *Clinical psychologists* deal with the study and treatment of sexual dysfunction and behavioral abnormalities.

Although *experimental psychologists* heavily rely on animals in their work and *physiological psychologists* primarily deal with neurophysiological processes, the psychological approaches to the study of sexuality tend to be different from those of the biologists. Whereas evolution is the key biological concept, **learning** is the cornerstone of psychological theories and practice.

Psychologists have made major contributions to the study of sex differences.[9] But until recently, they have shown much less interest in sexual behavior as such. The development of children has been observed and documented in exquisite detail by investigators such as Arnold Gesell, but massive volumes on child psychology barely touch the subject of sex.[10] Until recently it was only an exceptional

[7]Primates include monkeys (such as baboons and gibbons) and the apes (chimpanzees, gorillas, and orangutans), which are more closely related to humans than monkeys.

[8]For an overview of sex and evolution, see Le-Boeuf (1978). The evolution of sexuality is discussed more comprehensively in Symons (1979).

[9]Maccoby and Jacklin (1974).
[10]Mussen (1970).

investigator that addressed these issues.[11]

During the more recent past, there has been a tremendous upsurge of interest in sexuality within psychology. Both with respect to sexual behavior, as well as to related topics like gender, identity, and sex roles, there has been a great deal of research done by psychologists. Of all the specialties represented in the fields of sex research and sex education, psychologists probably now constitute the largest group. New methods of sex therapy that have been developed over the past two decades by Masters and Johnson and others largely depend on behavior modification techniques developed by psychologists.[12]

Sociological Approaches to Sexuality Sociologists have been greatly preoccupied with studying institutions like the family yet they have traditionally done little with respect to sexual behavior as such. The most famous sociological study of sex was conceived and carried out by Alfred C. Kinsey, who was not a sociologist but a biologist, even though his methods and interpretations were distinctively sociological in nature. This neglect has been subsequently rectified; currently there are many sociologists working in the field of human sexuality.[13]

Sociological contributions have enriched the study of sexuality in a number of important areas. Sociologists remain the foremost experts on marriage and the family in our society, which are the primary institutions for the expression and regulation of sexual interactions. Also, the survey, which is the most often used procedure to gather behavioral data about sex, is a sociological technique even though it is used by many others.

At a broader level, the concept of roles, which has been extensively studied by sociologists, has direct applications to the issue of sex roles and their relationship to sexual behavior. The concept of sexual deviance can likewise be formulated in terms of role theory, according to which some individuals take on the role of "deviant" for one reason or another. Sexual behaviors are seen as being determined by the sexual roles that we choose or which are more or less imposed on us by society. In other words, the sociological approach provides the larger societal framework within which to understand individual sexual behavior.

Anthropological Approaches to Sexuality
Given the critical role that culture plays in shaping human sexual behavior, the comparative study of sexuality across cultures is indispensable. Though the prospective gains from this approach have yet to be fully realized, much has been learned from it.

Sex does not have to be discovered by societies; its existence is one of the undeniable realities of life. But as a biological given, sex in human society is but a behavioral potential; it is only through socialization that it assumes form and meaning.[14] All cultures shape sexual behavior, but no two cultures do it exactly the same way. Since no single society can be regarded as representative of the human race as a whole, no serious understanding of sexuality is possible in an ethnocentric context. The only way to know the human family is to know something about its many members.[15]

Cultural anthropologists have traditionally studied societies markedly different from our own. Referred to by various labels ("primitive," "tribal," "preliterate," and so on), these societies have usually been relatively small, homogeneous, technologically less advanced, and changing at a slower pace; thus, easier to study.

Of all the social sciences, anthropology ("science of man") has shown the most

[11]Sears et al. (1957).

[12]For a review of behavioral approaches to sex therapy, see Caird and Wincze (1977).

[13]For a sample of readings in the sociology of sex, see Henslin and Sagarin (1978). Human sexuality textbooks that approach the subject from a primarily sociological perspective include Kando (1978); Gagnon (1977).

[14]Davenport (1977, p. 161).
[15]Ford and Beach (1951, p. 250).

longstanding interest in sexuality; but even at that sexual behavior has received less attention in the field than, for instance, kinship. Though this has changed over the past several decades, the cross-cultural study of sexual behavior is still in its formative stage. As Davenport sums it up,

> While some favorite assumptions and dated theories can now be shown to be either misconceptions or ethnocentric generalizations from our own culture, virtually no non-obvious and significant theoretical generalizations can yet be made. We are just beginning to sense the extent and limits of cultural variations. We do not know how and why these variations occur.[16]

People are naturally curious about each other, and it is to a society's advantage to know other cultures, be it for purposes of commerce, conquest, or conversion. Knowing about others also helps us to better understand our own culture. Since ethical and practical considerations preclude our conducting social experiments to study sexuality, the cross-cultural setting provides us with a natural laboratory where such experiments have already taken place through cultural diversity. We can thereby test the validity of our theories and the commonality of our sexual practices. Freud's claim of the universality of the Oedipus complex, for instance, was challenged by Malinowski's work in the Trobriand Islands (figure 1.3), where kinship is reckoned through mothers and the role of father as we understand it is fulfilled not by the child's biological father but the mother's brother.[17] Similarly, when Margaret Mead (1901–1978) reported in the 1920s that in Samoa premarital sex was part of a socially accepted pattern of adolescence, this called into question the belief that there was an incompatibility between

FIGURE 1.3 Decorated women of the Trobriand Islands photographed by Bronislaw Malinowski in the 1930s. (From *The Sexual Lives of Savages,* 1933.) Because of their exotic origins such figures could be publicly shown and helped liberalize sexual attitudes.

premarital sex and the harmonious functioning of a society.[18]

Evidence from another culture does not necessarily invalidate a theory nor does a different cultural experience mean that we are not entitled to our own values. But such contrasts do prompt us to consider if our way of doing things is not but one alternative and perhaps not the best. The cross-cultural perspective is thus conducive to questioning our cultural presuppositions, without necessarily subverting them.

The shortcomings of the anthropological approach to sexuality are in part inherent in the nature of sexual behavior. Since much

[16]Davenport (1977) pp. 1–2.

[17]Malinowski's *Sex and Repression in Savage Society* (1927) is one of the classics in sexual anthropology. Also see Malinowski (1929). The psychoanalytic counterarguments to Malinowski are put forth in Jones (1966).

[18]Mead (1929). Mead's views reached a wide professional and public audience and may have significantly influenced American attitudes toward adolescent sexuality. Freeman (1983) contends that Mead's research was faulty and the conclusions drawn from it with respect to cultural determinants in adolescent development were in error. For a critique of Freeman's view, see Marcus (1983).

of sexual behavior occurs in private, it precludes direct observation. The anthropologist must then rely on reports of such behavior, which are particularly liable to distortion due to the fear of disclosure or to distortion through exaggeration. Since talking about sex is itself a form of sexual behavior, people of conservative views or in sexually repressive settings are reluctant to reveal their sexual lives to outsiders. Thus, reports on such societies tend to portray their sexual values as more restrictive than may be the case. Similarly sexually liberal societies are likely to appear more liberal than they are in reality.

Particularly for those who are not anthropologists, the main body of ethnographical data in sexual behavior appears like a bewildering collection of tidbits.[19] Apart from the compelling impression that there is enormous variance with regard to all forms of sexual behavior, it is difficult to perceive more specific patterns with regard to a particular culture.

Especially when such material is lifted out of context, isolated behaviors make limited sense. For example, what are we to make of the information that women in Tikopia use bananas for self-stimulation or that Siriono men have a tendency to tug at their foreskin when preoccupied or worried? Having no clue about who these people are, we are not likely either to understand or to care much about what the behavior means, unless the acts are so exotic as to carry a "believe it or not" sort of fascination. One can of course go to the original studies from which the data on sexual behaviors have been taken, but that would be a formidable task; and even at

that there is no guarantee that the sexually relevant material will be found to be integrated and explained in the broader context of the culture. More recent anthropological reports do better in meeting these problems, both in terms of their more explicit attention to the sexual component as well as by their better analyses and syntheses.[20]

Humanistic Perspectives

The field of sexuality, as commonly defined, is dominated by the biomedical and the psychosexual perspectives. Yet there is a long and rich tradition of literary, artistic, and historical exploration of sexual themes that constitutes a source of deep insights into human sexuality. The behavioral sciences themselves owe their origins to the intellectual movement known as *Humanism* that flourished during the Renaissance (though the term was not coined until the 19th century). In the educational ideals of classical antiquity, which were revived during the Renaissance, the curriculum considered most suitable for the instruction of free and responsible persons was called the *studia humanitatis* ("the study of mankind," hence "the humanities").[21] Humanism held forth the possibility of human improvement through education and study. Based on this ideal, the spirit of rational inquiry into human intentions and actions gradually developed over the centuries into the systematic disciplines of the social sciences and by extension to the study of sexuality as a distinct field of inquiry.

The humanities currently include the study of literature, art, music, philosophy,

[19]Several decades ago Ford and Beach carried out an extensive review of cross-cultural research by abstracting data on sexual behavior from studies of about 200 societies (out of several thousand) described by anthropologists. This survey dealt with various cultures, not with separate geographical or political units. Of Western cultures, only that of the United States was included. The original reports from which the data were drawn were often quite dated. Most of these cultures have by now undergone profound changes, and some may be extinct. Nevertheless, these changes do not detract from the value of the original

data for our purposes (Ford and Beach, 1951). Similarly, Murdock has examined data on sexual behavior from a sample of 250 diverse societies ranging from primitive tribes to complex civilizations to determine the nature of social regulation of sexual behavior (Murdock, 1949).

[20]See for instance Marshall and Suggs (1971); Herdt (1981). For further discussion of anthropological perspectives on sexuality, see Davenport (1977); Fisher (1980).

[21]Chambers et al. (1979).

religion, and history. The contributions of these fields may be thought of as the earliest and broadest attempts to make sense of sexual and all other human experience.

The Erotic in Art   Erotic art (or what strikes us as erotic themes in art) provides the oldest and richest record of human sexual activity. Sexual representations, in naturalistic or symbolic form, are among the earliest surviving artifacts of human culture.[22]

Representations of coital postures, which one may associate with our permissive culture, can be found in the records of the most ancient cultures.

In its earliest forms, art was not meant to represent reality but to *be* reality.[23] When Paleolithic people fashioned statues of pregnant women or painted bisons pierced by arrows on cave walls, they were generating a new reality and enacting a ritual through which they hoped to increase fecundity and the food supply.

The magical and religious use of *sexual symbolism* has persisted through the ages, often in subtle and obscure forms. In a more secular sense, art has also provided a *pictorial record* of sexual behavior lending truth to the adage that one picture is worth a thousand words. Depictions of the body, levels of nudity, scenes of sexual interaction provide us with eyewitness accounts of times past, slices of life gone forever. When we look at Greek vase paintings that depict sexual scenes, we learn volumes about who these people were.

Instructive as the representational functions of art are, equally important is the artist's *interpretative* analysis of human sexual emotions and actions. Looking at Picasso's erotic drawings that depict the interactions of the artist with the model, one learns very little about sexual anatomy or coital postures. Yet the genius of the artist captures the mood of the moment, and explains the relationship of art and eroticism in ways that could not be accomplished otherwise.

The purpose of erotic art is not only to explore and explain but also to please the eye, to titillate, to arouse the viewer. Erotic imagery is among the most popular and powerful of aphrodisiacs, ranging from the priceless nudes gracing the walls of museums to the pornographic trash displayed in sex shops. Every society has to come to terms with erotic art. What a society does with it reveals a great deal about the society's sexual values and culture.[24]

The Erotic in Literature   Literary approaches to sex, like art, have served the purposes of *description, analysis,* and *erotic arousal.* The writer's approach to sexuality is closer to that of the clinician than that of the behavioral scientist. The emphasis is on individuals rather than groups, on keen observation and intuition rather than the systematic gathering and analysis of data. D. H. Lawrence's Lady Chatterly and James Joyce's Leopold Bloom may not be representative of men or women in general; they may not even be realistic portrayals of a particular woman or man, yet a writer may capture in a single imaginary character a universal element, the essence of some aspect of sex that no survey could ever hope to distill.

The earliest forms of literature with sexual themes are the *creation myths* of ancient cultures. Here again, as in art, we see sexual expression serving the most profound human strivings to make cosmic sense of our life. The Biblical story of creation with which we are most familiar is exceptional in its asexuality: God formed man from the "dust of the ground" and breathed life into his nostrils; He then took one of his ribs and fashioned his female companion (Gen. 2:7–3). In contrast, let us consider the Egyptian creation myths. The sun god Atum-Re created the first couple, Shu and Tefnut, by ejaculating into his hand, transferring the semen into his mouth and spitting it out again. A drop of semen fell into the waters and formed the first

---

[22]Field (1975).
[23]Eitner (1975).

[24]For a brief overview of the erotic in art, see Eitner (1975); for a more comprehensive presentation, see Webb (1975).

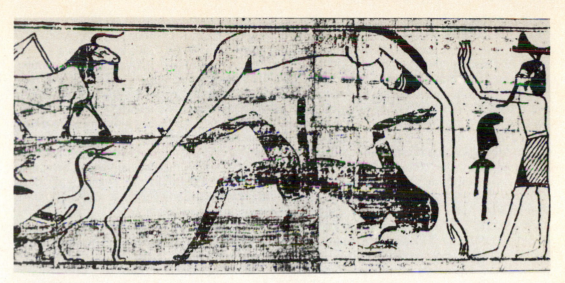

FIGURE 1.4 The Egyptian earth-god, Geb, and his sister the sky-god, Nut. From Papyrus of Tameniu, 1102–952 B.C. (The British Museum).

solid land.[25] Shu became the god of air, and his sister Tefnut became the goddess of moisture. They mated incestuously and gave birth to Geb, the earth god, and Nut the sky goddess. Geb and Nut were united in continuous coitus until wrenched apart by their father, who raised Nut to become the starry vault of heaven and out of reach of Geb's phallus, which continued to thrust skyward vainly trying to reach his sister-wife (figure 1.4). Bereft of her consort, Nut impregnated herself each night by swallowing the disk of the sun which then was reborn each morning from between her legs.[26]

Western erotic literature, like art, has mainly served secular purposes. It has taken many forms, with varying degrees of social acceptability. Much of literature dwells on the exploration of love relationships, but the sexual element in these works is often understated. More flagrantly erotic works have been labeled as *pornography,* although the standards by which to judge such works

have been hard to define and subject to dramatic changes.

The tendency of great writers to treat sexual themes subtly rather than flagrantly may be only in part due to the fear of social censure; erotic themes often carry a more powerful charge if some things are left to the imagination of the reader or audience. Most of erotic literature, at least of the more explicit variety, has been written by men and for men. Women are prominent as subjects but exist to fulfill male desires. One common literary type is that of woman as *erotic toy* typified by works in which the central characters are prostitutes.[27] Good examples are two 18th century English novels: *Moll Flanders* by Daniel Defoe; and *Memoirs of a Woman of Pleasure* (popularly known as *Fanny Hill*) by John Cleland. Alternately, women are cast as *demons* who victimize men: *Nana,* by Emile Zola; or as the *wily adulteress:* Boccaccio's *Decameron* of c. A.D. 1350; Chaucer's *Canterbury Tales* of c. A.D. 1375.

Though much of erotic literature dwells on heterosexual intercourse, all other varieties of sexual experience have also been

[25]There is an interesting similarity here to Japanese myths of creation in which the god Izanagi stirs the ocean with a "spear" and the water dripping from its tip forms the first island.

[26]Field (1975).

[27]This section is based on Purdy (1975a).

explored. Most important are works that deal with homosexual relationships. Even though this theme has had to be presented often in disguise, more explicit works have appeared from time to time (for instance, the earl of Rochester's *Sodom* of 1684), until the recent liberalization of social attitudes that has been followed by an outpouring of gay literature.[28]

The literature of sadomasochism is best exemplified by the 18th century works of the marquis de Sade. The theme of incest has attracted a great many literary talents: the poetry of Percy Bysshe Shelley (1792–1822), Thomas Mann's *Blood of the Walsungs* (1905), and William Faulkner's *The Sound and the Fury* explore brother-sister erotic love. Father-daughter incest is the subject of Shelley's *The Cenci* (1820) and F. Scott Fitzgerald's *Tender Is the Night* (1934). Mother-son incest was dealt with by the great Greek dramatists most notably in Sophocles' classic *Oedipus Rex*.

With the exception of Vladimir Nabokov's *Lolita* (1955), few serious authors have dealt with the theme of pedophilia. Likewise, masturbation is a topic that was generally avoided until recently, when works like Philip Roth's *Portnoy's Complaint* (1967) began to treat it openly.

These are but a few examples of the extensive literature on eroticism. If we were to expand our purview to include more general works on love, jealousy, infidelity, and other topics that have a bearing on sex, we would have to take on the bulk of the literature of the world since ancient times.[29]

The Erotic in Motion Pictures Compared with art and literature, the cinema is a very recent newcomer to the erotic scene. But what it lacks in longevity it makes up for by its pervasive presence. Over the past several decades, people have been exposed to movies and television to a far greater extent than to art and literature. Furthermore, romance and sex have been among the main staples of their scripts.

The cinema, like theater, is an art form, which has produced memorable explorations of erotic themes. Though moviemakers have mainly focused on romantic themes, even topics like rape have been explored sensitively by masters such as Ingmar Bergman (*The Virgin Spring*), Akiro Kurosawa (*Rashomon*), Luis Bunuel (*Viridiana*) and Vittorio de Sica (*Two Women*).[30] Television, by contrast, remains largely awash in soap operas.

Although the cinematic depiction of sexual scenes and the exploration of controversial subjects have become commonplace, there is still a substantial, though fuzzy, distinction between films produced for general consumption and the ***pornographic film*** ("blue film") of *soft core* (no genital exposure) and *hard core* varieties. This distinction is quite similar to that in literature.

Blue movies go back to the very origins of the film industry. Long an underground activity, the making of pornographic films mushroomed in the 1960s with the liberalization of censorship laws. Yet the value of the contributions of pornographic films to the exploration of sexuality has been so far minimal despite their explicitness and the fact that sex is their sole purpose.

At a technical level, there is a vast difference between the *stag films* of the past, which were mainly depictions of copulation clumsily acted and filmed, and more recent extravaganzas like *Behind the Green Door*. Yet basically this genre of film remains based on plots that are contrived and manifest neither much sense nor sensitivity.

Why is it that gifted artists have so far shied away from making erotic films, with few exceptions (Gustav Machaty's *Extase*, 1933; Bertolucci's *Last Tango in Paris*, 1972; Radley Metzer's *The Lickerish Quartet*, 1970)?

[28]Boswell (1980) explores this theme in detail from the beginning of the Christian era to the 19th century.

[29]For a brief overview of the erotic in literature, see Purdy (1975a), which also provides a selected bibliography. Legman (1963) and Atkins (1973, 1978) provide more extensive coverage. Marcus (1966) provides an excellent discussion of the nature of pornographic literature.

[30]Purdy (1975b).

The fear of social censure has been, and may still be, an important inhibiter. Yet beyond that, there may be a more intrinsic problem, an incompatibility between the effectiveness of artistic expression and relentless sexual exposure. It could also be that the issue is not one of explicitness but of fundamental purpose. The artist is interested in the exploration of sexual realities and the imaginative recreation of human erotic experience—sex is but one vehicle to understand the human condition. The pornographic work takes sex out of its living context, thereby dehumanizing it, and exploits it for all its worth by reenacting what are basically repressed infantile fantasies.

Pornographic films may be dreary and void of the breath of life, but there is an intrinsic force in sexual acts that asserts itself no matter how ineptly presented. In that sense, the unadorned depictions of genital organs and sexual activities serve a certain educational function. Films are also used now more effectively in sex research and education. Educational films depicting coitus, masturbation, and homosexual interactions began to be developed in the 1970s. Whether used in classrooms or with patients, there is a certain obvious merit to these films, but they also suffer from some of the same limitations ascribed to commercially made pornographic movies.

Historical Approaches   Historians have produced a stupendous amount of writing but very little of it has explicitly dealt with sexual behavior. And whatever attention has been paid to sex has tended to focus on the lives of those in power and socially privileged positions. Hence, we know very little about the sexual behaviors of the great mass of ordinary men and women in the past.

Bias is unavoidable in historical research since historians, like everyone else, have beliefs and prejudices that color their perceptions, often without their being aware of it. Such bias has been plentiful with respect to sexual matters. But beyond that, deliberate distortion has also compromised the historical record, especially with

socially problematic sexual behaviors. For instance, in Ovid's *Art of Love*, a "boy's love appealed to me less" was changed by a medieval scholar to read "a boy's love appealed to me not at all."[31] Likewise, by tampering with the gender pronouns the *Rubaiyat* of Omar Khayyam was made to sound like the heterosexual rather than the homosexual yearnings of the poet.

Another major source of bias is the almost exclusively male perspective in history. Thus, although female characters are present in great profusion, they are seen through the eyes of men, similar to the situation in much of erotic art and literature. This picture has been changing over the past two decades or so, as the sexual record from the past has begun to be cautiously reconstructed. Historians are now much more concerned with the "private side" of history. *Family history* and *women's history* are fields that are particularly concerned with these issues.[32]

Since there were no sex surveys of sexual behavior before the modern era, our understanding of sexual behavior in the past depends mainly on historical records. Historians also play a crucial role in elucidating the broader social context within which sexual relationships unfolded and in providing us with the thread of continuity running through civilization.

The Field of Human Sexuality

The *field of sexuality* (or *sexology*), as a specialized area of study, constitutes a more discrete entity than the multidisciplinary studies of sexuality we have discussed so far. Nevertheless, the issue remains unresolved as to what direction the study of

[31]A marginal note further informed the reader, "Thus you may be sure that Ovid was not a sodomite." For the original source of the quotation and other examples, see Boswell (1980, pp. 18–21).

[32]Freedman (1982). For a review of surveys of historical literature on sexuality, see Burnham (1972). Examples of the more specialized treatment of sexual topics by historians are Degler (1974); Boswell (1980); Robinson (1976); Foucault (1978). In a more popular vein, sex throughout history is presented by Tannahil (1980).

sexuality should take in the future: should sex be part of the concern of every relevant field, from biology to theology; or should there be a specialized discipline that concerns itself with various aspects of sexuality? These questions and issues currently engender considerable tension within the field of sexuality.[33]

History of Sexology

The term "sexology," though currently out of favor, has characterized the broad enterprise of sex research, education, and therapy (there is as yet no single term that could be accepted as an adequate substitute). The history of this enterprise can be divided into three phases: origins; establishment; revival.

Origins   The present-day study of sexuality can be traced to the broader concerns of Enlightenment figures, like Jean Jacques Rousseau (1712–1778), who addressed themselves to problems of sexual relationships and their proper place in society. Yet approached on such a general plane, the study of sexuality was merely part of larger intellectual inquiries into human nature and behavior.

Establishment   More focused attention to sexual behavior developed in central Europe, in the 19th century. This was primarily an effort on the part of physicians, mostly psychiatrists, to extend the benefits of scientific study to the area of sexuality. Some of the basic concepts of sexual pathology (like the notion of "degeneracy") had been expounded in France. But it was in Germany, in the second half of the 19th century, that the foundations of sexology were laid and then reached fruition at the turn of the 20th century.

The fact that physicians were in the vanguard of this movement is explicable on the grounds that sex constituted a natural extension of their concern with the human body, and they had the social standing and

FIGURE 1.5   Richard von Krafft-Ebing.

credibility to stand the heat of criticism (of which there was a great deal) for trespassing into this area.

Two early figures approached the area of sexuality as part of their broader professional concerns and were not directly involved with the establishment of sexology in the institutional sense. The first was *Richard von Krafft-Ebing* (1840–1902), a German neuropsychiatrist who systematized case studies of sexual aberrations into his *Psychopathia Sexualis*, first published in 1886 and revised through twelve editions (figure 1.5).[34]

Krafft-Ebing epitomizes the early medical-psychiatric approach to sexuality. In the tradition of clinical psychiatry, Krafft-Ebing used the case history as his investigative tool. Since his primary interest was in forensic medicine and psychopathology, he chose mostly extreme cases; hence, they make grim reading. Many of them, nonetheless, remain instructive to this day.

[33]See the exchange between Riess and Moser in Riess (1982), (1983); Moser (1983).

[34]The full English translation of this work has become available only recently. See Krafft-Ebing (1978).

The theoretical basis of much of what Krafft-Ebing had to say has turned out to be wrong. However, by drawing together, for the first time, materials on sexual behavior in a systematic and reasoned manner, he laid the foundations of sexology. By drawing this field into medicine, he also lent it a certain measure of respectability at a time of great intolerance to the candid discussion of sexuality. Even though Krafft-Ebing was a highly respected figure, he came under considerable criticism for stepping into forbidden territory. His self-justification is expressed in the following statement in the Preface to *Psychopathia Sexualis:*

> No physical or moral misery, no wound, no matter how infected, should frighten someone who has dedicated himself to the science of man; and the sacred ministry of the physician, while obliging him to see everything, also permits him to say everything.[35]

Sigmund Freud (1856–1939) also started his career as a neurologist, and his interest in sex was incidental to his more basic concern with personality development and psychopathology (figure 1.6). Yet sexuality came to dominate his theories as the driving force behind human motivation. The two most important concepts elaborated by Freud were those of the unconscious and of infantile sexuality.[36] Freud did not discover or invent either of these concepts, yet he articulated them with such persuasiveness that they became highly influential in the intellectual history of the modern Western world.

FIGURE 1.6   Sigmund Freud (Bettman Archives).

Freud had some dealings with the early sexologists but never quite joined their budding movement.[37] Nonetheless, his work has had a far greater impact in making sexuality a subject to be reckoned with than that of anyone else. Freud was wrong in a lot of ways, and some of his views are highly unpalatable to many. Yet his theories still constitute the most comprehensive and internally consistent system of sexual development and behavior, even though their validity is by no means firmly established.

The centrality of sex in the work of psychoanalytic theorists varies widely nonetheless. Some of Freud's major disciples, like Carl Jung and Alfred Adler, thought he put too much emphasis on sexuality. Others, like Wilhelm Reich, thought Freud did not do full justice to it.

Wilhelm Reich (1897–1957) was a Viennese physician who rose to prominence in the early psychoanalytic movement. Reich was also a Marxist who became increasingly dissatisfied with Freud's neglect of social and political issues. Reich thought Marx's concept of alienation should be extended to include sexuality, since the way of life imposed by the capitalist system

[35]Krafft-Ebing, 1978.

[36]Concepts of the unconscious antedating Freud are discussed in Whyte (1960). Freud's first major analysis of the unconscious is in the *Interpretation of Dreams* (1900). Five years later appeared his *Three Essays on the Theory of Sexuality* in which the theory of infantile sexuality was first formulated. For Freud's complete works, see *The Standard Edition of the Complete Psychological Works of Sigmund Freud,* edited by J. Strachey (1957–1964). There are many biographies of Freud; the most comprehensive is that of Jones (1953).

[37]Sulloway (1979). See pp. 277–319.

crippled the free and healthy expression of sexuality.[38] Furthermore, all neuroses and characterological problems resulted from the damming up of sexual energy because of inhibition of its full expression and release.

To rectify the neglect of political thinking in psychoanalysis and the neglect of sexuality in Marxism, Reich in 1929 organized the Socialist Society for Sexual Advice and Sexual Research. It proved an unworkable match; Reich was expelled both from the Communist and the psychoanalytic camps. In his later years (when he lived in the United States) Reich developed outlandish ideas about trapping "biological energy" in his "orgone box" and his work was discredited. Subsequently, he has been heralded as a champion of sexual and political freedom by radicals in the New Left.[39]

The establishment of sexology as a distinct discipline was mainly brought about by three other German physicians: Iwan Bloch, Albert Moll, and Magnus Hirschfeld.[40] Through the leadership of these men and several other collaborators the field of sexology was established at the turn of the century and thrived for a few decades until its suppression by the Nazis.

Iwan Bloch (1872–1922) was a dermatologist with a special interest in veneral diseases and a broadly educated man with an impressive fund of knowledge in the social sciences and humanities. Though he remained committed to the sexological theories of degeneracy that were espoused by the medical profession of his day, Block broke through the narrow confines of biology and introduced the perspectives of the social sciences (especially anthropology) into the study of sexuality. To reflect this concept of a multidisciplinary approach, in 1906 he coined the term *sexualwissenschaft*—the "science of sex" or *sexology*.

Albert Moll (1862–1939) was a neuro-psychiatrist. Though he lacked Bloch's erudition, Moll made up for it with his organizational skills. In 1913 he founded the *International Society for Sex Research* as well as the "Society for Experimental Psychology." Moll was a man with great respect for institutions and social acceptability, traits that often brought him into conflict with the more radical and iconoclastic Hirschfeld. His early contributions include monographs on homosexuality and the libido. His ideas on infantile sexuality (which may have influenced Freud) were given expression in his 1909 volume on the sexual life of the child.

Magnus Hirschfeld (1868–1935) was the most versatile figure of early sexology (figure 1.7). Aspiring first to be a writer, Hirschfeld then switched to medicine. Starting as a general practitioner, he developed an early interest in sexual matters and became a full-time specialist in sex research and treatment, as well as an activist organizer and crusader for sexual freedom.

Hirschfeld's primary interest was in homosexuality. As an avowed homosexual himself, he often took up the defense of other homosexuals in courts of law and tirelessly pursued the goal of reforming sex laws aimed at homosexuals. This earned him the reputation of being a special pleader with an ax to grind.[41]

Hirschfeld was just as enterprising as a researcher. In 1903 he surveyed 3,000 university students and then almost twice as many metalworkers, in an attempt to ascertain the prevalence of homosexuality (his response rates were better than many modern studies.)[42] In addition to a number of shorter works, Hirschfeld wrote a massive compendium of what was known about homosexuality in a series edited by Iwan Bloch. He coined the term, "transvestism"

---

[38]Reich (1969).

[39]For a discussion of Reich's work, see Robinson (1976).

[40]For short accounts of the history of sexology, see Haeberle (1982); Hoenig (1977).

[41]When Albert Moll was pressed to explain why he had not invited Hirschfeld to a congress in sexology, he said that "many serious researchers do not consider him to be an objective seeker of truth . . . he confuses science with propaganda" (quoted in Haeberle 1981, p. 272).

[42]Leseer (1967).

All of this and other promising developments (including the institution of sexological programs in two major universities) were stifled when the Nazis came to power. All three pioneers were Jewish, hence marked for persecution. Hirschfeld's ethnicity, homosexuality, and political radicalism proved to be a particularly calamitous combination; his institute was ransacked by a mob and its contents publicly burned in 1933—three months after Hitler came to power. (Hirschfeld was in France at the time, where he remained.)

The demise of sexology in Germany retarded similar developments elsewhere. *Auguste Forel* (1848–1931) in Switzerland and *Henry Havelock Ellis* (1859–1939) in England (figure 1.8) were the contemporaries of the German sexologists and had close ties to them. Yet in their respective countries they were unable to replicate the same organizational efforts.

Ellis nonetheless exerted a great deal of influence on the sexual attitudes of the English-speaking world. He did no original research, but gathered a vast amount of sexual materials in his *Psychology of Sex*, which he wrote and periodically revised

FIGURE 1.7 Magnus Hirschfeld (Bettman Archives).

and wrote the first systematic account of what we now call "transsexualism." In 1928 he published his culminating work—a five-volume presentation of the field of sexology.[43]

Hirschfeld's vision of sexology was an elaboration of Bloch's multifaceted view. In 1908 he used Bloch's term for sexology in the title of the first journal of sexology, (*Zeitschrift für Sexualwissenschaft*), which Hirschfeld edited. Along with Bloch, he was among the founders of the first sexological society (Medical Society for Sexology and Eugenics). In 1919 Hirschfeld realized his greatest ambition with the founding of the world's first *Institute for Sexology*. Housed in an elegant building in Berlin, the institute conducted research, provided clinical services (including premarital guidance) and medico-legal aid. It also served as a training center and housed a library with 20,000 volumes, 35,000 photographs, collections of objects of art, and some 40,000 biographical case materials.[44]

[43]Hoenig (1977).
[44]Haeberle (1982).

FIGURE 1.8 Havelock Ellis (Bettman Archives).

between 1896 and 1928.[45] It made his scholarly reputation, but he was subjected early on to a great deal of social opprobrium.

Ellis and his wife Edith were both active in the women's movement. They also had an unconventional marriage, which was emotionally close but precluded sex (each carried on affairs with other women, with mutual consent). Ellis' personal experiences and difficulties with the Victorian morality of the time made him a persistent advocate of tolerance for sexual diversity. In view of his positive views of sex, Brecher has called him the "first of the Yea-Sayers."[46]

FIGURE 1.9   Margaret Sanger. The Sophia Smith Collection, Smith College, Northhampton, Mass. Reprinted from *Margaret Sanger: An Autobiography* (New York: Norton, 1938). By permission.

[45]Ellis (1942).
[46]Brecher (1969). For Ellis' biography, see Grosskurth (1980); for his autobiography, Ellis (1940). Ellis' role in the modernization of sex is discussed in Robinson (1976).

Notably absent from this cast of characters are women. This is partly because the social climate of the Victorian era would not have tolerated their involvement in this field. But also, even those women who braved such criticism were usually social reformers who devoted their efforts to issues like birth control, which most directly affected women's lives. This is exemplified by the work of *Marie Stokes* (1880–1958) in England and *Margaret Sanger* (1883–1966) (figure 1.9) in the United States.

It is also possible that sex research undertaken by women has gone unnoticed. For example, in 1973 Carl Degler discovered in the Stanford archives an extensive survey conducted by Dr. Clelia Duel Mosher (1863–1940). Carried out between 1892 and 1920, it was a questionnaire study of the sexual attitudes of 45 women, 7% of them born before 1870. The findings of this work has helped place Victorian female sexuality in a new light (see box 18.9).[47]

Resurgence   Although the events in Germany of the 1930s were a serious setback for the further development of sexology, various efforts continued to study sexuality. In the United States, by the start of World War II, a number of behavioral investigations had been carried out as well as numerous clinical studies.[48] But the full resurgence of these efforts, which marked the beginning of the modern field of sex research, did not start until Kinsey's work in the 1940s.

Alfred C. Kinsey (1894–1956) was a zoologist at Indiana University. After twenty-five years of work with the gall wasp, he turned to the systematic statistical study of human sexual behavior. This was prompted by the need to find answers to the questions his students put to him in a course on marriage that Kinsey had been prevailed upon to participate in. From these early beginnings, Kinsey and his collaborators (Wardell B. Pomeroy, Clyde E. Martin, and Paul H. Gebhard) collected over

[47]Mahood and Wenburg (1980).
[48]For a review of this early work, see Kinsey et al. (1948).

FIGURE 1.10   Alfred C. Kinsey (Institute for Sex Research).

16,000 sex histories from people in all walks of life across the United States—a feat unprecedented and unequaled. Kinsey alone collected 7,000 such histories—an average of two a day for 10 years. He died, however, long before he could fulfill his goal of interviewing 100,000 individuals (figure 1.10).[49]

The Kinsey studies on the sexual behavior of the male and the female, despite the passage of almost four decades, remain the most comprehensive and systematic source of information on human sexual behavior. A number of studies have been done since that glibly compare themselves to Kinsey's work but come nowhere near it in scope and thoroughness.

The *Institute for Sex Research* founded by Kinsey (renamed recently as "The Kinsey Institute for Research in Sex, Gender,

and Reproduction") has pursued other extensive investigations of sexual behavior, including studies of sex offenders and homosexuality.[50] The institute library performs an outstanding archival and educational function with its collection of 60,000 books, 3,700 films, and extensive pictorial materials and artifacts relevant to sexuality.

Kinsey's work received such extensive public notice that in addition to describing the patterns of sexual behavior, it well may have changed them. The sex survey became a part of everyday life as other investigators, sexual entrepreneurs, and magazine editors kept probing and prodding people to reveal their sexual lives. Books dealing with sex became a substantial industry, although with rather modest contributions to serious advancement of our sexual knowledge.

The next threshold crossed in sex research was through laboratory investigations of human sexual physiology. Kinsey had anticipated the need for direct observation of sexual activity, but the progress from interviewing to observing did not take place until the 1960s, in the studies of *William Masters* (a gynecologist) and his research associate, *Virginia Johnson.*[51]

Working with 694 volunteer men and women, aged 18 to 89, these investigators observed, monitored, and filmed the responses of the body during 10,000 orgasms attained through masturbation or coitus. Their findings established at least a preliminary basis in the physiology of sex, a matter long neglected by experts in physiology and sexology. The subsequent work by Masters and Johnson with the treatment of sexual dysfunction established the modern field of sex therapy.[52]

Beyond these narrower confines of sex research, there have been enormous ad-

---

[49]Kinsey left no autobiography. There are many brief accounts of his career: see Brecher (1969); detailed biographies have been published by Christenson (1971) and Pomeroy (1972). Also see Robinson (1976).

[50]For a review of studies from the Kinsey Institute, see Weinberg (1976).

[51]Masters and Johnson (1966).

[52]See Brecher (1969) for a brief account of this work, and Robinson (1976) for a more critical analysis of the ideological aspects of the work of Masters and Johnson.

vances over the past three decades in the study of reproduction, contraception, and the treatment of the sexually transmitted diseases. Contraceptive devices like the birth control pill and antibiotics like penicillin have created new realities unprecedented in human history. Discoveries in molecular biology and manipulations of the reproductive process that have produced "test-tube babies" and embryo transfers are events that the sexologists earlier in the century could have barely imagined.

The Field of Sexuality at Present

Defined in its more circumscribed sense, the field of human sexuality today has three major components: research, education, and therapy. Since each of these areas is discussed in fair detail elsewhere in the text, we touch upon them here only briefly.

Sex Research   Except for a few institutes, most research in sexuality is undertaken by individuals in various university departments. Physicians continue to play an important role in this field but they no longer dominate it. The majority of investigators of sexual behavior are now behavioral scientists, many of them psychologists.

There are two primary journals devoted to sex research: *Archives of Sexual Behavior* and *Journal of Sex Research;* over a dozen other journals and newsletters also furnish useful information. Furthermore, given its multidisciplinary nature, important research on sexuality is just as likely to be reported in the scholarly publications of many other fields. Hence the serious student of sexuality must cast a wide net when investigating a given subject.[53] The methods of sex research are discussed in chapter 8.

Sex Education.   Under one or another guise, American educators have provided instruction in sexuality at least at the col-

lege level since the start of the century. Pioneers like Prince A. Morrow, who made efforts to develop more formal and effective programs, were primarily concerned with the prevention of venereal disease. In the period following World War II, courses in marriage and the family became popular, but it was not until the 1960s that human sexuality courses, as currently constituted, made their appearance.[54]

Although there are no comprehensive data on the number and nature of human sexuality courses in college, a 1973 survey of 213 institutions found that 42% had one or more courses in this area and an additional 46% offered courses that included some reference to sex; thus, close to 90% of these institutions made some provision for didactic instruction in human sexuality.[55]

The instructors of human sexuality courses come from a wide variety of disciplines but most commonly from departments of health education and psychology. Sex education courses in college have generally proven highly popular with students; they are generally perceived by the faculty as being academically marginal. Except for a few fledgling attempts to develop graduate level programs, this area remains a neglected field in higher education.

Sex education for adolescents and children has been fraught with controversy. Who should provide sexual instruction to youngsters, in what form, and for what purposes continue to be vexing problems (chapter 9).[56]

The organization that has been most prominent (and most actively attacked) in efforts to promote sex education is *SIECUS* (Sex Information Council of the United States) cofounded in 1960 and subse-

---

[53]Brewer and Wright (1979) provide extensive bibliographies. The Institute of Sex Research Catalogs are the most extensive sources for the literature of this still emerging field.

[54]For an overview of sex education in the United States, see Kirkendall (1981). The evolution of human sexuality courses in one major university is described in Katchadourian (1981). Also see McCary (1975); Anderson (1975); Sarrel and Coplin (1971).

[55]Sheppard (1974).

[56]For an account of the sex education battle in the Anaheim, California, school district, see Breasted (1970).

quently led by *Mary Steichen Calderone.* The concept of sexuality that underlies the organization's philosophy is as follows:

> The SIECUS concept of *sexuality* refers to the totality of being a person. It includes all of those aspects of the human being that relate specifically to being boy or girl, woman or man, and is an entity subject to life-long dynamic change. Sexuality reflects our human character, not solely our genital nature. As a function of the total personality it is concerned with the biological, psychological, sociological, spiritual, and cultural variables of life which, by their effects on personality development and interpersonal relations, can in turn affect social structure.[57]

The standard argument against sex education has been that it intrudes into the sexual innocence of children and fuels the flames of adolescent sexuality; thus, the argument goes, while attempting to inform, sex education wittingly or unwittingly encourages sexual experimentation with all its dire consequences. The advocates of sex education claim that it helps prevent problems like VD, unwanted pregnancy, and faulty attitudes toward sexuality: by providing the right information and instilling the right attitudes, it leads to more fulfilling sexual lives.

Ironically neither the advocates nor the opponents of sex education have any solid evidence to support their claims. Most of the attempts to assess the impact of sex education have involved college populations; in general, they show gains in sexual knowledge and shifts toward more liberal attitudes, but no clear evidence for behavioral change.[58] Not much is known about the effects of sex education in younger populations. The one extensive and careful evaluative study of adolescent sexuality education programs conducted so far has shown that, with very few exceptions, these courses have little impact. Students typically claim that the courses increased their factual knowledge and self-understanding. They and their parents are generally very positive about the quality of these courses. Yet objective pretest/posttest assessments show that these students learned very little, let alone experiencing major attitudinal and behavioral changes with respect to sexual behavior and the use of contraception. Only in one program, in which classroom education was combined with clinical services on campus, was there clearly increased use of contraceptives and decreased unintended pregnancy.[59]

It would be premature to conclude from this research that sex education is worthless; more properly, its conclusions point to the inadequacy of current programs, which often operate with limited funding, self-taught instructors who have to work under a cloud of potential social disapproval, and so on. It can be further argued that the existence of such programs should not be contingent on the rigorous demonstration of concrete change; after all, what is the evidence that social studies in high school make better citizens or learning about American history enhances patriotism? Yet schools teach these courses on the premise that they ought to be part of the background of every educated person. Why should sexuality not be given similar consideration?

Sex Therapy   Most people with sexual dysfunctions such as impotence and failure to reach orgasm are probably still treated by general practitioners, clinical psychologists, and other counselors. Yet a new field of sex therapy has also emerged over the past two decades with its own particular methods (chapter 15). It is unclear whether this *new sex therapy* will eventually evolve into a full-fledged specialty or whether its contributions will be absorbed into the mainstream of other established disciplines.

---

[57]Brown (1981), p. 252.

[58]For a review of studies assessing the effectiveness of sex education, see Kilmann et al. (1981).

[59]This study is described briefly in Kirby (1983). The full report of its findings has not yet been published.

The Status of the Field   Compared to the low visibility and shady image of sexology prior to the 1960s, its development within the last two decades has been quite striking. There are now hundreds of professionals of various disciplinary backgrounds who constitute the membership of organizations like the *American Association of Sex Educators, Counselors and Therapists* (AASECT), *Society for the Scientific Study of Sex* (SSSS), and *International Academy of Sex Research*. National meetings and international congresses on sexuality now attract the same sort of public notice as do other professional gatherings.

Yet in some ways the field of human sexuality remains ill-defined and marginal compared to other established disciplines in the behavioral and biomedical sciences. Its practitioners include highly trained professionals as well as people with very modest academic credentials. Reputations are all too often made through best-selling books rather than solid research or scholarly accomplishments.

The quality of sex research, though rapidly improving, is still largely substandard. College courses and texts typically carry light intellectual freight. There is as yet no sexuality department, graduate program, or institute affiliated with any of the more prestigious research universities. Most medical schools offer courses in sexuality but no well-known school has a department in this field. Even the most established academics who work in this area have their primary appointments in some other department.

None of this should be really surprising given the fact that the modern field of sexuality is barely two decades old. The issue is not over what does not yet exist but what is likely to come about over the next few decades. There can be no doubt that the study of sexuality will expand; the question is not whether that will happen but under what auspices its study and instruction will be conducted.

One possibility is that various disciplines will reverse the pattern of neglect accorded this topic and turn their specialized methods and tools to its study. Thus, sexual physiology will be studied by physiologists; the history of sexuality by historians, and so on. Much of this is already happening and is likely to continue. But if the area is thus parceled out, will there be any need for sexologists?

The advantage of this multidisciplinary approach is that various facets of sexuality will be studied by experts in established fields; this will result in higher quality work and greater acceptance of sex as a legitimate area of study. The disadvantage is that the subject will be fragmented. Furthermore, such dispersal will dilute if not vitiate altogether the thrust of sexology as a social movement. Given the political commitments of many people in this field, this could be a serious concern.

Alternatively, the field of sexuality, sexology, or whatever it may come to be called, could grow as a specialized discipline with its subspecialized approaches to various aspects of the subject. This would give the field a clear and coherent identity, and its subject of study, an integrated focus. It will also provide the field with a better chance to become a force for social reform or the advocacy of an enlightened sexual ideology. But as with any other new area, it is not easy for a field to lift itself up by its own bootstraps. As a result, it may take several generations for investigators and educators to establish a solid independent base in sexuality. And even then, should the field become identified with one or another ideology, it will be seen as the province of special pleaders, which will compromise its credibility.

Possibly, sexology could flourish as a discrete field as well as be an integral part of other fields. It could be an objective science and a force for social reform over a broad spectrum of sexual values.

In whatever manner these issues and choices are resolved, the field of sexuality is now poised to expand and occupy its rightful place as a subject worthy of study and instruction. After centuries of neglect and oppression, this vital topic deserves to be treated with the same honesty, rigor, and integrity that have been brought to bear on other aspects of human life.

Part I
Biology

Chapter 2
Sexual Anatomy

Praise be given to God, who has placed man's greatest pleasure in the natural parts of woman, and has destined the natural parts of man to afford the greatest enjoyment to woman.

Shaykh Nefzawi, *The Perfumed Garden*

The human body has few other parts as fascinating as the sexual organs. Venerated or vilified, concealed or exhibited, the human genitals have elicited a multitude of responses. They have been portrayed in every art form, praised and damned in poetry and prose, worshiped with religious fervor, and mutilated in insane frenzy.

Many of us combine a lively interest in the sex organs with an equally compelling tendency either to deny such interest or to be ashamed of it. There are men and women who have been married for years, who have engaged in sexual intercourse countless times, but who have never looked frankly and searchingly at each other's genitals. Some may argue that concealment promotes desire; but it also perpetuates ignorance.

Traditionally, men have been fascinated, aroused, and preoccupied by female genitals, but simultaneously they have feared and deprecated them. Males often have ambivalent views about their own genitals as well, alternately imbuing them with exaggerated importance and being anxious about their size, shape, and capacity for performance.

Female attitudes towards the genitalia, be it their own or those of males, have had much less occasion for public expression, but women too have combined feelings of pride and pleasure with shame and confusion with respect to their sex organs. Women also tend to be less knowledgeable than men about sexual anatomy because their own organs are more concealed and they have been expected or assumed not to be interested in the genitals of males. Attitudes toward male and female genitalia held by both sexes have recently become more open and accepting, but there is still much ambivalence in this regard.

Sexuality entails more than reproduction. Yet the structure of sex organs has evolved in ways to maximize their reproductive potential. Sexual anatomy is best understood therefore when viewed in reproductive terms.

The reproductive systems of both male and female are built on the same basic plan and fulfill similar functions: the production and transport of *germ cells* (*sperm* in the male; *ova* in the female); the production of *sex hormones* in both sexes which are secreted into the blood stream (discussed in chapter 4).

The reproductive system is housed in the *bony pelvis;* the main parts of each system, and their relationship to the bony pelvis are shown in figure 2.1. The male pelvis has a heavier bone structure while the female pelvis is broader. The bones of the pelvis consist of the triangular end of the vertebral column (*sacrum*), and a pair of *hip bones* that are attached to the sacrum behind and to each other in front, at the *symphysis pubis.*[1]

[1]Dienhart (1967).

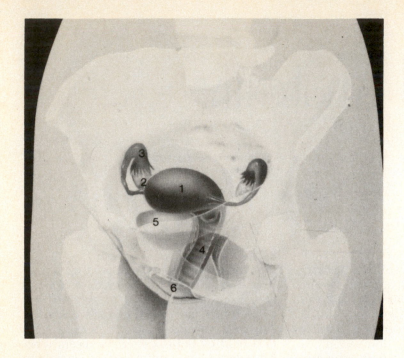

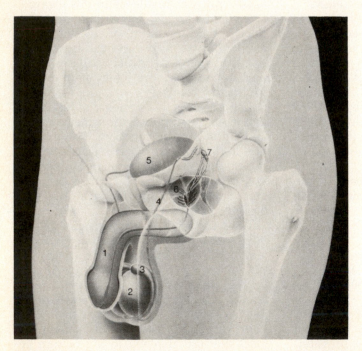

FIGURE 2.1   The reproductive system. (Top) Female reproductive organs. (1) uterus, (2) ovary, (3) fallopian tube, (4) vagina, (5) bladder, (6) labia majora and labia minora on the right side. (Bottom) Male reproductive organs. (1) penis, (2) testicle, (3) epididymis, (4) spermatic cord, (5) bladder, (6) prostate, (7) seminal vesicle. (From Nilsson, *A Child Is Born*. New York: Delacorte Press, 1977, pp. 17 and 21.)

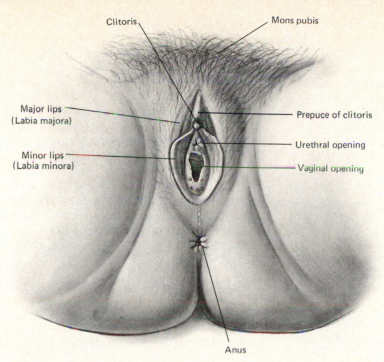

Clitoris   Mons pubis

Major lips
(Labia majora)   Prepuce of clitoris

                                  Urethral opening

Minor lips
(Labia minora)   Vaginal opening

                Anus

FIGURE 2.2 External female genitalia. (From Dienhart, *Basic Human Anatomy and Physiology*. Philadelphia: Saunders, 1967, p. 217.)

Female Sex Organs

External Sex Organs

The genitals of the female are collectively called the *vulva* ("covering").[2] They include the *mons pubis*, the *major* and *minor lips*, the *clitoris*, and the *vaginal introitus* or opening.

The Mons Pubis The mons pubis (or *mons veneris*, "mound of Venus") is the soft, rounded elevation of fatty tissue over the pubic symphysis. After it becomes covered with hair during puberty, the mons is the most visible part of the female genitals.

The Major Lips The major lips or *labia majora* are two elongated folds of skin that run down and back from the mons pubis. Their appearance varies a great deal: Some are flat and hardly visible behind thick pubic hair; others bulge prominently. Ordinarily they are close together, making the female genitals appear "closed."

The major lips are more distinct in front. Toward the anus they flatten out and merge with the surrounding tissues. The outer surfaces of the major lips are covered with skin of a darker color, from which hair grows at puberty. Their inner surfaces are smooth and hairless. Within these folds of skin are bundles of smooth muscle fibers, nerves, and vessels for blood and lymph. The space between the major lips is the *pudendal cleft;* it becomes visible only when the lips are parted (figure 2.2).

[2]The many colloquial names for the female genitals include: cunt, pussy, slit, box, quim, snatch, twat, beaver, and bearded clam. Ancient sex manuals like *The Perfumed Garden* have even more fanciful terms like: crusher, silent one, yearning one, glutton, bottomless, restless, biter, sucker, wasp, hedgehog, starling, hot one, delicious one, and so on (Nefzawi, 1964 ed.). Such designations reveal a good deal about cultural attitudes toward female sexuality.

The Minor Lips   The minor lips or *labia minora* are two lighter-colored hairless folds of skin located between the major lips. The space that they enclose is the *vaginal vestibule* into which open the vaginal and urethral orifices, as well as the ducts of Bartholin's glands. Near the clitoris the minor lips divide in two: The upper portions form a single fold of skin over the clitoris and are called the *prepuce of the clitoris;* the lower portions meet beneath the clitoris as a separate fold of skin called the *frenulum of the clitoris.*

From front to back the main structures enclosed by the minor lips are: the clitoris, the external urethral orifice, and the vaginal orifice. The anus, which is completely separate from the external genitals, lies farther back.

The Clitoris   The clitoris ("that which is closed in"; colloquial names include "clit,"

"button") consists of two masses of erectile spongy tissue called the *corpora cavernosa.* Most of the clitoral body is covered by the upper folds of the minor lips, but its free, rounded tip, called the **glans,** projects beyond it.

The clitoris becomes engorged with blood during sexual excitement. Because of the way it is attached, however, it does not become fully erect as the penis does. Functionally it corresponds more closely to the glans of the penis. It is richly endowed with nerves, highly sensitive, and a very important focus of sexual stimulation, which is its sole function.[3] In some cultures the clitoris has been subjected to ritual mutilation (see box 2.1).

[3]Given its importance for female sexual arousal, the clitoris has recently become the focus of more professional attention. For a detailed study of clitoral anatomy, function, and related considerations, see Lowry and Lowry (1976); Lowry (1978).

Box 2.1   Female Circumcision

The practice of female "circumcision" is far less known generally than its male counterpart. Yet the practice has been widespread in some cultures and continues to be practiced, mainly on the African continent where currently an estimated 30 million women have undergone one or another version of this mutilating procedure.[1]

To be precise, the term *female circumcision* should be restricted to the removal of the prepuce of the clitoris, but it is usually extended to other procedures such as the amputation of the clitoris by *clitoridectomy,* with or without excision of the labia. Sometimes the edges of the excised major lips are made to heal together through *infibulation* which obliterates access to the vaginal area (except for a small opening to let out urine and menstrual blood). This makes coitus impossible; when the woman is deemed entitled to engage in intercourse, the orifice is enlarged by cutting it open.

These practices, like male circumcision, have ancient roots. But while circumcision in no way incapacitates the male, it is a mutilating procedure in the female that seriously interferes with sexual function and poses serious health hazards. Yet the practice is defended on the grounds that cultures have the right to fashion their own rituals.[2]

The Western world has had its own version of female genital mutilation. Early in the 19th century, *declitorization* was a medical procedure for the treatment of female masturbation and "nymphomania" and was used both in Europe and the United States (discussed further in chapter 18).

[1]Remy (1979).
[2]For more detailed accounts of these practices, see Hayes (1975), Huelsman (1976), Paige (1978), and Taba (1979).

Urethral Opening   The external urethral orifice is a small, median slit with raised margins. The female urethra usually conveys urine and is totally independent of the reproductive system. Among some women there may be a discharge of fluid through the urethra during orgasm (chapter 3).

Vaginal Opening   The *vaginal introitus* or orifice is not a gaping hole but rather is visible only when the inner lips are parted. It is easily distinguishable from the urethral opening by its larger size and location. The appearance of the vaginal orifice depends to a large extent on the shape and condition of the **hymen**. This delicate membrane, which exists only in the human female, has no known physiological function, but its psychological and cultural significance, as a sign of virginity, has been enormous. It varies in shape and size and may surround the vaginal orifice, bridge it, or serve as a sievelike cover (figure 2.3).

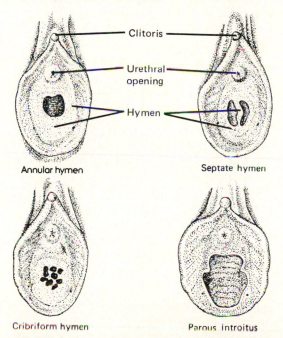

Clitoris

Urethral opening

Hymen

Annular hymen   Septate hymen

Cribriform hymen   Parous introitus

FIGURE 2.3   Types of hymens. The top two and the lower left figures represent intact hymens. The parous introitus shows the remnants of the hymen in a woman who has given birth. (From Netter, *The Reproductive System.* Ciba Pharmaceutical Company, 1965, p. 90.)

There is normally always some opening to the outside.[4]

Most hymens will permit passage of a finger (or sanitary tampon), but usually cannot accommodate an erect penis without tearing. However, a very flexible hymen will occasionally withstand intercourse. This, coupled with the fact that the hymen may be torn accidentally, makes the presence of the hymen unreliable as evidence for virginity. In childbirth the hymen is torn further and only fragments remain attached to the vaginal opening ("parous introitus" in figure 2.3).

When the major and minor lips are dissected away, the musculature of the vaginal area comes into view (figure 2.4). Several sets of **muscles** are important to sexual function in women: The *bulbocavernosus,* the *ischiocavernosus,* the *transverse perineal,* the *levator ani* (whose major inner portion is called the *pubococcygeus*), and the *involuntary smooth muscles* within the vaginal wall itself.

These muscles can be best thought of as constituting a muscular ring around the vagina. Such muscular rings that act to constrict bodily orifices are known as **sphincters.** Though the vaginal sphincter is not a single muscle or as highly developed as, for instance, the anal sphincter, women can voluntarily flex them or unconsciously tense them up, thus compressing the vagina and narrowing its opening. In normal function and in some pathological conditions, the level of control and tension exerted by these muscles is of prime importance (chapters 7 and 15).

Underneath the bulbocavernosus muscles are two elongated masses of erectile tissue called the **vestibular bulbs** (shown on the right in figure 2.4). These structures are connected at their upper ends with the clitoris and like that organ become con-

[4]In rare instances the hymen consists of a tough fibrous tissue that has no opening (*imperforate hymen*). This condition is usually detected after a girl begins to menstruate and the products of successive menstrual periods accumulate, swelling the vagina and uterus. It is corrected by surgery, with no aftereffects. The tearing of the hymen through coitus is referred to as *defloration* ("stripping of flowers").

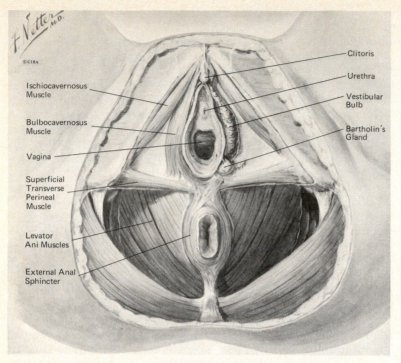

FIGURE 2.4 Musculature of the vaginal area (skin removed). (From Netter, *The Reproductive System.* Ciba Pharmaceutical Company, 1965, p. 92.)

gested with blood during sexual arousal. They too play an important function in the female sexual response cycle and, together with the vaginal sphincter, determine the size, tightness, and "feel" of the vagina (box 2.2).

Internal Sex Organs

The internal sex organs of the female consist of the paired *ovaries,* the uterine or *fallopian tubes,* the *uterus,* the *vagina,* and the paired *bulbourethral glands.*

The Ovaries The ovaries are the *gonads* or reproductive glands of the female. They produce *ova* ("eggs") and sex hormones (estrogens and progestins). In their natural position they lie vertically in the abdomen (figures 2.5 and 2.6) on each side of the uterus, and are about an inch and a half long. The ovaries are held in place by a number of folds and *ligaments;* these ligaments are solid cords and are not to be

confused with the fallopian tubes, which open into the uterine cavity.

The ovary has no tubes leading directly out of it. The ova leave the ovary by bursting through its wall and becoming caught in the fringed end of the fallopian tube. To permit the exit of ova, the surface of the ovarian capsule is quite thin. Before puberty the ovary has a smooth, glistening surface, while after the start of the ovarian cycle and the monthly exodus of ova, its surface becomes increasingly scarred and pitted. (The maturational cycle of the ovum is discussed later in this chapter.)

The Uterine (Fallopian) Tubes The two *uterine,* or *fallopian tubes* are about four inches long and extend between the ovaries and the uterus.[5] The ovarian end of the tube, called the *infundibulum* ("fun-

[5]Named after the 16th-century Italian anatomist, Gabriello Fallopio, who mistakenly thought the tubes were "ventilators" for the uterus.

Box 2.2   Size of the Vagina

The size of the vagina, like that of the penis, has been the object of interest and speculation. Popular notions differentiate between tight and lax vaginas, those that can actively "grasp" the penis and others that are passive. There has been no research to substantiate such notions, some of which sound more plausible than others.

Functionally it makes more sense to consider the vaginal introitus separately from the rest of its body. The vagina beyond the introitus is a soft and distensible organ; although it looks like a flat tube, it actually functions more as a balloon. Normally there is no such thing as a vagina that is permanently "too tight" or "too small." Properly stimulated, any adult vagina can, in principle, accommodate the largest penis.

The claim that some vaginas are too large is more tenable. A vagina may not return to normal size after childbirth, and lacerations produced during the process can weaken the vaginal walls. Even in these instances, however, the vagina expands only to the extent that the penis requires. When we add to these anatomical features the relative insensitivity of the vaginal walls, we can reasonably conclude that the main body of the vaginal cavity does not significantly influence the sexual experience of either partner. In short, most of the time there is no problem of anatomical fit between penis and vagina.

By contrast, the introitus is highly sensitive. The degree of congestion of the erectile tissues of the bulb of the vestibule and the level of tension of muscles that form the vaginal sphincter makes a great deal of difference in how relaxed or tight the vagina feels to the woman and her partner. If these muscles tense up they cause coital discomfort; if too lax, orgasm may not occur. To enhance her sexual experience, a woman can learn to relax or tighten her vaginal muscles and strengthen them with special exercises (chapter 15).

There is a long-standing controversy as to whether the penis can be "trapped" inside the vagina ("penis captivus"). The prevalent view is that this is a misconception arising from the observation of dogs, where this phenomenon does occur. (The penis of the dog expands into a knot inside the vagina and cannot be withdrawn until loss of erection.) But then, occasional reports continue to be published about such phenomena occurring in humans.[1]

[1]For a clinical account of such a case, see Melody (1977, p. 111).

---

nel"), is cone-shaped and fringed by irregular projections, or *fimbriae* ("fringe"), which cling to the ovary but are not attached to it (figure 2.6). After leaving the ovarian surface, the ovum must find its way into the opening of the fallopian tube; a remarkable feat considering that the ovum is about the size of a needle tip and the opening of the uterine tube is a slit about the size of a printed hyphen. The other parts of the fallopian tube are the *ampulla,* the *isthmus,* and the *uterine segment,* which runs within the wall of the uterus itself and opens into its cavity.

The cavity of the fallopian tube becomes progressively smaller between the ovarian end and the uterine end. Its lining is covered with numerous hairlike structures (*cilia*). The ovum, unlike the sperm, does not have its own means of mobility; its movement depends on the sweeping action of these cilia and the contractions of the tube.[6]

The fertilization of the ovum usually occurs in the infundibular third of the uter-

[6]If the ovum were the size of an orange, the cilia would be as big as eyelashes.

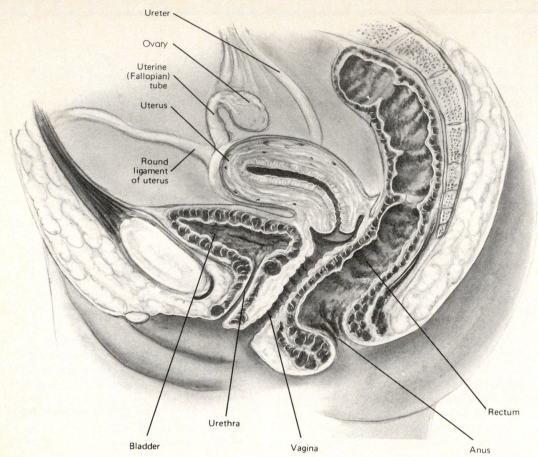

Ureter

Ovary

Uterine
(Fallopian)
tube

Uterus

Round
ligament
of uterus

Bladder

Urethra

Vagina

Anus

Rectum

FIGURE 2.5   The female reproductive system, side view. (From Dienhart, *Basic Human Anatomy and Physiology*. Philadelphia: Saunders, 1967, p. 213.)

ine tube. Although the fallopian tubes are surgically not as accessible as is the vas deferens of the male, they are still the most convenient targets for the sterilization of women.

The Uterus   The *uterus*, or *womb*, is a hollow, muscular organ in which the embryo (referred to as the fetus after the eighth week) is housed and nourished until birth.[7] It is shaped like an inverted pear and is usually tilted forward (anteverted) (figure 2.5).[8] The uterus is held, but not fixed, in place by various ligaments. Normally three inches long and three inches wide at the top, it expands greatly during pregnancy. There is no other body organ that ordinarily undergoes changes in size of comparable magnitude.

The uterus consists of four parts (figure 2.6): the *fundus* ("bottom") is the rounded portion that lies above the openings of the uterine tubes; the *body*, which is the main part; the narrow *isthmus* (not to be con-

[7]The Greek word for uterus is *hystera*, a term that supplies the root for words like "hysterectomy" (surgical removal of the uterus) and "hysteria" (a psychological condition believed by the ancient Greeks to result from the uterus wandering through the body in search of a child).

[8]Attempts at self-abortion or abortion by unqualified individuals often end in disaster because of this anatomical feature. When a probe or long needle is pushed blindly into the uterus, the instrument pierces the roof of the uterus and penetrates the abdominal cavity, causing infection.

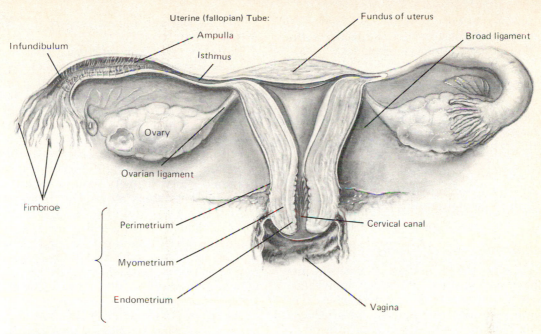

FIGURE 2.6   Internal female reproductive organs, front view. (From Dienhart, *Basic Human Anatomy and Physiology*. Philadelphia: Saunders, p. 215.)

fused with the isthmus of the fallopian tube); and the *cervix* ("neck"), the lower portion of which projects into the vagina.

The cavity of the uterus is widest at the level where the uterine tubes enter it, and narrows toward the isthmus. The *cervical canal* leads into the vagina; its opening is about the size of the lead in a pencil (figure 2.5). The uterus has three layers. The inner *mucosa*, or *endometrium*, consists of numerous glands and a rich network of blood vessels. Its structure varies with the period of life (prepubertal, reproductive, and postmenopausal) and the phases of the menstrual cycle (chapter 4). The second layer, the *myometrium*, is muscular. The third, the *perimetrium* or *serosa*, is the external cover.

The Vagina   The *vagina* ("sheath") is the female organ of copulation. Through it pass the menstrual discharge and the baby during birth, but it does not serve for the passage of urine.

The vagina is ordinarily a collapsed muscular tube, a potential, rather than permanent, space (box 2.2). Its side walls are quite narrow; hence it appears as a narrow slit in side views (figure 2.5). The vaginal canal is slanted downward and forward. Its upper end communicates with the cervical canal; its lower end opens into the vestibule between the minor lips.

The inner lining, or *vaginal mucosa*, resembles the skin inside of the mouth. In contrast to the uterine endometrium, it contains no glands, but during sexual excitement the vaginal mucosa exudes a clear lubricating fluid (chapter 3).

In adult premenopausal women the vaginal walls are corrugated, fleshy, and soft. Following menopause they become thinner and smoother. The vaginal walls are poorly supplied with nerves so that the vagina is a rather insensitive organ except for the area surrounding the vaginal opening, which is highly excitable.[9] There are,

[9]In gynecological examinations conducted for Kinsey, 98% of the subjects were able to perceive tactile stimulation of the clitoris; in contrast, less than 14% could detect being touched in the interior of the vagina (Kinsey et al., 1953).

however, recent claims that the anterior wall of the vagina has an erotically sensitive zone (the *Grafenberg spot*) (see box 3.3).

Bulbourethral Glands   The bulbourethral, or *Bartholin's glands,* are two small structures located behind the vestibular bulbs. Their ducts open on each side in the ridges between the edge of the hymen and the minor lips. These glands were formerly assumed to be the source of vaginal lubrication; they are now considered to play at most only a minor role in this process.

Breasts

The *breasts* are not sex organs, but they have reproductive and erotic significance. They are characteristic of the highest class of vertebrates (mammals), which suckle their young, and their milk-producing organs are called **mammary glands.** Among female primates, only women have large breasts even when not suckling.

Although we generally associate breasts with females, males also have breasts with basically the same structure but which are normally not as well developed. (If a male is given female hormones, he develops female-looking breasts.) The appearance of the female breast normally varies a good deal depending on age, body weight, and other factors. Its size and shape can also be altered through plastic surgery (figure 2.7).[10]

The adult female breast consist of about 15 to 20 lobes or clusters of glandular tissue, each with a separate duct opening on the nipple. The lobes are separated by loosely packed fibrous and fatty tissue, which gives the breast its shape and consistency (figure 2.8).

The *nipple* is the prominent tip of the breast into which the milk ducts open. It has smooth muscle fibers that make it erect in response to stimulation. The *areola* is the circular area around the nipple. The nipples, richly endowed with nerve fibers, are highly sensitive and play an important part in sexual arousal. In some cases, the nipple may be pushed inwards or inverted. This condition does not usually interfere with its normal functions.

The size and shape of breasts have no bearing on their erotic responsiveness. Furthermore, a smaller-breasted woman is

[10]Plastic surgery can be used to make breasts larger, smaller, and to correct asymmetries (that occur naturally) and deformities that may follow breast surgery. Techniques of breast enlargement that rely on liquid silicone injections or "inflatable" implants have led to numerous complications. A much safer approach utilizes soft silicone implants, whereby the materials introduced into the breast are encapsulated in an inert sac and do not come into direct contact with the breast tissue.

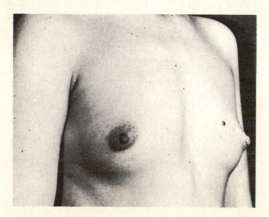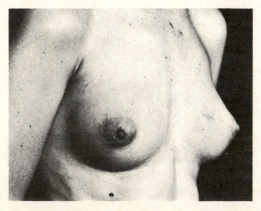

FIGURE 2.7   Breast augmentation through the implantation of silicone gel prosthesis. (Courtesy of Dr. Donald R. Laub.)

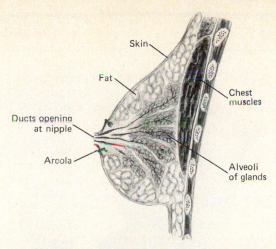

Skin

Fat

Chest
muscles

Ducts opening
at nipple

Areola

Alveoli
of glands

FIGURE 2.8 Internal structure of the breast. (From Dienhart, *Basic Human Anatomy and Physiology*. Philadelphia: Saunders, 1967, p. 217.)

not necessarily any less capable of breast-feeding than a larger-breasted woman. Many women, but not all, and some men, find stimulation of the breasts and nipples sexually arousing. In addition to personal idiosyncrasy, the sensitivity of the breasts has been shown to depend on the hormone levels, which fluctuate with the menstrual cycle and in pregnancy.

The size and shape of female breasts have no functional significance, yet such attributes tend to greatly influence their erotic appeal and a woman's aesthetic image of herself. Though such culturally shaped judgments are quite arbitrary, extremes in size in either direction generally tend to cause self-consciousness.

The female breasts develop during puberty (chapter 4), and sometimes one grows faster than the other. The resulting asymmetry may be disturbing, but eventually the two sides become approximately equal in size. With age, the breasts undergo other natural changes. As their supporting ligaments stretch, they tend to sag; following the menopause, they become smaller and less firm. Such changes, though physiologically normal, may have psychological repercussions for some. What to do, if anything, about these changes is a matter of personal choice. Exercises, creams, and

other popularly advertised methods do not demonstrably augment breast size, but plastic surgery can be quite effective.

Male Sex Organs

External Genitals

The external sex organs of the male consist of the *penis* and *scrotum* (figure 2.9). *Phallus* (the Greek name for penis) is often used in contexts where a less starkly anatomical designation is preferred.[11]

The Penis The penis ("tail" in Latin) is the male organ for copulation and urination. It contains three parallel cylinders of spongy tissue, through one of which runs the *urethra* that conveys both urine and semen (figure 2.9).

The three cylinders of the penis are structurally similar. Two of them are called the cavernous bodies (*corpora cavernosa*), and the third, which contains the urethra, the spongy body (*corpus spongiosum*) (figure 2.10). Each cylinder is wrapped in a fibrous coat, but the cavernous bodies have an additional common covering that makes them appear to be a single structure for most of their length. When the penis is flaccid, these bodies cannot be seen or felt as separate structures, but in erection the spongy body stands out as a distinct ridge on the underside of the penis.

As the terms "cavernous" and "spongy" suggest, the penis consists of an agglomeration of irregular cavities and spaces very much like a dense sponge. These tissues are served by a rich network of blood vessels and nerves. When the penis is flaccid, the cavities contain little blood. During sexual arousal they become engorged, and their constriction within their tough fi-

[11]Many colloquial terms exist for the penis, such as: prick, poker, pecker, rod, tool, cock, dick, dong, joy stick, boner, weenie (Haeberle, 1978, p. 491). *The Perfumed Garden* has more exotic descriptions like housebreaker, ransacker, rummager, pigeon, shamefaced one, the indomitable, and swimmer (Nefzawi, 1964 ed., pp. 156–157).

Box 2.4   Size of the Penis

The average penis is three to four inches when flaccid and about twice as long in erection. Its diameter in the relaxed state is about 1¼ inches with an increase of another ¼ inch in erection. Penises can, however, be consideraby smaller or larger.

Normal variation in size and shape among individuals is the rule for all parts of the human body. Nevertheless, the size and shape of the penis are often the cause of special curiosity and amusement as well as apprehension and concern. Representations of enormous penises can be found in numerous cultures, including some from remote antiquity (see accompanying figure). These anatomical exaggerations are generally not caricatures or monuments to male vanity but symbols of fertility and power. Symbolic representations of the penis have often been used for religious and magical functions (see chapters 17 and 18).

The size and shape of the penis, contrary to popular belief, are not related to a man's body build, race, virility, or ability to give and receive sexual satisfaction. Furthermore, variations in size tend to be less in the erect state: the smaller the flaccid penis, the proportionately larger it tends to become when erect. The penis does not grow larger through frequent use or "exercise."

Mochica pottery. Courtesy of William Dellenback, Institute for Sex Research.

brous coats causes the characteristic stiffness of the erect penis.

The smooth, rounded head of the penis is known as the **glans** ("acorn" in Latin). The glans is formed entirely by the free end of the spongy body, which expands to shelter the tips of the cavernous bodies. Like the clitoris, the glans penis has particular erotic importance. It is richly endowed with nerves and highly sensitive. At its underside, a thin strip of skin (*frenulum*) connects it to the adjoining body of the penis. At its rim (*corona*), the glans slightly overhangs the *neck* of the penis, which forms the boundary between the body of the penis and the glans. At the tip

of the glans is the longitudinal slit for the urethral opening (*meatus*).

The skin of the penis is hairless and unusually loose, which permits expansion during erection. Although the skin is fixed to the penis at its neck, some of it folds over and covers part of the glans (like the sleeve of an academic gown), forming the **prepuce,** or foreskin. Ordinarily the prepuce is retractable and the glans readily exposed. Circumcision is the excision of the prepuce. In the circumcised penis the glans is always totally exposed (see box 2.3).

There are small glands in the corona and the neck that produce a soft cheesy substance called *smegma* that has a distinctive

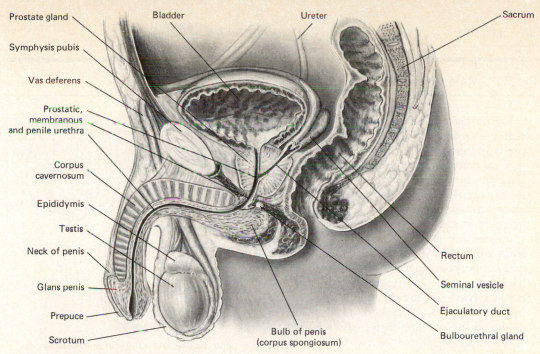

Prostate gland

Symphysis pubis

Vas deferens

Prostatic,
membranous
and penile urethra

Corpus
cavernosum

Epididymis

Testis

Neck of penis

Glans penis

Prepuce

Scrotum

Bladder

Ureter

Sacrum

Rectum

Seminal vesicle

Ejaculatory duct

Bulbourethral gland

Bulb of penis
(corpus spongiosum)

FIGURE 2.9   The male reproductive system. (From Dienhart, *Basic Human Anatomy and Physiology*. Philadelphia: Saunders, 1967, p. 207.)

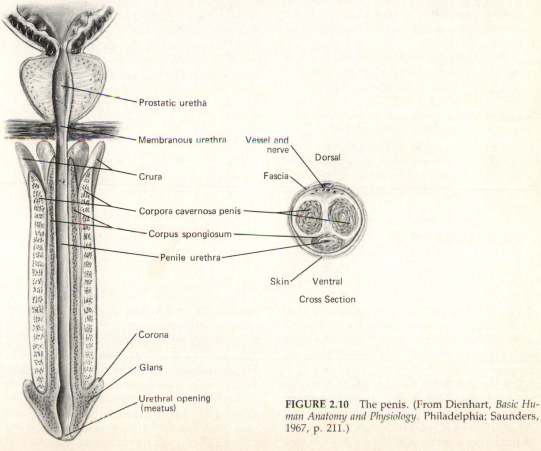

Prostatic urethra

Membranous urethra

Crura

Corpora cavernosa penis

Corpus spongiosum

Penile urethra

Corona

Glans

Urethral opening
(meatus)

Longitudinal Section

Vessel and
nerve

Dorsal

Fascia

Skin   Ventral

Cross Section

FIGURE 2.10   The penis. (From Dienhart, *Basic Human Anatomy and Physiology*. Philadelphia: Saunders, 1967, p. 211.)

Box 2.3   Male Circumcision

Circumcision ("cutting around") is the excision of the foreskin and is practiced around the world as a religious ritual as well as a medical measure. In the circumcised male the glans and the neck of the penis are completely exposed (see accompanying figure).

As a ritual, circumcision was performed in Egypt as early as 4,000 B.C. It thus long antedates the practice among Jews, Moslems, and other groups.[1] Circumcision for medical purposes in the United States dates back to the 19th century. Its original justification was to help combat masturbation. After this rationale was discredited, its advocates endorsed the practice on the grounds that circumcision prevented the accumulation of smegma under the prepuce. Further support for the practice came from reports that cancer of the penis appeared to be less frequent among circumcised men, and cancer of the cervix to be less common among their spouses.

The validity of these associations has now been called into question and physicians are currently divided as to the necessity of routine circumcision in infancy (see chapter 7). Circumcision remains a medical necessity however, whenever the foreskin is so tight that it cannot be easily retracted over the glans. But this condition, called *phimosis*, is rare and impossible to predict in infancy since it usually takes several years for the foreskin to become retractable.[2]

It is often assumed that because of his fully exposed glans penis, the circumcised male is more rapidly aroused during coitus and

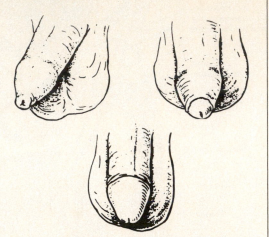

The penis before and after circumcision. (From Lauerson and Whitney, *It's Your Body*. Grosset and Dunlap, 1977.)

more likely to ejaculate prematurely. Current research has failed to support this belief: There seems to be no difference between the excitability of the circumcised and uncircumcised penis.[3] For some circumcised men, the absence of the prepuce seems to become a source of preoccupation. They therefore seek reconstruction of their foreskin through plastic surgery.[4]

[1]The basis for the practice among Jews is set forth in Genesis 17:9–15: "You shall circumcise the flesh of your foreskin, and it shall be the sign of the covenant between us."

[2]For a critical review and discussion of the function of circumcision, see Paige (1978b).

[3]Masters and Johnson (1966), p. 190.

[4]Mohl et al. (1981); Mohl et al. (1983).

smell. This is purely local secretion that accumulates under the prepuce. It has no known function and must not be confused with semen, which is discharged through the urethra.

The human penis (unlike that of the dog) has no bone. Nor does it have voluntary muscles within it. The *bulbocavernosus* and *ischiocavernosus* muscles surround the bulb and the crura externally (figure 2.11). Their function is primarily in connection with urination and ejaculation; they play no significant role in erection.

The Scrotum   The scrotum is a multilayered pouch. Its skin is darker than the rest of the body, has many sweat glands, and at puberty becomes sparsely covered with hair. Underneath it there is a layer of loosely organized muscle fibers (*dartos muscle*) and

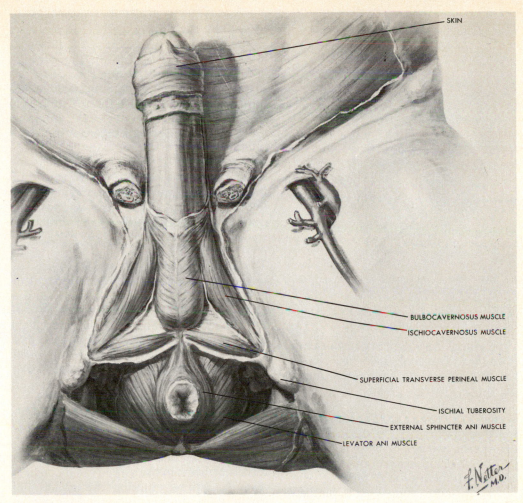

FIGURE 2.11 Male external genitalia (skin removed). (From Netter, *The Reproductive System*. Ciba Pharmaceutical Company, 1965, p. 11.)

fibrous tissue. These muscle fibers are not under voluntary control, but contract in response to cold, sexual excitement, and other stimuli, making the scrotum appear compact and heavily wrinkled. Otherwise the scrotum hangs loose and its surface is smooth. When the inner side of the thigh is stimulated, the dartos muscle contracts reflexively (*cremasteric reflex*).

The scrotal sac contains two separate compartments, each of which houses a testicle and its ***spermatic cord.*** The spermatic cord enters the abdominal cavity from the scrotal sac by traversing a region of the abdominal wall called the ***inguinal canal.***

The spermatic cord includes the vas deferens that carries sperm from the testicle. It also contains blood vessels, nerves, and muscle fibers. When these muscles contract, the spermatic cord shortens and pulls the testicle upward within the scrotal sac; an important feature in sexual arousal (chapter 3).

Internal Sex Organs

The internal sex organs of the male consist of a pair of ***testes*** or testicles, with their duct system for the storage and transport of sperm consisting of ***epididymis, vas def-***

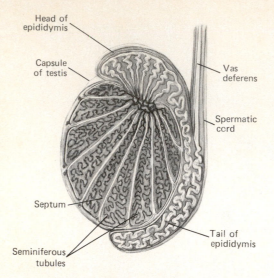

Labels: Head of epididymis; Capsule of testis; Septum; Seminiferous tubules; Vas deferens; Spermatic cord; Tail of epididymis

FIGURE 2.12 Testis and epididymis. (From Dienhart, *Basic Human Anatomy and Physiology*. Philadelphia: Saunders, 1967, p. 207.)

erens, and *ejaculatory duct* in pairs and the single *urethra.* The male also has paired *seminal vesicles* and *bulbourethral glands* and the single *prostate gland.*

The Testes   The testes are the gonads, or reproductive glands, of the male.[12] They produce sperm and sex hormones (androgens). The two testicles are about the same size (about two inches long), although the left one usually hangs somewhat lower than the right (a fact noted by classical sculptors). Testicles weigh about one ounce each and tend to shrink somewhat in old age.

Each testicle is enclosed in a tight, whitish fibrous sheath that thickens at the back of the organ and penetrates inside the testicle subdividing it into conical *lobes* (figure 2.12). Each lobe is packed with convoluted *seminiferous* ("sperm-bearing") *tubules.* These threadlike structures are the sites at which sperm are produced (figure 2.13). Each seminiferous tubule is one to three feet long, and the combined length of the tubules of both testes extends to over a quarter of a mile!

This elaborate system of tubules allows

[12]"Testis" is derived from the root for "witness"; based on the ancient custom of placing the hand on the genitals when taking an oath; hence "testify."

for the production and storage of billions of sperm. The process of *spermatogenesis,* or sperm production (discussed farther on), takes place exclusively within the seminiferous tubules, and the other major testicular function, the production of male sex hormone, occurs outside them. The testicular cells that produce the male hormone are found in the spaces between the seminiferous tubules, and are known as *interstitial cells* (or *Leydig's cells*). The cells responsible for the two primary functions of the testes are thus quite separate and not in physical contact (but the hormones produced by Leydig's cells are essential for spermatogenesis).

The Epididymis   The seminiferous tubules converge into an intricate maze of ducts that culminate in a single tube, the *epididymis.* The epididymis ("over the testis") is a remarkably long tube (about 20 feet) that is so tortuous and convoluted that it appears as a C-shaped structure not much longer than the testis to whose surface it adheres (figures 2.9 and 2.12).

The Vas Deferens   The vas deferens, or *ductus deferens* ("the vessel that brings down"), is the less tortuous and much shorter continuation of the epididymis. It travels upward in the scrotal sac for a short distance before entering the abdominal cavity; its portion in the scrotal sac can be felt as a firm cord. The fact that the vas is surgically so accessible makes it the most convenient target for sterilizing men (chapter 6).

The terminal portion of the vas deferens expands into the *ampulla* ("flask"). At the back of the urinary bladder, it narrows and joins the duct of the seminal vesicle to form the ejaculatory duct (figure 2.9). This duct is a very short (less than one inch), straight tube that runs its entire course within the prostate gland and opens into the prostatic portion of the urethra.

The Urethra   The urethra has a dual function in the male, conveying both semen and urine. (It must not be confused with the two *ureters,* each of which begins in one kidney and carries urine into the blad-

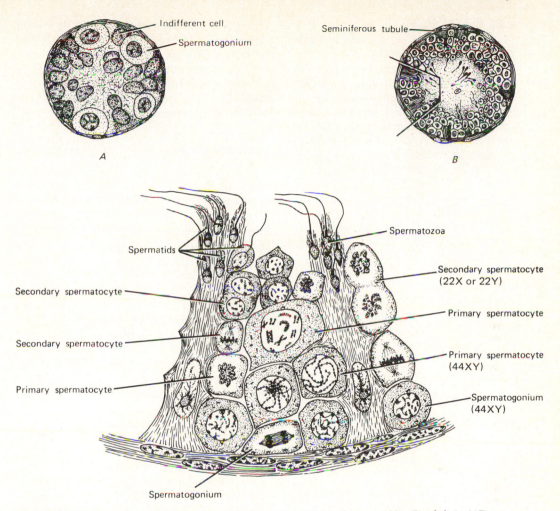

FIGURE 2.13 Human testis tubules, in transverse section. (*A*) Newborn (×400); (*B*) adult (×115); (*C*) detail of the area outlined in (*B*) (×900) (From Arey, *Developmental Anatomy,* 7th ed. Philadelphia: Saunders, 1965, p 41.)

der.) The urethra is about eight inches long and is subdivided into ***prostatic, membranous,*** and ***penile*** segments (figures 2.9 and 2.10). Into the prostatic portion of the urethra open the two ***ejaculatory ducts*** and the multiple sievelike openings of the ***ducts of the prostate gland*** (figure 2.10). It is here that the various components of semen coming from the testes, seminal vesicles, and the prostate gland are admixed prior to ejaculation.

The fact that semen and urine use the same urethral passage necessitates that the two not get mixed together. This is accom-

plished by the actions of two urethral sphincters: an *internal sphincter* located at the neck of the bladder (where the urethra enters the prostate), and an *external sphincter* that closely surrounds the membranous urethra (right below where the urethra exits from the prostate gland).

During urination, the internal sphincter and the external sphincter (which is under voluntary control after infancy) relax and the pressure of the bladder wall pushes urine out. During ejaculation, the internal sphincter remains closed, the external one open, thus allowing semen to flow out of the penis, instead of into the bladder. (This pattern is reversed in retrograde ejaculation, discussed in chapter 3.)

The penile part of the urethra pierces the bulb of the corpus spongiosum, traverses its whole length, and terminates at the tip of the glans in the external urethral opening or meatus. The external urethral opening has no sphincter, which is why at the end of urination the urine left in the penile portion must be squirted out through the contraction of the bulbocavernosus and ischiocavernosus muscles.

Accessory Organs Three accessory organs perform auxiliary functions in the male. They are the prostate gland, two seminal vesicles, and two bulbourethral glands.

The *prostate* is an encapsulated structure about the size and shape of a large chestnut located with its base against the bottom of the bladder (figure 2.9). It consists of three lobes that contain smooth muscle fibers and glandular tissue whose secretions account for much of the seminal fluid and its characteristic odor.

The *seminal vesicles* are two sacs, each about two inches long (figure 2.9). Each ends in a straight, narrow duct, which joins the tip of the vas deferens to form the ejaculatory duct. The function of the seminal vesicles was assumed to be the storage of sperm (each can hold about a teaspoon of fluid), but it is currently believed to be primarily involved in contributing fluids rich in nutrients that initiate the motility of sperm through the whiplike action of their tails.

The *bulbourethral glands* (*Cowper's glands*) are two pea-sized structures flanking the penile urethra, into which each empties through a tiny duct (figure 2.10). During sexual arousal these glands secrete a clear, sticky fluid that appears in droplets at the tip of the penis.[13] There usually is not enough of this secretion to serve as a coital lubricant. However, as it is alkaline, it may help to neutralize the acidity of the urethra (which is harmful to sperm) prior to the passage of semen. Although this secretion must not be confused with semen, it can contain stray sperm.

Gametogenesis

Sperm and ova are the germ cells (*gametes*) of the male and female, and constitute their reproductive elements.[14] The production of germ cells (*gametogenesis*) follows the same basic principles in both sexes, but the resultant sperm and egg are quite dissimilar cells, each with its own specialized function.[15]

Gametogenesis, like other processes of cell multiplication, relies on the process of *mitosis.* But in addition, germ cells undergo a special kind of reduction division called *meiosis* (Greek for "less"), through which their number of *chromosomes* is halved. All human cells (other than ova and sperm) have 46 chromosomes.[16] These consist of 22 pairs of *autosomes* that are similar in both sexes, plus a pair of *sex chromosomes* that differ: The cells of the female body have two X sex chromosomes, those of males one X and one Y sex chromosome. The genetic configuration (*genotype*) of female body cells is thus 44XX; that of male body cells, 44XY. Ova have $22 + X$; sperm have either $22 + X$ or $22 + Y$ chromosomes. Thus, when sperm and egg merge during fertil-

---

[13]Medieval theologians called it the "distillate of love." A Latin poet wrote the following epigram (quoted in Ellis, 1942, vol. 2, part I, p. 153):

> You see this organ . . . is humid
> This moisture is not dew nor drops of rain,
> It is the outcome of sweet memory
> Recalling thoughts of a complacent maid.

[14]Sperm were discovered by Anton van Leeuwenhoek (1632–1723), who conducted some of the first microscopic investigations in biology. He called them "spermatozoa" ("seed animals") since sperm looked like other life forms swimming in pond water. It took another century before the reproductive role of sperm was firmly established. The ovum was discovered in 1827 by Karl Ernst von Baer (1792–1876), the founder of modern embryology, who first identified the ovum in the ovaries of dogs (Stern, 1973).

[15]Sources for the following discussion are Moore (1982); Bloom and Fawcett (1975); Ham and Cormack (1979); and Arey (1974).

[16]Chromosomes are part of the nucleus of plant and animal cells. They carry genes, which are responsible for the transmission of hereditary characteristics.

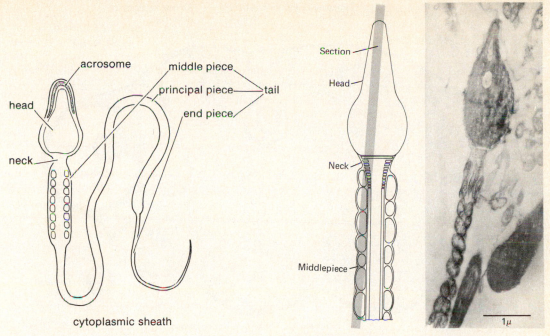

FIGURE 2.14   A human sperm. (Left) Component parts. (From Moore, *The Developing Human*, 3rd ed. Philadelphia: Saunders, 1982, p. 15.) (Right) Electronmicrograph of a section through head and middlepiece. (From Lord Rothschild, *British Medical Journal*, 1, 1958, p. 301.)

ization, the normal number of chromosomes is recreated instead of doubled (chapter 5).

Spermatogenesis

Sperm production (*spermatogenesis*) takes place within seminiferous tubules starting at puberty, prior to which the tubules consist of solid cords with dormant germ cells. A cross section of a tubule shows germ cells in various stages of development (figure 2.13). Sperm that are fully formed are released into the lumen of the tubule and transported to the epididymis, where they mature further while awaiting ejaculation. Spermatogenesis is in progress simultaneously in all seminiferous tubules; thus generations of sperm reach maturity in successive waves. The process takes 64 days and cycles follow each other uninterruptedly. The developing spermatozoa are provided support, protection, and nutrition by *Sertoli cells,* which are elongated structures interspersed among them.

Spermatogenesis consists of several phases. In the first phase, the earliest cell in the maturational chain, the *spermatogonium,* multiplies through mitosis and is transformed into the larger *primary spermatocyte,* but with no change in the number of chromosomes. In the second phase, each primary spermatocyte undergoes meiosis (through two specialized divisions called the first and second maturation divisions) giving rise to two *secondary spermatocytes,* which in turn each split into two *spermatids.* These cells now have half the normal complement of chromosomes: two are 22 + X; two are 22 + Y. The third phase (*spermiogenesis*) entails no further division, but involves an extensive process of differentiation that transforms spermatids into sperm.

Mature *sperm* consist of a *head, middle piece,* and *tail* (figure 2.14). The head contains the chromosomes and is the only part involved in fertilization. The tail is responsible for locomotion through its whiplash movements. Sperm are smaller than a tenth of a millimeter and therefore are not visible to the naked eye.

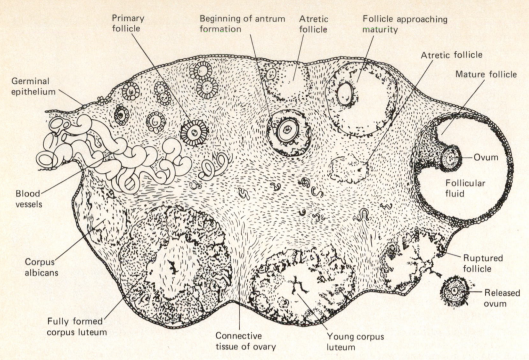

Primary
follicle

Beginning of antrum
formation

Atretic
follicle

Follicle approaching
maturity

Atretic follicle

Mature follicle

Germinal
epithelium

Ovum

Follicular
fluid

Blood
vessels

Ruptured
follicle

Released
ovum

Corpus
albicans

Fully formed
corpus luteum

Connective
tissue of ovary

Young corpus
luteum

FIGURE 2.15   Diagram of ovary, showing maturation and rupture of follicle, luteinization, and regression of the corpus luteum. Nonruptured follicles degenerate into atretic follicles. (From Patten, *Foundations of Embryology,* 3rd edition, New York: McGraw-Hill, 1974, p. 51.)

Development of the Ovum

Unlike men who produce billions of sperm during their lifetime, women are born with all the primordial ova they will ever produce. These number some 400,000, of which 40,000 will survive to puberty; of these, about 400 will reach maturity during a woman's reproductive lifetime, but only a few of which will ever get fertilized.

The first cell in the chain of the maturation of the ovum, (the process of *oogenesis*) is the **primary oocyte,** which with its surrounding cells constitutes the **primary follicle.** These primary follicles remain dormant until puberty when, each month, several of them begin to mature. Usually one of them gets ahead and becomes progressively larger (while the others regress), culminating in a *mature follicle* (*Graafian follicle*) (figure 2.15). This is a liquid-filled vesicle containing the ovum and *granulosa cells* that surround it and line the rest of the follicle wall (it is these granulosa

FIGURE 2.16   The human egg. The zona pellucida is the clear circle, the corona radiata the outer layer of cells. A human sperm is in the lower right corner, showing the relative size of the two cells. (From Arey, *Developmental Anatomy,* 7th ed. Philadelphia: Saunders, 1974.)

cells that produce estrogen). During ovulation the follicle bursts out ejecting the ovum (chapter 5). The rest of the follicle is then transformed into the **corpus luteum** ("yellow body"), which takes over the production of estrogen and starts producing progesterone (chapter 4). At the end of the ovarian cycle, the corpus luteum turns into scar tissue (the "white body" or *corpus albicans*).

The ovum is one of the largest cells in the body, next to which the sperm looks minuscule (figure 2.16). Nonetheless the ovum is still scarcely visible to the naked eye, having a diameter of about half a millimeter and a weight of one-twenty-millionth of an ounce.[17]

The *nucleus* of the ovum, which contains the genetic material, is surrounded by a large amount of *cytoplasm*. This is necessary for sustaining life after fertilization; the sperm has a minimal amount of cytoplasm, since to cover its long journey it must travel light. A clear-appearing area, the *zona pellucida*, is a protective layer around the ovum which in turn is surrounded by follicular cells that remain attached to it after ovulation and form the *corona radiata* ("radiating crown").

Development of the Reproductive System

The genital system of both sexes makes its appearance during the fifth to sixth week of intrauterine life, when the embryo has attained a length of 5 to 12 millimeters.[18] At this undifferentiated stage the embryo has a pair of **gonads,** two sets of **genital ducts,** and a **urogenital sinus**—a common opening to the outside for the genital ducts

and the urinary tract (figure 2.17). It also has the rudiments of external genitals (figure 2.18).

At this time one cannot reliably determine the sex of the embryo by either gross or microscopic examination; the gonads have not yet become either testes or ovaries, and the other structures are also undifferentiated. This lack of visible differentiation does not mean that the sex of the individual is still undecided; genetic sex is determined at the moment of fertilization and depends on the chromosome composition of the fertilizing sperm: If the fertilizing sperm carries a Y chromosome, the issue will be male, otherwise it will be female (chapter 5).

Differentiation of the Gonads

The gonad that is destined to develop into a testis gradually consolidates into a more compact organ under the influence of the Y chromosome. Testicular-differentiating *genes* on the Y chromosome play the dominant role in initiating fetal testicular differentiation, but genes on other chromosomes (especially the X chromosome) also play a part in this process.[19] In response, gonadal cells are organized into distinct strands (*testis cords*), the forerunners of the seminiferous tubules. By about the seventh week the organ is sufficiently differentiated to be recognizable as a developing testis. If by this time the basic architecture of the future testis is not discernible, it can be provisionally assumed that the undifferentiated gonad will develop into an ovary. More definitive evidence that the baby will be a girl comes at about the tenth week, when the forerunners of the follicles begin to become visibly organized.

Testosterone produced by the Leydig cells in the embryonal testis is necessary to promote the maturation of seminiferous tubules. The embryonal ovary also produces estrogenic hormones, but their role in the further development of ovaries is unclear.

[17]All the ova needed to repopulate the world would fill two gallons while all the sperm needed for the same purpose would fit into an aspirin tablet. The volume of DNA needed to produce the entire next generation of the world is less than one-tenth of an aspirin tablet (Stern, 1973).

[18]Discussion in this section is based on Langman (1981); Wilson et al. (1981); Gordon and Ruddle (1981); and Moore (1982). The modern understanding of the differentiations of the reproductive system was envisioned by Alfred Jost (1953).

[19]Bernstein (1981). For further discussion of the chromosomal basis of sex differentiation, see Gordon and Ruddle (1981); Haseltine and Ohno (1981).

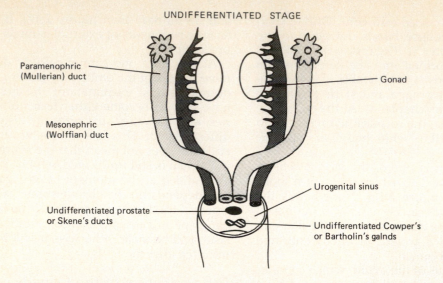

UNDIFFERENTIATED STAGE

Paramenophric
(Mullerian) duct

Gonad

Mesonephric
(Wolffian) duct

Urogenital sinus

Undifferentiated prostate
or Skene's ducts

Undifferentiated Cowper's
or Bartholin's galnds

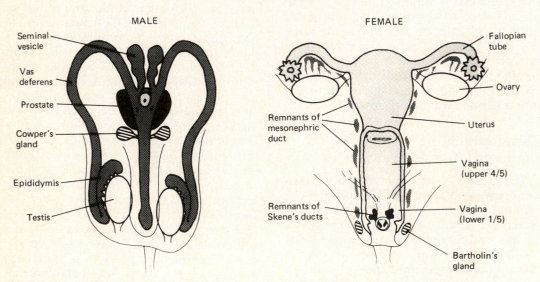

DIFFERENTIATED STAGE

MALE

FEMALE

Seminal
vesicle

Vas
deferens

Prostate

Cowper's
gland

Epididymis

Testis

Fallopian
tube

Ovary

Remnants of
mesonephric
duct

Uterus

Vagina
(upper 4/5)

Remnants of
Skene's ducts

Vagina
(lower 1/5)

Bartholin's
gland

FIGURE 2.17   Differentiation of the internal reproductive organs into male and female forms.

Differentiation of
the Genital Ducts

In the undifferentiated stage, the embryo has two sets of ducts: the *paramesonephric* or *Mullerian* (the potential female) ducts and the *mesonephric* or *Wolffian* (the potential male) ducts (figure 2.17).

Just as the Y chromosome directs the development of the undifferentiated gonad into a testis, the embryonal testes in turn determine the future development of the genital ducts. This is accomplished by two hormones: *testosterone,* produced by Leydig cells, which promotes the further differentiation of the Wolffian duct system,

UNDIFFERENTIATED STAGE

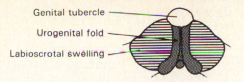

Genital tubercle
Urogenital fold
Labioscrotal swelling

PARTIALLY DIFFERENTIATED STAGE

MALE   FEMALE

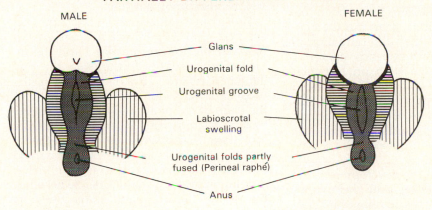

Glans

Urogenital fold

Urogenital groove

Labioscrotal
swelling

Urogenital folds partly
fused (Perineal raphé)

Anus

FULLY DIFFERENTIATED STAGE

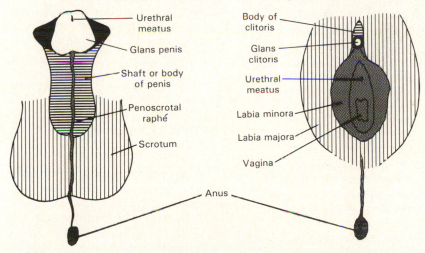

Urethral
meatus

Glans penis

Shaft or body
of penis

Penoscrotal
raphé

Scrotum

Body of
clitoris

Glans
clitoris

Urethral
meatus

Labia minora

Labia majora

Vagina

Anus

FIGURE 2.18 Differentiation of the external genitalia into male and female forms.

and a second substance called *Mullerian regression hormone* produced by Sertoli cells, which inhibits the further development of the Mullerian ducts. As a result the Wolffian duct on each side eventually becomes the epididymus, vas deferens, and seminal vesicle, while the Mullerian ducts degenerate.

In the absence of these testicular hormones (and without the need of ovarian hormones) the Wolffian ducts degenerate and the Mullerian ducts form the fallopian tubes, uterus, and the upper two-thirds of the vagina. The lower third of the vagina, the bulbourethral glands and the urethra in both sexes, and the prostate gland are derived from the urogenital sinus that is part of the embryonal urinary system.

The precise mechanisms involved in this process are not entirely clear, but the general principle holds that differentiation of the reproductive tract depends on the Y chromosome first and then on testicular hormones. Otherwise the undifferentiated system will develop into the female pattern. Thus, irrespective of genetic makeup, if the gonads are removed early in life, the reproductive tract will develop the female pattern.

Concomitant with the developments described, both the testes and the ovaries undergo gross changes in shape and position that are of significance. At first both of them are slender structures high up in the abdominal cavity. By the tenth week they have grown and shifted down to the level of the upper edge of the pelvis. There the ovaries remain until birth; they subsequently rotate and move farther down until they reach their adult positions in the pelvis. In the male this early internal migration is followed by the further descent of the testes into the scrotal sac. After the descent of the testes, the passage is obliterated. In about 2% of male births, one or both of the testes fail to descend into the scrotum before birth (*cryptorchism*). In most of these boys the testes do descend by puberty. Otherwise hormonal or surgical intervention becomes necessary since the higher temperature of the abdominal cavity would interfere with spermatogenesis, resulting in sterility. Undescended testes are also more likely to develop cancer.[20]

[20]Another problem arises when the passage traversed by the testes is not closed or reopens when the tissues become slack in old age, resulting in *inguinal hernia,* or *rupture,* which is readily corrected by surgery.

Differentiation of the External Genitals

The external genitals are also, at first, sexually undifferentiated (figure 2.18). Even after the gonads begin to be distinguishable by the second month of life, several more weeks are necessary for the more distinctive development of the external genitals; by four months the sex of the fetus becomes unmistakable.

The main components of the undifferentiated external genitalia are: the *genital tubercle;* the *urogenital folds;* and the *labioscrotal swellings.* In the male, the genital tubercle grows into the glans penis; the urogenital folds elongate and fuse to form the body of the penis and urethra; and the fusion of labioscrotal swellings forms the scrotal sac. The external genitalia in the female undergo relatively less marked changes in appearance during their further development: the genital tubercle becomes the clitoris; the urogenital folds, the labia minora; and the labioscrotal swellings, the labia majora.

The differentiation of the external genitalia also takes place under the influence of androgen. As with the development of the internal sex organs, the presence of androgen leads to the development of the male pattern; its absence results in the female pattern. However, while testosterone is responsible for the masculinization of the internal reproductive system, that of the external genitalia depends on one of its derivative hormones, called *dihydrotestosterone.*

Since the reproductive systems of male and female develop from the same embryonal origins, each part has its developmental counterpart, or **homologue** in the other sex. With the aid of figures 2.17 and 2.18 we can readily match the homologous pairs of organs in male and female. In principle it should be possible to identify the homologue to every part, even though the degenerated remnants of the Wolffian ducts in the female and the Mullerian ducts in the male are inconsequential structures.

The homologous pairs of organs and parts that are clearly functional in both sexes are as follows: testis—ovary; Bartholin's gland—Cowper's gland; glans penis—glans of clitoris; corpora cavernosa of penis—corpora cavernosa of clitoris; corpus spongiosum of penis—bulb of the vestibule; underside of penis—labia minora; scrotum—labia majora.[21] The homologue of the prostate, the urethral Skene's glands in the female, may be functional in some women, allowing them to "ejaculate" during orgasm as we discuss in the next chapter.

[21]A complete listing of all homologous pairs may be found in Moore (1973, p. 216). Colored drawings that more dramatically show the correspondence of the homologous pairs are shown in Netter (1965).

SUGGESTED READINGS

Netter, F. H. *Ciba Collection of Medical Illustrations, Volume 2: Reproductive System* (1965). Profusely illustrated with attractively executed colored plates.

Nilssen, L. *Behold Man* (1973). Superb photographs of reproductive organs in color.

Kessel, R. G., and Kardon, R. H. *Tissues and Organs: A Text-Atlas of Scanning Electron Microscopy* (1979). The reproductive system is covered in chapters 15 and 16.

Dickinson, R. L. *Human Sex Anatomy* (1949; first published in 1933). An extraordinary collection of sketches and measurements of the sex organs.

Langman, J. *Medical Embryology*, 4th edition (1981) and Moore, K. L. *The Developing Human*, 3rd edition (1982). The embryonal development and differentiation of the genital system is succinctly described and well illustrated.

Chapter 3
Sexual Physiology

Whatever the poetry and romance of sex and whatever the moral and social significance of human sexual behavior, sexual responses involve real and material changes in the physiological functioning of an animal.

Alfred C. Kinsey

This chapter is primarily concerned with physiological processes that mediate sexual activity. When discussing anatomy, we were concerned with *structure*, whereas with the physiological approach we will be dealing with *function*—how the sexual machinery and its control mechanisms operate. Ultimately our interest is in sexual *behavior*, which refers to the activities of the individual, or the organism, as a whole.

There is a regrettable, albeit understandable, tendency to make sharp distinctions between the *physiological* aspects of sex on the one hand and its *psychological* aspects on the other. It is as if mind and body each had a separate existence, or physiological and psychological factors acted as independent causes of sexual behavior. But as Frank Beach explains, that is not the case:

> Physiology and psychology relate to different levels of organization and not to different kinds of causal agents. At the physiological level we study the organization and interrelations of organs and organ systems; at the psychological level we concentrate upon functions of the total individual as he relates to his physical and social environment.[1]

It is common knowledge that a typical

episode of sexual experience begins with mounting excitement, culminates in orgasm, and is followed by sexual satiety. In more formal descriptions various stages have been ascribed to this process. At the turn of the century, Havelock Ellis envisaged two phases: tumescence and detumescence, based on the process of erection preceding orgasm and loss of erection following it. More recently, Masters and Johnson have proposed four phases: excitement, plateau, orgasm, and resolution. Such subdivisions are meant to facilitate description, not to obscure the basic continuity of sexual activity, where the manifestations of various phases overlap and merge into one another imperceptibly.

Sexual Stimulation

The ability to respond to sexual stimulation is a universal human characteristic. Although the nature of the stimulus varies widely, the basic physiological responses of the body are quite consistent. The varieties and intensity of sexual arousal that each person experiences during a lifetime are legion; virtually anything and everything can trigger sexual arousal. The stimuli may be sexual in the ordinary sense of the term, or they may involve factors that have no erotic appeal for most people.

This wide variability in the power of stimuli to elicit erotic responses testifies to

[1] Beach (1947, p. 15).

54

the importance of sexual *learning*. Individuals within a group are biologically not that different from each other, nor are intergroup differences that great. But the erotic significance of stimuli varies enormously between cultures and is subject to further individual idiosyncrasies within each culture. Sexual preferences—what turns us on or off—are thus largely the result of social conditioning. Nonetheless, there may also be certain commonalities and gender-specific patterns that characterize patterns of sexual arousal.

The Stimulus-Response Paradigm

A useful approach to the study of sexual behavior is to view it as an interaction between a *stimulus* and its behavioral *response*. This model permits us to distinguish more clearly between those variables that elicit sexual behavior and the components of those behavioral responses.[2]

Much of our sensory stimulation comes from the environment. In erotic terms, this may take the form of various sights, smells, and sounds that strike us as sexy. But in addition to such *external* sources of stimulation, there are also *internal* triggers for sexual arousal. These may be in the form of erotic thoughts—an idea, memory, or fantasy. Another category of internal stimuli are hormonal changes that may predispose us to having such thoughts or behaving sexually. As we discuss in chapter 4, the sexual behavior of lower mammals at least is very much under the control of such hormonal influences.

The stimulus-response model, though useful, is in itself insufficient to explain sexual behavior. Setting aside the complexities of human relations, let us consider how a pair of rats sexually interact. Hormonal changes (the stimulus) make a female rat behave in a sexually receptive manner (the response). Such female behavior now itself becomes a stimulus to which the male rat responds by mounting the female rat. These two sets of reactions

are temporally linked—one follows the other—but they need an intermediate stage to explain the manifest behaviors. To provide such an explanatory bridge, we invoke the concept of **sexual arousal.** The hormonal influence activates in the female rat a state of arousal that prompts her to act in a sexually receptive manner; her behavior, in turn, along with other factors, sexually arouses the male and leads to copulation.

The problem with the concept of sexual arousal is that it is an abstraction. We can measure the rise in hormones that stimulates the female rat; we can describe her behavioral response; and we can measure the physiological changes that accompany that response. But we cannot identify or measure "sexual arousal" directly—we can only infer its existence or intensity.[3]

When we extend these principles from rats to people, matters become even more complex. But then we also gain a tremendous advantage, because people, unlike rats, can express their feelings and thoughts. Thus, in addition to physiological measures of sexual arousal and behavioral observations of how sexually aroused people act, we have the subjective accounts of people of what it feels like to be sexually aroused.

Erotic Stimulation through Physical Means

There are five basic modalities of sensation—vision, hearing, taste, smell, and touch—and all of them transmit messages that may be interpreted by the brain as erotic. Though vision and hearing are especially important for humans in communicating verbal and nonverbal erotic messages, touch is the most potent physical mode of erotic stimulation. Tactile stimulation figures in almost all instances of sexual arousal leading to orgasm. It is, in fact, the only type of stimulation to which the body can respond reflexively, independent of higher psychic centers. Thus,

---

[2]The stimulus-response paradigm is an important aspect of learning theory which we discuss in chapter 8.

[3]For further elaboration of these concepts, see Beach (1965); Davidson (1977).

even if a man is unconscious or has a spinal cord injury that prevents any genital sensations from reaching the brain (but leaves sexual coordinating centers in the lower spinal cord intact), he may still have an erection when his genitals or inner thighs are caressed.

Erogenous Zones　The perception of touch is mediated through special nerve endings in the skin and deeper tissues. These end organs are distributed unevenly, which explains why some parts (like fingertips) are more sensitive than others (like the skin of the back): the more richly innervated the region, the greater is its potential for stimulation.

Some of the more sensitive areas of the body are believed to be especially susceptible to sexual arousal and are therefore called *erogenous zones.* They include the clitoris; the labia minora; the vaginal introitus; the glans penis (particularly the corona and the underside of the glans); the shaft of the penis; the area between the anus and the genitals; the anus itself; the buttocks; the inner surfaces of the thighs; the mouth, especially the lips; the ears, especially the lobes; and the breasts, especially the nipples.[4]

Although it is true that these areas are most effectively involved in sexual stimulation, one should not assume that they are the only ones. The neck, the palms and fingertips, the soles and toes, the abdomen, the groin, the center of the lower back, or any other part of the body may well be erotically sensitive to touch for someone or another. In unusual cases, women have been reported to reach orgasm by having their eyebrows stroked or through pressure applied to their teeth.[5]

The concept of erogenous zones is very old. Explicit or implicit references to them are plentiful in ancient love manuals. A knowledge of erogenous zones has been rightly assumed to greatly enhance one's effectiveness as a lover. However, the ultimate interpretation of all stimuli by the brain is profoundly affected by previous experience and the mood of the moment. One cannot therefore approach a sexual partner solely guided by an "erotic map" or in a mechanical, push-button manner and expect to elicit automatic sexual arousal. Subtle lovers seek out instead the unique erogenous pattern of their partners on each and every occasion of making love.

The frankly erotic component in being touched must be understood in the broader, more fundamental need for bodily contact that stems from our primate heritage and our infancy. A crucial component in infant care is the touching, caressing, fondling, and cuddling that is carried out by the adult caretakers. The dire effects that result from the deprivations of such care have been well documented for humans[6] and other primates.[7] Indeed, tactile communication plays such a major part in primate life that the practice of grooming has been called the "social cement of primates."[8] Thus, sexual arousal may well be embedded in the more fundamental need for security and affection.

Erotic Sights, Sounds, and Smells　Vision, hearing, smell, and taste are also important avenues of erotic stimulation. But it is generally believed that these modalities, in contrast to touch, do not operate reflexively: we learn to experience certain sights, sounds, and smells as erotic and others as neutral or offensive.

An alternative viewpoint suggests that although learning indeed helps shape our arousal responses, there are nonetheless innate mechanisms that help determine our response to particular sexual cues. Thus just as animals react to certain predetermined "sexual triggers," humans presumably respond likewise to visual, olfactory, and other cues in ways that are more or less consistent and genetically determined.[9] This issue is but one facet of the

---

[4]See Goldstein (1976) for further information on nongenital erotic zones.

[5]Kinsey et al. (1953, p. 50).

[6]Spitz and Wolf (1947).

[7]Harlow (1958). We discuss some of Harlow's work in chapter 8.

[8]Jolly (1972, p. 153).

[9]Morris (1977).

ongoing controversy over the biological versus cultural determinants of sexual behavior (which we discuss in chapter 8). Cross-cultural and cross-species comparisons are usually the sources of evidence examined in support of these views. Those who favor the concept that arousal patterns are learned point to the cultural diversity that exists in this regard; those who favor innate models dwell on the similarities across cultures and between humans and animals.

The sight of the female genitals is as nearly a universal source of erotic excitement for heterosexual males as any, but viewing the male genitals does not seem to excite females as intensely, possibly because of cultural inhibitions. Although there is great diversity in what sights are considered as erotic, and cultural standards differ and change, the impact of visual stimuli in sexual arousal can hardly be overestimated, as our preoccupation with physical attractiveness of the body, cosmetics, and dress indicate.

The effect of sound is perhaps less evident but nonetheless quite significant as a sexual stimulant. The tone and softness of voice (apart from the content of speech) as in whisperings of "sweet nothings" between lovers is a case in point. The sighs, groans, and moans uttered during sex can in themselves be highly arousing to the participants (and to others within earshot). Similarly, certain types of music with pulsating rhythms or repeated langourous sequences often serve as erotic stimuli, or serve to set the mood for sex.

The importance of the sense of smell has declined in humans relative to other species. Among other animals, chemical substances called *pheromones* act as powerful erotic stimulants through the sense of smell. An intriguing possibility exists that humans too secrete pheromones (see box 4.3 in chapter 4), but even without them, the use of scents and preoccupation with body odors in most cultures attest to the erotic importance of the sense of smell.[10]

We have dwelt so far on the positive ef-

[10]For further discussion of the role of olfaction, see Schneider (1971); Hopson (1979).

fect of sensory stimulation in generating erotic arousal. But sights, sounds, and smells can also exert an equally powerful inhibiting influence. We may also be simultaneously affected by conflicting signals: a person may look and sound very sexy but his or her erotic appeal may be compromised by a bad smell (in which case the person will look less sexy and sound less appealing). Effective sexual arousal thus depends equally on the signals we send as well as those we avoid sending to the potential sexual partner.

Psychological Means of Erotic Stimulation

Despite all the power of physical stimulation, the key to human sexual arousal remains locked in psychological processes. Sexual arousal is after all an emotional state and one greatly influenced by other emotional states. Stimulation through the senses will result in sexual arousal if, and only if, it is accompanied by the appropriate emotional concomitants. Feelings like affection and trust will enhance, and others like anxiety and fear will inhibit, erotic response in most cases.

Given our highly developed nervous system, we can also react sexually to purely mental images, which makes sexual fantasy the most common erotic stimulant. Our responsiveness is thus not based solely on the external characteristics of a given erotic situation, but also depends on the store of memory from past experience and thoughts that are projected into the future. What arouses us sexually at a given moment is the outcome of all of these influences.

As human beings we share common developmental experiences as well as a common biology. For example, we are all cared for as infants by adults who become the first and most significant influences in our lives. What was said earlier about erogenous zones is therefore applicable to the psychological realm as well. Just as most of us are likely to respond to gentle caressing on the inside of our thighs, we are also likely to respond positively to the expression of affectionate sexual interest in us. But as we are also all unique in our developmental histories so, in both cases, the response will vary with the persons and circumstances involved and is hardly an automatic or fixed reaction. The practical applications of how we respond to sexual stimulation are discussed in connection with the enhancement of sexual pleasure (chapter 15).

Gender Differences in Sexual Arousal

Do women and men react differently to various sexual stimuli? Are men turned on more easily than women? Are such differences, if they exist, primarily innate or culturally determined? Despite widespread interest in such questions, amazingly little research existed until recently concerning differences among males and females as to what they generally consider to be erotically stimulating.

Kinsey reported that men are generally more readily stimulated than women by viewing sexually explicit materials (such as pictures of nudes, genitals, or sexual scenes) but that women are as responsive as men in reaction to motion pictures and literary materials with romantic content. In the 1970s experimental rather than reported evidence (which was the basis for Kinsey's conclusion) began to appear in this connection. In one study, a group of students at Hamburg University were shown sexually explicit pictures under experimental conditions.[11] In general, men were more responsive than women to nude pictures. But when the scenes had an interpersonal or affectionate component (such as kissing couples) women were equally if not more responsive. Scenes of coitus were somewhat more arousing to men, but not much more so than to women. Sexual arousal in these contexts was reported by the women as genital sensations of warmth, itching, pulsations, and vaginal moistening; the men usually responded with erections.

A larger sample (128 males and 128 females) were shown films featuring male and female masturbation, petting, and co-

itus.[12] Among the women, 65% experienced genital sensations; among the men 31% had full erections and 55% had partial erections. About one in five men and women reported some masturbatory activity while viewing the film, and during the following 24 hours there was some increase in sexual activity for both sexes, especially masturbation. These findings tended to confirm the notion that males respond more readily than females to erotic visual material, especially of the more "hard-core" variety, but also that the gender differences reported by Kinsey had diminished.[13]

The early work on erotic responsiveness relied on verbal reports by the subjects. With the development of new technologies, special instruments have provided objective means of studying physiological arousal (box 3.1). This has added a whole new dimension to our understanding of these issues. For example, in one study subjects listened to tape-recorded stories with erotic and romantic themes. Their reactions were self-reported and monitored with instruments. Both sexes reacted similarly in finding the tape with explicit erotic content more stimulating than those with romantic or erotically neutral content. There was however an interesting discrepancy between the verbal reports of the women and their physiological reactions: only half of the women who were physiologically aroused reported this fact. The men on the other hand never failed to perceive their physiological arousal.[14]

There is no reason to believe that these female subjects were concealing the fact of their being aroused. Their relative failure to perceive sexual arousal may be explained by anatomical differences (it is harder to miss an erection than signs of female arousal) or by psychological mechanisms repressing such perception. Women

[11]Schmidt et al. (1970).

[12]Schmidt and Sigusch (1970). For a more comprehensive discussion of this research, see chapter 7 by these same authors in Zubin and Money (1973).

[13]Schmidt (1975); Athanasiou and Shaver (1971); Osborn and Pollack (1977).

[14]Heiman (1975).

Box 3.1:   The New Technology of Sex Research

To monitor and measure the physiological changes during the sexual response cycle, a number of new instruments have been developed. The objective data obtained from their use supplements, in important ways, the subjective reports of men and women undergoing the experiences of sexual arousal and orgasm. These instruments measure the two basic physiological processes that underlie the sexual response cycle—vasocongestion and increased muscular tension.

Genital vasocongestion in the male is assessed by measuring the degree of erection. This is easily done by a *penile strain gauge* which consists of a flexible rubber band that fits around the base of the penis and expands with the swelling penis (see accompanying figure). In the female, the *vaginal photoplethysmograph* essentially accom-

plishes the same purpose by relying on color changes that result from vasocongestion. It consists of a transparent acrylic cylinder that is placed within the vagina (see figure). The light within the cylinder illuminates the vaginal wall; the changes in color produced by the increased number of red blood cells are detected by the photoelectric cell.

Increases in muscular tension and the contractions of orgasm can be measured by a *perineometer*. When placed in the vagina, this instrument will register changes in pressure. *Electronic perineometers* are *myographs* which detect the electrical activity in muscle fibers being activated by nerves and measure changes in muscular tension within the vagina and anus (see figure).[1]

[1]See Ladas et al. (1982) for more details.

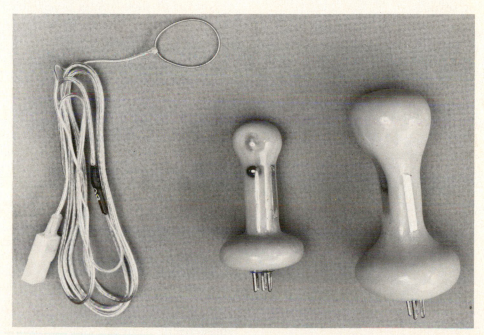

Penile strain gauge (left). The *anal probe* (center) and *vaginal probe* (right) combine the functions of a photoplethysmograph (note the photoelectric cell in the knob of the center figure) and those of a myograph (note the metal strip in the shaft of both instruments). They can therefore monitor sexual arousal by simultaneously measuring the effects of vasocongestion and myotomia. (Courtesy, Dr. Julian Davidson.)

have traditionally inhibited expressions of sexual arousal and interest because of their social unacceptability. While such attitudes are currently changing, their residual effects may still be operative. Current studies continue to report a stronger correspondence between subjective and physiological measures of arousal for males than females.[15]

The laboratory study of sexual responsiveness holds the promise of providing more objective data on the one hand, and on the other it poses the danger of trivializing what it sets out to investigate. For instance, a woman with a highly satisfying sexual life may find exposure to erotica under laboratory conditions quite uninteresting. But her failure to respond to contrived erotic cues would be of little real-life significance and hardly reflect her true sexual potential.[16] Even more misleading would be to draw conclusions from what can be observed publicly about the differential sexual responses of males and females to erotic themes. Since much of what passes for erotic and pornographic material in our mass culture (films, books, magazines and so on) are produced by men and for men, the fact that women may not care about them or find them offensive, says nothing about their capacity to respond to erotic visual cues in general.

Sexual Response

Sexual response to effective sexual stimulation can be examined in terms of the changes manifested by the sex organs, bodily responses more generally, and with regard to the subjective sensations experienced by the person.

The physiology of sexual function has been largely neglected in medical research. In the fourth century B.C., Aristotle observed that the testes are lifted up within the scrotal sac during sexual excitement and contractions of the anus accompany or-

gasm; more than 20 centuries passed before such observations were confirmed under laboratory conditions. There were very few investigations of the physiology of orgasm until the investigations conducted by Masters and Johnson in the 1960s.[17] (This research is discussed in chapter 8.) Since then, other studies have supplemented their work and sometimes amended it.[18]

We shall be dealing throughout this chapter with typical patterns of human sexual response. These are not, however, intended as standards of normality. There are many variations of these patterns, and they are also perfectly normal and healthy. The unity of the biological basis of sexual function does not mean that its manifestations are relentlessly uniform.

Response Patterns

We subjectively recognize sexual arousal and orgasm as highly pleasurable experiences, but it is difficult for us to be fully cognizant of their effects objectively. But when observed under laboratory conditions the human body is seen to exhibit distinct physiological patterns of response during sexual arousal and orgasm.

The sexual response patterns shown in figures 3.1 and 3.2 summarize observations by Masters and Johnson. The sexual response pattern for males (figure 3.1) and the three patterns for females (figure 3.2) include the same four phases: *excitement, plateau, orgasm,* and *resolution.* These patterns are generally independent of the type of stimulation or sexual activity that produces them. The basic physiology of orgasm is the same, regardless of whether it is brought about through masturbation, coitus, or some other activity.

The fundamental similarity of sexual responses in the two sexes notwithstanding, a number of differences between male and female responses must be noted. Some of these result from anatomical differences;

---

[15]Steinman et al. (1981). Also see Geer et al. (1974); Wincze et al. (1977); Henson et al. (1977); and Heiman (1977).

[16]Heiman (1980).

[17]Masters and Johnson (1966); (1979). For a summary of their work, see Brecher and Brecher (1966).

[18]See, for instance, Bohlen et al. (1980); Bohlen et al. (1981); Gillan and Brindley (1979); Campbell (1976); Petersen and Stener (1970).

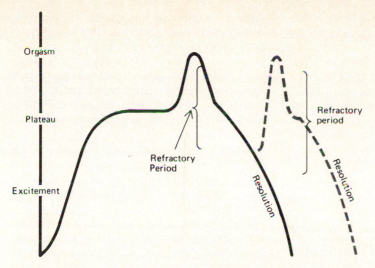

Figure 3.1   The male sexual response cycle. (From Masters and Johnson, *Human Sexual Response*. Boston: Little, Brown and Company, 1966, p. 5.) Reprinted by permission.

others cannot be explained structurally and possibly reflect variations in nervous system organization.

The first major difference is in the range of variability. As is apparent from figures 3.1 and 3.2, although a single sequence characterizes the basic male pattern, three alternatives are shown for females. The second sex difference involves the pres-

ence of a *refractory period* in the male cycle. This refers to the inability of the body to respond to further stimulation until an obligatory period of rest has elapsed after the preceding stimulation. The refractory period immediately follows orgasm and extends into the resolution phase. During this period, regardless of the nature and intensity of sexual stimulation, the male

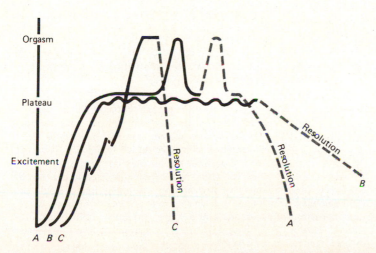

Figure 3.2   The female sexual response cycle. (From Masters and Johnson, *Human Sexual Response*. Boston: Little, Brown and Company, 1966, p. 5.) Reprinted by permission.

cannot achieve full erection and another orgasm; only after the refractory period has passed can he do so.[19]

Females do not have refractory periods (figure 3.2). Even in the pattern closest to that of the male (line A), soon after the first orgasm is over, the level of excitement can mount again to another climax. Women can thus have so-called multiple orgasms, that is, consecutive orgasms in quick succession. But this does not mean that all women, all of the time, can or want to be multiorgasmic.

Male capacity for multiple orgasm is very limited, because of the refractory period. Among the subjects studied by Masters and Johnson, only a few men seemed capable of repeated orgasm with ejaculation within minutes. More recent evidence suggests that multiple orgasm is not that rare if a man experiences orgasm without ejaculation.[20] We shall return to this issue shortly when we discuss orgasm in greater detail.

Apart from these differences, the basic physiological response patterns in the two sexes are the same (psychological response patterns are another matter). In males (figure 3.1) and females (figure 3.2, line A) excitement mounts in response to effective and sustained stimulation, which may be psychological (erotic thoughts and feelings) or physical, but usually involves both. Excitement may mount rapidly or more slowly, depending on various factors. If erotic stimulation is sustained, the level of excitement becomes stabilized at a high point, the plateau phase, until the man reaches the point of no return and orgasm follows. This abrupt release is succeeded by a gradual dissipation of pent-up excitement during the resolution phase.

The lengths of these phases can vary greatly, but in general the excitement and resolution phases are the longest. The plateau phase is usually relatively short, and orgasm usually is measured in seconds. Although the diagrams do not indicate it, there may be several peaks in the plateau phase, each followed by a return to a lower level of excitement. And of course not all episodes of arousal reach the point of orgasm. The overall time for one complete response cycle may range from a few minutes to hours.[21]

In alternative female patterns, the woman attains a high level of arousal during the plateau phase but instead of an orgasmic peak, there is a period of protracted orgasmic release (pattern B in figure 3.2).[22] Sexual tension is then gradually dissipated through a protracted period of resolution. In contrast, pattern C shows a more abrupt orgasmic response, which bypasses the plateau phase and is followed by a more precipitate resolution of sexual tension.

During the various phases of the sexual response cycle numerous physiological changes occur in the genitals and in other parts of the body (figures 3.3 and 3.4). As we describe these reactions we must keep in mind that not all of these responses continue through the entire cycle. Some, such as erection of the penis, set in right away and persist throughout the entire cycle, wherease others exist only during part of a given phase.

Physiological Mechanisms of Sexual Response

Two physiologic mechanisms explain how the body and its various organs respond to sexual stimulation: vasocongestion and myotonia.[23]

---

[19]The length of the refractory period seems to vary greatly among different males and with the same person on different occasions. It may last anywhere from a few minutes to several hours: The interval usually gets longer with age and with successive orgasms during a given sexual episode (Kolodny et al., 1979).

[20]Robbins and Jensen (1976).

[21]Among the Kinsey subjects, men reported reaching orgasm generally within four minutes, whereas it took women 10 to 20 minutes to do so. But when women relied on self-stimulation, they could reach orgasm as fast as the men.

[22]The term *status orgasmus* is used to describe the intensive orgasmic experience where a single prolonged orgasmic episode is superimposed on the plateau phase or a series of orgasms follow each other without discernable plateau-phase intervals (Masters and Johnson, 1966, p. 131).

[23]Masters and Johnson (1966); deGroat and Booth (1980).

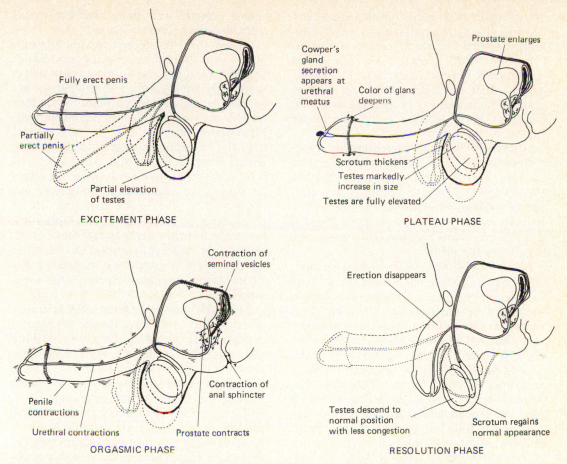

Figure 3.3   Changes in male sex organs during the sexual response cycle. (Based on Masters and Johnson, *Human Sexual Response*. Boston: Little, Brown and Company, 1966.)

Vasocongestion is the engorgement of blood vessels and tissues with blood. Ordinarily, the influx of blood through the arteries is matched by the outflow through the veins. When blood flow into a region exceeds the capacity of the veins to drain the area, vasocongestion will result. Blood flow is primarily controlled by the smaller arteries (arterioles), whose muscular walls constrict and dilate in response to nervous impulses and hormones.

Sexual excitement is accompanied by widespread vasocongestion involving superficial and deep tissues. Its most obvious manifestations is the erection of the penis. The physiology of erection has been the subject of considerable research, yet its precise mechanisms remain to be elucidated.[24] There is now general agreement that the primary cause of erection is increased arterial blood flow rather than impediments to venous outflow.[25] When the penis is flaccid, the blood coming in through the arteries is drained through the deep

[24]See Benson et al. (1981); Newman and Northup (1981); Krane and Siroky (1981); and Tarabulcy (1972).

[25]It used to be thought that venous outflow was impeded by pressure from the expanding penis, but that has turned out not to be the case (Shirai and Ishii, 1981). Similarly, the role of "polsters" (columns of smooth cells within the walls of blood vessels) in directing blood flow internally within penile blood vessels has now been called into question (Benson, 1981).

veins within the penis, thus largely by-passing the spongy tissues of the penis. During sexual arousal, the increased blood inflow is shunted into the penile tissue causing the corpora cavernosa to expand against their unyielding fascial cover thus making them rigid (just as a garden hose becomes stiff when filled with water under pressure).

Although women experience vaso-congestion in response to arousal just as men do, far less is known of female sexual physiology.[26] The assumption is that similar physiological processes are operative in both sexes. New research may reveal physiological mechanisms that are more specific to females.[27]

Myotonia is increased muscular tension. Even when a person is completely relaxed or asleep, the muscles maintain a certain degree of muscle tone. From this baseline, muscular tension increases during voluntary actions or involuntary contractions. During sexual activity myotonia is widespread. It affects both smooth (involuntary) and skeletal (voluntary) muscles. The muscles of the feet, in particular, tense up at the height of sexual arousal with extension of the toes (referred to as carpopedal spasm). Myotonia culminates in orgasmic contractions that involve muscles of both varieties. Although evidence of myotonia is present from the start of sexual excitement, it tends to lag behind vasocongestion and disappears shortly after orgasm.

Excitement and Plateau Phases

Physiologically the excitement and plateau phases have certain distinct characteristics, but basically they constitute one continuous process. Subjectively we also experience them as a sustained period of

sexual arousal. We shall, therefore, consider them together.

In response to effective sexual stimulation, a sensation of heightened arousal develops. Thoughts and attention turn to the sexual activity at hand, and the person becomes progressively oblivious to other stimuli and events in the environment. Most people attempt to exert some control over the intensity and tempo of mounting sexual tensions. They may try to suppress it by diverting attention to other matters, or they may deliberately enhance and heighten the feeling by dwelling on its pleasurable aspects. If circumstances are favorable to fuller expression, these erotic stirrings are difficult to ignore. On the other hand, anxiety or distraction may easily dissipate sexual arousal during the early stages.

Although excitement sometimes intensifies rapidly and relentlessly, it usually mounts more unevenly. In younger adults the progression is steeper, whereas in older people it tends to be more gradual. As the level of tension reaches the plateau phase, external distractions become less effective, and orgasm is more likely to occur. The prelude to orgasm is pleasurable in itself, and following a period of sustained excitement one may voluntarily forego the climax. But lingering tensions may also create irritability and restlessness if unrelieved by orgasm.

The behavioral and subjective manifestations of sexual excitement vary so widely that no one description can possibly encompass them all. With mild sexual excitement, relatively few reactions may be visible to the casual observer; on the other hand, in intense excitement behavioral changes are quite dramatic. The person in the grip of intense sexual arousal appears tense from head to toe. The skin becomes flushed, salivation increases, the nostrils flare, the heart pounds, breathing grows heavy, the face is flushed, the person feels, looks, and acts quite unlike his or her ordinary self. Sexual arousal may also result in more idiosyncratic manifestations: some stutterers speak more freely when sexually aroused; the gagging reflex may become

[26]An important factor for the lack of research in female sexual physiology is the difficulty in ascertaining sexual responses in female animals. Much of the experimental work in this area has been done with dogs where the male has clearly discernable erectile responses. This same consideration also applies to the exploration of nervous mechanisms to be discussed in the next section.

[27]Levin (1980).

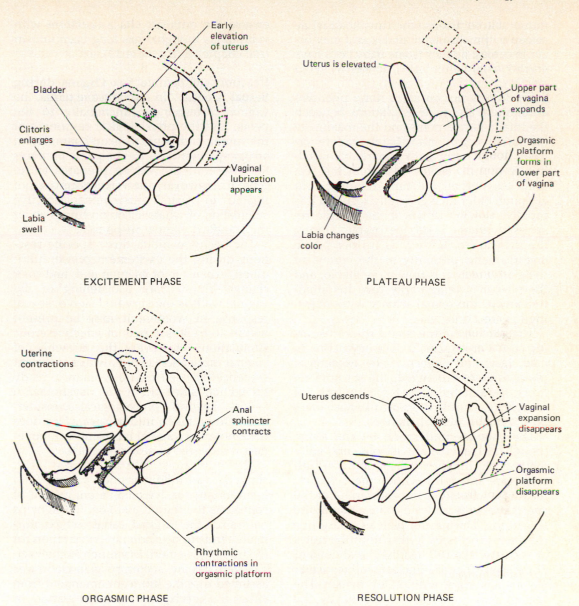

Figure 3.4   Changes in female sex organs during the sexual response cycle. (Based on Masters and Johnson, *Human Sexual Response.* Boston: Little, Brown and Company, 1966.)

less sensitive (which explains the ability of some people to take the penis deep into their mouths); persons afflicted with spastic paralysis coordinate their movements better; those suffering from hay fever may obtain temporary relief; bleeding from cuts decreases. The perception of pain is also blunted during sexual arousal (which may

explain the ability of people to withstand masochistic practices).

Reactions of Male Sexual Organs to Sexual Arousal   The *penis* undergoes striking changes during sexual excitement with erection as its most obvious manifestation (figure 3.3). Erection is experienced on innumerable occasions when awake or while asleep by practically all males and may occur in earliest infancy (some baby boys have erections right after birth), as well as in old age.

Erection is not an all-or-none phenomenon; there are many gradations between the totally flaccid penis and the maximally congested organ immediately before orgasm. It can occur with remarkable rapidity: in younger males, less than ten seconds may be all the time required. But although older men generally respond less rapidly, slowing down of the erectile response with age is not inexorable. During the plateau phase there is further penile engorgement, primarily in the corona of the glan. Erection is now more stable, and the man may temporarily turn his attention away from sexual activity without losing his erection.

The *scrotum* contracts and thickens during sexual arousal (as it does in reaction to cold, fear, and anger). If the excitement phase is quite prolonged, the scrotum relaxes, though the sexual cycle is not yet completed. During the plateau phase there are no further scrotal changes.

The changes undergone by the *testes*, though not as readily visible, are quite marked. During the excitement phase both testes are lifted up within the scrotum, mainly as a result of the shortening of the spermatic cords and the contraction of the scrotal sac. During the plateau phase this elevation progresses farther until the testes are actually pressed against the abdomen. For reasons that are unclear, full testicular elevation is a precondition for orgasm. The second major change undergone by the testes is a marked increase in size (about 50% in most instances) because of vasocongestion.

The *Cowper's glands* show no evidence of activity during the excitement phase. During the plateau phase, drops of clear fluid produced by these glands appear at the tip of the penis; some men produce enough of it to wet the glans or even dribble freely, others very little of it. The presence of this fluid is reliable evidence of a high level of sexual tension, but its absence does not ensure the opposite. Its association with voluptuous thoughts has long been known (see chapter 2, footnote 13).

Reactions of Female Sex Organs during Sexual Arousal   In reproductive terms, the vagina corresponds functionally to the penis, and the physiological responses of the vagina and penis to sexual stimulation are complementary. As the penis prepares for insertion, the vagina prepares to receive it. However, these reactions are not limited to sexual intercourse; effective stimulation of whatever origin brings about the standard vaginal responses.

The *vagina* exhibits three specific reactions during the excitement phase: lubrication, expansion of its inner end, and color change (figure 3.4). Moistening of the vaginal walls is the very first sign of sexual response in women; it may be present within 10 to 30 seconds of effective erotic stimulation. However, the presence of vaginal lubrication is not always sufficient evidence by itself that a woman is ready for coitus, since there are psychological factors beyond physiological considerations to be taken into account. Nor does the amount of vaginal lubrication alone reflect the degree of sexual arousal.

The lubricatory function of the clear, slippery, and mildly scented vaginal fluid is important for a woman's enjoyment of coitus. As the fluid is alkaline, it also helps neutralize the vaginal canal (which normally tends to be acidic) in preparation for the transit of semen. Inasmuch as the vaginal wall has no secretory glands, it was assumed that the lubricant emanated from either the cervix, the Bartholin glands, or both, until it was demonstrated that this fluid comes out directly from the vaginal walls (like perspiration from skin).[28]

Recently, considerable interest has been elicited by reports that there exists an erotically sensitive area palpable through the anterior wall of the vagina. Named the *Grafenberg spot* after the person who first described it, the exact physiology of this dime-sized region and its connection with

[28]Masters and Johnson (1966).

Box 3.2:   The Grafenberg Spot and Female Ejaculation

The question of whether women ejaculate has intrigued people for a long time. Until the 17th century the vaginal lubricant fluid was assumed to be a female "ejaculate" analogous to male semen and therefore essential for conception. This misunderstanding formed the basis of theological tolerance of female masturbation during coitus; if a woman could not reach orgasm during coitus it was permissible that she did so by manipulation, otherwise she would be unable to complement male's semen with her own ejaculate to make conception possible. When the nature of human reproduction became better understood this notion was discarded, and women were declared incapable of ejaculation. Yet the controversy lingered on.

Reports kept appearing of women experiencing emission of fluid at orgasm but there was much confusion about what that fluid consisted of. The alternatives included urine, fluid from Bartholin's glands, vaginal lubricant expelled forcibly during orgasmic contractions, and fluid from the urethra.

In 1950 Ernest Grafenberg published an article in which he made two claims.[1] First, there exists an "erotic zone" on the anterior wall of the vagina along the course of the urethra. This area becomes enlarged during sexual excitement and protrudes into the vaginal canal, and following orgasm, it reverts to its nonstimulated state. Second, "large quantities of a clear transparent fluid" gush out of the urethra during orgasm, at least in the case of some women.

Grafenberg explained both phenomena by the hypothesis that there persisted in the female the homologue of the erectile tissues surrounding the male urethra (the corpus spongiosum) and the homologue of the prostate gland (known as "Skene's glands"). The existence of the first would explain the vasocongestive swelling within the anterior vaginal wall. Similarly, the Skene's glands would function as a "female prostate," which would provide the fluid in female ejaculation (just as the ejaculatory fluid in the male is provided by the prostate).

Grafenberg's hypothesis was plausible on the grounds of embryonal development (chapter 2), but he offered no anatomical evidence or much objective clinical data to support his views. Recently, considerable attention has refocused on this issue.[2] The "erotic zone" in the anterior vaginal wall has been named the *Grafenberg spot* (G-spot), and detailed descriptions given of urethral ejaculation (including a film recording of the phenomenon in one woman). Based on this research, the G-spot is located in the anterior wall of the vagina, about two inches from the introitus; when properly stimulated, it swells and leads to orgasm (often a whole series of them) in many women. During orgasm, women emit a fluid through the urethra that is similar to the male ejaculate, except for lack of sperm; many women are embarrased about ejaculating because they believe they are losing urinary control, and hence try to suppress their orgasm.[3]

If these claims are confirmed, they would contradict established views that the inside of the vagina is largely insensitive, that women do not ejaculate, and that there are no physiological grounds for differentiating "vaginal" from "clitoral" orgasm (see box 3.4). Otherwise, yet one more sexual sensation would have been generated causing undue anxiety to women who would have been led to believe that they are somehow lacking sexually because they do not ejaculate or do not appear to be blessed with a G-spot.

[1]Grafenberg (1950).
[2]Sevely and Bennett (1978); Weisberg and Martin (1981).
[3]Addiego et al. (1981); Ladas et al. (1982).

"female ejaculation" are yet to be fully determined (box 3.2).

The second major vaginal change during the excitement phase is the lengthening and expansion of the inner two-thirds of the vagina ("tenting"), which creates a space where the ejaculate will be deposited. Finally, the vaginal walls take on a darker hue in response to sexual stimulation reflecting the progressive vasocongestion of the area.

During the plateau phase the outer third of the vagina becomes swollen forming the *orgasmic platform* and narrowing the vaginal introitus. Meanwhile the tenting effect at the inner end attains full vaginal expansion. Vaginal lubrication tends to slow down, and if the plateau phase is unduly protracted, further production of vaginal fluid may cease altogether.

The clitoris becomes markedly congested during the excitement phase. This response is more consistent for the clitoral glans and may result in the doubling of its diameter. This vasocongestive response is more rapid and more marked if the clitoris and adjoining areas of the mons are stimulated directly. It coincides with the vasocongestive response of the minor labia and comes quite late in the excitement phase (when the penis has been erect for some time and the vagina is fully lubricated). The clitoral glans, once tumescent, remains so throughout the sexual cycle.

During the plateau phase the entire clitoris is retracted under the clitoral hood receding to half its unstimulated length (which may be misinterpreted by the sexual partner as indicating loss of sexual tension). When excitement abates, the clitoris reemerges from under the hood. During protracted plateau phases there may be several repetitions of this clitoral retraction-emergence sequence.

The *labia* of women who have not given birth (nulliparous labia) respond somewhat differently from those of women who have (parous labia). Nulliparous major lips become flattened, thinner, and more widely parted during excitement revealing the congested moist tissue within. Parous major lips are larger and instead of flattening,

become markedly engorged, and may double or triple in size during arousal.

As the excitement phase progresses to the plateau level, the minor lips become severely engorged and double or even triple in size in both parous and nulliparous women. They also become pink, or even bright red in light-complexioned women. In parous women the resulting color is a more intense red or a deeper wine color. This vivid coloration of the minor lips has been called the *sex skin* of the sexually excited woman. Like full testicular elevation of the male, the presence of the sex skin heralds impending orgasm.

Bartholin's glands secrete a few drops of mucoid material rather late in the excitement phase, whose contribution to vaginal lubrication is relatively minor.

The *uterus* responds to sexual stimulation by elevation from its anteverted position, which pulls the cervix up and contributes to the tenting effect in the vagina. Full uterine elevation is achieved during the plateau phase and is maintained until resolution.

Orgasm

In physiological terms **orgasm** consists of the discharge of accumulated neuromuscular tension.[29] Subjectively, it is the high pitch of erotic tension constituting one of the most intense and profoundly satisfying human sensations; it lasts but a few moments, yet sometimes feels like an eternity.

The experience of orgasm varies somewhat with sex, age, physical condition, length of abstinence, and the context in which it occurs. The manifestations of orgasm may range from the subdued to the explosive. Typically, there is sustained tension or mild twitching of the extremities. The person may grimace or utter a muffled cry; rhythmic throbbing of sex organs accompanies pelvic thrusts. In more intense orgasms, the whole body becomes

[29]"Orgasm" is derived from the Greek *orgasmos* ("to swell," "to be lustful"). Colloquial terms include "to climax," "to come," "to spend."

rigid, the legs and feet are extended, the toes curl in or flare out, the abdomen becomes hard and spastic, the stiffened neck is thrust forward, the shoulders and arms are rigid and grasping, the mouth gasps for breath, the eyes bulge and stare vacantly or shut tightly. The whole body convulses in synchrony with the genital throbs or twitches uncontrollably. The person may moan, groan, scream, or utter fragmented and meaningless phrases. There may be uncontrollable laughing, talking, crying, or frenzied thrashing about.

Orgasm is felt by both men and women as an intensely pleasurable experience, but one that is difficult to describe adequately.[30] In adult males the sensations of orgasm are linked to ejaculation, which occurs in two stages. First, there is a sense that ejaculation is imminent, or "coming," and that one can do nothing to stop it. Second, there is a distinct awareness of the throbbing contractions at the base of the penis followed by the sensation of fluid moving out under pressure.

Orgasm in the female starts with a feeling of momentary suspension or "stoppage." Sensations of tingling and tension in the clitoris then reach a peak and spread to the vagina and pelvis. This stage varies in intensity and may also involve sensations of "falling," "opening up," or of emitting fluid. (Some women compare it to mild labor pains.) It is followed by a suffusion of warmth spreading from the pelvis through the rest of the body, culminating in the characteristic throbbing sensations in the pelvis.[31]

Are all orgasms the same? In principle no two sexual experiences of any kind are ever the same, but in practical terms the male orgasmic experience seems to follow a more standard pattern than the female. Women experience the same variations as men do, but there is also the possibility that women have qualitatively different orgasms based on different physiological mechanisms (see box 3.3). Other evidence suggests that the orgasmic experience in the two sexes is not that dissimilar; if brief descriptions of orgasm by men and women are submitted to a group of "judges," they are unable to identify the sex of the authors from the descriptions.[32]

Reactions of Male Sex Organs during Orgasm   The characteristic rhythmic muscular contractions during orgasm begin in the accessory sexual organs (prostate, seminal vesicles, and vas deferens) and extend to the penis (see figure 3.3) involving the entire length of the urethra, as well as the muscles covering the root of the penis. At first the contractions occur fairly regularly at intervals of approximately 0.8 second, but after the first several strong throbs they become weaker, irregular, and less frequent.

Studies subsequent to the work of Masters and Johnson have studied male orgasm as manifested in contractions of the anal sphincter. Masters and Johnson had reported that 2 to 4 anal contractions occurred during this period. Bohlen and his associates have found an average of 17 anal contractions (figure 3.5).[33] They occur over a period of 26 seconds, which may be taken as a close estimate of the length of male orgasm, although subject to considerable individual variation.

Ejaculation and orgasm are not synonymous. Orgasm is experienced in both sexes and at all ages. It consists of the neuromuscular discharge of accumulated sexual tensions. Ejaculation ("to throw out") entails the ejection of semen and is the most obvious manifestation of orgasm in males following puberty when the prostate and accessory glands have become functional.

Ejaculation consists of two distinct

---

[30]An interesting new way of conceptualizing the orgasmic experience is to view it as an "altered state of consciousness." See Davidson (1980).

[31]One woman describes it as follows: "There are a few faint sparks, coming up to orgasm, and then I suddenly realize that it is going to catch fire, and then I concentrate all my energies, both physical and mental, to quickly bring on the climax—which turns out to be a moment suspended in time, a hot rush—a sudden breathtaking dousing of all the nerves of my body in pleasure" (Hite, 1976, p. 73).

[32]Vance and Wagner (1976).
[33]Bohlen et al. (1980).

Box 3.3:   Varieties of Female Orgasm

There is an extensive literature on female orgasm, about which probably more has been written than any other aspect of human sexuality.[1] It has been a tenet of psychoanalytic theory that women experience two types of orgasm: clitoral and vaginal.[2] The term *clitoral orgasm* means orgasm attained exclusively through direct clitoral stimulation, usually by masturbation; *vaginal orgasm* means orgasm attained through vaginal stimulation, usually through coitus. This dual orgasm theory assumes that in young girls the clitoris is the primary site of sexual excitement. With psychosexual maturity the sexual focus is said to shift from the clitoris to the vagina, so that after puberty the vagina emerges as the dominant orgasmic zone (chapter 8).

Kinsey and his associates raised doubts about the concept of dual orgasm. Masters and Johnson confirmed these doubts and established the view that physiologically there is only one type of orgasm: orgasmic response to clitoral, vaginal, or any other form of stimulation is the same.[3]

Singer and Singer subsequently proposed that there are three types of female orgasm.[4] First is the *vulval orgasm,* characterized by involuntary rhythmic contractions of the vaginal introitus, which they consider to be the orgasm described by Masters and Johnson. Second, there is the *uterine orgasm,* which results from the repetitive displacement of the uterus, and unlike the vulval orgasm, depends on coitus or a close substitute. The third type is the *blended orgasm,* combining elements of the previous two.

Ladas, Whipple, and Perry suggest a *continuum of orgasmic response* that purports to integrate earlier orgasmic models with their own findings. At one end of this continuum is the *clitoral orgasm* (Singers' "vulval orgasm"). It is triggered by clitoral stimulation, involves contractions of the pubococcygeus muscle, and is felt primarily at the region of the orgasmic platform. At the other end is the *vaginal orgasm* (Singers' "uterine orgasm"). It is triggered by stimulation of the G-spot and involves contractions of the uterus, hence is experienced in the region of the pelvic organs. The vaginal orgasm is terminative: one is enough to attain sexual satiety.[5]

Part of the confusion is due to the blurring of distinctions between the *physiological* or objective manifestations of orgasm and the subjective or *psychological* aspects of the experience. Although the physiological reactions of orgasms may be similar, many women are reported to be able to distinguish one form of orgasm from another. Orgasm through clitoral stimulation is said to be more intense though shorter. It subsides quickly, but can be readily repeated ("multiple"). Orgasms through vaginal stimulation have a slower build-up, a more subdued peak, subside more slowly, result in a deeper and fuller sense of satisfaction, and are not apt to be repeated.[6]

Whatever experiential differences may exist in female orgasm, it is best to avoid evaluations of one form of orgasm as "better" than the other. Women have been needlessly burdened in the past because of the onus placed on clitoral orgasm. That notion has now been largely dispelled. Yet the expectations of multiple orgasm and ejaculatory orgasm may have taken over as the new standards by which to denigrate the sexual experiences of those women who do not measure up to these variations.

[1]Levin (1981).

[2]*New Introductory Lectures on Psychoanalysis* (1933), in Freud (1957–64), Vol. 22; Fenichel (1945); Deutsch (1945); Bonaparte (1963). Among modern psychoanalysts there is a shift away from the dual orgasm hypothesis. See Salzman (1968).

[3]Masters and Johnson (1966).

[4]Singer and Singer (1972).

[5]Ladas et al. (1982).

[6]Fisher (1973); Bentler et al. (1979); Ladas et al. (1982). Lessing (1962) gives a compelling description of the two types of orgasm.

phases. During the first or emission phase (*first-stage orgasm*) the prostate, seminal vesicles, and vas deferens pour their contents into the dilated urethral bulb. At this point the man feels ejaculatory pressure building up. In the next phase, ejaculation proper (*second-stage orgasm*), semen is expelled by the vigorous contractions of the muscles surrounding the root of the penis and the contractions of the genital ducts.

The amount of fluid and the force with which it is ejaculated are popularly associated with strength of desire, potency, and so on, though these beliefs are difficult to substantiate. There does seem, however, to be a valid association between the emission phase and the onset of the refractory period. Some men are able to inhibit the emission of semen while they experience the orgasmic contractions: in other words they have nonejaculatory orgasms. Such orgasms do not seem to be followed by a refractory period, thereby allowing these men to have consecutive or multiple orgasms like women. But should emission of semen with ejaculation occur, then a refractory period follows.[34]

However, the matter is even more complicated than that. The fact that no ejaculate comes out of the penis does not necessarily mean that emission did not take place. In cases of **retrograde ejaculation**, the flow of semen is reversed so that instead of flowing out of the urethra, semen is emptied into the urinary bladder. The sensation of orgasm in this condition is unchanged, and a refractory period presumably follows normally. This condition occurs in some illnesses and with the use of certain common tranquilizers as well as antihypertensive drugs (chapter 15).

In an attempt to integrate the various patterns of male and female orgasm into a coherent scheme, Julian Davidson has proposed a *bipolar hypothesis* of orgasm that postulates that when sexual excitement passes a critical threshhold, an orgasmic control area in the central nervous system triggers orgasm by sending neural impulses simultaneously in two directions: "upward" to higher brain areas in the cortex, which results in the intense subjective orgasmic experience; and "downward" to produce the physiological reactions in the genital-pelvic region.

This postulate can be extended into a basic orgasmic model that applies to both sexes. The subjective experience of orgasm with its euphoria and alterations in consciousness would be linked with orgasmic contractions. Contractions without ejaculation in the male and the "vulval" type of orgasm in the female would not induce a refractory period in either sex. On the other hand, a refractory period would follow if there was ejaculation in the male and uterine orgasmic contractions in the female ("uterine orgasm"). Especially if the latter can also be shown to be accompanied by female ejaculation, then male and female orgasm will be essentially identical phenomena.[35]

Reactions of Female Sex Organs during Orgasm   During female orgasm (see figure 3.4), the most visible effects occur in the orgasmic platform. This area contracts rhythmically and with decreasing intensity (initially at approximately 0.8-second intervals). The more frequent and stronger the contractions of the orgasmic platform, the more intense is the subjective experience of orgasm. At particularly high levels of excitement these rhythmic contractions are preceded by spastic (nonrhythmic) contractions of the orgasmic platform that last several seconds.

The clitoris remains hidden from view during orgasm. Orgasmic contractions in the uterus start at the fundus and spread downward; although these contractions occur simultaneously with those of the orgasmic platform, they are less distinct and more irregular. It has often been assumed that the contractions of the uterus during coitus cause the semen to be sucked into

---

[34]Ancient Chinese sex manuals have long extolled the virtues of nonejaculatory coitus (chapter 18). The Indian practice of *Karezza* or *coitus reservatus* similarly consist of protracted coitus without ejaculation (chapter 11).

[35]Davidson (1980).

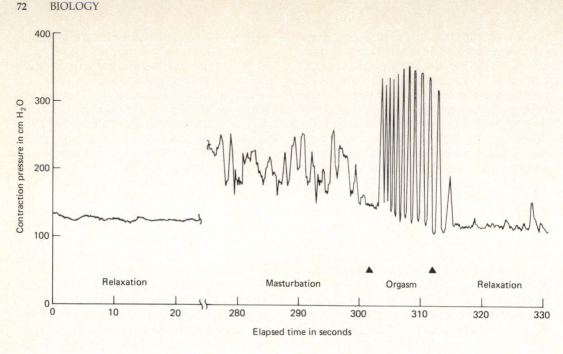

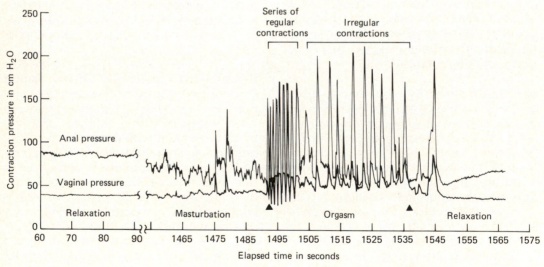

Figure 3.5 Computer-drawn plot of anal tension and contractions during male masturbation (above) and anal and vaginal pressures recorded during female masturbation (below). (From Bohlen, Held, and Sanderson, *Archives of Sexual Behavior*, vol. 9:6, 1980, and vol. 11:5, 1982.)

its cavity. Masters and Johnson found no evidence to support this, but others claim to have done so.[36]

[36]See Fox and Fox (1969); Fox and Fox (1967). The "gasping" and sucking action of the cervix at orgasm was first reported by Beek in 1874 (Levin, 1980, p. 219).

Through new laboratory studies, we now have more precise measures of the manifestation of orgasm. One such approach is to simultaneously record anal and vaginal pressures (which reflect the activity of contracting muscles) during orgasm. As shown in figure 3.5, pressures in both the anus and vagina erupt into regular and synchronized contractions with the onset of orgasm; in some but not all women, these are followed by irregular contractions to the end of the orgasm. The lengths of recorded orgasms range from 13 to 51 seconds; the period of signaled orgasm (that is, the interval between when the women say they started and stopped orgasm) ranges from about 7 to 107 seconds.[37] These investigators, unlike Masters and Johnson, found no correlation between the number of orgasmic contractions and level of satisfaction, perceived intensity, or sexual gratification.

Resolution Phase and the Aftereffects of Orgasm

Whereas the onset of orgasm is fairly distinct, its termination is more ambiguous. As the genital rhythmic throbs become progressively less intense and less frequent, neuromuscular tensions give way to a profound state of relaxation.

The manifestations of the postorgasmic phase are the opposite of those of the preorgasmic period. The entire musculature relaxes. The pounding heart and accelerated breathing revert to normal. Congested and swollen tissues and organs resume their usual color and size. As the body rests, the mind reverts to its ordinary state of consciousness.

The quiescent state of body and mind following sexual climax has given rise to the ancient belief that there is a feeling of sadness following orgasm.[38] Actually, the predominant sensation is one of profound gratification and peace. Facial expression conveys serenity as the eyes become luminous and languid, and a subtle flush lights the face.

The descent from the peak of orgasm may occur in one vertiginous sweep, or more gradually. Particularly at night, when profound postcoital relaxation compounds natural weariness, the person tends to fall asleep. Others feel alert or even exhilarated.

It is not unusual to feel thirsty or hungry following orgasm. Smokers may crave a cigarette. There may be a need to urinate, sometimes to move the bowels. Some feel numb, itch, or develop a headache.

Regardless of the immediate postorgasmic response, a healthy person recovers fully from the aftereffects of orgasm in a relatively short time. Protracted fatigue is often the result of activities that may have preceded or accompanied sex (drinking, drugs, lack of sleep), rather than of orgasm itself. When a person is in ill health, however, the experience itself may be quite taxing.

Reactions of Sex Organs in the Resolution Phase In the resolution phase the changes of the preceding stages are reversed (figures 3.3 and 3.4). In the male, erection is lost in two stages: a relatively rapid loss of tumescence, which reduces the organ to a semierect state, is followed by a more gradual decongestion in which the penis returns to its unstimulated size. In general, the longer the excitement and plateau phases (and the more marked the vasocongestion process), the longer the primary stage of resolution, which in turn delays the secondary stage.

After ejaculation, if the penis is kept in the vagina it remains tumescent longer. If a man withdraws, is distracted, or attempts to urinate, detumescence is more rapid. (A man cannot urinate with a fully erect penis because the internal urinary sphincter closes reflexively during full erection to prevent intermingling of urine and semen.)

In the female resolution phase, the orgasmic platform subsides rapidly. The inner vaginal walls return much more slowly to their usual form. With decongestion the color of the vaginal walls lightens over a period of 10 to 15 minutes. The process of lubrication may in rare instances continue into this phase, and such continuation indicates lingering or rekindled sexual tension. With sufficient stimulation, another orgasm may follow if the woman so desires.

---

[37]Bohlen et al. (1982).
[38]According to Galen (A.D. 130–200), "Every animal is sad after coitus except the human female and the rooster."

The labia return to their decongested state rapidly if orgasm has occurred. Otherwise the changes brought about during excitement take longer to dissipate. Following a protracted period of arousal, congestion may be so intense that the labia remain swollen for several hours after all sexual stimulation has ceased.

Following orgasm, the clitoris reemerges promptly (in five to ten seconds) from its retracted position. The rapidity and timing of this response are comparable to the first postorgasmic loss of penile erection. Final detumescence of the clitoris, like the second stage of penile erection, is much slower (usually taking five to ten minutes but sometimes as long as half an hour). When orgasm has not occurred, the engorgement of the clitoris may persist for hours.

Pelvic congestion unrelieved by orgasm can be a source of discomfort for both sexes. Men experience localized heaviness and tension in the testes ("blue balls"); women have a more diffuse sense of pelvic fullness with feelings of restlessness and irritability.

Extragenital Reactions

The Breast *Erection of the nipple* is the first response of the female breast in the excitement phase. Engorgement of blood vessels is responsible for the *swelling of the breasts* as a whole, including the areolae. In the plateau phase the engorgement of the areolae becomes more marked. As a result, the nipples appear relatively smaller. The breast swells further during this phase, particularly if it has never been suckled (it may increase by as much as a fourth of its unstimulated size). During orgasm there are no further changes. In the resolution phase the areolae become detumescent, and the nipples regain their fully erect appearance ("false erection"). Gradually breasts and nipples return to normal size. Changes in the male breast are inconsistent and restricted to nipple erection during the late excitement and plateau phases.

The Skin Sexual activity results in definite skin reactions, consisting of *flushing*, *temperature change*, and *perspiration*. The flushing response is more common and pronounced in women. It appears like a rash in the center of the lower chest in the excitement phase and spreads to the breasts, the rest of the chest wall, and the neck. This sexual flush reaches its peak in the late plateau phase and is an important component of the excited, straining, facial expression characteristic of the person about to experience the release of orgasm. During the resolution phase the sexual flush disappears quickly.

Although the temperature of the body generally remains stable, people frequently report feelings of pervasive warmth following orgasm; and there are popular references to sexual excitement as a "glow," "fever," or "fire." Superficial vasocongestion is the likely explanation of this sensation. Perspiration (apart from that caused by physical exertion) occurs frequently during the resolution phase. Among men this response is less consistent and may involve only the soles of the feet and the palms of the hands.

Cardiorespiratory Systems In the plateau phase the *heart rate* rises to 100 to 160 beats a minute (the normal resting heart rate of 60 to 80 beats a minute). The blood pressure also registers definite increases. These changes are comparable to levels reached by athletes exerting maximum effort. They entail considerable strain on the cardiovascular system, which is easily handled by most individuals. People with heart disease, however, require medical guidance in this regard.

Changes in *respiratory rate* lag behind those in heart rate. Faster and deeper breathing becomes apparent in the plateau phase, and during orgasm the respiratory rate may go up to 40 breaths a minute (the normal rate is about 15 a minute, inhalation and exhalation counting as one). Breathing, however, becomes irregular during orgasm, when some persons may momentarily hold their breath and then breathe rapidly.[39] Some of the panting and

---

[39]For a review and discussion of blood pressure and respiratory patterns during human coitus, see Fox and Fox (1969).

grunts uttered during orgasm result from involuntary contractions of the respiratory muscles, which force air through the spastic respiratory passages. Following orgasm the soft palate relaxes and the person may snore.[40]

Changes in Sexual Response with Aging

There are definite changes in the physiology of sexual response with aging.[41] However, the significance of these changes must neither be exaggerated nor denied. As with other bodily functions, the sexual responses and capabilities change over time. But that is not to say that the older person reacts in a novel or abnormal manner; some of the reactions of the sexual response cycle continue uninterrupted, although others are modified and a few cease altogether. Basically, older men and women respond as before and continue to be capable of orgasm.

In the Aging Male   The physiological changes that influence male sexual response are as follows. First, the reaction to erotic stimulation is slower; the older man needs more time to achieve erection, no matter how exciting the stimulation is. Second, after age 50 or so, psychological stimulation is often not enough in itself, and the male requires direct physical stimulation to achieve erection. However, once erection has been achieved, it can be maintained longer, perhaps because of better control based on experience or because of changes in physiological functioning. But should the older male lose his erection before orgasm, he will encounter greater difficulty in reviving it. During orgasm contractions die out after a few throbs, ejaculation is less vigorous, and the volume of ejaculate is reduced.

During the resolution phase whatever physical changes have occurred disappear with greater speed; in fact, some bodily responses disappear before they can be de-

tected. Penile detumescence occurs in a matter of seconds after orgasm, rather than in a lingering two-stage process. Few older men seek multiple orgasms, and even fewer can achieve them; but, given a chance to recharge during a longer refractory period, some men can and do go on to have additional orgasms.

In the Aging Female   The changes experienced by the aging male are very gradual; in the female, they are much more abrupt because of the *menopause* or the climacterium (discussed further in chapters 4, 7, and 15). The postmenopausal woman exhibits progressive and marked anatomical changes. The vaginal walls lose their thick, corrugated texture and elasticity, becoming thin and pale. Vaginal lubrication response takes longer (several minutes) and is less profuse; the tenting effect is limited and delayed; the orgasmic platform develops less fully. In the orgasmic and resolution phases the older vagina is comparable to the older penis: orgasmic contractions are fewer and less intense, and resolution is prompt and precipitate.

Despite these changes, there is even less physiological cause for cessation of female sexual activity than for male activity during late adulthood. Sexual desire does not decline during the menopause; in some cases it seems to be enhanced. The use of simple vaginal lubricants can fully compensate for the lack of natural lubrication. Most importantly, an active sexual life seems to significantly counteract the effects of aging and to maintain the sexual fitness of aging men and women. Time takes its toll, but it need not quench sexual desire or cripple its fulfillment.

Neurophysiological Control of Sexual Functions

Underlying the various patterns of response to sexual stimulation, there are complicated mechanisms of control. Some of these are hormonal (discussed in the next chapter). Others are neurophysiological and

---

[40]For details of general body responses, see Masters and Johnson (1966).

[41]Masters and Johnson (1966), chapters 15 and 16.

coordinated at two levels: in the spinal cord and in the brain.

Spinal Control of Sexual Function

Sexual functions are controlled at their most elemental level by spinal reflexes. *Reflexes* are the basic units of nervous organization and at their simplest have three components: a *receptor*, a *transmitter*, and an *effector*. The receptor can be any nerve that detects and conveys sensory information (such as touch, pressure, or pain) to the transmitter neurons in the spinal cord, which interpret the sensory input and convey appropriate responses to the third component, the end organs ("effectors"), thus completing the reflex arc. The end organ may be a muscle that contracts or a gland that secretes in response to such stimulation. An example of a simple reflex is withdrawal of the hand after touching a hot object. Reflexes are involuntary in the sense that their response is automatic and does not require a "decision" by the brain. Yet the brain is conscious of these responses and may be able to inhibit the reflexive response to a variable extent.

The examples of the reflex acts given above involve the *voluntary* nervous system. There are other reflexes, including those involved in sexual functions, that belong to the *autonomic* part of the nervous system. This system operates involuntarily to control many of the internal functions of the body. Its activities are normally carried out below the level of consciousness.

The autonomic nervous system has two main subdivisions, both of which are involved in sexual function: the *parasympathetic* and the *sympathetic.* Stimulation of one or inhibition of the other has the same overall effects. One of the basic functions of autonomic nerves is to control the flow of blood by constricting or dilating arteries. In the case of genital blood vessels the effect of parasympathetic stimulation is arterial *dilation* and the effect of sympathetic stimulation is arterial *constriction.* The reflexive processes in sexual functions have been better elucidated in the male. Hence we shall first discuss penile erection and ejaculation.

Mechanism of Erection and Ejaculation

The spinal components that control erection and ejaculation are easier to comprehend when considered separately. The first component in the process of erection consists of the *afferent nerves,* which convey a variety of stimuli from the genitalia to the spinal cord. The second division of the reflex mechanism consists of the spinal cord centers for erection and ejaculation. The afferent nerves convey sensory impulses to the S2 and S4 segments of the spinal cord, which contain the primary *parasympathetic erection center* activated reflexly by external or peripheral stimulation. A second spinal erection center located higher up (segments T11 to L2) is part of the sympathetic system, which also plays a role in the process of erection. This center is thought to mediate psychogenically or centrally induced stimulation from the brain.

The spinal *ejaculatory centers* also have dual locations. The main autonomic center is in the sympathetic segment of the spinal cord (T11 to L2). This is where the first phase of orgasm, seminal emission, is triggered. The second component, ejaculation proper, is controlled by the sacral portion of the cord (S2 to S4), but this time not by its autonomic component but rather by the somatic or voluntary part.

The third division of the reflex arcs consists of the effector or *efferent nerves.* The efferent or outgoing nerves to the tissues involve the parasympathetic, sympathetic, and somatic parts of the nervous system, all of which work in a closely integrated fashion. To clarify these variously overlapping and sequential activities, we shall next consider what transpires physiologically during an entire sexual orgasmic cycle.

When stimulated by appropriate incoming sensations, the parasympathetic reflex erection center in the spinal cord responds by sending out impulses through parasympathetic nerves to the arteries conveying blood to the spongy tissue of the penis. Ordinarily, cavernous and spongy bodies receive modest amounts of blood from the arteries, which are then drained by the veins. Under parasympathetic stimulation these arteries dilate flooding the penile tissues and causing erection. Loss of erection results from a reversal of this process: Sympathetic nerve fibers constrict the arteries, cutting down the inflow of blood, and drainage through the veins increases until the organ returns to a flaccid condition.

When sexual stimulation is effective to the point of triggering orgasm, then another reflexive mechanism comes into play. The first phase of orgasm, emission, is triggered by the sympathetic center (T11 to L2) through the sympathetic fibers in the hypogastric nerve. This results in the contraction of the vas deferens, seminal vesicles, and prostate gland, and the emptying of their contents into the urethra. In the next phase, impulses through somatic nerves to the muscles surrounding the penis (bulbospongiosus, ischiocavernosus, and other pelvic muscles) result in their contraction and the ejaculation of semen.

Emission and ejaculation are thus closely integrated yet separate physiological processes. Ordinarily they occur together, but under some conditions one or the other may be experienced separately. Thus emission may take place without ejaculatory contractions, in which case semen simply seeps out of the urethra. Or ejaculatory contractions occur without having been preceded by the emission of semen.[42]

The processes described so far are the reflexive aspects of penile erection. They are independent of the brain, in the sense that they can occur without assistance from higher centers. For instance, a man whose spinal cord has been severed above the level of the spinal reflex center may still be capable of erection. He will not "feel" the stimulation of his penis; for that matter, he may even be totally unconscious. But his penis will respond "blindly" as it were.

The independence of reflex centers does not mean that they cannot be influenced by the brain, however. Intricate networks link the brain to the reflex center in the spinal cord. Purely mental activity may thus trigger the mechanism of erection without

physical stimulation, or it may inhibit erection despite persistent physical stimulation. Usually the two components operate simultaneously and complement each other. Spurred by erotic thoughts, the man initiates physical stimulation of the genitals; conversely, physical excitement inspires erotic thoughts.

The instances in which erection seems to be nonsexual in origin involve tension of the pelvic muscles (as when lifting a heavy weight or straining during defecation). Irritation of the glans or a full bladder may have the same effect. Erections that occur in infancy are explained on a reflex basis also. An additional gruesome example is erection demonstrated by some men during execution by hanging.[43]

Reflexive Mechanisms in Women   It is generally assumed that there are spinal centers in women that correspond to the erection and ejaculatory centers of men. The reflexive centers in the female spinal cord are less well identified in part because the manifestation of orgasm is relatively more difficult to ascertain among female experimental animals.

Not only do we need to learn a good deal more about reflexive mechanisms of women, but our explanations of male reflexive controls are likely to be revised radically as we learn more about the precise role of *neurotransmitters* that mediate nerve impulses, including those that regulate sexual functions.

Brain Mechanisms

All sensory input must ultimately be interpreted in the brain before any sensation is felt by the person. No thought, however trivial, and no emotion, however ethereal, can exist in an empty skull. Therefore all sexual experience is represented at the level of cortical activity. This is not to say that the neurophysiological component is the most "important" or that such under-

---

[42]For further discussion of spinal sexual mechanism, see Tarabulcy (1972); Weiss (1973); and Hart (1978).

[43]Our knowledge of the neurophysiological mechanisms is quite fragmentary even where a "simple" process like erection is concerned. For more details, see Benson et al. (1981).

standing in itself will be an adequate substitute for other ways of conceptualizing human experience, such as in psychological or ethical terms. But neither can we ever dispense with the physical basis of these mental events if we aspire to comprehend them fully.

There are three general investigative approaches to the study of neurophysiological mechanisms in the brain. First is the use of *electrical stimulation*, whereby erection, ejaculation, and copulatory behavior are elicited by activating various areas in the brain. This approach has led to the discovery of "pleasure centers" in the brain (box 3.4). It is also possible to monitor the natural electrical activity of the brain through electroencephalography (box 3.5).

Box 3.4:   "Pleasure Centers" in the Brain

When microelectrodes are implanted in the brain of a rat and attached to a circuit, the rat can stimulate its brain electrically by pressing on a treadle. With the current off, rats will press the treadle several times in an hour as they explore their surroundings. But when activated electrodes stimulate certain portions of the brain located in the hypothalamus, thalamus, and elsewhere, rats will go on pressing the treadle thousands of times an hour, despite hunger or thirst, to the point of exhaustion. Since the stimulation of these locations appears to be so rewarding, Olds has called them "pleasure centers."[1]

The presence of pleasure centers has also been suggested among humans. When electrodes were implanted in the brain of a psychiatric patient for therapeutic purposes, pleasure of a clearly sexual nature was elicited when the septal region or the amygdala of the limbic system was stimulated. Furthermore, whenever the patient was sexually aroused, changes in brain waves emanating from the septal region could be detected. Such arousal could be elicited by external erotic stimuli (pictures, films), by having the subject fantasize, through the direct delivery of chemicals to the septal region, and through the administration of euphoria-producing drugs.

In addition to reporting feelings of sexual arousal and appearing to be aroused, the patient, just like the experimental rat, would stimulate himself incessantly (1,500 times over three hours on one occasion) and beg to continue whenever the stimulating apparatus was taken away from him.[2]

[1]Olds (1956).
[2]Heath (1972).

Self-stimulation circuit allowing the rat to excite "pleasure centers" in the brain electrically by pressing on a treadle. From James Olds, "Pleasure Centers in the Brain." Copyright 1956 by Scientific American, Inc. All rights reserved.

Box 3.5:   Brain Wave Manifestations of Orgasm

The electroencephalogram (EEG) is the record of the electrical activity of the brain. The record reflects the voltage fluctuations between two points on the scalp. The EEG tracings below are from the right and left parietal channels of a male subject. They successively show brain activity prior to sexual arousal (baseline), 30 seconds prior to orgasm, during orgasm, and two minutes following orgasm.

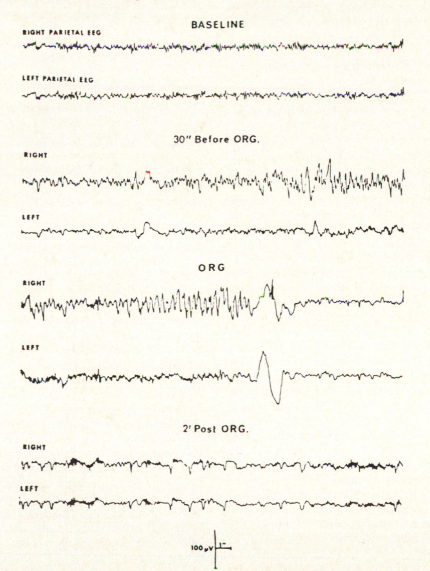

BASELINE

RIGHT PARIETAL EEG

LEFT PARIETAL EEG

30" Before ORG.

RIGHT

LEFT

ORG

RIGHT

LEFT

2' Post ORG.

RIGHT

LEFT

100 μV 1"

From H. D. Cohen, R. C. Rosen, and L. Goldstein, "Electroencephalographic Laterality Changes during Human Sexual Orgasm." *Archives of Sexual Behavior* 5:3 (1976), 189–199.

The second method of study is through *destruction of brain centers* that affect sexual function. It is possible for instance to eliminate male copulatory behavior in a variety of mammalian species by lesions in the medial preoptic region of the brain. Similarly, sexual receptivity may be eliminated in female mammals through hypothalamic lesions. If stimulation of a brain area results in a given sexual reaction, such as erection, and destruction of the region eliminates such response, then one can infer that the area in question is at least one link in the chain of brain mechanisms controlling that function.

A third and complementary approach is to identify brain areas that have an *inhibitory effect* on a given function. How we act is the outcome of the "push" to behave in a certain way as well as the inhibitory "pull" restraining us from doing so. The effect of inhibitory mechanisms becomes manifest by the activities that appear or are enhanced, instead of disappearing, as a result of the removal or destruction of the parts of the brain exerting the inhibitory effect. For example, removal of the temporal lobes in a monkey results in a striking increase in sexual activity involving autoerotic, heterosexual, and homosexual behavior. These monkeys often show penile erection even when sitting quietly. This phenomenon, known as the *Kluver-Bucy syndrome*, has been reproduced in humans by bilateral removal of the temporal lobes.[44]

In functional terms, the brain is organized hierarchically and its organization reflects the evolution of successively more complex nervous functions culminating in human thought. The central core of the brain is the oldest part in evolutionary terms and the portion involved in the most basic life-sustaining activities. Of particular interest to sexuality in this region is the *hypothalamus,* which exerts crucial nervous and hormonal influences (discussed in chapter 4).

Bordering the upper portion of the central core and lying deep within the innermost edge of the cerebral hemispheres are

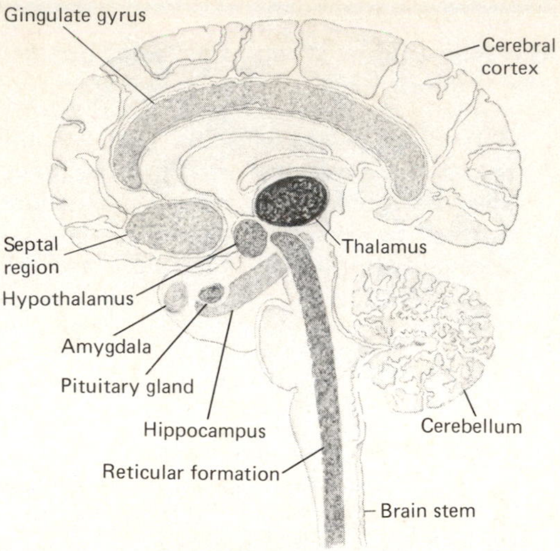

Figure 3.6   The limbic system of the human brain. The thalamus, hypothalamus, pituitary gland, and the reticular formation are interrelated with the limbic system but are not a part of it. (From Harlow, et al., *Psychology.* San Francisco: Albion Publishing Company, 1971, p. 156.) Reproduced by permission.

a set of discrete brain areas that belong to the *limbic system* (from Latin for "border") (figure 3.6). Included in this system are the hippocampus, the septum, the amygdala ("almond"), and the cingulate gyrus. Closely interconnected with the hypothalamus, the limbic system is involved in many "instinctive" activities of animals like feeding, attacking, fleeing from danger, and sex. Stimulation of the septal region elicits penile erection, mounting, and grooming in male animals.[45]

Although the localization of physiological functions in the brain provides very valuable information, the final answers to how the brain regulates sexual functions is not likely to come from the identification of "sex centers" but from a broader understanding of the integrated brain systems that are involved with sexuality and related processes. As Beach has stated,

> Instead of depending on one or more centers, sexual responsiveness and

[44]Terzian and Dalle-Ore (1955).

[45]For further discussion of brain mechanisms, see MacLean (1976).

performance are served by a net of neural subsystems including components from the cerebral cortex down to the sacral cord. Different subsystems act in concert, but tend to mediate different units or elements in the normally integrated patterns.[46]

We are still a long way from comprehending the enormous complexity and infinite subtlety of human sexual experiences in physiological terms, but the rewards of such understanding will be enormous.

SUGGESTED READINGS

Masters, W. H., and Johnson, V. E. *Human Sexual Response* (1966). The standard source on sexual physiology. For a brief and less technical summary of this work see *An Analysis of Human Sexual Response* by Ruth Brecher and Edward Brecher (1966).

Davidson, J. M. "The Psychobiology of Sexual Experience." (in Davidson, J. M., and Davidson, R., eds. *The Psychobiology of Consciousness*, 1980). Presents a comprehensive and clearly written overview of the role of neuroendocrine factors in sexual experience.

[46]Beach (1976), p. 216.

Chapter 4
Sex Hormones

We must examine the powers of humors, and how they are related to one another.

Hippocrates

Hormones are chemical substances that are secreted directly into the bloodstream by ductless *endocrine glands* and by specialized neurosecretory cells. Close to fifty hormones produced by some ten major endocrine sources are crucial for the development and sustenance of a vast range of vital physiological functions. Sex and reproduction in particular simply could not exist without hormones.

The hormonal system, like the nervous system (with which it is closely integrated), can be thought of as a vast communications network within the body. Instead of nerves, the hormonal system uses the bloodstream to reach *target organs* and tissues whose development, sustenance, and functions it controls.

The concept of chemical control of bodily functions and temperaments can be traced to the ancient concept of *bodily humors* (which were thought to be blood, mucus, yellow bile, and black bile). Yet the modern science of endocrinology is relatively quite young; the term *hormone* (from the Greek "to excite") only goes back to the turn of this century.

Hormonal Systems and Sexual Functions

A great many of the body's hormones have a bearing on sexual and reproductive development and functions. But we deal here only with those hormones that are specifically and directly linked to those functions. These are the hormones produced by the *gonads* (testes and ovaries), the *adrenal cortex,* the *pituitary gland,* and the *hypothalamus.*[1]

Gonadal Hormones

The hormones produced by the *testis* and *ovary* are commonly referred to as *sex hormones* because of their gonadal origins and the crucial roles they play in sexual physiology. It is also common to refer to the hormones mainly produced by the testis (*androgens*) and by the ovary (*estrogens* and *progestins*) respectively as "male" and "female" hormones. Such designations facilitate discussion but can also be misleading. Sex hormones are involved in functions other than sex; also, male and female hormones exist in both sexes (though in different concentrations) and are closely related chemically. These hormones, as well as the hormones produced by the cortex of the *adrenal glands* (located over the kidneys), belong to a family of chemical compounds called *steroids,* which has a common basic molecular structure.

Androgens *Androgen* is the generic name

[1]Gilman et al. (1980); Brobeck (1979).

82

for a class of compounds of which ***testosterone*** is the most important (hence the two terms are often used synonymously). Androgens are produced mainly by the testes, secreted by the interstitial *Leydig cells* located between the seminiferous tubules (chapter 2). Androgens are also produced in both sexes by the adrenal cortex, which is the main source of androgens in women.

Androgens are responsible for the sexual differentiation of the reproductive system in intrauterine life (see chapter 2). They bring about the sexual maturational changes of puberty in the male and contribute to the development of some of the secondary sexual characteristics of the female. They are also general *anabolic agents* that promote the building up of tissues. Testosterone is thought to be associated with the sexual drive in both sexes, and possibly with aggressive behavior.

The effects of *castration* (removal of testes) in humans and animals have been known since antiquity. The ancient Egyptians and the Chinese castrated boys to produce *eunuchs* (box 4.2). Similarly bulls are turned into tamer steers and stallions into geldings by castration. Some of the earliest experiments in endocrinology were done in this area. In 1771 John Hunter masculinized hens by transplanting testes from roosters. In 1849 Berthold showed that transplanting male gonads into castrated roosters prevented the typical effects of castration, thus providing the first formal experimental evidence for the existence of an endocrine gland. In the 1930s the testicular substance responsible for these effects was isolated and labeled testosterone.

Estrogens and Progestins   The two principal classes of "female" hormones are *estrogens* and *progestins*. Though estrogen is commonly referred to in the singular, there is no single hormone called "estrogen." Of the three main estrogens in humans *estradiol* is the most potent. Similarly *progesterone* is the most important of the progestins.

Our knowledge of female hormones has a short history. Because the ovaries are not as accessible as the testes, their experimental removal, while keeping the organism alive, was not possible in former times. The hormonal nature of the ovarian control of the female reproductive system was established in 1900, and since the early 1920s an enormous amount of research has been undertaken with these substances with respect to their critical roles in reproduction and contraception.

Estrogens and progestins are secreted by the *granulosa cells* of the ovarian follicles (chapter 2). Estrogens are mainly produced as the follicles are maturing; progestins are produced after ovulation when the follicle develops into the corpus luteum. During pregnancy the production of these hormones is taken over by the placenta.

Estrogens do not seem to play a critical role in the early differentiation of the female reproduction system in intrauterine life (see chapter 2). But they are responsible for most of the sexual maturational changes occurring in girls at puberty. Estrogens and progestins regulate the menstrual cycle and are essential for reproduction. Estrogens are necessary for the implantation of the fertilized ovum and the sustenance of embryonal life. Progestins in turn keep the uterine musculature quiescent, providing optimal conditions for implantation and the initial growth of the fertilized ovum (chapter 5). The relationship of these hormones to the female sexual drive and behavior is unclear. Small amounts of estrogens are produced in the male by the testes and adrenal cortex but fulfill no known function.[2]

In addition to the natural or endogenous forms of these gonadal hormones produced by the body, there are many synthetic products with androgenic, estrogenic, and progestational properties. Estrogens and progestins can be taken orally (as in birth control pills) or injected.

[2]Despite its reputation as the female hormone, estrogens have been extracted from the testes of bulls, and the grand champion of estrogen production among all animals is the stallion (Murad and Gilman, 1975, p. 1424).

Estrogens are also readily absorbed through the skin and mucous membranes and are used, for instance, in vaginal creams. Testosterone is much more effective when injected than when taken orally. Unlike the female hormones, its medical uses are far more limited. The inactivation of gonadal hormones takes place in the liver, and their metabolic by-products are excreted in the urine.

Pituitary Hormones

The **pituitary gland** (or the *hypophysis*) is a pea-sized organ located at the base of the brain whose multiple hormonal functions have earned it the grandiloquent titles of "master gland" and "conductor of the endocrine orchestra" (figure 4.1). The pituitary is in effect a composite of two glands: the *anterior pituitary* and the *posterior pitui-*

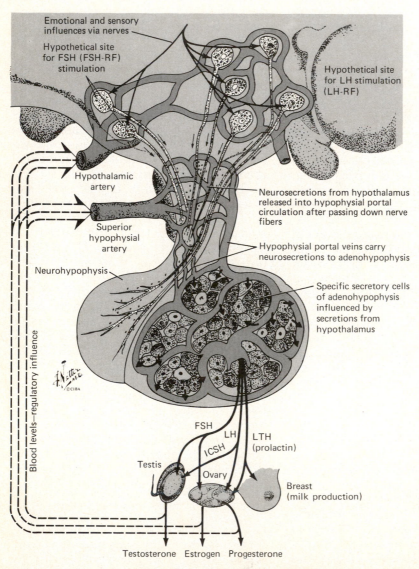

Figure 4.1 The pituitary gland. (From *The Reproductive System* by Frank H. Netter, 1953, Ciba.)

tary (figure 4.2). Our concern is mainly with two of the six hormones produced by the anterior pituitary known as *gonadotropins* because of their control of gonadal functions.

The two gonadotropins are the *follicle stimulating hormone* (FSH) and the *luteinizing hormone* (LH). Though named after ovarian structures (because they were first discovered in females), these two hormones are identical in males and females.[3] The placenta produces *human chorionic gonadotropin* (HCG) during pregnancy, which fulfills the same basic functions as its pituitary counterparts. (Its role in the detection and sustenance of pregnancy is discussed in chapter 5.)

Unlike the steroid hormones, which are relatively simple compounds, the gonadotropins are complex proteins. The names of the gonadotropins are descriptive of their functions in the female: FSH stimulates the maturation of ovarian follicles and LH promotes the formation of the corpus luteum. In the male, FSH is responsible for sustaining and regulating sperm production within the seminiferous tubules; LH stimulates the interstitial Leydig cells to produce testosterone.

Hypothalamic Hormones

The *hypothalamus* is a part of the brain closely linked to the pituitary. It functions both as a nervous center mediating various sexual functions (see chapter 3) and as an endocrine gland conveying its hormones through the portal venous system to the pituitary (figure 4.1). The hypothalamus itself is subject to complex influences from other parts of the brain. Much of our knowledge of hypothalamic hormones is very recent and in many respects is still quite tentative.[4]

The hypothalamic hormones that deal with sexual functions are known as *releasing factors* or *releasing hormones.* Earlier the hypothalamic hormones linked to the

[3]Gilman et al. (1980).

[4]For more detailed discussion of hypothalamic hormones, see Schally (1978); Guillemin and Burgus (1972); Gilman et al. (1980).

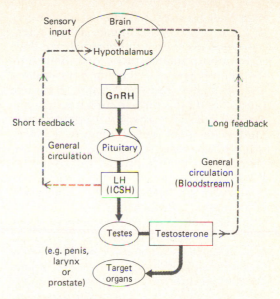

Figure 4.2 Diagrammatic representation of the neuroendocrine feedback mechanisms in the male.

gonadotropins were called FSH-releasing hormone (FSHRH) and LH-releasing hormone (LHRH). However, evidence for the existence of the former has not been conclusive, and what was called LHRH was shown to stimulate the release of both LH and FSH from the pituitary. Thus there is now a single hormone, *gonadotropin-releasing hormone* (GnRH), which is accepted as the active agent. Like pituitary hormones, the hypothalamic releasers are also made up of complex protein-like molecules, yet are relatively simpler structures (GnRH consists of a chain of 10 amino acids; LH has closer to 100).

The function of these hypothalamic hormones is to release the corresponding pituitary hormones. Thus the relationship of GnRH to the gonadotropins is functionally similar to the relationship of gonadotropins to gonadal hormones, in that one affects the output of the other.

Neuroendocrine Control Systems

From our discussion so far, one may get the impression of a hierarchic model in which gonadal hormones do the footwork in influencing the target organs while the hypothalamus controls the headquarters, and the pituitary hormones act as intermediaries. As shown in figure 4.1, the flow of hormones indeed proceeds down from the hypothalamus to the anterior pituitary and on to the ovary and testes. But then the gonadal hormones in turn travel to the anterior pituitary and the hypothalamus influencing their output. Figure 4.2 represents this same process in diagrammatic form using testosterone as the example. What is thus entailed is not a hierarchic but a *cybernetic system* where various parts of a system control each other.[5]

A basic principle in such cybernetic (from Greek for "helmsman") models is the concept of *feedback*. A common example is the way a household thermostat works: When the temperature in a room falls below the degree at which the thermostat has been set, a sensing device turns on the furnace; when the room temperature goes above the set point, the furnace is turned off. In this way, room temperature and furnace mutually control each other so that the level of heat is kept more or less constant around a predetermined level.

The concentration of hormones in the bloodstream is analogous to the temperature of the room and the endocrine glands to the furnace. As the body uses up hormones, their concentration in the bloodstream falls below the physiological set point and more hormones are produced; as the level catches up, hormone production is reduced. In the body, these processes are of course far more complicated. Thus, as shown in figure 4.2, several feedback loops link the endocrine centers together. *Long feedback* links up the gonads with the hypothalamus-pituitary complex. With testosterone as the example, we see that GnRH from the hypothalamus prompts the pituitary to release LH, which in turn increases the production of testosterone by the testes; higher levels of testosterone in the bloodstream then inhibit GnRH production through "negative feedback." When the level of testosterone falls, the

cycle repeats.[6] *Short feedback* similarly links up the hypothalamus with the pituitary. (For the female, we discuss further these and related mechanisms with respect to estrogen and progesterone regulation in connection with the menstrual cycle.) These self-regulatory activities are but one manifestation of the fundamental physiological process of **homeostasis** whereby the body maintains the constancy of its internal environment.

The target organs and tissues are outside of this feedback loop. They are the final recipients of the effects of the sex hormones, and the changes they undergo as a result are to a greater or lesser extent irreversible. But sexual organs, and presumably sexual behavior as well, are continuously dependent on appropriate levels of these hormones for optimal function. In chapter 2, we discussed the effects of sex hormones on the growth and differentiation of the reproductive systems during embryonal development. We now turn to the maturational changes that they bring about during puberty and to their effects on sexual behavior.

Reproductive Maturation at Puberty

During the second decade of life the child is transformed into an adult. **Puberty** (from Latin "to grow hair") refers to the biological aspects, **adolescence** ("to grow") to the psychosocial component of this developmental process. Maturation of the reproductive system and the development of secondary sexual characteristics constitute the core changes in puberty. Yet the changes manifested during this phase of development are so pervasive that there are hardly any tissues in the body that are not affected by it.

The changes that constitute puberty have been classified as follows: acceleration and then deceleration of skeletal growth (the

---

[5]For more detailed discussion of these control mechanisms see Hafez (1980); Root (1973).

[6]The suppression of FSH by the testes is actually due not to testosterone alone but to an associated testicular substance called *inhibin*, probably produced by Sertoli cells (Brobeck, 1979, pp. 7–108).

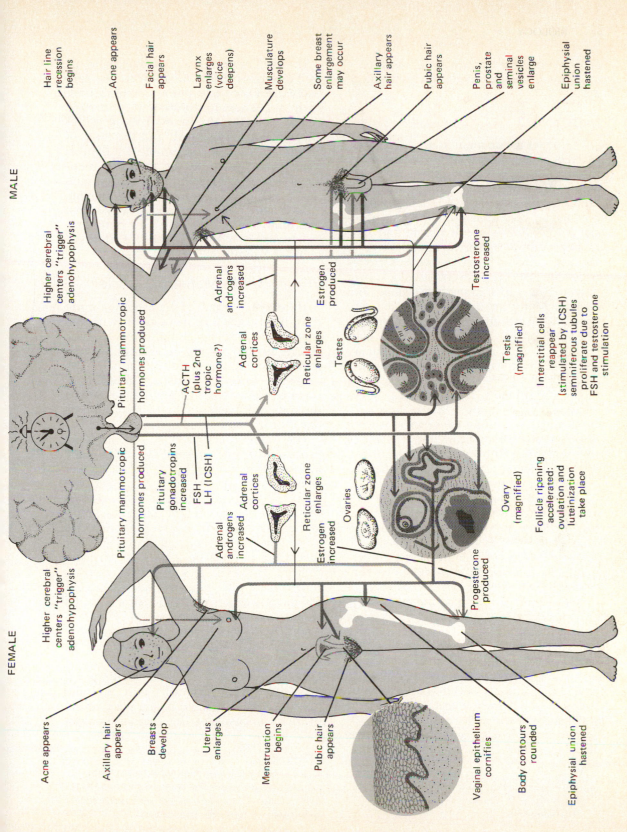

Figure 4.7 Effects of sex hormones on development at puberty. (From *The Reproductive System* by Frank H. Netter, 1953, Ciba.)

accompanies the ovarian cycle) exists only in women, female apes, and some monkeys.

Menarche

The onset of the menstrual cycles in puberty is called *menarche;* the cessation of the menstrual cycles in midlife is the *menopause* (discussed in chapter 7). Both are highly significant biological landmarks in a woman's life, with profound psychosocial implications. Many rituals and customs governing menarche have evolved in various cultures.[12] In Western societies menarche is a private event, usually known only to the girl's immediate family and friends. Many an uninformed pubescent has had the unsettling experience of suddenly bleeding from the vagina. Currently, however, most teenagers in our society receive at least some information and guidance about what to expect. Under these more enlightened circumstances, menarche, far from being a traumatic event, can be an exhilarating experience, since it signifies most dramatically the crossing of the threshold into young womanhood (psychological reactions to puberty are discussed in chapter 9).

The onset of menarche is determined by a variety of genetic and environmental factors. In the United States, menarche now occurs at the average of age 12.8 years with a normal range of 9 to 18 years.[13] Information from around the world shows considerable variation in this regard. For instance, among Cuban girls the median age at menarche is as early as 12.4 years while among the Bundi of New Guinea it is 18.8 years.[14] Among Western societies there has been a general trend over the past century or so whereby the average age at menarche has gradually declined from 17 years in 1840 to the current levels, which became stabilized a few decades ago. Teenage girls therefore now reach menarche at about the same ages that their mothers did.

The decline in the age of menarche, as well as cross-cultural differences, are generally attributed to environmental factors, such as better nutrition and improved health care. But the issue of racial differences is not fully settled, and the force of genetic factors is discernible in familial tendencies related to the onset of menarche. For instance, randomly chosen girls reach menarche differing on the average of 19 months; for sisters who are not twins, the difference is 13 months; for nonidentical twins, it is ten months; for identical twins, 2.8 months.[15]

Phases of the Menstrual Cycle

The average length of the menstrual cycle is 28 days (hence its association with the *lunar month* in folklore). But cycles that are shorter or longer by several days are also quite common and perfectly normal.[16] Because the onset of menstrual bleeding is more clean-cut than its gradual cessation, the time that bleeding starts is counted as day 1 of the menstrual cycle.

The menstrual cycle is a continuous process with one cycle following another. Following menarche it usually takes a few years before the cycles become regularized; they become irregular again during the menopause, before stopping altogether. During the intervening decades most women normally go through their menstrual periods on a fairly predictable rhythm. However, this process is often influenced by physiological and psychological factors, and irregularities in menstrual rhythm are quite common.

The menstrual cycle is controlled by complex neuroendocrine mechanisms involving the hypothalamic, pituitary, and gonadal hormones. Its manifestations involve the entire reproductive tract, with the most important changes taking place in the ovaries, uterus, and the vagina. Though

---

[12]Ford and Beach (1951) and Van Gennep (1960) provide numerous examples.

[13]Zacharias et al. (1976).

[14]Hiernaux (1968).

[15]Tanner (1978).

[16]The length of the ovarian cycle is specific for each species. It is approximately 36 days in the chimpanzee, 20 days in the cow, 16 days in sheep, and 5 days in mice. Dogs and cats are seasonal breeders and usually ovulate only twice a year.

the menstrual cycle is one continuous process, for descriptive purposes it is divided into four phases: *preovulatory phase; ovulation; postovulatory phase;* and *menstrual phase.*

Preovulatory Phase   The ***preovulatory phase*** starts as soon as menstrual bleeding from the previous cycle has ended. This period is also known as the *follicular phase* with reference to the development of the ovarian follicles in the ovary and as the *proliferative phase* with regard to the changes undergone by the uterine mucosa during this period.

As illustrated in figure 4.8, when menstrual bleeding dwindles away, the anterior pituitary begins to produce large amounts of FSH and small amounts of LH. Under the influence of FSH, a cluster of immature follicles in the ovary begins to mature, with one of them taking the lead.

The maturing follicles produce estrogens, of which the levels in the blood rise gradually. Increasing levels of estrogen cause thickening of the uterine endometrium with proliferation of its superficial blood vessels and uterine glands. The cervical glands produce a thick and cloudy mucous discharge, which gradually takes on a watery and viscous character. Simultaneously the lining of the vagina becomes thickened, and its component cells undergo characteristic changes.[17]

Increasing levels of estrogen in the bloodstream gradually inhibit, through negative feedback, the production of FSH. Meanwhile, the production of LH gradu-

[17]These changes are so distinctive that examination of vaginal cells, cervical mucus, and the endometrial lining is relied on in tests of ovarian hormonal function. For a description of these tests, see Williams (1981).

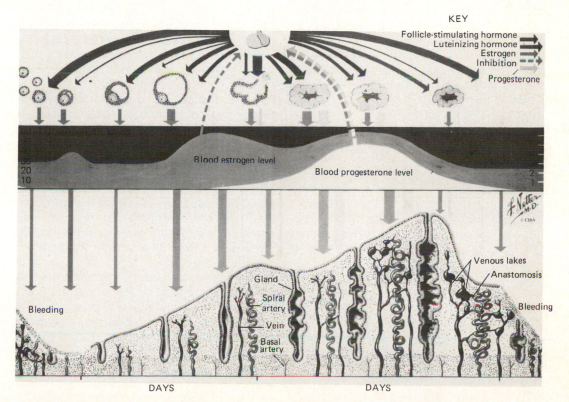

KEY
Follicle-stimulating hormone
Luteinizing hormone
Estrogen
Inhibition
Progesterone

Blood estrogen level

Blood progesterone level

Venous lakes
Anastomosis

Gland
Spiral artery
Vein
Basal artery

Bleeding

Bleeding

DAYS   DAYS

Figure 4.8   The menstrual cycle. (From *The Reproductive System* by Frank H. Netter, 1953, Ciba.)

ally   eases, in turn promoting proges-
ter   roduction.

O   ion   The climactic event in the
c   cycle is *ovulation,* which takes place
a   but the mid-point of a 28-day cycle.
C   ation is triggered by an upsurge in
    This in turn is thought to be caused
    rapid increases in levels of estrogen,
    ich act in a "positive feedback" rela-
    nship. Unlike the negative feedback
    odel discussed earlier, in this case in-
    easing levels of estrogen do not inhibit
    ut enhance the production of the pitui-
tary hormone until the critical point is
reached for the upsurge of LH, triggering
ovulation. As the discharged ovum starts
on its journey through the fallopian tube,
the menstrual cycle enters its postovulatory phase.

Some women experience pain in midcycle during ovulation (called "middle
pain," *Mittelschmerz* in German). This
symptom is most common in young women
and consists of intermittent cramping pains
on one or both sides of the lower abdomen
lasting for about a day.

Postovulatory Phase   Under the influence
of high levels of LH, the corpus luteum
begins to form, increasing the secretion of
progestins. Since it is now the corpus luteum rather than the developing follicle that
is the main source of hormones, the ***postovulatory phase*** is also known as the *luteal
phase*.

The sustained level of estrogens continues to act on the uterine endometrium,
making it progressively thicker. Under the
influence of progestins the uterine glands
now become active and secrete a nutrient
fluid; hence *secretory phase* as another name
for the postovulatory period. Secretions of
cervical mucus gradually thicken and regain their cloudy and sticky consistency.

High levels of estrogens and progestins
in the bloodstream inhibit gonadotropin
production by the anterior pituitary, which
results in gradually declining levels of FSH
and LH. This in turn means weaker stimulation of the corpus luteum and a reduction in the production of estrogens and

progestins. Since these hormones are being
constantly metabolized, their blood levels
decline. Without the sustenance of these
hormones the endometrial lining undergoes degenerative changes and sloughs off,
resulting in the menstrual discharge.

Menstrual Phase   Discharge of the uterine lining, which is manifested as vaginal
bleeding, constitutes **menstruation.** It lasts
four or five days, during which a woman
loses 50 to 200 ml (several large spoonfuls)
of blood; however, the amount varies a
good deal between individuals and in different cycles. Normally this blood loss is
rapidly replenished and there are no ill effects.[18]

The reduced blood levels of estrogens
and progestins during the menstrual phase
result in the loss of their inhibitory effect
on the anterior pituitary. As a result, FSH
production rapidly picks up, heralding the
start of the proliferative phase of the next
cycle.

This foregoing description presupposes
that the woman has not become pregnant.
If she does, the cells of the developing placenta produce *chorionic gonadotropin,* which
sustains the corpus luteum in maintaining
high levels of hormone production. Under
these conditions the uterine lining (which
now contains the embryo) does not slough
off and the pregnant woman misses her
period (chapter 5).

The process we have just described also
explains how birth control pills function.
The pill typically consists of various combinations of synthetic estrogens and progestins. These hormones act the same way
as their natural counterparts produced by
the ovary. Taken on a daily basis, they
maintain high blood levels of these hormones. Since the anterior pituitary has no
way of telling if these hormones are coming from the ovary or the pharmacy, its
production of gonadotropins is suppressed through negative feedback just as

---

[18]Symptoms of menstrual discomfort that affect
some women are discussed in chapter 7. For more
details on the various aspects of the menstrual cycle,
see Lein (1979); Dalton (1979).

it does naturally in the postovulatory phase. But since the birth control pill suppresses the pituitary prior to ovulation, there is no LH surge, no ovulation, and the woman cannot become pregnant. When she stops taking the pill, the hypertrophied endometrial lining sloughs off and menstruation follows. When she resumes taking the pill, another anovulatory cycle is repeated (discussed further in chapter 6).

Disorders of Sexual Development

Though our concerns are primarily with normal sexual development, some attention to abnormalities in sexual differentiation would be instructive, because pathological patterns shed light on normative processes and are of interest in and of themselves. This is particularly true with respect to the possible biological bases of gender identity and gender-related behavior. As we discuss later on, a good deal of the research in this area has been based on abnormal conditions caused either by endocrine disturbances or by exposure to hormones before birth. In order to understand the implications of this research, we need to briefly consider a number of representative examples of these conditions.

The embryonal differentiation of the reproductive system, which we described in chapter 2, and the maturational changes at puberty, which we reviewed earlier in this chapter, constitute the normal developmental patterns for the great majority of individuals. Yet rarely, things do go wrong at a variety of points. Abnormalities in sexual differentiation thus could occur at the *genetic* level, the *hormonal* level (hypothalamic, pituitary, or gonadal), or at the level of the *target tissues.*

The developmental abnormality may be one of timing, or faulty differentiation. When sexual development occurs abnormally early, the result is **precocious puberty** (box 4.1); if late, **delayed puberty.** Failure of puberty to occur altogether results in the inadequate development of sex organs and is called *sexual infantilism* (the male is called a *eunuch*—Box 4.2).

Faulty sexual differentiation creates incongruous combinations of male and female structures. Individuals with ambiguous or contradictory external genitals have long been known since antiquity as **hermaphrodites** (from *Hermes* and *Aphrodite*). They are now also referred to as **intersex.**

The *true hermaphrodite* has both male and female gonads, or admixtures of ovarian and testicular tissue in the gonads. Though usually a genetic female (XX), the external genitalia of such a person combine features of both sexes, with one or another form predominating (figure 4.9).

Where the sex chromosomes are correctly matched with male or female gonads but mismatched with external genitalia, the term *pseudohermaphrodite* is applied. Thus a female pseudohermaphrodite will have female sex chromosomes (XX), ovaries, fallopian tubes, uterus, but masculinized external genitals (figure 4.9). A male pseu-

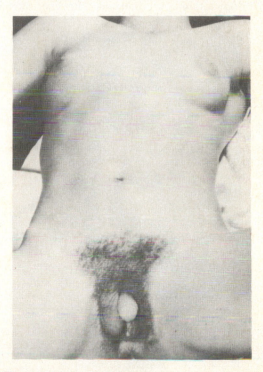

Figure 4.9   A true hermaphrodite with one ovary and one testis. A genetic female who has always lived as a male. (From Money, *Sex Errors of the Body*. Baltimore: Johns Hopkins, 1968, p. 121.) Reprinted by permission.

Box 4.1:   Precocious Puberty

When puberty begins before the age of eight in girls and nine in boys, the condition is called precocious puberty. Girls are twice as likely to have precocious puberty as boys. There are cases of menstruation beginning in the first year of life. The youngest known mother was a Peruvian girl who began menstruating at three years and gave birth (by cesarean section) to a baby boy at the age of five years and seven months (see accompanying picture).

There are cases in which penile development has begun at five months and spermatogenesis at five years (see accompanying figure). Boys as young as seven years of age have reportedly fathered children. In many cases, especially in girls, precocious puberty does not reflect any underlying pathology, but simply reflects natural variations in the body's timing mechanisms. However, 20% of girls and 60% of boys with precocious puberty have a serious organic disease. Tumors in the hypothalamic region of the brain, the gonads, or the adrenals can trigger the production of gonadotropins or sex hormones, which in turn leads to the early sexual maturation of the body. These cases clearly show that puberty depends on the maturation of the hormonal system and not of the body tissues; the reproductive system is ready to mature at any time.

In some cases, a child may undergo incomplete precocious puberty, with early breast development or early growth of pubic hair, but no other changes. These changes are less likely to be related to major organic disorders than is complete precocious puberty. Precocious puberty is usually not accompanied by the heightened sexual interest that typically occurs with normal puberty.

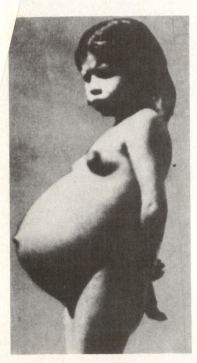 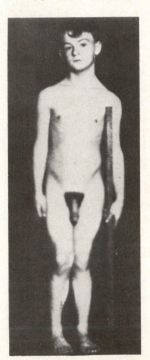

Lina Medina (left) is the youngest known mother in the medical literature. She gave birth in 1939 at the age of five years and seven months. A five-year-old boy with an unusual adrenal development (right) has the height of an 11.5-year-old and precocious penile development. (From Wilkins, Blizzard, and Migeon, *The Diagnosis and Treatment of Endocrine Disorders in Childhood and Adolescence,* 1965.)

Box 4.2:   Eunuchs

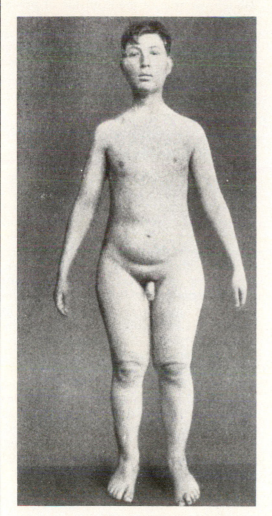

Twenty-two-year-old castrated man in a 1926 photograph (Bilder Lexicon).

Of the many causes of androgen insufficiency, the most dramatic is the removal of the testes or *castration*. If done before puberty, castration results in the failure of adequate genital development and the non-emergence of the secondary sexual characteristics. The person, known as a *eunuch,* will have bypassed most of the changes of puberty (see accompanying figure). He will have a high-pitched voice, poor muscular development, underdeveloped genitals, no beard, no pubic or axillary hair, and female-type subcutaneous fat deposits with partially developed breasts. Those castrated after puberty do not lose their secondary characteristics.

Castration has been practiced in many cultures since remote antiquity for various reasons but most notably to provide "safe" guardians for women (*eunoukhos* means "guardian of the bed" in Greek). Best known in connection with Islamic harems, the practice goes back to antiquity.

Some eunuchs wielded great power in Islamic and Chinese courts. With no family ties or offspring, their sole loyalty was to the ruler, and they often acted as members of his personal staff. At the peak of their influence, there were over 70,000 eunuchs in the service of the Ming dynasty. The eunuch system in China was not completely abolished until 1924, when the final 470 eunuchs were driven from the last emperor's palace.[1]

In Europe as late as the turn of the 19th century boys were castrated to maintain their soprano voices. These *castrati* sang in church choirs and performed female roles in opera, which excluded women. During this present century, castration has been used in Europe and the United States in efforts to modify various sexual behaviors such as "chronic" masturbation, homosexuality, exhibitionism, and child molestation. (Castration for legal reasons is discussed further in chapter 18.)

[1]Mitamura (1970).

dohermaphrodite will have male sex chromosomes, testes, male tubal structure but female external genitals or even a feminine body build. There are many variants of these abnormalities and numerous conditions that cause them. We shall restrict ourselves here to a few representative examples.

Disorders Caused by Sex Chromosomes

Instead of the normal complement of 46 chromosomes, including XX or XY chromosomes, some individuals are born with extra or missing sex chromosomes. Thus *Klinefelter's syndrome* results when a genetic male is born with an extra X chromosome (XXY); *Turner's syndrome* results when a genetic female has one X chromosome missing (XO).

In Klinefelter's syndrome masculinization is incomplete, resulting in a small penis, small testes, low testosterone production, and therefore incomplete development of secondary sex traits. Some show partial breast development at puberty. These men are infertile and may have problems with their gender identity.

In females with Turner's syndrome, ovaries are absent or present only in rudimentary form, incapable of producing ova or female hormones. Therefore, the XO female is infertile and does not undergo puberty unless treated with female sex hormones. These women are of short stature and may have congenital organ defects, webbing between the fingers and toes or between the neck and shoulders. Despite incomplete development of the female organs, XO females usually have no gender problems during psychosexual development.

Another chromosomal abnormality with intriguing behavioral consequences results from the presence of an extra Y chromosome in males. XYY individuals tend to be tall and markedly masculine in appearance even though they do not always fit this image. The XYY syndrome originally received a great deal of publicity because a high percentage of such men were reported to be found among prisoners who had committed violent crimes. Yet when a random sample of men outside of prison was examined, the association of violence did not appear to be higher in XYY than in XY males; the validity of the association between an extra Y chromosome and proneness to violence and antisocial sexual behavior is therefore doubtful.[19]

Disorders Caused by Abnormalities of Sex Hormones

Disorders of differentiation due to chromosomes usually exert their effects through resultant hormonal abnormalities. But developmental abnormalities may also result from hormonal dysfunction as such even where sex chromosomes are normal. Such hormonal abnormalities may entail overfunction or underfunction at any level of the hypothalamic-pituitary-gonadal system of the embryo. Or the abnormality in sexual differentiation may be caused by hormones produced or ingested by the mother during pregnancy.

In the female, hormonally caused pseudohermaphroditism most commonly is associated with *congenital adrenal hyperplasia* (CAH). The condition (also known as *adrenogenital syndrome*) results from a genetic defect that leads to the excessive production of androgen by the fetal adrenal cortex. Thus if the embryo is a genetic female her external genitalia will become masculinized. The clitoris is hypertrophied to look like a penis, and the labial folds fuse together giving the appearance of a scrotum. The condition can usually be treated by suppressing the excessive androgen production and surgically correcting the genital abnormality within the first few weeks of life.

The same end result occurs in a perfectly normal female embryo if the pregnant mother produces abnormal amounts of androgen (for example due to an androgen-

[19]Witkin et al. (1976).

producing tumor) or ingests androgenic hormones. Thus, some years ago when pregnant women were given certain synthetic steroids to prevent miscarriage (before the masculinizing effects of these compounds were known), they gave birth to masculinized female babies.

Male pseudohermaphroditism results if there is inadequate androgen production or if the hormones produced are not biologically potent. In the absence of normal male hormone, female genitalia will develop.

Disorders Due to Abnormalities in Tissue Response

Even where the hormonal system is quite normal, abnormalities will result if the bodily tissues do not respond normally to hormonal stimulation. Such a condition, known as the *testicular feminization syndrome,* is the most common cause of a male form of pseudohermaphroditism. In this case, a genetic male with androgen-producing testes will develop a female appearance with female genitalia (with clitoris, labia, vagina) and breasts, because of a failure on the part of these tissues to respond to testosterone stimulation (figure 4.10). The net effect is therefore the same as if no androgen were being produced. Since response to the Mullerian duct-inhibiting substance is not impaired, no uterus or fallopian tubes develop. Genetically male infants with this syndrome look like normal baby girls (unless the testes can be felt in the groin). The condition is usually diagnosed at puberty when menarche fails to occur.

Unlike other forms of pseudohermaphroditism that can be treated with surgery and hormonal treatment if detected early, cases of complete testicular feminization must be reared as female even if diagnosed early since neither surgery nor hormonal treatment will effectively reverse the established female pattern or create functional male genitals.

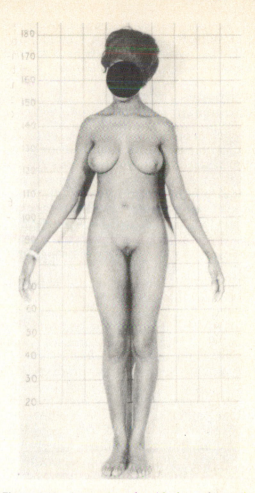

Figure 4.10 A genetic male with the androgen insensitivity syndrome. (From Money, *Sex Errors of the Body.* Baltimore: Johns Hopkins, 1968, p. 113.) Reprinted by permission.

Hormones and Sexual Behavior

Over the past half century a great deal of evidence has been gathered to demonstrate the critical influence that steroid hormones play in shaping and sustaining mammalian sexual behavior. The extent to which this mammalian influence persists among primates, particularly humans, is not yet fully established. But despite many uncertainties, it seems quite probable that sex hormones are significantly linked to at least some aspects of human sexual behavior. Equally important is the possible relationship of hormones to gender identity, and other gender-related behavioral and cognitive differences.

Hormones and Mammalian Behavior

Much of our knowledge of the effects of steroid hormones on sexual behavior comes from research on lower mammals (particularly rodents like the rat, hamster, and guinea pig—whimsically called the "ramstergig" by Beach).[20] These animals are convenient to study and have predictable, stereotyped, and gender-specific sexual behaviors that are under the strong influence of sex hormones.

The sexual behavior of the female is linked to the period of *estrus* (Latin, "frenzy") that coincides with ovulation and during which the animal is said to be "in heat." The estrous female arouses sexual interest in the male by physical changes in her genital region and the production of potent scent signals conveyed by *pheromones*. Whereas hormones affect the individual producing them, pheromones are secreted to the outside and the scents emanating from them influence the behavior of other animals of the same species (box 4.3). Unlike the female, male sexual interest is not cyclical; the male is always ready to copulate, provided there is a receptive female available. In effect, then, male sexual behavior is dependent on or controlled by female receptivity.

The copulatory behavior of both male and female mammals depends on sex hormones for its development and sustenance. Castration of sexually experienced rats is followed by decline in sexual behavior; subsequent administration of testosterone restores the sexual behavior to its earlier levels. The presence of testosterone during early life is necessary for adequate behavioral responses to testosterone during adulthood; reduced testosterone stimulation in infancy impairs sexual per-

formance in adulthood. Comparable conditions apply to female animals.[21] Furthermore, through experimental manipulation it has been possible to reverse the sexual behavioral patterns of male and female. Based on such findings, a model has been developed linking hormones to sexual behavior among mammals that may have possible applicability to humans.

The Organizational and Activational Effects of Hormones   The influence of sex hormones on behavior occurs at two levels. The *organizational* effects of hormones involve their influence on the development and differentiation of those portions of the brain that deal with sexual behavior. The *activational* effects of hormones pertain to the initiation and maintenance of sexual behavior after the organizational effects have taken place. These two effects are distinguishable in several ways.[22] The organizational effects are developmental in nature, typically occurring during a time-limited period of maximal susceptibility, or *critical phase*, either before birth (the *prenatal* period) or around the time of birth (the *perinatal* period). Their influence tends to be long-term and relatively permanent. By contrast, the activational effects are reversible, repeatable, and not limited to a critical phase of development. The organizational effects of hormones have been compared to the exposure of photographic film (which captures the image) and their activational effect to exposed film in developer fluid (which brings out the image). That this model operates at the level of the reproductive system is quite clear. Androgens exert their organizational effect in intrauterine life and shape undifferentiated structures into male or female patterns (chapter 2); but what is the evidence for sexual dimorphism at the level of the brain?

---

[20]Beach (1971).

[21]For more extensive discussion of animal research in this area see Goy and McEwen (1980); Bermant and Davidson (1974); Beach (1971); Gorski (1971); Goy and Goldfoot (1974); Feder (1981).

[22]Ehrhardt and Meyer-Bahlburg (1981).

Box 4.3:   Pheromones

The term *pheromone* (from the Greek, "to transfer excitement") was coined in 1959 to describe chemical sex attractants in insects, the existence of which had been known since the 19th century. These are remarkably potent substances; the minute amount of pheromone that exists in a single female gypsy moth is enough to sexually excite more than a billion males from as far as two miles away. In a number of insect species, the males in turn produce "aphrodisiacs" to induce the female to copulate. In the male cockroach this is an oily substance whose consumption induces the female into the coital posture.[1]

There are numerous examples of the influence of pheromones in mammalian reproductive and social behavior: housing female mice together inhibits their ovarian cycles; exposure to a male or just his urine will revive the cycles; the odor of male mice accelerates puberty of young female mice; the introduction of a male from a foreign colony suppresses the pregnancy of mice who have mated with their own males.

The menstrual cycles of women show a similar tendency to be influenced by the odors of other women without their being conscious of it. For example, women who live together in college dormitories or other close quarters often begin to menstruate close to the same time.[2] When a "donor" woman wears cotton pads under her arms for a 24-hour period and other women are then exposed to her armpit odor, the menstrual periods of the recipient women shift to become closer to the periods of the donor.[3] There is also some evidence of male influence. In one study, women who seldom dated had longer cycles; those who dated more often had shorter and more regular cycles.[2]

What are the substances that cause such curious effects? Some investigators have reported the presence of volatile fatty acids ("copulins") in vaginal secretions of rhesus monkeys in midcycle that stimulate male sexual interest, mounting, and ejaculation. Similar compounds have been identified in human vaginal secretions that peak at midcycle, but this does not seem to be the case for women on the pill.[4] But other investigators have failed to confirm these findings. Hence, the role of such human "copulins" is currently unclear.[5]

People are apparently able to detect differences in vaginal odors (without being told the source) and under experimental conditions most men and women find them rather unpleasant, but less so in the case of midcycle secretions.[6] On the other hand, erotic literature is full of testimonials to the rousing smell of genital secretions; perhaps people react to such odors when sexually excited in quite different ways.

Vaginal secretions moreover need not be the primary source of erotic scent signals. Our body is studded with odor-producing glands, and despite our scrupulous efforts to eliminate and conceal them, they well may still be part of the "silent language" of sex. One perfume company at least is counting on that by marketing a fragrance that contains *alpha androstenol*, a synthetic chemical similar to a substance in human perspiration (and a powerful pheromone produced by boars, which makes the sow adopt the mating posture) even though there is no scientific evidence of its sexual effect on people.[7]

[1]Hopson (1979).
[2]McClintock (1971).
[3]Russell et al. (1977).
[4]Michael and Keverne (1968); Michael et al. (1974); Michael et al. (1976); Bonsall and Michael (1978).
[5]Goldfoot et al. (1976).
[6]Doty et al. (1975).
[7]*Time*, June 29, 1981.

Sexual Dimorphism of the Brain　Evidence for sexual dimorphism in the mammalian brain comes from three areas. The first is *behavioral*. As we described in the case of the rat, mammalian sexual behaviors are typically sex-specific. Since such sex differences do not seem to be primarily acquired by learning, the inference is that behavioral dimorphism reflects brain dimorphism.

More direct evidence comes from *physiological* functions. In both sexes, the production of gonadal hormones is controlled by the same pituitary hormones (FSH and LH). Why is it then that the female produces gonadal hormones in a cyclical fashion, and the male does not? Experimental evidence points to a sex difference in the hypothalamic function rather than to pituitary or gonadal causes.[23]

The third type of evidence comes from the demonstration of *anatomical* differences between the male and female mammalian brains. A number of such structural differences have now been shown to exist, most notably in the *preoptic area* of the rat brain.[24] In the male, the medial part of the preoptic nucleus is strikingly larger than in the female. This is a significant finding, since the preoptic area is essential for the preovulatory surge of gonadotropin that occurs in the female but not in the male; it is also important for both male and female sexual behavior.[25]

The Role of Androgens　A good deal of evidence points to androgen as the key hormone in the sexual organization of the brain, as it is with respect to the sexual differentiation of the reproductive system. Depriving the male of testosterone during its species-specific critical time of brain differentiation results in a female pattern of sexual behavior; exposing the female mammal to androgen prenatally or perinatally leads to a male pattern of sexual behavior in adulthood.[26] These results are dependent on the organizational effect of androgen or its absence in early life, complemented by the activational effects of giving male or female hormones in adulthood.

Steroid hormones continue to exert an important influence on the sexual behavior of nonhuman primates, but their role is not as definitive as in the case of lower mammals. Thus, although removal of the ovaries in subprimate mammals virtually eliminates sexual receptivity, primate females appear to be somewhat less dependent on hormones for sexual responsiveness. Generally the same is true for the level of dependence of male primate sexual behavior on hormones.

Many questions still need to be answered even at the lower mammalian levels, and there are many complexities we have not dealt with here. For instance, androgens may not be the sole hormones exerting an organizational effect; progesterone may play a protective role against the effects of androgens on the brain; estrogens too, in low concentration, may exert an influence.[27]

With respect to activational effects, it would be misleading to think of hormones as a form of "erotic fuel" that ignites the sexual drive and keeps it going. The character and state of the body on which hormones act are important determinants in the behavioral outcomes that result from the action of hormones. Thus, in addition to the earlier organizational effects of hormones, age, conditions of rearing, experience, nutritional state, and the testing situation in experimental conditions all influence the behavioral outcome.

Hormones and Human Sexual Behavior

The task of determining the organizational and activational roles of steroid hormones among humans has proven to be far more difficult than among lower animals for the following reasons.

[23]For a summary of this research see Daly and Wilson (1978).

[24]Raisman and Field (1971). Also see Goy and McEwen (1980).

[25]Lisk (1967)

[26]Ehrhardt and Meyer-Bahlburg (1981).

[27]MacLusky and Naftolin (1981).

First, human sexual behavior is incomparably more complex than its animal counterpart. Cultural influences are so pervasive and varied in our lives that the subtlety and range of human sexual diversity is bewilderingly diverse. Thus in studying the relationship of hormones to sexual behavior we have no discrete and simple measure like mounting or lordosis to go by.

Second, the key experiments that have elucidated so much of animal sexual interactions cannot be replicated among humans. We are not about to castrate baby boys or give androgen to baby girls under experimentally controlled conditions to see what happens to their sexual development. Instead we must rely on studying the often haphazard consequences of such events when they occur because of pathological and unforeseen circumstances.

Third, it may be argued that the reason that we have difficulty in finding convincing evidence for the influence of hormones on human sexual behavior is perhaps because there is no such influence to be found. Beach has proposed that during the course of evolution, there has been a progressive "emancipation" of sexual behavior from hormonal control.[28] Particularly in humans, there is instead a far greater dependence on social learning and individual experiences. But although no one doubts the enormous influence of sociocultural factors on human sexual behavior, whether such behavior has become totally emancipated from the organizational and activational effects of hormones is yet to be settled.

Two decades ago, John Money was able to conclude, "The sex hormones, it appears, have no direct effect on the direction or content of erotic inclination in the human species."[29] Other investigators are now not so sure. The greater role of social learning in human behavior notwithstanding, there is sufficient evidence to suggest that biological factors influence human psychosexual differentiation. The subcort-

ical regions of the human brain are strikingly similar to those of subhuman mammals. However, no sex differences have been shown in the preoptic region of the human brain, as there have been in the rat. Preliminary observations suggest a sex difference in the shape and size of the human *corpus callosum*, but the significance of this is far from clear.[30]

The cyclicity in gonadal hormone secretion in women but not in men presupposes differences in brain function, just as it does among lower mammals. However, in order to understand the influence of steroid hormones, we need to ask more sophisticated questions than, "Do hormones have anything to do with sex?" They do and they do not. Hormones may have a significant impact on a particular facet of sexuality but not on another; they may be influential under certain conditions but not others. In short, hormonal influences, if they exist, only make sense in the broader context of human experience, as one factor among many, that predispose us or increase the likelihood of our acting in one manner or another, but not as some blind force that drives us into sex willy-nilly. We therefore need to consider separately the effect of hormones on various major facets of sexual behavior.[31]

Hormones and the Male Sexual Drive

Erotic desire that motivates us to act sexually is usually taken for granted as part of "human nature." But, as we discuss in chapter 8, the matter is far more enigmatic than that. One of the key questions in this respect is the extent to which biological factors, hormones in particular, are responsible for generating and sustaining sexual motivation, what is commonly referred to as the **sexual drive** or *libido*.

The most persuasive case for the effects of hormones on sexual drive is based on the role of androgen in the male, evidence for which comes from the effects of castra-

---

[28]Beach (1947).

[29]Money (1961), p. 1396.

[30]Lacoste-Utamsing and Holloway (1982). The issue of brain differences is further discussed in chapter 9 in connection with gender identity.

[31]Ehrhardt and Meyer-Bahlburg (1981).

tion (box 4.2). Despite past uncertainties in this regard, it is now generally accepted that castration is incompatible with normal sexuality in men, particularly where castration has been done prior to puberty.

There is little by the way of reliable and detailed information on the sexual capabilities and erotic life of castrates. Based on a review of the literature and the few cases he had seen, Kinsey concluded: "Human males who are castrated as adults are, in many but not in all cases, still capable of being aroused by tactile or psychogenic stimuli."[32] Subsequent surveys have reported similar retention of sexual function. In one instance, 50% of 38 men castrated 3 to 5 years previously, had erections in response to erotic films.[33] Other reports reveal complete loss of "sexual response" ranging from 37% to 90% of cases.[34] Contradictions abound because studies of castrates are rife with methodological weaknesses. Nonetheless, as Davidson sums it up, "It is clear that castration is incompatible with normal sexuality in men or males of any other mammalian species studied."[35]

Further confirmation of the importance of androgen for libido comes from clinical experience with the treatment of hypogonadal males. When patients with inadequate levels of androgen are treated with testosterone, there is a distinct improvement in their failing sexual functions (discussed in chapter 15). However, among normal males with no testosterone deficiency, levels of hormones do not relate significantly to levels of sexual desire, excitement, or frequency of coitus.[36] In other words, if a man has normal levels of androgen, giving that person more androgen has no effect (as with a full glass, the excess spills over).

The influence of hormones on male sexual behavior has an interesting corollary in the effect that testosterone may have on aggressive behavior, which in turn may have a significant bearing on sexual activity. Ordinarily there are no discernible differences in testosterone levels between less or more aggressive men. Although aggressive behavior is highly unlikely to be explained simply on hormonal grounds, it is possible that among individuals otherwise predisposed to develop such behavior, testosterone might be an additional facilitating factor. Consequently, just as testosterone has been used to treat people with low levels of sexual drive, *chemical castration* through the administration of antiandrogenic agents such as *cyproterone* and *medroxyprogesterone*, is now used to treat men with compulsively antisocial sexual behaviors (discussed in chapter 19).

Hormones and Female Sexual Drive   Unlike subhuman primates and other mammals, women have no period of estrus. Instead, like men, they are potentially sexually receptive at any time.[37] But whether or not female libido is completely free from hormonal influence remains an open question. Studies in this regard have been focused on the role of estrogens and progestins, as well as androgen.

The most obvious approach would be to look for fluctuations in sexual interest during the menstrual cycle, but attempts to do this have so far led to inconclusive results (box 4.4).[38] The cessation of ovarian function through removal of the ovaries or following the menopause represents the physiological counterpart of male castration. However, women, unlike men, generally do not show any serious deficiency in sexual drive following the loss of ovarian function; on the contrary, some report a heightening of sexual interest. This is not to say that the sexual relationships of many women are not hampered to various degrees by the menopause. But cultural attitudes and psychological factors are usu-

[32]Kinsey et al. (1953), p. 744.
[33]Heim and Hursch (1979); Heim (1981).
[34]Sturup (1979).
[35]Davidson (1980), p. 320.
[36]Persky (1983).

[37]The loss of estrus during human evolution is thought to have had important consequences on the development of sexual associations and institutions like the family, which we discuss in chapter 14.

[38]Persky (1983).

Box 4.4:   Sexual Responsiveness and the Menstrual Cycle

The regular fluctuation of estrogens and progestins during the menstrual cycle would seem to provide an excellent opportunity to study the relationship of hormones and the intensity of the sexual drive. The mammalian pattern is quite clear: the period of estrus and female receptivity overlaps with ovulation, which maximizes the likelihood that copulation will lead to fertilization.

Women have no period of estrus; but does that mean that they have no cyclic fluctuations whatsoever in the level of their sexual desire? Some 30 studies have so far failed to answer that with any certainty. Several studies have shown an increase in the likelihood of intercourse at the time or shortly after ovulation at midcycle.[1] A similar pattern has been found based on self-ratings of sexual arousal among young women during various menstrual phases.[2]

Other studies show the evidence of coitus to peak after menstruation with another peak at midcycle (see accompanying figure).[3] The peak after menstruation may be compensatory to the period of abstinence during menses. Women who engage in coitus during their period do not show such a rebound reaction.

The midcycle peak, when it occurs, may be in response to increased levels of estrogen, or it may be due to higher levels of androgen, which are also present at this time. The decline in sexual activity during the

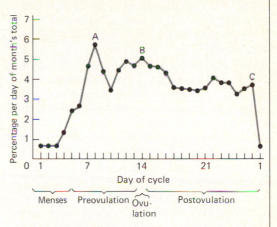

The distribution of intercourse for 52 women on 1,123 occasions during the menstrual cycle. Redrawn from McCance et al. (1952).

postovulatory period is consistent with the higher levels of progestogens, which are known to exert an inhibitory effect on libido.

The inconsistencies in this area of research result in part from differences in methodology. But in addition they attest to the social complexities of human sexual emotions and interaction. It is also possible that no consistent pattern can be found because female libido is truly emancipated from endocrine control.

[1]Udry and Morris (1968).
[2]Moos et al. (1969); McCance et al. (1952); James (1971); Gold and Adams (1978).

---

ally more relevant to such outcomes (see chapter 7). Current research shows some decrease in sexual response to erotic films by menopausal women as assessed by physiological measures. But it is not known whether the magnitude of change noted is sufficient to cause sexual dysfunction, and there is no comparable change in subjective ratings of sexual arousal between menopausal and premenopausal women.[39]

The administration of estrogens to a woman with nonfunctional or absent ovaries brings about a slowing or even some reversal of menopausal changes in the reproductive tract and sexual functions like the lubricatory response. These in turn may

[39]Morrell et al. (in press).

influence a woman's sexual interest and behavior. It is also claimed that the administration of estrogen more directly brings about an enhancement of sexual desire, enjoyment, and orgasmic frequency.[40]

The behavioral aspects of progestins are more enigmatic. Progesterone has been shown to inhibit sexual desire.[41] Yet young women who exhibit higher levels of activity in the luteal phase of their cycle also have higher progestins levels than women who do not.

Given the large numbers of women who have taken birth control pills, one would think we would have learned by now a great deal about the effect of steroid hormones on female libido. But studies of oral contraceptives in this respect have produced conflicting results. The impact, when there is an impact, seems to be in the direction of the birth control pill having an inhibitory effect on libido. But such a physiological effect can often be more than counteracted by the freedom from fear of pregnancy that results in greater sexual interest and responsiveness.

Another significant consideration is that the administration of sex hormones to individuals who already have normal levels of a given hormone is not a good test of whether or not hormones are responsible for sustaining the sexual drive. Even testosterone, whose function with respect to male sexuality is well established, has generally no effect on people who are not suffering from its lack.

It has been suggested that androgens are the hormones responsible for the female sexual drive as they are for the male. This may well be so, but both the experimental and clinical evidence for it are as yet inconclusive. For example, removal of the adrenal glands in women has been reported to result in reduction of sexual desire, responsiveness, and activity. But these cases have involved women with advanced stages of breast cancer; under such circumstances an objective assessment of sexual activity is hardly possible.[42] In other cases testosterone has been given to women for other medical reasons with reports of enhanced libido. But improvement here may be the result of the more general effects of testosterone on body tissue; or, since androgens cause hypertrophy of the clitoris, the effect may result from enhanced sensitivity of the genital tissues.[43] In light of all this, the nature of the relationship of female libido to sex hormones remains quite unsettled.[44]

Another area of uncertainty is the relationship between hormones and sexual orientation. Since this research has mainly dealt with the development of homosexuality, we discuss it in that connection (chapter 13).

Hormones and Gender-Related Behavior

Sexual drive and orientation are integral parts of sexual behavior. The gender-related behaviors we discuss in this section are somewhat more peripheral to sexual activity. Nonetheless, they have an important bearing on sexual behavior, and their importance extends to areas beyond the sexual.

Gender Identity

Dimorphism in sexual behavior—the different ways in which males and females behave sexually—is shared by humans and animals. But *gender identity* is a uniquely human phenomenon; it refers to our self-perceptions with respect to being *masculine* and *feminine*, the personality characteristics that are assumed to be associated with being male or female. Gender identity is a complex and controversial concept that cannot be addressed properly outside of its cultural context. We therefore deal with it in greater detail when we discuss sexual development (chapter 9). It would be convenient nonetheless to examine here the hormonal factors that could possibly be involved in defining gender identity, in the light of what we have discussed so far.

[40]Dennerstein et al. (1980).
[41]Bancroft (1980).
[42]Waxenburg et al. (1959)

[43]Gray and Gorzalka (1980)
[44]Persky (1983)

Gender identity is in part shaped by the person's awareness of his or her body and genital organs, a fact we become aware of as children. But beyond that, is there some other biological basis for our sense of being male or female? Do we start life with our brains "wired" one way or another? Does social learning merely amplify these differences, or is it the sole basis for it?

Much of the research undertaken to answer these questions has involved individuals born with the various hormonally induced abnormalities that we discussed earlier. Particularly instructive have been matched pairs of pseudohermaphrodites (or intersex) patients. Consider, for instance, two children who have the same genetic and gonadal sex, the same presumed prenatal hormonal environment, and the same degree of ambiguity of external sex organs. Based on medical opinion and parental preference, suppose it is decided to rear one child as male, the other as female. Through surgery and hormonal therapy the genital and physical appearance of each child is made to confirm to the gender according to which each child is to be reared, and then one child is treated as a boy, and the other a girl.

If gender is determined by the prenatal hormonal environment (the organizational effect), we would expect the two children to develop a similar gender identity, irrespective of their being reared differently; if the process of rearing is the decisive factor then one child will develop male gender identity, the other female gender identity, consistent with the gender basis for child rearing. The main body of evidence in this area of research supports the latter outcome.[45]

The preeminence of rearing in gender definition is also claimed to be attested by an unusual case where one of twin infant boys lost his penis through a surgical mishap during circumcision. Since it was easier to alter his genitals to make them look female than to reconstruct the penis, he was castrated, his genitals surgically re-shaped, and he was then treated with estrogen at puberty. Reared as a girl, the child is reported to have successfully developed an essentially female gender identity.[46]

Based on such evidence it is maintained that gender identity is not shaped by the effect of hormones on the brain but is a function of social learning. So instead of a theory of instinctive masculinity or femininity which is innate, the evidence of hermaphroditism suggests that gender is undifferentiated at birth and that it becomes differentiated as masculine or feminine in the course of growing up.

It also appears that there is a definite timetable according to which this development takes place. John Money and his associates place the critical period for the formation of gender identity at 18 to 36 months. The further one moves from this critical period the more difficult and problematic are the attempts to reassign gender.

The model that emerges from this perspective is one of *gender neutrality* at birth. So far as gender identity is concerned, life starts with a clean slate on which the text of gender identity is written by social learning. This model has not gone unchallenged. The counterargument is that an individual's biological heritage sets limits to the definition of gender identity and that it is within these limits that social forces exert their influence.[47] In this view, it would have made more sense to rear the boy with the lost penis as male, despite the problems he would have to confront as a result of it.[48]

The primacy of rearing and the existence of a critical period in childhood for the establishment of gender identity appear to be further contradicted by an intriguing phenomenon that has been reported from a remote rural area of the Dominican Republic. Over four generations, some 38 seemingly normal girls have undergone a

---

[45]Money, Hampson, and Hampson (1955); Money and Ehrhardt (1972); Ehrhardt and Meyer-Bahlburg (1981).

[46]Money and Ehrhardt (1972); Money and Tucker (1975). Recently, however, serious questions have been raised about the reported outcome of this case (Diamond, 1982).

[47]Diamond (1965); (1976); (1979).

[48]Diamond (1982).

mysterious transformation at puberty. Between the ages of 7 and 12, instead of developing breasts, they develop muscles; their voices deepen; the clitoris develops into a penis; testes descend; and ejaculation produces live sperm. The explanation for this phenomenon lies in an enzymatic defect, whereby fetal androgen cannot be converted into *dihydrotestosterone*, which is necessary for the sexual differentiation of the external genitalia. The result is a form of male pseudohermaphroditism. At puberty the increased level of testosterone is able to assert its influence causing the sex reversal. Along with these somatic changes, these "girls" are reportely able to shed their female identity in favor of a masculine one.[49] However, closer scrutiny of this study raises a host of methodological concerns that make its conclusion far from definitive.[50]

At present, the more widely held view is in support of the proposition that the sex of rearing, as established by parental and early social influences, is the primary determinant of gender identity.

Sex-Dimorphic Behavior

Gender identity is one component of being male or female. Another is **gender-role,** which refers to the various behavioral patterns that characterize men and women. Like gender identity, gender role is heavily laden with cultural expectations and strongly influenced by social learning (chapter 19). Nonetheless, there is evidence that sex-dimorphic behaviors related to gender role and temperamental sex differences appear to be modified by prenatal hormones.[51]

As with gender identity, the behaviors we are concerned with here are not sexual, in the sense of being erotic. However, they are intimately bound with sexuality by virtue of helping to define the interpersonal context of relationships between male and female, as well as members of the same sex.

These behaviors, which have been studied in children and adolescents (as well as nonhuman primates), cluster in the following areas: (1) Energy expenditure as manifested in active play and athletic skills. (2) Social aggression—physical and verbal fighting. (3) Parental rehearsal—doll play (as against playing with other toys like cars), playing "house," play-acting the role of mother or father, participating in infant care, fantasies about having children. (4) Preference of playmates by sex. (5) Gender role labeling—conformity and deviance from existing norms of masculine and feminine behavior; girls who are markedly different in this regard are labeled "tomboy," boys are called "sissy." (6) Grooming behavior—dress, make-up, hairdo, use of jewelry and the like.

Among nonhuman primates, there is good evidence linking prenatal hormonal influence with sex differences in behaviors like rough-and-tumble play, threat behavior and fighting, parenting, dominance behavior, sex-segregated play, and grooming. Studies among humans largely focused once again on individuals with unusual prenatal hormonal histories, to ascertain the behavioral consequences of exposure to unusually high or low levels of androgens, estrogens, and progestins.

Fetal Androgenization in Genetic Females and Males   Cases of congenital adrenal hyperplasia (CAH) have been particularly instructive in this area. Gender identity among CAH boys and girls typically agrees with their sex of rearing, even though CAH girls are prenatally exposed to high levels of androgen. But this hormonal influence seems to make a significant difference in the sex-dimorphic behavior of these girls: they engage in a great deal of active outdoor play, show increased association with male peers, tend to be identified as "tomboys" by themselves and others, show decreased parental rehearsing behavior such as doll play and baby care, and have a low interest in the roles of wife and mother, as

[49]Imperato-McGinley et al. (1974); (1979). These children are known locally as *guevodoces,* which means "penis at twelve." Now that the condition has been recognized, they are reared as boys from birth.

[50]For a critique of the Dominican case, see Rubin et al. (1981).

[51]Ehrhardt and Meyer-Bahlburg (1981).

against having career aspirations.[52] Although the concentration of such behavior among CAH girls is significant, the presence of such traits as such does not constitute evidence for hormonal abnormality. There are a great many perfectly normal girls who may act like tomboys, and such behavior is now widely acceptable in our culture. (This is an issue we discuss in greater detail in chapter 19.)

Since CAH girls have had their abnormally high levels of androgen controlled and their masculinized genitals surgically corrected soon after birth, their behavior is ascribed to the prenatal effects of androgen. Further confirmation comes from the study of females whose mothers were given masculinizing sex steroids during pregnancy.[53] These girls had no intrinsic hormonal abnormality, their masculinized genitals were corrected early, and they were reared as girls. Yet they too showed behavioral patterns similar to CAH girls.

Fetal Nonandrogenization in Genetic Males   The clinical syndrome in the male that corresponds to the CAH in women is the condition of testicular feminization. As discussed earlier, this results from the failure of male tissues to respond to the effects of testosterone. These genetic males are thus born and grow up with female bodies and external genitals (figure 4.10). Sex of rearing is female and in the majority of cases sex-dimorphic behavior is stereotypically feminine.

[52]Ehrhardt et al. (1968).
[53]Money and Ehrhardt (1972).

This outcome could be due to the failure of the brain to become androgenized, consistent with the insensitivity to testosterone shown by the rest of the body. These males may therefore develop "female brains" and behave accordingly. However, we cannot be certain of this. Unlike CAH females who grow up looking female and are reared as females, males with androgen insensitivity grow up looking not male but female and are reared not as males but as females. Hence, the influence of rearing cannot be separated in these cases from those of endocrine factors.

In this chapter, we have focused mainly on the role of hormones in shaping sexual and gender-dimorphic behaviors, because hormones are what we have been concerned with. But much more can be said about these issues from psychological and social perspectives. We address these matters at length in subsequent chapters.

SUGGESTED READINGS

Goodman and Gilman's The Pharmacological Basis of Therapeutics, sixth edition, edited by Gilman et al. (1980). An authoritative source for learning about specific hormones.

Hopson, J. L. *Scent Signals* (1979). An informative and entertaining account of pheromones.

Katchadourian, H. *The Biology of Adolescence* (1977), and Tanner, J. M., *Physical Growth from Conception to Maturity* (1978). Reproductive maturation and other changes in puberty are presented. Tanner is more comprehensive and technical.

Berman, G., and Davidson, J. M. (1974) *Biological Bases of Sexual Behavior*. Provides a fine overview of research related to the relationship of hormones and sexual behavior.

Chapter 5

Pregnancy and Childbirth

And God blessed them, and God said unto them, "Be fruitful and multiply."

Genesis 1:27–28

The generation of a new life is the most elemental, awesome, and wonderful process in nature. For a particular individual, reproduction may not be desirable or possible, yet for the human species, no event is more crucial. Though sex serves many additional purposes for men and women, no other aspect of sexual relations carries more important biological, psychological, and social implications and consequences. No decision could be more serious than that of wanting to become a parent.

Yet parenthood is not obligatory, and many individuals are voluntarily or involuntarily childless. About 10 to 15% of married couples in the United States are infertile. We shall consider some of the causes that lead to this condition toward the end of this chapter. The motivations for voluntary childlessness are discussed in the next chapter.

It is unclear when in human history it was that the causal connection was made between coitus and reproduction. The association between the two is by no means obvious and until fairly recent times remained unknown to people in isolated cultures. For instance, Trobriand Islanders in the South Pacific engaged in sex for plea-

sure while believing that conception followed the entry of a spirit embryo into the body of a woman through the vagina or the head; the Kiwai of New Guinea ascribed pregnancy to something the woman ate.

There are many compelling reasons why people want children, apart from the sexual enjoyment entailed in generating them. Some of these reasons are socioeconomic, others more personal and psychological. Traditionally, children have been the family's primary economic assets, and the cycle of generations has depended on parents looking after their children and then the children looking after their parents. Similarly children are a contribution parents make to the clan, ethnic group, and nation to be in turn accorded recognition as mature, full-fledged adult members of society.

Though in modern industrialized societies children are much less of an insurance against the future, there is still considerable social pressure on young adults to become parents, both from their own parents and from society at large.[1]

But beyond these more external consid-

[1]Pohlman (1969); Fawcett (1970).

erations, there are powerful psychological forces that motivate us to have children. The notion of a "parental instinct," though hard to define and even harder to substantiate, aptly conveys the sense of a deep and elemental urge to procreate even when the rewards are uncertain or the costs prohibitive. To love and to be loved by a child, to share that love with others, and the sheer enjoyment that children provide are matched by few other human experiences.

Yet the burdens and penalties of parenthood are no less significant than the rewards. Half of all pregnancies in the United States are unplanned (chapter 6), and many pregnancies that are planned are motivated by the wrong reasons: to cement an uncertain or faltering marriage; to entrap another into a more permanent relationship; to assert one's sense of manhood or womanhood; to enhance self-esteem; as a shortcut to adulthood; to have a child for want of something better to do, and other manipulative and self-serving reasons. Though we shall be primarily concerned here with the biological aspects of reproduction, clearly there is much more to reproduction than its physiology.

Conception

For a fertile and sexually active woman trying to have a baby, it takes on the average 5.3 months to get pregnant: 25% conceive after one month of unprotected intercourse; 63% in six months; 80% by the end of the year.[2] Some women get pregnant after the first act of coitus, but others may take more than a decade to achieve it.

The Journey of the Sperm

Sperm that develop within the walls of the seminiferous tubules (chapter 2) are led through the testicular duct system in a current of fluid to the epididymis where they attain full maturity and become motile.[3] They then move through the vas deferens

to the ejaculatory duct. During emission (chapter 3) sperm are mixed with the secretions of the seminal vesicles and the prostate gland, which provide nutrients and substances that enhance the sperm's motility.[4] The resultant semen is then ejaculated through the urethra. Up to this point, the movement of sperm has been assisted by the contractions of the tubal system; from here on they must move on their own.

The spermatic fluid in a normal ejaculation has a volume of 2 to 3 milliliters (approximately a teaspoonful), containing 300 to 500 million sperm. The bulk of the fluid of the ejaculate comes from the seminal vesicles and the prostate (whose secretions are also responsible for its odor). Semen is whitish and semigelatinous, but gets more watery after repeated ejaculations. It coagulates on exposure to air.

Sperm deposited in the vagina begin to make their way into the uterus, leaving most of the fluid of the ejaculate behind. If intercourse has taken place while the woman was lying on her back, and if she has remained in that position, there is a greater chance that more of the sperm will reach the cervix and enter the uterus. To further enhance the chances of impregnation the penis should be kept for a while in the vagina following orgasm without thrusting. If the secretions of the cervix and vagina are strongly acidic, sperm are destroyed quickly; even in a mildly acidic environment the movement of sperm ceases.

Sperm swim by the lashing of their tails, moving at a rate of one to two centimeters (less than one inch) an hour, but once they are through the cervix they are aided in their journey by muscular contractions of the uterus and fallopian tubes. Whether orgasmic uterine contractions have a "suction" effect pulling in sperm is unclear (chapter 3). Once through the uterus and into the fallopian tube, sperm complete the final two inches or so of their journey by swimming against the current generated by the hairlike structures (*cilia*) that line

---

[2]Shane et al. (1976).

[3]Discussion in this section is primarily based on Moore (1982) and Langman (1981).

[4]Prostatic secretions are rich in acid phosphatase whose presence in the vagina is used as presumptive evidence in legal cases that coitus has occurred.

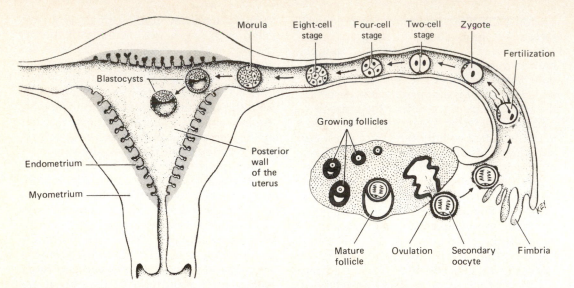

Figure 5.1 Diagrammatic summary of the ovarian cycle, fertilization, and human development during the first week. Developmental stage 1 begins with fertilization and ends when the zygote forms. Stage 2 (days 2 to 3) comprises the early stages of cleavage (from 2 to about 16 cells or the morula). Stage 3 (days 4 to 5) consists of the free unattached blastocyst. Stage 4 (days 5 to 6) is represented by the blastocyst attaching to the center of the posterior wall of the uterus, the usual site of implantation. (From Moore, K. L., *The Developing Human*, 3rd ed. W. B. Saunders, 1982, p. 15.)

the tube and which are helping to propel the ovum from the opposite direction.

The journey ends about an hour after ejaculation when sperm reach their fertilization site, usually at the outer third of the fallopian tube (figure 5.1). Of the several hundred million sperm that start this journey, only a few hundred make it to their final destination, and then only one sperm actually gets to unite with the egg.

Ovulation and Migration of the Egg

At the time of ovulation, the graafian follicle is 10 to 15 millimeters in diameter and protrudes from the surface of the ovary. It is filled with fluid, and its wall has become very thin. The egg has become detached within the follicle and is floating in its fluid. At ovulation, the follicle wall bursts through the ovarian surface, and the ovum is carried in a stream of fluid into the fallopian tube by the sweeping movements of its fringed end.

Once the egg has entered the fallopian tube it begins a leisurely journey toward the uterus, taking about three days to move three to five inches. The egg, in contrast to the sperm, has no means of self-propulsion and is carried along by the current of the cilia lining the tube. If the egg is not fertilized, it disintegrates during its journey through the fallopian tube.

Fertilization and Implantation

Fertilization consists of the fusion of sperm and ovum (figure 5.1). The egg is usually fertilized within 12 hours after ovulation and cannot live longer than a day; sperm can stay alive for about 3 days following intravaginal ejaculation.[5] For a woman to get pregnant, intercourse must therefore take place within less than 3 days before ovulation or within a day or less

[5]Frozen sperm is said to survive for up to ten years. The use of such sperm for artificial insemination is discussed in chapter 6.

Box 5.1.   Boy or Girl

The sex of the baby is determined at the time of fertilization by the type of sperm that fertilizes the egg: if the sperm carries an X chromosome the child will be a girl; if it carries a Y chromosome the child will be a boy.

Attempts to predetermine the sex of offspring are as old as recorded history. Aristotle recommended having intercourse in a north wind if boys were desired. A more recent formula is to engage in intercourse two to three days before ovulation and then to abstain, if a girl is desired (since X-bearing sperm are heavier, they will travel more slowly and get to the ovum at the right time). To conceive a boy, the couple must limit intercourse to the time of ovulation (Y-bearing sperm are lighter and hence move faster). This timetable is combined with an acid douche, shallow penetration, and no orgasm if the woman is to conceive a girl, or an alkaline douche, deep penetration, and orgasm to conceive a boy. The rationales behind these recommendations are not all clear and their results often conflicting.[1]

Another approach is to separate X-containing from Y-containing sperm by centrifugation. Since the heavier X-bearing sperm go to the bottom of the tube that fraction can be used in artificial insemination if a girl is desired, sperm taken from the top layer if a boy is desired. This approach too has not been notably successful.

A study of 5,981 married women in the United States indicates that if sex selection were practiced by this sample, the long-term effect on the ratio of male to female births (currently 105 male to 100 female) would be negligible. However, there was a strong preference by these women for their first child to be male, so that the short-term effect would be a preponderance of male births, followed several years later by a preponderance of female births. When asked about the desirability of the practice of sex preselection, 47% of the women in the sample were opposed, 39% were in favor, and 14% were indifferent.[2]

[1]For further discussion of these methods, see Shettles (1972) and Guerrero (1975).
[2]Westoff and Rindfuss (1974).

following it. Rhythm methods of contraception rely on avoiding this fertile period, but the timing of ovulation is not easy to determine accurately (chapter 6).

When the fertilizing sperm meets the egg, it needs to undergo further changes before it can penetrate the ovum's protective layers. Through the process of *capacitation* the coating that covers the head of the sperm is removed, releasing enzymes which then digest a path through the ovum's barriers. Normally, once a sperm has made its way into the ovum, no other sperm can get through because of an immediate inhibitory reaction developed by the egg.

The head of the sperm that has successfully penetrated the ovum detaches from the rest of its body. The nuclei of the sperm and the egg merge, intermingling their 23 chromosomes and restoring the full complement of 46 chromosomes typical of human cells (chapter 2). This new combination of genes from the parents determines the genetic makeup of the new organism, including its gender (box 5.1).

The fertilized egg, which is called a *zygote*, continues the journey toward the uterus. After some 30 hours, the zygote divides into two cells, these two become four, the four become eight, and so on. Though there is no significant change in volume during these first few days, the zygote has become a round mass of cells called a *morula* (Latin "mulberry") by the time it reaches the uterus in three days. The cells of the morula arrange themselves around the outside of the sphere, leaving a fluid-filled cavity in the center. This structure, called a **blastocyst,** floats about in the uterine cavity, and sometime between the fifth and seventh days after ovulation attaches itself to the uterine lining (*implantation*) initiating the pregnancy. Though the normal site of implantation is in the uterus (usually in the back wall), one in 250 pregnancies is extrauterine or *ectopic,* most often involving the fallopian tube. Should the ectopic pregnancy progress, the tube will rupture, causing the death of the embryo and endangering the mother's health.

Two ova that are fertilized simultaneously give rise to fraternal or *dizygotic twins.* When a single fertilized egg subdivides before implantation, identical or *monozygotic twins* develop. Genetically, fraternal twins are like ordinary siblings. Identical twins have the same genetic make-up, hence they are always of the same sex and look alike.[6]

Pregnancy

It is common practice to divide the nine months (266 days) of pregnancy into three-month periods or **trimesters.**[7] During this period the growing organism is called an *embryo* (Greek, "to swell") for the first 8 weeks and *fetus* (Latin, "offspring") thereafter.[8]

---

[6]Twins occur in 1 out of 90 births; two out of three sets of twins are fraternal. Triplets occur in about 1 of 5,000 births; quadruplets, in 1 of 500,000 births. Births of more than four children are extremely rare. The first quintuplets known to survive were the Dionne sisters born in Canada on May 28, 1934.

[7]The average length of pregnancy is about 266 days after coitus. Authenticiated lower and upper limits are 230 and 349 days respectively. (Haynes, 1982, p. 362.) In extended pregnancies the length of period assumes legal importance in establishing the legitimacy of a child when the presumed father has been away for more than 10 months. In the United States the longest pregnancy upheld by the courts as legitimate lasted 355 days.

[8]The following discussion of embryonal development is based on Langman (1981) and Moore (1982) Obstetrical aspects based on Pritchard and MacDonald (1980) and Danforth (1982).

The Embryo in the First Trimester

At the time the blastocyst is implanted, the uterine endometrium is at the peak of the secretory phase of the menstrual cycle (chapter 4). The blastocyst burrows in, as its enzymes digest the outer surface of the uterine lining, permitting the developing organism to reach the blood vessels and nutrients below. By the 10th to 12th day after ovulation, the blastocyst is firmly implanted in the uterine wall; but the woman does not yet know that she is pregnant, for her menstrual period, which she is going to miss, is not due for several more days.

This is a critical time for the embryo's survival. **Trophoblast cells** (which will become part of the placenta) start producing *chorionic gonadotropin,* that stimulates the corpus luteum to maintain its output of estrogens and progestins. Otherwise if the level of these hormones drops, the uterine lining will slough off and the embryo will be lost without the woman even realizing that she was pregnant (chapter 4).

In the early stages of development, the dividing cells form a hollow ball within which a layer of cells forms across the center of the blastocyst. From this **embryonic disk** will eventually grow all of the parts of the embryo. The remaining cells develop into the *placenta* and the *fetal membranes* that form a sac filled with *amniotic fluid* within which the fetus floats, cushioned and protected until birth (figures 5.2 and 5.3).

The embryonic disk becomes elongated and ovoid by the end of the second week after fertilization. The embryo is still barely visible to the naked eye at this stage, as depicted in figure 5.2, which shows the actual sizes of the embryo during the first 7 weeks. During the third week the embryonic disc differentiates into its three distinctive layers, **endoderm, mesoderm, ectoderm.** From the inner endodermal layer will develop the internal organs; from the middle mesodermal layer will come muscles, skeleton, and blood; from the ectodermal outer layer will emerge the central nervous system, skin, and other tissues.

The head end of the embryo develops

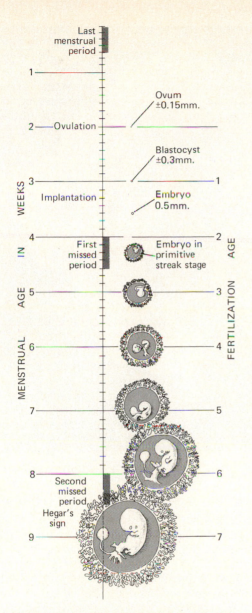

Figure 5.2 Actual sizes of embryos and their membranes during the first 7 weeks of life. (From Patten, B. M., *Human Embryology*, 3rd ed., McGraw-Hill, Inc., 1968, p. 145.)

embryo has a primitive heart and the beginnings of a digestive system. In the fifth week, precursors of arms and legs become visible; fingers and toes begin to appear between the sixth and eighth weeks. Bones are beginning to ossify, and the intestines are forming. By the seventh week the gonads are present, but cannot yet be clearly distinguished as male or female.

By the end of 8 weeks, when the embryonal periods ends, the rudiments of all essential structures are present, and by 4 months the major organs are fairly well formed. The clearly defined external features now mark the fetus as a tiny human-looking figure (figure 5.3). From here on fetal development is mainly a matter of the further growth and differentiation (figure 5.4). This process of embryonal development, where a simple cluster of cells develops into a complex organism, has been a source of much wonder, speculation, and study throughout history (box 5.2).

The Development of the Placenta   During embryonal life, the fetal lifeline to the mother is through the *placenta.* The placenta develops from both fetal and maternal tissues, eventually growing to a sizable organ weighing about a pound (450 grams). It sustains the life of the fetus by conveying to it nutrients and oxygen and carrying away its waste products. This transport takes place through the blood vessels of the *umbilical cord,* which connects the fetal and maternal circulatory systems (figure 5.3). The placenta also functions as an endocrine organ, first producing chorionic gonadotropin to keep the corpus luteum active and then itself taking over the production of estrogens and progestins.

The Mother in the First Trimester

Pregnancy is an experience of unique significance to women, and no other issue has had a more profound effect on women's lives than their capacity to bear children. Though pregnancy is a normal phys-

faster than the rest of the body during these early stages. By the end of the third week the beginnings of eyes and ears are visible, and the brain and other portions of the central nervous system are beginning to form. By the end of the first month, the

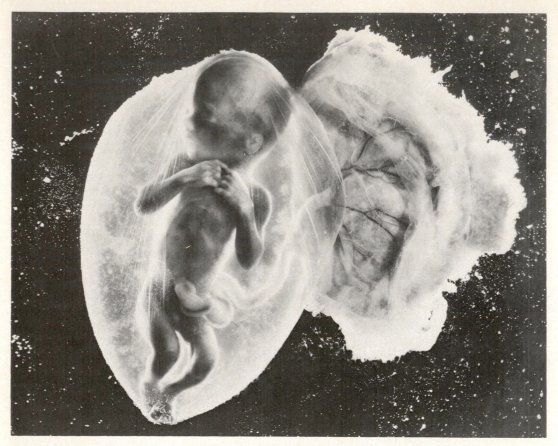

Figure 5.3 Fetus about 4 months, 8 inches long. (From Nilsson et al., *A Child is Born*. Boston: Delacorte Press/Seymour Lawrence, Inc., 1965, pp. 116–117.) Reprinted by permission.

iological process and not a disease, it does constitute a period of increased physical and psychosocial vulnerability and risk. Before the advent of modern obstetrics, countless women suffered and died from complications of childbirth; pregnancy nowadays carries minimal risk for most women who are under proper prenatal care.[9]

[9]Prior to the middle of the 19th century, *puerperal* or *childbed fever* caused the death of 1 in 10 women giving birth in hospitals. In 1847, Ignaz Semmelweiss established the infectious origin of this illness and the simple expedient of physicians disinfecting their hands before delivery dropped the mortality rate dramatically. Currently the *maternal death rate* in the United States is less than 10 in 10,000 births (Thompson, 1982, p. 286).

Pregnancy is manifested by a number of signs and subjective experiences.[10] Most common in early pregnancy are feelings of *fatigue* and *drowsiness* (though some women experience a sense of heightened energy and well-being). The mood of the woman who knows she is pregnant largely depends on the particular circumstances. If it is an event that is genuinely wanted, she may have a sense of deep satisfaction and anticipation. If for one reason or another the pregnancy feels wrong, then her experience may be colored by feelings of worry, anger, and depression.

[10]Signs and symptoms of pregnancy are discussed in detail in all obstetrics texts. See Danforth (1982); Pritchard and MacDonald (1980).

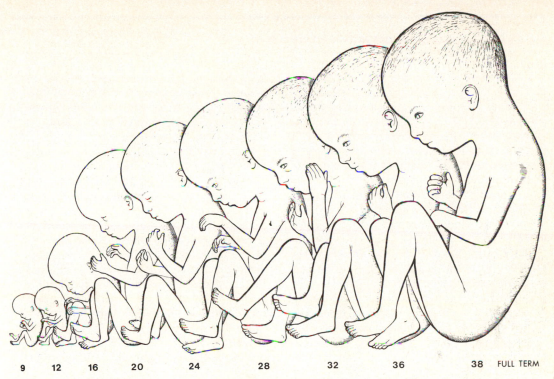

9   12   16   20   24   28   32   36   38   FULL TERM

Figure 5.4   The fetal period, extending from the ninth week until birth (fetuses shown about one third actual size) (From Moore, K. L., *The Developing Human*, 3rd ed., W. B. Saunders, 1982, p. 5.)

The sign commonly associated with pregnancy is a missed menstrual period. While all pregnant women eventually do fail to menstruate, there are many other reasons for missed periods, which include illnesses, emotional factors, and sometimes no apparent reason. This is particularly true for women younger than 20 and older than 40. Conversely, the presence of a vaginal bloody discharge does not rule out pregnancy, since about 20% of pregnant women have *spotting*, a short period of slight bleeding connected with implantation. Such bleeding is usually harmless but may also be an early sign of miscarriage.

Other physical signs of early pregnancy are swelling and tenderness of the breasts, frequent urination, irregular bowel movements, and increased vaginal secretion. A particularly bothersome symptom experienced by many women during the first 6 to 8 weeks of pregnancy is *morning sickness.* This consists of queasy sensations in the stomach upon awakening, accompanied by an aversion to food or even to the odors of certain foods. In more serious cases, nausea may be accompanied by vomiting and reluctance to be near food. Some women experience these symptoms in the evening. About 25% of pregnant women experience no vomiting at all, while in about 1 in 200 cases vomiting is so severe that the woman must be hospitalized. This condition (known as *hyperemesis gravidarum*) may have serious consequences, including malnutrition, if it is not properly treated.[11]

[11]Some husbands identify with their pregnant wives so closely that they too suffer from morning sickness and other signs of early pregnancy. In some preliterate societies, men actually went through "labor" in concert with their wives (*couvade*) in an effort to distract evil spirits away from the mother and baby (Davenport, 1977).

Box 5.2.   Theories of Embryonal Development

The discovery that coitus results in pregnancy, momentous as it was, did not resolve the mystery of how a complex being develops from formless elements during embryonal life. People have speculated about the nature of this process until the modern science of embryology was able to elucidate its underlying mechanisms.

Aristotle believed that the human embryo originated from the union of semen and menstrual blood. This same notion has also occurred to preliterate societies. For instance, the Venda tribe of East Africa believed that "red elements" like muscle and blood were derived from the mother's menses (which, they explained, ceased during pregnancy because the menstrual blood was being absorbed by the developing fetus); the "white elements"—like skin, bone, and nerves—developed from the father's semen.[1]

Following the invention of the microscope in the 17th century, sperm and ovum could be visualized and their role in reproduction correctly surmised. Yet since neither looked remotely "human," how could a child develop from them? Two schools of thought emerged to explain the mystery. The *ovists* claimed that a minuscule, but fully formed, baby was contained in the egg and that the sperm functioned only to activate the growth of the preformed baby. The *homunculists* held the opposite view, that the preformed baby resided inside the head of the sperm but did not begin to develop until it arrived in the fertile uterine environment. Looking through their crude early microscopes, some homunculists claimed to have actually seen a homunculus ("little man") inside the sperm. (See accompanying figure.)

Ovists and homunculists were both *preformationists,* presuming that a fully formed miniature being simply grew bigger during development. This necessitated the further presumptions that all generations of humans, past and future, were stacked inside each other, like so many dolls-within-dolls.

Curiously enough, our modern concepts of genetics provide some vindication to this fantastic notion in the sense that our genetic endowment is transmitted in a continuous line from one to another generation.

Preformationist theories died hard. Aristotle had posited the alternative to preformation by considering the possibility that differentiation rather than simple growth characterized development. However, it was not until the 19th century, that the theory of *epigenesis,* which embodies this doctrine, became established. Final proof for it was provided by Driesch in 1900, when he showed that in many life forms divided cells of a fertilized egg, when separated, will develop into complete embryos.[2] This could only mean that the development of simpler components gave rise to complex parts in a continuous sequence of growth and differentiation.[2]

[1]Meyer (1939).
[2]Arey (1966).

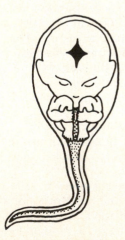

A homunculus as drawn by Niklaas Hartsoeker in 1694.

More objective evidence for early pregnancy can be gathered through gynecological examination. The vagina shows a purplish coloration (*Chadwick's sign*), as does the cervix. By the sixth week of pregnancy (that is, a month after missing a period) a soft and compressible area between the cervix and body of the uterus can be felt on examination (*Hegar's sign*). Occasionally, a woman develops some of the symptoms of pregnancy without being pregnant ("false pregnancy" or *pseudocyesis*). These women, who are usually intensely desirous of having a child, stop menstruating and develop symptoms of morning sickness, breast tenderness, a sense of fullness in the pelvis, and the sensation of fetal movements in the abdomen (caused by contractions of the abdominal muscles). The absence of objective signs and negative pregnancy tests reveal the true nature of their condition.

Pregnancy Tests   The manifestations listed above, including the objective signs, constitute only presumptive evidence for the presence of pregnancy. More positive confirmation comes from laboratory tests that indicate the presence of *human chorionic gonadotropin* (hCG) in blood or urine; this substance is normally found only in pregnant women.[12]

In earlier versions of these tests, urine or serum was injected into female mice or rabbits: the presence of hCG made the animals ovulate (or, injected into male frogs, made them ejaculate). At present, hCG is detected by immunologic tests that are simpler, less expensive, and 90 to 95% accurate: the woman's blood or urine sample is mixed with specific chemicals; if hCG is present, even in very small amounts, it can be detected within minutes. The test for hCG using urine is now the most common test for determining pregnancy. It can detect hCG as early as 2 weeks after conception. The blood test for hCG is used less frequently, in part because it is more ex-

pensive, though it can detect hCG within 6 to 8 days after fertilization. The most sensitive test for hCG is through radioimmunoassay methods that are virtually 100% accurate but not yet widely available in clinical laboratories. They can detect the hCG present 1 day after implantation.

Another version of the hCG test is available in pharmacies in the form of kits for home use. If used correctly, it can detect hCG in the urine of a pregnant woman as early as 9 days after a missed period. Recent studies have found that only in 3% of cases, home tests indicated pregnancy when the woman was not pregnant. But in 20% of tests giving negative results, the women were actually pregnant. When the women with negative results performed a second test 8 days later, the accuracy of the test increased to 91%.[13]

With regard to all pregnancy tests, there always exists the possibility of such *false negative* results when the woman is pregnant and *false positive* results when she is not. Thus, absolute confirmation of pregnancy can be established only by one of three means: (a) verification of the fetal heartbeat, (b) photographic demonstration of the fetal skeleton, (c) observation of fetal movements. Until recent developments in technology, none of these signs of pregnancy could be verified until well into the second trimester.

Using a conventional stethoscope a physician can hear the fetal heartbeat by the fifth month. (The fetal heart rate is 120 to 140 beats per minute, and thus can be differentiated from the mother's heartbeat, which is usually 70 to 80 per minute.) A fetal *pulse detector* now commercially available can detect the fetal heartbeat as early as 9 weeks, quite reliably after 12 weeks.

Photographic evidence is obtained through X rays and *ultrasonography*. Variations in the echo from an ultrasonic pulse reflecting off the fetal skeleton are converted to a photographic image that is far more distinct than a conventional X ray; it is also safer than X rays since it does not pose radiation hazards to the fetus. Nei-

---

[12]The following discussion of pregnancy tests is based on Hale (1980). Also see Haynes (1982, p. 374) for additional references.

[13]McQuarrie and Flanagan (1978).

Box 5.3.   Amniocentesis

Amniocentesis is the analysis of fetal cells floating in amniotic fluid in order to obtain genetic information on the fetus. The procedure is as follows. First, sound waves are used to make an image (an *ultrasonic scan*) of the fetus in the womb. Once the position of the fetus is known, it is possible to insert a long needle through the abdominal wall into the uterus and draw a small amount of amniotic fluid (see accompanying figure). Since the fetus floats in this fluid some of its cells are shed into it. These cells are received from the fluid and analyzed. The condition and number of the chromosomes of these cells as well as the presence of XX or XY sex chromosomes, hence the sex of the child, can be determined with a high degree of accuracy.

Amniocentesis is an accurate and relatively safe procedure for both mother and fetus, and is usually performed after the 14th week. Among the abnormalities that can be detected by amniocentesis are Down's syndrome (mongolism), neural tube closure defects (open spine), Tay-Sachs disease, cystic fibrosis, and Rh incompatibility. When abnormalities are diagnosed, women can choose to have an abortion—although by the time the chromosome analysis is complete (the 20th week to 24th week of pregnancy) abortion is no longer a simple, risk-free operation (see chapter 6).

Because a woman in her forties is 100 times more likely than a woman in her twenties to have ova with abnormal chromosomes, amniocentesis is currently recommended for all women who become pregnant after the age of 35, and for women who have previously had a child with a chromosome abnormality or other problems that can be detected by analysis of fetal cells. It is also recommended for couples with family histories of certain hereditary problems. Since about 3% of newborns in the United States suffer an obvious defect, the application of procedures like amniocentesis is of relevance to large numbers of individuals.

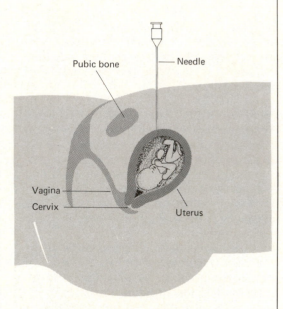

Amniocentesis involves removing a few milliliters of amniotic fluid to test for cellular abnormalities. (Redrawn from *Resident & Staff Physician*, 1977.) By permission.

ther of these techniques is used routinely, but each is of value in verifying suspected complications such as fetal head size being larger than the pelvic opening, gross malformation, and multiple fetuses. Further information about the fetus can be obtained through *amniocentesis* (box 5.3).

Fetal movements in the abdomen can usually be felt by the end of the fourth month ("quickening"). These are described as being similar to the fluttering of a bird in the hand. The observation of fetal movements not only confirms pregnancy but indicates that the fetus is alive.

Expected Date of Delivery Once pregnancy has been confirmed, the next question usually is, "When is the baby due?" The expected delivery date (*EDC*, or *expected date of confinement*) can be calculated by the following formula: Add 1 week to the first day of the last menstrual period, subtract 3 months, then add 1 year. Thus, if the last menstrual period began on January 8, 1980, adding 1 week (to January 15), subtracting 3 months (to October 15), and adding 1 year gives an expected delivery date of October 15, 1980. In fact, only about 4% of births occur on the day predicted by this formula; but 60% occur within 5 days of it.

The Fetus in the Second Trimester

In addition to further maturation of internal organ systems during the second trimester, there is a substantial increase in the size of the fetus. By the end of the sixth month the fetus weighs about 2 pounds and is some 14 inches long. Among the changes during this period is the development of a temporary fine coat of soft hair (*lanugo*) on the scalp. The fetus at first has a wrinkled appearance but with the seventh month layers of subcutaneous fat begin to build up eventually making the baby look more chubby.

By the end of this period the eyes can open and the fetus moves its arms and legs spontaneously and vigorously. The fetus alternates between periods of wakefulness and sleep. Though the uterus is a very sheltered environment, a loud noise near the uterus or a rapid change in the position of the mother can disturb the tranquility of the womb and provoke a vigorous movement by the fetus. Changes in outside temperature are not perceived, since the intrauterine temperature is maintained by internal mechanisms at a constant level just slightly above the temperature of the rest of the mother's body.

Normally, fetuses are assumed to become *viable*, that is, able to survive outside of the uterus, 28 weeks after the last menstrual period of the mother. Fetuses typically weigh about 1,000 grams at this stage.

Both weight and length of gestation are important in determining viability. If delivered at the end of 6 months, the fetus may live for a few hours to a few days. With heroic efforts 5 to 10% of babies weighing less than 2 pounds (about 900 grams) may live. A few infants born at less than 24 weeks of gestation have survived, as have a few others weighing less than 600 grams; but there is no reliable case of survival of any infant born at less than 24 weeks and weighing less than 600 grams.[14]

Since fetuses can now be legally aborted up to 24 weeks, it is possible that some of them could survive with extraordinary efforts. Continuing advances in medical technology will probably make it possible to sustain life for increasing numbers of such very prematurely born infants, which further complicates the ongoing controversy on abortion (see chapters 6 and 8).

The Mother in the Second Trimester

For many women, the second trimester is the happiest time of pregnancy. The nausea, lethargy, and other troublesome symptoms of the earlier period have usually subsided and the concerns about delivery are still in the future.

The pregnancy now feels secure and is publicly visible because of the expanding abdomen and bustline that may necessitate maternity clothes. The experience is also made much more real because late in the fourth month the mother can feel the fetal movements more strongly and the "kicking" of the fetus becomes outwardly visible on the abdomen. The mother can be now more directly aware of the new life growing within her, and begin to relate to it. Being more comfortable than earlier, and not yet as burdened as she will be later, more women are able to keep active during this time, be it at work, at home, or in sports and leisure; there is also heightened sexual interest (to be discussed shortly).

[14]Tietze (1983).

Nevertheless, the second trimester has its own discomforts. Pressure from the expanding uterus may lead to indigestion and constipation. Varicose veins and hemorrhoids may appear or get worse. Fluid retention causes swelling of the feet and ankles (*edema*), and the woman may begin to gain excessive amounts of weight.

The Fetus in the Third Trimester

This last phase of development is mainly a period of further maturation. The fetus is now well formed with improved chances of survival; by the middle of the last trimester the prematurely born baby has a 70% chance of survival.

Early in the trimester the fetus is in an upright or *breech* position. But it keeps shifting around in its increasingly cramped quarters and in 97% of cases assumes a head-down position by the time it reaches full term.

During the ninth month the fetus gains more than 2 pounds (0.9 kg), and essential organs like the lungs reach a state of maturity that makes them readily compatible with life in the outside world. At full term the average baby weighs 7.5 pounds (3.4 kg) and is 20 inches long (50.8 cm).[15] Ninety-nine percent of full-term babies born alive in the United States now survive, a figure that could be improved even further if all expectant mothers and newborn babies received proper care.

The Mother in the Third Trimester

The relative comfort and tranquility of the second trimester gradually give way to the special discomforts of the last phase of pregnancy. Most of these have to do with the increased activity and size of the fetus, which displaces and presses upon the mother's organs (figure 5.5). What were occasional fetal movements now turn to periods of seemingly perpetual kicking, tossing, and turning, which may keep the mother awake at night.

[15]There is great variation in birth weights, ranging usually from five to nine pounds. The largest baby known to have survived weighed 15.5 pounds at birth.

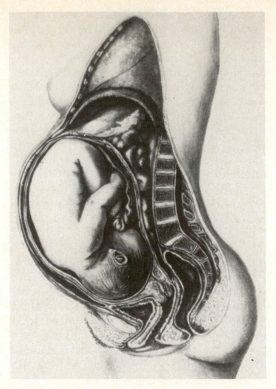

Figure 5.5   Full-term fetus. (From Cherry, S. H., *For Women of All Ages*, Macmillan Publishing Co., 1979, p. 186.)

The woman's weight, if not controlled up to this point, may become a problem. The optimal weight gain during pregnancy is 24 to 27 pounds (11 to 13 kg). The average infant at 9 months weighs about 7.5 pounds (3.4 kg). The rest of the weight gain is accounted for as follows: the placenta, about 1 pound (0.4 kg); amniotic fluid, 2 pounds (0.9 kg); enlargement of the uterus, 2 pounds (0.9 kg); enlargement of the breasts, about 1.5 pounds (0.7 kg); and the retained fluid and fat accumulated by the mother, the balance of 10 pounds (4.5 kg). Excessive weight gain is associated with a higher incidence of medical complications during pregnancy, such as strain on the heart and high blood pressure.

By the ninth month of pregnancy a woman is usually anxious to have it over with. There is speculation about the sex of the child and concerns about whether or not the baby is going to be all right. There may also be some anxiety about the process of delivery. But the properly prepared woman in good health can also count the days with much pleasure and anticipation before she finally meets her baby face to face.

Sexual Interest and Activity During Pregnancy

Given the marked physiological and psychological changes during pregnancy, sexual interest and behavior can be quite understandably affected. But there is much variation in this regard among women and between the different phases of pregnancy. In general, sex in pregnancy is quite safe for the mother and the fetus. Due precaution and abstention are necessary in certain circumstances, but how a couple's sex life changes during pregnancy is usually more a matter of personal choice than medical necessity.

Some studies show a steady decline in the frequency of coitus throughout the pregnancy. Other studies show a different pattern. In the first trimester, women show a highly variable response. There may be heightening of sexual feelings and behavior, no change, or a decline in interest often associated with the symptoms of morning sickness, fatigue, fear of disrupting the pregnancy or resulting from high progesterone levels. But in the second trimester, some 80% of women report an increase in sexual desire and responsiveness.[16] In the last trimester there is a consistent drop in the frequency of coitus linked with physical awkwardness and discomfort, a feeling of loss of sexual attractiveness by the women (a perception not always shared by their husbands), and concerns about hurting the baby.[17]

The health concerns in this area have to do with the prospects of orgasm causing miscarriage or premature labor (through uterine contractions) and the risk of infection through coitus. The risk is realistic for women with a history of miscarriage or who are showing early signs of it (like vaginal bleeding). Otherwise there seems to be no correlation between experiencing orgasm and given birth prematurely.[18] The risk of infection becomes a serious consideration if the membranes are ruptured.

Some of these problems related to sex during pregnancy are easily resolved. Rear entry coitus obviates much of the physical awkwardness in later months and is less taxing on the woman. Use of condoms will help prevent infection. Noncoital sexual activities like masturbation or mouth-genital stimulation provide alternatives to coitus. But these activities carry their own risks, as the effect of orgasm is the same with regard to miscarriage, however orgasm is reached. The practice of blowing air forcefully into the vagina (which some people do during cunnilingus) is very dangerous in pregnancy because it may introduce air bubbles into the woman's bloodstream, causing air embolism.

Health considerations with respect to sex during pregnancy are important but by no means the sole factors to be considered. Because a woman is pregnant her sexual needs as well as those of her sexual partner cannot be simply ignored. Of the 79 husbands of pregnant women interviewed by Masters and Johnson, 12 had turned to extramarital sex.[19] Discovery of such an occurrence is likely to be particularly aggravating for a pregnant wife.

Childbirth

As the end of pregnancy approaches, the expectant mother experiences contractions of her uterus at irregular intervals referred to as *false labor.* Three or four weeks be-

[16]Masters and Johnson (1966). Also see Tolor and DiGrazia (1976); Calhoun et al. (1981).

[17]Such concerns are variously shared by other cultures, some of which prohibit coitus during all or part of pregnancy, while others permit it all the way up to onset of labor (Ford and Beach, 1951, pp. 213–220).

[18]Perkins (1979).

[19]Masters and Johnson (1966). For further sources on the influence of pregnancy on sexuality, see Calhoun et al. (1981); White and Reamy (1982).

fore delivery the fetus *drops* to a lower position in the abdomen. The next major step is the softening and dilation of the cervix. Usually just before labor begins there is a small, slightly bloody discharge (*bloody show*) that represents the expelled plug of mucus occluding the cervix. *Labor* follows this event within a few hours or as long as several days.

Labor

Labor consists of the regular and rhythmic uterine contractions or *labor pains* that dilate the cervix and culminate in the delivery of the baby, the placenta, and the fetal membranes. The exact mechanism that initiates labor is not fully understood, but a number of hormonal factors are known to be involved.

Current research indicates that it is the fetus rather than the mother that actually triggers labor.[20] The fetal adrenal gland produces hormones that act upon the placenta and uterus to increase the secretion of *prostaglandins* which stimulate the muscles of the uterus, thereby causing labor to begin. *Oxytocin,* a hormone produced by the posterior pituitary gland, is also released in the late stages of labor and stimulates the more powerful contractions required to expel the fetus.

Labor is divided into three stages. The *first stage* is the longest, extending from the onset of regular contractions until the cervix is fully dilated to about 4 inches (or 10 centimeters) in diameter. This stage lasts about 15 hours in first pregnancies and about 8 hours in subsequent ones. (Deliveries after the first child are generally easier in all respects.) Uterine contractions begin at intervals as far apart as 15 or 20 minutes, occurring more frequently and with greater intensity and regularity in time.

The *second stage* starts with the fully dilated cervix and ends with the delivery of the baby (see figure 5.6). It may last for only a few minutes or up to several hours in particularly difficult births. At some point during the first two stages, there is a rupture of the fetal membranes and a gushing

out of amniotic fluid. In 10% of cases, a premature rupture of the membranes initiates the process of labor. In other instances the membranes may be ruptured deliberately to accelerate the progress of labor. After the baby is born and starts breathing, the umbilical cord is cut, thus severing the last physical link to the mother.

In the *third stage* the placenta separates from the uterine wall and is discharged as the *afterbirth* (placenta and fetal membranes). The uterus contracts considerably during this stage, and there is a variable degree of bleeding. The third stage of labor lasts about an hour, during which time the mother and baby are carefully examined for signs of trauma during birth and other problems.

Methods of Childbirth

The child plays no active part in the birth process but is passively propelled out of the birth canal. Yet in earlier times it was believed that babies fought their way out into the world.

Societies have devised various ways of assisting women in childbirth. In preliterate cultures, sitting, squatting, or kneeling postures for the mother were usually considered optimal. But women have also been "assisted" by being suspended, sat upon, tossed about in blankets (to shake the baby loose), and having smoke blown into their vagina.

Traditionally, women have been assisted in childbirth by a *midwife* (from "with wife").[21] Men were for a long time strictly excluded from attending to the birth process: In 1552, a Hamburg physician called Wertt was burned at the stake for posing as a woman to attend a delivery. As late as the 18th century, physicians considered it beneath their dignity to care for pregnant woman, and notions of modesty precluded their assisting at labor.[22]

---

[20]Daly and Wilson (1978, p. 192).

[21]The Old Testament has several references to midwives, as in Genesis 35:15, Rachel's delivery of Benjamin: "While her pains were upon her, the midwife said, Do not be afraid, this is another son for you.'" Also see Genesis 38:27 for a midwife delivering twins.

[22]Speert (1982).

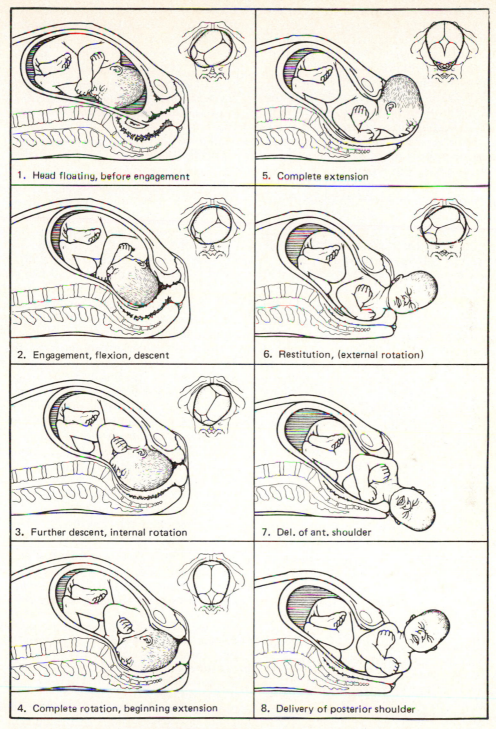

1. Head floating, before engagement

2. Engagement, flexion, descent

3. Further descent, internal rotation

4. Complete rotation, beginning extension

5. Complete extension

6. Restitution, (external rotation)

7. Del. of ant. shoulder

8. Delivery of posterior shoulder

Figure 5.6 Principal movements in the mechanism of labor and delivery, left occiput position. (From Pritchard and McDonald, *Williams Obstetrics*, 16th ed., Appleton-Century-Crofts, 1980.)

With the advent of modern medicine the delivery of children in industrialized countries was taken over by physicians, and became almost exclusively a hospital procedure. This has recently elicited objections from both within and outside of the medical profession on the grounds that the hospital setting turns childbirth into an impersonal, expensive, and needlessly medicalized procedure.[23] As a result, a number of alternative ways of *natural childbirth* have been advocated, and women have chosen to deliver at home or in *birthing centers* that provide the advantages of an institutional setting without some of the objectionable aspects of hospitals. Meanwhile, hospitals have modified their routines to counter some of the criticisms aimed at them.

A safe choice exists between delivery in a hospital or elsewhere only in cases where the pregnancy has progressed normally and there are no anticipated complications. Childbirth can lead very quickly to acute emergencies that can endanger the life of mother and child, and these can be best handled in a hospital. But since the majority of deliveries are quite normal, many women in fact do have a choice. They should be free in exercising that choice after being properly informed and should be guided by their own preferences instead of being pressured either by convention or swayed by prevailing fads.

Hospital Delivery   The main advantage of a hospital delivery is the security it provides against unforeseen complications. It also relieves members of the family or friends of the responsibility of caring for the new mother and baby right away. A woman may also prefer the privacy provided by the hospital setting.

The experience of giving birth is basically the same wherever it takes place; yet a number of significant differences are likely to exist, depending on the setting. In the hospital setting, the woman who is to give birth is placed in a *labor room* where she is prepared for labor through various procedures (such has having her pubic area

shaved and cleansed, her bowels voided, and so on). The actual birth takes place in the *delivery room* (which is like an operating room). The patient is usually given an **analgesic** to combat pain and anxiety or some form of **anesthetic.** Especially with first deliveries, an **episiotomy** or incision is made in the perineum (the area between vagina and anus) to facilitate the passage of the head and to avoid jagged tears.[24]

General anesthesia is now used much less than in earlier decades since it entails considerable risk to the mother, slows down labor, and may interfere with the newborn's respiration. It also deprives the mother of assisting and witnessing the birth of the baby. To avoid some of these ill effects yet still provide relief from pain, **local anesthesia** (which may be *caudal* or *spinal*) is currently popular. The loss of sensation and paralysis that result involve the body only below the waist, leaving the mother quite awake.

Hospital deliveries can allow for the **induction** of labor through infusion of *oxytocin* when the mother is near the end of her term or is overdue. The procedure is not without risk, but in select cases the advantages warrant it. The same is true for deliveries by **cesarean section,** a major surgical procedure whereby the fetus is delivered through an incision in the abdominal wall and uterus.[25]

Some 15% of births in the United States currently take place through cesarean section.[26] This practice has tripled over the past decade and strikes some as another

[24]Chloroform anesthesia was introduced in the 19th century, but its use was resisted because it was "unnatural" and appeared to nullify the judgment to Eve in Genesis 3:16 to bear children in "labor" and "groaning." But the practice became quite popular in England after Queen Victoria was delivered of her eighth child under chloroform in 1853.

[25]Although the procedure is named after Julius Caesar, it is highly unlikely that he was in fact delivered in this fashion. The Roman practice was to remove the entire uterus to save the child if the mother was dying. Caesar's mother survived his birth. The first authentic cesarean delivery on a living patient was performed in 1610 (Speert, 1982).

[23]Arms (1975).

[26]Kolaka (1980).

instance of the penchant of physicians to resort to needless surgery. Among its legitimate indications are discrepancies between the size of the baby's head and the mother's pelvis (making passage difficult if not impossible); complications endangering the fetus (such as abnormal positioning or premature separation of the placenta); and other conditions interfering with normal childbirth. Though cesarean section results in only one death in 10,000 cases, this is still twice the rate of maternal deaths in vaginal deliveries. The fact that a woman was delivered by surgery does not preclude her giving birth vaginally on subsequent occasions.

Prepared Childbirth   *Natural childbirth* refers to a set of attitudes and practices that seek to free the process of giving birth from unnecessary pain and anxiety while simultaneously dissociating it from excessive medical intervention. To some, the term simply means giving birth at home, with the help of a midwife, without drugs and doctors. To others, it signifies one of several methods to facilitate the process of birth through natural means rather than through the use of anesthesia, instrumentation, and so on.

The term "natural childbirth" was coined by the English physician Grantly Dick-Read in 1932 in his book, *Childbirth Without Fear*. Dick-Read postulated that the pain of childbirth is primarily related to fear and consequent muscular tension during labor. So he sought to educate women about the birth process in order to break the cycle of fear, tension, and pain.

The *Lamaze* form of natural childbirth originated in Russia, then was popularized by the French physician Bernard Lamaze in 1970. It entails teaching women how to relax their muscles and breath properly during labor, coupled with various massage techniques.[27]

Methods of prepared childbirth currently in use in the United States incorporate some of these techniques from Dick-Read and Lamaze, as well as other sources.

[27]Lamaze (1970).

An expectant mother and her partner (usually husband, friend, or relative) attend prenatal classes for 6 to 10 weeks before the birth. These classes are designed to inform prospective parents about each step of labor, to answer questions, and dispel anxiety. The women are also taught a variety of exercises that increase muscle control. Part of the pain that women experience in childbirth results from their tightening of abdominal and perineal muscles, thereby making it harder for the baby to emerge. By learning to relax these and other muscles, women can allow the baby to pass through the birth canal more easily and thus decrease pain, while helping to push the baby through. Other techniques such as massaging the abdomen in special patterns or concentrating on visual targets work by distracting attention from the experience of pain.

The woman's partner plays an important part in these approaches. He or she is present at the delivery not only to provide psychological comfort and security to the mother, but to help her to divert her attention from the pain, to breathe properly, and keep her abdominal muscles relaxed; in short, to implement the various techniques they have been taught together.

Another method of childbirth that has become popular in recent years was expounded by the French physician Frederick Leboyer in *Birth Without Violence* (1975). This method is mainly concerned with protecting the infant from needless pain and trauma during childbirth. It is a slow, quiet delivery, often done in hospitals where everything is aimed at protecting the infant's delicate senses from shock; hence it takes place in a warm room where the lights are kept low, unnecessary noises avoided, and so on. Following birth, the baby is gently settled onto the mother's abdomen where he or she is allowed to adjust to its new environment for several minutes while still attached to the umbilical cord. Only after breathing starts spontaneously is the cord cut and the baby gently lowered into a warm bath. This gentle entry into the world is a far cry from the image of the kicking and screaming infant being dangled from the feet. But whether Leboyer's method results in more relaxed and healthy children in the long term remains to be substantiated.[28] Nonetheless natural childbirth approaches have already had an impact on hospital practices. Increasingly hospitals have modified their regulations to allow husbands to be present at delivery and to allow *rooming-in* whereby newborn babies live in the same room with their mothers rather than in a separate nursery.

Early-Mother-Infant Interaction Another area of recent interest is the interaction between mother and infant soon after birth. Among animals there are *critical periods* during which mother and offspring become *bonded.* The mother's willingness to care for the newborn or even recognize it as its own is contingent on such a bond being established. Similarly the young animal becomes attached to the mother through *imprinting.* Imprinting has been found in a number of species but is most clearly developed in birds that are able to walk or swim right after birth. For example, a newly hatched duckling that has been incubated artificially and not seen its mother will become imprinted on any moving object (a wooden decoy or human being) that it sees after birth. From then on the duckling will follow the imprinted object around, preferring it even to a live duck.

By inference from such work, as well as on independent grounds, similar processes are assumed to operate among humans. It is hypothesized that, "the entire range of problems from mild maternal anxiety to child abuse may result largely from separation and other unusual circumstances which occur in the early newborn period."[29]

To enhance the process of early attachment, the newborn is shown to the parents right after birth and then placed next to the mother face-to-face to facilitate touching and eye contact. Cuddling, fondling, cooing, and other forms of tender and affectionate contact are encouraged during early infancy. The baby actively participates in this process by responding to parentally initiated interactions, best exemplified by the *smiling response* that accords recognition to the parent's affectionate attention, thus reinforcing it.[30] Such practices may well enhance attachment between infant and parents and be beneficial to subsequent psychological development. However, there is no convincing evidence that they work on the same bonding and imprinting principles that are operative among animals.

The Postpartum Period

The *postpartum period* (or the *puerperium*) is the period of 6 to 8 weeks that follows delivery and ends with the resumption of ovulatory menstrual cycles. It is an important period for the mother, as well as for the baby and the father, from physiological, psychological, and practical perspectives.[31]

Physiological Aspects

We tend to take for granted the adaptation of the female body to the burden of pregnancy, losing sight of the tremendous changes it entails. In the postpartum period, tissues, organs, and physiological systems face the considerable task of reverting to their previous state, a process that is usually accomplished quite successfully.

After delivery, the uterus shrinks markedly (*involution*) and gradually regains its prepregnancy size; its weight drops from about 2 pounds to 2 ounces over a 6-week period. The cervix, which is stretched and flabby after delivery, regains its tone and tightness within a week. For several weeks there is a uterine discharge called *lochia* that gradually turns from reddish-brown to yellowish-white.

Psychological Aspects

A woman now usually leaves the hospital 2 or 3 days after an uncomplicated

[28]Nelson et al. (1980).
[29]Klaus and Kennell (1976).
[30]Scharfman (1977).
[31]Easterling and Herbert (1982).

delivery. The first week at home may be quite taxing to the new mother who is trying to cope with the many needs of the baby (especially around-the-clock feedings). Fatigue may be a major complaint at this point, along with a general "let-down" feeling.

About two-thirds of women experience transient episodes of sadness and crying sometime during the first 10 days after delivery, a phenomenon known as *baby blues* or in more serious cases as *postpartum depression*. In one to two cases out of 1,000 births, the disturbance is severe enough to lead to *postpartum psychosis* with hallucinations, profound depression, and suicidal and infanticidal impulses. This is a condition that requires urgent psychiatric attention.[32]

These problems are quite puzzling to all concerned, particularly since they occur when the mother is expected to be especially happy over the arrival of the baby. Yet the emotional upheaval of the postpartum period is often characterized by considerable maternal ambivalence towards the infant and fears that she may hurt the baby. Doubts about one's competence as a mother, fatigue, feelings of rejection or neglect by the father, as well as the biochemical changes that occur during this time are among the causative factors likely to be involved in these conditions.[33] Circumstances where the baby is unwanted, defective, or to be given up for adoption, inevitably compound the problems of the postpartum period. On the other hand, under more usual circumstances, the postpartum period is often a time of great joy, and deep satisfaction as both parents revel in the presence of the fledgling human being they have brought into the world.

Nursing

The sucking of milk at the breast among animals is referred to as *suckling*, and among humans, as *nursing* or *breastfeeding*. Immediately following delivery the breasts do not contain milk but **colostrum**.[34] Milk production (*lactation*) begins 2 to 3 days after childbirth and is accompanied by a feeling of engorgement in the breasts. Two pituitary hormones are involved in the physiology of milk production. **Prolactin** (from the anterior pituitary) stimulates the production of milk by the mammary glands, and **oxytocin** (from the posterior pituitary) causes the flow of the milk from the breast to the nipple in response to the stimulus of suckling by the baby. After weaning, or if a women does not breastfeed, the breast is no longer stimulated by the nursing infant and lactation ceases.

Nursing has been the universal method of feeding infants during most of human history and remains so in many parts of the world. Factors that have led to a decline in the prevalence of breastfeeding have been mainly the results of urbanization and industrialization. For mothers working outside the home regular breastfeeding is not practical. Those who work in the home have been influenced by the advice of health workers and manufacturers of milk substitutes to rely on alternative methods of feeding their babies. Women have been further discouraged from nursing because of concerns over how it might affect the shape of their breasts. Nonetheless breastfeeding is currently becoming more popular in the United States. It is an experience that can be emotionally quite satisfying and pleasurable. It is also not unusual for women to be sexually aroused, or in some cases to reach orgasm while nursing. Some enjoy the experience, others react to it with anxiety and guilt.

For the baby, there is no question of the superiority of human milk over cow's milk or commercial formulas. Human milk contains the ideal mixture of nutrients and antibodies that protect the infant from certain infectious diseases; it is free of bacteria; it

[34]Colostrum has more protein, less fat, and comparable levels of sugar as normal milk. It is rich in antibodies that may help provide immunity to the infant.

[32]Yalom et al. (1968).
[33]Simons and Pardes (1977, pp. 41–42).

is always at the right temperature; and it costs nothing. On the other hand, breast-feeding may be inconvenient or highly impractical for some women. Others do not produce sufficient milk. Drugs taken by the mother are usually secreted in the milk. For these and other reasons a mother may well decide not to nurse her infant, and she need not feel burdened by that decision.

Resumption of Ovulation and Menstruation

If a woman does not nurse her child, the menstrual cycle will usually be reinstated within a few months after delivery, though this may take as long as 18 months. The menses will usually not appear as long as the child is suckled. Lactation inhibits ovulation in the majority of women, but this is not a reliable method of contraception, even though it has been widely used in many parts of the world (chapter 6).

It should be emphasized that ovulation can occur before the first postpartum menstrual cycle; consequently, a woman can become pregnant without having had a menstrual period after the birth of her baby. The first few periods after pregnancy may be somewhat irregular in length and flow, but become regularized in time. Also, women who have painful periods may find that they suffer less from such discomfort after they have had a child.

Sex in the Postpartum Period

There is considerable variation in sexual activity after delivery. Fatigue, physical discomfort, level of sexual interest, and the obstetrician's injunctions play an important part in determining when a woman resumes sexual relations after childbirth. Women who nurse their babies reportedly have a higher sexual interest than those who do not; the underlying physiological or psychological reasons for this are not quite clear.

Doctors used to advise women to refrain from intercourse for "6 weeks before, 6 weeks after" delivery. But this rule does not hold anymore.[35] There is no medical reason why a healthy woman cannot have vaginal intercourse as soon as the episiotomy scars or other lacerations of the perineum have healed and the flow of lochia has ended, which usually takes about 3 weeks. The only medical risk at this time is the possibility of infection through coitus, so couples who practice sexual activities other than vaginal intercourse early in the postpartum period need not be hampered by this concern.

Particularly with the first baby, the intimate relationship between the parents must make room for a third person whose helplessness inevitably takes precedence over their more personal needs. A child can provide a tremendous boost to the bonds of affection linking a couple, with concomitant enhancement of sexual interest and satisfaction. But a baby can also be an intruder and a competitor for the affections of the parents. An admixture of all these considerations probably operates in most instances, making it necessary for a couple to go through a period of adjustment.

Prenatal Care and Disorders of Pregnancy

In most cases, pregnancy should not seriously interfere with a woman's ability to continue leading a normal life with regard to her work, social activities, and sexual relations. Some adjustments may need to be made but there is no reason why a pregnant woman should be treated like an invalid. However, for the pregnancy to progress normally, appropriate *prenatal* ("before birth") care is necessary to safeguard the health of the mother and the baby. Teenage mothers (below 15) and older mothers (over 35) are especially vulnerable to complications of pregnancy and need to be watched with particular care for signs of complications.

[35]Easterling and Herbert (1982).

Nutrition and Exercise

During pregnancy, a woman does not need to eat for two, as conventional wisdom claims, but she does need about 200 calories above her normal daily intake (a grand total of 40,000 calories throughout pregnancy).[36] Her diet should be rich in proteins, supplemented by vitamins and minerals. Calcium, in particular, is necessary for the growth of the fetal skeleton, iron and niacin for the prevention of anemia.

Pregnancy is no time to indulge in dietary fads. Poor nutrition may endanger the mother's health; it will also seriously compromise the development of the fetus, leading to premature or low-weight babies who are subject to higher death rates and more vulnerable to brain damage and mental retardation. The pregnant woman thus walks a thin line with regard to food intake. On the one hand she must make sure she gets adequate nutrition; on the other she must watch for excessive weight gain, which carries its own penalties.

A similar situation exists with regard to physical activity. Ample exercise is as essential as adequate rest and sleep. Being pregnant is an added source of stress for the mother's body and allowance must be made for it. The fatigue and lethargy that mark certain periods of pregnancy are adaptive reactions to induce the mother to rest more. How much to exert oneself is a matter of individual judgment. Some women are in better physical shape than others and thus can do more. Common sense would exclude more hazardous sports, but activities like swimming and walking are most beneficial to women with normal pregnancies.

Effects of Smoking, Alcohol, and Drugs

The fetus is vulnerable to all harmful substances that cross the placental barrier. Thus every chemical that is ingested or otherwise introduced into the mother's body must be considered with that risk in mind.

Cigarette smoking during pregnancy is harmful above and beyond the risks it ordinarily entails. Fetuses carried by women who smoke while pregnant have an increased risk of prematurity, lower birth weight, and death in the *perinatal* (before and soon after birth) period.[37] It has also been shown that children born to heavy smokers carry almost twice the risk of the children of nonsmokers to develop impulsive behavioral disturbances in childhood, as well as showing lower IQs, and poorer motor skills.[38] A woman who is a smoker but quits no later than by midpregnancy, does not cause similar risks to the fetus.

Heavy *drinking* by a pregnant woman can cause serious abnormalities in the child (*fetal alcohol syndrome*) including birth defects and mental retardation.[39] Even moderate drinking (one or two drinks a day) or a single binge may be enough to cause damage. Though abstinence from alcohol is clearly the safest course, an occasional drink may not be harmful. The risk of damage to the fetus is highest at the time of implantation.

The active ingredients in **marijuana** cross the placenta and have been shown to cause fetal damage in some animals.[40] Though the influence of smoking marijuana in pregnant women remains unclear, it would be safer to avoid its use during pregnancy and nursing.

In principle, any chemical substance or *drug* can be harmful to the fetus and therefore must be used only under medical supervision.[41] It is not only substances like *Thalidomide* (which causes dramatic malformations of the limbs) but also very commonly used items like aspirin and Valium

---

[36]Moghissi (1982).

[37]Meyer and Tonascia (1977); Coleman et al. (1979); Haynes (1982).

[38]Dunn et al. (1977).

[39]Haynes (1982).

[40]Harbison and Mantilla-Plata (1972).

[41]Chemicals and other factors that cause birth defects are called *teratogens* (from the Greek word for "monster"), and the field of study that deals with their effects is *teratology*.

that pose risks. Similarly, hormones can have serious effects on fetal development (chapter 4) that may be manifested years later. The use of *narcotics* by the mother exposes the fetus to very high risks, including addiction of the infant, which necessitates a gradual withdrawal after birth.

Complications of Pregnancy

The most serious problem in the first trimester is miscarriage or **spontaneous abortion,** which accounts for the termination of 10 to 15% of pregnancies (75% occur before the sixth week). In about half of these cases a defect in the fetus is the probable cause of the miscarriage. But in 15% of cases the miscarriage is due to some condition in the mother such as illness, malnutrition, or trauma.

Another complication of pregnancy is a condition called **toxemia** (from Latin for "poison"). The cause of toxemia is unknown; presumably a toxin produced by the body causes the symptoms of high blood pressure, headaches, protein in the urine, and the retention of fluids causing swelling in feet, ankles, and other tissues. In the early stages the illness is called *preeclampsia;* when it progresses to coma and convulsions, it becomes *eclampsia.*

Toxemia occurs only in pregnant women, usually in the last trimester, affecting some 6% of pregnancies, but as many as 20% of those not under prenatal care. Uncontrolled toxemia and hemorrhage are the major causes of maternal mortality and account for 9.4 maternal deaths in 100,000 births or 1 death in 10,000 pregnancies in the United States (as against 70 per 10,000 in Bangladesh).[42]

During the third trimester, **premature birth** is the most serious concern. Since the date of conception is not always accurately estimated and because the age and weight of the fetus are highly correlated, prematurity is defined by weight rather than age. Thus, an infant who weighs less than 5 pounds and 8 ounces (2.5 kg) at birth is considered *premature.* The mortality rate among premature infants is directly related to size: the smaller the infant, the poorer are its chances for survival.

An estimated 7% of births in the United States are premature. Prematurity may be associated with various maternal illnesses (such as high blood pressure, heart disease, and syphilis) as well as heavy cigarette smoking, and multiple pregnancies; half the time the cause of prematurity is unknown.

Not all women are equally vulnerable to the complications of pregnancy. A woman's general health, age, and the quality of prenatal care are all significant factors. Pregnancies before age 17 run a higher risk of toxemia, prematurity, and infant mortality. These problems are often compounded by ignorance, negligence, and poor care.

Women who wish to become pregnant after 35 face other special problems. To start with, they are less likely to succeed because of reduced fertility. If they do get pregnant, they run a higher risk of miscarriage and have a greater likelihood of having children with certain birth defects. But many older women are also quite capable of bearing perfectly normal children.

Conditions Affecting the Fetus

The **infant mortality rate** is the number of infants who die within the first year of life per 1,000 live births; the current rate in the United States is 14.0. The **neonatal mortality rate** is the number of live-born infants who die within 28 days after birth, currently 9.8.

A variety of maternal ailments adversely affect the fetus but developmental problems also arise independent of the mother. Abnormalities of structure or function present at birth are called **congenital malformations** or *birth defects.* They affect 3% of all live births. In 70% of cases there is no identifiable cause; about 20% are due to genetic factors inherited from the parents; 10% are due to chemicals (often drugs and alcohol used by the mother), radiation, infections, and other causes.[43]

---

[42]Thompson (1982, p. 287).

[43]Oakley (1978).

One example of a chromosomal disorder is *Down's syndrome*, which results in severe mental retardation and defective internal organs. It affects one in 1,500 births but the incidence rises sharply with maternal age to one in 300 births at age 35, and 1 in about 40 births at age 45. For men older than 55, age is similarly linked to a higher incidence of this defect in the offspring.

Among infectious conditions, particularly damaging are certain viruses. German measles (*rubella*), for instance, if contracted during pregnancy, in 50% of cases causes serious defects of hearing, vision, damage to the heart, and mental retardation.

Congenital disorders other than birth defects include diseases, such as syphilis, which are contracted by the fetus from the mother (chapter 7). Another serious condition whose symptoms become manifest soon after birth involves destruction of the newborn's red blood cells because of *Rh incompatibility* with the mother. The Rh factor in blood is present (positive) or absent (negative) in various men and women. The problem arises when the mother is Rh-negative and the baby Rh-positive. In this case, the mother's body produces antibodies against the positive Rh-factor, which destroys the red blood cells of the fetus, causing anemia, jaundice, and sometimes death. The development of Rh sensibility can be prevented in Rh-negative pregnant women by medications that neutralize antibody formation. The condition is treated in the newborn by special transfusions.[44]

Infertility

In our preoccupation with contraception, it is all too easy to lose sight of the opposite problem of the inability to reproduce. *Infertility* means failure to conceive or to impregnate over a given period of time (usually 1 year); *sterility* entails permanent infertility.

The world may have more people than it needs, yet for an individual couple the failure to have offspring can be deeply frustrating. As we said earlier, about 1 in 10 marriages in the United States are involuntarily childless after attempts to conceive for a year or longer. Another 1 in 10 couples would like to have more children than they do but cannot.[45]

Causes of Infertility

Reproductive failure occurs in the male partner in about the same proportion of cases as in the female (each 40%); in 20%, the problem is due to both. The most frequent cause of *male infertility* is a *low sperm count*; when the ejaculate contains fewer than 20 million sperm per milliliter, impregnation becomes highly unlikely. Additional causes are *defects* in a large proportion of sperm and *blockage* somewhere in the tubal system. These problems may be due to developmental abnormalities in the testes (including undescended testes), infections (including VD), exposure to radiation or chemical agents; severe malnutrition, and general debilitating conditions. The administration of steroid hormones also inhibits spermatogenesis by suppressing gonadotropin production.

The most common causes of *female infertility* are the *failure to ovulate* and *blockage of fallopian tubes*. A large variety of causes may in turn account for these problems, prominent among which are defects of reproductive organs, hormonal disorders, diseases of the ovary, severe malnutrition, chronic ailments, drug addiction, scarring of the fallopian tubes by pelvic infections (including gonorrhea); in rare cases the problem may be psychologically caused.[46]

Treatment of Infertility

The diagnostic investigation of infertility involves both the female and male. Once the problem has been pinpointed, it can often be corrected by the appropriate treatment. This includes the use of *drugs* like *clomiphene*, which induces the pituitary to produce LH and FSH, or failing that, *HMG* (human menopausal gonadotropin), which

---

[44]Vaughn et al. (1979).

[45]Menning (1977).
[46]For more details, see Moghissi and Evans (1982).

acts directly on the ovary. In a high proportion of cases these induce ovulation and make pregnancy possible. (In fact, the ovaries are often overstimulated, resulting in multiple pregnancies.) Also, *microsurgery* may succeed in opening blocked tubes.

Despite the often successful treatment of infertility, in a certain proportion of cases none of the ordinary measures work. Faced with this fact, many couples give up the idea of having children or adopt a child. But some now go to greater lengths to conceive by resorting to artificial insemination and surrogate mothers.

Artificial insemination, consists of the placement of semen in the vagina by means other than sexual intercourse. When the source of the semen is the *husband,* this approach raises no unusual problems; it is simply a means of pooling sperm from several ejaculations from a husband who has a low sperm count and depositing it in the wife's vagina. But when the semen comes from a *donor* because the husband is sterile, then many concerns arise in psychological, social, and legal terms. Otherwise, there are no serious technical problems; 75% of women get pregnant if artificially inseminated with fresh donor sperm.

The donor is carefully selected with respect to race, blood type compatibility, physical appearance, general health, and genetic background. His identity is kept secret; he never sees the child, nor even knows if the artificial insemination worked. The procedure entails the deposition of the donor semen (obtained through masturbation) into the cervical canal and the region around it.

Artificial insemination is almost always done with fresh semen. But frozen sperm has also been used, though with a lower rate of success.[47] Frozen sperm is kept in *sperm banks* as "fertility insurance" by some men who are to undergo vasectomy (discussed in chapter 6).

When it is the wife who is sterile, the male counterpart of artificial insemination is to have another woman get impregnated artificially with the semen of the husband. This *surrogate mother* carries and then relinquishes the baby to the couple as stipulated by a contractual agreement. The comparison with artificial insemination is, however, not quite apt, since for the surrogate mother to give up her child is a much more wrenching experience than for the sperm donor to suspect that he may be the biological father of a child somewhere.

In 1978 a more astounding event, long a science fiction fantasy, became reality with the birth of Louise Brown, the first *test-tube baby* in England. The procedure was to induce ovulation in the mother (who had defective fallopian tubes), then recover a mature ovum from her ovary and fertilize it with the father's sperm in a laboratory dish containing appropriate nutrients. After the fertilized egg had reached the blastocyst stage it was transferred to the mother's uterus, which had meanwhile been primed with hormones. From then on the embryo developed as in any ordinary pregnancy, making it possible for a woman to conceive who otherwise could not have done so. Finally, in cases where a woman cannot ovulate, it is now possible for her to become pregnant by *embryo transfer*. This is achieved by having a donor woman artificially impregnated with the sperm of the infertile woman's husband. The developing embryo is then flushed out of the donor's uterus and implanted in the infertile woman's uterus. Or the embryo is frozen and implantation attempted at a later time.

Through these various methods, hundreds of children have been born. These technological developments open up unprecedented possibilities and problems. Some see these as a ray of hope for thousands of sterile women who wish to experience motherhood; for others it signifies the heralding of an Orwellian world of assembly-line produced humanity. We return to the social and legal implications of some of these novel means of conception and parenthood in chapter 19.

---

[47]For women who "qualify" (by having high IQ; professional achievements; good health), *Germinal Choice* in California will provide frozen sperm from Nobel laureates, only one of whom, William Shockley, has so far identified himself as a donor.

SUGGESTED READINGS

Pritchard, J. A. and MacDonald, P. C., eds. *Williams Obstetrics*, 16th ed. (1980). The course of normal pregnancy is discussed fully in this standard obstetrical text. Hughey, L. and Weber, M., *Woman Care*, 1982 present a less technical account.

Wertz, R. W. and Wertz, D. C., *Lying In* (1979). A history of childbirth in America.

Fetal development is covered in embryology texts (listed in readings for chapter 2). There are superb photographs of fetal development and a concise text in *A Child Is Born* by Lennart Nillson (1977).

Chapter 6

Contraception and Abortion

Birth control affects nearly everybody—people either have used it, will use it, or, at the very least, are against it.

Carl Djerassi

Contraception has been the object of greater attention during the past several decades than any other aspect of sexuality, and no other issue is likely to have a potentially greater impact on our sexual attitudes and behavior in the future. Our ability and willingness to use contraceptive devices constitutes one of the most critical of all problems in view of the potentially catastrophic growth of the world population.

There is nothing new about the interest on the part of individuals in avoiding the reproductive consequences of sexual intercourse (box 6.1). What is distinctive about contemporary contraceptive practice is the ability to decouple sex from reproduction in highly effective and reasonably safe ways, and extending this choice to large populations as a matter of public policy.[1]

Contraception (against conception) and *birth control* (a term coined by Margaret Sanger in 1914) encompass attempts at avoiding reproduction or attaining *fertility control. Family planning* refers to the spacing of children through birth control.

Abstinence from coitus also avoids pregnancy, but it does so by shunning sex altogether rather than making it contraceptively safe (just as refusing to ride in a car will avoid getting into car accidents but cannot be said to make driving safer). Similarly, there is a valid distinction between preventing pregnancy through contraception and terminating it through *abortion,* as there is between abortion and *infanticide* (killing or letting a newborn baby die).[2]

Our primary interest in this chapter is with the biological and behavioral aspects of contraception and abortion. But these are also matters of enormous psychological and social significance; we deal with their moral and legal aspects in chapters 19 and 20.

[1] For overviews of social considerations on contraception, see Noonan (1967); Djerassi (1981).

[2] Infanticide has been practiced in some cultures living under harsh economic conditions (such as among the Eskimo), to ease population pressure (as in Polynesia), to eliminate abnormal offspring, or the issue of illicit relations (like incest). The ritual of firstborn sacrifice has also been known since Biblical times as a means of offering to the deity one's most precious possession (Genesis 22).

Box 6.1: Early Contraception

People have been attempting to prevent unwanted pregnancies for millenia.[1] With the exception of the use of hormones, most of the current contraceptive methods have their ancient precursors, which were based on the right ideas but were subject to the technological limitations of their time.

Probably the oldest and most common contraceptive practice has been withdrawal before ejaculation, as attested in the Old Testament. The oldest known medical recipes to prevent conception are contained in the Egyptian Petri papyrus of 1850 B.C. Ancient Egyptians used vaginal pastes containing crocodile dung, honey, and sodium carbonate to act both as a barrier and spermicidal agent. They also relied on vaginal douches containing mixtures of wine, garlic, and fennel.

In ancient Greece and Rome, there was much concern over contraception and reliance on absorbent materials, root and herb potions, pessaries, and more permanent means of sterilization. Soranus, a physician who practiced in Rome in the 2nd century A.D., gives a clear account of the fertility control practices of the period, distinguishing between contraception and abortion-inducing agents (recommending reliance on the former). Through Islamic physicians, this classical knowledge passed on to medieval Europe and formed the basis of contraceptive practice up to the end of the 17th century.

Much of early contraception relied on barrier methods, and many materials were used in various cultures to provide obstacles to the passage of sperm.[2] Some native American groups blocked the cervix with soft clay; the Japanese inserted paper; the French, balls of silk. Various types of sponge and cotton balls were also used, sometimes in conjunction with medicated mineral oil or mild acid ointments to serve as spermicides.

Intracervical pessaries made of gold, silver, or rubber have also been used to plug the cervix (see accompanying figure). They were inserted by a physician at the end of one menstrual period and removed at the onset of the next period. Intrauterine pessaries were extended into the uterus to decrease the likelihood of their slipping out of place.

Along with these rational, even if not very effective, methods, countless magical means have been resorted to especially in preliterate societies. Among some North African tribes, for example, water that had been used for washing a dead person was secretly given to a woman to drink to make her infertile. In another group, a woman would eat bread into which had been ground a honeycomb containing a few dead bees. (The magical association between death and sterility was presumably the basis for these practices.)

Other contraceptive attempts sound even more farfetched. Moroccan men would turn a special ring on a finger from one side to another following coitus. Papuan women of New Guinea who did not want a child, tied a rope very tightly around their waist during coitus. But they also washed carefully following it, which probably exerted enough beneficial influences to keep the two practices alive in combination.

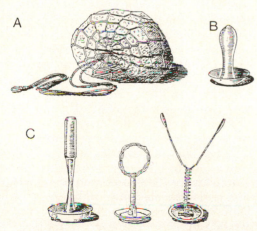

A sponge in a silk net has a string to facilitate withdrawal (A). An intracervical pessary used to plug the cervix (B). Intrauterine pessaries extended into the uterus (C).

[1] For a comprehensive history of contraception, see Himes (1963). Also see Suitters (1967); Draper (1976). Dawson et al. (1980) discuss fertility control in the United States before modern contraceptives were developed.
[2] Cooper (1928).

Patterns of Contraceptive Usage

Just as there are numerous reasons for wanting children (chapter 5), there are also many reasons for not wanting them, or more often, not wanting them at a particular time. Thus, for many couples, it is the wish for family planning rather than the avoidance of parenthood altogether that prompts their use of contraception.

Reasons for Contraceptive Use

One set of reasons for avoiding parenthood is based on the nature of the couple's relationship. Although women occasionally choose to become single parents, most people wait till they are married or have a stable relationship before having children. Another is the general tendency to limit family size, especially among the middle classes. Unlike earlier times, when infant mortality rates were high, parents now can be fairly certain that their children will grow to maturity, so there is no reason to have many children to ensure that at least a few will survive to adulthood. And rearing children being as expensive as it is, not many parents can afford to have large families without financial strain. Other reasons for avoiding or postponing pregnancy include health considerations and career aspirations.

Traditionally, childlessness has been due to the renunciation of sexual relationships or the inability to bear children. With the advent of effective contraception, sexually active individuals now have the choice not only to postpone childbearing but avoid it altogether if they consider the personal and social advantages of not having children to outweigh the benefits of parenthood. To express the positive component in their choice, these couples are sometimes called *childfree*, or following a "childfree lifestyle."[3] Some couples arrive at the decision not to have children through a series of postponements, others decide it relatively early in life and enter marriage with that understanding.[4] The proportion of women who express their intention early to remain childless is quite small (about 6%) and constitutes a group that has not yet been well studied. It has been suggested that family backgrounds that encourage the development of traits like independence and achievement motivation among women make the voluntary choice of childlessness more likely. In addition to counteracting the various personal motivation for having children, these individuals have to contend with the expectations and pressures of *pronatalist* society, which views parenthood as a normative and desirable part of adult life.[5]

Although relatively few people limit their family primarily out of concern for the world's expanding population, this is a matter of great importance for social agencies and governments, particularly in developing countries with the most explosive rates of population growth. The alarm over the dire consequences of population explosion was sounded by the English clergyman and economist Thomas R. Malthus (1766–1834). Malthus argued in *An Essay on the Principle of Population* (1798) that population grows exponentially by doubling, while the means of subsistence expands by simple increments. To avoid the inevitable prospects of famine and war that would result from this discrepancy, he advocated abstinence and late marriage. His successors shifted the emphasis to contraception instigating the *population control* movement that attained great visibility in the 1960s.

From the beginnings of prehistory, it took until 1830 for the population of the world to reach 1 billion. During the next century this figure doubled; and it is likely to double again in less than half that time. The problem, though, is highly regional. Where the population of Brazil may double in two decades, that of Denmark will take a century to do so. The *birth rate* (the number of live births for 100,000 women per year) has dropped very significantly in the United States within the past 2 decades (15 per 100,000 in 1976) but even if couples restricted themselves to two children, the

[3]Cooper et al. (1978).
[4]Veevers (1974).

[5]Houseknecht (1978).

population will continue to grow until 2040, when it would stabilize at about 280 million.[6] When births equal deaths in a population, we have a state of *zero population growth.*

Prevalence of Contraceptive Use

Pressed by all these considerations, there is currently extensive reliance on birth control in the United States. In 1976 an estimated 13.5 million people, half of all married couples where the wives were aged 15 to 24, used contraception; and another 5.1 million (18.6%) had been surgically sterilized. Thus, 67.7% of this population (or 18.6 million people) were using birth control. Of the remaining 32.3% of couples who did not use birth control, 13.3% were pregnant, postpartum, or trying to get pregnant; 11.4% were sterile for reasons other than contraception; and 7.6% were not using contraception for other reasons.[7]

Contraceptive devices may be categorized by their modes of action. There are *hormonal methods* (as exemplified by the birth control pill); *intrauterine devices* (IUDs); *barrier methods* (condom, cervical cap, and diaphragm); *spermicides* (foam, jelly, cream); *withdrawal;* and methods which rely on *periods of infertility.* In addition, there are various agents that induce abortion (*abortifacients*) and more permanent forms of achieving infertility through *sterilization.*

In the United States, the birth control pill is the single most frequently used device for females. Condoms and withdrawal are the only exclusively male methods; sterilization is shared by both sexes; the rest are all female methods. The main burden of contraceptive responsibility clearly falls on women. This may well be another instance of men burdening women with what should be a shared concern. But there are also important physiological reasons why it has been more feasible so far to develop female methods of contraception, especially with respect to hormonal

methods.[8] Technological breakthroughs and changes in social attitudes may alter this picture, but because it is only women who can get pregnant, contraception is likely to remain a matter of more concern to them.[9]

Under ideal conditions, the only women who would get pregnant would be those who wished to do so. Instead, about half of all pregnancies in the United States (including about a third of all births to married women) are currently unplanned.[10] This is not to say that half of all fertile women do not use contraceptives; rather, it is a relatively small proportion of women who repeatedly fail in this regard. Furthermore, not all unplanned pregnancies result in unwanted children. Fewer than half of unplanned pregnancies are aborted. But while some people may be perfectly happy with an unplanned pregnancy after the fact, not all of those who do not get abortions can be assumed to feel similarly, since getting an abortion may not be an easy matter of them.[11] But, as with population problems on a global scale, these women are often the ones that can least afford the burden.

An especially serious situation exists with regard to teenagers, who account for an estimated 1 million pregnancies a year—or one teenage pregnancy every 30 seconds. Of these teenagers, 30,000 are younger than 15.[12] Half of teenage women aged 15 to 19 report having had premarital coitus; of these 27% do so without ever using contraception and 62% of this group consequently get pregnant. For the sexually active teenage population as a whole (including contraception users and nonusers) the pregnancy rate is 16%. Recent trends show a somewhat greater use of contraception by young women. For instance, the percentage of those who used contraception at first intercourse has gone up from 38% percent in 1976 to 49% in 1979.[13] Yet this has been mainly due to an increased reliance on

[8]Djerassi (1981).
[9]*Boston Women's Health Book Collective* (1976).
[10]Munson (1977).
[11]Diamond et al. (1973).
[12]Hatcher et al. (1982).
[13]Zelnik and Kantner (1980).

[6]Ehrlich and Ehrlich (1972); Westoff and Westoff (1971).
[7]Mosher (1981).

condoms and withdrawal rather than the more effective contraceptive devices.

In addition to age, the prevalence and types of contraceptive use are correlated with marital status, socioeconomic class, ethnicity, religious affiliation and other factors.[14] Younger, unmarried women rely more often on the pill, less on IUDs, and a great deal on their partners using condoms or withdrawal. These trends are quite fluid, especially among well-educated younger women. For example, on one University of California campus, between 1974 and 1978, the percentage of women using the pill fell from 89% to 63%, diaphragm usage went from 6% to 33%, while IUD use remained steady at 8%.[15]

Obstacles to Contraceptive Usage

There are many reasons why contraception is not used more effectively. The opposition of the Catholic church has attracted much public discussion (see chapter 20) but relatively few women shun contraception because of it. Among married Catholic American women, every two out of three now use a contraceptive device other than rhythm; among those married between 1970 and 1975, the proportion of Catholic couples using contraception is 90.5%.[16]

Factors that inhibit effective use of birth control measures are more often not the dictates of conscience but rather, the result of ignorance, lack of access to contraceptives, and various psychological and social factors. A great deal of sexual misinformation exists, particularly among teenagers, even sexually experienced ones. Many think they are "too young" to get pregnant.[17] Others assume that it takes repeated intercourse to lead to conception; that if they do not reach orgasm they will be protected; or that they are in a safe period of the month.[18] Some think they must be sterile because coitus in the past did not result in pregnancy. Men have a tendency to view contraception as a "woman's problem" and to assume that if she says nothing, she must be safe.

Some couples seemingly forego contraception for nothing more serious than because it is "too messy" or "too much of a hassle." The unpredictable association between sex and pregnancy in itself generates a false sense of security. In fact, there is only a 2 to 4% chance that any single act of coitus will lead to pregnancy. But the odds go up to 30% if coitus takes place on the day before ovulation, to 50% on the day of ovulation, and to 90% over the period of a year.[19] The more often one has sex, and the more often that happens in high-risk times, the greater the chances of getting pregnant. As in any gamble, the odds sooner or later catch up with the unwary or risk-taking player.

The willingness to gamble in unprotected coitus may sometimes be based on feelings of guilt. It is not unusual to court punishment to alleviate guilt or to vindicate oneself. If you have sex when you think you should not, and you get away with it, then it must have been all right; if you get pregnant (or catch VD, or get caught) then it is only fair that you suffer.

A surprising number of sophisticated men and women risk pregnancy because contraceptive measures imply forethought; because such precautions are thought to rob sex of romance and spontaneity and turn it into a cold and calculated affair. To come prepared appears exploitative on the man's part, and promiscuous on the part of the woman.

Some people get pregnant even when it seems to others that they should not, because they want to. Among one group of sexually active teenagers, 7% said they wanted to get pregnant and another 9% would not mind if they did get pregnant.

---

[14]Anderson (1981).

[15]Harvey (1980).

[16]Westoff and Jones (1977).

[17]There is in fact a period of relative infertility in puberty before the ovarian cycle becomes well established (see chapter 4). But that does not amount to contraceptive security.

[18]Evans et al. (1976). Among sexually active teenage girls, 70% of whom had sex education courses, only one-third knew correctly the period of highest risk (Zelnick 1979).

[19]Luker (1975).

Many more may have similar motivations without quite being conscious of it. Even under the most inauspicious circumstances, having a baby carries enormous psychological significance. It may represent a teenager's desperate chance to be someone, to have someone to love, and to receive a measure of affection and attention herself. Pregnancy is also a powerful tool to hold on to another, to induce marriage, to embarrass families, to punish others, and to assert oneself. For some men impregnating a woman is the ultimate sign of "machismo."

Additional obstacles to contraceptive usage are created by the practical problems of availability, cost, and access to contraceptive information and devices. Pharmacies are full of condoms, but a man needs to go up and ask for them; not an easy matter for many, particularly the young. Planned Parenthood clinics will give a woman whatever contraceptives she needs but she must be willing to reveal her sexual intentions by going to them first, which is also not an easy matter for many. If society makes it mandatory that a teenager's parents be notified of her receiving contraceptive help, the effect on her willingness to seek such help is not hard to imagine.

Concern over the side effects of contraceptive agents is quite legitimate. But such concern may get exaggerated to the point of inhibiting its use. The risks in contraception are minimal and should be put in perspective. The chances of death in a year for a woman on the pill is 1 in 63,000 (1 in 16,000 for smokers); the fatality rate for the IUD is 1 in 100,000, which is one tenth the death rate in pregnancy. The chances of death for drivers of automobiles is 1 in 6,000 in a year; 1 in 200 for those smoke 1 pack of cigarettes a day; 1 in 50, for motorcyclists.[20]

Guidelines for Effective Contraception

The first step in making a contraceptive decision is coming to terms with one's sex-uality in general and in regard to a particular relationship, and on a specific occasion. There are many profound and practical issues to ponder in this regard but ultimately the question is whether or not one will use birth control, not in principle, but in the concrete context of a given sexual act. There are many cogent reasons to engage or not to engage in sex, but there are never any good reasons, only bad ones, for getting pregnant unless the circumstances are propitious. Effective contraception starts with that very conviction.

Of equal importance is knowledge of one's procreative vulnerabilities. A woman needs to monitor her high-risk periods, and be aware of the stages of increased conceptive vulnerability in her life. These stages tend to cluster around periods of transition: the time of becoming fertile during adolescence, and infertile during the menopause; at the start of a new sexual relationship; early in marriage; following childbirth; or abortion. There is also heightened contraceptive vulnerability during periods of stress such as during or following separation and divorce or breakups and quarrels. Moving away from home, involvement in a new circle of friends, experimenting with a different lifestyle are other situations where sexual activity may take place before a consistent contraceptive practice is established or reestablished.[21]

Contraceptive Methods

The ultimate purpose for all contraceptives is the same, but the methods for achieving that aim vary widely in efficacy, facility of use, the nature of side effects, and cost. The ideal contraceptive would be socially acceptable, usable by either sex, fail-safe, free of side effects, aesthetically inoffensive, and readily and cheaply available. No such contraceptive now exists or is likely to.

What we have instead is a set of alternatives with a varying mix of assets and

[20]Hatcher et al. (1982, p. 7).

[21]Miller (1973).

liabilities. Each device must be evaluated not just in the abstract but with regard to the needs of a particular individual at a given time. No contraceptive method, at present, is absolutely reliable or foolproof in preventing pregnancy (though some methods in combination come pretty close to that).

The only way to avoid pregnancy with absolute certainty is to avoid sexual intercourse or to be sterilized. Those who are not willing to do either must be prepared to take certain contraceptive risks, just as they take risks every time they step into a car or cross the street. The sensible course is to take *calculated risks;* to know what benefits one is likely to get for what probable cost, rather than simply to stumble into situations and hope for the best.

Box 6.2 shows in summary form the various aspects of commonly used methods of contraception. It is presented here, before we discuss these individual methods, so that it may serve as a preview and

Box 6.2   Summary of Contraceptive Methods and First-year Failure Rates

Method	Lowest Observed Failure Rate	Failure Rate in Typical Users	Disadvantages
Tubal ligation	0.04	0.04	Possible surgical, medical, and psychological complications
Vasectomy	0.15	0.15	Same as above
Injectable progestin	0.25	0.25	Continual cost
Birth control Pills	0.5	2	Requires daily attention; discontinuation rate 50–60% in first year
Progestin only pill	1	2.5	Same as above
IUD	1.5	4	Side effects, particularly increased bleeding, possible expulsion
Condom	2	10	Continual expense; interruption of sexual activity and possible impairment of gratification
Diaphragm w/cream or jelly	2	10	Repeated insertion and removal; possible aesthetic objections
Cervical cap	2	13	Same as above
Vaginal foam	3–5	15	Continual expense
Withdrawal	16	23	Possible sense of frustration
Rhythm, BBT, cervical mucus method	2–20	20–30	Requires significant motivation, cooperation, and intelligence; useless with irregular cycles and during postpartum period; requires careful observation
Douche	—	40	Inconvenient; possibly irritating
Chance	90	90	

(From Hatcher et al., *Contraceptive Technology 1982–83* and *1980–81,* Irvington Publishers.)

help keep various approaches in perspective. Note that the effectiveness of contraceptive methods is measured in terms of *failure rates:* the number of women out of a hundred who get pregnant while using the method over a year. A further distinction is made between the *ideal* or the *lowest* rates of failure with optimal use and the *actual* failure rate, or the failure rate of *typical* users. Thus, it is the element of human error that accounts for the discrepancy between the two rates.

Hormonal Methods

Hormonal methods turn off the reproductive process at its source. They are the most effective of all reversible contraceptive methods discovered to date. Properly used, the most effective hormonal methods provide over 99% protection (box 6.2). However, hormonal intervention is also physiologically the most intrusive and its side effects more serious than is generally the case with other methods. But then, in addition to contraception these hormones have additional health benefits, which other methods lack.

Hormones can be introduced into the body orally, by injection, and through other means like implantation under the skin. So far oral use has proven to be by far the most practical method, and that is what we focus on here.

Oral contraception became commercially available in the 1960s, but its use spread so widely that it has come to be known simply as *the Pill.*[22] Some 60 million women worldwide and 10 million in the United States now rely on the Pill, making it the most extensively used effective contraceptive measure in the world. It accounts for 40% of contraceptive methods used by never-married American women and 33% by those married.[23]

The development of the Pill was a natural consequence of our modern understanding of the ovarian cycle. In the 1930s it was established that progesterone inhibited ovulation and hence could prevent pregnancy. Simultaneously, the estrogenic hormone was chemically isolated and used to treat certain menstrual disorders. The first oral contraceptive chemical substance (*norethindrone*) was synthesized in 1951 at the Syntex laboratories in Mexico City followed by wide-scale clinical trials in Puerto Rico in 1956. The first Pill (*Enovid*) became commercially available on the market in the 1960s.[24] Over the next two decades, oral contraceptives were studied more extensively than any other drug, since they are used by millions of healthy women for a critically important reason.[25]

Birth control pills contain various combinations of **synthetic steroids** with *estrogenic* and *progestational* actions, and like the natural hormones, they inhibit the secretion of LH and FSH by the anterior pituitary, thus preventing ovulation (chapter 4). They also interfere with the development of the lining of the uterus, making implantation difficult, and cause thickening and increased acidity of the cervical mucus, which thereby becomes a more effective barrier to the passage of sperm.

The Combination Pill   Oral contraceptives come in several forms. The most commonly used is the *combination Pill,* which consists mainly of progesterone and a small amount of estrogen (the precise amounts vary depending on the manufacturer). The combination Pill is taken once a day for 21 days beginning on the fifth day of the menstrual cycle (that is, 4 days after the onset of menstrual bleeding). The woman then waits for 7 days, during which time withdrawal bleeding simulates the normal menstrual flow; she then resumes taking the Pill.

To help the user stay on schedule, manufacturers add seven inactive or vitamin pills colored differently to be taken on

[22]It was Aldous Huxley who first referred to "the Pill" in *Brave New World Revisited* (1958).

[23]Ory et al. (1980). Percentages of use from Ford (1978) and Zelnik and Kantner (1980).

[24]For a history of the pill, see Pincus (1965). For an account of its chemical development, see Djerassi (1981).

[25]Djerassi (1981). For an assessment of this research over two decades, see Ory et al. (1980).

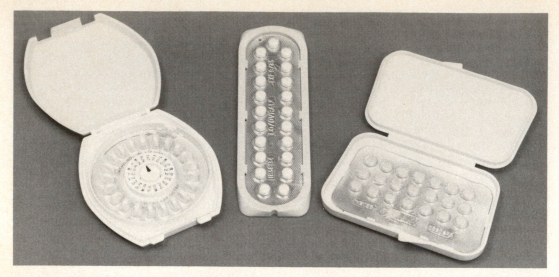

Figure 6.1   Various brands of oral contraceptives.

the off days; thus every day the woman takes a pill of one or the other variety (figure 6.1). If a birth control pill is missed for a day, two are taken the following day; if two are missed, then both should be added to the pill of the third day. If the Pill is missed for three days or longer, the method is no longer reliable for the rest of that cycle, and some other form of contraception must be used until the woman gets back on track following the next menstrual period.

The combination Pill is currently the most effective form of reversible contraception in wide use. Under ideal conditions no more than 1 in 200 (0.5%) sexually active women risks getting pregnant over a period of a year; the actual failure rate is 2% (box 6.2).

The Pill has the major advantage of freeing the couple from having to take any measures just before, during, or right after sexual intercourse; it is thus the least intrusive of all birth control methods. More than any other contraceptive device, the Pill puts a woman in full and effective charge of her reproductive process.

Any drug may manifest a number of secondary or **side effects** that are undesirable. Birth control pills show a wide variety of such side effects that complicate their use to varying degrees. We shall restrict our-

selves here to the more common and the more serious risks entailed.[26]

Among the minor side effects of the Pill are symptoms of nausea, breast tenderness, constipation, skin rashes (such as brown spots on the face), weight gain, vaginal discharge, and headaches; these effects are quite similar to the symptoms of early pregnancy and are usually transient.

The more serious risks caused by birth control pills relate to the *cardiovascular system.* There is increased risk of heart attack, stroke, hypertension, and the formation of blood clots. These clots may cause only local discomfort, but if as venous emboli they lodge in the brain or in the lungs, resultant damage could be serious. Older women and smokers are particularly susceptible to these serious side effects: for instance, women in their thirties who smoke and use the pill are seven times more likely to suffer a heart attack than nonsmokers.

The prospect that has caused the greatest alarm is that birth control pills may cause *cancer* of the uterus or the breast; yet no

[26]For more detailed discussion of side effects of birth control pills, see Isung et al. (1979); Hatcher et al. (1982).

such effect has been substantiated despite a great deal of research. It is not possible to assert that such an effect will never appear in the future, but as more time passes beyond the two decades the Pill has been around, it becomes less and less likely that any significant risk of cancer remains undetected.[27]

Similarly, there is no evidence that birth control pills cause *birth defects* unless they are taken by the mother when she is already pregnant (therefore a woman should only start taking the Pill right at the end of the menstrual period). Nor is there evidence that the use of oral contraceptives has any long-term effect on subsequent fertility even though there may be a delay in returning to normal fertility levels for several months. The use of the Pill has no effect on the onset of the menopause. Its influence on sexual interest follows no consistent pattern because of the multiplicity of interacting factors (chapter 4).

Frightened by its negative prospects, which were often played up in the popular press, many women stopped using the Pill. But comprehensive recent reviews of the evidence show that not only are the fears with regard to cancer unfounded, the Pill appears actually to protect women against certain forms of cancer: the risks of getting cancer of the endometrium and cancer of the ovary are reduced by about 50% among users of the Pill. Women on the Pill are also protected to some extent from various other ailments including pelvic inflammations, rheumatoid arthritis (the incidence of which the Pill reduces by half), anemia, ovarian cysts, and menstrual irregularities and tension.[28]

Nonetheless, the decision to use oral contraceptives should be undertaken by a woman carefully, with medical advice, and with due attention to the circumstances of her life. The risks entailed should neither be minimized nor needlessly exaggerated. For many women who are young and who do not smoke, oral contraceptives may well represent the best possible choice. Yet for others they pose sufficient risks that they are best avoided. These include older women, smokers, and those suffering from cardiovascular disease, liver disease, diabetes, cancer of the breast, cancer of the reproductive organs, and other conditions that are likely to be aggravated.[29] There is also some evidence that pills that are high in progesterone and low in estrogen lead to higher cholesterol levels, which may predispose women to heart attacks.[30] Among women younger than 30 (except Pill-takers who smoke), the risk of death associated with the Pill or other major methods of contraception is 1 to 2 per 100,000 women per year. Beyond age 30 the risk to life is closely linked to the method used and increases rapidly for Pill-takers (especially smokers). After 40 the Pill poses a higher risk to life than other methods or avoiding contraception altogether, while methods other than the Pill remain safer than pregnancy.[31]

With regard to these contraceptive risks, one must bear in mind that the alternative of getting pregnant poses a considerably higher threat to health; particularly in pregnancies that occur before age of 15 when there is an increased likelihood of toxemia, stillbirth, postpartum hemorrhage and infection (chapter 5). No less serious is the daunting prospect of bringing an unwanted child into the world.

The Minipill   There have been many attempts to minimize the undesirable side effects of the Pill while retaining its effectiveness by using progressively smaller doses of the hormones, especially the estrogenic component, which is held responsible for most of the side effects; as a result, the estrogenic content of most Pills is now a fraction of the quantities used earlier.

The *Minipill* contains only a small amount of progestin and no estrogen. It is taken

---

[27]*Population Reports* (1982).

[28]Altman (1982). Also see MacLeod (1979); Droegemueller and Bressler (1980).

[29]For a comprehensive list of contraindications to pill use, see Hatcher et al. (1982, pp. 22–23).

[30]Wahl et al. (1983).

[31]Tietze and Lewit (1979).

every day including during menstruation. Though it does have fewer side effects, it may lead to more irregular bleeding. The Minipill is also somewhat less effective in suppressing ovulation than the combination Pill. It has an ideal failure rate of 1% as against 0.5% for the combination Pill, and an actual failure rate of 2.5%.

Biphasic and Triphasic Pills   In order to more closely imitate the hormonal sequence of the natural menstrual cycle, a *sequential Pill* was developed some years ago which permitted a woman to take estrogen for 15 days followed by a combination of estrogen and progesterone for the next 5. Though these Pills were effective, they had a higher risk of serious side effects and were withdrawn from the market. This same intent of imitating the hormonal pattern of the menstrual cycle is now better served by *biphasic* and *triphasic* Pills. Since the daily physiological levels of the hormones are not constant, this allows lower daily doses in the early part of the cycle, thus reducing the total doses of hormone taken per cycle to less than that in most conventional oral contraceptives.

The biphasic Pill has a small amount of progestin and higher level of estrogen for the first 10 days, followed by another Pill that has a much higher level of progestin and the same level of estrogen as before, which is used for days 11 to 21. The triphasic formulation uses three combinations of progestins and estrogens in varying doses over successive phases of the cycle. These new formulations appear as effective as the more standard forms of the Pill.[32]

The Post-coital Pill   A pill that works postcoitally (hence the term, "the morning after pill") would have the great advantage of providing hormonal protection only when coitus necessitates it. Its action would depend on interfering with implantation of the fertilized ovum, instead of preventing ovulation.

An early version of such a pill contains such a potent estrogen, *diethylstilbestrol,* (DES) which prevents pregnancy if given to a woman twice a day for five days beginning within 72 hours (or preferably within 24 hours) of unprotected intercourse. However, the estrogen content of the DES pill is 500 times the level of estrogen in ordinary birth control pills and causes severe nausea and vomiting. DES has also been linked with birth defects and increased risk of cancer in female offspring many years later if the drug is taken when the woman is already pregnant.[33]

A more current and safer version of the postcoital pill contains another form of estrogen (*ethinyl estradiol*) singly or in combination with a progestin (*norgestrel*). The first dose is taken as soon as possible after coitus, and a second dose in 12 hours (or sometime within 72 hours after coitus). Since the hormone content of these pills is not that much higher than in the regular pills, problems with nausea and vomiting are not so severe.

In a group of women exposed to single acts of unprotected coitus followed by the above regimen, only 7 pregnancies occurred, instead of the 30 to 34 that would have been expected with no such intervention.[34] The postcoital pill has proved most useful in cases of rape or other unusual circumstances but is not yet commonly used as a regular form of contraception. Studies are underway to explore further its broader uses along with other means of using hormones, such as in long-acting preparations that we discuss later.

The Intrauterine Device (IUD)

The *intrauterine device* is a variously shaped object about the size of a paperclip, now usually made of plastic, which is inserted in the uterus to provide contraceptive protection. Some 25 different forms of the IUD have been used, some of which are shown in figure 6.2. The origin of modern IUD goes back to the early 1930s when

[32]*Population Reports* (1982).

[33]Herbst (1981).

[34]*Population Reports* (1982).

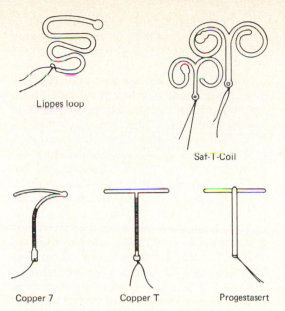

Lippes loop

Saf-T-Coil

Copper 7   Copper T   Progestasert

Figure 6.2 IUDs approved by the Federal Drug Administration.

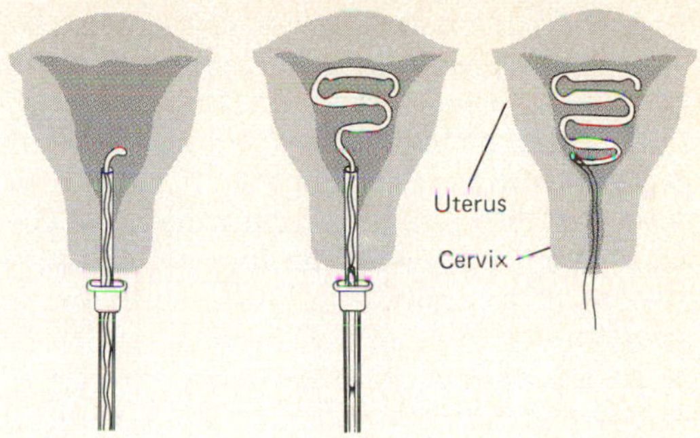

Figure 6.3 Insertion of the Lippes Loop. (Courtesy of Ortho Pharmaceutical Corporation.)

Grafenberg in Germany and Ota in Japan experimented with metal rings inserted into the uterus as a contraceptive device. Its widespread use came in the 1960s.

The IUD is a close second to the pill in effectiveness and prevalence of usage. Some 50 million women throughout the world rely on it, including 35 million in China (where one type of IUD is called the "Flower of Canton").[35] It accounts for 9% of contraceptive methods used by married women in the United States but only 2% for those never married.

The precise mechanism of how the IUD works is unclear. The most widely accepted explanation is that it causes cellular and biochemical reactions in the uterine endometrium that interfere with implantation. It may also act by dislodging the implanted blastocyst, in which case it would be more properly viewed as inducing early abortion rather than providing contraception.

Early IUDs were made from various metals; IUDs currently in use are usually made of a flexible plastic that can be stretched out in the inserter tube but once released in the uterus returns to its usual shape. (Figure 6.3 illustrates this for the *Lippes Loop,* one of the most commonly used devices.) A nylon thread or "tail" attached to the lower end of the IUD trails out of the cervix into the vagina enabling the woman to check that the device is in place. Small amounts of barium may be incorporated in IUDs to make them visible on X rays.

The IUD has an ideal failure rate of 1.5% and an actual failure rate of 4%. The failure rate tends to decline after the first year as longer-term users tend to be those who are able to better tolerate it. To further their effectiveness, some IUDs rely on chemical measures. Some have copper filament wrappings that slowly dissolve in the uterus. Another type called the *Progestasert* (figure 6.2) releases progesterone at a slow rate, which although insufficient to inhibit ovulation, brings about additional changes in the lining of the uterus that further interfere with implantation.

While the pill's effectiveness largely depends on the woman's use of it, a greater share of the responsibility to make the IUD work well rests with the health professional who inserts it, since proper placement greatly influences the effectiveness of the device. Once the IUD is in place, the woman need not do anything further except occasionally check the thread to make sure the device is still in the uterus. Unless there are complications, IUDs can be left

[35]Djerassi (1981).

in place for up to 4 years (progesterone-releasing types must be replaced every year).

IUDs have also been experimented with as postcoital contraceptives. When inserted within five days of unprotected coitus, the copper-containing IUDs have shown to be quite effective in preventing pregnancy. Further research needs to be done to confirm their utility in this regard and to ascertain better if any complications are entailed.[36]

Side Effects   Since the IUD is just a mechanical device rather than a potent chemical like hormones, one would think that it would be free of side effects, but that is not the case.[37] The two most common side effects are *irregular bleeding* and *pelvic pain*. Mild bleeding or "spotting" may occur at various times in the menstrual cycle, and the menstrual periods of women with IUDs tend to be heavier than usual (with smaller type IUDs this is less likely). Pelvic pain is caused by uterine cramps. It affects some 10 to 20% of users but tends to disappear after several months. In 10% of women, pain or bleeding are severe enough to necessitate removal of the IUD. Side effects are said to be less of a problem with IUDs containing progesterone. In 5 to 15% of users the IUD is expelled spontaneously (which is why women need to check the thread).[38]

Less common, but more serious, complications include perforation of the uterus, pelvic infection, and problems related to pregnancy, should it occur despite the IUD. The danger of *uterine perforation* is about 1 in 1,000 insertions (and requires surgery to treat it). IUD use results in a 5- to 10-fold increase in the risk of developing *pelvic inflammatory disease* (PID). This is because the tail acts as a conduit for bacteria from the vagina into the uterus and the device in-

terferes with uterine protective mechanisms. In view of this, women with acute or recurrent pelvic infections should not use IUDs.

Should a pregnancy occur while an IUD is in place there is a threefold increase in the risk of *spontaneous abortion*, a tenfold increase in the risk of *ectopic pregnancy* (5% of all IUD failure pregnancies are ectopic), and increased risk of *infection* during the pregnancy.[39] But there is no danger that the IUD will cause cancer or birth defects.

Despite the possibility of these complications, the IUD is an effective and safe device. Generally the IUD does not interfere with sexual activity. Some women experience pain during orgasm when the contractions of the uterus press on the IUD. Though the string is not even detected by many men, a few complain of penile irritation caused by its friction.

Barrier Methods

The principle of barrier methods of contraception is simple: to mechanically prevent sperm from reaching ova. This is normally done by either covering the penis with a sheath or **condom,** or obstructing the mouth of the cervix with a **diaphragm, cervical cap,** or **contraceptive sponge.** Before the advent of the Pill and the IUD, barrier methods were the most reliable and most frequently used contraceptives. They remain in wide use today.[40]

The main advantage of barrier contraceptives is their freedom from side effects combined with high levels of effectiveness when used correctly and in conjunction with spermicidal substances. Their main disadvantage is the likelihood of their being used incorrectly, hence their higher actual failure rates, such as 10% for the diaphragm and 10% for the condom (box 6.2). Some people dislike them because they interrupt sexual activity, may not be available when needed unexpectedly, or on aesthetic grounds. Since there are distinctive features for each of these devices we further discuss them separately.

---

[36]Yuzpe (1979).

[37]For detailed discussion of the side effects of IUDs, see Osser et al. (1980). Contraindications to IUD use are listed in full in Hatcher et al. (1982, p. 74).

[38]Sparks et al. (1981).

[39]Mishell (1979); Droegemueller and Bressler (1980).
[40]Connell (1979).

The Diaphragm   Blocking the cervical opening for contraceptive purposes has been practiced for a long time (box 6.1). The modern *diaphragm* was invented in 1882 in Germany, and like the modern condom, was made possible by advances in the manufacture of rubber. It is now used by 2.3 million women throughout the world, and accounts for about 4% of all methods used by American married women, and a somewhat lower proportion for those never married. When used with spermicide, the diaphragm has an ideal failure rate of 2% but a typical failure rate of 10% (box 6.2).[41]

Diaphragms in current use consist of a thin rubber dome attached to a flexible, rubber-covered metal ring. They are inserted in the vagina in a way that will cover the cervix and prevent passage of sperm into the cervical canal (figure 6.4). The inner surface of the diaphragm (that is, the surface in contact with the cervix) is coated with a layer of contraceptive jelly or cream before insertion; without these spermicidal substances, its effectiveness is greatly reduced.

[41]For the use of the diaphragm among college women see Hagen and Beach (1980).

Diaphragms come in various sizes (most are around 3 inches in diameter) and must be individually fitted by a physician or other health professional. A woman must be re-fitted following pregnancy, major changes in body weight, and other circumstances that may alter the size and shape of her vaginal canal.

The diaphragm must be inserted no more than 6 hours before intercourse in order for the spermicide to retain its effectiveness. If a woman has intercourse more than two hours after insertion of the diaphragm, she should leave it in place and add more spermicidal jelly or cream into the vagina. Similarly, more cream or jelly should be placed in the vagina (not the diaphragm) every time intercourse is repeated. The diaphragm is removed 6 to 8 hours after the last act of intercourse; though it can be left in place for as long as 24 hours.

Unless a woman knows in advance that she is going to have intercourse, she has to stop and insert the diaphragm in the midst of making love, or risk pregnancy. This is distracting to some; but others incorporate the procedure into foreplay, thus making a virtue out of necessity. When the

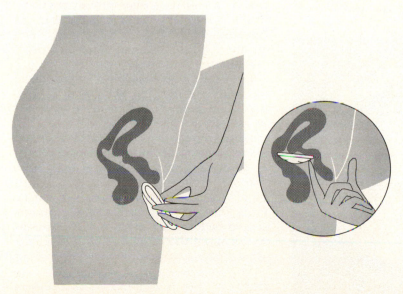

Figure 6.4   Insertion and placement of diaphragm. (Courtesy of Ortho Pharmaceutical Corporation.)

diaphragm has been properly inserted, neither partner is aware of its presence during intercourse. They also are likely to be unaware if the diaphragm becomes dislodged. This is more likely to happen if it fits loosely, or if the partners engage in vigorous coital movements.

The Cervical Cap   Though it works on the same general principel as the diaphragm, the *cervical cap* fits snugly around the cervical tip by suction. Like the diaphragm, it must be individually fitted and is comparably effective. One of the most popular forms of contraception in the 19th century, the cervical cap has been almost completely displaced by newer methods. Since there is not enough known about its safety and effectiveness it has not yet been approved by the Food and Drug Administration. But it is under active investigation and can be prescribed by those engaged in such work.[42]

[42]Fairbanks and Scharfman (1980); Smith (1980).

The Contraceptive Sponge   A contraceptive device approved in 1983 by the Food and Drug Administration is the *vaginal sponge* (figure 6.5). Made of polyurethane,

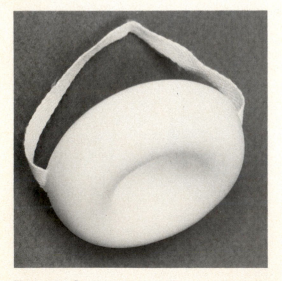

Figure 6.5   Contraceptive sponge.

the circular sponge has a diameter of two inches and is permeated with a spermicide (nonoxynol-9), which is the active substance in other creams and gels. Inserted into the upper vagina, the sponge blocks the cervical opening and traps sperm. The barrier and spermicidal attributes of the sponge retain their effectiveness over 24 hours (hence the trade name, *Today*), irrespective of the number of acts of coitus.

The actual failure rate of the sponge is about 15%, which is comparable to that of the diaphragm. Its failures are ascribed mainly to improper use. So far, few side effects have been reported, mainly allergic reactions to the spermicide. The sponge does not require a prescription, does not need to be specially fitted, and is disposable. It remains to be seen how widely it will be used.

The Condom   The *condom* is the only male contraceptive device in widespread use that is acceptably reliable. The rubber condom has been in use since the middle of the 19th century, when it was more often relied on to prevent venereal disease than pregnancy. Also known as *prophylactic, rubber*, and other fanciful names (such as *French letters*) its earlier prototypes were made of linen.[43] Casanova referred to the condom as the "English riding coat," and as "the English vestment" that puts one's mind at rest.[44]

The major manufacturers of condoms in the world are the U.S. and Japan, supplying over a billion condoms used in a year worldwide by 15 to 20 million men. The modern condom is a cylindrical sheath made of very thin latex rubber with a ring of harder rubber at the open end and a "nipple" at the closed end (figure 6.6). The nipple acts as a receptacle for the ejaculate (otherwise half an inch of the condom should be left loose at the end). Condoms are all about the same size, and stretch to

[43]The first known description of the condom was published in 1564 by the Italian anatomist Fallopius (Himes, 1970, p. 188).
[44]Himes (1970, p. 195).

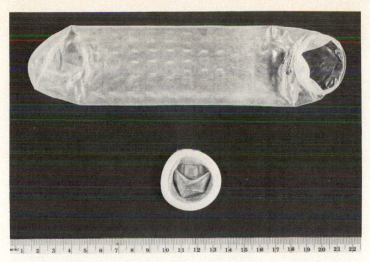

Figure 6.6   Condom, rolled and unrolled. (Scale is in centimeters.)

fit most penises.[45] They are sold rolled up in individual packages; some are lubricated to facilitate their use. Fancier versions are made in different colors, with ribbed elevations, or other features to convey greater vaginal stimulation. Condoms are sold over the counter in drugstores, by mail, and in coin-operated dispensers.

Condoms have the same effectiveness as the diaphragm (ideal failure rate of 2%). Being manufactured to stringent specifications, they are unlikely to burst. Nonetheless, their effectiveness is compromised through improper use, which results in a 10% actual failure rate.

The condom should be put on before intromission, or any other contact between the penis and the female genitalia. However, putting on the condom too early in foreplay may lead to its getting damaged. To avoid leakage of semen, the penis should be withdrawn from the vagina before loss of erection. To engage in coitus again, the used condom should be discarded, the penis washed or at least wiped clean and a new condom put on. These devices are not reusable even if they are washed and dried.[46]

The condom is virtually free of side effects though rarely a person may be allergic to the rubber or the lubricant. Women are generally not troubled by it especially since it relieves them from the contraceptive burden, but some miss the sensation of ejaculation in the vagina. More often, men complain that it lessens their pleasure ("like taking a shower with a raincoat on"). Those with potency problems find that the distraction of putting on the condom makes them lose their erections. But premature ejaculators may find condoms helpful. In view of the fair protection it also offers against some forms of sexually communicable diseases, the condom is probably the most useful contraceptive device for engaging in sex with occasional partners. It can also render good service as a back-up system in more stable sexual relationships.

[45]Barbara Seaman has suggested that just as women buy brassieres in different cup sizes, it would help condom efficacy if they too were packaged in different sizes and labeled "jumbo," "colossal," and "supercolossal" so that men don't have to ask for the "small" (quoted in Djerassi, 1981, p. 17).

[46]Unopened condoms remain good for two years if stored away from heat. They should not come into contact with vaseline or other petroleum-based products. Spermicidal agents, K-Y jelly, or other water-soluble lubricants do not affect them.

Spermicides

Spermicides are chemical agents that kill sperm. They also provide a physical obstacle to the passage of sperm into the cervix, so they can also be considered as another form of barrier contraception. Spermicides come in the form of *creams, jellies, foam, tablets,* and *vaginal suppositories.* They are all placed directly in the vagina but nonetheless they vary in their effectiveness and method of application.[47] As a group, they have an ideal failure rate of 3 to 5% and actual failure rate of 15%.

There are no firmly established serious side effects attached to the use of spermicidal substances. There is a reported correlation between the use of spermicides and a slightly higher rate of birth defects when the woman gets pregnant while using them (2.2% for spermicidal users against 1% for nonusers). Presumably these defects result from damaged sperm that then fertilize the egg. There is a similarly slight increase in spontaneous abortion in spermicide users. Though these findings are as yet quite tentative, it is safer for women to stop using spermicides as soon as they suspect they may be pregnant.[48]

Other drawbacks include occasional complaints of burning sensations or allergic reactions. The overly fastidious may find them too messy or too "sloshy." Suppositories must be in place for 10 minutes or so before they become effective. Foam works right away, but only for half an hour.

In the light of these considerations, spermicidal contraceptives have an important place, not so much as individual agents but in their use with barrier contraceptives. Without cream or jelly, the diaphragm would not be safe enough, hence it is now standard practice to combine their use. It is much less common to combine the use of foam with a condom though the combination would make them close to 100% effective. In cases where pregnancy must be avoided at all cost, spermacides may be combined with both diaphragm and condom used together. There is no reason to settle for just one method, especially when there is no risk of cumulative side effects.

Douching, washing out the vagina after coitus, can be thought of as a spermicidal method of sorts, but it is so ineffective (actual failure rate of 40%) that it hardly deserves to be discussed as a contraceptive measure. This is true no matter what douching fluid is used and however fast the woman rushes to the bathroom to use it following coitus.

Fertility Awareness Techniques

By definition, all attempts at avoiding pregnancy while engaging in coitus represent a form of contraceptive practice. Yet distinctions are made by some between contraceptive methods that are "natural," from others that are not. The Catholic church, for instance, opposes the use of all active methods to avoid pregnancy. But it considers as morally permissible the taking advantage of a woman's transient periods of infertility during the menstrual cycle to avoid conception. Others are interested in fertility awareness techniques not for moral reasons but because they may provide a method of birth control that needs no contraceptive devices, is free of their intrusion and side effects, and costs nothing.

Such an approach, popularly known as the *rhythm method*, depends on *fertility awareness*. As such, this knowledge can be used not only for avoiding pregnancy but also for increasing the chances of bringing it about. Currently there are several ways of determining the "unsafe" periods (that is those periods when a woman is fertile).[49] Their ideal failure rate ranges from 2 to 20% and their actual failure rate from 20 to 30%.

Calendar Method   This is the original rhythm method and the least reliable alternative (failure rates estimated by some

---

[47]There are explicit instructions for use enclosed with these products, which are available on open pharmacy counters and require no prescription.

[48]Jick et al. (1981).

[49]Hatcher et al. (1982) have explicit instructions for using these methods.

to be as high as 45%).[50] The calendar method avoids conception based on the facts that ovulation occurs on day 14 (give or take 2 days) in a 28-day cycle; sperm remains viable 2 to 3 days; the ovum survives 24 hours after ovulation. Since women do not all have 28-day cycles, the first task is to construct a personal *menstrual calendar* by noting the length of each menstrual cycle over 8 months (counting the first day of bleeding in each cycle as day 1). The earliest day in which the woman is likely to be fertile is calculated by subtracting 18 days from the length of her shortest cycle, and the latest day she is likely to be fertile is obtained by subtracting 11 days from the length of her longest cycle. For example, if a woman has regular 28-day cycles, ovulation will take place on day 14 or within 2 days earlier or later, which makes days 12 to 16 unsafe. Sperm deposited on the 10th or 11th day may live to day 12, making that day unsafe. The egg may still be alive on day 17 if ovulation took place on the 16th, so cross off one more day. This means that the period from day 10 to day 17 is unsafe. If the couple also refrains from sex during the several days of menstrua-

[50]Ross and Piotrow (1974).

tion (though there is no physiological reason to do so) that still leaves sixteen days or so that are safe to engage in coitus. This is all very logical, but in fact it does not work well because menstrual cycles do not function like clockwork. They are influenced by many physiological and psychological factors that throw off the predicted pattern. In an extreme case, a woman with a very short cycle but a long period of bleeding may ovulate while she is still menstruating; so unprotected coitus even during menstruation is not entirely safe. To better predict ovulation one can look for more objective evidence than simply rely on the calendar. The following two methods attempt to do that.

The Basic Body Temperature (BBT) Method   This method (also called the *sympto-thermal method*) depends on the fact that ovulation is accompanied by a discernible rise in the *basal body temperature* (the lowest temperature during waking hours). In order to pinpoint the time of ovulation, a woman must take her temperature by a special BBT thermometer, immediately upon awakening every morning, before getting out of bed (or doing anything in bed). An increase of 0.4°F

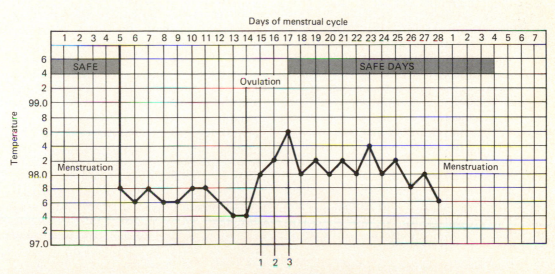

Figure 6.7   Basal body temperature variations during a model menstrual cycle. (From Hatcher et al., *Contraceptive Technology 1982–83*, Irvington Publishers, 1982, p. 131.)

(0.2°C) or more over the temperatures of the preceding 5 days, for 3 days, indicates that ovulation has occurred. Sometimes the temperature may also drop before it begins to rise (figure 6.7). One abstains from intercourse from the end of the menstrual period until 3 days after the time of ovulation as revealed by the temperature rise. To have coitus anytime before there is clear evidence for ovulation is risky.

This method is subject to considerable error because so many factors other than ovulation can also cause a rise in body temperature. Furthermore, in 20% of cycles, no temperature rise accompanies ovulation; then that entire month must be considered unsafe. If women do scrupulously follow the BBT chart and engage in coitus only after the requisite 3 days following unequivocal evidence of temperature rise, success rates as reported by some are extremely high (99.7%).[51] Others, however, regard this method as far less reliable.[52]

The Cervical Mucus Method   This approach, which is also known as the *Billings method*, relies on changes in *cervical mucus* to predict ovulation. To use it a woman must learn how to identify correctly the changes in the amount and consistency of cervical mucus.[53] For a few days before and shortly after menstruation, many women have "dry" periods during which no noticeable cervical discharge is present and there is a sensation of dryness in the vagina. These "dry" days are considered relatively safe for intercourse.

Next, the cervix produces a thick, sticky mucus discharge that may be white or cloudy. Gradually the mucus becomes more watery, slippery, and clear (like egg white). This is called the *peak symptom* and usually lasts for 1 or 2 days. Generally a woman ovulates about 24 hours after the last peak symptom day. The mucus then changes back to being cloudy and sticky. Intercourse must not occur from the first day in which mucus is present to until 4 days

after the last peak symptom day. To help women make these determinations, an *ovutimer* has now been developed in the form of a plastic device that when inserted into the vagina measures the stickiness of cervical mucus.

A World Health Organization review in 1978 concluded that even combinations of these rhythm methods were relatively ineffective. But they ascribed failures more to risk taking (that is engaging in coitus) during fertile periods than difficulties in interpreting BBT or cervical mucus changes. But people being what they are the effectiveness of any method must be undertaken in the light of that reality.

Prolonged Nursing

It has been long observed that nursing an infant protects the mother from getting pregnant (chapter 5). In developing countries more pregnancies are thought to be prevented by breastfeeding than by any other contraceptive method. Yet breastfeeding is widely regarded in the West as an unpredictable and unreliable form of contraception. How do we reconcile these conflicting observations?

Suckling does inhibit postpartum resumption of ovarian activity, including ovulation and menstruation. The endocrine mechanisms underlying this effect are not fully understood. They originate in the sensory nerve endings of the nipple, from which impulses reach the brain through the spinal cord and inhibit the hypothalamus. This depresses the secretion of LH from the pituitary, which in turn inhibits ovulation.

However, this mechanism only works if breastfeeding is done consistently and around the clock. Among certain nomadic tribes in the Kalahari desert in Africa, women on the average conceive at 4-year intervals without relying on any other contraceptive. These mothers nurse their infants up to 60 times a day. The practice of giving the breast to the infant on demand has been and remains quite common in many other developing countries. Even if there is no milk transfer at every feeding,

[51]Doring (1967) in Hatcher et al. (1982).
[52]Bauman (1981).
[53]Billings and Billings (1974).

infants derive comfort from sucking and such nipple stimulation keeps the mother's contraceptive mechanisms active.

In recent years, breastfeeding has been rapidly losing its contraceptive function in the third world, as it has in the West, because of the introduction of powdered milk and other food supplements that reduce the frequency of nursing. The Western practice of bottle feeding, and the use of "pacifiers" in the form of rubber teats has compromised the efficacy of the suckling mechanism even when women do nurse their infants. Yet this innate mechanism is by no means entirely lost. A study of nursing women in Scotland has shown that if breast feeding is carried out with more than five feeds per 24 hours, including a nighttime feed, ovulation can be inhibited for up to a year or more.[54] However, once menstruation returns nursing exerts no further effect in preventing pregnancy. Even the amenorrheic period is not completely reliable, since 80% of breastfeeding women ovulate before their first period. Thus, all things considered, breastfeeding, as currently practiced, is not a highly reliable method of contraception.

Withdrawal   Another ancient and very widely used means of avoiding pregnancy is to interrupt coitus by *withdrawal* of the penis from the vagina prior to ejaculation; hence the term, *coitus interruptus.* The major problem with coitus interruptus is that it requires a great deal of will power just at the moment when a man is most likely to throw caution to the winds. Nevertheless, this method costs nothing, requires no device, and has no physiological side effects—although some people find it psychologically frustrating or socially unacceptable.

When withdrawal is the only contraceptive measure taken, the lowest observed failure rate is 16% and the actual failure rate 23%. This is mainly because the male does not always withdraw quickly enough or because some sperm may have already seeped out before ejaculation. Though

withdrawal is not a reliable method, nonetheless it may be still useful as a last resort. The failure rate of withdrawal is admittedly poor, but that of not withdrawing is surely worse.

Sterilization

Sterilization is the induction of permanent (but not necessarily irreversible) infertility through surgery. It is the most effective method of contraception in both sexes and is now the most widely used method of birth control in the world. The estimated number of couples who rely on it rose from 20 million in 1970 to 80 million in 1977. By 1975 one partner had been sterilized in a third of all American married couples using contraception.[55] There are now well over 10 million Americans who have been undergone this procedure and a million more are added each year. Until a few years ago women underwent sterilization in larger numbers, but more recently the proportion has been shifting towards a more equal distribution with men.[56] Since 3 out of 4 women now have all the children they want by age 30, that leaves a period of another 15 to 20 years during which they remain vulnerable to unwanted pregnancy. Sterilization provides such women a reliable, safe, and simple way to be free from contraceptive concerns once and for all. Similar considerations apply to men when they reach a point where fatherhood is no longer a likely or desirable prospect.

Male Sterilization   The operation that is most often used to sterilize the male is *vasectomy,* a simple procedure that can be done in a doctor's office in about 15 minutes.[57] Under local anesthesia a small incision is made on each side of the scrotum to reach

---

[54]Short (1979, p. 361).

[55]Mishell (1982). This is not true for all subgroups. The sex ratio among Blacks and Mexican-Americans is still heavily in favor of women who seek sterilization.

[56]Droegemueller and Bressler (1980).

[57]For a general review, see Ackman et al. (1979); Samuel and Rose (1980).

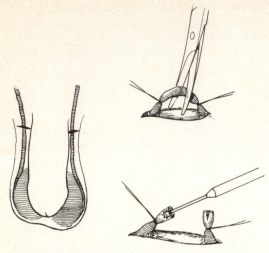

Figure 6.8 Vasectomy.

the vas deferens (figure 6.8). Each vas is then tied in two places and the segment between is removed or the ends cauterized in order to prevent them from growing together again. Sperm continues to be produced but it now accumulates in the testes and epididymis, then is broken down and reabsorbed.

No changes in hormonal function occur as a result of vasectomy. The testes continue to secrete testosterone into the blood stream in normal amounts. Ejaculation remains intact. Since the contribution by the testes to semen accounts for less than 10% of its volume, most men cannot detect any change in their ejaculate. The only difference is that the semen is now free of sperm.

Sperm may still be found in the ejaculate two or three months after a vasectomy due to their presence in the duct system beyond the site of the vasectomy. These can be flushed out at the time of the vasectomy, or more simply, some other contraceptive can be used for the next 3 months or until two successive ejaculations are sperm free. From then on, the man is sterile (Latin, "unfruitful").

The rare cases of failure (box 6.2) are due to unprotected coitus in the postvasectomy period before full sterility has been achieved. Very rarely the cut ends of the vas have reunited while the wound is still fresh; current techniques make this virtually impossible.

The main disadvantage of vasectomy is its permanency. Despite many advances in microsurgery that permit reuniting the vas (*vasovasostomy*), there is only a 50% chance that a vasectomized man can be made fertile again (depending on the methods used, the rates actually vary from 5 to 70%). About one in 500 vasectomized men seek vasovasostomies, usually following a new marriage.

Before undergoing vasectomy, some men have a sample of their sperm frozen and deposited in a *sperm bank*, which then can be used for artificial insemination. This practice attracted considerable attention a decade ago but enthusiasm for it has diminished more recently. Although some pregnancies have been achieved with sperm that has been kept for several years, the success rate is much lower than with fresh sperm. Furthermore, the possibility that damage to sperm stored over a decade or so will cause genetic defects cannot be ruled out. Sperm banks, therefore, are not reliable fertility insurance for men who elect to undergo vasectomy.[58]

There are no serious side effects to vasectomy. The local discomfort after surgery is minimal and lasts only a few days with minimal risk of complications. One lingering concern has to do with the development of *autoimmune reactions* that result from the reabsorption of sperm into the body.[59] This in fact is partly responsible for the lowered chance of fertility after vasovasostomy. It has also been found that vasectomized monkeys develop blood vessel changes (atherosclerosis), but this has not been confirmed for humans.[60] Therefore, although no long-term effects have been shown to follow vasectomy, that prospect cannot be ruled out entirely given the relative recency of our experience with this procedure.[61]

[58]Ansbacher (1978).
[59]Shahani and Hattikudur (1981).
[60]Clarkson and Alexander, (1980).
[61]Hussey (1981).

On purely psychological grounds, vasectomy may interfere with sexual performance because a man may feel less virile, or "damaged." Though there is no evidence that vasectomy as such causes impotence, men who suffer from problems of potency or are likely to have their sense of masculinity threatened are better off avoiding this procedure. Resentment and conflict may also arise if a man feels pressured to have a vasectomy to relieve his wife from the burden of birth control; just as a woman would feel resentful if pushed to such a choice. These occasional concerns notwithstanding, the more typical response to vasectomy is a sense of freedom and relief that leads to greater interest in sex.

Female Sterilization   The most common surgical procedure for sterilizing women is *tubal ligation.*[62] Tying or severing the fallopian tubes prevents the meeting of eggs and sperm. The eggs that continue to be released are simply reabsorbed by the body. The ovaries continue to produce their hormones normally and the menstrual cycle is not interfered with, nor is sexual interest diminished.

Female sterilization used to be a major surgical procedure, but now there are effective and inexpensive procedures involving over a hundred techniques for cutting, closing, or tying the tubes (figure 6.9). Although in some situations sterilization is

[62]Rioux (1979).

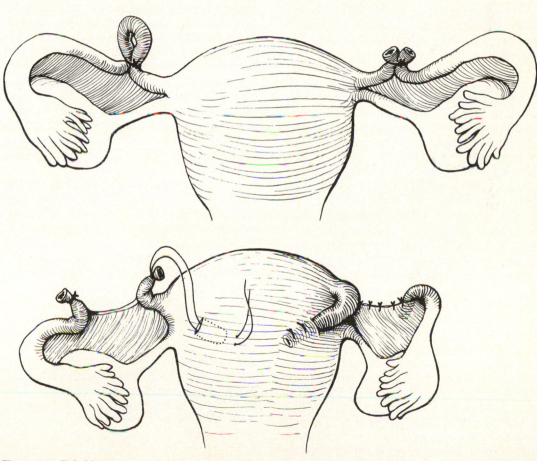

Figure 6.9   Tubal ligation.

still accomplished by major surgery (such as by removal of the uterus), the current trend is to use outpatient procedures performed under local anesthesia. It is possible to approach the fallopian tubes through a small incision in the vagina or in the abdominal wall and to perform the sterilization with the help of a *laparoscope* (a tube with a self-contained optical system that allows the physician to see inside the abdominal cavity). Chemicals that solidify in the tubes, and lasers that destroy a portion of the tubes are among new sterilization techniques under investigation.

Most methods currently used to sterilized women are virtually 100% effective. The occasional failures in the past have been due to improper procedures. More often, the woman is already pregnant when the tubes are tied but that fact is not known. Female sterilization, like its male counterpart, must be approached as a permanent procedure. The chances of becoming fertile again after reuniting the tubes is 10 to 50%, depending on the techniques used.

In the past, women have been generally more willing than men to undergo sterilization. The fact that women will become sterile anyway at the menopause has been one factor facilitating their decision. More importantly, if given no other choice, women have been more willing to undergo the procedure, since it is they who usually carry the burden of contraception and the risk of pregnancy. But female sterilization is a more difficult procedure, and many couples' greater sense of shared responsibility has now made more men more willing to undergo sterilization.

There are certain risks of complications with sterilization procedures, as with any other type of surgery, but serious consequences are rare. The fatality rate for female sterilization is 25 deaths per one million cases. Adverse psychological reactions to sterilization are less common among women than men. Relief from the fear of pregnancy and freedom from the trouble and side effects of contraceptives generally make women sexually more interested and responsive. Only occasionally does a sterilized woman (especially after hysterec-

tomy) feel "less feminine," or defective, and develop sexual problems.

Future Methods for Birth Control

Despite the enormous expansion of contraceptive technology and usage over the past two decades, we are still in the "horse and buggy days" of effective contraception, according to Alan Guttmacher, one of the pioneers in this field. The future prospects of contraception will depend on technological advances and changes in social attitudes influencing the use of these devices.

Advances in Male Contraception Currently there are only two reliable contraceptive methods available to men: the condom and vasectomy. The advances to be made for better condom use are mainly attitudinal, not technological. With vasectomy research, prospects are aimed mainly at improving reversibility. This will entail both the reestablishment of the patency of the vas and counteracting the antifertility effects of autoimmune reactions. In addition to various removable clips and plugs, there have been attempts at installing blocking mechanisms in the vas that can be turned on and off from the outside through the use of magnets.

A major breakthrough would be the development of a *male pill* that would interfere with spermatogenesis, or somehow neutralize the ability of sperm to fertilize. There are interesting leads in this area as well as formidable problems in controlling millions of sperm without damaging other cells, causing genetic mutations, or loss of libido.

Some drugs have shown promising results in reducing male fertility.[63] *Danazol* is a synthetic hormone with a structure similar to the androgens, which prevents the release of FSH and LH from the pituitary. In preliminary studies, men who were given Danazol in conjunction with testosterone for 6 months at a time, had their sperm counts drop to 0.5 to 5% of normal levels.

[63]Shapiro (1977).

Furthermore these men did not lose sexual function, and they regained normal fertility within 5 months after the end of treatment.

The Chinese have reported success with a method that they claim is 99.8% effective and has no serious side effects. It had been noted that men showed decreased fertility in certain parts of China where unrefined cottonseed oil is part of the daily diet. The substance in cottonseed oil that interferes with sperm production and mobility is *gossypol*, and thousands of Chinese men have been using it now for birth control, with apparent success. They are reported to regain fertility readily after they stop taking gossypol.[64]

Gossypol comes in tablet form, hence it would be the closest thing to a male pill (though it is not a hormone). However, this contraceptive method is not likely to be rapidly adopted in the United States because gossypol is known to have several toxic effects (including weakness, decrease in libido, change in appetite, nausea), hence extensive testing would be needed first to establish its safety. The prospects of having on the market a male pill therefore appear to be many years away.

Advances in Female Contraception Much of the research in female contraception is aimed at the improvement of existing methods, with the emphasis on disrupting the hormonal cycle necessary for reproduction. Thus the same hormones that constitute the Pill are put to use through intravaginal and intrauterine devices and in longer acting forms to provide sustained protection against fertilization.[65]

Contraceptive vaginal rings have been shown to be 98% effective in preventing ovulation. These rings have an inert core of plastic surrounded by a layer of steroid hormones.[66] They are slightly smaller than a regular diaphragm and more easily inserted. They are placed in the vagina on the fifth day of the menstrual period and left in place for the next 21 days. Menstruation begins a few days after the ring is removed. Each ring can be used for up to 6 months. The estrogens and progestins in the rings are released slowly and absorbed through the vaginal wall into the woman's bloodstream. The hormones prevent ovulation, cause changes in the endometrium, and thicken the cervical mucus. There are few side effects, and the ring does not interfere with intercourse.

Instead of taking a pill every day, long-acting injections of hormones have been developed for contraceptive purposes. One such preparation is already being used in many parts of the world but has not yet been approved for use in the United States.[67] The preparation, marketed as *Depo-Provera*, contains a synthetic form of progesterone that prevents ovulation for 90 days. It is very highly effective but side effects include weight gain, loss of sex drive, menstrual irregularities, and an unpredictable period of infertility following cessation of the shots. It has been linked with cancer in certain animals but not in women.[68]

The IUD is also being tested for further improvements. *Tailless IUDs*, for instance, may greatly reduce the risk of infection. Though this would make it more difficult to check that the IUD is in place, new technologies using ultrasound may make such checks easier. The *progestin-releasing IUD*, which now needs to be replaced on a yearly basis, may be further refined to slowly release a powerful progestin, like norgestral, over a period of 6 to 10 years.

If an *antipregnancy vaccine* could be developed, it would be activated in the event of conception and terminate it. Antibodies to human chorionic gonadotropin (hCG) are being developed that allow the female body to be immunized to its own hCG. Women with antibodies to hCG no longer respond to the mechanism that signals the body to begin the hormonal developments

---

[64]Hatcher et al. (1982). For recent developments in family planning in China see Kaufman et al. (1981).

[65]Diczfalusy (1979). For a series of articles, see October 1981 issue of *Contraception*.

[66]Shapiro (1977); Gonzalez (1980).

[67]Aitken (1979); Hatcher et al. (1982).

[68]Diczfalusy (1979).

of pregnancy (see chapter 5). Immunized women will thus menstruate and disrupt implantation.

All of these interesting developments notwithstanding, it is estimated that women now in their early twenties are likely to reach the menopause before any major breakthroughs occur in male or female contraceptives. Hence the birth control methods available at the end of the century are most likely to be indistinguishable from those we have today.[69] Our governmental regulatory controls are currently such that a scientific breakthrough made today would take 12 to 15 years before becoming a practical device in general use. Meanwhile, 350,000 babies are born into the world every day, but only 200,000 persons die.[70]

Abortion

In medical terminology, **abortion** refers to the termination of an established pregnancy before the fetus has attained the ability to survive, with appropriate life support, outside of the uterus. Typically such viability is attained 28 weeks after the last menstrual period, at which time the fetus typically weighs about 1,000 grams (chapter 5).

Abortions may be *spontaneous* (also known as *miscarriage*) or they may be *induced*, that is, initiated voluntarily either as an *elective abortion* or as a *therapeutic abortion*. The latter is done for medical reasons affecting the health of the mother; the former may be done for a variety of other considerations that lead to the wish not to have the child. Induced abortion may be *legal* or *illegal* depending on the particular laws in effect. The complex legal components of abortion are dealt with in chapter 19. Our concern here is mainly with its physiological and some of its psychological aspects.

Following the liberalization of abortion laws in the United States in 1973, the number of abortions went up from 744,600 to 1,553,900 in 1980.[71] In 1980 a quarter of all pregnancies, and about half of all unintended ones, were terminated by elective abortion (30 abortions per 1,000 live births). This rate has been slowing down and is probably stabilized by now.[72]

Simultaneously the number of deaths associated with abortion have declined dramatically from 364 deaths annually during 1958–62 to 14 deaths during 1980, because of the reduction in illegal abortions that tend to be carried out under more hazardous circumstances. These trends have been replicated in the experiences of other countries.[73]

Methods of Abortion

The method used for abortion is usually determined by the length of the pregnancy. During the first trimester, abortion is performed by mechanically removing the contents of the uterus through the cervix. Evacuation is sometimes used as late as the 20th week, but during the second trimester, abortion is usually performed by stimulating the uterus to expel its contents—in effect, inducing labor.

Although abortion does present certain risks, the overall death rate associated with legal abortions is very low, especially when performed during the first trimester (when the great majority of abortions are done). Maternal deaths in pregnancy are 1 in 10,000. In legal abortions performed before 9 weeks, the death rate is 1 in 400,000; between 9 and 12 weeks, 1 in 100,000; between 13 and 16 weeks, 1 in 25,000, and after 16 weeks, 1 in 10,000. By contrast, 1

---

[69]For further examples and details in the development of new contraception, see Djerassi (1981); Hatcher et al. (1982).

[70]Djerassi (1981, p. 225).

[71]Tietze (1983, p. 33). An estimated 55 million abortions are reported worldwide (30 abortions per 100 pregnancies).

[72]Henshaw et al. (1982).

[73]For instance, mortality rates declined 56% and 38% in Czechoslovakia and Hungary, respectively, after their abortion laws were liberalized in the mid-1950s. The opposite effect occurs when abortion laws become more restrictive. When this happened in Romania in 1966, there was a sevenfold increase in deaths due to abortion (Tietze, 1983, p. 102).

in 3,000 women die during illegal abortions.[74]

Vacuum aspiration   *Vacuum aspiration* or *vacuum curettage* is the preferred method for first trimester abortions because it can be performed on an outpatient basis, quickly, and at a relatively low cost. Prior to 8 weeks of pregnancy, minimal or no anesthesia may be required, and it is often unnecessary to dilate the cervix mechanically.

As shown in figure 6.10, a *suction curette* is passed through the cervical opening into the uterus and its contents are extracted. Complications of vacuum aspiration are relatively uncommon, but may be serious. They include perforation of the uterus, hemorrhage, uterine infection, and cervical lacerations.

Dilation and Curettage (D and C)   Before the advent of vacuum aspiration in the 1960s, the most common method of abortion was dilation of the cervix and curettage (scraping) of the uterine lining.[75] The first step, *cervical dilation*, can be accomplished by passing a series of progressively larger metallic dilators through the cervical opening, but in recent years a less painful but slower method has become popular using *laminaria sticks*. Made of compressed seaweed, these are inserted into the cervix where they absorb cervical secretions and expand to five times their dry size in about a day. When the cervix is enlarged sufficiently, a *curette* (a bluntly serrated metal instrument) is inserted to scrape off the uterine contents. The complications of the D-and-C method are the same as those of vacuum aspiration but occur more often, which is why D and C has now been largely superseded by vacuum aspiration.

Dilation and Evacuation (D and E)   After

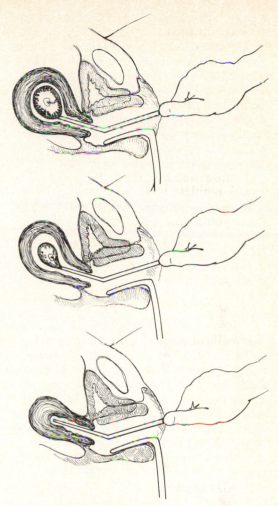

Figure 6.10   Abortion by vacuum aspiration. Suction curette is inserted through the cervical canal; uterine contents aspirated; uterus contracted after completion of evacuation. (From Shapiro, *The Birth Control Book,* St. Martin's Press, 1975, p. 165.)

the 12th week of pregnancy, abortion becomes an increasingly serious surgical procedure, and the rate of complications increases. One abortion technique that is often used between the 13th and 20th weeks of pregnancy is *dilation and evacuation*. D and E is similar to the two previous methods, but the fetus now being larger, it is not as easily removed. After adequate cervical dilation is achieved, suction, forceps, and curettage are used to remove the uterine contents.

[74]Hatcher et al. (1982, p. 7).

[75]D and C is also employed for the diagnosis and treatment of a number of pelvic disorders, so if a woman has undergone this procedure, it does not necessarily mean she has had an abortion.

Saline Abortions During the second trimester, abortion is commonly induced by the injection of a concentrated salt (or urea) solution into the uterine cavity. Known as *saline abortion*, this is a technically more difficult procedure if done earlier. The fourth month is a particularly difficult time: The pregnancy is too far along to allow for a safe, simple aspiration; but the uterus is not yet large enough to allow the physician to easily locate the proper place in the abdominal wall for the saline injection.

Following the injection of saline solution into the uterine cavity, contractions usually begin within 12 to 24 hours, and about 80% of women deliver the fetus and placenta within 48 hours of the injection. Some physicians also administer oxytocin to stimulate more vigorous uterine contractions.

Uncommon but severe complications are known to occur with saline abortion. Most serious is *hypernatremia* ("salt poisoning"), which can result in high blood pressure, brain damage, and death. Other complications include intrauterine infection and hemorrhage. Delayed hemorrhage (days or weeks after the abortion) occurs most commonly with a retained placenta—that is, in instances where the fetus is successfully aborted but part or all of the placenta remains behind.

Prostaglandin abortions *Prostaglandins* cause uterine contractions, hence they can be used to induce abortion. They are usually injected directly into the uterus but can also be injected into the bloodstream or intramuscularly. Laminaria sticks are sometimes used to assist the process.

Complications with prostaglandin abortion include nausea, vomiting, and headache; at least 50% of women experience one or more of these side effects, which are however temporary and rarely serious. Hemorrhage, infection, and uterine rupture (a danger with all types of second-trimester abortion) are infrequent with prostaglandin abortion. The proportion of live births is higher with prostaglandin than with saline abortion, especially after the 20th week, which makes some physicians reluctant to perform abortions by this method after the 20th week.

Psychological Aspects of Abortion

The psychological reactions of couples, and of women in particular, faced with the prospect of abortion are likely to be complex and varied. Much depends on their perceptions of abortion in moral and psychological terms: For some, abortion is tantamount to murder; to others it is a matter of disruption of a physiological process whose continuation is not in their best interest.

Reactions to Unwanted Pregnancy For most people the reality of an unwanted pregnancy and the prospect of abortion elicit strong feelings. A typical initial reaction is disbelief or denial, which may lead, especially among teenagers, to delays in confirming the pregnancy and to coming to terms with it. A woman may also be shocked to discover that when her pregnancy is confirmed, she is further into it than she assumed based on her missed period (by which time she was already several weeks pregnant). Denial gives way to a cluster of negative feelings. Distress may be compounded by guilt and mutual recrimination. There is anger and outrage, especially if there is the perception that seduction, deception, or pressure were used in the sexual interaction that led to the pregnancy.

Yet simultaneously there may be a deep sense of satisfaction at being pregnant and a most reluctant acquiescence to terminating it, however necessary or inevitable that may be. Nor are men immune to these yearnings of parenthood; their own feelings of anguish, guilt, and anger often receive insufficient recognition. Men are also likely to feel helpless in these situations, since it is usually the woman who eventually determines what is to happen to the pregnancy.

The most trying time is often between the discovery of the pregnancy and the decision to abort. Once there is the prospect of action, at least the burden of uncertainty is lifted. But often doubts persist and women continue to feel anxious and depressed.

Deciding between Abortion and Its Alternatives No two cases of unwanted pregnancy are the same. A mature woman in a secure marriage who fully intended to have a child at some future date but got pregnant unexpectedly and the teenage girl who is not even sure who got her pregnant would be obviously facing the abortion decision from quite different perspectives.

A single woman who gets pregnant unintentionally faces four choices. Traditionally, the most desirable alternative has been *marriage*. This continues to be the case if all else about the relationship indicates a good marital choice; in other words, if it would make good sense for the woman to marry the man at this time even if she were not pregnant by him. Otherwise, marriage under these circumstances may turn out to be a good short-term or face-saving solution, but a long-term liability. The prospects of an unhappy marriage, or a divorce, with a child in the picture, deserve serious thought.

The next alternative is *single parenthood*. Many more women decide on this alternative now than in the past because there is far less social ostracism entailed and women are more independent. But raising a child on one's own is still not easy. Hence, the prospects of help from the man, other sources, the impact of single parenthood on career choices, and other related considerations need to be taken into account.

Another solution is to give up the child for *adoption*. This avoids the prospect of abortion, which may be unacceptable on moral or psychological grounds, as well as the need for dealing with the child. However, the decision to give up the child is not easy, and a woman may not be able to make up her mind until the baby is born. If there are no adoptive parents available, the child can live in a foster home. Before the baby can be placed for adoption, the father of the baby has the right to have his views heard at a legal hearing.

If the above alternatives are unavailable or undesirable, then the choice is abortion—which is what 4 out of 10 pregnant teenagers choose to do. The earlier the decision to abort is made, the better it is from a health perspective. This is why it is most important for women to keep close watch over their menstrual cycles. A period that is late by a day or two usually means nothing, but if a woman loses track of when her period should come, then she may not know soon enough when to become concerned and get a pregnancy test. Delay in getting confirmation of pregnancy is the major difference between women who receive first trimester abortions and second trimester abortions.

Though time is of the essence, a woman should not be rushed or pressured into an abortion decision. All women faced with this situation need information and support and should be provided with an opportunity for counseling. The discussions should review her moral values, her life situation, aspirations for the future, feelings about her partner, and whatever conflicts she may be experiencing in making a choice. In addition to traditional sources of guidance, such as physicians, organizations such as Planned Parenthood provide such services in many communities.

Psychological Reactions to Abortion A woman's reaction to the abortion procedure itself depends on the length of the pregnancy and the sensitivity with which the woman is treated. The later the abortion, the more upsetting it is likely to be. Women who have had children are startled to discover that the induced uterine contractions feel just like labor pains, and the recollection of those memories can be disturbing.

These emotions are much influenced by various psychological and practical circumstances. Having an abortion is by no means a source of social ostracism in all groups.[76] A woman's husband or lover, family, and friends can provide much

---

[76]See chapter 11 in *Boston Women's Health Book Collective* (1976).

emotional support. If there is ready access to appropriate medical facilities the prospect of abortion can be taken in stride by many women rather than being experienced as a catastrophe.[77]

The aftermath of an abortion is usually followed by a sense of relief. The mechanisms of denial and repression help to bury the experience in the past. Yet painful affects may also linger in the form of doubts about the decision. Sooner or later these feelings begin to recede: by the end of several months, many of these women consider the matter resolved. In some cases women require more extensive counseling, and a few may require psychiatric help.[78]

The law and large segments of professional and public opinion now endorse the right of each woman to make decisions regarding her own pregnancy. Since the fetus is part of the woman's body, it is argued that she ought to exercise her right of ownership as to what can and cannot be done with it. But this view may be unfair to the man who is the father, provided he is known, is present, and cares about the outcome of the pregnancy; after all, he has a biological investment in the fetus and society expects him to have a vital interest in the future of the new life he has helped create. Does his fatherhood, with all its rights and responsibilities, start only after the moment of birth? Yet to allow the father to have a decisive role in the abortion decision may place an intolerable burden on the woman. Is she to bear a child that she does not want and then be primarily responsible for rearing that child as well?

There are no ready answers to these dilemmas, as we shall see when we discuss them in greater detail from a societal perspective (chapter 19). The need for abortion may never be eliminated entirely, but the effective use of contraception would certainly spare millions of people the experience of abortion and save them much grief.

[77]Freeman (1978).

[78]For further discussion on the emotional impact of abortion, see David (1978); Nadelson (1978).

SUGGESTED READINGS

Djerassi, C. *The Politics of Contraception* (1981). Provides an authoritative, broad, and well-written account of the background and current status of contraception and the prospects for the future.

Hatcher, R. A. *Contraceptive Technology 1982–83* (1982). Gives detailed and clear descriptions of all contraceptive methods with instructions for their use. Hatcher, R. A. *It's Your Choice* (1982), offers detailed instructions for choosing a contraceptive method and using it safely and effectively.

In-depth reports and review articles are published by various population programs on contraceptive technology policy and statistics. See, for instance, *Population Reports* published by the Population Information Program of Johns Hopkins University; *Population Council Fact Books* published by the Population Council, New York.

All books dealing with women's health issues have chapters on contraception and abortion. See references among the suggested readings for chapter 7.

Chapter 7

Diseases of Sex Organs

"If I got it again I'll kill him," she said, her small, childish face, despite her bitterness and tears, still somehow smiling . . . she was only sixteen years old. Two and a half years ago we had treated her for syphilis. Now she had come because her boyfriend . . . had told her to have a checkup too.

"I know the girl who gave it to him," she informed me.

"What's the name of your boy friend?" I asked.

"I don't know," she said laughingly, "I always call him Tony."

Basile Yanovsky, *The Dark Fields of Venus*

The body suffers inevitably from wear and tear throughout life, and there are illnesses we cannot do very much about. But there are also ailments that can be avoided entirely or the risk of contracting them minimized. Our focus in this chapter is on these latter conditions and on staying sexually healthy.

Sexual disorders can take many forms. The reproductive system is subject to the same range of ailments as the other systems of the body, and in addition, conditions such as disorders of reproduction and sexual dysfunction are specific to it. Of particular importance are infections that are acquired through sexual contact and hence are referred to as *sexually transmitted diseases* (STD).

Diseases of sexual organs do not necessarily interfere with sexual functions. A man may have syphilis but no difficulty in having an erection; a woman may have gonorrhea but have no problem reaching orgasm. In other circumstances, physical ailments interfere with sexual function, as do a wide variety of psychological and interpersonal problems. The resultant disturbances in sexual performance or satisfaction are referred to as *sexual dysfunction*

("dys" for difficult or faulty). We deal with sexual dysfunction in chapter 14.

Finally, there are disorders involving *sexual behavior* that are known as *paraphilias.* These conditions usually entail behaviors that are considered socially unacceptable, such as the sexual abuse of children. Although these are generally treated as psychiatric conditions, some social scientists object to the use of medical models of illness to categorize such activities. The paraphilias are discussed in chapter 10.

Maintaining Sexual Health

The human body is a superbly designed biological system, but it requires proper care and maintenance to work well. Sexual functions are handicapped in this regard because of the secrecy and embarrassment that shroud them. This puts an additional responsibility on individuals to look after their own sexual health.

Sexual Hygiene

Cleanliness is the first requirement in the care of genital organs, coupled with an awareness of the early signs of illness that could affect them. The genitalia of the male are clearly in view, but those of the female are more hidden. Women, therefore, tend to be less informed about their genital structures than men. In addition to learning about the genitals through books and pictures, a woman can explore her genital region by holding a mirror to it.

Cleanliness   Apart from hygienic considerations, a clean and fresh body enhances sexual attractiveness. Yet it is not necessary to equate cleanliness with the elimination or covering up of all natural odors. Stale and offensive smells result from the action of skin bacteria and other microorganisms on accumulated body secretions. Cleanliness of the genital region is especially important since the skin in this area is full of folds and wrinkles, covered by thick hair, and rich in various glands. Another source of strong odor is the armpit where similar conditions exist. Such odors can easily be prevented from becoming offensive by regular, careful washing of the genital region with soap and water several times a week or more, each time one has sexual intercourse, and preferably before intercourse. Men who are not circumcised need to retract the prepuce and thoroughly wash the glans where smegma tends to accumulate. The rest of the genital organs and surrounding area (including the anal region) can be best cleansed with a wash cloth.

Women should avoid vigorous rubbing of their genital organs with soap or wash cloth.[2] It is best to lather the pubic hair, then spread the soap bubbles over the vulva and use the fingers to clean the folds between the labia without getting soap into the vagina. On the toilet, the anus must be wiped backward and away from the vaginal opening to avoid fecal contamination. Following urination, dabbing the area with toilet paper or a wet cotton ball will lessen the chances of lingering odors (especially when wearing panty hose on warm days). It is better to keep the genital area clean rather than to try to mask smells with spray deodorants, to which some women are allergic.

Although the vagina is a self-cleansing organ, some women like to wash it by *douching* (French for "shower") after menstruation or coitus. This is usually not necessary for hygienic purposes, except where contraceptive creams, foams, and the like leave residues that may cause odors. Douching can be done by using disposable kits, but this entails needless expense, and some of the chemical solutions used may be irritating. One can instead buy a douche bag and use plain lukewarm tap water or make it mildly acidic by adding two tablespoons of vinegar to a quart.

Menstrual Hygiene   Women in most societies are no longer burdened with menstrual taboos or think of menstruation as the "curse" (box 7.1), but some are still embarrassed by menstrual blood and hence excessively preoccupied with concealing its presence. To absorb the menstrual flow women use *tampons* (absorbent cylinders inserted into the vagina) or *sanitary napkins* (absorbent pads placed against the vaginal opening). In various forms, both devices have been in use since ancient times.

Teenage girls learn the use of these materials when they reach menarche, but boys are often mystified by them. Tampons come in various sizes and can be used by women of all ages. Junior-size tampons can be inserted into the vagina without damage to the hymen. It is preferable that the size of the tampon used on a given day be appropriate to the amount of flow expected. Small ones will not be able to handle heavy flow; large ones will be insufficiently saturated by scanty flow to be taken out easily and may cause vaginal irritation. When placed in properly (they should be pushed as far back as possible), a tampon can hardly be felt and does not interfere with the freedom of movement. Menstrual blood develops an odor mainly after exposure to air and the action of bacteria.

---

[2]The following discussion is based on Budoff (1980). For further practical advice, see pp. 229–235.

Box 7.1:   Menstrual Taboos

In various societies over many centuries, women have been burdened by pervasive attitudes that have regarded menstruation as rendering them *ritually unclean*. Menstruation is but one form of pollutant among other physiological processes, but occurring as it does only among women, it has constituted a significant social handicap for them.

The origins of *menstrual taboos* are ancient. The Old Testament and the Quran are replete with regulations concerning the proper conduct of the menstruating woman. Not only are women considered impure during this period, they are also deemed dangerous by being more likely to harbor and transmit evil spirits (chapter 20).

In cultures with such taboos, menstruating women were segregated from others (especially from those who were ill and women in labor). They were subject to various restrictions in the home, such as having to sleep on the floor, avoiding sexual relations and all other physical contact with their husbands (not even touching his bed or preparing meals).[1] At the end of her period the woman would become clean again following a prescribed ritual bath.

Among Native Americans of the southeast, girls were secluded during their first menstruation and from then on women either absented themselves from the household during their period or were subjected to restrictive taboos. Among the monastic orders of India, nuns were relegated to secondary status because of their being periodically "polluted" and thus having to be barred from certain key rituals.

Even in societies with no specific menstrual taboos, sexual intercourse during menses has been often avoided or prohibited. In many instances this is done to avoid harm to the man: The Lepcha believed that if a man had sex with a menstruating women he would get ill; Thonga men who had yielded to such temptation would tremble before battle or fail to fight; Mataco men developed headaches.[2]

The Roman historian Pliny writing in 77 A.D. described the effects of exposure to menstrual blood as follows:

> Contact with it turns new wine sour, crops touched by it become barren, grafts die, seeds in gardens are dried up, the fruit of trees falls off . . . the edge of steel and the gleam of ivory are dulled, hives of bees die, even bronze and iron are at once seized by rust, and a horrible smell fills the air, to taste it drives dogs mad and infects their bites with an incurable poison.[3]

We no longer believe in such effects. Yet more subtle fears of the menstrual "curse" may well continue to linger on in our attitudes toward the menstruating woman.

[1]Shiloh (1976).
[2]Ford and Beach (1951).
[3]Quoted in Delaney (1977), which is a good source on the cultural history of menstruation.

Toxic Shock Syndrome   A great deal of concern has been generated since 1980 by a newly recognized condition associated with tampon use called *toxic shock syndrome* (TSS). It usually occurs in younger women during or immediately following menstruation and is thought to be caused by toxins produced by certain bacteria (*staphylococcus aureus*) that are present in many people but tend to accumulate in and around highly absorbent tampons. It is possible that a virus is responsible for triggering these bacteria to produce their toxin. The early symptoms of toxic shock are sudden high fever accompanied by sore throat, rash, vomiting, diarrhea, dizziness, and abdominal pain. The condition is very rare, occurring in less than 10 per

100,000 menstruating women (of whom there are 50 million), but it can be fatal; 88 deaths were ascribed to it between 1978 and mid-1982. The particular brand of tampon (*Rely*) that was found to be most frequently associated with this condition has been taken off the market.[3] But there has been no significant drop in the number of cases (about 40 to 65 per month).

Though there are many other causes for the TSS-like symptoms, if a woman experiences them while wearing tampons, she should seek immediate medical care. To help reduce the risk, tampons should be changed at least every 6 hours, hands washed prior to insertion, sanitary napkins substituted at night, excessively absorbent types ("super" or "superplus") used only with heavy flow, and avoided by adolescents altogether.

Menstruation and Sex   There is no health reason why a woman should refrain from coitus while menstruating, yet many do. If a woman does not feel well during menstruation that is good enough reason to abstain. Otherwise it is a matter of the aesthetics of the situation. Washing before coitus and placing a towel on the bed will minimize the problem of soiling; inserting a diaphragm will solve it.

Menstruation and Menopause

The menstrual cycle, whose physiology we discussed earlier (chapter 4), is subject to a variety of disturbances. There are normal fluctuations in its timing, duration (from 3 to 7 days), and amount of flow. But there can also be absence of menstruation (*amenorrhea*), increased amount or duration of menstrual bleeding (*menorrhagia*), and painful periods (*dysmorrhea*).

There are many physical conditions that account for such menstrual disorders. The menstrual cycle is also subject to the influence of emotional factors. Sometimes the experience of going to college or going back home from college is enough to disrupt temporarily the menstrual rhythm. Fol-

lowing unprotected intercourse the fear of pregnancy is a well-known cause for delayed menstruation. This is explained by the simultaneous involvement of the hypothalamus in the regulation of emotions and endocrine functions.[4]

Menstruation is a normal bodily process, which most of the time does not interfere with physical or psychological functions. Its cessation during the menopause is part of the normal process of aging. Yet these events may also be accompanied by physical and psychological manifestations that cause variable degrees of distress without constituting an illness.

Menstrual Distress

For a significant number of women menstruation entails some discomfort. This is usually quite mild, but in 10% of cases (usually among younger women) it is severe enough to necessitate bed rest.

The symptoms of menstrual-cycle distress are categorized under the *premenstrual tension syndrome* and *dysmenorrhea* (painful menstruation).[5] Premenstrual symptoms are manifested during the week preceding the menstrual flow; the symptoms of dysmenorrhea accompany menstruation itself. Though there may be some overlap in the discomfort caused by these two conditions, they should be dealt with separately since their symptoms, causes, and treatments differ.

Dysmenorrhea   Typically, menstrual discomfort is manifested by cramps in the lower abdomen, backache, and aches in the thighs. Less often, there may be nausea, vomiting, diarrhea, headache, and loss of appetite. When these symptoms become severe enough to interfere with work or school, and last 2 days or more, the woman can be said to suffer from dysmenorrhea.[6] Menstrual pain can be due to various pelvic diseases, in which case it is called ***secondary dysmenorrhea.*** Where there is no

[3]Paige (1978).

[4]Jewelewics (1983).

[5]Some 150 symptoms have been linked to the menstrual cycle. For details, see Moos (1969).

[6]Jones and Wentz (1982).

apparent illness causing it, then it is *primary dysmenorrhea,* which is what we are concerned with here.

The menstrual cramps of primary dysmenorrhea are attributed to *spasms* of the uterine musculature. This is caused by *prostaglandins* that are normally released from the endometrial lining shed during menstruation (chapter 4). Prostaglandins are a family of compounds, widely distributed in various tissues of the body, with a broad range of actions. These actions include: spasmodic contraction of the smooth muscles of the uterus (hence their use in inducing labor); various effects on the gastrointestinal tract (causing diarrhea, cramps, nausea, and vomiting); dilation of blood vessels (causing hot flashes); and other changes in the endocrine and nervous systems of the body.[7]

The time-honored, though not very effective, remedy for dysmenorrhea is aspirin with bed rest and warm fluids. More recently a wider choice of drugs, many of which are *antiprostaglandins,* have proved quite effective in preventing as well as treating dysmenorrhea by inhibiting the synthesis or the actions of the prostaglandins. Women on the Pill are also likely to get relief because steroid hormones reduce the amount of endometrial sloughing and hence the level of prostaglandins released in the process. The use of the Pill for this purpose only is not advocated.

A similar caution applies to *menstrual extraction,* especially when practiced by self-help groups. This procedure uses the same technique described for the vacuum aspiration method for abortion (chapter 6), namely suctioning out the uterine contents. This so-called "5-minute period" is not a safe procedure, particularly for women with a history of uterine infections, tumors, and a number of other conditions.

Premenstrual Tension Syndrome   The symptoms of the *premenstrual tension syndrome* (PMTs) are more varied and less discrete than those of dysmenorrhea. And as yet none of the explanations offered for its cause has been generally accepted.

The symptoms of premenstrual tension appear 2 to 10 days before the onset of menstrual flow and usually abate by the time the menses start. In some cases, these symptoms merge with those of dysmenorrhea, but the latter can frequently occur without being preceded by menstrual tension.

The more common symptoms of premenstrual tension can be subsumed under three categories. First, are symptoms associated with *edema* (swelling) consisting of a bloated feeling in the abdomen, swelling of the fingers and legs, swelling and tenderness of the breasts, and weight gain. Second, is *headache.* Third, is *emotional instability,* including anxiety, irritability, depression, lethargy, insomnia, and cognitive changes such as difficulty in concentration and forgetfulness. More idiosyncratic reactions include changes in eating habits (such as a craving for sweets), excessive thirst, and shifts in sex drive (increase for some, decrease for others).

There has been much confusion in the assessment of these symptoms and the ascertainment of their causes. To begin with, it is unclear what proportion of women actually suffer from such symptoms: estimates range from 30% to 90%. Moreover, these symptoms tend to be highly subjective: though generally perceived as unpleasant, some women experience premenstrual tension as bursts of energy and creative activity. The symptoms of edema are ascribed to fluid retention, but recent studies have failed to document significant weight gain in the premenstrual period (in which case the swelling of body parts must be explained by internal shifts of fluid from one part of the body to another). Finally, in studies where women with premenstrual tension believed that their mental or physical performance was impaired, tests and other objective measures failed to substantiate these self-perceptions.

A variety of physiological causes have been proposed to explain premenstrual

---

[7]Gilman et al. (1980). Early studies of prostaglandins were based on the effects of semen on uterine muscle strips. Since much of the seminal fluid comes from the prostate gland, prostaglandins were named after it.

symptoms: The drop in *progesterone* level late in the postovulatory phase; *vitamin B-complex* deficiencies causing decreased liver metabolism, hence higher levels of estrogen; increased *aldosterone* (a steroid hormone secreted by the adrenal cortex) secretion; and abnormalities in *endorphin* production causing estrogen-progesterone imbalances.[8]

The likelihood of experiencing menstrual distress also seems to be linked to purely psychological or cultural factors, such as religious background. Among college women, Catholics and Orthodox Jews tend to have a higher prevalence of menstrual distress.[9] Negative attitudes toward menstruation and the expectation of menstrual discomfort may also create a self-fulfilling prophecy. For instance, in one study, women who were led to believe that their menstrual periods were due in a few days were more likely to experience premenstrual distress than other women who had been persuaded by the experimenter not to expect their period until quite a bit later.[10] Other attempts to establish a psychic origin for premenstrual tension have linked it to personality types and various life circumstances.[11] Negative associations attached by society to the menstrual experience as well as rejection by the individual of this characteristic experience of women, have received special attention.[12]

Premenstrual tension may be helped by a *low-salt diet* during the week before the period, or by taking *diuretics*, both of which can counter fluid retention. Coffee, tea, cola drinks, and sweets tend to aggravate the condition for some women and are best avoided. If symptoms of physical discomfort and headache are severe enough, *analgesics* (medications to counteract pain) may be helpful. One must be careful, however, not to substitute one problem for another by becoming overly dependent on drugs

in dealing with the ordinary problems of menstrual distress.

There have also been claims of good therapeutic results through the administration of *progesterone* and *vitamin B-complex*, to correct their deficiency that presumably causes the premenstrual symptoms. Yet such results have not been generally replicable. Also, these symptoms seem to be relieved in a considerable proportion of cases by *placebos* (chemically inert substances), which depend on a psychological mechanism for their apparent efficacy. Finally, some women find that *orgasm* (attained through whatever means) is quite effective in attenuating the discomfort of menstrual distress both with respect to premenstrual tension and menstrual cramps.[13]

Behavioral Changes during the Paramenstruum   Katharina Dalton, who has been studying various forms of menstrual distress for over 3 decades, has focused particular attention on the emotional turmoil and its behavioral impact associated with the menstrual cycle (which she ascribes to progesterone withdrawal).

The 4 days before and the 4 days after the onset of menses, which Dalton calls the ***paramenstruum,*** appear to be particularly stressful times for women.[14] For example, higher rates of accidents, psychiatric admissions for acute illness, and attempted suicides have been reported to occur during the paramenstruum. Similarly, this period has been associated with significantly higher rates of behaviors leading to imprisonment and absenteeism from work.[15] However, studies that relate female behavior to stages of the menstrual

---

[8]Debrovner (1983).

[9]Page (1973).

[10]Ruble (1977).

[11]Kinch (1979).

[12]For further sources on the causes of PMTs, see Rubin et al. (1981).

[13]For more detailed discussion on the relief of menstrual distress, see Budoff (1980); Holt and Weber (1982).

[14]Dalton (1979); Dalton (1969). Queen Victoria reportedly suffered from fierce tempers during her premenstrual period. ''Even (Lord) Melbourne, a past master at dealing with women, had on one occasion quavered and feared to sit down as the fire blazed in the eyes of the eighteen-year-old queen.'' Quoted in Dalton (1979), p. xiii.

[15]Dalton (1979).

cycle, often without measuring hormone levels, have been criticized on methodological and conceptual grounds. More systematic attempts to relate levels of estrogens and progestins to moods and daily activities have failed to show any correlation.[16]

Antedating attempts to document systematically the association of the paramenstruum with behavioral disturbances, allowances have been made by social institutions in recognition of it. For instance, in France menstrual distress is considered a mitigating circumstance in violent crimes, and in Britain women who have committed murder while in the throes of menstrual distress have been merely put on probation.

This poses a dilemma. One the one hand, it can be argued that the paramenstruum should be recognized as a time of higher risk and distress for women. Therefore, albeit part of a normal physiological process, serious attention should be paid to anticipation of stress and the relief and management of its physical and psychological manifestations. On the other hand, it is countered that while some women may well be adversely affected during the paramenstruum, its behavioral aberrations have been highly exaggerated. There is no good evidence that the mental and physical capabilities of women are generally compromised during the paramenstruum (even if some of the affected women themselves may think so). To recognize diminished responsibility for actions committed by women during this period and make allowances casts an aura of unreliability on all women, and creates one more excuse to discriminate against women in various occupational and social settings.

The Menopause

Just as we do not know what initiates ovarian function at puberty, so we do not understand why ovarian function comes to an end between the ages of 45 and 55 during the *menopause* ("stopping of men-

struation"). The alternative term *climacteric* (from Greek for "crisis") more broadly encompasses the many biological and psychological changes characteristic of this period in addition to the cessation of menstruation. (Another common term is "change of life.")

In physiological terms, the source of failure is in the ovary and not in the pituitary, which continues producing gonadotropins. For some reason, the ovarian follicles fail to respond to gonadotropin stimulation anymore, and in turn fail to produce estrogen. Infertility and other manifestations of the menopause are then the direct result of estrogen deprivation.

Menopausal Symptoms   The most common of the *physical symptoms* of the menopause are *hot flashes* (or flushes), which are present in close to 70% of menopausal women.[17] They are experienced as a sensation of warmth rising to the face from the upper body, with or without perspiration and chilliness. These come at about hourly intervals over a period of a few months or several years. Other physical symptoms include dizziness, headaches, palpitations, and joint pains. The bones tend to become more porous (*osteoporosis*), making postmenopausal women more liable to suffer fractures.

Among the *psychological symptoms* of the menopause, *sadness* is the most prominent (experienced by about 40%) and may range from mild moodiness to severe depression. Other symptoms are tiredness, irritability, and forgetfulness.

With respect to *sexual functions*, the most important menopausal changes are those affecting the reproductive system. They include loss of elasticity of the vagina and thinning of the vaginal lining, with marked decrease of the lubricatory response during sexual excitement; atrophy of the uterine endometrium, with decline in cervical mucus secretion; and shrinkage of the breasts. The vaginal changes may lead to painful coitus but the problem is easily remedied by hormones or more simply by

---

[16]McCauley and Ehrhardt (1976); Abplanalp et al. (1979).

[17]Jones and Wentz (1982).

the use of lubricants. Older women also develop a tendency toward a burning sensation during urination after intercourse because of irritation of the bladder and urethra by the penis, since the thinner vaginal wall gives inadequate protection to the bladder. Discomfort may persist for several days following coitus. It does not follow orgasm through masturbation. Women who remain sexually active seem less likely to manifest some of these changes.

Following menopause, all women become infertile, but sterility does not set in abruptly. As with menarche, there is a period of several years of menstrual irregularity and relative infertility. Pregnancies are rare beyond age 47 though some have been medically documented for as late as 61 years. The effect of the menopause on sexual behavior is dealt with in chapter 15.

Treatment of Menopausal Symptoms

Much of what we said earlier about menstrual distress would apply equally well to the menopause. The menopause as such is also not a disease. Many women experience no significant ill effects. Among those that do, only 1 out of 10 is markedly inconvenienced by them. The lingering stereotype that the menopause is a very difficult time for all women is not true, yet one must not deny the real distress experienced by some and take appropriate measures to deal with them. Furthermore, the fact that there are important physiological changes present does not necessarily mean that they are the cause of all the problems that may be manifested. Cultural and psychological factors have a great deal to do with how this period is experienced. The fear of the loss of attractiveness and the assumption that a woman has passed her prime are likely to have a greater impact on a woman's self-esteem and feeling of well-being than the hormonal changes of the menopause.

An effective *treatment* for some of the symptoms of the menopause is *estrogen replacement*. When such therapy became available several decades ago it was enthusiastically embraced as a panacea and then, with the cancer scare in the 1970s, estrogen use fell into disrepute. Present opinion is that estrogen replacement is a legitimate form of therapy when needed, but it should be used in the lowest effective doses, for the shortest possible period of time, and under medical supervision.[18] As with other drugs, the judgment to use them must be based on a cost-benefit assessment in each individual case.

Estrogen is most helpful in attenuating the hot flushes and counteracting the vaginal atrophy and dryness. Its effectiveness in preventing osteoporosis is more open to question, as is its ability to prevent wrinkles or keep a woman youthful.

The side effects of estrogen use are the same as those for birth control pills (chapter 6), but a more serious threat in the case of menopausal usage is a five-fold higher risk of endometrial cancer.[19] This is mitigated by the relative rarity of endometrial cancer, which causes only 1% of all female deaths from cancer. Women with relatives who have had cancer of the reproductive system are at higher risk. There are also higher risks if estrogen use is associated with smoking, vascular disease, and certain pelvic disorders.[20]

Men typically do not suffer a comparably rapid loss of testicular function at midlife, but show a gradual, albeit variable, decline in testicular function and potency with aging. Rarely, a man may undergo a true male climacteric, with hot flushes, depression, and other signs similar to the female menopause, when testosterone production declines precipitously.[21] However, periods of psychosexual distress in middle-aged men (the "midlife crisis") are quite common and can be disruptive to sexual function (chapter 15).

[18]For details of estrogen replacement therapy, see Jones and Wentz (1982); Mosher and Whelan (1981); Dalton (1979).

[19]Jick et al. (1980).

[20]For comprehensive discussions on many aspects of the menopause, see Rose (1977); Weideger (1976); Jones and Wentz (1982).

[21]Bermant and Davidson (1974).

Common Ailments of the Reproductive System

Genitourinary Infection

The most serious infections of the genital and urinary tract are due to sexually transmitted diseases, which we discuss separately. A number of other conditions deserve brief mention.

Genital Discharge   Normally, men should pass only urine and semen through the urethra; any other discharge or the presence of blood is always abnormal. But women, in addition to menstrual bleeding, secrete vaginal mucus during their monthly cycle (apart from vaginal fluid generated during sexual excitement). When vaginal discharge is excessive it is called *leukorrhea*. Almost every woman experiences leukorrhea at some time in her life. This is not a discrete disease entity but rather a condition that can be caused by infections, chemicals, and physical changes. For example, irritating chemicals in commercial douche preparations may cause excess vaginal discharges, as also may foreign bodies (such as contraceptive devices) and alterations in hormone balance (as during pregnancy or menopause).[22]

Cystitis   The proximity of the urethral opening to the vagina and anus predisposes women to urinary bladder infections (*cystitis*). Sometimes bladder irritation is caused by frequent coitus ("honeymoon cystitis"). More often it is due to the thinning of the vaginal wall following the menopause, which makes the bladder more liable to irritation.

The primary symptom of cystitis is frequent urination accompanied by pain and a burning sensation. Though it may subside spontaneously in a few days, it is advisable to receive proper antibiotic treatment because untreated infections may spread from the bladder to the kidneys. Other measures that may help prevent cystitis are drinking lots of fluids, urinating after coitus, wearing cotton underpants, and maintaining good general hygiene. Condoms can also help reduce infection.[23]

Prostatitis   A common problem among men is inflammation of the prostate gland (*prostatitis*) manifested by increased frequency and burning on urination, and painful ejaculation. Often there is no apparent bacterial cause, but a curious association exists between prostatitis and infrequent or irregular sexual activity—long periods of abstinence followed by bouts of intensive sex (hence, "sailor's disease").[24]

Cancer of Reproductive Organs and the Breast

Cancer of the Breast   Cancer of the breast is the most common form of cancer in women, accounting for 25% of all female malignancies. Though very rare before age 25, its occurrence increases steadily in each decade thereafter. For women in the 40 to 44 age groups it is the most common cause of death. About 5% of American women develop breast cancer (100,000 new cases a year). Women who are at higher risk include those over 50, women who have a family history of breast cancer, those who experience a late menopause, and those who have never had children. Men very rarely develop breast cancer.

The primary symptom of breast cancer is a painless mass in the breast; much less common are dimples on the breast surface or discharge from the nipple. The cancerous lump will not make itself felt: hence regular and systematic breast self-examination is of great importance. It should be done by every woman, once a month, about a week after the end of the menses or on a set day each month (box 7.2). Most breast masses, especially in younger women, are due to other causes, not cancer. So when a mass is felt, one should not panic but seek prompt medical attention.

---

[22]For a general discussion of abnormal vaginal discharge, see Capraro et al. (1983).

[23]Hatcher et al. (1982).
[24]Silber (1981, p. 104).

Box 7.2   How To Do a Breast Self-exam

1. In the shower. Examine your breasts during your bath or shower, since hands glide more easily over wet skin. Hold your fingers flat and move them gently over every part of each breast. Use the right hand to examine the left breast and the left hand for the right breast. Check for any lump, hard knot, or thickening.

2. Before a mirror. Inspect your breasts with arms at your sides. Next, raise your arms high overhead. Look for any changes in the contour of each breast: a swelling, dimpling of skin, or changes in the nipple.

Then rest your palms on your hips and press down firmly to flex your chest muscles. Left and right breast may not exactly match—few women's breasts do. Again, look for changes and irregularities. Regular inspection shows what is normal for you and will give you confidence in your examination.

The basic treatment for breast cancer is the surgical removal of the breast (*mastectomy*). Though current surgical procedures try to cause as little disfigurement as possible, mastectomy may still pose a problem in creating doubts about femininity and sexual attractiveness. Following such surgery, some women give up sex altogether. However, treatment procedures and reconstruction techniques are now sophisticated enough that disfigurement need not be as much of a problem.

Cancer of the breast can be rapidly fatal if it spreads to vital organs, but with early diagnosis and treatment the prognosis is much more favorable. About 65% of patients with cancer of the breast now remain alive 5 years after the initial diagnosis.

Cancer of the Cervix   Cancer of the cervix is the second most common type of cancer in women. About 2% of all women ultimately develop it (60,000 new cases a year). It is very rare before age 20, but the incidence rises over the next several decades. The average age of women with cancer of the cervix is 45. This disease is more common in women who have had large numbers of sexual contacts and who have borne children. A study of 13,000 Canadian nuns failed to reveal a single case of cervical cancer. The disease is also rare among Jewish women, which suggests that perhaps circumcision in the sexual partner is somehow linked to the disease. But in India there is no difference in rates between Muslim and Hindu women, though the husbands of the former but not the latter are circum-

Box 7.2   How To Do a Breast Self-exam *(Continued)*

3. Lying down. To examine your right breast, put a pillow or folded towel under your right shoulder. Place your right hand behind your head; this distributes breast tissue more evenly on the chest. With the left hand, fingers flat, press gently in small circular motions around an imaginary clock face. Begin at outermost top of your right breast for 12 o'clock, then move to 1 o'clock, and so on around the circle back to 12. (A ridge of firm tissue in the lower curve of each breast is normal.) Then move 1 inch inward, toward the nipple, and keep circling to ex-

amine every part of your breast, including the nipple. This requires at least three more circles. Now slowly repeat the procedure on your left breast with a pillow under your left shoulder and left hand behind your head. Notice how your breast structure feels.

Finally, squeeze the nipple of each breast gently between the thumb and index finger. Any discharge, clear or bloody, should be reported to your doctor immediately.

(From American Cancer Society pamphlet. Used with permission.)

cised.[25] (Also see comments below about the association of circumcision and cancer of the penis).

Cancer of the cervix may present no symptoms for 5 or 10 years; if detected early, treatment during this period is highly successful. The well-known Papanicolaou or **Pap smear** test is the best means now available for identifying cancer of the cervix in these early stages. It should be done annually beginning at age 20 (or earlier if a woman is sexually active). The procedure is simple: a wooden spatula lightly scrapes the cervix, transferring cells to a glass slide that is then stained and examined for the presence of abnormal cells.

As cancer of the cervix begins to invade

surrounding tissues, irregular vaginal bleeding or a chronic bloody vaginal discharge develops. Treatment is less successful when the cancer has reached this stage. If treatment by *surgery, radiation,* or both is instituted before the cancer spreads beyond the cervix, the 5-year survival rate is about 80%, but it drops precipitously as the disease reaches other organs in the pelvis. The overall 5-year survival rate for cancer of the cervix (including all stages of the disease) is about 58%.

Cancer of the Endometrium   Cancer of the lining of the uterus is less common than cancer of the cervix, affecting about 1% of women. It usually occurs in women over 35, most commonly those between 50 to 64. Many but not all cases are detected by

[25]Novak et al. (1970, p. 212).

the Pap smear test. Thus, in addition to having a regular Pap test each year, women over 35 should watch for any abnormal vaginal bleeding. The 5-year survival rate for endometrial cancer is 77%.

Cancer of the endometrium has been linked with estrogen replacement therapy, as we discussed in connection with the menopause. This is not true for birth control pills, which, on the contrary, seem to exert a protective influence (chapter 6).[26]

Cancer of Male Sex Organs   In the male, *the prostate* is the reproductive organ that most frequently causes trouble. It tends to grow larger in **benign hypertrophy** as men get older, obstructing the neck of the urethra and interfering with normal urination. As a result the man has to go to the bathroom frequently, getting up several times at night.

Cancer of the prostate is the most common form of cancer of male sex organs (and the third most common form of cancer in men). But it is not a very significant cause of death because it is uncommon before age 60 and grows very slowly. Thus, those who have it are more likely to die of other causes, such as heart disease.

The initial symptoms of prostatic cancer are similar to those of benign enlargement of the prostate, resulting from partial obstruction of the urethra. Early in the course of the disease, sexual interest may increase because of frequent erections caused by local changes. Later on there is usually a loss of genital functioning.

A tentative diagnosis of cancer of the prostate can usually be made on the basis of a *rectal examination* (palpation of the prostate through the rectum), the history of symptoms, and laboratory tests. A prostate examination should therefore be part of an annual physical checkup for any man over 50, since as with other cancers, the

prognosis is much more optimistic when it is diagnosed and treated early. The cause of prostatic cancer remains unknown despite efforts to link it with hormonal factors, infectious agents, excessive sexual activity, or abstinence.

Unlike most other cancers, which strike later in life, *cancer of the testes* affects young men. It is the most common cancer in males 29 to 35, yet it accounts for only about 1% of all cancers in males. Males who have undescended testes or whose testes descended after age six are at greater risk for developing testicular cancer (between 11 and 15% of these males will develop it).

If testicular cancer is detected early, it is curable. Otherwise it can spread to other parts of the body and cause death. To check for early evidence of testicular cancer a man should examine his testes periodically. This is best done when the scrotum is lax after a warm shower or bath. Each testicle is examined separately by holding it between thumb and two fingers and feeling the surface for lumps (which may be no bigger than a bean).

Treatment involves the removal of the affected testicle, which does not affect sexual activity or fertility. A prosthetic implant is inserted into the scrotum for cosmetic purposes. Nonetheless, some men may still feel that lacking a testicle interferes with their self-image of masculinity.

Cancer of the penis is rare in the United States, accounting for about 2% of all cancers in males. Cancer of the penis almost never occurs among Jews, who undergo circumcision in infancy, and is also rare among Muslims who get circumcised before puberty. Yet in areas of the world where circumcision is not common, cancer of the penis is much more prevalent. For instance, it accounts for about 18% of all malignancies in Far Eastern countries. The usual explanation, though unconfirmed, is that circumcision prevents accumulation of potentially carcinogenic secretions (possibly a virus) around the rim of the penis, which is the usual site of this tumor.[27]

---

[26]For further study, there are good sources of easily understood information with regard to female cancers, in addition to standard medical texts. See for example Lauersen and Whitney (1977); Cherry (1979); Holt and Weber (1982).

[27]See Silber (1981) for more detailed discussion of common male genital ailments.

Health and Sexual Activity

Illness in general, and diseases affecting sex organs in particular, can significantly influence sexual function. But what is the effect of sexual activity on general health? Despite much speculation over the centuries there is no evidence that in a person of average health, even high levels of sexual activity lead to debility, premature aging, or ill health. The opposite view—that regular sexual outlets are essential to good health—has also been popular but also remains to be demonstrated.[28] Nevertheless it is reasonable to assume that a satisfactory sex life contributes to psychological contentment and well-being, which in turn affects the health of the body. By the same token, sexual dysfunction causes frustration and conflict and may contribute to psychological tension and psychosomatic disorders.

For those with debilitating ailments or serious heart conditions, orgasm places additional strain on the cardiovascular system. Thus, people who would react adversely to any kind of physical exertion may also suffer from the physiological strain of sexual activity. Occasionally death occurs as a result of heart attack or stroke during coitus (as it does during running) among those predisposed to such conditions, more often males. But with caution, medical guidance, and common sense, people with heart conditions can enjoy sex without adverse effects and will benefit from it as they would from moderate exercise. Even heart-transplant patients have been able to resume sexual activity only months after surgery.[29]

There is not very much known about how to keep sexual functions themselves healthy. Men in particular have been much preoccupied with resisting the effects of aging, and innumerable substances have been tried to enhance sexual desire and performance (chapter 15). In addition to being in good physical and psychological health and having a positive and accepting attitude towards sex, the one specific factor that seems most consistently linked to a vigorous and active sex life is steady sexual activity itself for both men and women.[30]

Sexually Transmitted Diseases

Sexually transmitted diseases (STD) are infections caused by various microorganisms that are primarily acquired through sexual contact. They were formerly called, and still are popularly known as, *venereal diseases* (VD) in reference to Venus, the goddess of love).[31] Though primarily affecting the sex organs, STD are by no means localized ailments; they affect many other parts of the body as well, with potentially grave consequences.

The STD constitute one of the most serious of all public health problems, affecting millions of individuals in near-epidemic proportions. Many of these diseases could be controlled. As shown in figures 7.1 and 7.3, the tremendous increase in their prevalence is a phenomenon of the past 2 decades. Prior to the 1960s these diseases were much less common, and even those rates could have been further reduced with more rigorous public health measures.

As matters now stand, there are more than ten million cases of STD a year in the United States. Younger people are particularly vulnerable; the preponderance of cases are in the 15 to 29 age group (figures 7.2 and 7.4). It is estimated that half of the country's youth now contracts a sexually transmitted disease by age 25 (which does not mean that every other person has or will get VD; in some groups almost everyone, in others almost none will have it.

---

[28]In the Kinsey sample there was one man who had had only one orgasm in 20 years; another had averaged 20 orgasms a week for the same period. Both were ostensibly equally healthy (Kinsey, 1948).

[29]For the effects of sex on various ailments, see Kolodny et al. (1979). Specifically with regard to heart disease, see Brenton (1968).

[30]Masters and Johnson (1970).

[31]The term "venereal disease" is ascribed to Jacques de Bethencourt (1527). Its replacement with STD in the professional literature was in part because of the social condemnation that has come to be associated with "venereal disease."

In 1979 there were over one million reported cases of gonorrhea and presumably twice as many unreported cases.[32] Currently, even gonorrhea has been surpassed by genital herpes, which now affects an estimated 20 million Americans. Unlike gonorrhea and syphilis, which are ancient diseases (box 7.3) and which can be treated readily, we know much less about herpes and as yet lack an effective cure for it.

The rate of gonorrhea may be 50 times higher among young, single, inner-city poor than among older people who are married and financially well-off. There is also a direct correlation between the number of sexual partners one has and the chance of getting infected. The soaring rates of STD in the past decade have been clearly linked to the liberalization of premarital sex.[33] The pattern of multiple sexual contacts characteristic of some male homosexuals makes them especially vulnerable: half of all male patients with syphilis name other men as their sexual contacts.[34] Yet it is important to realize that anyone can have a sexually communicable disease irrespective of age, sex, marital status, education, affluence, or social status.[35]

Gonorrhea

Gonorrhea (Greek, "flow of seed") is an infection caused by the bacterium *Neisseria gonorrhoeae.* This microorganism only infects humans and cannot survive for long outside the living conditions (temperature,

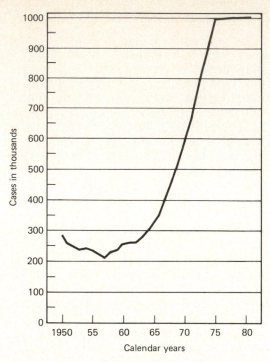

Figure 7.1   Reported cases of gonorrhea, United States, calendar years 1950–1979. (From *STD Fact Sheet,* Edition Thirty-five, HHS Publications, No. 81-1895, p. 28.)

moisture, and so on) provided by the human body.[36]

Gonorrhea is transmitted from one person to another during intimate contact with infected mucous membranes of the genitalia, throat, or rectum. Women run a higher risk of infection following exposure than men: A woman has an 80 to 90% chance of getting gonorrhea after having intercourse once with an infected man; a man in a comparable situation runs a 20 to 25% chance.[37] This is presumably because the penis is exposed to the infectious female discharge only during coitus, whereas the infected ejaculate is deposited deep within the vagina where it remains after withdrawal of the penis.

[32]*STD Fact Sheet* (1981). Although public health codes require physicians to report all cases of venereal disease to public health authorities, in actual practice probably less than one-third are reported (Kolata, 1976).

[33]NIAID Study Group (1980).

[34]Sparling (1982).

[35]Those who call the *VD Hotline* are 83% white, 88% heterosexual, and 26% married. A third of them have bachelor's degrees and a quarter earn over $25,000 a year (Hoffman, 1981). But these are also the sort of people more likely to use such a service and hardly representative of the general population.

[36]There is now evidence that these bacteria can survive for up to 2 hours on toilet seats and wet toilet paper (Gilbaugh and Fuchs, 1979).

[37]Rein (1977).

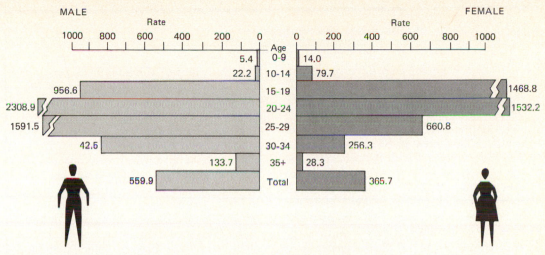

MALE   FEMALE

	Male Rate	Age	Female Rate	
	5.4	0-9	14.0	
	22.2	10-14	79.7	
956.6		15-19		1468.8
2308.9		20-24		1532.2
1591.5		25-29		660.8
	42.5	30-34	256.3	
	133.7	35+	28.3	
	559.9	Total	365.7	

Figure 7.2   Gonorrhea: Age-specific case rates (per 100,000 population) by sex, United States, calendar year 1979. (From *STD Fact Sheet*, Edition Thirty-five, HHS Publications, No. 81-1895, p. 30.)

The diagnosis of gonorrhea is suggested by the symptoms and signs of the illness. But it must be confirmed by identification of the causative organism in the genital discharges. If microscopic study is non-conclusive, then the organism must be cultured (grown on special nutrients) and then its presence confirmed.

Symptoms in Men   In males the primary symptom of gonorrhea, also known as the *clap* (from "clapoir," French for brothel) is a purulent, yellowish urethral discharge.[38] The usual site of infection is the urethra, resulting in *gonorrheal urethritis.* Most infected males have symptoms, but 10% are asymptomatic (yet infectious). A discharge from the tip of the penis appears usually within 2 to 10 days after infection. It is often accompanied by burning during urination and a sensation of itching within the urethra.

The inflammation may subside within two or three weeks without treatment, or it may persist in chronic form, in which case it tends to spread to the rest of the genitourinary tract to involve the prostate gland, seminal vesicles, bladder, and kidneys. In rare cases, the disease spreads to the joints of the knees, ankles, wrists, or elbows, causing *gonorrheal arthritis,* a very painful condition.

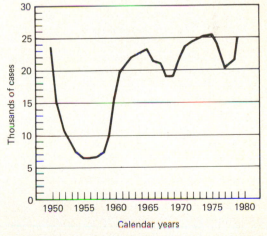

Figure 7.3   Primary and secondary syphilis, reported cases in the United States, calendar years 1950–1979. (From *STD Fact Sheet*, Edition Thirty-five, HHS Publications, No. 81-1895, p. 32.)

[38]Detailed discussion of these disease entities can be found in standard medical texts such as Wyngaarden and Smith (1982); and in more specialized sources and literature reviews as that of Olin (1981).

Box 7.3   History of Gonorrhea and Syphilis

Ancient Chinese and Egyptian manuscripts refer to a contagious urethral discharge that was probably gonorrhea. The ancient Jews and Greeks thought that the discharge represented an involuntary loss of semen. The Greek physician Galen (A.D. 130–201) is credited with having coined the term *gonorrhea* from the Greek words for "seed" and "to flow." For centuries gonorrhea and syphilis were believed to be the same disease, but by the 19th century it had been demonstrated that they were two separate entities. In 1879 A. L. S. Neisser identified the bacterium that causes gonorrhea (which now bears his name). Not until 1905 was the microorganism that causes syphilis identified by Fritz Richard Schaudinn, who named it *Spirochaeta pallidum*, because of its "slender, very pale, corkscrew-like" appearance.

Some believe that syphilis was present in the Old World since ancient times but became more virulent in the late 15th century. But it is more commonly conjectured that syphilis was brought to Europe by Columbus's crew since within a few years after Columbus's return in 1493 from his first voyage to the West Indies, epidemics of syphilis spread across Europe with devastating effects. (Columbus himself died in 1506 with symptoms of advanced syphilis.)

It is further assumed that the Spaniards introduced the disease to the Italians while fighting beside the troops of Alfonso II of Naples. Then in 1495 an army of mercenaries fighting for Charles VIII of France conquered Naples. As they returned home through France, Germany, Switzerland, Austria, and England, they spread the disease along their route. By 1496 syphilis was rampant in Paris, leading to the passage of strict laws banishing from the city anyone suffering from it. In 1497 all syphilitics in Edinburgh were banished to an island near Leith. In 1498 Vasco de Gama and his Portuguese crew carried the disease to India, and from there it spread to China; the first epidemic in that country was reported in 1505.

The New World origin of syphilis is also suggested by the discovery of syphilitic lesions in the bones of Indians from the pre-Columbian period in the Americas. There are no comparable findings in the bones of ancient Egyptians, nor are there any clear descriptions of the disease in the medical literature of the Old World before Columbus.

In the early 16th century the Spaniards called syphilis the "disease of Espanola" (present-day Haiti). The Italians called it "the Spanish disease"; and the French called it "The Neapolitan disease." As it spread to many countries it acquired the name *morbus Gallicus*, the "French sickness," a name that persisted for about a century. The term *syphilis* was introduced in 1530 by the Italian physician Girolamo Fracastoro, who wrote a poem in Latin about a shepherd boy named Syphilius (from the Greek for "crippled" or "maimed"), who was stricken with the disease as a punishment for having insulted Apollo. As this name did not indict any particular country as its origin, it gradually became accepted as the term for the dread disease.

More than 90% of cases of gonorrhea clear up promptly with **penicillin** treatment. The discharge often disappears within 12 hours, although a thin flow may persist for a few more days in some patients. Since 1958 new strains of gonorrhea have appeared that are resistant to penicillin. Higher doses or alternative antibiotics are required in these cases. There is also a strain that destroys penicillin ("superclap") but can be effectively treated with **spectinomycin.** Such penicillinase-producing strains of gonorrhea account for only a very small proportion of cases.

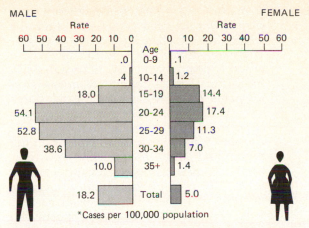

MALE   FEMALE

Figure 7.4 Primary and secondary syphilis, age-specific case rates (per 100,000 population) by sex, United States, calendar year 1979. (From *STD Fact Sheet*, Edition Thirty-five, HHS Publications, No. 81-1895, p. 33.)

Symptoms in Women The symptoms of gonorrhea in 80% of women are mild or altogether absent in the early stages, a major factor leading to complications and the unwitting spread of the disease to others. The primary site of infection is usually the cervix, causing *cervicitis.* The only early symptom may be a yellowish vaginal discharge that may easily escape notice. Microscopic examination and bacterial culture of the discharge are required for definitive diagnosis. There is no routine blood test that can detect gonorrhea in either sex. This is a major reason why identification of asymptomatic gonorrhea has been much less successful than identification of asymptomatic syphilis. Treatment with the same antibiotics as used for men is quite effective if the disease is recognized and treated promptly.

If left untreated, the infection may spread to the uterus, involve the fallopian tubes and other pelvic organs, causing *pelvic inflammatory disease* (PID). Often this occurs during menstruation, when the uterine cavity is more susceptible to gonorrheal invasion. Acute symptoms—severe pelvic pain, abdominal distension and tenderness, vomiting, and fever—may then appear during or just after menstruation. Chronic inflammation of the uterine tubes (*chronic salpingitis*) results in formation of scar tissue that obstructs the tubes; a com-

mon cause of infertility. The disease can also be disseminated to more distant parts of the body, as in the male.

In earlier years, a common cause of blindness in children was gonorrheal infection of the eyes acquired during passage through the mother's infected organs. Instilling penicillin ointment or silver nitrate drops into the eyes of all newborn babies is now compulsory and has eradicated this disease.

Nongenital Gonorrhea Nongenital forms of gonorrhea are most common among male homosexuals. *Pharyngeal gonorrhea* is an infection of the throat that is transmitted most commonly during fellatio (oral stimulation of the penis). Cunnilingus (oral stimulation of the vulva) and kissing do not usually cause pharyngeal gonorrhea. The primary symptom is a sore throat, but there may also be fever and enlarged lymph nodes in the neck. In some cases there are no symptoms, though the person remains contagious.[39]

Rectal gonorrhea is an infection of the rectum usually transmitted during anal intercourse. In women with gonorrhea, the infection sometimes spreads to the rectum through the infected vaginal discharge. The symptoms are itching associated with a

[39]Fiumara (1971); Wiesner (1975).

rectal discharge. Many cases are mild or asymptomatic. Treatment of rectal or pharyngeal gonorrhea is the same as for gonorrheal urethritis.[40]

Syphilis

Syphilis is an infection caused by a cork-screw-shaped microorganism called ***Treponema pallidum*** (also called *spirochete*). It is usually transmitted through intimate sexual contact including kissing, but the spirochete can penetrate through all mucosal surfaces and through minor abrasions in the skin. Syphilis (also called *lues*) is the most serious of the STD, and until the advent of antibiotics posed a formidable medical challenge. Syphilis currently ranks third (after gonorrhea and chicken pox) in the United States among communicable diseases that must be reported.[41]

Primary Stage Syphilis   The clinical course of untreated syphilis is divided into three stages. The first or *primary stage* of syphilis

[40]Schroeter (1972).
[41]*STD Fact Sheet* (1981).

is marked by a skin lesion known as a ***chancre*** (pronounced "shank-er") at the site where the spirochete has entered the body. The chancre is a hard, round ulcer with raised edges, which is usually painless. In the male, it most commonly appears on the penis, the scrotum, or in the pubic area (see figure 7.5). In the female it usually appears on the external genitals (see figure 7.6), but it may be in the vagina or on the cervix and thus not readily apparent. Chancres may also occur on the lips, the mouth, the rectum, the nipple, a fingertip; in short, anywhere the organism has entered the body. The chancre appears 2 to 4 weeks after infection and, if not treated, usually disappears in several weeks, giving the false impression that the individual has recovered.

The presence of syphilis is strongly suggested by *blood tests*.[42] But its definitive

[42]Serological tests, like the *VDRL*, or the *fluorescent treponemal antibody-absorption* (FTA-ABS) provide effective screening to detect syphilis by identifying antibodies produced against the spirochete. But there may be *false-positive* results in the absence of the disease or *false-negative* results in its presence.

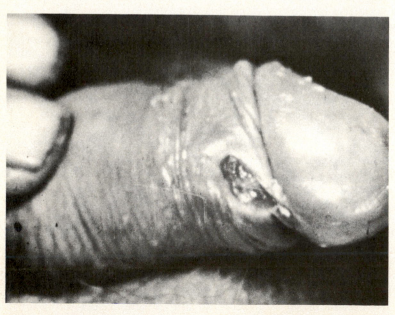

Figure 7.5   Chancre of the penis. Note the raised, hard appearance of the ulcer. (From Dodson and Hill, *Synopsis of Genitourinary Disease*, 7th ed. St. Louis: Mosby 1962, p. 201.)

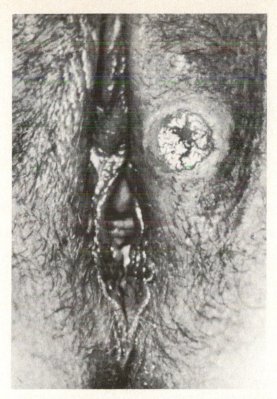

Figure 7.6 A large chancre on the labia majora. (Primary syphilis in the female is not usually this obvious.) From Weiss and Joseph, *Syphilis*. Baltimore: Williams & Wilkins, 1951, p. 73.)

diagnosis is made of identifying spirochetes from within the lesions through the special techniques of *darkfield microscopy*. Treatment with penicillin or other antibiotics at this stage cures most cases promptly.

Secondary Stage Syphilis When syphilis is untreated, the *secondary stage* becomes manifest anywhere from several weeks to several months after healing of the chancre. There is usually a generalized skin rash, which is transient and may or may not be accompanied by such symptoms as headache, fever, indigestion, sore throat, and muscle or joint pain. Many people do not associate these symptoms with the primary chancre as being part of the same disease.

Latency Phase The first two stages end within a year, during which the person is highly infectious. Following stage two, there is a period of *latency* that may last 2 to 20 years. During this period the person has no symptoms and is not infectious. But the spirochetes continue to cause internal damage by burrowing into blood vessels, bones, and the central nervous system. The resultant symptoms constitute the third stage.

Tertiary Stage Syphilis About 50% of untreated cases reach the final or *tertiary stage* of syphilis, in which heart failure, ruptured blood vessels, loss of muscular control, disturbances in the sense of balance, blindness, deafness, and severe mental disturbances occur: mainly affected are the cardiovascular and central nervous systems. Ultimately the disease can be fatal, but treatment with penicillin even at late stages may be beneficial, depending on the extent to which vital organs have already been damaged.

Congenital Syphilis Syphilis can be transmitted to the fetus through the placenta from the infected mother, hence the need for mandatory blood tests to identify untreated cases before marriage and before the birth of a child. Nine out of ten pregnant women who have untreated syphilis either miscarry, bear stillborn children, or give birth to living children with congenital syphilis. Treatment with penicillin during the first half of pregnancy can prevent congenital syphilis in the child.

Herpes

Genital herpes is manifested by painful blisters and sores on or around the sex organs. It is caused by a virus (*Herpes Simplex Virus Type 2* or HSV-2) which belongs to a family of viruses that infect humans more than any other viral group. A similar virus (*Herpes Simplex Virus Type 1*; HSV-1) causes *oral herpes* (also called "cold sores") with lesions on the lips, mouth, or face. Occasionally, HSV-1 causes genital herpes, and HSV-2 may lead to oral and labial lesions.

These usually occur through mouth-genital contact in oral sex. The genitals and lips are not the only vulnerable parts. One may get herpes on any body surface by touching a lesion (including one's own).

Prevalence and Symptoms   Though known for some time, genital herpes did not become highly prevalent in the United States until between 1966 and 1979.[43] The number of cases seen by physicians increased almost 100-fold during this period and an estimated one in five American adults now has genital herpes. Each year, another half-million cases are added to the 20 or so million already infected, making herpes a major focus of medical and public concern.

Herpes is typically transmitted by contact with infected areas. In the case of genital herpes, such contact is most often sexual. This includes coitus, mouth-genital contact, anal intercourse, and any other form of physical intimacy that can bring one into contact with herpes lesions. All sexual activity must therefore be avoided during the active phase; for two weeks after the blisters have healed; and when there are early signs of a recurrence, usually manifested by itching, burning, and tingling sensation at the site where blisters are to appear.

The chances of infection are very high during these active periods when the virus is being shed. Risk of infection is lower during recurrences, and believed to be absent during asymptomatic periods.[44] Hence, the fact that a person has had herpes does not mean that he or she can never again be a safe sexual partner. When in doubt, the use of condoms, and spermicides, followed by careful washing after coitus can help to further minimize the risk.

The herpes lesions consist of small, fluid-filled blisters surrounded by inflamed tissue (figure 7.7), which usually appear 2 to 20 days after exposure to the virus. The most common sites of herpes infections in women are the surface of the cervix, the

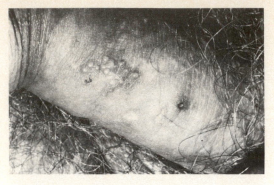

Figure 7.7   Herpes lesion on penis. (Courtesy Dr. Eugene Farber.)

clitoral prepuce, the major and minor lips; in males, the foreskin and glans. When the blisters are internal, such as on the cervix, the condition may go undetected. In some instances, males may also be asymptomatic (yet infectious).

Herpes blisters cause painful sensations of burning and itching. When they break open, the area may become infected with bacteria. Otherwise the blisters clear up spontaneously after a few weeks or sooner with treatment. However, once the virus has gotten into the body, it retreats along nerve fibers into nerve clusters near the sacral spinal cord (or nerve cells lying in the cheek, in the case of oral herpes) where the virus hides from the body's immune system. Henceforth, it may cause no further symptoms or may cause recurrences as often as twice a month or as rarely as once a decade. The diagnosis of herpes is suggested by the clinical picture but can only be confirmed by culturing and identifying the herpes virus.

The Role of Triggers   Recurrence of the herpes is linked up to a number of *triggers* that appear to lower the body's resistance to the virus. Trauma, marked physical exertion, prolonged exposure to the sun (attacks are more common during the summer), smoking, certain foods (such as nuts, seeds, and onions), debilitating diseases, menstruation, stress, and sexual activity

[43]*Morbidity and Mortality Weekly Report* (1982).
[44]Judson (1983).

(apart from reinfection) seem to lead to flare-ups of the condition.[45]

These triggers are also deemed important in determining whether or not one will manifest the disease in the first place. Since the herpes virus exists in the bodies of many more people than those who have been exposed to it sexually, in at least some cases the disease may be precipitated by a combination of the above triggers without sexual contact with an infected person. Hence, not all individuals with genital herpes (and few of those with oral herpes) can be said to be suffering from a sexually transmitted disease.

It is possible that the disease may also be transmitted through accidental contamination, since the virus has been shown to survive for short periods outside the body such as on infected drinking glasses, washbasins, and toilet seats. But there is no evidence such possibilities represent significant risks in practical terms. Hence, there is no cause for concern that we are wallowing in an environment contaminated with herpes.

Complications   The complications of herpes are rare but may be quite serious. Transferring the virus into the eyes after touching an infected area may cause *herpes keratitis* and possible damage to the eye. Herpes on the lips may in 5 to 8% of cases lead to *viral encephalitis.*

The virus can also be transferred to babies during birth if the mother has active lesions (*neonatal herpes*) a condition that affects 100 to 200 babies a year. A pregnant woman who has an active outbreak at the time of delivery runs a 50% risk that her newborn child will contract the disease and may be seriously damaged or die. To avoid this the baby is delivered by cesarean section. Fondling and kissing of infants by persons with active oral lesions create similar risks. Finally there is the possibility that HSV-2 increases the risk of cervical cancer five to eight times; so it is recommended that women who have herpes get Pap smear tests every 6 months.

Treatment   Though people have resorted to all sorts of remedies, the only proved benefits for genital herpes have been achieved with **acyclovir,** which is commonly used locally as an ointment but can also be taken orally or intravenously. It is thought to work by interrupting the multiplication of the herpes virus. If used in the initial period, this drug can alleviate symptoms and shorten the duration of the attack, but it will not cure the disease. **Lithium succinate,** a nonprescription ointment available at many health food stores, appears to have an effect similar to acyclovir's.

Bathing with soap and water and the use of soothing agents may also give some relief. Personal hygiene is very important since one can reinfect oneself by touching the sores and other parts of the body. Drugs that can prevent recurrences and vaccines that can prevent acquiring the disease are actively being sought.

Herpes is a benign condition, not a serious threat to health in most cases, unpleasant and painful as it may be. Yet psychological reactions and fears to having the disease may be quite severe. Even people with no reasonable cause to worry, fret about herpes or become obsessed with it. Its threat is reportedly having widespread effects on patterns of sexual behavior, specially among singles.[46]

Acquired Immune Deficiency Syndrome (AIDS)

In early 1981 it was noted in several medical centers that small groups of sexually active gay young men were suffering from an unusual and deadly form of pneumonia (*pneumocystis carinii*) and a rare, lethal skin cancer (*Kaposi's sarcoma*) usually seen only in older men.

After the connection was established between these conditions and a homosexual life style with multiple partners, a new group—mostly heterosexual drug abusers—appeared to be coming down with AIDS (*Acquired Immune Deficiency Syn-*

[45]Wickett (1982).

[46]*Time,* August 2, 1982.

drome). This was followed by the identification among AIDS victims of several hemophiliacs (who receive many blood transfusions), other transfusion recipients, female partners of drug users, and a few children of those with the disease. Especially baffling was a similar vulnerability among immigrants from Haiti.[47]

Male homosexuals with multiple sexual contacts have continued to constitute the preponderance (over 70%) of reported cases. Of these, 2,640 had been reported by late 1983; 1,092 of whom had died.[48] Reports of new cases officially diagnosed as AIDS are still doubling every 6 months. The reported cases are said to be the tip of the iceberg.

Though the cause of AIDS remains a mystery, its primary manifestations have become fairly well established. AIDS is not a discrete disease entity but a pathological condition of the body's immune system that predisposes its victims to a variety of other ailments. In the early stages the common symptoms are recurrent fever, weight loss, severe fatigue, diarrhea, chronic cough, and swelling of lymph nodes (such symptoms are also caused, of course, by many other ailments). There is a tendency to get sick more frequently, more severely, and for longer periods. The full-blown symptoms of pneumocystis carinii pneumonia or Kaposi's sarcoma then follow. Some persons may develop the preliminary symptoms and enlarged lymph nodes but no further symptoms, thus representing perhaps a milder variant of AIDS.[49]

Two theories have emerged to explain AIDS. One is that the condition is caused by a specific agent, probably a virus. This agent is then transmitted whenever body fluids are exchanged. Sexual transmission occurs through infected semen entering the mouth, anus, or vagina of another person; but women do not seem to pass it to others. Transmission through blood takes place when infected needles are shared in drug abuse and blood from an AIDS case is given to another.

The alternative theory is that the immune system of AIDS victims is overpowered by exposure to repeated infections. Sex with multiple partners among male homosexuals and the living conditions of drug addicts exposes them to repeated and multiple infectious agents; after battling against these endlessly, the immune system presumably simply gives up.

The mystery, ominous outcome, and sexual associations of AIDS, along with the threat that it may engulf larger segments of the population, have brought about strong public reactions. Health agencies have mounted a major effort to track down its causes and find means of curing and preventing the condition. There is optimism that these tasks will be accomplished; but because of the apparently long incubation period of the illness, many more cases will develop even if a preventative measure were found right away.

Public interest in AIDS has been reflected in and fanned by the attention paid to it in the media.[50] This preoccupation has led to the development of a plague mentality, with all sorts of irrational fears and misperceptions.[51] Its association with homosexuality has further led to the issue becoming politicized.[52] Though words like "epidemic" are often bandied about, AIDS is not highly contagious. It requires intimate and repeated exposure to the body fluids of an infected person. One does not get AIDS by touching its victims or being in the same room with them; even having sex will not necessarily lead to infection. Otherwise, there would be many more cases than exist.

Nonetheless, the threat of AIDS is very real to gay men with multiple sexual partners. There are reports of significant

[47]It has been found that a quarter of these men may have worked as gay prostitutes. Nonetheless, the Haitian connection remains puzzling. There is no evidence that AIDS originated in Haiti (Bazell, 1983).

[48]James Curran (director of federal AIDS task force) quoted in Perlman (1983).

[49]For details see Harris et al. (1983); Small et al. (1983); Gottlieb et al. (1981); Center for Disease Control Task Force (1982).

[50]See, for example, *Time,* July 4, 1983.
[51]Bazell (1983).
[52]Krauthammer (1983).

changes in gay lifestyles; bathhouses and bars that function as meeting places for casual sex are said to be losing their clients. Yet others persist in the very behaviors that are known to lead to higher risks of acquiring AIDS.[53]

Nonspecific Urethritis and Vaginitis

Nonspecific Urethritis   Cases of urethritis that are not due to gonorrhea are designated as *nongonococcal urethritis* (NGU) or *nonspecific urethritis* (NSU). Because NGU was only recently identified as a sexually transmitted disease, and because it is often asymptomatic (especially in women), many cases have gone undiagnosed, resulting in its wider spread. The condition is now recognized as being highly prevalent. With some three million cases a year, it is the most common form of urethritis in men seen by physicians in private practice and in college student health centers. The organism most often responsible is *chlamydia trachomatis.*[54]

In the male, NGU is manifested by symptoms of mild urethritis such as burning on urination and a purulent discharge. In women the condition causes inflammation of the cervix and may become generalized into pelvic inflammatory disease (PID) with potentially serious consequences (such as sterility). NGU can be effectively treated with tetracycline.

Nonspecific Vaginitis   This condition is said to be present when women have vaginitis (inflammation of the vagina) but the organism responsible for the infection cannot be isolated or identified. Nonspecific vaginitis is often caused by chlamydia trachomatis but *hemophilus vaginalis* is another common cause. Because it can often be recovered from the sexual partners of infected women, it is possible that it too is sexually transmitted.

The prevalence of chlamydia infections has increased to epidemic proportions, afflicting perhaps as many as 10 million persons (making it far more common than gonorrhea). In both sexes, it may result in sterility and other ailments. Infants born to mothers who harbor the organism may develop eye infections and pneumonia.

Other Sexually Transmitted Diseases

Sexually transmitted diseases that are rarely encountered in the United States are *chancroid* ("soft chancre"), which is manifested by a chancre, but unlike that of syphilis, this chancre is soft and painful. *Lymphogranuloma venereum* produces enlarged lymph nodes in the groin. *Granuloma inguinale* results in ulcerating skin lesions in the genital area. All three conditions are caused by infectious agents and are treatable with antibiotics.[55] Far more common are less serious infections with *trichomonas, candida,* and other agents which we shall discuss briefly.

Trichomoniasis   Trichomoniasis is a vaginal infection caused by protozoa called *Trichomonas vaginalis.* Close to a million women a year are seen by doctors for this condition.[56] It is characterized by a smelly, foamy, yellowish discharge that irritates the vulva, producing itching and burning.[57]

A man too may harbor the organism in the urethra or prostate gland without symptoms, or he may have a slight urethral discharge. Since sexual partners usually share the infection, it is customary to treat both of them simultaneously with *metronidazole* (*Flagyl*) to prevent reinfection. Though usually transmitted sexually, trichomoniasis occasionally may spread nonsexually by genital contact with the washcloth or wet bathing suit of an infected person.

Candidiasis   Another common vaginal infection is *candidiasis* (or *moniliasis*). It is

---

[53]Shilts (1983). For further details on STD among gay men see Ostrow et al. (1983).
[54]Wong (1983).

[55]For further details, see Sparling (1982).
[56]*STD Fact Sheet* (1981).
[57]Rein and Chapel (1975).

caused by a yeastlike fungus called **Candida albicans.** The thick, white discharge it produces causes itching and discomfort, which may be quite severe. The organism is present in the vagina of a substantial number of women, but it produces problems only when it multiplies excessively.

Candidiasis is more commonly seen in women using oral contraceptives, diabetics, during the course of pregnancy, or with prolonged antibiotic therapy. The condition responds to treatment with nystatin (*Mycostatin*) suppositories or miconazole cream, but it is difficult to completely eradicate it.

Though much less common, the same infection occurs in about 15% of the male sexual partners of women with moniliasis. They usually have no symptoms but there may be marked inflammation of the glans, especially under the foreskin.

Genital Warts   Genital warts (*condylomata acuminata*) are commonly, but not always, transmitted by sexual contact. Their prevalence has been rising rapidly. They are caused by a virus and usually appear within 3 to 8 months of infection. Genital warts most often appear in women on the vulva, less frequently inside the vagina. In males they are usually seen on the surface of the glans, below the rin of the corona, and around the anus. Warts are unsightly and cause itching. They may last from a few weeks to several years. They may be treated surgically, with electrodes, cryosurgery (freezing), or with the direct application of various drugs; none are particularly effective or free from causing scarring.

Pubic Lice   *Pediculosis pubis* ("crabs") is an infestation of pubic hair by lice (*Phtirus pubis*). Transmittal is usually sexual, but occasionally it is acquired from infected bedding, towels, or clothing. The primary symptom is intense itching. Cream, lotion, or shampoo preparations of benzene hexachloride (*Kwell*) are effective in eliminating both adult lice (about the size of a pinhead) and their eggs ("nits") that cling to pubic hair.

Scabies   Scabies is a contagious skin infection that is caused by the itch mite (*Sarcoptes scabei*). The mites provoke intense itching and can be spread by close personal contact, sexual contact, and exposure to infested clothing or bedding. Scabies is commonly found in the genital areas and buttocks as well as between the fingers. The female itch mite burrows into the skin and lays eggs along the burrow. The larvae hatch within a few days. The mites are more active and cause the most itching at night. The treatment is the same as for pubic lice.

Conditions Associated with Anal Sex   The rectum is liable to the same infections as the urethra and vagina. But there are other conditions that are more specific to sexual activities involving the anus such as anal intercourse and oral-anal contact. As a result, a number of diseases are transmitted sexually even though these are not venereal diseases as traditionally defined.

Men who have coitus following anal intercourse without carefully washing the penis in between, risk infecting the vagina with fecal organisms. Transmission of such microorganisms can also occur in oral-anal sex or indirectly when hands touch the anus, or the fecally contaminated penis, and then the mouth.

Since anal sex is far more frequently engaged in during homosexual activities, such conditions are typically seen among gay men. The more important among these diseases are *viral hepatitis*, enteric infections like *amebiasis, shigellosis, giardiasis;* and various urethral and prostatic infections by organisms that commonly occur in the rectum (like *E. coli*).

The rectum is also vulnerable to injury during anal intercourse or through the insertion of objects during anal masturbation. Because of the "pull" of the anal sphincter, objects (some as large as a Coke bottle) that are inserted in the anus can be inadvertently drawn into the rectum. Repeated stretching of the anal sphincter may also lead to fecal incontinence.[58]

[58]For further health problems associated with homosexual practices, see Rowan and Gillette (1978).

Prevention of Sexually Transmitted Diseases

Within two decades of the establishment of the People's Republic of China, sexually transmitted diseases were virtually wiped out among the country's enormous population. This was accomplished by suppressing prostitution, treating all known cases and contacts with STD, and public educational campaigns. Western democracies do not exercise comparable levels of social control over their citizenry to hope to replicate this outcome. On the other hand, a good deal more could be done if our approach to sexually transmitted diseases were not rooted in "ignorance, apathy, and neglect."[59]

Moral considerations form a crucial part of sexuality (chapter 20), but they are best kept out of the realm of illness. There are no moral or immoral diseases: Whether an act is moral or immoral must be determined on ethical, not medical, grounds. Illness is not a punishment for sin, nor health a reward for virtue. To claim otherwise makes a mockery out of human suffering. The one ethical concern that is legitimate in this regard is in acting responsibly so one does not hurt someone else's health and safeguards one's own.

If a person has the slightest suspicion of having a STD, he or she should not engage in sex—be it with spouse, friend, or prostitute. At the very least, the sexual partner must be forewarned so that he or she can make an informed decision, and, having done so, watch for its possible consequences and seek treatment should the need arise. Once a case of STD has been identified, all sexual contacts must be promptly checked and treated if necessary.

There are several ways to help protect oneself. Knowing the sexual partner well is reassuring, but no guarantee for safety. Sophisticated prostitutes examine their clients (including "milking" the penis to check for discharge). If one cannot go that far, it may still be possible to undertake a

Figure 7.8   Swedish condom symbol used in public education campaign.

reasonably careful genital inspection during foreplay and look for at least telltale signs of infection.

A condom is the single most practical device to avoid infection. It will give a man more protection against some conditions (like gonorrhea) than others (like syphilis or herpes). Equally important, it will protect the woman should the man be infectious. Increased use of condoms is held responsible for the 50% drop of gonorrhea rates in Sweden between 1969 and 1976.[60] This was done through public education campaigns (figure 7.8) and by making condoms more readily available in public locations. Despite the blatant exploitation of sex in advertising in the United States, our sense of public decency still inhibits wide media exposure of condom use.

The use of spermicidal substances (chapter 6) gives a woman some additional protection against conditions like gonor-

[59]NIAID Study Group (1980, p. 10).   [60]Kolata (1976).

rhea. Since these devices also function as contraceptives, they provide an additional safeguard and a tactful excuse for their use without raising the issue of infection. The suggestion or request to use condoms or foam should be made by the woman as freely as by the man.

Thorough washing of the genitals with bactericidal or even plain soap as soon after sexual exposure as possible, is another measure that may help. Even simpler is urinating in forceful spurts to wash out the urethra. Prophylactic administration of penicillin or antibiotics is used when people are likely to be exposed to high risk (like soldiers going on leave in high-risk areas). If all else fails, vigilance for early signs is the key to getting prompt treatment.

Admittedly, all such preventive measures are a nuisance; they are embarrassing, distasteful, and antithetical to all that is erotic and romantic, but they are well worth it, when circumstances warrant them. Men and women in truly monogamous relationships who start off healthy have no need for such measures. The extent to which they are called for in other circumstances is a matter of individual judgment. Much of what was said about contracep-tive risk taking also applies here. The level of risk determines the right course of action with regard to the stringency of preventive measures called for. The issue of infection is but one of the variables that must be taken into consideration before making a sexual decision. To repeat a phrase we used in discussing contraception, there are many good and many bad reasons to engage in sex, but there are only bad reasons for contracting or transmitting a sexually communicable disease.[61]

SUGGESTED READINGS

Holt L. H. and Weber M. *Woman Care* (1982); N. Lauersen *It's Your Body* (1977). Both discuss conditions affecting the sexual health of women.

Budoff P. W. *No More Menstrual Cramps and Other Good News*; L. Rose, ed. *The Menopause Book* (1977). These books are more specific to the subjects of their titles.

Silber S. S. *The Male* (1981); R. Rowan and P. J. Gillette *The Gay Health Guide* (1978). For conditions affecting men.

[61]One of the simplest ways of getting information and advice on venereal disease is to call the VD National Hotline, which can be done anonymously and toll free. The number is 800-982-5883.

Part II

Behavior

Chapter 8
Sexual Behavior

Sex isn't the best thing in the world, or the worst thing in the world—but there's nothing else quite like it.

W. C. Fields

Beginning with this chapter, we shift our emphasis from the biological to the psychological and social aspects of human sexuality. We must bear in mind that what is shifting is the perspective of study, not its subject; it is not a different kind of sex that we are discussing, but another facet of the same entity. All of these facets of sex are closely integrated: Without biological underpinnings, there would be no sexual behavior; by the same token, sexual behavior looked at in social isolation would make no sense.

The sexual machinery in our bodies makes sexual activity possible, but such activity is only given meaning as human experience in its psychological and social contexts. We see genital organs like the penis and vagina not only for what they are in a *naturalistic* sense but for what they represent in a *symbolic* sense. They are imbued with feelings of power, dominance, beauty, and pride as well as shame, guilt, and ugliness. These feelings are determined by our culture, which generates and manipulates symbols through language and imagery. A relentlessly biological perspective reduces us to reproductive machines; an insistently sociocultural approach transforms us into disembodied abstractions.

The Nature of Sexual Behavior

If asked, what is sexual behavior, we are likely to think of one or another form of sexual activity, usually sexual intercourse (figure 8.1). If pressed further as to what makes these behaviors sexual, we may point to their reproductive consequences, pleasurable aspects, and the like, but the essence of the question is likely to elude us; or at least, we may find it hard to communicate the answer to others even though we feel that we know what it is.

One of the reasons why definitions of sexual behavior seem so elusive is because its *essence* is hard to separate from its *associations*. Sexuality touches virtually every aspect of life at one level or another. But that does not entitle us to equate sex with all of life, anymore than it allows us to relegate it to some dark corner of life.

Components of Sexual Behavior

Of the myriad associations of sex, four areas are especially important: *reproduction, pleasure, gender,* and *affection.* Sexual behavior overlaps with all four but cannot be equated with any one; conversely there is more to each than its association with sex.

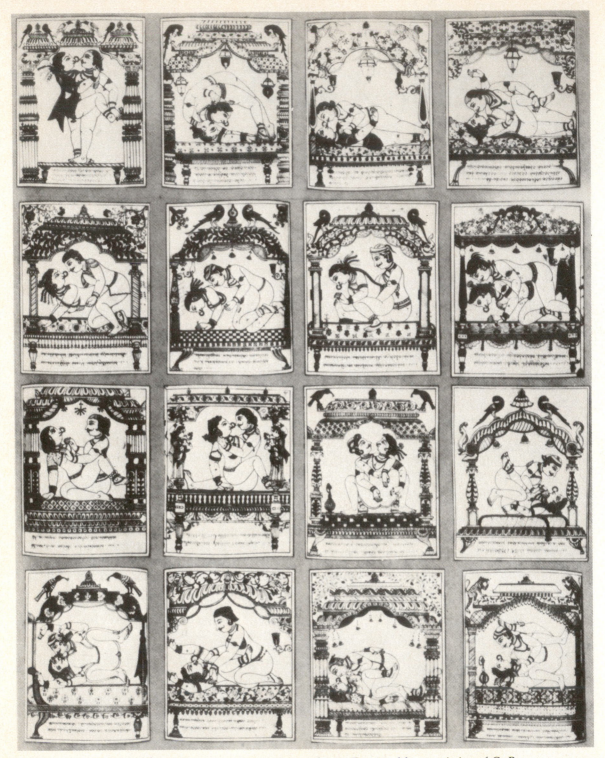

Figure 8.1 Pages from a 19th-century Orissan posture-book. (Reprinted by permission of G. P. Putnam's Sons and Weidenfeld and Nicholson Ltd., from *Erotic Art of the East*, Philip Rawson, ed., p. 43.)

Definitions of sexual behavior flounder if we substitute *normative* and ethical criteria (such as, "love and sex should not be separated") for *descriptive* ones. It is important that we not confuse what sex is in terms of how people generally behave or with how they ought to behave, though that is of great importance in its own right.

A parsimonious definition of sexual behavior should not be encumbered by its common associations. Reproduction and sex are closely linked, yet a great deal of sex is engaged in without reproductive intent or outcome. Reproduction itself is now possible without sex: an artificially inseminated woman has not engaged in sexual behavior, in any reasonable sense of the term.

Gender is defined by characteristics that are linked to being male or female. Gender differences are of great importance in shaping sexual behavior, but gender behavior, as such, is not sexual behavior.[1] Similarly, emotions like love are of crucial significance to sex (chapter 12), but a great deal of sexual activity has nothing to do with such emotions; conversely many forms of love are not erotic. Finally, while sensual pleasure is a key component of sex, not all sexual experiences are pleasurable nor are all pleasurable experiences erotic.

If reproduction, gender, pleasure, and affection are stripped away, then what is left of sex? Does sex have an independent existence apart from these components? If so, what does that consist of? There are no clear answers to these questions, but we shall assume that sex does constitute a separate entity and we shall define sexual behavior as *erotic behavior,* that is, any activity that is sexually arousing.[2]

What defines a behavior as erotic is a combination of *subjective* effects experienced by the individual and *objective* manifestations linked to that experience. Hence, to qualify as erotic, a given experience must have the following components: a set of unique physiological responses that become manifest during sexual arousal (chapter 3) and distinctive, subjective experiences commonly designated as erotic pleasure. In this scheme a fleeting erotic fantasy without objective signs of arousal would not qualify as sexual behavior since it would lack the necessary physiological punch; by the same token physiological signs of arousal, such as reflexively caused erections in the newborn, would not count as erotic because of the absence of (or lack of evidence for) the subjective component of sexual arousal.

We shall settle on this definition being aware that in one sense it is too narrow, in another, too broad. By imposing a physiological criterion, it excludes from the realm of the erotic all manner of sexual thoughts and imagery that would seem manifestly sexual; by settling for physiological reactions short of orgasm, it makes sexual behavior so ubiquitous and global that it virtually precludes any quantification of the experience.

All behavior, including sexual behavior, has several main characteristics.[3] First, it is integrated and *indivisible.* When we describe behavior as determined by factors that are conscious or unconscious, innate or learned, we are referring to different aspects of the same behavior, not to different behaviors. Second, all behavior expresses the actions of the *total organism,* the personality as a whole. Third, all behavior is part of a lifelong *developmental sequence* and can be understood only as links in a chain of events. Finally, all behavior is *multidetermined* or shaped by multiple forces; each and every act has biological, psychological, and social determinants. We never think, feel, or do anything for a single reason.

The purpose in pointing out all these ambiguities in meaning is not to induce a sense of despair over our ever being able to define what sexual behavior is. Rather, it is to guard against the all too common tendency to discuss sexual behavior in simplistic terms. The ignorance, uneasi-

---

[1]Maccoby and Jacklin (1974).

[2]The term is derived from *Eros,* the Greek god of love, but what we understand by it is different from the Greek conception of it (see chapter 17).

[3]Engle (1962, p. 4).

ness, and bias that infuse the subject of sex is compounded by the simplemindedness of many who write on the subject (including the authors of textbooks).

Varieties of Sexual Behavior

If variety is the spice of life, then sexual behavior is spicy indeed. Throughout history people have resorted to every imaginable type of activity in pursuit of sexual gratification. They have coupled with members of the opposite and the same sex, with animals and inanimate objects. They have had sex in solitude, in pairs, and in groups. Every orifice that can possibly be penetrated has been; every inch of skin has been caressed, scratched, tickled, pinched, licked, and bitten. Orgasms have come amid murmurs of tender affection and screams of pain, when fully awake and while fast asleep.

Some of the observed differences in sexual behavior between people are undoubtedly real. Variation is a cardinal rule of all biological phenomena, and cultures are inevitably diverse. Other presumed differences in sexual behavior are based on ignorance and concealment. If we could be open about what we do in our sexual lives, a great many apparent differences would be revealed to be considerably less sharp.

Finally, the magnitude of these differences may be a function of the particular methods of observation and interpretation used. Among psychoanalysts, for example, what is considered of erotic significance is much broader than among psychologists or sociologists. When investigators use different methods of study, disparities emerge independent of whatever actual variation exists in the groups under study.

Even when defined fairly restrictively, a large number of behaviors are commonly considered sexual. To impose some order on such variety, it is useful to subsume sexual behaviors into various categories. One such division is based on the *direction* of the sexual drive; whether it is aimed "inward" or "outward." Another distinc-

tion is based on the *magnitude* or amount of sexual behavior.[4]

Direction of Sexual Behavior   When the sexual drive is directed inward, or at the self, the sexual behavior is called *autoerotic* ("auto" from Greek for "self"). The key autoerotic activities are: *sexual fantasy, nocturnal emission,* and *masturbation* (chapter 10).

When the sexual drive is directed outward, or at others, the sexual behavior is considered *sociosexual* (a term coined by Kinsey). It includes a variety of sexual interactions, all of which depend upon some confluence of the capacities, interests, and desires of individuals and on the willingness of each to adjust to the other."[5]

The distinction between autoerotic and sociosexual behaviors has a certain face validity but becomes untenable if pressed too far. For instance, what would we call the behavior of a group of adolescents masturbating together but not touching each other; or that of a person engaged in coitus yet so absorbed in fantasies as to be virtually oblivious to the partner? Thus it is more realistic to think of all sexual behaviors as having a combination of "inner" and "outer" components.[6]

The concept of direction of sexual drive was central to Freud's categorization of sexual behaviors. It has two components: choice of *sexual object* (the person or entity one finds sexually attractive); choice of *sexual aim* (what one wishes to do with the sexual object).[7]

Freud's paradigm for mature sexual behavior is *coitus,* where the object is an adult of the opposite sex (who is not a close relative) and the aim is genital intercourse.

---

[4]Katchadourian and Martin (1979) discuss this approach in greater detail.

[5]Kinsey et al. (1953, p. 250).

[6]Harry Stack Sullivan (1953) used a continuum based on "primary need" to gauge the level of involvement between the person and the partner.

[7]"Object" in this context means an individual who fulfills essential functions in the gratification of the needs of others (see Engle, 1962, p. 17). It has nothing to do with what is conveyed in expressions like, "You are treating me like an object."

Departures from this entail choices of other objects or means. Alternative object choices could be an adult of the same sex (*homosexuality*); a child (*pedophilia*); a close relative (*incest*); an animal (*zoophilia*); an inanimate object (*fetishism*); or a dead person (*necrophilia*). Alternative aims would entail the preference to watch others in erotic contexts (*voyeurism*); to expose oneself (*exhibitionism*); to inflict pain (*sadism*); and to suffer pain (*masochism*), for erotic gratification.[8]

Currently the term **sexual orientation** is used to differentiate between heterosexual and homosexual object choices. **Sexual preference** is a broader term to indicate the particular mode of sexual gratification sought by a person through various choices of objects and means.

Despite the wide variety involved in these sexual activities, we are in effect dealing not with so many discrete behaviors but with different methods of invoking sexual arousal. The differentiating feature of these various behaviors is in the substitution of an unusual sexual stimulus for the more usual ones. Otherwise, once past the point of sexual arousal, most sexual activities culminate in masturbation or coitus.

Magnitude of Sexual Behavior   The concept of magnitude of sexual experience was central to Kinsey's attempts at quantifying sexual behavior. Just as direction of impulse tells us *how* a person wants to behave sexually, magnitude reveals *how much* of a behavior actually occurs.

Kinsey used *orgasm* as the unit to measure sexual experience, since it is an unmistakable sexual event and dramatic enough to be remembered and counted. To estimate how much sex a person was having over a period of time Kinsey added orgasms from whatever source and considered it as the person's "total sexual outlet." Such an approach would seem to reduce sex to its lowest common physiological denominator, hence robbing it of its subtlety and emotional attributes. These attempts at

---

[8]Freud's formulation of these issues appeared first in 1905 in his *Three Essays on the Theory of Sexuality*.

---

making assessments of sexual behavior do have their shortcomings, but they are nonetheless useful in providing us with a fuller picture of the sexual lives of groups and individuals.

Judgments of Sexual Behavior

What we discussed so far pertains to attempts at classifying sexual behavior in *descriptive* terms. More problematic are distinctions made on *judgmental* grounds. Such judgments are commonly made on the basis of four criteria, or sets of norms: the **statistical** (How common is the behavior?); the **medical** (Is it healthy?); the **moral** (Is it ethical?); and the **legal** (Is it legitimate?). In chapter 16, we deal with the societal aspects of such judgments; at this point, we are concerned mainly with assessments of the normality of sexual behaviors from the perspective of psychological health.

The Problem of Deviance   The core issue that all judgments of behavior deal with is that of **deviance** (from the Latin, "away from the road")—departures from the established *norms* or socially accepted standards of behavior. Unlike illnesses that may affect the sexual organs (such as syphilis) or sexual functions (such as impotence), problematic sexual behaviors are harder to define.

Some behaviors are socially intolerable on grounds other than the nature of the sexual activity itself. The best example is rape: no one argues that coitus as such is pathological, but to force someone to have sex is clearly unacceptable. The same argument would apply to any behavior that exploits and victimizes others, such as in pedophilia and exhibitionism. Yet in other behaviors, like homosexuality or fetishism, there is no victim, yet these activities have often been deemed socially offensive nonetheless.

When judgments of sexual behavior were made primarily on religious grounds, those with aberrant sexual behaviors were considered *sinners*. With the secularization of society, they came to be seen by the law as a class of criminals called *sex offenders*. Emphasis then shifted from the ethical and legal to the medical approach, whereby sexual deviants came to be viewed as *sick;* they could not help being what they were because of organic, psychological, or social causes. Hence what was called for was not damnation or punishment but treatment.

In current psychiatric classification, behaviors that were labeled *perversions* and then as *sexual deviations* are now called **paraphilias** and are subsumed under *psychosexual disorders* (chapter 14). Paraphilia is a neutral descriptive term that merely states that what the person likes ("philia") is other than, or "beside," the usual ("para"). The behaviors included under this category are: *fetishism, transvestism, zoophilia, pedophilia, exhibitionism, voyeurism, sexual masochism,* and *sexual sadism.*[9] In a major departure from past practice, *homosexuality* was excluded, in the mid-1970s, from this group of behaviors.[10]

Diagnostic judgments in these areas are problematic on conceptual as well as practical grounds. It is argued that where there are no complaints by individuals engaging in a given behavior, no evidence of resultant disability, and no victim, ostensibly medical judgments may become fronts for perpetrating dominant societal values and prejudices; what is called "sick" may merely represent unconventional behavior that offends social sensibilities. Therefore, there is a division between those who use an *illness model* and others who view individuals who engage in such sexual activities as following *alternative sexual life styles,* or belonging to *sexual minorities.*[11]

Difficult as they may be, judgments of sexual behavior are just as appropriate as are judgments of other forms of human activity. Not all sexual behaviors are equally adaptive, healthy, socially desirable, moral, and so on. Such judgments, however, ought to be made with great care. Experience shows how arbitrary they can be and how easy it is to inflict unnecessary hardship on people whose main offense may consist merely in being different.

Methods of Sex Research

In chapter 1 we reviewed the various disciplinary approaches to the study of sexuality and the historical background of sex research. We shall now examine in more detail the various research methods that are used to study sexual behavior.

Methodological Considerations

The study of an entity as complex as sexual behavior requires that we approach it with a variety of research methods. Each method has its advantages and shortcomings, but a number of basic considerations apply to all of them. We address these issues first before we turn to the specifics of each method.

The Purpose and Perspective of Study Every study must have a definite purpose; a haphazard collection of data is of little value. There are so many facets and wrinkles to even the simplest sexual act that unless investigators set out with a clear idea of what they are investigating, they are likely to get lost.

This does not mean that one sets out to prove a fixed idea, disregarding whatever information that contradicts it. Instead, the starting point is a *hypothesis,* a tentative assertion that is subject to verification, restatement, or negation. The use of hypotheses helps to focus the investigation and makes it easier to determine its outcome; to decide if the idea tested was right or wrong. There are innumerable studies that report that so many men do this and so many women do that. Although all carefully gathered data are interesting, it is often hard to know what to make of such findings, which stand by themselves.

---

[9]American Psychiatric Association (1980).

[10]Only one of its forms ("ego-dystonic homosexuality") is treated as a category of psychosexual disorder; in these cases the person is unhappy with his or her homosexual inclination and behavior and would rather be heterosexually oriented (chapter 13).

[11]Gagnon (1977) discusses under this heading sadomasochism, animal contacts, fetishisms, transvestism, and transsexualism.

The Problem of Bias   Sex researchers are not detached intellects wandering around looking for the truth. They are members of a profession, heirs to intellectual traditions, and human beings with personal needs and aspirations. They have their own sexual values (to which they are entitled) and their prejudices (which they try to keep out of their work). They may also be either unaware of their biases or bent on furthering them through their work.

In these respects, sex researchers are not basically different from other behavioral scientists except that the study of sexual behavior poses more pitfalls in that there may be more axes to grind in this field than in others. In the light of this, a healthy skepticism (but not cynicism) is a good starting point in evaluating any piece of sex research.

Ethical Considerations   Investigators have generally been quite conscious of their responsibility not to harm their subjects. Currently more formal and stringent regulations are imposed by institutions and governmental agencies to ensure that human subjects are protected from harm (animal research is also regulated, but there is a far greater latitude in what can be done with animals).

The protection from harm has several components. The study must not expose its subjects to pain and seriously distressing conditions that are likely to cause harm. We would not, for example, give pregnant women testosterone to study its androgenizing effects on the fetus. Nor would we expose children to sexual experiences with adults to see how that would affect their sexual development.

The issue of *confidentiality* is particularly important in survey research and clinical studies of sexual behavior. Kinsey and his collaborators devised an elaborate coding system that kept inviolate the identity of thousands of subjects (figure 8.2). Clinical records are likewise considered privileged information. There is a need for protection since the revelation of the sexual details of a person's life could lead to scandal, social ostracism, marital strife, professional jeopardy, and legal action.

Even if the research is carefully designed to avoid harm, its consequences for subjects cannot be entirely predicted. This is why it is essential that every participant give his or her *informed consent* after being fully informed about what the experience will entail. To further protect them, subjects are *debriefed* after the experiment is over. This gives the investigator an opportunity to clarify the true nature of the study (if there was deception involved) and to help overcome whatever disturbing thoughts or feelings the subject may have as a result of the experience.

The use of *deception* raises its own special problems. There are experiments that are based on the subjects not knowing the true nature of the situations they are put in. With groups that are unlikely to admit outsiders, some investigators gain entry by pretending to be persons of similar sexual interest. For example, one anthropologist and his wife associated themselves with "swingers" to study group sex. They managed to stay sexually uninvolved through various excuses ("It's my wife's period") while sustaining the false assumption that they were active swingers themselves.[12]

Though such investigators do not intend to harm their subjects, the surreptitious manner in which such work is conducted, the absence of both informed consent and of debriefing raise questions about their propriety as research endeavors. The counterargument is that without such methods, certain areas of sexuality would be impossible to explore. Besides, our society permits their use by the police and investigative journalists.

The Subjects Studied   A basic aim of all behavioral research is to discover both the common patterns, or the universals, and the variant forms of human activities. We want to be able to make *generalizations,* statements that apply not to a few individuals but to entire populations. For instance, we would want to know how col-

[12]Bartell (1971).

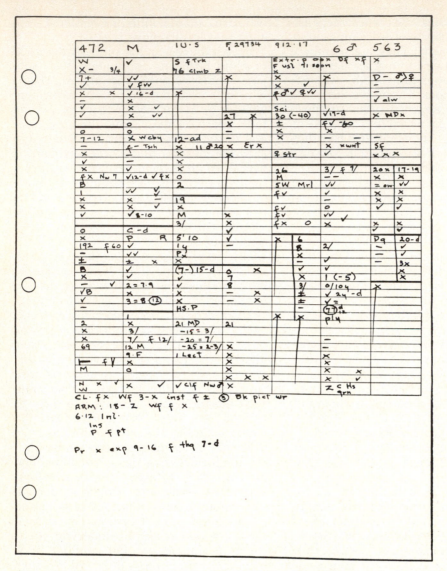

Figure 8.2 Sample of Kinsey study coded sex history. (From Kinsey et al., *Sexual Behavior in the Human Male*, 1948, p. 72. Courtesy of the Institute for Sex Research.)

lege students behave not only in a given institution but more generally; how a particular sexual activity varies with age, marital status, socioeconomic class, or some other variable.

Information about the sexual behavior of a few individuals is not enough, but studying entire populations is usually not feasible. As a result, investigators must select a *sample,* a subset of the population they are interested in. If this sample is *representative* or typical of the larger population, then the findings from the sample can be generalized to the rest of the population.

There are many methods of sampling. A common method is *random sampling* whereby every member of the population has an equal likelihood to be selected by chance. Thus, to find out how often the members of a freshman class masturbate, the question is asked to a subgroup of freshmen who have been selected by lot (the proper size of the sample for a given purpose is determined by a number of statistical considerations).

Random samples are not always adequate. In the above example, there may be too few women, blacks, or Catholics in the freshman class chosen randomly; if we expect these variables to make a significant difference, then we cannot rely on chance alone to yield appropriate ratios of freshmen from the three subgroups. We could better achieve such representation by relying on a *stratified sample* whereby members from each subgroup are randomly selected and thus are certain to be present in appropriate proportions within the overall sample.

Virtually all general studies of sexual behavior suffer from serious *sampling bias*. The nature of such bias varies with the type of study and often is due to the unwillingness of chosen subjects to participate. Thus, even if the sample is chosen randomly, the low response rate makes it into a *self-selected* sample and hence, most probably, less likely to be representative.

As a result of these and similar considerations, no general study of sexual behavior has yet been conducted whose findings could be extended to large segments of the population, let alone the country as a whole. Since people with conservative sexual values, and hence presumably more restrictive sexual patterns, are less likely to participate in a sexual survey or volunteer to be subjects in a sexual study, most studies convey an overly liberal picture. On the other hand, socially problematic or unconventional behaviors tend to be underreported to the extent that subjects refuse to reveal them.

Clinical Research

Investigations that study patients, or those who are under care, are referred to as *clinical research* (from Greek for "bedside"). Traditionally this has been the domain of physicians, but more recently clin-

ical psychologists, social workers, marriage and family counselors have enriched the clinical field with their own special perspectives.

Clinical research may consist of a single *case study*, or it may be based on a series of cases. The research component in clinical work may be part of the broader treatment procedure or conducted separately. This tradition of single case study continues;[13] but clinical research in sexuality is now more typically conducted on larger groups of patients and with more emphasis on control groups and the quantitative analysis of data.

The strength of the clinical approach is threefold. First, clinical investigators tend to be well-trained professionals. Any enterprising editor is at liberty to conduct a magazine survey. But to be permitted to treat patients and clients, formal training and certification procedures are required. Second, clinical research has a higher likelihood of focusing on significant issues. People seek treatment because they are hurting, hence clinicians deal with people with real problems; and are less likely to fiddle with contrived issues. Third, the clinical setting allows for the most in-depth investigations. A patient in psychoanalysis will see the therapist over many years; a research subject is not motivated to go through a comparable process. Even the most short-term form of psychological counseling usually entails a greater intensity of contact than the typical research interview or questionnaire approach.

The shortcomings of the clinical approach are the inevitable consequences, or the obverse, of its advantages. Thus, the fact that clinicians usually see their research subjects personally and over extended periods of time means that they can only deal with a limited number of such subjects (though cases from many sources may be pooled together). These subjects are self-selected and therefore unlikely to be representative of the general population. Furthermore, the case studies that get

[13]For a current example of a case study, see Stoller (1979).

published often involve dramatic and unusual situations which are atypical.

These shortcomings are further compounded by the fact that subjects of clinical research are in distress and therefore may behave in special ways. The clinical context is not neutral nor geared for dispassionate inquiry. The primary purpose of the encounter is treatment; thus both investigator and subject have a personal stake in the process that may well interfere with the objectivity with which issues should be studied. Therapists may only want to hear what fits with their theoretical bias, and patients may oblige by saying what the therapists want to hear.

In sum, the clinical approach provides the most in-depth information on the smallest numbers of people with regard to problems that are real life concerns, but where considerations of pathology and cure are most likely to color the behaviors being observed.

The Research Interview

The research interview has much in common with taking a *case history*, which is the basic clinical tool in elucidating the story of the patient's illness. In taking a *sex history* the investigator similarly inquires into various aspects of sexual behavior or focuses on a particular type of activity.

In the **structured interview** approach, a predetermined set of issues are explored and the same questions put to all subjects. In the **open-ended interview,** the subject is encouraged to reveal spontaneously his or her sexual life with minimal guidance or prodding from the interviewer. The former approach yields more systematic and uniform data; the latter results in more spontaneous and free revelations. Hence the two methods are complementary.

Unlike clinical research that deals with patients, the research interview method uses a broader, randomly chosen group of subjects. This advantage, however, is often not realized because only some of the subjects selected agree to participate, which results in a self-selected sample. Another drawback is the relatively superficial contact between interviewer and subject. Particularly in a sensitive area like sexual behavior, a person may not reveal intimate details of his or her life to a virtual stranger who, unlike the clinician, is not there to help the person. If one adds to these concerns the inevitable failings of memory and the tendency to conceal or exaggerate sexual experiences, then what we can learn from research interviews becomes further compromised.

Yet in the hands of a skillful researcher like Kinsey, the research interview can produce a wealth of information over a matter of hours. As a good representative of this approach, and because of its historical importance (chapter 1), we consider the Kinsey studies in some detail (box 8.1). Though other studies have used the interview method, none has come close to the Kinsey effort in magnitude. Typically, interviews are conducted to supplement a larger questionnaire study. For example, in the Hunt survey, which we discuss below, subjects were interviewed in addition to the 2,000 questionnaire respondents. In other cases, interviews have provided a convenient device to obtain titillating vignettes, to dress up the more drab data obtained by other means.

Questionnaire Methods

The *questionnaire method* is a further extension of the structured interview whereby the subject responds to a set of written questions instead of voicing answers to an interviewer. Just as it takes skill and training to be a good interviewer, the construction of an effective questionnaire is a difficult task that requires special expertise.

The main strengths of the questionnaire method are in the large numbers of subjects that it can easily reach, and the consistency with which it presents its questions. It took heroic effort for the Kinsey group to interview some 16,000 subjects; far less skillful investigators with much less effort have gotten responses from 10 times as many respondents.

On the other hand, the questionnaire method suffers from poor response rates.

Box 8.1   The Kinsey Surveys

The studies of sexual behavior by Alfred Kinsey and his associates were the most ambitious efforts at sex research based on interviews; after four decades they remain unsurpassed in their scope, level of sophistication, and social impact.

Kinsey and his associates undertook extensive training as interviewers. They relied on a standard set of questions to obtain comparable answers and the hundreds of questions they put to their subjects included numerous checks and safeguards (such as comparing answers given by husbands and wives) to ensure the reliability of the responses.[1] These investigators also memorized their codes so that the answers were recorded directly in code allowing for greater accuracy and confidentiality (figure 8.2).

Between 1938 and 1950 these researchers interviewed more than 16,000 persons; Kinsey personally obtained 7,000 sex histories—an average of two a day for ten years.[2] But the samples used as the bases of the reports consisted of 5,940 female and 5,300 male residents of the United States.

Kinsey was aware that it would be very difficult to get a random sample of the population to participate in his study; given the topic, many were likely to refuse, thus compromising the representativeness of the group he would end up with. So he chose a *group sampling* procedure and selected a wide variety of groups (such as churches, prisons, colleges) and then tried to persuade everyone within these groups to be interviewed. In a quarter of instances, he managed to obtain *hundred percent samples* in which a sex history was obtained for every member of a given group. Since these groups were drawn from many sectors of society, they provided a greater measure of representativeness.

The population of 11,240 people whose histories were analyzed included individuals of a wide variety of ages, marital status, educational level, occupation, geographical location, religious denomination, and so on. Since all these groups were represented in the sample population in sufficient numbers to permit comparison, the Kinsey sample was *stratified*. But it was not *representative*, since each group in the sample population was not proportionate to the size of that group in the American population at large. Lower educational levels, in particular, and rural groups were underrepresented. Some groups were, in fact, not represented at all; for example, all of the subjects whose histories were used were white; Kinsey collected histories from blacks, but not enough to permit statistical analysis. Three quarters of the women subjects had been to college. Many of the lower-class males were ex-convicts. Thus his subjects were mainly representative of the white, urban, Protestant, college-educated population of the northeastern United States. Kinsey was quite aware of the problems posed by these sampling problems and made no claims to the contrary. He called it a study of sexual behavior within certain groups of the human species, "not a study of the sexual behavior of all cultures and of all races."[3]

[1] See Kinsey et al. (1948, pp. 47–59) for the interviewing devices they used to get to sensitive information.
[2] Brecher (1969, p. 116).
[3] Kinsey et al. (1953, p. 4). For other critiques of the Kinsey survey, see Chall (1955) and Cochran et al. (1954).

Whether or not people respond more truthfully to a person or to a questionnaire is hard to settle. It is possible that under the cloak of anonymity, respondents are more truthful in response to a questionnaire. But the presence of an interviewer enhances the chances that the subject understands the questions and responds seriously. The interviewer can also *probe* a particular area by asking more questions and pursuing whatever leads a subject may provide.

Questionnaires are more objective tools than the methods of case study, but they are by no means free of bias. The way that the questions are phrased may affect the answers. Apart from such ''loading,'' subjects are likely to respond, just as they do in other contexts, in the direction of *social desirability*; that is, they give the answers that they think are compatible with dominant social values or the expectations of the investigator.

The questionnaire approach thus represents the opposite extreme from the case study method by providing more superficial information for larger numbers of people over a short period of time. Many questionnaire studies have been conducted over the past two decades. We describe some of the better known ones by way of illustration (box 8.2)

The examples of research we have cited so far have all dealt with general studies of sexual behavior. These same methods have also be used to focus on a more specific form of behavior (such as premarital sex) or a special facet of a given behavior (such as the interpersonal aspects of homosexual interactions). As we discuss particular sexual behaviors, we shall have occasion to refer to these more specialized research efforts.

Direct Observation and Experimentation

In the study of sexual behavior, direct observation and experimentation have had very limited scope so far. Among the reasons for this are the requirements of privacy that shroud sexual experience in most cultures. The most significant application of direct observation in sex research has been so far in the **laboratory studies** conducted by Masters and Johnson. It must be noted, however, that in this work the primary focus was on the physiological aspects of sexual function. If Masters and Johnson had simply sat back and observed people engaging in various sexual activities, perhaps more serious questions would have been raised about the propriety of their research.

The subjects studied by Masters and Johnson were 694 normally functioning volunteers of both sexes between the ages of 18 and 89 years. This was neither a random sample nor were these subjects specifically selected for their sexual prowess; the only requirement was that they be sexually responsive under laboratory conditions.

The research procedure was to observe, monitor, and sometimes film the responses of the body as a whole and the sex organs in particular to sexual stimulation and during orgasm attained by masturbation and sexual intercourse. Over a decade (beginning in 1954) some 10,000 orgasms were monitored. Because more of the subjects were women, and because females registered more orgasms than males, about three-quarters of these orgasms were experienced by women. (The findings of this research are discussed in chapter 3.)

There is a long tradition of **participant observation** in anthropology but very little of it has been applied to studying sexual behavior. It is easier, of course for an anthropologist to witness a marriage ceremony than to be present when the sexual union is consummated. Thus, what anthropologists frequently describe are inferences from informants' self-reports and from various observations rather than eyewitness accounts of sexual behavior. In other words, anthropologists use the case study or research interview method rather than imitate the approach of direct observation under laboratory conditions. But their observations are more naturalistic (that is, they occur where the subjects normally live) than is the case under the more artificial settings of clinical work.

The issue of **experimentation** is even more ticklish when applied to sex research because it entails the most active form of intrusion on the part of the researcher. Direct observations may involve no more than simply watching what people would be doing anyway, such as engaging in coitus. But in an experimental set-up, the subjects are induced to perform certain acts in prescribed ways to fulfill the experimenter's purposes. The increasingly stringent

Box 8.2   Questionnaire Surveys of Sexual Behavior

In the early 1970s, the Playboy Foundation sponsored a national sex survey that was carried out by an independent market survey and behavioral research organization. The sample consisted of 2,026 adults randomly chosen from the phone directories of 24 cities in the United States. Its 982 males and 1,044 females paralleled closely the U.S. population of persons 18 years old and over. However, not everyone is listed in a phone directory; therefore the sample cannot be said to be representative, especially in light of the fact that 80% of those contacted refused to participate.

The data were gathered by means of an extensive self-administered questionnaire. An additional sample of 100 men and 100 women were selected for more intensive investigation through interviews by Morton Hunt and his wife, Bernice Kohn, both professional writers. Hunt wrote the report, hence we shall shall hereon refer to this research as the *Hunt survey*.[1] Though hardly in the same class as the Kinsey survey, the findings of the Hunt report are nonetheless more readily comparable to the Kinsey findings than those of any other study; this makes the Hunt survey useful as a point of comparison with the Kinsey results two decades earlier.

In the late 1970s a questionnaire study of female sexuality by Shere Hite attracted wide public notice.[2] Its respondents were reached mainly through various women's groups (such as chapters of the National Organization of Women, abortion rights groups, and university women's centers). To broaden the study's reach, subjects were also recruited through various magazines (such as *Penthouse*) and church newsletters. After all was said and done, only 3,000 persons responded out of the 100,000 who had received questionnaires—a response rate of 3%. A similar study was done with 8,000 men (with a response rate of 6%).[3] The *Hite reports* of female and male sexuality, although of negligible statistical significance, have been quite instructive because of the wealth of anecdotal comment provided by the informants.

The ultimate in the mass gathering of questionnaire data is the *magazine survey*. This is done by printing the questionnaire in a magazine and inviting its readership to respond. Those who respond constitute the study sample. Given the large readership of some magazines, even a 1% response may yield thousands of responses. The results obtained cannot be generalized even to the readership of the magazines, which itself is hardly representative of the population at large. Sex surveys of this sort have been conducted by *Psychology Today*,[4] *Redbook*,[5] *Ladies Home Journal*,[6] *Cosmopolitan*,[7] and *Playboy*.[8] If seen within their limitations, such information can be of some interest, and we occasionally refer to it in subsequent discussions.

[1]Hunt (1974).
[2]Hite (1976).
[3]Hite (1981).
[4]Athanasiou et al. (1970).
[5]Tavris and Sadd (1977).
[6]Schultz (1980).
[7]Wolfe (1981).
[8]Peterson et al. (1983).

standards and controls that have been instituted in recent years to ensure the safety of experimental subjects may be particularly difficult to satisfy in sexual experiments. Nevertheless a number of such approaches has been used. One example is the exposure of subjects to sexually arousing materials while special instruments monitor physiological responses (chapter 3). More typically, investigators must still rely on the effects of so-called *natural experiments*, such as involving the effects of exposure to prenatal adrogens on subsequent general development (chapter 4).

STATISTICS OF SEXUAL BEHAVIOR

Investigators gather, organize, and interpret *quantitative data* through **statistical methods.** Statistics are a powerful tool in helping us to understand sexual behavior. But they can be abused by conveying a spurious sense of accuracy and reliability. Statistics can be especially misleading when the sample on which they are based is highly selective. Thus, in a study with a response rate of a few percent, whether 10% or 90% of the respondents report one thing or another may say little with regard to the actual prevalence of the practice in the general population. Yet such figures are constantly bandied about (including in this text).

Statistical Methods

Statistical tools are only as good as the data that they manipulate. Given an adequate data base, they are indispensible to make sense out of masses of information. Kinsey's background in entomology was responsible for his awareness that large populations of people would have to be studied to establish a *taxonomic* base, behavioral standards based on groups against which to compare individuals. Given the wide variation in sexual behavior, it is especially important that we have such general bases for comparison; otherwise the fact that a person has one orgasm per day, month, year, means little if one does not know what experiences others have

Perhaps the greatest benefit to be derived from the study of sexual statisitics is the realization that the concept of sexual behavior, as such, is an abstraction: in reality, many different groups and individuals within them behave in many different ways; there is not one but many patterns of sexual behavior. This extreme variability of sexual experience is good reason for caution. Even when careful study has uncovered the patterns of behavior in a given group, this information cannot be automatically applied to all of its members. *Averages* tell us about clusters of people, not individual human beings. Under-

standing each person requires a study of that person. But individual study can yield more meaning if we know the range of variance for the group. The study of individuals and of groups is therefore meant to complement each other, not substitute for one another.

The concept of the average is expressed through a variety of ways. The *mean* is the familiar arithmatical average. It is most useful in answering the question, "How often does a given activity occur within a group?" For instance, if a group of 250 high school seniors has a total of 1,000 orgasms during a week, the mean will be four orgasms per week.

The *median* is the midpoint that divides the distribution of scores into two equal halves when the scores are arranged in linear fashion in increasing order. It best answers the question, "How often does the average person engage in a given activity?" That is because the median, unlike the mean, is not influenced by extremely high or low figures in the group. In the foregoing example, 10 seniors each with 30 orgasms or more a week will inflate the mean but not affect the median. The *mode* is the score that occurs most frequently.

The process of averaging may obscure a lot of useful information. A group of women with an average of five orgasms a week may include a women with one orgasm a week and another with 20; the average figures will not truly reflect the experience of either. *Measures of variance* are thus necessary for a more accurate representation of behavior. Futhermore, even the same total figure may not mean the same thing: one couple may engage in coitus frequently over weekends, another couple may spread such engagements over the entire week: the monthly totals of these couples may be the same but their patterns of sexual activity are clearly different.

Statistical measures are not necessary to all approaches in understanding sexual behavior. The study of a single case may be highly revealing. Writers and artists to not work with representative samples yet can attain profound insights into human sexual behavior. Statistics are but one tool and they should be neither uncritically accepted nor blindly rejected.

In subsequent chapters, we refer to various statistics on sexual behavior. At this point, it might be instructive to see how the use of statistics, in general, might help us to understand some of the broader aspects of sexual behavior, such as frequency of sexual activity and the prevalence of components of total sexual outlet. We will do this by using illustrations from the Kinsey survey, even though the specific figure may or may not be generally valid because of sampling problems and the passage of several decades since these data were collected. Nonetheless, the Kinsey figures remain the best models to illustrate the nature and use of sexual statistics.

Frequency of Sexual Activity

How much sex do people have? Although each culture apparently assumes a range of permissible or desirable sexual activity for those with legitimate access to it, that range is rarely, if ever, spelled out. Nonetheless, there has been a longstanding fascination with individuals with high levels of sexual activity.

The frequency of sexual behavior has health implications in the minds of many people. Not infrequently there are disagreements between couples over how often they should make love. These conflicts are usually assessed against what is thought to be "average" and "normal."

Frequency must be measured over specific periods of time. As our lives are generally organized around weekly schedules, Kinsey found the week to be a useful temporal unit for sexual behavior. Furthermore, since the level of sexual activity changes over the years, Kinsey examined behavior over 5-year periods. Thus, he defined *total sexual outlet* as the number of orgasms achieved during an average week in a given 5-year period of a person's lifetime.

In the Male   The mean frequency of total sexual outlet for white males between ado-

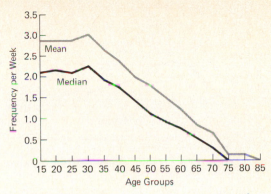

Figure 8.3   Frequency of total sexual outlet for males, by age groups. (From Kinsey et al., *Sexual Behavior in the Human Male*, 1948, p. 220. Courtesy of the Institute for Sex Research.)

lescence and age 85 in the Kinsey sample is nearly three orgasms in an average week. This does not mean that the "average man" has this many weekly orgasms, or that such frequencies constitute the standard of normality, health, virility, and so on.[14] The frequency of total outlet is related to a variety of factors, of which age is the most important. As shown in figure 8.3, the frequency of orgasms steadily declines following age 30. This pattern runs contrary to the popular notion that sexuality is awakened during adolescence, gradually reaches its peak during the "prime of life" and then dies down. Instead, sexual outlet is already at near peak level in the youngest age group.

The curve for the mean is higher than that for the median at every point. The peak mean (for the group between 26 and 30 years old) is three orgasms a week, whereas the median is only a little more than two. This means that the group is averaging three orgasms a week but the average person within the group is having about two orgasms each week.

In the Female   Orgasm is a less satisfactory measure of female than of male sexuality since it is more difficult to recognize and count female orgasm than male eja-

---

[14]The statistics to be cited here on are rounded figures based on Kinsey et al. (1948).

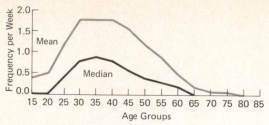

Figure 8.4   Frequency of total sexual outlet for females, by age groups. (From Kinsey et al., *Sexual Behavior in the Human Female*, 1953, p. 548. Courtesy of the Institute for Sex Research.)

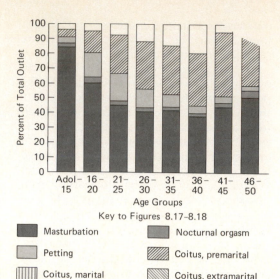

Key to Figures 8.17–8.18

■ Masturbation   ▨ Nocturnal orgasm

▢ Petting   ▧ Coitus, premarital

▥ Coitus, marital   ▨ Coitus, extramarital

▤ Coitus, postmarital   ▢ Homosexual

Figure 8.5   Percentages of total outlet: sources of orgasm among single females. (From Kinsey et al., *Sexual Behavior in the Human Female*, 1953, p. 562. Courtesy of the Institute for Sex Research.)

culation. Figure 8.4 shows the mean and the median frequencies of total sexual outlet among female age groups. Comparison of the female with the male pattern reveals the frequencies of total sexual outlet for females to be lower than those for males in all age groups, but the differences between the two sexes are not that striking. The wider disparity between the female means and medians indicates wider variability. There must be some women whose levels of activity are so much higher than the rest that their performances unduly inflate the mean values for the group as a whole.

Whereas men achieve high levels of sexual activity right after puberty, for women, the period between 16 and 30 years is still one of gradually increasing activity. From then on, instead of declining, these rates are sustained for a decade. It is unclear if these sex differences continue to persist at the same level at present despite changed social circumstances.

Components of Total Sexual Outlet

An important contribution of large-scale statistical studies is that they reveal broad patterns of behavior for a given population. Though the means of obtaining sexual satisfaction are legion, the Kinsey survey showed that most orgasms are achieved through six means, some much more prominent than others. Furthermore, the contributions of these components of total outlet vary with age and other variables like education (reflecting social class) and marital status.

Female Patterns   Figures 8.4 and 8.5 show how marital status is correlated with sources of orgasm in the Kinsey female sample. For single women (figure 8.5) in the youngest age group, masturbation is the primary orgasmic outlet (84% of total outlet). In successive age groups, this proportion declines, then it rises again in the two oldest age groups. Coitus is the second most important source of orgasm for single women. Beginning with the youngest age group, in which premarital coitus accounts for 6% of orgasms, this outlet reaches its peak between the ages of 41 and 45, when it provides 43% of total outlets. Homosexual contacts account for a steadily increasing proportion of orgasms for single women up to the age span of 36 to 40 years, when almost one out of five orgasms is reached by this means. In contrast, married women show much more uniform sex lives (figure 8.6). At all age levels they register a primary reliance on coitus as a source of no fewer than four out of five orgasms (some attained through extramarital intercourse). Masturbation provides a small but steady proportion (about 10%) of total sexual outlet, with a slight increase with age. Homo-

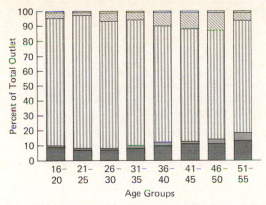

Figure 8.6 Percentage of total outlet: sources of orgasm among married females. (From Kinsey et al., *Sexual Behavior in the Human Female*, 1953, p. 562. Courtesy of the Institute for Sex Research.)

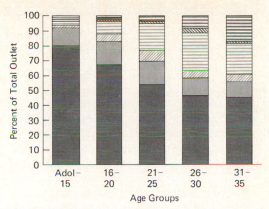

Figure 8.8 Sources of orgasm for single males with some college education, by age groups. (From Kinsey et al., *Sexual Behavior in the Human Male*, 1948, p. 491. Courtesy of the Institute for Sex Research.)

sexual contacts are a negligible source of orgasm for married women.

Male Patterns   Among single males with elementary school educations (figure 8.7) masturbation accounts for a little more than half the total sexual outlets even at the youngest age levels, and becomes even less significant with increasing age. Homosex-

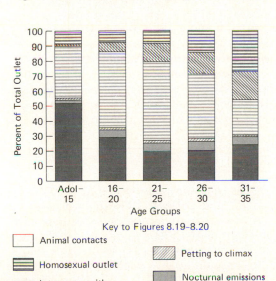

Key to Figures 8.19–8.20

☐ Animal contacts

▤ Homosexual outlet

▨ Intercourse with prostitutes

▥ Intercourse with companions

▨ Petting to climax

▤ Nocturnal emissions

▨ Masturbation

Figure 8.7 Sources of orgasm for single males with elementary school educations, by age groups. (From Kinsey et al., *Sexual Behavior in the Human Male*, 1948, p. 490. Courtesy of the Institute for Sex Research.)

ual contacts also increase somewhat with age so that between the ages 31 and 35 they account for one-fourth of all orgasms.

In contrast, single males with at least some college education rely heavily on masturbation (figure 8.8). In the youngest age group it accounts for 80% of orgasms. In successively higher age groups the ratio declines, but not much below 50%. The better-educated single male's reliance on masturbation is at the expense of coitus. Petting and nocturnal emissions are more substantial outlets in this group, but homosexual contacts are less important than for less-educated males. There is, however, a similar increase in reliance on such contacts with age. These class-based distinctions appear to have been greatly attenuated in more recent studies.

THEORIES OF SEXUAL BEHAVIOR

There are no behavioral *theories* (Greek for "contemplation") that are specific to sexuality; theoretical approaches to sexual behavior are the application of more general theories of human behavior to the sexual realm. Thus our understanding of sexual behavior eventually depends on the solution of the more basic riddle of human behavior in general.

Theories are useful to the extent that they can be tested and verified by empirical evi-

Box 8.3   The Nature versus Nurture Controversy

The question as to whether behavior is preordained by nature or acquired by nurture has preoccupied thoughtful observers since antiquity. In the 18th century the Enlightenment instituted a critical examination of these issues. Especially influential were the ideas of the English philosopher *John Locke* (1632–1704). Locke argued that our minds at birth are like blank tablets (*tabula rasa*); hence, everything that defines us as individuals is learned.[1]

In the middle of the 19th century, Darwin's theory of evolution by natural selection generated a momentous challenge. Human beings were now seen as but one link in a long chain of organisms, and the biological factors shaped by evolution determined the nature of individuals and societies. Darwin himself said relatively little about these matters, but *Francis Galton* remorselessly applied the concept of natural selection to all aspects of human character and history. It was Galton who in 1874 coined the phrase "nature and nurture" to separate "under two distinct heads the innumerable elements of which personality is compared. Nature is all that a man brings with him into the world; nurture is every influence from without that affects him after birth."[2]

Galton had no doubt that nature was by far the more important determinative force, and devoted his life to the furtherance of that proposition. Under his influence, there developed a school of thought in biology that basically discounted the importance of cultural influences. The application of these principles in turn led to the *eugenics* movement aimed at "race improvement." These ideas and other excesses of *social Darwinism* were then applied to justify colonialism, racism, and the dominance of those in power over the socially disadvantaged in Victorian society.

Not all biologists adhered to these views. T. H. Huxley, for instance, explicitly recognized the relevance of cultural processes in the "evolution of society," which was "a process essentially different from that which brought about the evolution of species, in the state of nature." Yet by the early part of the 20th century, the more extreme view of biological determinism had gained the upper hand.

The opposition to the extreme evolutionists was led by *Franz Boas* (1858–1942), the founder of American anthropology and his students: Alfred Kroeber, Robert Lowie, and later on Ruth Benedict and Margaret Mead, among others. The central issue that concerned Boas was the ways in which cultures shaped and sometimes shackled the lives of individuals. He began his career at a time when evolutionary determinism reigned supreme not only in biology but in anthropol-

---

dence; theories that pertain to sexual behavior are handicapped in this regard since naturally occurring sexual behavior is generally concealed from observation, and sexual experimentation is socially unacceptable. Theories of sexual behavior, therefore, tend to remain highly speculative.

Theories that pertain to sexuality are usually of two kinds: **motivational theories** that attempt to explain why people behave sexually the way they do, and **developmental theories** (often called *theories of psychosexual development*) that trace the origins and patterns of sexual behavior.[15] We discuss motivational theories here and developmental theories of sexual behavior in the next chapter. These theories share the common goal of explaining the nature

[15]*Motivate* literally means to "cause to move." Psychologists use the concept of motivation to refer to those factors that energize behavior and give it direction. For example, a rat is said to be motivated by hunger to seek food purposefully and energetically.

Box 8.3   The Nature versus Nurture Controversy *(continued)*

ogy as well. E. B. Tyler represented the view that cultural phenomenon were just as subordinate to the laws of evolution as were natural phenomena: "Our thoughts, will and actions" accorded with "laws as definite as those that govern the motion of waves . . . and the growth of plant and animals." And these laws were primarily biologically determined. Against this view, Boas put forth his conviction that culture "is an expression of the achievements of the mind and shows the cumulative effects of the activities of many minds." These cultural activities entailed phenomena to which of the laws of biology did not apply.

Attempts at rational discourse between these two intellectual camps proved increasingly futile as the evolutionists and eugenicists (now led by Charles Davenport in the United States) pressed their cause with uncompromising rigidity and mounting moral fervor. In response, the proponents of cultural determinism undertook to go their separate way, declaring their complete independence: an "eternal chasm" (in the words of Kroeber) now separated cultural anthropology from biology. This schism was reflected in subsequent conceptions of culture that saw it as quite independent of evolutionary antecedents or origins. Culture was "not a link in any chain, not step in any path, but a leap to another plane." The laws

of biology and evolution had nothing to do with it.

The revolt of the anthropologists against biological determinism was reinforced by the work of other social scientists. In sociology, the doctrine of *Emile Durkheim* (1858–1917) that society is "a thing in itself," and the behaviorist theory in psychology of *J. B. Watson* (1878–1958) were highly compatible with the cultural determinism approach. These developments, combined with new discoveries in genetics and the revulsion generated by the incorporation of eugenics principles in Nazi ideology in the 1930s, ensured the victory and the subsequent dominance of principles of cultural determinism in our understanding of human behavior and society.

The pendulum began to swing back over the past several decades as advances in genetics, the study of the social behavior of animals by ethologists, and major fossil discoveries by paleontologists shed new light on human evolution. The application of evolutionary principles to understanding human behavior through sociobiology, has now once again generated a great deal of interest and controversy over these issues.

[1]Russell (1945).
[2]The following discussions and quotes are from Freeman (1983, pp. 3–50).

and manifestations of human sexual behavior. That they come up with such diverse explanations results from differences in their intellectual assumptions, methodology, and the historical factors pertaining to the growth of their underlying discipline.

Theories seek to explain objective reality but they themselves are part of the human intellectual enterprise and thus subject to historical and cultural influences. Just as sexual behavior is profoundly influenced

by its social content, theorizing about such behavior also reflects the particular intellectual temper of the time and must be understood in that setting. One of the key conflicts between theories of human behavior has been, and remains, over the respective role of biological and cultural determinants, historically known as the *nature versus nurture* controversy (box 8.3). Even though almost everyone now recognizes the importance of both sets of factors, the matter is by no means resolved.

Evolutionary Perspectives on Sexuality

In their reproductive functions, humans clearly represent evolutionary extensions of primate and mammalian patterns. To what extent could the same be said about sexual behaviors more generally? Have we irrevocably broken off the behavioral patterns of our evolutionary forebears, or, underneath our cultural refinements and complexities, do we still behave in ways that continue to serve the same basic evolutionary purposes?

The basic premise of the evolutionary approach is that our sexual structures and behaviors have evolved through two processes: natural selection and sexual selection. *Natural selection* favors survival and depends on the success of both sexes in staying alive within whatever conditions of life prevail at a given time. *Sexual selection* has to do with all of the physical and behavioral characteristics of an animal that enhance its being chosen by a mate.

Primate reproduction depends on male-female interaction, hence evolutionary approaches to sexual relations focus on male-female differences and the effects of those differences on sexual interactions. Darwin noted in *The Descent of Man and Selection in Relation to Sex* (1871), the striking sex differences in structure and behavior among many species of animals: males were larger, stronger, more aggressive; they took the initiative in courting and fought off their rivals.[16] Darwin attributed these differences to sexual selection, the evolutionary process that favored those individuals that had a reproductive advantage over others of the same sex within the species.

Sexual selection has two components: *intrasexual selection*, which entails the competition of males with males for females whereby some males breed more than others; *female choice*, which results in the female selecting the males who will father her offspring. The purpose of both

processes is *reproductive success:* to produce as many offspring with as high a chance of survival as possible, to contribute maximally to the gene pool of the next generation.

The recent formulations of sociobiology are not exclusively based on the biological determinism of sexual behaviors. Nonetheless they have proven to be quite controversal. As popularly perceived, sociobiology renders human beings into biological puppets. And by linking certain behaviors (like male dominance and sexual promiscuity) to evolutionay antecedents, it appears to lend legitimacy to sexually prejudicial attitudes and practices (chapter 12).

The process of learning is essential for explaining the wide repertory of our sexual behaviors, no matter how one looks at it. Sociobiologists do not deny that. But in their view, learning is *biologically directed,* which predisposes the individual to learn certain things more readily than others and, hence, to behave in certain ways more often than others. As E. O. Wilson states, "Only small parts of the brain resemble a tabula rasa . . . the remainder is more like an exposed negative waiting to be dipped into developer fluid."[17] In this view, society is the "developer fluid"; it has the crucial task of bringing the image out of the negative, but it is not the inscriber of the image or the author of our sexual "script."

Based on these assumptions and on evidence from various quarters, sociobiologists formulate explanations of why and how we behave sexually. Since these have mainly to do with various *reproductive strategies* used by men and women, we also discuss the sociobiological perspective and various objections to it when we deal with sex and interpersonal relations (chapter 12).

Theory of Instincts

Prior to the theory of evolution, animals were seen as categorically different from humans. Without benefit of reason or soul, animal behavior was ascribed to *instincts*

---

[16]This is by no means a universal pattern. There are many species in which the sexes are similar in size or the females are more aggressive.

[17]Wilson (1975, p. 156).

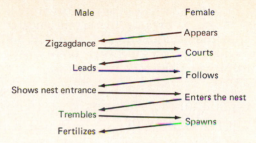

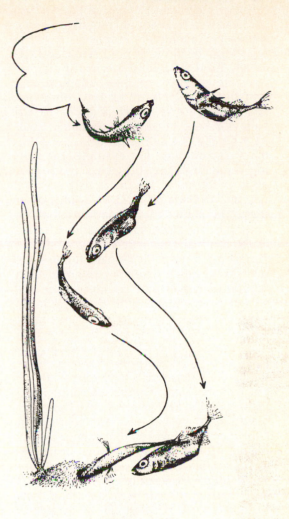

(Latin, *incite*), which were conceived as biological forces that compelled the organism to act in characteristic ways without the necessity of learning those behaviors. Animals copulated, reproduced, and looked after their offspring because that was part of their "animal nature"; they did not need to be taught how to perform any of these functions. In the context of evolutionary theory, the concept of instincts was extended to humans as well. By the turn of the century, this idea had caught on so firmly that psychologists like William McDougall sought to explain human behavior through an ever-expanding list of instincts.[18]

By attempting to explain everything instincts explained nothing; so the term gradually fell into disrepute among social scientists. But the fundamental notion of a biologically determined *predisposition*, or tendency, to act in certain ways persisted. The term "instinct," however, is now used to refer to processes far more complex than previously understood.[19] To what extent instinctual processes, as currently defined, apply to humans, is, however, still not quite clear. Yet efforts to extend insights from animal to human sexual behavior continue through naturalistic observations in the wild as well as experimental work in the laboratory.

[18]McDougall (1908).

[19]Tinbergen (1951) has defined instinct as "a hierarchically organized nervous mechanism which is susceptible to certain priming, releasing, and directing impulses of internal as well as external origin, and which responds to these impulses by coordinated movements that contribute to the maintenance of the individual and species."

Figure 8.9 The mating ritual of the three-spined stickleback. Each step in this behavioral sequence is a fixed-action pattern that serves as a releasing stimulus for a fixed-action pattern by the mate. (From H. Gleitman, *Basic Psychology*, 1981; based on Tinbergen, 1951.)

The sexual behavior of animals generally represents involuntary responses to external stimuli resulting in predictable and relatively fixed behavior patterns. These patterns are inherited, complex, adaptive, and remain fairly stable under environmental changes. This gives much of animal sexual behavior a *stereotypical*, "automatic" quality whereby one fish or rat behaves pretty much like another. For example, the mating behavior of the stickleback progresses quite predictably: The male fish builds a nest and defends his territory against other males; he allows the egg-laden female to enter it (distinguishing her from males because the male has a red belly while the female does not); the female follows the male to the nest where the pair go through additional moves and countermoves; the sequence culminates with the male fertilizing the eggs laid by the female (figure 8.9).

In these instances animals exhibit specific, sterotyped behaviors believed to be mediated through neurophysiological mechanisms called **innate releasing mechanisms** (IRM). The external cues that stimulate these mechanisms are specific *cue stimuli* or *social releasers* that trigger the behavior. The applicability of such principles to human sexual behavior is by no means self-evident. Nonetheless, the assumption that there are innate patterns underlying human sexual behavior is an important element in all biologically oriented theories of sexual theory.

The terms *innate* ("present at birth") is often used in contrast with *acquired:* The former is said to represent *nature,* or heredity; the latter, *nurture,* or the influence of the the environment through learning. While there is a certain face validity to this distinction, it readily leads to the false dichotomy of whether one or another behavior is genetically or environmentally shaped: in effect all behavior, like all bodily structures (the *phenotype*), is the product of genetic endowment (*genotype*) and environment.[20] The continuing search

for biological factors in sexual behavior is not meant to exclude the significance of environmental influences but to put the interactional patterns between both sets of variables in a biosocial perspective.

Psychoanalytic Theories of Sexual Behavior

The concept of a *sexual instinct* has been central in psychoanalytic theory. Freud viewed sex as a psychophysiological process (like hunger) with both physical and mental manifestations. By *libido* (Latin for *lust*) he meant the psychological aspects, the erotic longing induced by the sexual instinct. His libido theory evolved into a broad conceptual scheme that purported to explain the nature and manifestations of sexuality throughout psychosexual development. Freud initially formulated a *single-instinct* theory, whereby the sexual interest served to sustain life and reproduction. He then shifted to a *dual-instinct* concept where *eros* represented the libidinal or "life instinct" and *thanatos* stood for aggression or the "death instinct."[21]

Freud's critics have attacked him for unduly stretching the meaning of sexuality. Some of his followers, notably Carl Jung, deemphasized the erotic in favor of equating sexuality with a more general *life force*. By contrast, Wilhelm Reich expanded the meaning of sexuality, eroticizing virtually the entire cosmos (chapter 1).[22]

Freud did not divest sex of its commonly understood sense of pleasure through orgasm. Rather he extended the term to other pleasurable experiences ordinarily considered to be nonsexual. This was comparable to his broadening of the concept of "mental" from its ordinary sense of conscious experience to include the entire range of unconscious mental activity as well.[23] Sexuality, in this broader sense, is the central theme of psychoanalytic theory, and in-

[20]For further dicussion of the various meanings of "innate," see Symons (1979); Lehrman (1970).

[21]Freud, *Beyond the Pleasure Principle,* in vol. 18 of *Collected Works.*

[22]Reich (1942). For discussions of Reich's work, see Marcuse (1955) and Robinson (1976).

[23]Jones (1953, vol. 2, p. 284).

fantile sexuality is the cornerstone of psychological development and mental health (chapter 9).

The key to Freud's conception of the mind is the **unconscious.** Freud neither discovered nor invented the concept of the unconscious.[24] But he did designate it as the pivotal point in his theory of mental functioning and the primary location of basic human motivations. Though we often speak of the unconscious as if it were a part of the brain, it is only meant to represent a class of mental functions. Thoughts (as well as feelings) can thus be placed in three categories: *conscious* thoughts, of which we are aware at any given time; *preconscious* thoughts, which can be brought into awareness more or less at will (as when we recall a telephone number); and *unconscious* thoughts, which cannot be ordinarily brought into consciousness. Unconscious thoughts can, however, become conscious in a number of ways: they spill into dreams and slips of the tongue; they appear in neuroses in distorted or camouflaged forms; and they may be coaxed out through *free association* during the psychoanalytic process. Unconscious thoughts and feelings profoundly influence how we feel, what we think, and how we act, specially with regard to sexual matters. To explain these processes, Freud proposed a hypothetical model of the mind consisting of the id, the ego, and the superego, which, once again, are not parts of the brain but categories of mental processes.

The *id* embodies the psychic representations of biological instincts and its contents are entirely unconscious. Only its derivatives may come out in veiled form, passing through the censorship of the ego. The id strives for expression through pleasure and immediate gratification regardless of reality considerations or social concerns.

The *ego* has a dual nature. Part of it (the *autonomous ego*) consists of mental processes like perception and memory. The ego also mediates between the demands of the id and the constraints of the superego.

Through various *defense mechanisms* the ego blocks recognition of the raw demands of the id (repression and denial), rechannels them (sublimation and displacement), or distorts them (projection). The ego thus manages a working compromise between drive gratification and social constraints and limitations. The ego is partly conscious, which accounts for our faculty of self-awareness, but its defensive functions against the id are unconscious (since the most effective way of hiding something involves hiding the very fact of doing so).

The *superego* is popularly equated with the conscious. Part of its function is to maintain an awareness of the social acceptability or morality of one's conduct. But the superego, like the ego, is also partly unconscious, for it too must work to block the id. The superego embodies the moral and value judgments of those who have helped to shape one's life. It may be inconsistent and corruptible or excessively harsh and punitive. When the expectations of the superego are not met, we experience guilt and shame.

Freud's concepts have enormously influenced contemporary Western notions of sexuality (chapter 1). Over the past several decades, however, the influence of psychoanalytic theory has declined within the behavioral sciences because many of its basic assumptions cannot be subjected to empirical verification. Freud's view of human nature, his assumed primacy of biological over cultural determinants of behavior, and his contention that personality and sexual development are for all purposes fixed by age five or six, have been widely criticized. More recently, Freud's male-centered conception of female sexuality and personality has come under critism both within the field[25] and from the outside, specially from feminists.[26]

On the other hand, psychoanalysis has itself evolved in new directions, becoming less dogmatic and more diverse in its viewpoints. Yet Freud remains the dominant intellectual force within psychoanalysis, and

[24]Whyte (1960).

[25]See Salzman in Marmor (1968); Blos (1980).
[26]Miller (1973); Mitchell (1974).

the sheer scope of his all-embracing vision of sexuality continues to make itself felt in our culture, however much it may be modified and reinterpreted.

Drive Theory

While biologists pursued the study of innately determined behaviors and psychoanalysts elaborated on sexual instincts, psychologists, by the 1920s, came to favor the concept of drives. A *drive* is a state of psychological arousal elicited by some physiological need (such as for food and water) or the avoidance of pain that motivates behavior to satisfy the need.

The concept of drives is no less hypothetical than that of instinct; its existence can only be inferred from certain behaviors. Yet by being more circumscribed, the concept of drives has proved more useful and compatible with the broader physiological principle of *homeostasis* ("equal state"), which refers to the tendency of the body to maintain a constant internal environment with respect to the chemical composition of its internal fluids like blood, its temperature, and so on.

A drive can be thought of as the psychological representation of a physiological need created by disturbances of the homeostatic balance, such as hunger or thirst. According to Clark L. Hull's *drive reduction theory,* formulated some 40 years ago, the basic motives for behavior emanate from the need to reduce such internal bodily tensions. In this homeostatic model, sexual arousal generates tension that disturbs the state of inner calm. The impetus to achieve orgasm is then due to the need to reduce sexual tension and regain internal calm.[27]

Drive theory as applied to sexual behavior has several shortcomings. Unlike hunger or thirst, there is no evidence that sexual interest is generated by a physiological lack or imbalance; nor is sex essential to sustain individual life. Yet because sexual behavior has a "driven" quality to it and is linked with specific physiological processes, it remains convenient to refer to the *sex drive* as a motivating force.

Another problem is the reliance of drive theory on tension reduction in both physiological and psychological terms. But tension reduction is clearly not a universal goal of human behavior; on the contrary there are many occasions where we seek tension through excitement; particularly in sexual terms, we go to great lengths to arouse ourselves. Furthermore, organisms can be "pushed" into behavior by internal drives, while behavior can also be elicited by the "pull" of environmental incentives. For instance, one eats when hungry as well as when enticed by the offer of something one likes to eat; the same is true for sex.

In view of these considerations, *incentive theory* is now seen as a more satisfactory explanation of sexual behavior. The *positive incentive* in sex is the pleasure one anticipates in the experience; one does not necessarily have to be sexually hungry to be aroused (although the element of deprivation can clearly enhance sexual desire). A *negative incentive* on the other hand has an inhibiting effect because of the threat of pain or anxiety. If a person is taught that masturbation has ill effects, he or she will tend to shun it. The interplay of positive and negative incentives thus motivates the organism to behave in certain ways under given conditions.

Theories of Learning

Psychologists define *learning* as a relatively permanent change in behavior that occurs as a result of practice. But not all behavioral changes are due to learning. They may be due to *maturation*. For instance, infants begin to walk at about 15 months because at that time their neuromuscular system reaches the necessary level of maturity. Behavioral change may also come about through temporary conditions such as fatigue or the influence of drugs.[28] The limits of what we learn and how we learn it are determined by the nature of our brain. But the substance of what we

[27]Hull (1943).

[28]Hilgard (1956).

learn and the methods of learning are culturally determined. For example, our mental capacity to use language allows us to speak, but in order to speak a particular language, we need to learn it. Presumably a similar situation prevails with respect to sexual learning.

Human learning is so complex that psychologists have had to study the mechanisms of learning in simpler situations using experimental animals like rats and pigeons. Therefore, our knowledge of how sexuality is learned is based on inferences from general *theories of learning* or (*behavior theory*). These theories focus on the acquisition of behavior patterns through various forms of learning rather than through the influence of biological determinants. And unlike psychoanalytic theory, they focus on conscious, cognitive processes, rather than unconscious motives.

Learning theory is the dominant intellectual orientation in psychology, but its adherents come from other disciplines as well. For instance, Kinsey, despite being a biologist, viewed most aspects of human sexual behavior as the product of learning whereby exposure to various sexual attitudes and experiences were assumed to shape an individual's sexual preferences and behavior. Masters and Johnson also base their therapeutic approaches on behavior modification techniques derived from learning theory (chapter 15).

Learning theories include a variety of approaches that can be subsumed under three main categories: classical conditioning, operant conditioning, and observational learning. None of these processes are specific to sexuality, but we shall explain them by using examples of sexual learning.

Classical Conditioning   The original experiments of *classical conditioning* were conducted by Ivan Petrovich Pavlov in the 1920s.[29] This is a form of learning by association. Pavlov noted that dogs reflex-ively salivated when food was placed in their mouth. By association, dogs became conditioned to salivate from experiencing a host of other stimuli such as the sight of food or even the dish in which it was served. If a buzzer was sounded before the food was served, the dog would eventually salivate in response to the sound of the buzzer alone. Pavlov called the dog's reflexive salivation to food in its mouth an *unconditioned reflex* that was based on the connection between an *unconditioned stimulus* (UCS) (food-in-the-mouth) and *unconditioned response* (UCR) (salivation). By contrast, the buzzer represented a *conditioned stimulus* (CS), salivation in response to it a *conditioned response* (CR), and the association constituted a learned or *conditioned reflex*.

The scope of learning by classical conditioning is expanded through *stimulus generalization* whereby conditioned stimuli similar to the original one will also elicit the conditioned response (for instance, a sound similar to the original buzzer works just as well). But learning is not permanent: the CR may disappear (or become *extinguished*) if food ceases to follow the buzzer. Yet the CR may reappear (*spontaneous recovery*) after a rest period during which the animal is not exposed to these stimuli.

This learning process is assumed to apply to a large array of human experiences, including sexual behavior. But these applications are largely inferred. One of the basic problems is to find an unconditioned stimulus of an erotic nature which would reflexively elicit the unconditioned response of sexual arousal. Physical stimulation of the genitalia would qualify since the sexual arousal response to it has a demonstrably reflexive basis. But beyond that, the sights, sounds, and smells that are erotic have no demonstrably reflexive basis among humans (chapter 3).

If the condition of a reflexive basis for such learning is set aside then greater possibilities open up for sexual learning by association. For instance, if sexual arousal happens to coincide with a particular scent (such as bodily odor or a perfume), sound

[29]Pavlov (1927).

(utterances during coitus, a piece of music), or sights (nudity, undergarments), then these sources of stimuli may become conditioned stimuli (or *eroticized*) and may elicit sexual arousal on their own. How or where such learning takes place must often be taken on faith since rarely can we trace a learned sexual trigger or preference to its original associational source. For example, a man may find women with long legs sexy and a woman may find men with slim buttocks to be erotically appealing but they would be hard pressed to trace back these preferences to when they were learned.

Instrumental and Operant Conditioning
A learning mechanism related to classical conditioning is *instrumental conditioning*, where an animal learns to perform an act to obtain a reward. For example, a seal will balance a ball on its snout to get a fish from the trainer. Yet these two conditioning processes are not the same. The salivation of Pavlov's dog is merely anticipatory to getting food, whereas the performance of the seal is "instrumental" (that is, it acts like a tool) in getting to the sought-after goal.

Classical conditioning presumes a stimulus response linkage, called an *S–R bond;* where there is no stimulus, there is no response. This is a rather passive way of learning, when in reality animals manifest a great deal of spontaneous activity. To differentiate such spontaneous behavior from activity that involves actions elicited by specific external stimuli, the term *operant behavior* is used (since the animal "operates" or acts on environment) and the acquisition or modification of spontaneous behavior is called *operant conditioning*.[30] The concept of operant behavior thus provides a much more versatile and comprehensive framework than does classical conditioning for understanding sexual behavior.

In operant conditioning, learning occurs because of the consequences of a given behavior. When the outcome of an act is rewarding, the behavior is *reinforced*. Such reinforcement may be obtained by offering

the animal a positive reward (such as food) or the elimination of something bad (like electric shock). As with classical conditioning, the processes of generalization and extinction also function in operant conditioning. Furthermore, the reinforcers need not always be *primary*, that is, responsive to some basic biological need. Stimuli acquire *secondary* reinforcing potential by being paired with primary reinforcers. Thus, a chimpanzee will work for a token after learning that the token can be used to get food.

We can envisage numerous examples of how operant conditioning would work with respect to sexual behavior. Thus, when a child becomes sexually aroused while pressing against the mattress or fondling the genitals, the experience is found to be pleasurable, hence it is repeated until one day it is carried to orgasm: the child has now learned to masturbate. Furthermore, as the person continues to explore ways to enhance sexual pleasure, some approaches prove successful and hence get repeated and learned; others do not. Once sexual arousal and orgasm become established as sought-after goals, a person will go through considerable trouble to obtain the sexual rewards.

Social Learning   The principles of classical and operant conditioning have been developed under rigorous experimental conditions. But many behavioral scientists think that these learning models in themselves are inadequate to explain the process of *socialization* whereby the child acquires the behavioral patterns characteristic of his or her society. To allow for these more subtle and complex forms of learning, *social learning theory* offers additional modalities of learning.

One such modality is **observational learning** or learning by *imitation* from the behavior of others who act as *models*.[31] Such learning does not require the repeated pairing of stimuli with reinforcement, thus eliminating the necessity of direct personal experience. It thus opens up vast possibil-

---

[30]Skinner (1938).

[31]Bandura (1969); Mischel (1981).

ities for learning sexual behaviors through observing people engage in the act, listening to them describe it, or reading about it, seeing it portrayed on TV, and so on. That a great deal of such learning goes on all the time seems self-evident. So the questions have to do with the mechanisms whereby such learning takes place.

Classical and operant conditioning theorists view socialization as a process of molding the individual. In social learning, the individual is seen as being shaped not through simple learning devices but by the entire force of culturally defined interactional patterns.

To this end, social learning theory emphasizes the importance of *cognitive processes* that allow us to think through and foresee the likely consequences of our actions. Our actions can thus be governed by anticipated circumstances rather than being merely reactive to immediate stimuli. And through *symbolic thinking*, the scope of our psychological experiences is greatly expanded; even the most farfetched experience or object can acquire erotic significance, as amply illustrated by the behavior of paraphiliacs. In chapter 9, we discuss

the significance of cognitive processes and the acquisition of "sexual scripts" through social learning during sexual development.

All of the theoretical approaches we have discussed offer useful insights into one or another aspect of sexual behavior. But there is as yet no generally acceptable theory that in a comprehensive and internally consistent manner explains why we behave sexually the way we do.

SUGGESTED READINGS

The Kinsey volumes are quite long and rather technical, yet there is no better way to grasp the problems and intricacies of conducting large-scale sex research. Particularly instructive as introductions are chapter 2 of Alfred Kinsey et al., *Sexual Behavior in the Male* (1948), and chapter 1 in Kinsey et al., *Sexual Behavior in the Female* (1953).

Bullough, V., ed. *The Frontiers of Sex Research* (1979). Brings together a series of essays on various aspects of sex research.

Brecker, E. M. *The Sex Researchers* (1969). Reviews the work of major sex researchers and some of their key findings.

Robinson, P. *The Modernization of Sex* (1976). A historical analysis and intellectual critique of the work of Havelock Ellis, Kinsey, and Masters and Johnson.

Chapter 9

Sexual Development

It is meaningless to speak of a human child as if it were an animal in the process of domestication.

Erik Erikson

A thorough understanding of any behavior requires some knowledge of its development. In this chapter, we examine the genesis of sexual behavior through childhood and adolescence and follow its further maturation in adulthood and old age.

Much of what we have to say is descriptive, for as yet there is no comprehensive theory of psychosexual development that is generally acceptable. We have many facts, but only some can be explained; we have many hypotheses, but only parts of them are supported by facts.

Sexuality in Childhood

In the Western world, childhood came to be recognized in the 17th century as a discrete phase of life.[1] From then on, the protection of the rights and welfare of children gradually led to strict prohibitions against their sexual abuse and simultaneously to the inhibition of any overt form of sexual expression in childhood.

Modern recognition and acceptance of childhood sexuality is to a significant extent due to the influence of Freud. His views were not substantiated by systematic empirical data on sexual behavior in childhood and such empirical data are still largely lacking. This is in part because of the social constraints placed on such study but also because childhood sexual behavior tends to be sporadic, which makes it difficult to be observed in research settings.

Responsive Capacity

The capacity for sexual arousal is present at birth; ultrasound studies suggest that reflex erections may even be present for several months before birth.[3] Erections in childhood, long noted by caretakers of infants, have also been formally reported by investigators.[4] Physiological evidence for sexual arousal, as manifested by vaginal lubrication, has been similarly noted in female infants.[5]

During erections, infants show a greater tendency for thumb-sucking and restless

---

[1]Aries (1962).

[2]There have been others who have contributed to attracting attention to childhood sexuality. For instance, Albert Moll wrote in 1912, "When we see a child lying with moist, widely opened eyes, and exhibiting all the other signs of sexual behavior as we are accustomed to observe in adults, we are justified in assuming that the child is experiencing a voluptuous sensation."

[3]Masters (1980).

[4]In one study of nine baby boys aged 3 to 20 weeks, erections were noted in seven of them to occur from 5 to 40 times a day (Conn and Kanner, 1940).

[5]Martinson (1976). For futher discussin of female genital excitability at age two, see Galenson and Rolphe (1976).

behavior, including stretching and limb rigidity, fretting, and crying; erections in 2- and 3-day old males have been noted to be more common during periods of crying. There is also an apparent association between alimentary functions and the incidence of erections. Feeding, sucking, fullness of the bladder and bowels, urination, and defecation often seem to prompt erections.[6] Infants experience erections during sleep even more commonly than adults.[7]

Reflexive Response   Given the contexts in which such erections commonly occur and the psychological immaturity of infants, it is assumed that such erections are *reflexive*. But they are not erotically neutral events (like a knee jerk) since there is some expression of pleasure that accompanies them: stimulation of the genital area of children of both sexes aged 3 to 4 months elicits smiles and cooing sounds (frequently accompanied by erection in males).[8] Thus while erections and corresponding female responses may start as reflexive response to stimulation in the genital region, they are soon endowed with pleasurable potential, even though we do not know when and how the reflexive responses become *eroticized*.

Sources of Arousal   Sexual arousal in childhood and early adolescence is often part of a more generalized state of *emotional excitement*. A wide variety of sexual as well as nonsexual sources of stimulation have been reported to elicit erotic responses among boys (no comparable data are available for girls). Sexual sources of arousal involve nudity and sight of genitals. The nonsexual sources may be mainly physical (such as friction with clothing) or emotional (exciting situations like watching fires, fighting, accidents, or wild animals).[9] Such indiscriminate erotic reac-

tions gradually give way to more selective responses. By the late teens sexual response is by and large limited to direct stimulation of the genitals or to obviously erotic situations.

Kinsey and others have reported experiences highly suggestive of orgasm among infant boys and girls.[10] But we do not know what proportion of infants and children have such experiences or how often. Although such activity may not be an everyday occurence, more of it probably goes on than is observed by adults; even when present, adults ignore it or fail to recognize its sexual significance.

Sexual Behavior in Childhood

Sexual behavior, as with many other forms of childhood activity, takes the form of *play*. And as with adult sexual behaviors, such play can be *autoerotic* or *sociosexual*. Such characterizations are useful for descriptive purposes but we should not impute adult meanings to such experiences.

Autoerotic Play   *Self-exploration* and *self-manipulation* are the most common forms of sex play in children. Such activity has been observed starting between 6 and 12 months.[11] Genital play starts somewhat earlier among boys (6 to 7 months) than girls (10 to 11 months).[12] *Infantile masturbation* has been noted among 61% of 1-year old infants in a residential treatment center, and 91% of infants from "superior home environments."[13] Such activity is thought to be discovered accidentally; as the infant explores his or her body in a random manner, contact with the genitals proves pleasurable, so it tends to be repeated. As the child grows older, self-exploration becomes more focused and its sexual intent more obvious (see figure 9.1).[14]

[6]Halverson (1940); Sears, Maccoby, and Levin (1957).

[7]Karacan, Marens, and Barnet (1978); Korner (1969).

[8]Martinson (1980).

[9]The full list is quite extensive. See Kinsey et al. (1948, pp. 164–165).

[10]Kinsey et al. (1948); (1953); Bakwin (1973, p. 52).

[11]Spitz (1949).

[12]Galenson and Rolphe (1974).

[13]Spitz (1949).

[14]Kleeman (1975).

Figure 9.1 Infantile masturbation. (From *The Erotic Drawings of Mihaly Zichy*, 1969, plate 24.)

Among the Kinsey subjects, an occasional individual could remember masturbating, sometimes to the point of orgasm, as early as at 3 years of age. Autoerotic activities in infancy appear more prevalent among boys. During the first year of life, males have been noted to engage in autoerotic activity about twice as frequently as female infants.[15] A higher prevalence for this activity among males has also been reported for ages 1 and 2 years,[16] and 4 through 14 years.[17]

Fondling the penis and manual stimulation of the clitoris are the most common autoerotic techniques.[18] Children also discover the erotic potential of thigh friction by rubbing against their beds, toys, and the like.[19] They also learn to masturbate

from seeing someone else do it or by being stimulated by another person; these interactions are more common among boys than girls.

Infantile masturbation may or may not be continuous with *childhood masturbation.* Most male children seem to relearn to masturbate from one another. In the Kinsey sample nearly all males reported have heard about the practice before trying it themselves, and quite a few had watched companions doing it. Fewer than one in three boys reported discovering this outlet by himself, and fewer than 1 in 10 was led to it through homosexual contact.

By contrast, two out of three females in the Kinsey sample had learned to masturbate primarily through accidental discovery. Sometimes this discovery did not take the place until adulthood. At other times it was not the act itself but awareness of its significance that came late; occasionally a woman would have masturbated for years before she realized the nature of her behavior.

Though masturbation is now considered a normal part of childhood, the practice is by no means universal. Based on data from

[15]Spitz (1949).

[16]Kleeman (1975).

[17]Gebhard and Elias (1969).

[18]Spitz (1949) subsumes autoerotic activities during the first year of life under three categories: self-stimulatory rocking, stimulation of the genital area, and manipulation of the penis.

[19]Bakwin (1973) describes a young girl who would lie down on top of her rag doll and rhythmically press her body against it until she "felt satisfied."

Box 9.1   Cross-Cultural Comparisons in Childhood Sexuality

Ford and Beach have categorized cultures according to their levels of sexual permissiveness as being restrictive, semirestrictive, and permissive.[1]

Restrictive societies vary in the severity of their control of preadolescent sexuality, but as a rule they attempt to keep children from learning about sexual matters and inhibit their spontaneous sexual activities. For instance, both the Apinaye (a primitive, peaceful, monogamous people in Brazil) and the Ashanti (a complex polygynous society in Ghana) expressly forbade children to masturbate from an early age. In New Guinea a Kwoma woman who saw a boy with an erection would strike his penis with a stick; these boys soon learned not to touch their penises even while urinating.

The secrecy in restrictive societies often extends to reproductive functions. Even animals giving birth may not be watched by children, and fictitious explanations of where babies come from are offered. Special precautions were taken to keep children from surprising adults engaged in coitus; children might even be placed in separate sleeping quarters at an early age. Yet the expression of sexuality in childhood could not be entirely checked. Children still engaged in sex play when they could, even though they may have done so in fear or shame.

Messenger provides an account of a sexually repressive island folk community in Ireland, which he calls Inis Beag:

> The seeds of repression are planted early in childhood by parents and kin through instruction supplemented by rewards and punishments, conscious imitation, and unconscious internalization. Although mothers bestow considerable affection and attention on their offspring, especially on their sons, physical love as manifested in intimate handling and kissing is rare in Inis Beag. Even breast feeding is uncommon because of its sexual connotation, and verbal affection comes to replace contact affection by late infancy. Any form of direct or indirect sexual expression—such as masturbation, mutual exploration of bodies, use of either standard or slang words relating to sex, and often urination and defecation—is severely punished by word or deed.[2]

Permissiveness toward sexual behavior during the developing years does not mean complete absence of rules and regulations. Relative to the cultures already discussed, however, permissive societies show remarkable laxity toward the sexual activity of youngsters.

Among various Pacific islanders and other permissive cultures, children of both sexes freely and openly engaged in autoerotic and sex play, including oral-genital contacts and imitative coitus. These children were either instructed in sex or allowed to observe adult sexuality activity. In a few societies adults actually initiated children sexually. Siriono (polygynous nomads of Bolivia) and Hopi parents masturbated their youngsters. On Mangaia island in the South Pacific women orally stimulated the penises of little boys.

In these permissive cultures sex play became progressively more sophisticated and gradually merged with adult forms of sexual activity. With the exception of incest, youngsters were generally free to gratify their growing and changing sexual needs. Sexual activity was actually encouraged early in life. The Chewa (polygynous advanced agriculturalists of central Africa) believed that children should be sexually active if they wished to ensure fertility in the future. The Lepcha (monogamous agriculturalists of the Himalayas) believed that sexual activity was necessary if girls are to grow up: at 11 or 12 years most girls were engaging in coitus, often with adult males. Trobriand boys of 10 to 12 years and girls of 6 to 8 years were initiated into coitus under adult tutelage.

[1]Ford and Beach (1951).
[2]Messenger (1971, p. 29).

as they now come in contact with children from diverse backgrounds. Their ideas of where babies come from (a matter of curiosity by age four) and what sex is all about range from enlightened views to utterly baffled or grossly misinformed notions. Children also have a rich repertory of "forbidden" riddles, songs, verses, and games that are suffused with sexual themes.[23]

Elementary school generates new opportunities for sexual learning as well as imposing greater constraints on sexual behavior. Attempts at sex play are now more susceptible to discovery and disapproval. Beyond the home and the school, children are subjected to a wide variety of sexual influences. *Television* is a major source of such influence since more time is spent watching TV than in any other leisure activity. The nature of the medium, the cumulative effect from different programs, the characters with whom children identify themselves, all add up to a persistent set of sex- and gender-linked messages that are part of the process of sexual socialization.[24]

Sexual Socialization

Sexual interactions between adults and children take the place in two broad contexts: In the first, adults act as the *socializing agents;* in the second, adults themselves become the *sexual partners* of children.

Parents as Socializing Agents   *Socialization* is the process whereby children learn the norms of their society and acquire its values, beliefs, and culture. Children are socialized with regard to most developmental tasks through such institutions as the family and schools. Sexuality is the only major facet of life that is by and large left out of the educational process, be it in school or in the home. The prevailing view in our culture has tended to regard the best sexual teaching as that which teaches

without provoking further questions or stimulating sexual behavior (chapter 16).

Even though relatively few parents actively and openly instruct their children in sexual matters, all of them inevitably influence their sexual development by communicating, one way or another, their own attitudes and sexual values. Parents have been described as exercising three main types of information control concerning sexual matters. First, through *unambiguous labeling*, parents point out and label certain behaviors as wrong but with no explanation as to why. A child may simply be told, "That's not nice" or "Good kids do not do that." Second, through *nonlabeling*, parents avoid the sexual issue by attempting to distract the child or shifting focus to a less sensitive area. A girl who asks how mommy got pregnant, may be told that mommy and daddy "were in love," leaving the child no wiser. Third, through *mislabeling*, a sexual activity is condemned not for what it is but for some spurious reason; a boy many be told not to play with his penis to avoid "getting germs."[25] Parents may also avoid labeling sex altogether or refer to the genitalia and sexual acts in ambiguous terms. The net effect of these parental manuevers is to effectively ban meaningful discussion of sexual issues and to compound ignorance with shame and anxiety.

All of these are examples of *verbal communication*. Equally important are the messages communicated to the child through *nonverbal communication*—looks, gestures, and actions. When a parent reacts to a sexual question or situation by blushing, fidgeting, or freezing up, the child will pick up the negative message no matter what the parent may say subsequently.[26] The openly repressive and punitive attitudes that characterized the sexual socialization of children in the past have been largely overcome. Yet many parents still lack adequate information and feel uncomfortable

---

[23]Extensive studies have been conducted in Germany in this area, which are not yet translated into English. Borneman (1983).

[24]Himmelweit and Bell (1980).

[25]Gagnon (1968); Sears et al. (1957).

[26]For futher discussion of sexual communication between parents and children, see Kahn and Kline (1980).

discussing the subject; even sexually sophisticated parents may feel uncertain as to what to encourage and what to discourage in their children when it comes to sexual behavior.

As a result, very little by way of a meaningful and positive sex education takes place in most homes. A study involving 1,400 parents of 3- to 11-year-old children, conducted in the mid-1970s in the Cleveland area, showed that the vast majority of children received little or no direct sexual information or advice from their parents. Fathers particularly were uninvolved in the sexual education of their children; both boys and girls directed their questions to the mothers (pregnancy and birth being the most talked about topics). Futhermore, when parents responded to sexual questions, they usually dealt with them as one-time issues rather than as part of an ongoing dialogue.[27]

Children will become aware of sexuality whether parents like it or not. It is precisely because awareness of sexuality is unavoidable that appropriate overt and covert communication about it is important. Knowing that sex is there, the child is puzzled and made anxious if he or she is left in the dark.[28]

Parental anxieties in dealing with sexual questions may arise from their own recollections of unhappy childhood sexual experiences. Or evidence of the awakening of sexuality in their children raises the threat of incestuous wishes. Even though very few parents act on these impulses, even talking about sex may be too close for comfort.[29] Finally, parents do not know what to tell their children because they themselves are uncertain of their sexual values.

All of these problems notwithstanding, some parents manage to give their children sensible sexual guidance. This is best done in small installments. Very young children need no more than learning the proper names of their genital organs—in the same manner that they are taught about the rest of their body. When children become curious about the differences between male and female bodies, they need simple, straightforward answers. If a 5-year-old asks where babies come from, it will suffice to say "from their mommy's tummy." "How does the baby get there?" may be dealt with a simple explanation how daddy puts the sperm that are in his penis in mommy's vagina to join with her egg.[30] Much of what to say and how to phrase it must be left to parental discretion.[31]

Psychiatrists have written extensively on the effects of children witnessing or overhearing parental coitus (Freud called it the *primal scene.*). Whether or not all such experience is traumatic to the child is open to question. The crowded conditions under which the great mass of the world's population lives must make such experiences fairly common. Perhaps it is not the act itself but the emotional context within which it is seen, and not knowing what it means, that makes the difference.[32]

Adults as Sexual Partners The earliest and most crucial physical relationship is between the infant and the adult caretaker, typically the mother. It may strike some as offensive to think of such relationships in sexual terms, but in effect it has distinctive erotic components that are inseparable from the affection and tenderness that we rightly associate with parental love. For example, though not every mother may experience it, recognize it as erotic, or enjoy it as such, nursing an infant may bring forth certain unmistakably sexual physiological and subjective responses (chapter 5). The infant also finds the experience highly grat-

[27]Roberts et al. (1978). Roberts (1980).

[28]For further discussion of the parents as sex educators, see Kelly (1981); Gilbert and Bailis (1980). Snyder and Gordon (1984) present an annotated bibliography.

[29]Rosenfeld et al. (1982).

[30]Finch (1982).

[31]The Sex Information and Educational Council of the United States (SIECUS) provides a book list for the use of parents. Scanzoni (1982) provides guidance to Christian parents in answering their children' sexual questions. Pomeroy (1976) provides more general guidance to parents.

[32]Borneman (1983).

ifying because of the alleviation of hunger and the pleasure obtained through sucking. Some infants show erections during breast feeding but whether that is due to sexual arousal or some other physiological process is not clear.[33] Mother and infant go through many affectionate exchanges during nursing involving nuzzling, snuggling, patting, kissing, cooing, and smiling.[34] But nursing is not the only occasion for such intimacy, nor the mother the only adult to behave in this manner. The father, grandparents, siblings and others close to the child also interact in similar ways.

We should neither view all such behavior as relentlessly and openly sexual nor deny the subtle eroticism that may characterize some aspects of it. Not only is such intimate interaction normal, but it is essential for the formation of early affectionate ties ("bonding") and subsequent psychosexual development.[35] In studies of human infants and experimental work with primate infants, deprivation of such "contact comfort" early in life has been shown to lead to severe disruption of the developmental process[36] and sexual function,[37] as we discuss shortly.

Direct and deliberate sexual interactions between adults and children constitute *sexual abuse* or *child molestation.* In most cases, such encounters involve persons known or close to the child such as an older sibling, relative, neighbor, or babysitter. Among Kinsey's female subjects, about one out of four had been sexually approached in childhood by someone 5 or more years older (80% had experienced such contacts only once each; 5% nine or more times). In Hunt's sample, 14% of males, prior to adulthood, had had sexual contact with a relative (13% heterosexual; 0.5% homosexual). The corresponding figure for females was 9% (8% heterosexual and 1% homosexual). Two-thirds of the male and two-fifths of the female experiences were with cousins; over a quarter of the males and over half of the females went no further than light petting. Sexual contacts within the nuclear family mainly involved siblings: 3.8% of the males reported contact with a sister; of the female sujects 3.6% reported contact with brothers and 0.7% with sisters. Contact with parents was restricted to fathers with daughters (0.5%).[38] In the *Psychology Today* survey sample, 1% of women and 2% of men had had their first coitus with a relative.[39]

Incestuous relations with parents are of particular concern. Though long thought to be quite rare, such involvement may be more common than previously assumed. Among cases of child abuse, 12% involve sexual abuse of the child as well.[40] The prevalence of incest in the general population is estimated at 5 cases per 1,000 per year.[41] Given the likelihood of underreporting, the true figures may be much higher. The majority of cases that come to public attention involve father-daughter incest.

The impact of precocious sexual experience on young children is generally believed to be quite detrimental, specially in incestuous settings. Not only are girls more often abused than boys but they also seem to be more adversely affected. The ill effects on the child may consist of immediate reactions of anxiety, guilt, depression, and behavioral disorders. Longer-term consequences involve disturbances with sexual function, sexual identity, and marital relationships.[42] Various studies have found increased incidences of prostitution and drug abuse among women with incestual experience, especially father-daughter incest.[43] They also show a higher rate of running away, suicide attempts, and adolescent pregnancy.[44] Incest abuse by close relatives increases the child's likelihood of

---

[33]Goodman and Goodman (1976).
[34]Halverson (1940).
[35]References on bonding.
[36]Spitz (1949), Spitz (1946); Bowlby (1969, 1973).
[37]Harlow et al. (1971).

[38]Hunt (1974).
[39]Athanasiou et al. (1970).
[40]Brown and Holder (1979).
[41]Nakashima and Zakus (1979).
[42]LaBarbera and Dozier (1981); Meiselman (1978).
[43]Giarretto (1976).
[44]Herman and Hirschman (1981).

being sexually exploited by others, especially if the child runs away from home.[45]

These consequences of incest are present in cases which come to the attention of the public agencies or are seen in treatment. It is possible that other cases of incest that do not have such deleterious effects remain largely hidden. There are studies that report that some cases of father-daughter incest the daughters showed no evidence of psychopathology.[46] The reason that so few cases of mother-son incest are known may be due to the fact that such interactions are more likely to occur in nonthreatening and affectionate settings. Such experiences are now assumed to exist in thousands of cases.[47] In societies where sexual interaction with children by adults is socially acceptable, there seem to be no serious psychological problems. The cultural and interpersonal context of incest may therefore significantly influence the outcome. Where fear, guilt, intimidation, or violence mark the encounter, the experience can be terrifying and traumatic, but under the opposite circumstances the child may retain positive memories of the experience. Nonetheless, since children are dependent on adults, they are vulnerable to exploitation. Nor can they be expected to enter into sexual relationships with a reasonable understanding of what it entails.

Dealing with the Sexually Abused Child
The discovery that a child has been sexually abused, especially by a member of the family, is a highly traumatic experience for all concerned and a situation that is all too easy to mismanage.

Children are often afraid or reluctant to report such experiences, hence parents and clinicians must be alert to signs of disturbance that may be caused by it. If there is sufficient reason to warrant it, the child must be gently coaxed to reveal the truth.

On the other hand, just as some children may conceal such experiences, they may conjure up fantasies of sexual interaction, make false accusations (or accuse the wrong person), or misinterpret signs of affection for sexual advances.[48]

When children are subjected to incestuous experiences, it is not just the nature of the act but the manner in which the problem is dealt with that may determine the psychological impact. Parental dismay, anger and worry, intrusion of the police, and court appearances may well add up to a more traumatic experience for the child than the actual sexual encounter itself.[49] There are cases where the child needs to be separated from the abusing parent but in such cases, incarceration of the offender or other punitive measures may do more harm than good (chapter 14).[50]

Adolescent Sexuality

The importance of sexuality in childhood notwithstanding, it is during adolescence that adult patterns of sexuality emerge. There is a close interaction between the biological changes of *puberty* (chapter 4) and psychosocial development in *adolescence*.[51] Yet the two processes are not the same, and there are many uncertainties about the respective roles of biological and social factors in shaping adolescent sexual behavior. In contemporary industrial societies, the aquisition of adult status takes longer than the attainment of reproductive maturity. Adolescence thus entails a contradictory state when one is simultaneously "mature" (biologically) and "immature" (psychosocially). This is one

[45]James and Meyerding (1977).

[46]LaBarbera and Dozier (1981). The Kinsey survey and other general surveys have tended to substantiate this.

[47]Sarrel and Masters (1982).

[48]Rosenfeld et al. (1980).

[49]For a review of the literature on child abuse, see Price and Valdisseri (1981).

[50]Rosenfeld (1979) For further discussion on incest, see Meiselman (1978).

[51]"Puberty" is derived from the Latin *pubescere* ("to be covered with hair"), "adolescence" from *adolescere* ("to grow up"). *Teenager* is an American expression refering to ages 13 to 19. "Youth" may refer to younger persons in general or to a stage of life between adolescence and adulthood. *Nubile* means marriageable.

of the main reasons why adolescent sexuality is so often problematic.

Biological Aspects

Hormones and Sexual Drive   The intensification of sexual interest and activity during adolescence has been traditionally ascribed to an upsurge of the *sexual drive*, which in turn has been assumed to be due to the hormonal changes of puberty. As shown in figure 9.5, the increase in the level of testosterone appears to correlate with several significant sexual behaviors. Yet there is no firm evidence that such behavioral changes in adolescence are a direct outcome of higher hormone levels. For instance, in precocious puberty, higher hormonal levels are generally not accompanied by a concomitant precocity in dating, romantic, or sexual experiences despite an early awakening of sexual fantasies.[52] For normal adolescents, the onset of dating for a given teenager is influenced more by social factors (such as parental and peer expectations) than by hormone levels.[53]

[52]Money and Ehrhardt (1972).
[53]Dornbusch et al. (1981). Also see Peterson and Taylor (1980).

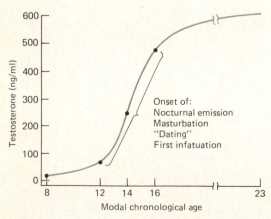

Figure 9.5   Amount of testosterone in the plasma of human males at different ages. (From Beach, in *Reproductive Behavior*, ed. Montagna and Sadler, 1974.)

Orgasmic Experience   Among males in the Kinsey sample, the earliest ejaculation remembered was at about 8 years and the latest first ejaculation at 21 years of age; the majority (about 90%) of all males had had this experience between the ages of 11 and 15 years (mean age close to 14).[54] In the lowest educational group the mean age for first ejaculation was 14.6 years, almost a year later than in the most educated group. The first ejaculation resulted most often from masturbation (in 2 out of 3 instances), nocturnal emissions (1 out of 8 instances), and homosexual contacts (1 out of 20 instances).

Psychosocial Aspects

Psychological Reactions to Puberty   The dramatic transformations of puberty require adjustments in body image and the concept of the self. There is a great deal of pride and pleasure in watching oneself grow into manhood and womanhood; a source of satisfaction shared by the parents. But there is also the potential for tension and distress. As the parts of the body grow at different rates, there is concern about one's ultimate shape and size. The breasts among girls, and the genitalia among boys, are especially likely to be sources of preoccupation.[55] Long hours in front of the mirror are not just to satisfy adolescent vanity (though there is some of that) but to monitor the transformations of one's face and body (figures 9.6 and 9.7).

Menarche used to be a stressful period for girls, but most teenagers are now well prepared to face it and welcome it; some parents participate in its celebration by sending their daughter flowers, gifts, and taking them out to dinner to mark the occasion.

Sex and Adult Status   An adolescent may engage in sex because he or she feels sufficiently adult, or because engaging in sex-

[54]Kinsey et al. (1948, pp. 185–187).
[55]Betancourt (1983a) and (1983b) addresses these issues for those 9 to 14 years old.

Figure 9.6 Paul Burlin, *Young Man Alone with His Face*, 1944. (Collection of Whitney Museum of American Art, New York.)

Figure 9.7 Pablo Picasso: *Girl Before a Mirror*, 1932. (Collection of The Museum of Modern Art, New York. Gift of Mrs. Simon Guggenheim.)

ual behavior will generate such recognition among adults.

Unlike many preliterate cultures, industrialized societies do not have commonly accepted **rites of passage** that signal to the individual and to the community that a shift in social status has taken place.

There are, nonetheless, strong social expectations of how adolescents should behave socially. As measured against prevailing standards, *lack* of sexual interest, *excessive* sexual interest, or the *wrong* kind of sexual interest, alarms parents and other significant adults. Parents may be concerned over teenage pregnancy, sexually transmitted illness, and the consequence of their children consorting with unsavory characters. So they feel a rightful responsibility to guide and protect their offspring. But parents may also try to enforce behavioral codes on their children that they themselves neither believe in nor follow; they may be more concerned with the family "image" than the intrinsic merits or demerits of their children's sexual behavior.

Other parents simply stay out of their teenager's lives as long as they do not get into trouble. Adolescents in these situations enjoy their freedom but they may yearn for guidance and the setting of limits, since such measures show that their parents care enough to be concerned.

Adolescent Sexual Behavior

Sexual behavior in adolescence is in some ways continuous with its childhood antecedents. But the transition is by no means a simple progression from sexual play to the "real thing." The meaning of various sexual acts in childhood may be quite different from their counterparts experienced later on. Adolescence itself is a time of sexual exploration, and patterns of sexual behavior are bound to change over time.

Heterosexual interactions during adolescence usually take place between peers in contexts like *dating*, where the sexual and the more purely social elements coexist. What two teenagers do and at what age,

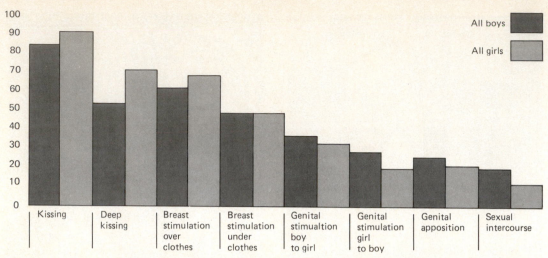

Figure 9.8 The incidence of eight sexual activities shown in percentages for boys and girls. (From Schofield, *The Sexual Behavior of Young People*, 1965, p. 31.)

varies with class background and other social factors. Generally, there is a gradual progression in the level of intimacy that follows a predictable order. Schofield, in a study of British teenagers, aged 15 to 19, has described this process in terms of certain "landmark" activities (figure 9.8).

Premarital Sex There are a few reliable studies of adolescent sexual behavior among high school populations. In a statewide sample in Illinois studied in 1972, fewer than 1 in 10 boys and girls have had coitus by age 15; one in five had done so by their 17th birthday.[56] A sample from a small Western city, studied in 1972, revealed that less than 10% of those younger than 15 had had coital experience, but one third of the boys and over half of the girls had done so by their senior year.[57] Vener and Stewart studied a sample of adolescents in 1970 and another group in 1973, in the same nonmetropolitan community: among 15-year-olds 20% of boys and 13% of girls in 1970, and 38% of boys and 24% of girls in

1973 reported to have engaged in sexual intercourse.[58]

Comparable figures showing the same trend towards increasing prevalence of premarital sex come from the studies of Kantner and Zelnick based on national probability samples.[59] As shown in figure 9.9, for women aged between 15 and 19, the more recent percentages are higher: whereas 27% of this age group had engaged in premarital sex in 1971, 35% did so in 1976. The median age at first intercourse declined during this interval by about 4 months (from 16.5 to 16.2 years).

These data also show significant differences between blacks and whites that are not due to race, as such, but reflect socioeconomic and subcultural factors. The rates for black women between 15 and 19 years of age are much higher than for white women: 51.2% against 21.4% for whites in 1971 and 62.7% versus 30.8% for whites in

[56]Miller and Simon (1974).

[57]Jessor and Jessor (1974).

[58]Vener and Stewart (1974). A study by Sorenson (1973) based on a flawed sample of 13- to 19-year-olds found 52% to have had coitus (boys 59%; girls 45%).

[59]Zelnick and Kantner (1977); Kantner and Zelnick (1972). These are among the most reliable studies conducted in the field of sexual behavior.

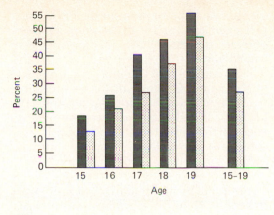

1976 ■   1971 ▨

Figure 9.9   Percent of never-married women aged 15 to 19 who have ever had intercourse, by age, 1976, 1971. (From Zelnik and Kantern, *Family Planning Perspectives*, 9:2, 1977, p. 56.)

1976. The higher ratio for blacks is true for each age group for both studies, but the relative differences are smaller in 1976, showing that the rates are rising more steeply for whites.

Although about a third of young women, and a somewhat higher proportion of men, have already had premarital sex before coming to college, for many it is during the college years that initiation into coitus takes place. This is also the period when the patterns for more intimate interpersonal relationships are rehearsed and established.

Premarital sex among college students became more prevalent during the 1970s. Though various studies report different rates, on average 3 out of 5 men and 2 out of 5 women attending college have reportedly engaged in premarital coitus.[60] Women in this group have registered the sharper increases, whereas the male rates have remained more stable; as a result, the sex discrepancy between premarital rates has diminished. This is especially apparent in studies that reveal higher rates of premarital sex: 73% for both sexes reported by Baumann and Wilson[61]; 82% for males and 85% for females, among seniors, reported by Jessor and Jessor.[62] When lower rates are reported, the percentage for males (60%) is twice as high as that for females (29%), as in the study by Lewis and Burr.[63] Yet large differences may exist from one college to another. For instance, in one institution, 90% of the women were still virgins at age 16.[64] In more conservative settings where such studies would not even be tolerated, the proportion could be even higher.

The Experience of First Coitus   Premarital sex constitutes a landmark in sexual development for those who experience it, but it is not an obligatory or normative stage for everyone. While many teenagers voluntarily choose to have sex, others are rushed and pushed by peers before they are psychologically ready or feel at peace about it. In the Hunt sample, only 4 out of 10 young women found their first coital experience "very pleasurable." More than a third of young males and close to two-thirds of young females felt regret and worried afterward. Their concerns were generally related to emotional and moral conflicts, fear of STD and pregnancy, and worry about adequacy of performance.[65]

In a more recent study of a group of women, the average age at first coitus was 17 (62% had it between 16 and 18; one as early as 9).[66] The male partners were about 2 years older; in 2 out of 3 cases the couple was engaged or going steady; in 13% of cases the man was a casual acquaintance; in 2% the woman was raped. Circumstances were described as spontaneous by 60%, and planned by 40%; in only half of the cases was any contraception used. The emotional reactions to first coitus were characterized as highly pleasurable by a third of these young women; an equal proportion reported high levels of anxiety and guilt; the balance had a mixture of pleas-

[60]Hopkins (1977).
[61]Baumann and Wilson (1974).

[62]Jessor and Jessor (1974).
[63]Lewis and Burr (1975).
[64]Jackson and Polkay (1973).
[65]Hunt (1974).
[66]Weis (1983).

ure, guilt and anxiety. The most important factor was the quality of the male partner: those seen as loving, tender, and considerate elicited the most positive reaction; those who acted in a rough and inconsiderate manner left the women feeling anxious, guilty, and exploited. These reactions prevailed irrespective of the nature of the relationship (such as being engaged). The experience also tended to be more positive if the woman was older, had extensive dating and petting experience, and had known her partner for a longer period.

In physical terms, first coitus need not be a traumatic event. The tearing of the hymen (chapter 2) entails minimal pain and bleeding, yet, despite more liberal sexual attitudes, the first coital experience remains problematic to a significant number of women, as well as to some proportion of men. If too much fuss was made in the past over first coitus, too cavalier and thoughtless a manner may now prevail in dealing with this important step in our sexual lives.

The Interpersonal Context of Premarital Sex   The increased prevalence and openness of premarital sex notwithstanding, there seems to be a remarkable constancy in the nature of the interpersonal relationships within which premarital sexual interactions occur. Kinsey and Hunt noted that premarital coitus more often occurred between people who intended to marry or were in a committed relationship. Even among the sexually liberal respondents of the *Cosmopolitan* survey, 64% women listed their first sexual partner as a steady boyfriend, 6% as their fiancé, 5% as their husband, and only 16% as a casual acquaintance.[67]

For many young people, sex requires that there be affection and caring, what Ira Reiss has called "permissiveness with affection."[68] But this is more of a precondition

for women than for men. In the Hunt sample, even among 18- to 24-year olds, sex outside an affectionate relationship was considered wrong for a man by 53% of the women and 29% of the men; for a woman in the same situation the corresponding percentages were 71% for women and 44% for men. Whether this reflects the persistence of the sexual double standard or is somehow determined by more intrinsic gender differences, is an open question.

Consequences of Premarital Sex in Adolescence   There are significant differences in how cultures view sexual behavior during adolescence (box 9.2). In our culture, societal concerns over premarital sex relate to unwanted pregnancy, sexually transmitted diseases, damage to reputation, and potential consequences to subsequent marital happiness.[69] Adolescent girls are at a higher risk with respect to most of these dangers, though boys are by no means immune to their effects.

Contraception should make pregnancies unlikely, in principle, yet in practice a million teenagers a year get pregnant (chapter 6). Effective public health measures should keep sexually transmitted diseases in check, yet their prevalence is highest among youth (chapter 7). Judgments of premarital sex have become far less harsh but promiscuity continues to be disapproved by most people. On the positive side, there are adolescents who engage in sex without suffering any of the above-mentioned ill effects. On the contrary, their lives seem to be enriched by the experience and their ability enhanced to make intelligent sexual decisions in adulthood.

Despite a good deal of preoccupation with it, the question of the impact of premarital sex on subsequent marital happiness remains unanswered. This is hardly surprising in view of the varied reasons for which people engage in premarital sex and the numerous variables that determine marital satisfaction.

Those who value sexual exclusivity in a monogamous bond argue that premarital

---

[67]Wolfe (1981, p. 270).

[68]The issues of premarital sexual standards and the social context of premarital sexual permissiveness have been studied extensively by Reiss (1960); (1967); (1977); (1980).

[69]Miller and Simon (1980).

sexual experiences lays the groundwork for future extramarital involvements. In fact, those with extensive premarital sexual experience do report more extensive extramarital activity.[70] Various studies have shown that marriage tends to be rated as more successful when premarital chastity has been maintained (the correlation is positive but not a strong one).[71]

Others find such evidence unpersuasive. Sexually conservative persons who shun premarital sex may also accept their marriages as "happy" because they are supposed to be; some respondents who rate their marriages "very happy" rank below average in their measures of coital frequency and sexual enjoyment.[72] Despite their correlation, there may not be a *causal* connection between engaging in premarital and extramarital coitus; both may be the result of an extraneous variable such as "sexual liberalism."[73] One could also argue on commonsense grounds that the best way to find out if a partner is sexually compatible would be to test it before marriage.

Cross-cultural evidence with respect to premarital practice suggests that its effect on marriage may be a function of the congruence between socially held values and practice (box 9.2). So the question of whether or not premarital sex will help or hinder marital happiness cannot be answered in the abstract. Much depends on whether such behavior creates a tension between personal values and behavior, and between individual choice and generally held cultural values with respect to it.[74]

Sex Education in Adolescence   The sexual education of adolescents is in some ways simpler, in other ways more difficult, than that of children. Courses in high school present many more opportunities to incorporate the teaching of sexuality. On the other hand, adolescents are not satisfied to learn about where babies come from: they want to know about sex in straightforward terms.

The proponents of such instruction point the urgent need to enlighten adolescents (whether sexually experienced or not) so they can make intelligent sexual decisions and act responsibly. To this end, courses in health education and marriage have existed for many decades.[75] What is new is an emphasis on the positive aspects of sexual relationships, not just the prevention of pregnancy and sexually transmitted diseases.

The opponents of sex education claim that such instruction incites sexual behavior by providing it with tacit approval if not downright encouragement. Teaching about contraception is claimed not to reduce the frequency of pregnancy but to make it more prevalent.

Parents may have a hard time with teenage sexual behavior because it tends to become entangled in the adolescent drive for emancipation and parental efforts at control. But not all adolescents go through such rebellion or make sex a central part of it. Many young people and their parents now go through this phase of life in relative peace with the assurance that adolescents have their freedom of action but use such freedom responsibly.[76] We have more to say about the societal aspects of adolescent sexuality and sex education in chapter 16.

Sexual Identity

The term *identity* (from the Latin for "sameness") refers to the persistent individuality and sameness of a person or thing over time and under different circumstances. *Sexuality identity* is that component of a person's identity that pertains to his or her sexuality. Defined in this global sense, sexual identity has several components: the basic perception of oneself as

---

[70]Athanasiou and Sarkin (1974).

[71]See Shope and Broderick (1967); Reiss (1966).

[72]Athanasiou et al. (1970).

[73]Reiss (1960).

[74]Christensen (1966) applies this argument in her analysis of contemporary shifts in sexual values.

[75]For a collection of essays in sex education, see Brown (1981).

[76]Numerous books exist to guide sexual behavior during adolescence. See, for instance, Pomeroy (1968), (1969); Comfort and Comfort (1979).

Box 9.2   Cross-Cultural Perspectives on Premarital Sex

In about half of preliterate societies, women are allowed to have premarital coitus. If societies that overtly condemn but covertly condone such behavior are included, the figure rises to 70%. Where they have free access to each other, young men and women appear to engage in sex almost daily. But the choice of partners is not indiscriminate.

Marshall has provided a detailed account of a sexually permissive society of Polynesians living on the island of Mangaia (one of the Cook Islands in the South Pacific).[1] These people seem to have mastered the art of "looking the other way" where sexual matters are concerned.

Mangian boys may experiment with coitus prior to being formally initiated into manhood, but these contacts usually involve sexually experienced older women or widows. To have access to their peers or the more desirable "good girls," a boy must first undergo the initiation rite of superincision, whereby the skin at the upper surface of the penis is cut along the midline.

The standard pattern for premarital sex is "sleep crawling" (*motoro*), which appears to be a general Polynesian cultural practice. But as the following description indicates, for youth to get together in Mangaia is not that easy, which again shows that even sexual permissiveness does not imply free and indiscriminate access to sex.

A youth bent on courting must first slip out of his home and then avoid village policemen who are out searching for violators of the nine o'clock curfew. He must then get to the girl's house before other suitors. His effort may be made at sometime between ten o'clock in the evening and midnight; if successful, he may remain until three o'clock in the morning. If the girl has invited him to visit her, or if he is a regular visitor, his task is much easier. If not, he is faced with the alternative of "sweet talking" her (with the risk of waking the family) or of smiting her or gagging her mouth with his hand or a towel until he has made his penetration. Mangaian boys believe that most girls will not call the family if the suitor talks to her but that

---

male or female, as a sexual being ("sexy," "not sexy," "oversexed," "undersexed"); sexual orientation ("heterosexual," "homosexual"); sex values ("permissive," "liberal," "conservative"); and gender identity ("masculine," "feminine").

The genetic, hormonal, and anatomical characteristics of the person form the biological groundwork for sexual identity. The resultant *maleness* or *femaleness* is the *sex* of the person. On this biological bedrock are constructed the psychosocially defined gender characteristics whereby the fact of being a biological female or male is given psychological and cultural meaning. Thus, beyond being male or female, the person develops a sense of being *masculine* or *feminine*, which constitutes the individual's *gender identity* (box 9.3). Social expectations of gender-linked behavior are, in turn, known as *sex roles*.

Psychologists also use the term *sex-typed behavior* for "role behavior appropriate to a child's gender" and *sex-typing* for "the developmental process by which behavioral components of one or another gender role are established."[77] Sex-typed behavior is thus basically synonymous with gender behavior.

The consolidation of a sense of *ego identity*, the answer to the question "Who am I?" is the key task of adolescence according to Erik Erikson (discussed further on). Sex

[77]Sears (1965). For further terminological consideration of sex and gender see Katchadourian (1979).

Box 9.2   Cross-Cultural Perspectives on Premarital Sex *(Continued)*

many a girl will call out to her father if the boy tries force. In any event, once penetration is made, most resistance is gone; "She has no voice to call." Less aggressive boys or . . . less socially desirable youths may take up to a year of "sweet talk" to win the girl they desire; others may speed their suit by serenades. Most of them simply take what they want, when they want it.[2]

The majority of cultures surveyed by Ford and Beach were characterized less by differences in sexual codes for juveniles than by the lack of vigor in enforcing existing prohibitions.[3] For example, the Alorese (polygynous agriculturalists of Indonesia), the Andamanese (monogamous seminomadic gatherers, hunters, and fishers), and the Huichol (monogamous Mexican people who live by animal husbandry and agriculture) all formally disapproved of premarital intercourse, but made little effort to check it. As long as the practice was not flaunted and did not result in pregnancy, it was generally tolerated. When pregnancy occurred the cou-

ple was pressured into marriage. Contraceptive measures were therefore frequently used by youngsters in these cultures. Methods included coitus interruptus, ejaculation between the girl's legs, placing a pad in the vagina and washing the vagina after coitus. A pregnant girl might also have resorted to abortion.

In sexually restrictive societies, female virginity at the time of marriage is highly valued and society goes to great pains to enforce prohibitions on preadolescent sexual behavior. Chaperoning, and segregation are some preventative measures. Disgrace, punishment, and even death are invoked when prevention fails. Among the Chagga (advanced polygynous agriculturalists of Tanzania), for instance, if a boy was caught in an illicit sex act, he and his partner were placed one on top of the other and staked to the ground.

[1]In Marshall and Suggs (1971, pp. 103–162).
[2]Donald Marshall, "Sexual behavior in Mangaia," in Marshall and Suggs (1971, p.129). Reproduced by permission.
[3]Ford and Beach (1951).

plays an important role in defining who one is. Sexuality leads to establishing intimacy with others and the capacity for intimacy in turn allows the expression of sexuality. Adolescents thus use sex to get to know each other as well as to define themselves. Such interactions lead to close associations outside the family circle and eventually to the establishment of one's own family.

Gender identity and sex roles are particularly important in the shaping of *interpersonal relationships*. The significance of gender is integral to all sociosexual behavior; it is not possible to act sexually without simultaneously behaving either as a male or female. So when adolescents behave sexually, it is not possible to always tell

whether they are fulfilling sexual needs or trying to live up to social expectations of what it means to be a man or woman.

The Study of Sex Differences

Researchers in the 1920s began to study personality traits that would differentiate between men and woman.[78] These studies yielded correlations between gender and various traits (such as assertiveness or passivity) but did not explain why and how these sex differences came about. Furthermore, these *masculinity-femininity* tests revealed no personality characteristics that

[78]A good example is the "Masculinity-Femininity Test" by Terman and Miles (1936).

Box 9.3   The Concepts of Gender Identity and Sex Roles

The term *gender identity* is recent, yet preoccupation with masculinity and femininity long antedates contemporary scholarly interest in it. In Chinese philosophy and religion, the masculine principle called *Yang* and the feminine principle called *Yin* were attributed not only to men and women, but to everything in existence, including inanimate objects, spirits, and events.[1]

Although the concept of Yang and Yin cannot be simply equated with sexual identity, some contemporary writers have used it or some similar idea to represent the male and female elements as present in different proportions in every individual of either sex. The Jungian concept of the animus and anima is a case in point.

Jung conceived of the anima and the animus as archetypal images that are rooted in the collective unconscious (chapter 8). The *anima* represents the feminine personality element within the male and the image that the male has of feminine nature in general— in other words, it is the archetype of the feminine. Likewise, the *animus* represents the masculine personality components of the female and her concept of masculine nature.

Normally, both masculine and feminine characteristics are present in every individual, but the person expresses outwardly only the set of characteristics that are considered socially appropriate to his or her sex, and are therefore not disturbing to the ideal self-image.[2]

The word "gender" is derived from the Latin *genus* meaning birth or origin, and hence representing a class of a given entity. The term *gender identity* was introduced by Robert Stoller since the use of the word "sexual" in *sexual identity* was confusing. Stoller explains it as follows:

the word sexual will have connotations of anatomy and physiology. This obviously leaves tremendous areas of behavior, feelings, thoughts, and fantasies that are related to the sexes and yet do not have primarily biological connotation. It is for some of these psychological phenomena that the term gender will be used: One can speak of the male sex or the female sex, but one can also talk about masculinity and femininity and not necessarily be im-

were exclusively male or female; rather, various traits were unevenly distributed among males and females.

Although these gender tests did not claim that the perceived differences between men and women were innate, immutable, or socially desirable, they nonetheless reinforced common perceptions of what was considered to be "natural" behavior for male and female. They also perpetrated a gender model that was *unidimensional* and *bipolar*: gender was viewed as a single personality dimension with masculinity and feminity at the opposite poles of a continum; if you scored high in masculinity, you had to score low in feminity, and vice versa.

This approach to the study of gender

differences is now considered inadequate.[79] Instead of viewing them as mutually exclusive entities, gender traits are now seen as characteristics that coexist in various combinations. Thus, a man or woman can be assertive or dependent in one context and the contrary in another without being type-cast as being less or more masculine or feminine.

In a review of over 1,400 studies, Maccoby and Jacklin found only four sex differences that have been well established: girls show higher *verbal ability*; boys do better in *visual-spatial tests*, show higher *mathematical performance*, and are more *aggres-*

[79]Constantinople (1973); Spence and Helmreich (1978).

Box 9.3   The Concepts of Gender Identity and Sex Roles *(Continued)*

plying anything about anatomy or physiology. Thus, while sex and gender seem to common sense to be practically synonymous, and in everyday life to be inextricably bound together, the realms (sex and gender) are not at all inevitably bound in anything like a one-to-one relationship, but each may go in its quite independent way.[3]

A concept closely related to gender identity is that of *sex role,* or *gender role.* The term as well as the notion of role originated in the theater. The Latin word *rotula* means a small wooden roller. The parchment containing the actor's script was rolled around this rod and hence was referred to as the roll (*rowle*). The actor's role was thus defined by the script to be enacted in the play.

Sociologists define role in terms of an individual's *position,* by which they mean the location of an individual in a system of social relationships. Such a position is independent of any particular person and entails a more or less explicit set of responsibilities and prerogatives. For instance, many different persons could be a "teacher" or a "par-

ent," and so on. A role is a set of social expectations as to how someone in a given position should behave toward others in other positions. Roles are in this sense another variety of norms or shared rules about behavior.

Sex roles are those roles determined by sex. They pertain to many aspects of our lives. Some of these define societal expectations of, for instance, what women and men are supposed to do vocationally. Other sex roles pertain more directly to how males and females are supposed to behave sexually. Such social expectations are enormously influential in how we behave sexually. Thus, in addition to determining, for instance, whether or not a man or a woman or both should take the initiative in sexual encounters, the differential functions and statuses to which the two sexes are relegated even in ostensibly totally nonsexual contexts may significantly affect their sexual lives.

[1]Gulik (1974).
[2]Jung (1969).
[3]Stoller (1968, pp. vii–ix).

sive. However, these reviewers left entirely from consideration gender differences in sexual behavior while acknowledging that "sexual behavior per se may of course be the sphere of behavior most affected by the biology of sex."[80]

An Interactional Model of Gender Identity Differentiation

It is now commonly agreed that the development of gender identity can be best

understood as the result of the interaction of biological and social variables.[81] But there is disagreement about the relative importance of biological influences and the extent to which they predispose the development of gender differences. So the question is whether societies merely reinforce biologically determined gender traits or generate such gender models and then ascribe biological roots to them. In any case, it is clear that children acquire their gender identity at a young age and act according

[80]Maccoby and Jacklin (1974). Also see Astin et al. (1975); Ortner and Whitehead (1981).

[81]For overviews of various influences on sexual identity, see contributions in Katchadourian (1979).

Figure 9.10 Girls and boys playing at their gender roles. (John Rees, figure above; Billy Barnes, right figure. Black Star.)

to the dictates of sex roles quite consistently (figure 9.10)

Money and Ehrhardt have proposed an *interactional model* of psychosexual differentiation that is integrative and attempts to avoid dichotomies between "nature versus nurture, the genetic versus the environmental, the innate versus the acquired, and biological versus the psychological, or the instinctive versus the learned." They define gender identity as follows: "Gender identity: The sameness, unity and persistence of one's individuality as male, female, or ambivalent, in greater or lesser degree, especially as it is experienced in self-awareness and behavior; gender identity is the private experience of gender role, and gender role is the public expression of gender identity."[82]

As shown in figure 9.11, gender differentiation is conceived as a continuous process starting at conception and culminating in the emergence of adult gender identity. The succesive stages in sexual differentiation follow each other as in a relay

race; the "program" of instructions of sexual differentiation is initially carried by the *chromosomes* and handed on to the undifferentiated *gonad*, which in turn passes it on through its *hormones* to various tissues. Bodily and genital differences that result from genetic and hormonal influences lead to *genital dimorphism* (chapter 4). The perception of ourselves in these physical terms leads to our *body image*.

Although the initial processes in this sequence are biological, there are profound and pervasive social factors at work that also influence early gender development. From the moment of birth, boys and girls are treated differently. Such differential treatment of male and female persists throughout life, constantly reinforcing culturally defined gender models and stereotypes. The influences that are exerted by the reactions of others (*other's behavior*) contribute to the emergence of *juvenile gender identity*. In the next major developmental stage, changes brought about by *pubertal hormones* set apart male and female even further in terms of bodily and sexual structures (*pubertal morphology*) and sexual functions (*pubertal eroticism*). The conflu-

[82]Money and Ehrhardt (1972).

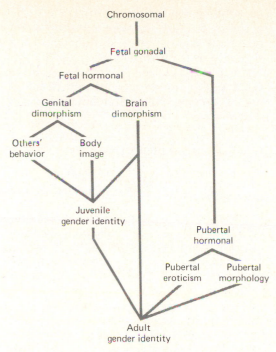

Chromosomal

Fetal gonadal

Fetal hormonal

Genital dimorphism   Brain dimorphism

Others' behavior   Body image

Juvenile gender identity

Pubertal hormonal

Pubertal eroticism   Pubertal morphology

Adult gender identity

Figure 9.11   The sequential and interactional components of gender-identity differentiation. (Money and Ehrhardt, *Man and Woman, Boy and Girl,* 1972, p. 3.) Reproduced by permission.

activities are predominantly performed by men (such as hunting and metal working) others by women (cooking; grinding grain; repair of clothing); and yet others by either sex (planting; making leather products).[83] Industrialization has attenuated such role specialization but even in settings, such as the Israeli kibbutz, with an ideological commitment to sexual equality, there is a tendency for women to gravitate into traditional housekeeping and childcare activities.

Some see this as evidence for the assertion of biological factors despite cultural attempts to tamper with them. They would therefore like to sustain these distinctions by raising children to follow discrete and mutually exclusive gender-linked personality patterns and roles. Others object to the perpetuation of these patterns, which have traditionally severely restricted the vocational choices available to women. It is further argued that sharp distinctions between what is considered masculine (ambitiousness, self-reliance, independence) and feminine (being affectionate, gentle, nurturant) create restricted personality types. Instead, an ***androgynous*** ("andro," man; "gyne," woman) ideal integrates both sets of characteristics within men and women, whereby one is tough or gentle, not by virtue of being male or female, but in response to what would be most appropriate for a given situation.[84] Though recently there has been a significant shift toward the attenuation of gender differences be it in dress, behavior, or sense of identity, we are still far from living in a *unisex* society.

Children with ***atypical gender behaviors,*** such as in their manner of dress, choice of toys, roles taken in games, peer group compositions, and participation in sport, continue to be labeled "tomboys" and "sissies," and may experience considerably difficulty, especially in the latter case.[85]

ence of all these developments leads to the emergence of *adult gender identity.*

In addition to the two main avenues of influence by hormones, there is a third path shown in figure 9.11 that leads to *brain dimorphism.* What this means is that just as the genitals are shaped differently by hormones, so presumably are the male and female brains. This would in turn lead to differences in gender-related behaviors, including some aspects of sexual behaviors. If true, this is where endogenous differences in gender would come in, even though they would still be modified by social learning. Such hormonally induced modification of the brain is not assumed to dictate specific behavioral patterns, but to predispose the individual to elect a particular set of gender roles, or to facilitate the development of a male or female gender identity. The research underlying these conclusions, which are as yet not firmly established, was discussed in chapter 4.

Cross-cultural studies show that certain

[83]D'Andrade (1966).

[84]See Bem (1974) for the measurement of psychological androgeny.

[85]For systematic descriptions of children with atypical sex-typed behaviors, see Green et al. (1982); Green (1976).

When these problems are serious enough, they are referred to as *gender identity disorders* in childhood, *transsexualism* in adulthood (box 9.4)

There is an important distinction, however, between these gender disturbances and androgyny. Children with gender disorders and transsexuals are often deeply unhappy about their biological sex: they feel trapped in the wrong body. What they want is to become a member of the opposite sex. Androgynous persons do not reject their biological sex nor the traits that typically go with it. Instead they say they wish to have the best of both worlds.

Sexuality in Adulthood

Traditional concepts of development have largely focused on childhood and adolescence. In these contexts, adulthood has represented a developmental plateau leading to the final changes of old age. Common perceptions of sexuality fit into a similar paradigm: sexual stirrings in childhood, awakening in adolescence, peak performance in early adulthood, gradual decline in middle age, and extinction of sexual desire and performance in old age.

Over the past two decades, adulthood has become a more focal point of study and is now seen as a dynamic period of continuing change and growth. This extension of the developmental perspective to the entire life span was present in the works of some of the early psychoanalysts, like Jung, but it has been only recently that this approach has attracted wider attention through the publication of longitudinal studies of adult life.[86]

Yet so far, this life-cycle literature has not specifically focused on the sexual aspects of adult development. We thus have a lot of theorizing on childhood sexual development, but little data, and a lot of data

on adult sexual behavior but little by way of theory to make developmental sense out of it. Since most of this text deals with adult sexuality, our purpose here is not to reiterate or summarize that material but to point out a number of salient developmental aspects of sexual behavior beyond adolescence.

Early Adult Transition

Following the end of adolescence, the person goes through a transitional period that leads to adult status. The precise ages for this period depend on what occupational path the person takes and other factors. Levinson assigns ages 17 to 22 to this phase; hence it roughly corresponds to the college years, although not everyone, of course, spends those years in college.[87]

College-age students are biologically fully adult, but in sociosexual terms they range from late adolescents to young adults. They have also been characterized as belonging to the stage of *youth,* a phase of life that has come about because of the lengthy period of vocational and social maturation required in highly industrialized societies like the United States.[88]

Based on their work with college students, Philip and Lorna Sarrel have characterized the interaction of biological, psychological, and social factors in sexual development as a process of *sexual unfolding* that they view as the central developmental task in moving toward adult sexuality.[89] Through dating experiences, young men and women develop sexual competence, the skills for sexual negotiation; they also come to terms with issues of intimacy, commitment, and fidelity.

Contrary to the stereotypical image of carefree sexuality, it is not unusual for open and sophisticated young men and women to have sexual problems during this stage of their lives. Sometimes all that is required in helping them is to provide relief from the social pressures to act sexually, thus allowing them to develop at their own

[86]See, for example, Block (1971) and Vaillant (1977). Other studies of adulthood include Levinson (1978); Gould (1978). For general overviews see Neugarten (1968); Erikson (1978); Smelser and Erikson (1980); Brim and Kagan (1980); Steinberg (1981). For an introduction to adulthood as a stage of life, see Katchadourian (1984).

[87]Levinson (1978).
[88]Erikson (1963).
[89]Sarrel and Sarrel (1979).

rate. At other times, more specific sex therapy techniques are called for.

Midlife Transition and Aging

The sexual changes and challenges that accompany the transition into middle age and subsequently into late adulthood have been mainly studied from the perspective of declining sexual activity and sexual dysfunction. This is unfortunate since it suggests that the process of growing old is pathological, which of course it is not.

However, since we do not as yet have a systematic picture of the normative changes in sexual behavior during middle and late adulthood, we have to deal with these behaviors in the more general context of the chapters that follow and then focus on the issue of age-related sexual change and decline when we deal with sexual dysfunction (chapter 15).

Theories of Psychosexual Development

Theories of *psychosexual development* are attempts to explain the development of sexual behavior through childhood and adolescence and their relation to adult sexuality. As the term "psychosexual" implies, sexual and psychic processes are intimately interrelated, and it is not possible to understand one apart from the other.

There is as yet no generally accepted theory of psychosexual development; certain aspects of sexual development are best explained by one theory and other aspects by another. These developmental models are closely linked to the broader theories of sexual behaviors that we discussed earlier (chapter 8) and should be approached as their extensions.

Primate Models of Psychosexual Development

Based on studies of the development of affectional interractions in monkeys, Harlow has discribed five basic *love systems:* maternal love, paternal love, infant-mother love, peer or age-mate love, and heterosexual love.[90]

[90]Harlow et al. (1971).

The Maternal-Love System   This affectional phase begins with the stage of care and comfort as exemplified by the feeding, protection, and physical *contact comfort* provided by the monkey mother to her clinging infant. As the infant grows older, the mother becomes less solicitious; she may even reject the infant to foster independence.

The reciprocal love of the infant for the mother (*infant-mother love*) is initially one of "organic affection" based on unlearned patterns that prompt the infant to feed at the breast or seek contact comfort. As infants feel more secure, they tend to wander off, gradually becoming less dependent on the mother.

Paternal Love   The attachment of monkey fathers to infants seems to be less compelling, although monkey fathers do protect infants that they live with. Conversely, the infant's affectional tie to the father is less evident than is its attachment to the mother, although infants occasionally seek out adult males for comfort when no females are available.

Peer Affectional System   Peer attachment among monkeys may well be the most important of all affectional systems. The rudiments of love appear in early infancy and persist throughout childhood and adolescence. The primary vehicle for the development and expression of peer love is *play*, through which earlier affectional systems are integrated and the love systems that follow are anticipated.

Heterosexual Affectional System   Rooted in earlier love systems, this phase emerges as a distinct entity at the time of puberty, matures during adolescence, and operates as the primary sociosexual affectional system for most adults.

The heterosexual love system develops through three subsystems: mechanical, hormonal, romantic. The *mechanical subsystem* depends on the proper function of the sexual organs, physiological reflexes, appropriate body postures and movements. Monkeys have a basic adult copulatory posture (see figure 11.4). This posture arises out of earlier responses during infant play

Box 9.4   Gender Identity Disorders

To warrant the diagnoses of gender identity disorder, a condition must entail more than nonconventional gender behavior; there must be a strong and persistent desire to be a member of the opposite sex. Hence, what is at issue here is not a rebellion against social convention, but a rejection of one's own physical sex. Therefore, departures from traditional gender behaviors, as such, do not constitute gender disorders, nor do alternative sexual orientations like homosexuality. In short, the essential feature of gender identity disorders is "an incongruence between anatomic sex and gender identity."[1]

Gender identity disorder of childhood is characterized by a persistent rejection by a child of his or her anatomic sex, accompanied by the desire to be, or the insistence that one is, a member of the opposite sex. Thus girls may seek male peer groups, show an avid interest in aggressive sports, rough-and-tumble play, and a lack of interest in playing with dolls or playing house (unless they play the role of the father). Occasionally a girl will claim that she will not grow into a woman but will become a man, and that she will grow (or already has) a penis. Some of these children have additional disturbances (like nightmares and phobias).[2]

These tendencies become manifest early in childhood but most girls yield to social pressure by adolescence and give up their attachment to male activities and attire. A minority retain a strong masculine identification and some develop a homosexual orientation. Boys with this disorder show a preference in dressing in female clothes (three-fourths begin to cross-dress before age four) and a compelling interest in girls as playmates and their games (such as playing with dolls, being the mommy when playing house). Rough-and-tumble play and other boyish activities are avoided, making these children subject to teasing and rejection by

other boys. These boys express disgust at their penis and wish that it would disappear, or they may believe that they will grow up to become women.

By age seven or eight, these boys may become seriously ostracized by their peers and may refuse to go to school ("sissies" have a harder time in this regard than "tomboys"). Under social pressure, such behavior may lessen in adolescence. But as many as half become aware of homosexual inclinations at this time.

In order to be diagnosed as a *transsexual* (or suffering from the *gender dysphoria syndrome*) a person must manifest persistent and serious discomfort with respect to his or her genital organs, wish to be rid of them, and want to live as a member of the opposite sex. The person should be free of hermaphroditic abnormalities and must not be psychotic. The condition must have been continuous over at least 2 years.

Superficially, there is much overlap in the behavior of transsexuals and transvestites. A male transsexual will dress up in female clothing as will a male transvestite. But unlike the latter, the transsexual does not derive sexual pleasure from dressing up as a female—so far as he is concerned, he is merely putting on the appropriate clothing for his true gender, which he believes to be female. Similarly, the transsexual male, like the homosexual, will be sexually attracted to other men. But unlike the homosexual who is seeking a partner of the same sex, the transsexual male feels like a woman would toward a man. Conversely, neither transvestite nor homosexual males think they are female. The same distinctions hold true for transsexual women, transvestites, and lesbians.

The world became widely aware of transsexualism in 1953 when an American ex-marine underwent surgery in Denmark to

Box 9.4   **Gender Identity Disorders** *(Continued)*

become Christine Jorgensen. Since then, about a thousand transsexual men and women have had sex changes through hormones and transsexual surgery, some of them gaining international prominence.[3]

The cause of transsexualism is unknown and the dilemma over biological versus psychosocial causation remains unresolved despite a good deal of research.[6]

Responsible programs deal with the treatment of transsexuals in a comprehensive psychological, social, hormonal, and surgical approach.[7] Following careful selection, the candidate receives psychological counseling, practical guidance (with matters like grooming, vocational choice, and legal issues) as well as a hormonal therapy to bring about changes in secondary sexual characteristics (most prominent of which are the growth of

breasts in males and facial hair in women). The transformation that these individuals undergo can be quite spectacular (see accompanying figures).

[1]American Psychiatric Association (1981).

[2]For therapeutic approaches to these childhood problems, see Green (1980).

[3]Among the more notable cases: Dr. Richard Raskin, an ophthalmologist and accomplished tennis player who became Dr. Renee Richards and began to compete in women's tennis tournaments with some controversy. John Morris, noted British journalist and writer became Jan Morris, whose autobiography, *Conundrum,* is an insightful and moving account of her experience as a transsexual.

[5]American Psychiatric Association (1980).

[6]Stoller (1972, 1975); Green (1974); Sorensen and Hertopt (1982).

[7]Laub and Gandy (1973); Laub and Dubin (1979).

Female to male transsexual change (courtesy, Dr. Donald Laub).

Figure 9.12   Basic presexual position. (Courtesy of Wisconsin Regional Primate Research Center.)

that involve interactions that establish dominance relationships. When there is a confrontation between two monkeys, the matter is usually settled by *threat displays* rather than actual fighting. The female monkey that usually yields takes a submissive posture by presenting its hindquarters to the more assertive male animal (figure 9.12) who may subsequently mount it (figure 9.13). At this stage, there is no apparent sexual intent in the act for either animal. But when monkeys reach puberty and are ready to copulate, this gender-linked behavior, which is already in place, leads to sexual intercourse.

The *hormonal subsystem* consists of the endocrine factors that we discussed with respect to puberty (chapter 4). It may sound strange to refer to a *romantic subsystem* among monkeys, but nonetheless, their sexual relationships are not engaged in a random fashion, and expressions of special interest and attachments of varying transience can be readily discerned among consort relationships (chapter 12).

These heterosexual affectional subsystems vary in their vulnerability to disruption. For example, social isolation early in life does not disturb hormonal functioning but may seriously disrupt the romantic subsystem. When monkeys are reared in social isolation, the socially deprived animals develop normally at puberty. But the effects of social isolation on sociosexual behavior is disastrous. These males may get

Figure 9.13   The first attempts at mounting. (Courtesy of Wisconsin Regional Primate Research Center.)

Figure 9.14 Inappropriate response by socially deprived male to receptive female. (Courtesy of Wisconsin Regional Primate Research Center.)

visibly aroused in the presence of receptive females, but stand puzzled, not knowing what to do (figure 9.14); they grope aimlessly, act clumsily and sometimes assault the female. The impact of social isolation on females is somewhat less damaging. Although they mistrust physical contact and will flee or attack males, they can be induced in time to endure at least partial sexual contact with normal males; figure 9.15 shows the inadequate sexual posturing of a socially deprived female and the sexual uninterest of a normal male in response to her.

The effects of isolation on monkeys have generally proven resistant to efforts at reversing them through aversive conditioning methods. Far more successful has been the use of "therapist" monkeys. These are normally reared monkeys who are younger than their socially deprived "patients" and who are able to initiate social contact with them in a nonthreatening manner. Within 6 months of such interaction, the disturbed behavior of the socially deprived monkeys has given way to normal age-appropriate social and play behaviors.[91]

Monkeys clearly do not "automatically" learn how to copulate; their ability to do

so depends on a long sequence of antecedent interactions with caretaker adults, as well as with peers. As Harlow has put it: "Sex secretions may create sex sensations, but it is social sensitivity that produces sensational sex."[92]

Psychoanalytic Theory of Psychosexual Development

Psychoanalytic formulations of psychosexual development assume that the newborn child is endowed with a certain libidinal "capital" (chapter 8). Psychosexual development is then the process by which this diffuse and labile sexual energy is invested in certain pleasurable zones of the body (mouth, anus, genitals) at successive stages.[93]

Stages of Psychosexual Development The oral zone is the first site of libidinal investment, and during the *oral stage* the mode of gratification is through "taking in," or *incorporation*. Erotic gratification through

[92]Harlow et al. (1971), p. 86.

[93]Freud's theory of infantile sexuality was first formulated in *Three Essays on the Theory of Sexuality*. For a clear and comprehensive review of the psychoanalytic model of psychosexual development, see Meissner et al. (1975).

[91]Suomi et al. (1972).

Figure 9.15   Inadequate sexual presentation by socially deprived female. (Courtesy of Wisconsin Regional Primate Research Center.)

the mouth persists through life and is obtained through kissing, mouth-genital contact, and so on.

The erogenous zone invested with libido during the second stage of development is the anus, and during the **anal stage** the primary modes of gratification are *retention* and *elimination*. The conflict over toilet training (retaining or expelling the feces at will) leads to *ambivalence*, a love-hate relationship with the parents that serves as a model for all subsequent interactions whereby we neither exclusively love or hate anyone. *Anal erotism* refers to the sexual pleasure in anal functioning. This sets the stage of obtaining erotic gratification through rectal stimulation such as during anal intercourse. The oral and anal stages generally extend through the first 3 years. As the genitals have not yet been invested with libido, these two stages constitute the *pre-genital* phase of psychosexual development and are identical for both sexes.

Between the ages of 3 to 5, the zones invested with libido are the *penis* in the male and the *clitoris* in the female. During the **phallic stage,** behavior is dominated by the *intrusive mode*. The main issue at this time is the development of the *oedipus complex* whereby the child forms an erotic attachment to the parent of the opposite sex

and feelings of hostility towards the same-sexed parent. Since the child also loves the rival parent he or she suffers guilt. A boy imagines that he will be punished by *castration* (by which Freud meant loss of the penis, not the testes). A girl's reaction to the discovery that boys have penises is to conclude that she has already been "castrated," and to envy the male's genitals ("penis envy").[94]

The oedipal conflict is resolved by the child giving up the opposite-sex parent as an erotic object and identifying with the parent of the same sex. This results in a *heterosexual* orientation. If the child identifies with the parent of the opposite sex and chooses the parent of the same sex, a *homosexual* orientation develops. In this, every child has the potential of developing in either direction. To ensure that these unacceptable wishes never become known, the entire conflict is buried in the unconscious. Oedipal wishes nevertheless continue to influence sexual behavior and their derivatives occasionally seep into consciousness.

The resolution of the Oedipus complex

[94]See footnote 27 in chapter 8 for the male-centered aspect of psychoanalytic views with regard to female sexual development

signals the end of the phallic stage. During the next phase of development (*latency*), the sexual drive is relatively quiescent while intellectual growth and social maturation are given greater impetus. With adolescence, sexuality awakens. To the extent that past conflicts have been satisfactorily resolved, the person is free to initiate sexual interaction along adult lines, attaining *genital maturity* whereby the young man ceases to be preoccupied with his penis and reacts sexually with his whole self; the young woman likewises incorporates the "inceptive" and "inclusive" modes into her psychosexual structure, finding sexual fulfillment in the prospect of motherhood. The successful outcome of this period leads to the dissolution of the dependent relationship on parents. The person is then ready to establish a mature, nonincestuous heterosexual relationship with another adult and pursue the aims of life that Freud characterized as *work* and *love*.

The demands of pregenital impulses give infantile sexuality an undifferentiated or *polymorphously perverse* character. The child represents at this time the aggregate of all possible sexual interests; heterosexual, homosexual, and paraphiliac (chapter 14). Normally these unbridled sexual wishes become *repressed* and *sublimated* into socially acceptable behaviors along with the emergence of heterosexual object choices. Yet some of the residue of these early wishes continue to break through in normal adult life in the form of dreams, fantasies, and other symbolic forms. By and large, early memories of sexual thoughts and behaviors are thoroughly forgotten through the process of *infantile amnesia*.

The Eriksonian Approach to Psychosocial Development

Building on psychoanalytic concepts, Erik Erikson has developed a scheme of psychosocial development that encompasses the entire life span rather than only the early years of life.[95] Erikson characterizes

the *life cycle*—the entire life span from birth to death—by eight phases of psychosocial development. Each phase is defined by the primary accomplishment of a *phase-specific task*, even though the resolutions are generally prepared in preceding phases and worked out further in subsequent ones.[96]

Erikson borrows from embryology the *epigenetic* principle, "that anything that grows has a ground plan, and that out of this ground plan the parts arise, each part having its time of special ascendency, until all parts have arisen to form a functioning whole."[97] Viewed in this light, the child is not simply a miniature adult, and sexuality is a potential that develops only if permitted and assisted.

Sexuality is an important factor in the task of *identity formation* (phase 5), the concept for which Erikson is best known. Biological features are the primary "givens," but they do not necessarily determine the individual's own definitions of himself or herself as masculine or feminine. It is up to each individual to clarify and consolidate his or her own sexual character as part of the larger task of identity formation. Cultures that provide clear and consistent models and guidelines facilitate this task for their members.

Despite the upsurge in sexual activity during adolescence, only after identity formation has been fairly well consolidated does true intimacy with the opposite sex (or with anyone else) become possible. Adolescent sexuality is often experimental and part of the search for identity; young people try to find themselves through each other.

The task of resolving the conflict of *intimacy versus isolation* belongs to young adulthood (phase 6). The young adult works out a compromise between these two polar opposites. Intimacy requires expressing, exposing, giving and sharing of oneself. Isolation ("distantiation") is also

---

[95]For Erikson's reformulation of psychosexual development, see *Childhood and Society* (1963), chapter 2.

[96]Erikson's scheme of psychosocial development is briefly stated in chapter 7 of the same work, but it is more fully elaborated in *Identity and the Life Cycle* (1959), Part 2, and *Identity: Youth and Crisis* (1968), chapter 3.

[97]Erikson (1968, p. 92).

necessary since no matter how satisfying a relationship, a certain distance helps people to keep each other in proper focus. When intimacy fails or is exploited, self-preservation requires that one repudiate the relationship and sever such destructive ties.

Cognitive Developmental Models

A fundamental principle underlying development theories is the concept of *maturation,* an innately governed pattern of growth that does not depend on particular environmental events, although the requisite amount of environmental stimulation and support is necessary for its unfolding. Developmental theories also postulate certain *stages* through which the individual progresses; closely related is the concept of *critical periods* during which key events must take place during circumscribed periods of time (chapter 4).

Freud's theory of psychosexual development and Erikson's stages of psychosocial maturation are classic examples of this conceptual paradigm. Other theorists like Jean Piaget and Lawrence Kohlberg have proposed stage-based maturational theories based on stages of *cognitive development* or intellectual growth.

Piaget argues that mental development is characterized by qualitative changes whereby the child's mind is able to gradually shift its exclusive focus from the concrete realities of the here and now to a conception of the world in abstract and symbolic terms.[98]

As the mind passes through successive stages of cognitive development, there are qualitative differences in the way the child comprehends the patterns of interpersonal and sexual relationships and the moral rules that govern them. Based on this premise, Kohlberg has formulated three levels of development in *moral judgment,* each of which is subdivided into two stages. In *Level 1,* moral values are sustained to avoid punishment and gain rewards. In *Level 2* moral values reside in fulfilling the proper duties in order to please others and avoid social

disapproval (being a "good" boy or girl). In *Level 3* shared moral values are internalized. Duties are defined in terms of contractual agreements and commitments that must be fulfilled. Finally, conscience takes over to regulate behavior on the basis of moral principle.[99]

At first glance, these cognitive-developmental approaches would seem to have little bearing on psychosexual development. But in effect, sexual behavior, like any other form of activity, is eventually dependent on cognitive functions. How we perceive and interpret sexual stimuli, comprehend the methods and rules of sexual interaction, generate and manipulate sexual symbols are all manifestations of cognitive functions. But we have as yet not integrated these cognitive-developmental perspectives into an explicit and coherent view of sexual development.

Social Learning Models of Sexual Development

Some behavioral scientists have used the metaphor of "scripts" to conceptualize the shaping and expression of sexual behavior. Through childhood and adolescence every individual develops a *sexual script* that acts as a record of past sexual activities, a standard for present behavior, and a plan for the future.[100] Like a blueprint, the script regulates the individual's sexual behavior with regard to five key variables: whom one has sex with; what one does sexually; when sex is appropriate (both with regard to time of life as well as in a more immediate sense); where is the proper setting for sex; and why one has sex.[101]

The theory of sexual scripts challenges other schemes of psychosexual development on several grounds.[102] First, it denies the necessity of postulating a biologically

---

[98]Gleitman (1983); Piaget (1972); (1952).

[99]Kohlberg (1969a); (1969b).

[100]Laws and Schwartz (1977) define sexual scripts as a "repertoire of acts and statuses that are recognized by a social group, together with the rules, expectations, and sanctions governing these acts and statuses" (p. 2).

[101]Gagnon (1977); Libby (1976).

[102]Gagnon (1973); Gagnon and Simon (1973).

based sexual drive. Sexual behavior is accepted as socially scripted behavior and not the expression of some primordial drive. We are born with the capacity to behave sexually, but we are not driven to behave sexually in one or another fashion. It is sexual learning and social contingencies that shape our erotic life.

Second, unlike the stage theories that trace development along a continuum, it is argued that sexual experiences are far more discontinuous. The infant fondling his penis is not masturbating in an adult sense, but merely engaging in an act that is diffusely pleasurable. Childhood and adolescent behaviors, therefore, do not necessarily have a developmental linkage. Much of what may appear as continuous from the past may be a function of our interpreting and reconstructing the past in order to bring it into harmony with our present identities, roles, and needs. This approach contradicts the psychoanalytic and other developmentalist views that childhood experiences in large part shape adult behavior.

Finally, the scripting theory rejects sharply drawn distinctions between what is "sexual" and what is "nonsexual." Beyond reproductive functions, what is defined as sexual is culturally determined and socially scripted. In short, sex is what society says it is. As a result, one encounters a wide diversity of signals, symbols, and behaviors that cultures define as "sexual." In this context, sexual activity need not necessarily be associated with erotic arousal, just as the capacity to experience erotic arousal need not necessarily accompany behavior that might conventionally be described as sexual.

The reason that culturally shared themes and symbols tend to be fairly uniform is because they are based on common childhood experiences, whereas interpersonal scripts tend to be more idiosyncratic. But both are shaped by social learning.

Patterns of sexual behavior are intimately linked with gender-role learning. Of all the experiences of the child, such learning exerts the greatest influence in shaping sexual behavior. In other words, by learning to be male or female, the person will act in certain sexually prescribed ways even though what has been learned in childhood are the gender roles and not the attendant sexual behaviors.

The child starts life with no sexual scripts. Typically, bits and pieces of sexual information are picked up that traditionally have been laden with prohibitions and guilt. Some children may become involved in adult-initiated sexual behaviors. But in the absence of the appropriate scripts to give meaning to the experiences, they remain unassimilated.

During adolescent sexual development, intrapsychic and social demands are often poorly matched; what the person wants to do may not be what society expects him or her to do. The conflict may be between acting or not acting sexually, or choosing one form of sexual behavior as against another. The adolescent is thus forced to come to terms with satisfying his sexual needs and peer expectations without getting into trouble with the adult world. If the sexual values of the adult culture are reasonable and clearly defined, there is less potential for conflict; if they are unduly restrictive or chaotically loose, the adolescent has a harder transition to make into adulthood.[103]

SUGGESTED READINGS

Chilman, C. S. *Adolescent Sexuality in a Changing American Society* (1979). Presents an extensive annotated bibliographic review of the literature on adolescent sexuality.

Sarrel, L., and Sarrel, P. *Sexual Unfolding* (1979). The development of sexuality in the adolescent and young adult. Especially pertinent for college youth.

Money, J., and Ehrhardt, A. A. *Man and Woman Boy and Girl* (1972). The complex and often contradictory research on gender identity is presented in a straightforward, authoritative, and highly readable manner.

Katchadourian, H. A. (ed.) *Human Sexuality: A Comparative and Developmental Perspective* (1979). Brings together a broad range of views by behavioral and biological scientists.

[103]For further discussion of adolescent sexual development from this perspective see Miller and Simon (1980).

Chapter 10

Autoerotism

Michelangelo said to Pope Julius II, "Self-negation is noble, self-culture is beneficient, self-possession is manly, but to the truly grand and inspiring soul they are poor and tame compared to self-abuse."

Mark Twain

At the turn of the century, Havelock Ellis coined the term *autoerotism* to characterize episodes of sexual arousal that were "spontaneous"; that is, they emanated from within the individual.[1] Although a precise definition of what constitutes autoerotic behavior is not easy, it is still useful to distinguish such activities from *sociosexual behaviors* that involve direct sexual interactions between people (chapter 8).

One could say that autoerotic activities bear the same relationship to sociosexual behaviors as talking to oneself does to talking to others. What we say to ourselves and what we do by ourselves is often a rehash, rehearsal, and substitute for what we say to others and do with others. Conversely, while ostensibly talking to others, we are sometimes really talking to ourselves just as having a sexual partner can merely be a front for a sexual monologue. Autoerotic and sociosexual behaviors, thus, are the two sides of the same coin of sexual behavior.

Erotic Fantasy

There is a whole world of sexual activity that is confined to the mind; fleeting erotic images, intricately woven fantasies, fading sexual memories, and fresh hopes are moving in and out of our consciousness a good deal of the time. Erotic fantasies are clearly the most ubiquitous of all sexual phenomena. It is difficult to imagine anyone who does not have them.

Fantasies may be the prelude to other sexual activity, such as masturbation and sexual intercourse, but more often they exist on their own. Medieval theologians called sexual reveries *delectatio morosa* and differentiated such dallying with erotic images from simple sexual desire or intent to engage in coitus.[2]

The propensity of adolescents to *daydream* is well known (figure 10.1), especially when they are in love. In the private and safe theaters of their minds, they endlessly rehearse their favorite fantasies—sometimes in the form of continuing stories, but more often as variations on a theme. Boys say they think about sex more often than girls do. College students spent 17% of their conversational time talking about sex.[3]

Sexual fantasies continue through adulthood, as do other forms of autoerotic ac-

---

[1]Ellis (1942), vol 1, p. 161. Another term current in the 19th century was *onanism*. Ellis rightly objected to it, pointing out that the Biblical passage in Gen. 8:8–11, from which the term is derived, did not refer to masturbation.

[2]A medieval penitential assigned the following penances for this offense: for a deacon 25 days, for a monk 30 days, for a priest 40 days, for a bishop 50 days. Quoted in Ellis (1942), vol. 2, p. 184.

[3]Cameron (1967).

Figure 10.1 *Young Man Fantasizing* (artist unknown).

tivity. The frequency of erotic contemplation, however, does decline with age. Whereas those 18 to 22 years old say they think about sex 20% of the time, those 28 to 35 years old do so 8%, and those 60 years and older 1% of the time.[4] Even people with satisfactory sex lives ruminate about past experiences (particularly missed opportunities) as well as future prospects. Psychoanalysts believe these thoughts are often sustained by unconscious associations with ungratified childhood wishes that have been repressed.

The cast of characters in daydreams varies, but the day-dreamer usually remains the central character. The objects of these fantasies may be acquaintances or people that the fantasizer knows only from afar, or they may be fictitious figures with vague and changing features. The imaginary activities are countless and are determined by the dreamer's past experiences, unconscious wishes, and his or her imaginative gifts.

The Nature of Sexual Fantasies

By far the most extensive examples of sexual fantasy are in literature and art. At their most explicit, the artistic and literary expression of sexual fantasies are considered *pornography*.[5] Such works are by no means the most effective means of sexual arousal; more subtle expressions of eroticism are often more compelling.

A second source of information on fantasy is clinical. Psychoanalytic case studies, in particular, dwell at great length on the erotic fantasies of patients and the analyst's interpretation of them.[6] Because there are greater social constraints on what we can or cannot do sexually than on what we can imagine, erotic fantasies may be a more accurate reflection of our true sexual proclivities than our sexual behaviors.

The third, and so far the most meager source, is sex research. Kinsey, who set the tone of current efforts in this field, was preoccupied with quantifying concrete behaviors that led to orgasm (chapter 8). The will-o-the-wisp world of fantasy did not figure prominently in his work, nor has it in the work of most other survey researchers. This gap has been partly filled by journalists and writers who have gathered sexual fantasies from people willing to tell it to them. They make fascinating reading but their manner of collection and presentation makes serious analysis difficult (boxes 10.1 and 10.2).[7]

Masturbatory Fantasy Unlike pure erotic fantasy, masturbatory fantasies are accompanied by various forms of self-stimulation frequently leading to orgasm. Although it is difficult to conceive that one could mas-

---

[4]Cameron (1967). Also see Verwoerdt et al. (1969).

[5]Problems in the definition of pornography and shifts in attitudes toward it are discussed in chapter 19.

[6]See, for example, Stoller (1979).

[7]Friday (1973); (1975); (1980). A more systematic effort at studying fantasies has been made by Shanor (1977).

Box 10.1   Female Erotic Fantasies[1]

"In the first sexual fantasy that I can remember, I thought of myself undressing while a boy I liked watched me. That became one of my most common fantasies when I was a teen-age girl."

*

"I'm at this hair store . . . I've just had a facial, so I've got this mask on, and there are cool cotton pads on my eyes. I can't see a thing. Not that I could see what's going on anyway, because there's a white silk curtain that falls from the ceiling down to my waist, then on down to the floor. No one can see me from the waist down. Neither can I . . . Over there, on the other side, is a young man, actually, lots of them, a row of young, big, strapping types, half nude . . . They are there to service us . . . My particular guy is dark, good-looking in a hard, impersonal sort of way . . . He crouches there between my legs, and with the greatest expertise in the world, he goes down on me."

*

"In my lesbian fantasies, I can never put an identity to my partner. She is no one I know and has no face or personality. In my dreams she is just a female body who takes most of the initiative, while I am merely passive and just lie there as she makes love to me. I fantasize that she plays with my breasts and sucks them while masturbating herself. Then she performs cunnilingus on me. We do not kiss, and I do not touch her genitals in these fantasies; however, I do play with her breasts. I often engage in this fantasy while making it with my husband, particularly when he performs cunnilingus on me."

*

"I especially enjoy the thought of being watched by someone who is not aware that I know he is watching me. Or I imagine that I am making love to someone, perhaps a close friend of the family, and my husband comes in and watches us, as prearranged between my husband and myself without the knowledge of the other man. It would be equally intriguing to walk in and catch my husband with another woman, also by prearrangement."

*

"The doorbell rings and it is this father and his son. The father has been a lover of mine . . . comes in and says, 'This is my fourteen-year-old son, and I want him to be as adept as I am, as I think I am, and I want you to teach him everything you know . . . So the son and I begin, the father sitting there watching as I undress the boy, caress him, totally initiate him. But it's not the boy that excites me in this fantasy, it isn't the idea of having a young boy, it's the idea of being watched by the father."

*

"During the last phase of intercourse is when I fantasize. I pretend I have changed into a very beautiful and glamorous woman (in real life I know I'm somewhat plain), and that my husband and I are in bed in very luxurious surroundings, usually in a hotel, far away from where we live. I can see the bottle of wine in a silver bucket waiting for us when we finish. I think of the people walking along outside our room in the coridor who are unaware of what we are doing only a few feet away from them, and how they'd envy us if they did know."

[1]From Nancy Friday, *My Secret Garden* (Pocket Books, 1973).

turbate with a blank mind, not everyone does in fact fantasize during masturbation. Some, for instance, focus on the bodily sensations elicited by sexual arousal.

During masturbation, the person usually evokes past memories or standby erotic themes, but especially among better-educated people, erotic photographs or literature may also be used as sources of stimulation. In the Hunt sample, about half

Box 10.2   Male Erotic Fantasies[1]

"I usually imagine that my girl friend or some girl that I've seen recently and I are making love, in various positions and places. Sometimes I'm lying in a bed, and she'll come into the room and start taking her clothes off without saying anything, and we'll start running our hands over each other, then start having sex. Sometimes I'll come up behind her (she knows and wants it, but doesn't respond until I touch her) and enter her from the rear. Sometimes we'll be in the shower, outside, etc. Sometimes she's passive, sometimes I am, all depending on my mood. We'll have oral sex, genital, you name it, I'll be on top, whatever."

*

"The fantasy begins as my mistress and I are together in a motel room. She is angry with me and will not let me kiss her. She makes me take off my clothes. After I am nude, she ties my hands behind my back. She then lifts her skirt and removes her panties and makes me step into them. As she pulls them up to my crotch she tells me not to get a hard-on or I'll get a spanking. All the while she is fondling my penis and balls. Naturally I get an erection. She slaps my hard penis and says, "You disobeyed me . . . now you are going to get your fanny warmed." She sits on the edge of the bed and forces me across her legs. She pulls down the panties and caresses my buttocks, running her finger down the crack. Then she begins to

spank me with her open hand. The spanking lasts a long time and my buttocks get hotter and hotter, the pain gets stronger until I plead with her to stop . . . promising to do anything she wants."

*

"This young male asks me to come to a party at his house, so I go, and he and I are the only ones who are gay. The rest of them are regular. Well, about when the party is over, we're going into the bedroom. We undress each other. Then he goes down on me and I go down on him. I would like this to happen, but not now."

*

"I am twenty-three and have been married for three years. I love my wife very much, and we share fantasies occasionally. I find that when I am fucking Jane (my wife), I can enjoy it much more if I know she is imagining being screwed by some other man. Sometimes she'll talk out loud of her fantasy and tell me that I am better than this other guy and that she would rather screw me. She will not, however, give the other man a specific identity. (He is not anybody we know). I suspect that this is because she does not like me to fantasize about specific other women. She is particularly jealous of a former girl friend of mine; and once when I mentioned her in a fantasy, she became upset."

[1]From Nancy Friday, *Men in Love* (Dell 1980).

---

of the men and a third of the women indicated that exposure to erotic pictures or movies increased their desire to masturbate; this was true for married as well as single persons. Reading erotic literature was found to be even more potent than visual material, especially among women.[8]

Kinsey reported an interesting sex dis-

[8]Hunt (1974).

crepancy with respect to the role of fantasy in masturbation: among males, 72% almost always fantasized while masturbating, 17% did so sometimes, and 11% did not fantasize at all. The corresponding percentages for females were 50%, 14%, and 36%. This pattern seems to run counter to the popular notion that it is women rather than men who are more attracted to the psychological aspects of sexual activities.

Women are now more responsive or more willing to admit interest in erotic materials and fantasies than those of a generation earlier.

In the Hunt survey, the fantasy most commonly mentioned involved sexual intercourse with a loved one (reported by three-quarters of all men and four-fifths of all women). But nearly half of the males and more than a fifth of the females also fantasized sex with more casual acquaintances in various forms of sexual encounter. Masturbatory fantasies clearly provide "safe" expression for a variety of sexual interests whose implementation would be impossible or unacceptable to the individual. A higher percentage of men than women reported fantasies of having intercourse with strangers, engaging in group sex, or forcing someone to have sex.

In Sorensen's sample of 13- to 19-year-olds, 57% of boys versus 46% of girls reported fantasizing "most of the time" while masturbating; but almost twice as many girls as boys said they combined masturbation with fantasy "some of the time," and 20% of boys but only 10% of girls rarely or never fantasized while masturbating.[9] Boys reported fantasies involving sex with someone who was forced to submit (including themselves), sex with more than one woman, as well as oral, anal, and group sex. The girls reported fantasies of sex with someone they liked, having to submit to several males, and inflicting mild pain on the sexual partner.

Coital Fantasy   As with masturbation, fantasy seems to be an integral part of the coital experience for many.[10] In one study, the most common themes imagined by women during coitus involved a person other than the partner; being coerced into sex; some forbidden activity; being in a different place while having sex; recalling a previous sexual experience; multiple partners; or scenes of oneself or others having sex.[11] Another group of women imagined

a former sexual encounter (79%); a different sexual partner (79%); a scene from a sexually exciting movie (71%); fondling male genitals (64%); a man stimulating her genitals (64%); a romantic scene (61%); or being seduced (54%). Less common were fantasies of having sex with several men, watching others have sex, being tied up while being stimulated, being humiliated and beaten, initiating a boy into sex, and having sex with an animal.[12]

A comparison of the fantasies of small samples of heterosexual and homosexual men and women has been reported as part of a broader investigation of homosexuality by Masters and Johnson.[13] The female subjects in this study reported more fantasies than males in both heterosexual and homosexual groups. Lesbians had the highest frequencies of the four study groups; bisexual ("ambisexual") subjects showed the lowest frequencies of fantasizing.

There was a high incidence of fantasies of "cross-preference" whereby heterosexual men and women contemplated homosexual activities, and vice versa. Also noteworthy were fantasy substitutions of someone other than the sexual partner and the use of coercion. In the case of heterosexual males, the fantasized alternative to the actual partner was most frequently identified as a specific woman. Heterosexual women most often thought of another man they could identify but did not know personally. Homosexual males likewise tended to favor a fantasy substitute. Only lesbians brought their current partners into their fantasies with any significant frequency, and they did so by invoking idealized situations.

Forced sexual encounters were the most common theme for lesbians and the second most common theme for homosexual males and heterosexuals of both sexes. Homosexual male fantasies were more violent than those of heterosexual males; lesbians more typically relied on psychological coercion than physical force and frequently switched between the roles of

---

[9]Sorensen (1973).
[10]In the *Cosmopolitan* survey, over a third of the women said they had sexual fantasies during coitus (Wolfe, 1981).
[11]Hariton and Singer (1974).

[12]Crepault et al. (1977).
[13]Masters and Johnson (1979, Chapter 9).

forcing sexual compliance and being forced into it. For the heterosexual male, the fantasy of being forced into sex was somewhat more frequent than his using force. In the former case he was usually set upon by a group of unidentified women. When he took the active role, there would usually be just one identified woman involved. Heterosexual females usually thought of themselves as victims of unidentified males singly or in groups, but there was little violence entailed beyond being rendered helpless.

The Meaning of Violence in Erotic Fantasy   Although the subjects of none of these studies can be taken as representative of the general population, the themes of sexual coercion and violence occur with sufficient frequency to give one pause. People do not commonly fantasize being robbed or beaten up; why then do they choose to imagine being coerced into sex?

Freud claimed that people are driven to debase their sexual partners. This admixture of sex and aggression has since been a salient theme in the psychoanalytic literature. Robert Stoller has further expanded on this association by claiming that *hostility*, the overt or covert desire to harm another, is central to generating *sexual excitement*, not just by sadomasochists, but by everyone. Adults reenact childhood hurts and fears, and in order to master them, they retaliate by hurting their sexual partners; this colors the process of sexual arousal: "Triumph, rage, revenge, fear, anxiety, risk are all condensed into one complex buzz called sexual excitement."[14]

Such theoretical explanations, or just the confronting of some of these fantasies in ourselves or others, may be disquieting. Are these the sorts of things that deep down we crave? Do these wishes reflect our true selves? Those who abhor violence and the subjugation of women are particularly appalled by the presence of violent fantasies and the interpretations attached to them. Susan Brownmiller, for example, rejects the psychoanalytic view that women may have *unconscious* rape fantasies. She accepts the fact that some women have *conscious* rape fantasies but ascribes these to the imposition of male rape fantasies on women. She states, "Fantasies *are* important to the enjoyment of sex, I think, but it is a rare woman who can successfully fight the culture and come up with her own non-exploitation, non-sadomasochistic, non-power-drive imaginative thrust." So, "when women do fantasize about sex, the fantasies are usually the product of male conditioning and cannot be otherwise."[15]

An important aspect of coercive fantasies that is often lost sight of is that fantasies are like play; they are a form of make-believe. To imagine being coerced into sex is not the same as wanting to be raped. No sane person would ever want to suffer the agony of a true rape experience.

Then why have such fantasies? Among other reasons, because they absolve us of responsibility. Though much has changed in this regard, sex is still laden with much shame and guilt, especially where unconventional activities are concerned. If you imagine that you are being held down and having sexual enjoyment forced on you, then you can permit yourself to enjoy it because you did not seek the pleasure; it was imposed on you. If women are more prone to fantasize about being in such "victim" roles, maybe it is because women have been socially more constrained from enjoying sex freely and need an externally imposed justification. Hence, in order to understand sex fantasies, we need to understand not just what they consist of, but the purposes they serve.

Functions of Sexual Fantasies

Erotic fantasies fulfill many functions. They are a source of pleasure to which everyone has ready access. They are also substitutes for action: as temporary satisfactions while awaiting more concrete ones (tonight's date; a honeymoon in the future), or as compensations for unattainable goals. They allow partial, tolerable expression of forbidden wishes and to some degree relieve sexual frustrations. Fantasies

---

[14]Stoller (1979)

[15]Brownmiller (1975, p. 360).

cannot yield full gratification of sexual desires, but they ease the frustration of unfulfilled wishes.

Fantasies that revolve around future events can be of help in coping with real-life situations. As one anticipates problems, plans for contingencies, and mentally rehearses alternative modes of action, a person develops better sense of control that reduces anxiety and helps master novel situations. There is thus an important difference between fantasies that substitute for action and those that prepare for it.[16]

As we have noted, an important function of fantasy involves its use as an erotic stimulant, or psychological "aphrodisiac." For many individuals it is as an integral part of autoerotic and sociosexual activities. But the practice of using fantasies during coitus elicits mixed reactions. Some see it as beneficial and others as detrimental to a relationship. The first consideration in this regard is whether or not the fantasies are shared. If they are not, they may distract the person from interacting directly with the sexual partner. But although one hopes for the undivided attention of one's partner during such intimate moments, what if the partner can switch back and forth between reality and fantasy, thereby intensifying the sexual interaction to the benefit of both?

If the fantasy is shared, there is a better chance of enhancing mutual arousal. The ability to reveal one's thoughts, however idiosyncratic, presupposes true intimacy and may further strengthen the willingness and capacity to be open with others. But the revelation of sexual fantasies can also have an offensive and disruptive influence. Especially if extraneous characters are involved, the exclusivity of the relationship is diluted. Those who can use fantasy effectively in these situations are able to safely isolate its erotic enhancement function from its wish-fulfillment components. Also, the more "fantastic" the fantasy, the less threatening it will be; to

imagine making love to an unlikely or unattainable figure (such as a movie star) is less threatening to the partner than if the next-door neighbor is the subject of the fantasy.

Pathological Fantasies

Sexual fantasizing is part of normal mental life. But does it have pathological as well as healthy forms? Fantasizing beyond a certain point about sex, or anything else, could be wasteful and maladaptive. Particularly during adolescence, erotic fantasies may become excessively absorbing. If they interfere with everyday tasks or if they regularly take the place of interacting with others, they may intefere with normal psychosocial development.

Can fantasies be evaluated by content? If we had a commonly accepted set of criteria to separate healthy from unhealthy forms of sexual behavior, then it would also apply to fantasies. But we have no such clear yardstick of normality (chapter 8). Furthermore, subjective reactions are highly variable: some people are disturbed by thoughts of rather innocuous activities, whereas others may be unaffected by the most bizarre fantasies.

Fantasies involving unconventional or socially unacceptable behaviors are very common. But they do not "define" us as adults. And it bears repeating that most such fantasies are rarely acted upon. What really matters in this regard is, therefore, not what we think or feel but what we do.

Unpleasant or disturbing fantasies become more of a problem when they become persistent. Conscious attempts to dispel them often simply cause us to focus on them more strongly. So it is better to take them for what they are: isolated thoughts that do not mean very much in practical everyday terms. But since fantasies reflect our thoughts and feelings, they can also mirror problems that may exist regarding sex. For instance, fearful fantasies about intercourse may interfere with sexual enjoyment. Women may be concerned about injury to their genitals; men may fear that their penis will be trapped

---

[16]For further consideration of the role of fantasy in sex, see Sullivan (1969); see Singer (1975) for a more general discussion of daydreaming.

or damaged by the vagina. There are reports of men imagining (but without truly believing) vaginas to be full of razor blades, ground glass, or armed with teeth.[17]

Fantasies are clearly pathological when they become *delusional* whereby the person firmly believes what is patently untrue. Paranoid women may misinterpret innocent comments as obscene propositions; men are more likely to have delusions of homosexual assault or being accused of being gay. In both sexes, the persecutor is a male, but women typically suspect acquaintances while men fear strangers.

In rare instances a person may correctly realize that he or she is losing control and is likely to commit a seriously antisocial act. Such a person must seek help, or should be prevailed upon to do so. In such situations it is better to find psychiatric or other professional assistance than to turn to family or friends; the latter of course can help secure such help.

Sexual Dreams and Nocturnal Orgasm

People have sought in their dreams the key to the future and a window into the past. Freud called dreams "the royal road to the unconscious." More recently, inquiries into the neurophysiology of sleep and dreaming have yielded new and intriguing findings, some of which are particularly relevant to the study of sexuality.

Like all other dreams, sexual dreams are usually fragmentary and difficult to describe. Patently sexual content may be accompanied by strong erotic feeling (figure 10.2); or imagery and affect may seem contradictory: a person may dream of a flagrantly sexual activity without feeling aroused; or may feel intense excitement while dreaming of an apparently nonsexual situation like climbing stairs, flying in

the air, and so on. Psychoanalysts explain such dreams in terms of their symbolic meanings and make the interpretation of dreams an important part of their work.

Psychoanalytic Concepts of Dreaming

Freud thought that the function of dreams was to protect sleep. During sleep, when the vigilance of the ego relaxes, repressed (often sexual) wishes threaten to break through into consciousness and disrupt sleep. These wishes are permitted partial expression in disguised form as dreams. The process of transforming unconscious wishes into a dream imagery is known as *dream work*. A dream therefore has two components: the *manifest content* (usually woven around some actual, recent experience, "the day residue") and the *latent content* (which carries the real message in disguised and symbolic form).[18]

In psychoanalytic reckoning, a great many dream symbols have sexual significance (the same would be true for symbols in myth, jokes, art, and literature). For example, objects like sticks, tree trunks, knives, daggers and nail files (because they rub back and forth) typically symbolize the penis and are referred to as *phallic symbols*. By the same token, boxes, chests, cupboards, ovens, rooms, ships—in fact any enclosed space or hollow object—usually represent the female genitals.

Actions involving such objects or places likewise carry sexual significance: The opening of an umbrella may represent erection; going in and out of a room would signify intercourse, and whether or not the door is locked could have special meaning. Walking up and down steps, ladders, and staircases also stands for coitus (for as in orgasm, we reach the apex through rhythmic movements and with panting breath).

Sexual dreams, particularly those culminating in orgasms, are intensely plea-

---

[17]The belief that the vagina has teeth (*vagina dentata*) represents in psychoanalytic theory an unconscious equation of the vagina with the mouth and may be associated with fears of the penis being "devoured" or castrated during coitus (Fenichel, 1945).

[18]The psychoanalytic literature on dream interpretation is very extensive. The classic source is Freud's *Interpretation of Dreams*.

Figure 10.2 *The Shepherd's Dream.* Roman. (Museum of Fine Arts, Boston).

surable but can also be quite bewildering. It has been suggested that *nightmares* too have a sexual basis: that the fear, anxiety, and the feeling of helpless paralysis that characterize nightmares are suggestive of the experience of being raped (figure 10.3).[19]

Fantasies and dreams are closely related (hence, "daydreams"). In both, there is a transient relaxation of conscious restraints that allows repressed wishes partial expression. Daydreams generally remain subject to the dictates of logic and reality (*secondary-process thinking*); whereas in dreams, the vigilance of the ego is lessened to the point at which deeply buried material emerges, without much regard for ordinary logical processes and the restraints of reality (*primary-process thinking*).

Carl Jung believed that some symbols originated from a deeper level of the unconscious—the **collective unconscious.** The symbols in this *objective psyche* are derived from the collective experience of the human race, not just individual experience;

[19]Jones (1949).

Figure 10.3   Henry Fuseli, *The Nightmare* (1781). (The Detroit Institute of Art.)

they thus represent *archetypes* of our modes of thought.[20]

Most behavioral scientists are by no means persuaded of the validity of these concepts. Moreover, dreaming is not the sort of behavior that lends itself easily to the standard methods of psychological investigation. However, over the past several decades, a new approach to the study of dream phenomena has attracted much interest. Rather than dealing with issues of content and symbolism, this approach studies the neurophysiological basis of the dream process.

[20]Jung's best-known work on dreams is his *General Aspects of Dream Analysis*. Also see *Archetypes of the Collective Unconscious*.

The Neurophysiology of Sleep and Dreaming

Sleep is not the uniform state that it appears to be, nor are dreams erratic events that punctuate it unpredictably. Rather, there is a definite *sleep-dream cycle* that recurs nightly. Brain waves as shown in *electroencephalogram* (EEG) tracings show four distinct sleep patterns. One of these is characterized by a fast rhythm and bursts of *rapid eye movements* (REM). A person who is awakened during REM sleep will report experiencing vivid dreams. During the other sleep phases dreaming is erratic and less vivid.

During an ordinary night's sleep there are four or five active **REM periods**, accounting for almost a quarter of total sleeping time: The first occurs 60 to 90 minutes after falling asleep, and the rest at approximately 90-minute intervals. These dream periods are generally quite constant, which means that everyone dreams every night, regardless of whether or not the dreams are remembered in the morning.[21]

REM periods are times of intense physiological and sexual activity; in a remarkably high number of instances (85 to 95%) partial or full erections have been observed during them even among infants and elderly men. These erections are not necessarily accompanied by sexual dreams, and their full significance remains unclear.[22] The presence or absence of erection is an important diagnostic test to separate organic from psychologically caused cases of impotence (chapter 15).

Evidence of sexual arousal during sleep has been more difficult to detect among women than it is among men. But the association of REM sleep with female sexual response has now been convincingly demonstrated; women manifest cyclic episodes of vascular engorgement during REM periods equivalent to erection in men and at the same high frequency (95%).[23]

Nocturnal Orgasm

"Visitations by the angel of the night" is what the pioneer sexologist, Paolo Mantegazza called *nocturnal emissions.* Babylonians believed in a "maid of the night" who visited men in their sleep and a "little night man" that slept with women.[24] Such imaginary beings became more prominent in medieval times in the form of demons who would lie upon women (*incubus*) (figure 10.3) and under men (*succubus*). The West African Yoruba believed in a single versatile being who could act either as male or female and visit members of either sex in their sleep.[25]

Although "nocturnal emission" (*wet dreams*) or "nocturnal orgasm" are commonly used to designate this experience, they are not quite accurate since the experience does not necessarily occur during nighttime and women who have such orgasms do not ejaculate.

Orgasms during sleep (nocturnal or daytime) did not constitute an important proportion of total sexual outlets (2 to 3% for females, 2 to 8% for males in the Kinsey sample), but nevertheless substantial numbers of people experienced them; by age 45 almost 40% of females and more than 90% of males have had such experiences at least once. Among men they occurred most frequently in late adolescence and the early twenties, among women between the ages of 30 and 50. Nocturnal emissions were seven times more frequent among the college-educated Kinsey men than in the lower educational group.

As involuntary acts, nocturnal orgasms have been less subject to moral censure, and as a "natural" outlet they have been supposed to act as a safety valve for accumulated sexual tensions. Men who are single, divorced, or in jail do show a somewhat higher rate of nocturnal emission than married men. But the compensatory function of this outlet is at best a partial one. When orgasms through other outlets become less frequent, nocturnal orgasms do not become significantly more frequent.

The manifest content of wet dreams may or may not be clearly erotic, but there is always a subjective sensation of sexual excitement. Pleasure may, however, be tinged with apprehension and remorse, especially among adolescent boys who are uninformed about the nature of the experience. Women also may be disturbed by such involuntary orgasms.[26] Occasionally the dreamer will sleep right through the experience and will have only vague recollections of the dream in the morning

[21]Dement (1972); (1965).

[22]See Fisher et al. (1965), Gulevich and Zarcone (1969); Kleitman (1963); Dement (1965); and Karacan et al. (1976).

[23]Fisher et al (1983).

[24]Jastrow, *Religion of Babylonia,* quoted in H. Ellis (1942), vol. 1, part 1, p. 198.

[25]H. Ellis (1942), vol. 1, part 1, p. 188.

[26]Truxal (1983).

(which does not make the mess in the bed any less awkward).

Nocturnal orgasms may also occur independently of erotic dreams. For instance, they are present among paraplegic men in whom nerve impulses from the brain (where dreams are enacted) cannot reach the spinal cord centers. Similarly, animals like cats, dogs, bulls, and horses have spontaneous emissions in their sleep. It is assumed that in these cases, perhaps in men as well, nocturnal emissions serve a reproductive purpose: the periodic release of sperm improves its quality and enhances its reproductive potential.[27]

Masturbation

Masturbation refers to any sexual activity that involves physical self-stimulation.[28] Like fantasy, masturbation is usually carried on in solitude and constitutes one of the most common sexual activities practiced in all cultures (box 10.3), and is discernible among other nonhuman primates and mammals (box 10.4).

Methods of Masturbation

The more common forms of masturbation are: *manual stimulation, genital friction* against objects, and the use of *special devices.* These methods are not mutually exclusive. Just as most people do not use a single coital position, nor do they rely on just one method of masturbation. Whatever the method used, masturbation proceeds along a fairly standard sequence.

[27]Levin (1975).

[28]The origins of the Latin verb *masturbari* are obscure. It may be derived from the words for "hand" (*manus*) and "to defile" (*stuprare*) or the Greek for "virile member" (*mazdo*) and Latin for "disturbance" (*turba*). Pejorative terms like "self-abuse," "self-pollution," "defilement of flesh," "solitary vice" have long been dropped. Current vernacular expressions for masturbation include: "jerking off," "jacking off," "whacking off," "rubbing off," "beating off," "frigging," "hand job," and so on (Haeberle, 1978, p. 492). Some sex therapists refer to it as *self-pleasuring*.

There is first a period of psychological interest and mild arousal that leads to the decision to masturbate. Then follows a period of preliminary play, during which physical arousal becomes manifest, which is then usually pushed to the point of orgasm.

The preliminary techniques of arousal, just as in foreplay, are slower, gentler, more indirect. There may be stroking of the thighs, fondling of the breasts (by women), and caressing of various other parts of the body. These then lead to more direct forms of genital stimulation by hand or by friction against objects that becomes progressively more vigorous, rhythmic, and faster, triggering orgasm.

Manual Stimulation Manual techniques are most common for both sexes. The most frequent form for males involves stroking and rubbing the penis and moving the hand over it firmly to and fro or in a "milking" motion (figure 10.4). Women also rely primarily on genital manipulation.[29] The clitoris and the labia minora are the structures most commonly stroked, pressed, and rhythmically stimulated (see figure 10.5). Since they are the most sensitive parts of the female genitals, the motions are usually quite gentle and deliberate. Women usually avoid the glans of the clitoris; instead, they concentrate on the clitoral shaft, which they can stimulate on either side. If too much pressure is applied or manipulation is prolonged over an area, the site may become less sensitive. Switching hands or moving the fingers about is therefore quite common.

Men commonly think that women insert their fingers or objects into their vaginas when they masturbate, but only one in five women in the Kinsey sample reported such practice, and often these consisted only of slight penetrations of the vaginal intro-

[29]In the Hite sample almost 80% of the women exclusively relied on manual stimulation of the clitoral/vulval area. The majority did this while lying on their back.

Figure 10.4   Masturbating figure on a Greek black-figure cup by the Amasis Painter, 530–520 B.C. (Courtesy, Museum of Fine Arts, Boston.)

itus.[30] Some women, however, do derive additional pleasure from deep penetration, perhaps because it simulates coitus or because vaginal stimulation may yield special sensations. When insertions do occur, fingers are most often used.

Particularly among homosexual men, stimulation of the anus with or without digital penetration may be quite arousing. As an occasional experiment, heterosexual males and women may also try out anal masturbation. Rarely, the urethra may be similarly explored with various objects.

Friction against Objects   Those who do not wish to touch their genitals may resort to friction against objects.[31] The possibilities are many: a pillow, towel, nightclothes tucked between the legs, the bed cover, or the mattress itself may provide a conven-

ient surface to rub and press against (figure 10.6). A jet of warm water such as from a hand-shower directed at the clitoral region can also be stimulating (but it is not safe to aim it inside the vagina).

Muscular Tension   Associated with friction against objects is the use of thigh pressure. Muscular tension is necessary whatever the technique used, but some women are able to reach orgasm through this method alone.[32] When a woman's legs are crossed or pressed together, steady and rhythmic pressure can be applied to the whole genital area, a method that combines the advantages of direct stimulation and muscular tension.

Havelock Ellis describes an episode where he observed a young woman stimulate herself to orgasm by swinging her legs

[30]Among the Hite subjects only 1.5% of women said they used this method exclusively. But vaginal insertion was used more often in conjunction with other techniques such as clitoral stimulation.

[31]Among the Hite subjects 4% of women relied on

"pressing and thrusting" the clitoral/vulval area against a soft object.

[32]Among the Hite subjects 3% of women reported masturbating through rhythmic thigh pressure.

Figure 10.5 Illustration by Caylus in *Therese Philosophe*.

while waiting in a train station.[33] There are also descriptions of how in 19th-century dress factories that were equipped with treadle-operated sewing instruments, shop stewards would listen for the occasional uncontrollable acceleration of a machine as its female operator went through mounting excitement and orgasm.[34]

Masturbatory Devices Most people make do with whatever is around in stimulating themselves but others will go to greater lengths. For example, they may avail themselves of hand lotions, scented oils, special kinds of underwear or other objects of fetishistic significance.

Masturbatory aids can be quite specialized. Artificial penises in many cultures have been fashioned from gold, silver, ivory, ebony, horn, glass, wax, wood, and stuffed leather, ranging from crude specimens to products of fine craftsmanship.[35] Their common name in English is *dildo* (from the Italian *diletto*); the French form *godemiche* is less often used. Aristophanes refers to dildos in his plays and Herodas has a play (*The Private Conversation*) in which two women friends discuss the fine workmanship of a particular cobbler in fashioning such devices. Despite the numerous references to the use of dildos in pornographic novels, their use is in fact not common among women and more often it is done to entertain men who find the spectacle sexually arousing. Dildos are probably more often used by gay men for anal penetration than women vaginally.

There are other ingenious devices. One is the Japanese *rin-no-tama* or *ben-wa*, consisting of two metal balls: one is hollow and introduced first into the vagina; the other contains a smaller metal ball, lead pellets, or mercury, and is inserted next. The metal balls are then made to vibrate by movements, especially by swaying in a hammock, swing, or rocking chair.

Over the last two decades, hand-held vibrators have become quite popular. Some are mechanized dildos, others have vibrating tips of various shapes or are attached to the back of the hand through

[33]H. Ellis (1942), vol. 1, part 1, p. 180. A modern description of the same practice comes from one of the women respondents to the *Cosmopolitan* survey: "I have a beautiful way of masturbating. By rubbing my legs together. *No one* can tell. I used to do it all the time in grade school—at my desk—especially when I was under pressure taking a test. (Now you know where I got my 3.5 to 4.0 average in college.)" (Wolfe, 1981, p. 119).

[34]Pouillet (1897).

[35]Cucumbers, bananas, sausages, candles, bottles, pencils, and countless other convenient objects have been pressed into similar service. (Some of these devices can cause injury as discussed in chapter 7.) A poem in the *Arabian Nights* extolls the erotic virtues of bananas: "O bananas, of soft and smooth skins, which dilate the eyes of young girls . . . you alone among fruits are endowed with a pitying heart, O consolers of widows and divorced women." (Quoted in Ellis 1942, vol. 1, part 1, p. 171.)

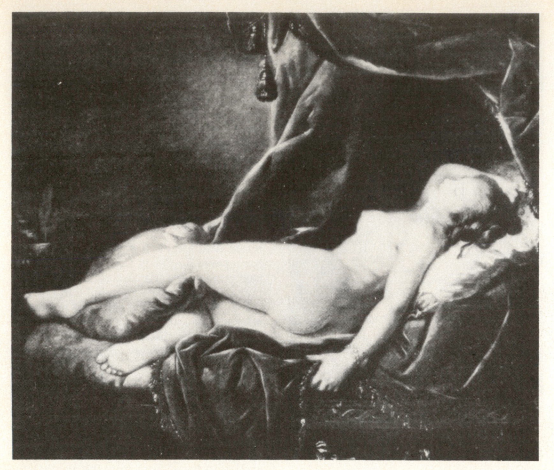

Figure 10.6   Jean Honoré Fragonard, *Sleeping Girl*.

which they transmit their vibrations. They may be sold openly for what they are or coyly advertised (for "soothing massage"). Vibrators are intensely stimulating and may lead to orgasm even in cases where a woman may have difficulty attaining it otherwise (their use in the treatment of sexual dysfunction is discussed in chapter 15).

There are no reliable statistics pertaining to the use of such devices. Among the *Redbook* survey respondents, 21% of women reported having used some form of gadget ("vibrators, oils, feathers, and dildos").[36] Among the younger women in the Mas-

ters and Johnson sample, half of those who masturbated had tried a vibrator at least once; a quarter preferred it to other forms of self-stimulation, making it the second most popular method next to manual stimulation of the genitals. But given the self-selected nature of all these subjects these figures are probably quite inflated relative to the general population.

Those who favor the use of vibrators see them as effective and harmless devices to intensify autoerotic pleasure, a form of "super-masturbation." Those who object to the sexual dependency of women on men see in vibrators a liberating device with which no penis could compete. For women who cannot reach orgasm any other way, it is a necessary path to sexual satisfaction.

[36]Tavris and Sadd (1978, p. 76.)

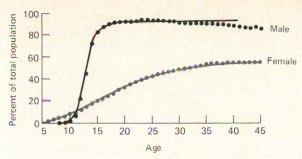

Figure 10.7 Accumulative incidence curves for masturbation. From Kinsey et al., *Sexual Behavior in the Human Male*, (1948), p. 502, and from Kinsey et al., *Sexual Behavior in the Human Female*, (1953), p. 141. Courtesy of the Institute for Sex Research.

Those who object to the practice of using vibrators and similar gadgets see it as further evidence of the mechanization of our lives, the alienation within human relationships, and the trivialization of sexuality.[37] Some therapists are concerned that vibrators may become sexual crutches that deflect attention from underlying problems. Ultimately, vibrators must be judged like any other artificial device (such as alcohol) that influences human experience: not in "good-or-bad" terms but whether they serve the user or the user serves them.

There are also some rare methods of self-stimulation that can be detrimental or even highly dangerous. In chapter 7, we discussed the health hazards of inserting foreign objects into the rectum and urethra. A more elaborate method of heightening the masturbatory climax is through partial asphyxiation, usually by tightening a rope around the neck. Occasionally the person (usually a man) doing this inadvertently hangs himself.[38] It is estimated that hundreds of deaths a year occur from this, particularly among young men, which are mistaken for suicides.

The Statistics of Masturbation

Prevalence of Masturbation   In the Kinsey sample, 92% of males and 58% of females were found to have masturbated to orgasm at some time in their lives (an additional 4% of women had masturbated without reaching orgasm). Figure 10.7 shows the percentages at various ages in the Kinsey samples with such experience. These graphs (known as *accumulative incidence curves*) answer the question: "How many people ever have such experiences by a given age?" A person who masturbates only once and one who does so many times are counted in the same way. Between ages 10 and 15 the male curve climbs dramatically and then levels off as it approaches age 20. Practically every man who is ever going to masturbate would have already done so by this time. Nonetheless, the curve does not go beyond 92%; 8 out of every 100 males do not ever masturbate.

The female curve peaks at 62% and it climbs only gradually to this high point. Until the age of 45 increasing numbers of women are still discovering this outlet.[39] More current data show women discovering masturbation earlier in life. Among the *Cosmopolitan* survey respondents, 54% had started masturbating by age 15; only

---

[37]These concerns especially bear on the use of "masturbatory dolls" that are life-size rubber or vinyl inflatable bags shaped like a woman with open mouth, vagina, and anus for receiving the penis. More sophisticated versions have vibrating and sucking devices. There are also separate artificial "vaginas" and "anuses," which simplify matters even further.

[38]In 1978 a woman in Texas found her husband hanging dead from their bedroom door dressed in her wig, bra, and panties. Through an arrangement

of pulleys, the man intended to control the tightness of the rope around his neck. But he had slipped, placing his full weight on the noose and hanging himself. (*San Francisco Chronicle*, October 29, 1981, p. 29.)

[39]Kinsey et al. (1948); (1953).

14% had started the practice when older than 20.[40] Yet the discrepancy between male and female prevalence figures persists: In the Hunt survey 94% of men but only 63% of women had masturbated;[41] in the college sample, 89% of men and 61% of women had done so.[42] Other studies also from the mid-1970s show somewhat higher ratios for women: 78% in one case;[43] 82% in another.[44] The *Redbook* survey showed 3 out of 4 married women to have masturbated.[45] Among the *Cosmopolitan* survey women, 89% had masturbated.[46] But it is quite likely that the higher female rates in these magazine surveys is at least in part an artifact of their samples, since other studies report much lower rates.[47]

Frequency of Masturbation   How frequently do people masturbate? Various factors, like age and marital status, make a good deal of difference. Among males in the Kinsey sample, this practice was at its peak during adolescence when the mean frequency of masturbation was twice a week. In the total "active population" (that is, counting only those who actually had masturbated sometime or other) the weekly mean was 2.4 times. Frequencies decreased steadily with age: unmarried men 45 to 60 years old averaged fewer than one orgasm every 2 weeks; the figure for married men was even smaller.

The Hunt sample did not reveal a significant change in how often men masturbate. But there was an increase in the frequency with which women masturbate: Single women aged 18 to 24 in the Kinsey sample were masturbating about 21 times a year; in the Hunt sample, they did so 37 times a year.

Frequency figures are even higher in the *Cosmopolitan* sample: 35% of the women who masturbate do so rarely; 37% several

times a month; 25% several times a week; 3% daily.[48] The average frequencies for the active female sample were quite uniform at various age levels (up to the mid-fifties) and did not show the steady decline with age that was characteristic of males. The average unmarried woman, if she masturbated at all, did so about once every 2 or 3 weeks; the median for her married counterpart was about once a month.

The frequency of masturbation among college students has been found to range quite widely. For example, 1% of freshmen males and a smaller proportion of females report masturbating daily; 17% of freshmen males and 25% of freshmen females masturbate less than once a month.[49] In another college sample, daily rates were about 6% for the active population of both sexes (some masturbating "several times a day"); 76% averaged "a few times a week"; another third or so, "a few times a month."[50]

Social Correlates   The better-educated person in the Kinsey sample was more likely to masturbate: 89% of males with only grade-school educations masturbated some time in their lives, but for those with college educations, the figure was 96%. The corresponding figures for females were 34% and 60%. For the college-educated males, masturbation constituted not only the chief source (60%) of orgasm before marriage but also almost 10% of orgasms following marriage.

The significance of educational level as a correlate of masturbation is with respect to its reflecting social class differences. The fact that the higher-educated person relied more on masturbation does not mean that this is something one learns in college. Rather, the better-educated person is more likely to come from a middle-class background and hence share the sexual values of that group. For example, better-educated people are less fearful of masturbation as a health hazard. In their social cir-

[40]Wolfe (1981, p. 114).
[41]Hunt (1975).
[42]Arafat and Cotton (1974).
[43]Miller and Lief (1976).
[44]Hite (1976).
[45]Tavris and Sadd (1975).
[46]Wolfe (1981).
[47]Fisher (1973).

[48]Wolfe (1981).
[49]Gagnon (1977, p. 157).
[50]Arafat and Cotton (1974).

Box 10.3   Masturbation in Cross-Cultural Perspective

The anthropologist Paolo Mantegazza (who was prominent in the 1930s) called Europeans a "race of masturbators." He reasoned that Western civilization simultaneously stimulates and represses sexuality and that restrictions on nonmarital coitus compel people to masturbate instead. But masturbation has been documented for many ancient cultures, including the Babylonian, Egyptian, Hebrew, and Indian. Greeks and Romans believed that Mercury had invented the act to console Pan after he had lost his mistress Echo. Zeus himself was known to indulge occasionally. Aristophanes, Aristotle, Herodas, and Petronius refer to masturbation.[1] Classical attitudes, nevertheless, seem to have been ambivalent: Demosthenes was condemned for the practice, whereas Diogenes was praised for doing it openly in the marketplace.[2]

Masturbation is reported by Ford and Beach for about 40 primitive cultures.[3] It is thought to be less prevalent in societies that are permissive toward nonmarital coitus, and rare in many preliterate groups.[4] Most groups seem to disapprove of it for adults. For instance, Trukese men (monogamous fishermen of the Caroline Islands) were said to masturbate in secret while watching women bathe. Men of Tikopia (Pacific island agriculturists) and of Dahomey (West African agriculturists and fishermen) masturbated occasionally, even though both cultures permitted polygamy.

Female masturbation has been reported less often and has been generally disapproved. Vaginal insertions seem more common than clitoral stimulation among some primitive people: African Azande women used wooden dildos (and were severely beaten if caught by their husbands); the Chukchee of Siberia used the calf muscles of reindeer; Tikopia women relied on roots and bananas; Crow women used their fingers, and so did the Aranda of Australia.

Among the Lesu (polygamous tribesmen of New Ireland) female masturbation was condoned. Powdermaker has reported:

> Masturbation . . . is practiced frequently at Lesu and regarded as normal behavior. A woman will masturbate if she is sexually excited and there is no man to satisfy her. A couple may be having intercourse in the same house, or near enough for her to see them, and she may thus become aroused. She then sits down and bends her right leg so that her heel presses against her genitalia. Even young girls of about 6 years may do this quite casually as they sit on the ground. The women and men talk about it freely, and there is no shame attached to it. It is a customary position for women to take and they learn it in childhood. They never use their hands for manipulation.[5]

[1]Aristophanes finds it unmanly but acceptable for women, children, slaves, and feeble old men.
[2]H. Ellis (1942), vol. 1, part 1, p. 277.
[3]Ford and Beach (1951).
[4]Gebhard, in Marshall and Suggs (1971, p. 208).
[5]Powdermaker, quoted in Ford and Beach (1951, p. 158).

cles masturbation may be openly condoned or at least not seriously condemned. At least at Kinsey's time, there were stronger taboos against masturbation among the less educated. An additional factor may be the difference in class attitudes toward premarital coitus. Although better-educated people have considerably relaxed restrictions against premarital relations, they are probably still less permissive than their counterparts in the lower socioeconomic classes. Their greater reliance on masturbation may then arise partly from less access to coitus.

Box 10.4   Masturbation Among Animals

Among males of many animal species masturbation to the point of orgasm has been well documented: It is less certain for female animals whose capacity to attain orgasm appears to be less developed so far as can be determined.[1]

Self-stimulation can be observed among subprimates. A sexually aroused porcupine, for instance, holds one forepaw to its genitals while walking about on the other three paws. It may also rub its sex organs against various objects; one captive male was observed hopping about with a stick between its legs, showing obvious signs of arousal.

Young monkey masturbating. From Harlow, "Lust, Latency and Love: Simian Secrets of Successful Sex," in D. Byrne, and L. Byrne, eds., *Exploring Human Sexuality.* New York: Crowell, 1977, p. 166. Reproduced by permission.

Dogs and cats lick their penises before and after coitus; elephants use their trunks; captive dolphins rub their erect penises against the floor of the tank, and one male has been observed stimulating itself by holding its penis in the jet of the water intake. Comparable activities have been noted among rats, rabbits, horses, bulls, and other animals.

Self-stimulation is quite common among primates in captivity (see accompanying figure). Male apes and monkeys manipulate their penises by hand or foot and also take them into their mouths or rub them against the floor. Male baboons in one African park have reportedly developed erotic responses to automobiles and are in the habit of jumping on the hoods of cars and ejaculating at the windshield.[2]

Masturbation is not restricted to captivity; free spider monkeys manipulate their genitals with their tails, and adult male rhesus monkeys have been observed masturbating to the point of orgasm in the presence of receptive females. But this does not apply, for example, to wild gorillas.

Female mammals of subprimate species apparently masturbate only rarely. Masturbation is also less frequent among female primates than among males, both in the wild and in captivity. Females can be seen fingering or rubbing their genitals but often only perfunctorily. Even when such activity is clearly autoerotic, it does not seem to lead to observable orgasm.

[1]Ford and Beach (1951, pp. 159–161).
[2]Bates (1967, p. 70).

In the Hunt sample, additional differences toward masturbation between blue-collar and white-collar workers, as well as between noncollege and college-educated individuals were clearly discernable and even seemed to widen. Among all groups, however, masturbation was viewed with more tolerance than before.

With respect to marital status, there is an increase in the prevalence of masturbation among both single and married women in the Hunt sample compared to the Kinsey sample. The same is true for married men but not single men.[51] These data when viewed in the larger context of the Hunt study do not suggest that increased masturbation is in compensation for frustrations in marital coitus. Rather, modern married persons, especially those who are younger, appear relatively freer to rely on masturbation as an auxiliary outlet for their sexual needs.

Kinsey found religion to have a greater effect on females than males. A man may feel more guilty about masturbating, but sooner or later more than nine out of ten indulge in the practice. There is a difference, however, in how often they do so. The more religious men (particularly Orthodox Jews and practicing Roman Catholics) in the Kinsey sample masturbated somewhat less often. The highest frequencies were among religiously inactive Protestants. Among women, masturbation was definitely less widespread among the more devout (41%) than among the nondevout (67%). The degrees of devoutness seemed more important than the particular religious denomination.

In the Hunt sample, a higher proportion of Jewish men and women than non-Jews reported masturbating. Catholics and Protestants had similar patterns of activity, but Catholics who did masturbate did so more often than Protestants. Hunt reports religious devoutness to continue to have a significant influence on the practice of masturbation: the nonreligious are more likely to masturbate; they start doing so at a younger age; and they are more likely to continue it into adult life.[52] But these differences between the devout and the nondevout are operating less effectively among the young. The effect of religion continues to be generally much more marked among women than men.

Functions of Masturbation

The primary functions of masturbation are similar to those of sexual fantasy. Like fantasy, masturbation is an activity that is indulged in for its own sake, as a vehicle, or as a substitute.

Masturbation plays an important role in psychosexual development (chapter 9). Starting with self-exploration, the child discovers the pleasurable potential of the genitals, which in turn becomes the vehicle for further learning and sexual maturation. In adolescence, masturbation continues to fulfill a developmental function in self-exploration as well as providing the primary outlet for sexual release and gratification. As a vehicle for learning about the sexual aspects of one's body and one's self, masturbation continues to play a useful role throughout adulthood.[53]

Probably the most common function of masturbation in adulthood is by way of a temporary substitute for coitus or other sociosexual outlets. The relief of sexual tension is the most frequently mentioned reason by adults for masturbating (claimed by four out of five men and two out of three women in the Hunt survey).[54] The need for such release is often to compensate for the lack of a sexual partner or temporary unavailability of one because of illness, absence, or some other reason. Of all the alternatives to marital coitus (other than abstinence and nocturnal orgasm) masturbation is the least threatening to the marital relationship.

In instances where there is a lack of sociosexual opportunities, masturbation becomes, by necessity, the primary sexual

---

[51]Hunt (1974, p. 75). Women respondents in the *Redbook* survey say that since their marriage, 16% have masturbated often; 51% occasionally; 7% once; 26% never (Tavris and Sadd, 1975).

[52]For denominational differences in this regard see Hunt (1974, pp. 87–88).

[53]For a series of contributions on the significance of masturbation in human development from infancy through adulthood, see Marcus and Francis (1975).

[54]In the Arafat and Cotton (1974) sample of college students, "feeling horny" was the reason given by 48% of the men and 39% of the women; an additional 21% of men and 24% of women cited pleasure as the motivation. In the *Redbook* survey close to 40% of married women said they masturbated when their husband was absent; 18% when coitus had not been satisfying (Tavris and Sadd, 1975).

outlet. There are countless people who because of their personality, physical attributes, health condition, age, social position, and other reasons seem unable to link up sexually with others who appeal to them. Masturbation to them is one outlet they can always count on.

There are other reasons for preferring masturbation that are intrinsic to the activity. Masturbatory orgasm can be more intense (truer for women than men). It allows the use of special devices like vibrators that may provide the person with a whole range of new sexual sensations. There is a freer reign for sexual fantasy and the use of erotic materials. One may reach orgasm whenever and however one prefers; there is no need to wait for a slow partner or fear being left behind by a fast one. No effort is necessary to be charming, no need to worry over sexually transmitted diseases or pregnancy; and it costs nothing.

Masturbation has currently attained particular importance for those women who see it as a means toward female sexual autonomy. Given that women generally learn about their own sexuality through male initiative, masturbation allows them to become free of such dependence.[55] In this manner women are able to discover their own eroticism and their own sexual responses without male dominance.

Finally, masturbation, like other sexual behaviors, can serve nonsexual needs. Some 12% of college men and 16% of women say "loneliness" is the reasons for masturbating; and about 10% of each sex indulge in it to combat psychological frustration and mental strain.[56] It is not uncommon to masturbate to help fall asleep, especially if one feels burdened with stressful thoughts.

Given the high proportion of people who masturbate, it is easy to lose sight of the fact with not everyone indulges in the practice. College students who refrain (11% of men, 39% of women) cite lack of desire in the majority of cases (76% of women, 56% of men). Others think it is a waste of

energy or cite shame ("makes me feel cheap"), guilt, religious prohibition, and other objections against the practice.[57]

Masturbation, Health, and Society

"There is really no end to the list of real or supposed symptoms and results of masturbation," wrote Havelock Ellis, who was himself ambivalent about the practice. An incredible list of ailments were supposed to have been caused by masturbation at one time or another. These included insanity, epilepsy, headaches, and "strange sensations at the top of the head"; dilated pupils, dark rings around the eyes, "eyes directed upward and sideways"; intermittent deafness; redness of nose and nosebleeds; hallucinations of smell and hearing; "morbid changes in the nose"; hypertrophy and tenderness of the breasts; afflictions of the ovaries, uterus, and vagina (including painful menstruation and "acidity of the vagina"); pains of various kinds, specifically "tenderness of the skin in the lower dorsal region"; asthma; heart murmurs ("masturbator's heart"); and skin ailments ranging from acne to wounds, pale and discolored skin, and "an undesirable odor of the skin in women."[58]

There is absolutely no evidence to support any of these or any other claims of physical harm resulting from masturbation as such. One can flatly say that, in physical terms, whenever sexual intercourse is not harmful, neither is masturbation.[59]

Masturbation and Mental Health Assessments of psychological health are more difficult because it is harder to define mental than physical health. It is important to distinguish here between the possibility of masturbation causing psychological maladjustment from masturbation that may be a manifestation of it. This has been often

---

[55]Dodson (1974).
[56]Arafat and Cotton (1974).

[57]Arafat and Cotton (1974).
[58]H. Ellis (1942); Vol. 1, part 1, p. 249.
[59]Some physicians may be still ambivalent about the practice being carried "to excess," though no one is able to draw the lines of its "normal" limits. For a range of medical opinions, see *Medical Aspects of Human Sexuality* (April 1973, pp. 12–24).

confused in the past; physicians would see patients in psychiatric wards openly masturbating and would conclude that chronic masturbation had driven them mad; whereas in effect, indulging in such behavior in public was just one more sign of their mental disorder. (The same would be true if they urinated or defecated in public.) We must also distinguish between the effects of masturbation as such and the impact of the negative feelings it may engender.

Masturbation can be a sign of psychological disturbance, but so can coitus, or eating, or any other behavior. It is not the behavior as such that determines this but its underlying motivation, and the purposes to which it is put. Masturbation as such does not cause mental illness. It is necessary to state this, even in this day and age, because it seems there still are some who harbor fears of becoming insane as a result of it.[60]

A proper evaluation of masturbatory behavior, as with any other type of behavior, eventually has to be made in the context of the individual's broader life. In this regard, it is conceivable that masturbation can become a liability if it is relied on compulsively at the expense of ultimately more rewarding interpersonal encounters. It is a convenient shortcut to erotic pleasure that carries the potential of shortchanging deeper means of sexual gratification. The problem in these cases is not primarily masturbation but other, more fundamental psychological conflicts. It must be recognized, however, that even in these pathological conditions, masturbation may provide one of the few readily available forms of sexual release and psychological solace to the disturbed individual.

Masturbation, Guilt, and Shame   There have been a great many changes since the Victorian era when lives were warped by needless concerns over masturbation (chapter 18). Yet despite the liberalization of sexual attitudes, especially among younger people, masturbation continues to generate for some reactions of guilt, shame, sadness, and loneliness.

Sorenson reports that of his adolescent subjects, only 19% claimed never to have felt guilty (32% did so rarely, 32% sometimes, and 17% often).[61] Since masturbation implied that one is not mature enough, attractive enough, or sophisticated enough to have a sexual partner, there was a concurrent feeling of shame. Yet paradoxically, masturbation was more common among those also engaging in coitus than those who were not.

Similar attitudes have been reported for college students. In one study, guilt feelings were reported by 42%;[62] in another, 13% of men and 10% of women reported feelings of guilt, and 11% and 25% respectively reported feeling depressed following masturbation.[63] Nor are adults immune from the negative reactions engendered by masturbation. In the Hunt report, people generally felt ashamed and secretive about the practice: "almost no adults, not even the very liberated can bring themselves to tell friends, lovers, or mates that they still occasionally masturbate."[64] One of the *Cosmopolitan* survey respondents, a 39-year-old divorced woman (who had had 25 lovers) wrote, "I have been masturbating since I was 18 years old, but it's something I've never admitted to anyone except my present lover."[65] Though a remarkable percentage of people still feel that masturbation is wrong, there is a clear association between the prevalence of such attitudes and age. In the 55 and older age group, 29% of males and 36% of females agree that "masturbation is wrong." These percentages steadily decrease so that in the 18 to 24-year-old bracket only 15% of males and 14% of females still agree with this statement. These figures indicate not only a relative change in attitude but the disappearance of sex discrepancy in such attitudes.

---

[60]Arafat and Cotton (1974).

[61]Sorensen (1973, p. 143).
[62]Greenberg (1972).
[63]Arafat and Cotton (1974).
[64]Hunt (1974, p. 55).
[65]Wolfe (1981, p. 110).

Box 10.5   Masturbation in Literature

Philip Roth, *Portnoy's Complaint*. New York: Random House, 1967, pp 17–19.

Then came adolescence—half my waking life spent locked behind the bathroom door, firing my wad down the toilet bowl, or into the soiled clothes in the laundry hamper, or splat up against my medicine-chest mirror, before which I stood in my dropped drawers so I could see how it looked coming out. Or else I was doubled over my flying fist, eyes pressed closed but mouth wide open, to take that sticky sauce of buttermilk and Clorox on my own tongue and teeth—though not infrequently, in my blindness and ecstasy, I got it all in the pompadour, like a blast of Wildroot Cream Oil. Through a world of matted handkerchiefs and crumpled Kleenex and stained pajamas, I moved my raw and swollen penis, perpetually in dread that my loathsomeness would be discovered by someone stealing upon me just as I was in the frenzy of dropping my load. Nevertheless, I was wholly incapable of keeping my paws from my dong once it started the climb up my belly. In the middle of a class I would raise a hand to be excused, rush down the corridor to the lavatory, and with ten or fifteen savage strokes, beat off standing up into a urinal. At the Saturday afternoon movie I would leave my friends to go off to the candy machine—and wind up in a distant balcony seat, squirting my seed into the empty wrapper from a Mounds bar. On an outing of our family association, I once cored an apple, saw to my astonishment (and with the aid of my obsession) what it looked like, and ran off into the woods to fall upon the orifice of the fruit, pretending that the cool and mealy hole was actually between the legs of that mythical being who always called me Big Boy when she pleaded for what no girl in all recorded history had ever had. "Oh shove it in me, Big Boy," cried the cored apple that I banged silly on that picnic. "Big Boy, Big Boy, oh give me all you've got," begged the empty milk bottle that I kept hidden in our storage bin in the basement, to drive

---

Though masturbation is still not quite respectable, the social acceptability of the practice has been clearly on the rise as judged by explicit discussions of masturbation in popular and literary works (box 10.5). The polite tolerance of earlier marriage manuals toward the practice has also now given way to unabashed endorsement by writers of popular sex manuals and advocates of "liberating masturbation."[66]

Mark Twain seems to have anticipated this turn of events in a satirical address, "Some Remarks on the Science of Onanism," which includes the following memorable lines:

Homer in the second book of the *Iliad*, says with fine enthusiasm, "Give me masturbation or give me death!" Caesar, in his *Commentaries*, says, "To the lonely it is company; to the forsaken it is a friend; to the aged and to the impotent it is a benefactor; they that are pennyless are yet rich, in that they still have this majestic diversion." In another place this experienced observer has said, "There are times when I prefer it to sodomy."

[66]Dodson (1974), "J" (1969, pp. 38–52); "M" (1971, pp. 60–68). For additional positive views on masturbation, see DeMartino (1979). For a review of the theme of masturbation in literature, see Kiell (1975); (1976).

Box 10.5   Masturbation in Literature (*Continued*)

wild after school with my vaselined upright. "Come, Big Boy, come," screamed
the maddened piece of liver that, in my
own insanity, I bought one afternoon
at a butcher shop and, believe it or not,
violated behind a billboard on the way
to a bar mitzvah lesson."

. . .

Erica Jong, *Fear of Flying.* New York: Holt,
Rinehart and Winston, 1973, p. 121.

She is lying beside him very still. She
touches herself to prove she's not dead.
She thinks of the first two weeks of her
broken leg. She used to masturbate
constantly then to convince herself that
she could feel something besides pain.
Pain was a religion then. A total commitment.

She runs her hands down her belly.
Her right forefinger touches the clitoris
while the left forefinger goes deep inside her, pretending to be a penis. What
does a penis feel, surrounded by those
soft, collapsing caves of flesh? Her finger is too small. She puts in two and

spreads them. But her nails are too long.
They scratch.

What if he wakes up?

Maybe she wants him to wake up and
see how lonely she is.

Lonely, lonely, lonely. She moves her
fingers to that rhythm, feeling the two
inside get creamy and the clitoris get
hard and red. Can you feel colors in
your finger tips? This is what red feels
like. The inner cave feels purple. Royal
purple. As if the blood down there were
blue.

"Who do you think of when you
masturbate?" her German analyst asked,
"Who do you sink of?" I sink therefore
I am. She thinks of no one really, and
of everyone. Of her analyst and of her
father. No, not her father. She cannot
think of her father. Of a man on a train.
A man under the bed. A man with no
face. His face is blank. His penis has
one eye. It weeps.

She feels the convulsions of the orgasm suck violently around her fingers.
Her hand falls to her side and then she
sinks into a dead sleep.

Robinson Crusoe says, "I cannot
describe what I owe to this gentle art."
Queen Elizabeth said, "It is the bul-
wark of Virginity." Cetewayo, the Zulu
hero, remarked, "A jerk in the hand
is worth two in the bush." The immortal Franklin has said, "Masturbation is the mother of invention." He
also said, "Masturbation is the best
policy." Michelangelo and all the other
old masters—Old Masters, I will re-
mark, is an abbreviation, a contraction—have used similar language. Michelangelo said to Pope Julius II, "Self-
negation is noble, self-culture is be-
neficent, self-possession is manly, but
to the truly grand and inspiring soul

they are poor and tame compared to
self-abuse."[67]

SUGGESTED READINGS

Stoller, R.J. *Sexual Excitement* (1979). A fascinating account of the psychoanalytic treatment of a woman
with an unusual sexual fantasy.
May, R. *Sex and Fantasy* (1980). Explanations of gender differences through comparison of male and
female patterns of fantasy.
Friday, N. *My Secret Garden* (1975); *Forbidden Flowers*
(1975). Narrative accounts of female erotic fantasies. *Men in Love* (1980), describes male sexual fantasies.
Marcus, I.M., and Francis, J.J. *Masturbation from Infancy to Senescence* (1975). DeMartino, M.F., *Human
Autoerotic Practices* (1979). Collections of essays on
various aspects of masturbation.

[67]Twain (1976, pp. 23–24).

Chapter 11

Sexual Intercourse

Those who know the art of love are like those who can mix five different flavors in the cooking pot to produce a good meal.

Ancient Chinese saying

For most adults, most of the time, heterosexual intercourse is what sex is all about. Animals *copulate* (Latin: to link together) primarily for reproductive purposes. But men and women *make love* in fulfillment of various other psychological and social needs as well. Sexual intercourse is a complex interaction between two individuals that can be only understood within the context of their relationship.

Sexual intercourse refers to penile-vaginal intercourse and is called *coitus* or *coition* (Latin: to come together).[1] More broadly defined, sexual intercourse encompasses all erotic activities between men and women that may or may not culminate in coitus. There is thus more to sexual intercourse than genital activity.

Coitus may not be the preferred sexual activity for every individual but for the species as a whole it is clearly the most important sexual act, since reproduction depends on it. It is no wonder that coitus represents the key sexual behavior in all cultures.

The great majority of men and women engage in sexual intercourse at least sometime during their life. As we discussed in chapter 8, coitus accounts for the majority of sexual orgasms in the total sexual outlets of both sexes. However, the prevalence and frequency of coitus are not uniform; they are influenced by age, marital status, level of education, and other factors (see figures 8.5 to 8.8).

In this chapter, we are primarily concerned with the descriptive aspects of heterosexual intercourse and the enhancement of coital pleasure. In the next chapter we deal with the interpersonal and institutional context of heterosexual relationships; hence, these two chapters should be seen as a unit. In subsequent discussions, we take up the social issues (chapter 16), legal considerations (chapter 19), and moral concerns (chapter 20) that relate to sexual intercourse.

Foreplay

In physical terms, masturbation, petting, and foreplay consist of the same basic activities aimed at sexual arousal: in masturbation, sexual stimulation is applied to one's own self; in *petting*, two individuals arouse each other, sometimes to the point of orgasm; when petting acts as a prelude to coitus, we refer to it as *foreplay*. But in

---

[1]Numerous colloquial expressions for coitus include: To make love, have sex, go to bed, sleep together; to ball, bang, hump, fuck, score, screw; to lay, get laid; to get into somebody's pants; to get a piece, a piece of ass, a piece of tail; to make, to make out; to roll in the hay, rub bellies, have the banana peeled (Haeberle, 1978, p. 492).

sociosexual terms, masturbation, petting, and foreplay have more distinctive meanings; caressing your own thighs and having them caressed by someone else are quite different experiences. Similarly, whether the activity is an end in itself or a way station to coitus carries different meanings.

Kissing

Touching or caressing with the lips is widely used as a sign of affection, respect, and greeting. When used to express and excite sexual passion, it becomes an **erotic kiss** (figure 11.1). Erotic kissing is a common component of foreplay in our culture but is by no means ubiquitous. There are important differences within various social classes: In the Kinsey sample, 77% of the college-educated males but only 40% of

those with only grade-school educations reported practicing the erotic, or *deep kiss*. Simple kissing was much more frequent, being reported by almost all (99.4%) of the married male Kinsey respondents.[2]

Erotic kissing may start with simple lip contact. A light, stroking motion (French, *effleurage*) may be used in alternation with tentative tongue caresses and gentle nibbling of the more accessible lower lip. Gradually, as the tongue becomes bolder, it can range freely in the mouth of the partner, who may suck on it (French, *maraichinage*). The use of the tongue in erotic kissing is sometimes referred to as *French kissing*, but this is a common form of foreplay whose merits have been widely known

[2]Kinsey et al. (1948).

Figure 11.1   The erotic kiss. *The Lickerish Quartet*. (Courtesy of Audubon Films, Inc.).

Figure 11.2 Mochica pottery. (Courtesy of William Dellenback and the Institute for Sex Research.)

for a long time (figure 11.2). "A humid kiss," says an ancient Arabian proverb, "is better than a hurried coitus."

Kissing need not be restricted to the sexual partner's lips. Any part of the body may be similarly stimulated. However, the erogenous zones such as the breasts, nipples, genitals, inside of the thighs, the neck, and earlobes have a higher erotic potential (chapter 3).

Kissing is unknown in some non-Western societies, at least in a form familiar to us.[3] Tinquians (monogamous Philippine agriculturists and headhunters) and the Balinese bring their faces together and smell each other ("rubbing noses"). The Thonga of Mozambique (primitive agriculturists) are revolted by the Western kiss, saying: "Look

at them—they eat each other's saliva and dirt."[4]

Behavior resembling human kissing can also be observed in nonhuman primates and other animals. Chimpanzees occasionally press their lips together as a form of greeting as well as a prelude to further sexual exploration (figure 12.1). During coitus a male chimpanzee may occasionally suck vigorously on the female's lower lip.[5] Oral contact during sexual activity also occurs in other animals: for example, male mice lick the female's mouth, sea lions rub mouths, and male elephants insert the tips of their trunks into their partners' mouths.[6]

Touching and Caressing

Caressing, like kissing, is a means of expressing affection that may or may not be erotic in either intent or effect. Other forms of erotic touching are *hugging, fondling,* and gentle *scratching.* Such activities are perceived as erotic when the erogenous zones are being stimulated or because of the manner and context in which it is done.

To describe the use of touch in conveying sensory and eventually erotic pleasure, sex therapists have introduced the rather awkward term, **sensate focus** (chapter 15). Another term that has gained currency is **pleasuring;** it subsumes all of the touching and feeling activities that are meant to provide the partner with sensuous pleasure and excitement.

Erotic stimulation through touch is part of a much larger pattern of interaction that relies on touch to convey feelings of affection (holding hands, putting an arm over the shoulder), comforting (hugging), greeting (shaking hands, embracing), appeasement (touching, patting) and so on. The animal counterpart of these activities is **grooming** and can be observed in a wide variety of species. Grooming acts as an an-

---

[3]For a discussion of cross-cultural aspects of kissing, see Opler (1969).

[4]Ford and Beach (1951, p. 49).

[5]Bingham (1928), in Ford and Beach (1951, p. 49).

[6]The origin of kissing has been traced back to the transfer of chewed food from the mother to the infant, a practice that has persisted in some preliterate societies. Wickler (1972).

tidote to aggression (the one groomed is usually the dominant animal) but also occurs in intimate relationships such as between mothers and infants and between sexual consorts.

Breast Stimulation The practice of fondling the female breasts is well known in many cultures. In the Kinsey sample 98% of respondents reported manual stimulation and 93% oral stimulation of female breasts by males. But despite such widespread acceptance of the practice, one should not assume that all women necessarily find it enjoyable: almost one-third of heterosexual women queried by Masters and Johnson reported that their breasts were not especially pleasurable erogenous zones.[7] Even women who generally do enjoy such stimulation may find the experience uncomfortable during certain phases of the menstrual cycle and nursing, when the breasts are congested.

Breast stimulation can be enhanced by using both the mouth and the hands. The nipples, though the most responsive parts, need not be the exclusive focus of attention. Women may also enjoy more general fondling, mouthing, and caressing of their breasts. The size and shape of the breast has no relation to its sensitivity.

Breast stimulation is an important part of lesbian sex, and women are usually more adept at it than men. Stimulation of the male nipple occurs mainly in homosexual encounters. Masters and Johnson observed nipple stimulation in almost three-quarters of "committed homosexual male couples," whereas no more than 3 or 4 among 100 married men were similarly stimulated by their wives.

Genital Stimulation Tactile stimulation of the genitals is highly exciting for most people. In the Kinsey sample 95% of males and 91% of females reported manually stimulating the genitals of their sexual partners. Males more often take the initiative to stimulate their female partners. Though women have become more active

in this regard, one of the biggest complaints of heterosexual men in the Hite survey was that women did not touch them enough, or if they did, it was not done right—either because women were not interested enough or lacked the expertise.[8]

The techniques of genital stimulation during foreplay are similar to those employed in autoerotic activities; there is, therefore, much that men and women can learn from each other's autoerotic practices. It is also instructive for heterosexuals to know how homosexuals stimulate each other, since they have the gender advantage of knowing about each other's bodies. For instance, in lesbian genital play, the labia, mons pubis, inner thighs, and the vaginal opening are usually approached prior to direct clitoral stimulation. Likewise homosexual males dwell on the thighs, lower abdomen, scrotum, and the anal region before turning to the penis.

Oral-Genital Stimulation

A highly effective erotic practice is oral stimulation of the genitalia. *Cunnilingus* ("to lick the vulva") and *fellatio* ("to suck") refer to stimulation by the mouth of the female and male genitalia respectively; it is the organ stimulated not the gender of the stimulator that defines the practice.[9]

The Hunt survey found 80% of single males and females between the ages of 25 and 34 years, and about 90% of those married and under 25 years, to have engaged in mouth-genital stimulation during the preceding year.[10] The practice was more prevalent among the better educated and less prevalent among older individuals.[11]

Evidence from the *Redbook* and Hite surveys further attests to the growing accep-

[7]Masters and Johnson (1979, p. 67).

[8]Hite (1981).
[9]The Dutch gynecologist Van de Velde boldly (for his time in 1926) favored this practice. It was he who coined the term *genital kiss* in preference to the technical terms cunnilingus and fellatio. (Van de Velde, 1965, pp. 154–156.) Colloquial terms for oral sex include: "frenching," "eating," "going down," "sucking," and "blow job" (for fellatio).
[10]Hunt (1974).

Figure 11.3 Chinese print (? 19th century). (Courtesy of William Dellenback and the Institute for Sex Research.)

tance of these practices, especially among the sexually more liberal. This evidence also highlights the residual ambivalence and negative reactions some people have in this regard (see boxes 11.1 and 11.2). Among the *Redbook* wives, about 40% had "often" practiced cunnilingus and 40% fellatio. Another 48% and 45%, respectively, had "occasionally" engaged in these activities. Only 7% of the women had never had their genitalia orally stimulated, and 9% have never stimulated a male in similar fashion. Although women engaged in both practices with about equal frequency, they found cunnilingus "very enjoyable" in 62% of the cases as against 34% for fellatio. Only a small percentage found these activities unpleasant or repulsive (6% for cunnilingus, 15% for fellatio).[12]

There are no standard procedures for oral-genital contact. Usually the partner gently licks the minor lips, the area of the introitus, and the clitoris, along with sucking and nibbling at them. In the male, the glans is the primary focus of excitement: gentle stroking of the frenulum with the tongue and lips, mouthing and sucking the glans while firmly holding the body of the

penis and grasping and pulling gently at the scrotal sac are some of the means of stimulation. These activities must be conducted with skill, tact, and tenderness, since not many people appreciate a tooth-and-nail assault.[13]

A couple may engage in mutual oral-genital contact (called *sixty-nine* because of the positioning of the numerals: 69) as a prelude to coitus or for its own sake (see figure 11.3). When oral sex leads to ejaculation into the mouth, it can be thought of as *oral intercourse* (discussed below).

Oral-genital stimulation is repulsive to some people. If the genitals are clean, objections are difficult to support on hygienic grounds. A person is not obliged to perform or submit to such activity, of course, if it does not appeal to him or her, but to carry judgments beyond personal preference and to make legal or moral issues out of them is another matter.

Other Means of Foreplay

Anal stimulation through stroking or pressing on the anal orifice and inserting fingers into the rectum is appreciated by some and rejected by others on aesthetic and other grounds. The same applies to the oral stimulation of the anus called *an-*

[11]Oral-genital contact for both sexes is also reported in other societies. Oral stimulation of female genitals is quite common among primates, and females reportedly respond with obvious pleasure. Stimulation of the male organ, however, is less well documented.

[12]Tavris and Sadd (1978).

[13]Currently popular books on sex provide explicit instructions on oral-genital stimulation. See, for example, "J" (1969, pp. 119–123); "M" (1971, pp. 110–116); Comfort (1972, pp. 137–140). Also see Hite (1976, pp. 244–245).

Box 11.1   Cunnilingus: Current Attitudes

Female views (From Hite, S., *The Hite Report*. Macmillan, 1976, pp. 233 and 245.)

"I lie on my back with my partner between my legs, flicking his tongue very gently over the same area, over and over. I like not doing anything else except concentrating on the sensation until I orgasm."

"I dislike it when my partner's tongue digs too close to the clitoral nerve inside the hood—it really hurts."

"My husband asks me to let him frequently but I shudder at the thought. This is an issue that causes arguments. My husband says that all women do it and like it. I say he's crazy."

"Nibbling and nuzzling on the clitoris, like simulated chewing, is good—but gently and tenderly."

"I like a slow, steady rhythm, very gentle and circular in motion, right at the front part of my private parts, then moving down to my opening, with a deep penetration of his tongue just before I come."

Male views (From Hite, S. *The Hite Report on Male Sexuality*. Balantine, 1981, p. 709.)

"I feel that the genital kiss given by a man to a woman is one of the most intimate expressions of love that there is. I often have dreams involving sex, most of which do not end with my orgasm or ejaculation but do include a protracted period of my kissing the woman's genitals."

"I love the way a woman's genitals look when I am close enough to see all of the features. When a woman begins to respond to the act of cunnilingus, her vagina opens like the petals of a flower—her lips fill and enlarge, just as a man's penis does. It tastes like ambrosia that only a woman's body could exude."

"I like cunnilingus sometimes, but I think its pleasures are overrated. I think men feel like they have to say they like it, because it's the macho thing to do—like eating raw eggs or raw meat, or drinking a pint of whisky."

"It's ghastly. I tried it once and barfed all over my wife."

"She has gorgeous legs even at sixty-one. I usually start it. I could wish she would ask, or tell me to eat it, but she is bashful. I massage her with my tongue and she directs me. We have done this for forty-one years, and more frequently the last two years."

ilingus (or *rimming*).[14] Pulling at the buttocks will stretch the anal sphincter causing mild stimulation that may be quite pleasurable, without actually touching the anal orifice.

The erotic potential of **mild pain** is well known; bites and scratches can be quite stimulating. Tickling is another activity to which people react differently.[15]

How long does foreplay usually last? Preferences seem to vary widely among in-dividual and cultures. Western sex manuals devote much attention to the necessity of foreplay in preparing the woman for intercourse. Some even specify various time periods as irreducible minimums.[16] Among married couples in the Kinsey sample the lengths of time involved in foreplay were 3 minutes or less in 11%, 4 to 10 minutes in 36%, 11 to 20 minutes in 31%, and beyond 20 minutes in 22%. Some couples spent several hours each day in erotic play.[17] In the Hunt sample, single

[14]Unlike oral-genital sex, oral contact with the anus (even if it is clean), carries certain health risks (see Chapter 7).

[15]For a discussion of the sexual significance of tickling, see Kepecs (1969).

[16]For a range of opinions, see Kroop and Schneidman (1977).

[17]Kinsey et al. (1953, p. 364).

Box 11.2   Fellatio: Current Attitudes

Male views: (From Hite, S. *The Hite Report on Male Sexuality.* Balantine, 1981, p. 538).

"I believe every man's dream is to have a woman who would, if you'll forgive the vulgarism, suck him off. If I could find the woman who would suck me off in the morning to wake me up, I would lay my life in the mud at her feet, for she would be one woman in a million."

"I enjoy very much to have fellatio performed on me. However, I have yet to find a woman who can do it properly. I have orgasmed like this only once, and one other time with the help of masturbation. Many times I feel I could orgasm this way, but something in the way she does it turns me off, either it's getting too repetitive, I hear a sigh, she bites, or I can tell her heart's not in it."

"My fear is that I will orgasm in my partner's mouth and choke her, so I retreat from the situation."

"I have never had fellatio. I have no desire to have it done, it is repulsive to me."

"Coming in a woman's mouth and having her swallow it is something special. I don't associate fellatio with the degrading of woman's aspects that I have heard about it. A man in love loves his woman's cunt and a woman in love loves her man's cock and the oral caresses are just a magnificent way of expressing it."

Female views. (From Tavris and Sadd, *The Redbook Report on Female Sexuality.* Dell, 1978).

"Anyone, including so-called doctors, that advocate or engage in oral-genital sex is unsure of whom or what they are. Oral-genital sex is abnormal and immoral. Believe me, everybody doesn't do it! Our mouths are for eating, drinking, speaking, and kissing those we love—not for using on someone's genitals.

"I feel that sex is something beautiful and good, but oral sex makes it seem dirty. Animals go around licking each other. Human beings are supposed to be more intelligent."

"Women should not object to or feel disgusted by fellatio and cunnilingus. If two people are clean, as they should be for any sexual activity, those practices are as sanitary as sucking one's thumb or french kissing. Orgasms for both parties are usually as intense and pleasurable as coitus, sometimes more so. Of most importance, there is absolutely no fear of unwanted pregnancies; abortion can't exist. Oral procedures, if generally accepted and practiced worldwide, would solve the population explosion in a hurry. With that grave problem under control, hunger, starvation, and low standards of living could largely disappear."

---

people under 25 typically spent about 15 minutes in foreplay.[18]

Enhancement of Pleasure in Foreplay

A simplistic search for bodily levers and push-buttons leads to mechanical sex since the energy that charges the erotic circuits is emotion. Nonetheless, effective love-making requires some knowledge of *sexual technique.* As with any other art or skill (which is what "technique" means in Greek) natural aptitude may not always be an adequate substitute for knowledge and practice. There is nothing unromantic about technical expertise as long as one does not act like a "pro" who does this sort of thing for a living.

[18]Hunt (1974, p. 160).

Most people find it best to start slowly and with the less sensitive areas. Some may crave rough handling, but many others do not. As no one objects to tenderness, it would seem preferable to start gently and to become more forceful as appropriate, rather than to startle and offend at the beginning. Particularly effective is a carefully timed *teasing* approach, whereby the partner is helped to reach peak excitement in successive waves before moving on to coital activity.

Sensitive surfaces respond best to gentle stimulation, whereas larger muscle masses require firmer handling. Once an area has been singled out, it should be attended to long enough to elicit arousal. Frantic shifting from one part of the body to another can be as ineffective as monotonous and endless perseverance. But if stimulation is to be effective, it must be steady and persistent so that sexual tension is not dissipated. Erotic arousal requires patience and is not always free of tedium.

It seems best to leave the length of foreplay for each couple to decide. Aside from personal preferences, different circumstances require different amounts of preparation. Physiologically, full erection and adequate vaginal lubrical indicate readiness. Emotionally, a couple is only ready when both partners feel a mutual urge for sexual union.

A common source of difficulty is the male tendency to treat foreplay as an obligation. This approach is not likely to be effective no matter how fervently a man may kiss, caress, lick, suck, twist, and turn. The woman may sense in this utilitarian approach the genitally oriented cast of his mind and will resent it; so as he rushes her, she will drag her feet.

This conflict of interest during foreplay is part of a broader discrepancy in how men and women approach coitus: the former are much more genitally oriented and focused on orgasm; the latter are more attuned to the totality of the experience of which orgasm is but one part. This difference also explains why in the aftermath of coitus, women more often than men want to cuddle up, talk, and extend the experience.

To break this deadlock, one needs to understand the other's needs and be accommodating. A man may approach foreplay pretending he is engaged in petting, as if that is as far as things are going to go; this might make him want to extend the experience. In response, the female partner may want to press on to coitus.

Similarly a woman must realize that her male partner has a problem that she does not; as foreplay stretches out, the male faces the prospect of ejaculating extravaginally if he is young, or losing his erection if he is older. So when a woman senses that either prospect is becoming imminent, she may have to yield to his tempo. They may then start again or he may help her reach orgasm by some other means.

Intromission marks the start of coitus proper but it need not terminate foreplay. In face to face coitus, the couple can continue to kiss, in rear entry coitus the breasts and clitoris are readily accessible to manual stimulation.

Coitus

Making love has been compared to a game of chess: The moves are many, but the ending is the same. Coitus proper starts with the introduction of the penis into the vagina. It can be achieved from a variety of postures, and intercourse involves various types of movements. Even though there is more to making love than coital mechanics, understanding the physical basis of the act is helpful. The relationship between male and female sexual organs during face-to-face intercourse is illustrated in figure 15.2 (a theme also treated by Leonardo da Vinci five centuries ago: see figure 1.2).

Intromission

Clumsy lovers act as if there was nothing more to intromission than sticking the penis into the vagina. Subtle lovers recognize it as an important step in lovemaking on both physical and psychological grounds. Foreplay prepares the couple for coitus; intromission sets the tone for it.

Physical Considerations To initiate coitus gently, the glans is placed firmly against the introitus, until the vaginal orifice relaxes. A skilled lover will not penetrate even then but keep moving the glans within the introitus as well as in the clitoral region until the woman shows unmistakable signs of wanting deeper penetration.

By taking the initiative herself, the woman may gauge even better the best time for penetration to occur. Here, it is perfectly appropriate and sometimes preferable that the woman introduce the penis into her vagina herself. But in doing so she must in turn be mindful to the man's level of arousal. If the penis has not attained a certain firmness when intromission is attempted, a sense of failure may cause some men to lose their erection entirely.

If the vagina is insufficiently lubricated, intromission must not be attempted. Either futher stimulation or an artifical lubricant is called for.[19] But the fact that there is adequate lubrication does not necessarily mean the woman is ready for coitus. The final criterion is psychological readiness.

Psychological Considerations Coitus has close associations with dominance and aggression. To *penetrate* and to *be penetrated* are therefore laden with symbolic meanings. Between affectionate, trusting, and self-confident lovers, it matters little who penetrates whom. And once the penis and vagina are joined they become shared organs. However, if a man uses penetration as a form of assertiveness, the woman may experience it as a put-down. Or if a woman exhibits excessive preoccupation over the process, then the man may feel drawn into a power struggle. Neither case is conducive to mutual sexual enhancement. It takes sometimes very little to make the differ-

[19]Water soluble lubricants like K-Y jelly or contraceptive jellies, creams, and foam work best. Oily substances are best avoided; Vaseline in particular will harm rubber condoms and diaphragms (chapter 6). Saliva works reasonably well if there is a lot of it (such as following mouth-genital stimulation); merely moistening the glans will usually not work.

ence between the feeling of making love and "getting fucked."

Coital Postures

Coitus among animals follows species-specific standard patterns. Among most primates, the male mounts the female from behind (figure 11.4); face-to-face coitus is rare though it does occur occasionally, for example, among captive gorillas and orangutans. Men and women are far more versatile in their coital behavior. Pictoral representation of coitus going back to ancient Egypt show all varieties of coital postures. Similar depictions can be found in the art of virtually all cultures (figure 11.5).

Discussions of the mechanics of coitus may suggest that their main contribution is to enhance the physical aspects of intercourse, yet the benefits are just as much psychological. Some of us approach sex with uncertainty and a tendency to hurry. Through experimenting with various approaches, we become more deliberate,

Figure 11.4 Basic adult copulatory posture in monkeys. (Courtesy of Wisconsin Regional Primate Research Center.)

Figure 11.5 A carved wooden figure from the Luba (or BaLuba) tribe of central Africa. The sculpture shows the two figures in a coital position. (Courtesy of the Musee Royal de L'Afrique Centrale, Tervuren, Belgium.)

controlled, and purposeful. This effort implies care and concern for the partner, so that intercourse literally fits the meaning of the term: an interaction between two people.

Variety is the second advantage obtained from coital postures. Married couples, especially, sooner or later tend to find sex a bit monotonous. As Honoré de Balzac wrote, "Marriage must continually vanquish a monster that devours everything: the monster of habit."[20] The imaginative use of different positions can be an antidote to such habituation.

There is nothing magical about coital positions. Experimentation can show one approach to be more exciting at one time and another more suited to other circumstances. There is no position to end all positions, and the search for mechanical perfection is endless and pointless. The cookbook approach to lovemaking may rob the act of the very spontaneity that ought to be its hallmark.

Despite innumerable variations, the basic coital approaches are relatively few: The couple may stand, sit, or lie down; the partners may face each other, or the woman can turn her back to the man; one can be on top of the other, or they can lie side by side. Figures 11.6 and 11.10 illustrate some of the more common coital postures, which obviates the need for detailed descriptions. Instead, we shall examine briefly the relative advantages of these various alternatives.

Face to Face The primary advantage of face-to-face approaches is the opportunity for direct interaction. The partners can see each other's faces, gaze into each other's eyes, and communicate their feelings through facial expressions. They can also kiss and caress most easily.

The traditional and probably most commonly used variant (but not necessarily the most enjoyable for everyone), is the *man-above* position where the woman lies on her back with the man on top of her (figure 11.6).[21] Among Kinsey subjects, 70% had tried no other coital approach. This is also the usual pattern in many societies other than our own.[22] The man-above position is helpful when the erectile capacity of the male is relatively weak. Penetration is shallow unless the woman lifts up her legs. Placing a pillow under her lower back makes this alternative more comfortable. Entwining her legs around the man's provides the woman with good leverage for

[20]Quoted in Van de Velde (1965, p. 45).

[21]Western missionaries were observed by natives to always engage in coitus in this manner; so they called it the "missionary position."

[22]Ford and Beach (1951).

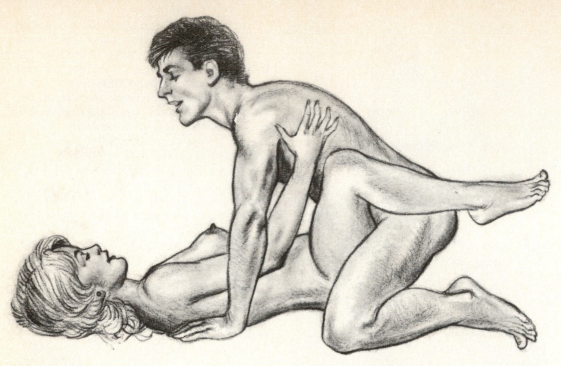

Figure 11.6 Man-above position.

Figure 11.7 Woman-above position.

Figure 11.8   Side-by-side position.

moving her pelvis to meet his thrusts or for simply sustaining muscular tension.

This position is the most likely to lead to pregnancy (but there are no positions that provide contraceptive protection). In order to maximize the chance of conception, the woman must maintain this posture for a while after coitus, and the man should not withdraw abruptly (chapter 3).

There are several drawbacks. The man's weight can be a problem: he can support himself on his elbows, but a really heavy man will still be a considerably burden on the woman. Furthermore, in this position the man's hands are not free to stimulate his partner. A more serious problem is the restriction that this position imposes on the woman's movements. Her pelvis remains largely immobilized or can be moved only with considerable effort.

A commonly used alternative is the *woman-above* position where the woman sits astride or lies on top of the man (figure 11.7). Among the Hunt sample, three-quarters of married couples used this approach at least occasionally.[23] The wom-

an's weight is usually less of a problem for the man, and she has the opportunity to regulate the speed and vigor of her movements and the depth of penetration. Many women find it easier to achieve orgasm in this manner (chapter 15).

A variety of other choices is available in *side-by-side* postures (figure 11.8). Penetration is somewhat more difficult, hence partners may effect intromission in some other position then roll onto their sides. The primary advantage of this approach is the ease and comfort it provides to both partners by eliminating weight bearings (and the issue of "who is on top"). Coitus tends to be prolonged and leisurely. Ovid commended this approach above others saying: "Of love's thousand ways, a simple way and with least labor, this is: to lie on the right side, and half supine."[24] It is also recommended by Masters and Johnson in their treatment of certain forms of sexual dysfunction, and subsequently adopted by three-quarters of the couples who tried it. More than half of Hunt's sample also reported using it.

[23]Hunt (1974, p. 202).

[24]Quoted in Van de Velde (1965, p. 227).

In *seated intercourse* the man sits in a chair or on the edge of a bed. The woman stands in front of him and lowers herself onto his erect penis (figure 11.9). By keeping her feet in the ground she can control her pelvic movements; she can also achieve deeper penetration by straddling him and locking her feet behind the chair. The man's pelvic movements are restricted, but his hands are free.

Penetration is difficult when the couple is standing face to face; if, however, the woman lifts one leg high enough, it can be done. She may then wrap her legs around his waist while he supports her buttocks (if she is fairly light and he has a strong back). Ancient manuals call this position "climbing the tree."

Rear Entry   Although the dominant form of coitus among animals is from the rear, it is not so for most people.[25] In addition to providing variety, the rear entry positions make it easy for the man to fondle the woman's breasts and stimulate her clitoris and genitals manually.

Rear-entry positions somewhat isolate the partners, since they cannot conveniently see each other. On the other hand, close and comfortable body contact is easy to achieve—no other position allows a woman to curl up as snugly in a man's lap. (Men also enjoy being cuddled in this "spoon position.")

Rear entry is possible with the woman lying down, sitting in the man's lap, leaning over, and on her hands and knees (figure 11.10). When the woman is lying on her abdomen, penetration is shallow. (Coitus in the manner may be difficult for obese people.) If the two lie on their sides, intromission is easy even with a relatively weak erection. This method is particularly restful and is suitable for coitus during pregnancy or ill health, when exertion is to be avoided.

[25]In the Kinsey sample only about 15% reported having tried it, but in the Hunt sample 20% in the 18- to 24 age group reported using this position often.

Figure 11.9   Sitting posture.

Coital Movements   Generally the initial coital thrusts are slow, deliberate, then getting progressively deeper. The man may thrust and withdraw; or the woman may do the same. Both may thrust and withdraw simultaneously. These movements may be rhythmic, fast or slow, with shallow or deep penetration. They may follow an in-and-out or circular "grinding" motion ("pestle in mortar"). It is possible for a movement to be executed by only one person while the other remains relatively passive. Some ancient manuals recommend a nine-stroke rhythm: nine shallow thrusts, then one full lunge, or nine full lunges followed by nine shallow thrusts.

In all these movements, variation and steady effort must be artfully combined with an element of surprise. Frantically rushing through pattern after pattern is as bad as monotonous concentration on one (unless

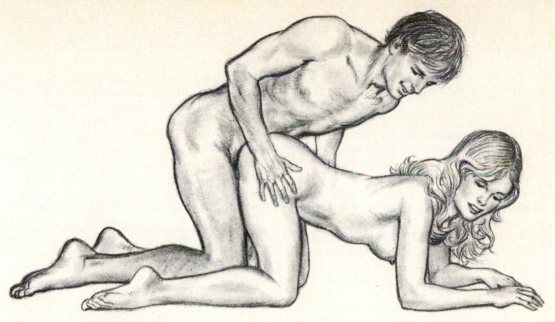

Figure 11.10   The rear approach.

both of the partners prefer it that way).[26] The most important dictum in this connection is that control of coital movements should be as much a prerogative and responsibility of the woman as of the man.

Variant Forms of Heterosexual Intercourse

Defined broadly, heterosexual intercourse would encompass activities other than coitus whereby a man and women reach orgasm through various means such as manual stimulation to orgasm or ejaculation into the mouth or anus. Though some of these activities might just as easily be characterized as *mutual masturbation*, we include them here because these activities are not autoerotic and go beyond what most people undertake during petting. They provide convenient ways of culminating a sexual encounter where something more than petting and less than coitus is desired.

[26]For anecdotal accounts of what women and men like about coitus, see Hite (1976); (1982).

These behaviors are, however, also engaged in by people who have ready access to vaginal coitus because they are unusual, provide novel physical sensations, manifest sexual daring, and provide opportunities to expect and offer something "special" to a sexual partner. For most people, these outlets are experimental or occasional variants rather than mainstays of their sexual lives.

Coitus Interruptus   Coitus that culminates with extravaginal ejaculation is called *coitus interruptus* or *withdrawal*. It is mainly used as a contraceptive device and not a very reliable one at that (chapter 6). The man may use a towel or whatever else is handy to catch the ejaculate or let it spurt against the woman's body.

The interruption of coitus at its climax is likely to be frustrating for the couple. Nonetheless, the practice has been widespread until the advent of modern contraception, and is still quite common (the French call it "watering the bushes").

Manual Orgasm   Manual stimulation usually accompanies fellatio and cunnilingus but may also be undertaken alone. When the female partner is being stimulated to orgasm, the practice is called *digital intercourse* ("finger fucking").

Ejaculation against Body Surfaces   Coitus can be simulated by *apposition* of the genital regions without intromission. Or the genital region of one partner may be pressed and rubbed against a part of the body of the other. Typically this is done by placing the thigh between the legs of the partner. The male may also ejaculate between the thighs of his partner (*intercrural coitus*) or between her breasts (*intermammary coitus*).

Oral Intercourse   Fellatio which culminates in ejaculation in the mouth may be considered as *oral intercourse*. Usually, the person whose penis is being stimulated remains fairly passive. But when a man forcefully thrusts his penis into the partner's mouth, it is called *irrumation*. The practice can be uncomfortable to the woman unless she is "constructed to withstand it, excited enough to enjoy it, or willing to use teeth, if necessary, to stop it."[27]

Ejaculation in the mouth is an innocuous practice from a health perspective: healthy semen is free of germs and harmless. So whether or not to engage in it is a matter of personal preference. The physical sensation of such an orgasm can be highly pleasurable to a man; he may also feel flattered that a woman will allow it or even welcome it (not unlike the reaction of a woman whose lover performs cunnilingus when she is menstruating).

A woman may likewise find the experience exciting and novel, or may simply tolerate it because of what it means to her partner. What to do with the ejaculate presents something of a problem. Some women swallow it, others discreetly dispose of it.

Anal Intercourse   Anal intercourse is usually thought of as a male homosexual practice, but a substantial number of heterosexual couples indulge in it as well. According to the Hunt survey, almost half of married men and more than a quarter of married women between the ages of 18 and 35 had tried anal intercourse at least once. The rates for singles and older married couples were lower.[28]

As discussed earlier, the anus is highly sensitive and its stimulation can be sexually quite arousing (chapter 3). Such stimulation may occur in both sexes by the insertion of fingers, objects, or by anal intercourse. Among the self-selected male respondents of the Hite survey, 31% of heterosexual men had experienced anal penetration by a finger (either during masturbation or by a partner) and another 12% had been penetrated by a penis or object of similar size (often as a teenager) as against 85% of homosexual men who had had similar experiences.[29]

Anal intercourse is enthusiastically endorsed by some and thought to be vulgar and offensive by others. Women may allow it as a concession to their partners; or because they themselves enjoy it and may reach orgasm through it. The attraction of anal sex is based partly on the physical sensations it can yield and partly on its being a novel, forbidden, unusual activity. It represents for some people the ultimate in sexual openness and receptivity when a woman allows a man to enter the most private and guarded part of her body.

Anal sex often entails some discomfort for the woman even in the best of circumstances; if clumsily done, it can be quite painful. To make it pleasurable (or even feasible) ample lubrication is necessary. The anus must be relaxed and gently stretched first with a finger, then the glans is pushed in carefully and not much beyond the anal sphincter. Movements following penetration must be slow and restrained to avoid injury to the delicate rectal mucosa. Anal

[27]Offit (1981, p. 66).

[28]Hunt (1974). Also see Masters and Johnson (1979, pp. 83–86).

[29]Hite (1982, p. 517).

sex can safely follow vaginal coitus but the reverse must be avoided unless the penis (or finger or vibrator) is carefully washed in between: otherwise bacteria that exist naturally in the rectum can cause vaginal infection.[30]

Orgasmic Control

Since orgasm effectively ends coitus for most men, its control and timing are matters of considerable significance. Generally, men are worried about reaching orgasm too soon; women have the opposite concern of reaching orgasm not soon enough or not at all.

Prolongation of Coitus Extending the period of coitus may seem desirable on the grounds that what feels good over 10 minutes should feel twice as good over 20. Yet sexual intercourse is not an endurance contest, and the only criterion is the optimal gratification of both partners, whatever the time it takes.

Women can remain actively engaged in coitus for longer periods of time; they do not have to worry about maintaining erection. Second, orgasm does not incapacitate women, as it does men, from continuing with coitus; they can persist with or without attaining subsequent orgasms (see chapter 3). The burden is thus more on the man to sustain his erection and control the timing of orgasm.

Control of orgasm involves modulating sexual arousal through physiological and psychological means. Of the two physiological processes underlying sexual response—vasocongestion and myotonia (chapter 3)—the first is not under voluntary control. But muscular tension can be influenced by various means. The more vigorously a man thrusts, the faster is the build-up to orgasm. Hence, to delay that moment, the man (and the woman) need to slow down or stop all coital thrusts. He also needs to relax his muscles, particularly those at the root of his penis (chapter

2). This can be done more easily if he is lying next to or under the female partner. A balance must be maintained by stop-and-go alterations of coital thrusting and other means to keep up erection without pushing on to ejaculation (other techniques to achieve this end are discussed in chapter 15).

Equally important is mental control over the level of arousal. During sexual excitement there is constant stimulation through all the senses as well as in terms of mental imagery. Though there is not much that a man can do to block sensory input he can effectively distract himself by thinking about a nonerotic topic to reduce his level of arousal. But he needs to do this without losing touch with his partner. To be most effective, all of these measures require the partner's cooperation and considerable practice on the man's part. The ultimate in ejaculatory control is *coitus reservatus* where a man engages in prolonged coitus without ejaculation (box 11.3).

Mutual Orgasm For a man and woman to reach orgasm simultaneously conveys an enhanced sense of shared pleasure. There is also a certain sense of pride in being able to synchronize one's climax with that of the partner. This is fine as long as mutual orgasm does not become the overriding criterion of a successful sexual encounter; such preoccupations can mar sexual enjoyment. It should be just as acceptable to reach orgasm separately, preferably the woman first. If she wishes to experience orgasm several times, the last may be made to coincide with the man's ejaculation. Then there is no risk that she will be left high and dry should the joint maneuver fail.

The Aftermath

If the proof of the pudding is in the eating, then the quality of intercourse is best judged by the mood that follows orgasm. Coitus does not end with orgasm. In its final phase the physiological effects of excitement recede and the participants regather their wits. (The physical manifes-

[30]Persons with ailments like hemorrhoids and fissures should avoid anal sex. For other complications see chapter 7.

Box 11.3   Coitus Reservatus

At least as far back as 2,000 years ago, Chinese Taoist physicians claimed that during intercourse, the male absorbed the enriching female essence that prolonged his life; ejaculation, however, depleted the man's own vital fluids (chapter 15). Hence, the ideal sexual practice was to engage in prolonged coitus without ejaculation, except at carefully regulated intervals. Such caution was especially called for once a man reached middle age.[1]

Foregoing the explosive, yet brief, pleasure of ejaculation provided a man with a more enduring and healthful form of sexual satisfaction, as well as offering his female partner ample opportunity for similar gratification. As one Taoist sage put it, "Although the man seems to have denied himself an ejaculatory sensation at times, his love for his woman will greatly increase. It is as if he could never have enough of her.[2]

The ancient Chinese clearly understood the distinction between ejaculatory and nonejaculatory orgasm (chapter 3). Those who mastered the art of achieving the latter obviated the refractory period, which made it possible for them to engage in prolonged coitus with multiple nonejaculatory orgasms.

In the West, the term *coitus reservatus* came to be applied to this technique several centuries ago.[3] Among others, the practice was adopted by the utopian Oneida community, in the 19th century, in upstate New York. Founded by John Humprey Noyes, the male and female members in this groups could freely engage in sex, but every man was not at liberty to have children with every woman. Hence, nonejaculatory coitus became the standard method of birth control and was called *male continence.*

Early in this century, Alice Stockham advocated this form of coitus in her book *Karezza* and her views were given a cautious endorsement by Marie Stopes. The practice called for prolonged caressing followed by very subdued coitus without ejaculation. The benefits to be derived from it were stated in terms of men conserving their vital energies, women obtaining a sense of union and soothing of the nerves, and both participants being spiritually enriched. The practice did not quite catch on.

Recently, there has been a resurgence of interest, not so much with the more esoteric aspects of these practices, as in their more general intent. Namely, that heterosexual intercourse need not be relentlessly genital and orgasmic in orientation. There is much erotic pleasure to be obtained through tender and affectionate intimacies, which may be engaged in as an end in themselves as well as a prelude to coitus.

[1]Chang (1977).
[2]Chang (1977), p. 23.
[3]The term has been applied to *nonejaculatory orgasm* as well as *retrograde ejaculation.* Though their outward manifestations are the same (namely, no semen is ejaculated), the underlying physiology of these two processes is different (chapter 3).

tations during this stage of detumescence and its behavioral components are described in chapter 3.)

Laboratory observation seems to confirm gender differences in postcoital behavior. Irrespective of sexual orientation or the manner of reaching orgasm, males tend to rest briefly and then disengage. Women usually desire to remain quiet longer and maintain the embrace. Even before such documentation was present, marriage manuals devoted considerable attention to this phase. Men were warned not to dismount abruptly, turn their backs, and go to sleep; rather they were told to disengage gradually with endearments and tender caresses, and remain attentive to their partner's needs for quiet intimacy.[31]

There is a good deal to be said for these admonitions, although the gender-linked differences should not be exaggerated. A man needs and may want as much tenderness as a woman, unless he is made to feel "unmasculine" or "weak" if he should reveal such inclinations.

The aftermath of coitus can be a time of quiet reflection. Thoughts meander; earlier experiences, sexual or otherwise, float into consciousness. There must be room for parallel solitude. But the partners should not drift away emotionally. One may share feelings and thoughts at this time with some discretion. If coitus was not all that it could have been, tenderness restores confidence and makes amends. But this is no time for clinical postmortems or statistical tallies. Even if orgasm has been highly pleasurable, the experience may be partially ruined if the feeling is conveyed that a victory was scored. Likewise, there may be a sinking feeling that one was seduced or manipulated. What transpires following coitus is said to reveal more about the relationship than the preceding sexual activity.[32] Even in the best of circumstances, one may experience a certain feeling of letdown, lassitude tinged with sadness, in the postorgasmic phase.

The aftermath of coitus may signify the end of an episode of sexual activity or it may simply be an interlude between two acts of intercourse, where afterplay imperceptibly merges into foreplay again. Even nonconsecutive acts of coitus are to some extent linked together through memory.

Enhancement of Coital Pleasure

It should be clear by now that coital satisfaction is a function of a host of physical, psychological, interpersonal, and social factors, and that none of these can be neglected or placed in hierarchies of importance. It is not enough to be in love to enjoy sex fully, just as all the refinements of erotic technique will be of little avail in an unhappy relationship; to glorify the emotional, one need not denigrate the physical aspects of sex, and vice versa.

We have so far dealt with various means of sexual enhancement in connection with specific facets of coitus. We shall now turn to more general means of making intercourse more satisfying. Our purpose here, as in the rest of this text, is to inform and stimulate reflection. For "how-to" instruction on sexual enhancement one can turn to *sex manuals* (box 11.4). A similar function is served by magazine articles and advice columnists. More pervasive though indirect are the effects of movies and television that portray screen lovers. Sexually explicit films may also fulfill an instructional role, but more focused in this regard are films made especially for purposes of teaching and sex therapy. Such devices are meant to be useful both by imparting information as well as by removing sexual inhibitions and negative attitudes through "desensitization."[33] Finally, trial-and-error is probably the primary method of sexual instruction for most people, be it for better or for worse.

Physical Factors

How we look, sound, smell, and even taste, are important to our sexual appeal. These sensory modalities have obvious physiological bases (chapter 3), although their erotic characteristics are largely learned and culturally determined.

It is tempting to equate the prospect of sexual satisfaction with *sexiness*, but past the point of arousal the popular verdict is that the two do not necessarily go together. Nonetheless, there is enormous emphasis placed in our culture (and in many others as well) on enhancing physical attractiveness for sex appeal. A good deal of the effort to keep fit, dress up, use makeup, and so on, are geared to this aim, whether or not we admit it or even rec-

---

[31]Van de Velde (1965).
[32]Crain (1978).

[33]Major outlets for such films include Focus International (333 West 52nd Street, New York) and Multimedia Resource Center (1525 Franklin Street, San Francisco).

Box 11.4   Sex Manuals

People may associate sex manuals with contemporary preoccupation with sexual enhancement in Western culture, but in fact, such books are quite ancient and come from the East. Chinese handbooks of sex existed 2,000 years ago and were widely studied until about the 13th century. These manuals instructed men how to conduct their relations with their women.[1] Other illustrated works known as *pillow books* gave more specific instructions on coital techniques.

A similar literature in India was brought to the attention of the Western world through the efforts of Sir Richard Burton and F. F. Arbuthnot, who founded in 1882 the Kama Shastra Society in London to publish translations of Hindu erotic works. Thus came to light the great Indian love manual, the *Kama Sutra*, written about the 3rd century A.D., by Vatsyayana.[2] It was not an original work but the summation of existing wisdom; as the earliest and most profound discussion of love and sex written in India, it remained the classic treatise in this field for a thousand years.[3]

Two other works from this same Hindu tradition are also noteworthy. One is the *Koka Shastra* composed in the 12th century by a poet named Kokkoka;[4] the other is the *Ananga Ranga*, written in the 16th century by Kalyana Malla.[5] These books were not merely compilations of erotic techniques, though they contain much of that, but texts with a broader scope, which influenced art and literature as well as the popular culture.

Persian and Arabic erotic works of this genre, though not as ancient, are equally extensive. The classic Arabian treatise on physical love (and another of Burton's finds) is the *Perfumed Garden*, written in the 16th century by Umar ibn Muhammed al-Nefzawi.[6]

The Western tradition is virtually unique among the great cultures in lacking a literature of this kind. It was only early in the 20th century that the *marriage manual* appeared on the scene. As exemplified by Theodore H. Van de Velde's *Ideal Marriage* (1926), these books were written by physicians and cast in the rather arid language of science, in contrast to the more humanistic and celebratory manner of the ancient love manuals.

Van de Velde's book remained the standard manual for almost four decades. Though quite positive in tone and daring (for its time) it did not survive the freewheeling 1960s, which saw the immergence of a new style of sex manual—explicit, unabashedly enthusiastic, and addressed to all, married or not.

---

ognize it to ourselves. But it would be wrong to assume that, in every individual case, such activity is erotically motivated. A woman may wear lipstick or a man will grow a mustache simply because the culture defines these as gender-appropriate means of looking respectable.

The sexual image we present is the outcome of our physical endowments and what we do with them; rarely is one so attractive or so homely that efforts at self-improvement are either irrelevant or hopeless. There are numerous ways of enhancing our physical endowment, ranging from the healthy to the misguided. It is one thing for women to exude a robust sexuality through exercise and another to deform their ribcages through corsets and cripple their feet through foot-binding. Similarly, a lot of bad cosmetic practices must be differentiated from artful ways of correcting physical flaws and enhancing attractive features (whether it concerns women or men). The trick is to project a distinctive image without appearing eccentric; to capitalize on the culturally recognized sexual

Box 11.4   Sex Manuals (*Continued*)

The more sophisticated of current sex manuals are Alex Comfort's *Joy of Sex*[7] and *More Joy*.[8] Unlike Comfort, who is a reputable scientist and scholar, most authors of the more typical popular sex manuals are writers with modest credentials in the field. They reiterate commonplace and common-sense information and advice in a light-hearted and titillating style. When Van de Velde wrote of the "genital kiss," he cautioned his readers that from the sublime to the ridiculous is but a step. His modern counterparts are less troubled by such distinctions. Note, for instance, the following "recipe" for fellatio from the *Sensuous Woman* (which went through more than 10 printings in 1970).

The Whipped Cream Wriggle
   If you have a sweet tooth, this is the one for you. Take some freshly whipped cream, to which you have added a dash of vanilla and a couple of teaspoons of powdered sugar and spread the concoction evenly on the penis so that the whole area is covered with a quarter-inch layer of cream. As a finishing touch, sprinkle on a little shredded coconut and/or chocolate. Then lap it all up with your tongue. He'll wriggle with delight

and you'll have the fun of an extra dessert. If you have a weight problem, use one of the many artificial whipped creams now on the market (available in boxes, plastic containers and aerosol cans) and forego the coconut and chocolate.[9]

These books clearly fulfill a need since they sell in the millions. But their informational content is meagre. What they provide people with is reassurance and permission to act more freely in their sexual lives. They also provide a certain measure of erotic titillation and vicarious gratification that makes them a form of soft-core pornography.

These modern sex manuals are also very much an American product, part of the quintessentially American self-help literature that tells us how to cook, diet, exercise, play, work, and get ahead in life.

[1]Gulik (1974).
[2]Vatsyayana (1966).
[3]W.G. Archer in Comfort (1965).
[4]Malla (1964).
[5]Comfort (1965).
[6]Nefzawi (1964).
[7]Comfort (1972).
[8]Comfort (1974).
[9]"J" (1969).

symbols (if not harmful) without being engulfed in the vulgar images of what passes for sexiness.

Nudity is a key feature in sexual enhancement, though cultural and personal preferences vary. The degree of bodily exposure permitted in a culture is a good index of its level of sexual permissiveness. But nudity does not always enhance eroticism. On the contrary, artful concealment may provoke desire and sexually permissive cultures may go to great lengths to devise clothing and ornaments whose

function is to make the body more erotically appealing.[34]

Men have been generally thought to derive more pleasure from female nudity than vice versa (chapter 3). But even if that is true, this need not apply to every woman. Furthermore, apart from visual appeal, the tendency is for both sexes to want coitus in the nude to enhance pleasure from skin

[34]For an illustrated historical overview of eroticism in dress, see Glynn (1982). Body ornamentation is discussed in Morris (1977).

contact and to convey a greater sense of intimacy.

Pleasurable sex requires attention to many practical details, beauty of mind and soul notwithstanding. All that was discussed with regard to sexual hygiene (chapter 7) is therefore highly pertinent to the enhancement of sexual pleasure. The trick here is to combine a taste for cleanliness with an earthy reveling in the messiness of sex; one must be prepared to take a fair amount of secretions, noises, and odors in stride. Rumbling sounds in the abdomen can be distracting; belching and farting are awkward; the noise made by the expelling of air sucked into the vagina may be mystifying. One can ignore these noises or laugh at them together, but if handled clumsily, they become sources of embarrassment.

Time and Place

Where and when people make love help determine how pleasurable the experience will be. The time of day, the level of fatigue, the temperature in the room, the sights, sounds, smells, even the hardness of the bed, may have a bearing on the quality of the sexual experience. Such factors help set the emotional tone and ambiance of the surroundings and hence influence the mood of the participants. People generally do best under circumstances that are restful, safe, and secure on the one hand and exciting, novel, and romantic on the other (a bit like sleeping in a warm bed while a cold wind is howling outside).

Animals prefer certain settings over others and some can mate only under given conditions. For instance, male mammals may restrict their contacts to their own territories. There is a general tendency to seek sex where it has been enjoyed before. Copulation occurs during the active period of the day; primates have sex in broad daylight. Our shift to more nocturnal sexual patterns is probably the result of cultural influences and work schedules.

There is an almost universal human preference for *privacy* during coitus. This too would seem to be culturally deter-

mined, since animals generally do not seem to mind copulating in front of one another (although they are particularly vulnerable to attack when so preoccupied). Coitus occurs in various cultures in or out of doors, depending on which affords more privacy. In our society, the bedroom is where sex usually takes place, but shifting to another room (especially the bathroom) may add novelty and excitement. More adventuresome couples seek the great outdoors (secluded beaches are great if you can keep the sand out). Cars are widely used for amorous purposes and there has been an ongoing romance between sex and the automobile from its rumble-seat origins.[35] The very sense of novelty, sometimes even danger, may endow a location with a sense of sexual excitement.[36]

Bedrooms are usually furnished with some thought to creating a romantic and erotic atmosphere through the choice of colors and lighting. Soft lights convey a mellow mood and are kind to physical blemishes; candles are particularly effective as they cast their languorous rays on the intertwined limbs and flushed faces of lovers. Darkness may appeal to those who feel self-conscious, others may prefer brighter lights. Music is another common mood enhancer, as we discuss shortly.

There is no end to the refinements that can be added to enhance the erotic potential of the physical setting. Mirrors may be positioned strategically so that the partners can observe themselves (but rarely is a couple bold enough to place a mirror on the ceiling where its exclusively sexual function would be obvious). Such erotic devices work at least for some people some of the time. But it is more difficult to decide whether these sights, sounds, and smells awaken the "sexual animal" within us or are mainly cultural artifacts to whose symbolic meanings we are conditioned to respond.

---

[35]Lewis (1980).

[36]New York helicopter police once arrested a teenage couple having coitus on top of a 500-foot tower of the Williamsburg Bridge over the East River. (*San Francisco Chronicle*, June 30, 1981, p. 2.)

The preferred *times for coitus* in different cultures are determined by practical as well as psychological considerations. Most couples in our culture make love at night because it fits most conveniently with the routine of everyday life; the lassitude that follows orgasm also nicely dovetails with sleep. But it is unwise to always relegate coitus to the last waking moments of the day, since a certain amount of stamina and alertness are essential for greatest pleasure. At least an occasional shift to the morning or the middle of the day may freshen up coital routines.

Should intercourse be planned or spontaneous? One may say that coitus requires preparation, as does any other activity. Besides, the anticipation that goes with planning heightens excitement. Others find such planning to be cold-blooded and detracting from the romantic spontaneity of the occasion; there are young men and women who cannot even bring themselves to take contraceptive measures because that would imply forethought (chapter 6).

Many an exciting sexual encounter may be the gift of the moment and occur in the least likely of places. Nevertheless, as with everything else in life, virtuosity in sex requires that spontaneity be supplemented by preparation and practice. One practical way to combine the best of both worlds is for one partner to plan and to surprise the other.

Erotic Aids and Exotic Practices

Sexual arousal can be further enhanced by a variety of devices and practices that titillate the senses of sight, hearing, smell, and taste. In order to be effective, such erotic aids must be both novel and exciting but not so foreign or far out that they induce anxiety.

Scents   Pleasant scents are among the most popular erotic enhancers. Men may use aftershave lotion to freshen up, but it is female perfumes that are really supposed to do the job; a great deal of advertising is banked on that promise. Yet there is no evidence that any scent, natural or artificial, acts as a pheromone (see box 4.2).

Music   The enormous and universal appeal of music may be at least partly due to its erotic qualities. Shakespeare calls music "the food of love."[37] At its most obvious, the pulsating beat and suggestive lyrics of certain forms of rock music leave little to the imagination. More sentimental songs elicit sexual enhancement by creating a romantic mood.[38] But it is not only popular music that has erotogenic potential—virtually any form of classical music can exert a similar effect if it falls on the right ears, under the right circumstances. For some reason, Ravel's *Bolero* has developed a reputation as the ultimate musical aphrodisiac. It has been cited so often in this regard that it probably works by now as a self-fulfilling prophecy. People can be conditioned of course to regard even a dirge as erotic.

Clothing   The use of sexy female *underwear* is another widely used turn-on. Even panties and bras for everyday use are often designed with a hint of erotic appeal. Other items range from subtly suggestive bedroom apparel sold at exclusive stores to undergarments peddled through sex catalogues, fit for professional strippers.

Lotions   During recent years, the use of *lotions* and erotic *massage* has become quite popular. Tactile stimulation can be greatly enhanced if surfaces are moist. Massage adds many dimensions to lovemaking. It can be used to stimulate the skin, induce deep muscular relaxation, generate a sense of restful contentment, and convey a sense of tenderness and caring.

Vaginal lubrication was discussed earlier. Lubricant need not be applied just prior to intromission but could be used earlier during foreplay to enhance tactile stimu-

---

[37] *Twelfth Night* act 1, scene 1.

[38] In one survey of a popular music station in Connecticut in 1977, about half of the songs pertained to romantic or erotic love (Tennov, 1979, p. 82). An analysis of the lyrics of popular songs in the 1950s revealed that 83% concerned *love* (Horton, 1957). *Sex* became a prominent theme in such songs later in the decade (Carey, 1979).

lation of the clitoris and the adjoining structures.

Bathing   Combining sex with *bathing* is an ancient device. Here again several factors contribute its beneficial effect: nudity, the relaxing effect of the warm water, the gentle mood induced by soaping and caressing each other. To actually engage in coitus while showering or in the bath tub may not be easy since erections are difficult to maintain in warm or cold water. This may necessitate that intromission precede submersion or take place while the man is not exposed to the jet of water.

Special Equipment   More esoteric forms of lovemaking often rely on *special equipment*. These may be no more complicated than condoms of different colors and textures, or they may involve specially constructed chairs, beds, and the like. Water beds must owe part of their popularity to their contribution to lovemaking. The use of vibrators and other mechanical gadgets was discussed in connection with masturbation (chapter 10); some couples use them in foreplay or as a substitute for coitus.

Combining *motion* and sex offers many possibilities. Speeding trains with their rhythmic noises are much recommended, as are gently rolling ships and boats. Where life comes cheap, moving cars and motorcycles may intrigue some; transport animals like horses, camels, and elephants have all been used for the same purpose. Eastern erotic works even show couples precariously balancing upside down on galloping ponies, which is probably pure fantasy.

Lovemaking, like eating, offers more pleasure with more variety, but it is a matter of personal preference as to what one chooses from a menu. In sex, too, there are both gluttons and people with discriminating palates. Eventually, sex must fit with and make sense in the larger context of our everyday lives. Relatively few of us have the energy, the need, or the means to stage a five-course dinner every night; nor can we mount a comparable production each time we engage in sex. On the other hand, far too many people settle for the sexual

equivalent of fast food and reduce making love to the level of a routine, as regular and uninspiring as going to the bathroom; a welcome relief, and not without its satisfactions, but nothing to particularly gladden one's heart.

Aphrodisiacs   Finally, there is the ancient quest for *aphrodisiacs* (from *Aphrodite*, goddess of love), various substances that are said to enhance libido, sexual performance, and pleasure. The popular use of aphrodisiacs is discussed in box 11.5; its medical applications are dealt with in connection with the drug treatment of sexual dysfunctions (chapter 15).

Psychological and Relational Factors

It is a truism that personality factors and the quality of the relationship are the most important determinants in experiencing and providing sexual enjoyment. Yet it is not entirely clear what sorts of people make the best lovers. Much of what we discuss with regard to sex and interpersonal relationships (chapter 12) is pertinent here but our more immediate concerns are with those aspects of personality and styles of relating that are more immediately conducive to high levels of satisfaction in sexual relations. Though we use a heterosexual model in this discussion, much of what is said here is just as applicable to homosexual relationships.

A close correlation between the quality of the sexual and other aspects of a relationship is often taken for granted, yet the two do not always seem to match: sexually accomplished lovers do not necessarily make the best friends or spouses and vice versa. That there are no established personality profiles of effective lovers is hardly surprising, given the complexity of the factors entailed and the wide range of variability in who gets turned on by whom.

Gender Considerations   Traditionally, sexual attractiveness has been strongly influenced by gender characteristics: the more "masculine" a man or the more "feminine" a woman, the sexier they have been

Box 11.5   Aphrodisiacs

The search for substances to enhance the sexual desire and experience is as ancient, and so far as futile, as the search for the fountain of youth.

Erotic potions are described in medical writings from ancient Egypt (c. 2000 B.C.). Among various societies since that time all the following and many more preparations have been recommended: pine nuts, the blood of bats mixed with donkey's milk, root of valerian plant, dried salamander, cyclamen, menstrual fluid, tulip bulbs, fat of camel's hump, parsnips, salted crocodile, pollen of date palm, the powdered tooth of a corpse, wings of bees, jasmine, turtles' eggs, groundcrickets, spiders, ants, the genitals of hedgehogs, rhinoceros horn, the blood of executed criminals, artichokes, honey compounded with camel's milk, swallows' hearts, vineyard snails, certain bones of the toad, sulfurous waters, powdered stag's horn, and so on.[1]

Presently, people still refer to bananas and oysters, half-jokingly, as sexual enhancers. As with some of the items listed above, the association is probably due to their shapes, linked by a primitive logic; since a banana is like an erect penis, if a man eats it his penis will become erect. However, more typically, alcohol and other drugs now dominate as substances with aphrodisiac potential.

Alcohol is a central nervous depressant that actually has a dampening effect on sexual response or performance. But in small amounts, it lifts inhibitions and prompts some people to be sexually more open and forward. Alcohol is also touted as a tool for seduction ("Candy is dandy but liquor is quicker") although its effects are mainly on those who are already quite inclined to be seduced; or who use it as an excuse for yielding.

The influence of *marijuana*, like that of alcohol, appears to be secondary. It is not as such a sexual stimulant but through its influence on mood, it alters the perception of the sexual experience, which some interpret subjectively as an enhancement of erotic pleasure.[2] Psychedelic drugs, such as LSD, may have similar effects.[3]

Cocaine and *amphetamines* are claimed by users to increase sexual desire. This may be part of their more general euphoric effect, or the result of direct brain stimulation.[4] Chronic use of these substances causes dryness and inflammation of the vagina, anxiety, and exhaustion, all of which are inimical to sexual enjoyment. *Heroin* depresses sexual desire. Drug addicts may be driven to prostitution to support their habit, not because they enjoy sex more under the influence of drugs.

Amyl nitrite causes rapid dilation of blood vessels; hence its use to relieve chest pains resulting from coronary spasm (angina). It comes in small vials that are popped open (hence, "poppers") and inhaled. It is claimed that inhaling the substance just before orgasm prolongs and intensifies the experience. A common side effect is severe headache. Occasionally, death may result from the sharp drop of blood pressure.[5] Other drugs with aphrodisiac potential that are used in the treatment of sexual dysfunction are discussed in that connection (chapter 15).

In the light of these ambiguities, whether or not there is such a thing as an aphrodisiac is a matter of definition. If one means by it a substance that directly increases sexual drive or libido under vigorously controlled conditions (such as a double-blind study), then no such substance has yet been shown to exist. If, however, the definition is extended to include various effects of subjective erotic pleasure, however such effects may be achieved, then it is quite plausible to characterize a number of substances as aphrodisiacs.[6]

[1]For historical overviews on aphrodisiacs, see Himes (1970; Licht (1969).
[2]Goode (1970).
[3]Grof (1975).
[4]Gay et al. (1975).
[5]Hollister (1974).
[6]Gawin (1978).

taken to be. Stereotyping provides enough stability to such images to lend them public credibility. Movie stars are the most common embodiments of these images in the mass culture.

Whatever the differences in the way the two sexes express their sexual attractiveness, it is now generally assumed that men and women ought to be equal partners in sexual relations.[39] It should be irrelevant, for instance, as to who takes the initiative or who does what to whom. When lovemaking becomes a shared and reciprocal pleasure, then sex ceases to be something you do *to* someone and becomes something you do *with* someone.

The Self and the Other We can take it for granted that obtaining sexual satisfaction for yourself should not be separated from providing such satisfaction to the partner and vice versa; to enhance your own sexual gratification at your lover's expense would be selfish; to provide pleasure at your own expense, unfair to yourself. This is not to say that at any given moment each and everyone must be getting the same share of gratification as the other; attempts at such mindless parity ruin the experience for all concerned. Sexually happy couples compensate for such discrepancies by extending to each other ample "credit." They go out of their way to please the other, knowing that they will be paid back sooner or later. Such confidence eliminates the need for sexual bookkeeping, as the satisfaction in giving pleasure complements the gratification of receiving such pleasure oneself.

Being Yourself Those who get high marks as lovers seem to share a number of traits.

First and foremost is the ability to *be yourself*. Sex exposes us psychologically, as it does physically. There is, therefore, a strong temptation to cover up and to pretend to be what you are not. This obtains some immediate gains but robs the experience of its authenticity by reducing it to playacting. It is similarly important to accept the other person as he or she is. What is called for in these situations is not stark realism but placing the partner in the best possible light, within the bounds of honesty. Without deluding oneself or lying to the other, one can emphasize the positive aspects of the other. The shortcomings of the other person must be recognized, but lovemaking is not the appropriate occasion to criticize or reform another.

Freedom from Anxiety If the condition of being yourself is fulfilled, half the battle is won, since it is attempts at pretence and the fear of discovery that often cause *anxiety*. Freedom from anxiety also presupposes freedom from *guilt* and *shame*. If you believe what you are doing is wrong or unworthy of yourself, you are not likely to get very much joy out of it. Similarly, to the extent that sex gets contaminated by hidden motives like dominance and aggression, the more likely it is to generate fear and anxiety. And as we discuss later on, sexual enjoyment is particularly vulnerable to anxiously watching over your own performance in order to meet some preconceived notion of how it all should feel and function.

There is also the matter of having confidence in your *desirabiliy* and *competence* as a lover. This is not a matter of showing off how expert you are but exuding the calm assurance of the person who knows what he or she is doing. Yet effective lovers do not come across as cool and collected operators. On the contrary, a certain sense of vulnerability is a most endearing quality. It is as if one should simultaneously project an image of the parent (strong, knowledgable, and trustworthy) as well as the child (helpless, innocent, and lovable). It is the ability to handle this paradox that characterizes master lovers.

---

[39]There is considerable cross-cultural variation about which sex takes the more active role. Some primitive peoples are quite similar to those of the West in this respect. The Chiricahua Apaches, for instance, even shared the Victorian notion that a woman, besides being passive, should display no emotion during coitus. On the other hand, the Hopi, Trukese, and Trobrianders considered an inactive female apathetic and a poor sexual partner (Ford and Beach, 1951).

Affection and Trust   In relational terms, trust and commitment are basic requirements for a satisfactory sexual relationship. *Love* is a powerful erotic force, but for many people it is not a precondition to enjoy sex. A more realistic expectation is a level of mutual *trust* that is commensurate with the extent of sexual involvement of the moment; there should be enough security that one is not going to get "screwed" (which means different things under different circumstances). This is also where *honesty* comes in, which is not an obligation to bare one's soul to everyone but to tell the other person what it would be in his or her legitimate interest to know under the circumstances.

Honesty is not enough. There must be a sense of caring and sharing if *intimacy* is to develop. This sense of closeness, the coming together of one's innermost self with that of another is the core of a sexually gratifying experience. True intimacy requires the maturing of a relationship over time but the passage of time alone will not do it. Some couples married for decades may be no more intimate than others who have spent an afternoon in bed together.

Communication   Intimacy allows *communication*, which in turn generates greater intimacy. Nothing could be more appropriate when making love than expressions of affection, and few measures are more effective in arousing a partner. The same is true for the less articulate means of conveying sexual arousal. The sighs and moans, the grunts and groans that we utter on these occasions are in their own way also songs of love.

Even verbal communications during lovemaking have their own cryptic style. At its most elementary level, communication allows informing the partner what to do and what not to do. Such messages can be conveyed verbally as well as nonverbally; an especially effective technique is *hand-riding*, whereby one guides the partner's hand to where one wants it to be. Countless couples suffer in frustrated silence because they are unable or unwilling to communicate their feelings and desires

because of shyness, the fear of ridicule, the need to avoid incurring an obligation, or out of the hope that their partners will somehow read their mind.

There is inevitably some friction in all sexual relationships, which causes trouble if it is allowed to build up. Communication is the grease that cuts down such friction.[40] It is also the basis of negotiating for what you want. *Sexual negotiation* is not an easy matter for a lot of people. Even within a marital context, sex is not always available just for the asking. It is especially hard if one wants to engage in practices that are unusual or socially disapproved. Negotiating is not the same as *bargaining* or *pressuring*, and people with healthy relationships are able to say "yes," "no," or "not now" without rejecting or slighting the other.

How one relates during coitus is part of the larger pattern of communication between a couple. Successful relationships are based on appropriate means of expressing feelings, getting one's views across, and listening to the other's views. Blame, generalizing ("always," "never"), arguing rather than discussing, are all inimical to intimate relationships.[41]

It may be necessary once in a while to air a major issue (or even have a good fight). But most everyday needs and problems are best dealt with in smaller segments without overloading the channels of communication. Perhaps most destructive is the "silent treatment" that cuts off any opportunity of resolving differences.[42]

Communication is essential, but there is nothing magical about it: it does not resolve all sexual conflicts. Many couples who complain of poor communication actually communicate very well. The problem stems from the fact that what is coming across loud and clear is the other's dissatisfac-

[40]Among the subjects of the *Redbook* survey, the more often wives could discuss their sexual feelings and wishes with their husbands, the more likely they were to describe the sexual aspect of their marriage as "good/very good" (Levin, 1975).

[41]Baucom and Hoffman (1983).

[42]James (1983).

tion, anger, and hatred, which the person does not want to hear.[43]

In our preoccupation with affection and other serious sentiments, we should not neglect the light-heartedness and *humor* that infuse fun into sex. Wit takes the sting out of awkward situations; an open and infectious smile is hard to beat as a source of reassurance. Teasing, bawdy humor, and "dirty talk" can also work as sexual stimulants for some. By allowing sex to become a form of adult play we can regain the carefree ability of our childhood to pretend and play-act without losing touch with reality or becoming deceitful. Such play-acting revitalizes sex by allowing our fantasies safe and partial fulfillment. Some people engage in "baby talk." Others pretend to be strangers, romantic characters, or even antagonists in coercive scenes. Whatever the scenario, both partners must find it pleasurable and comfortable; there is no justification for forcing such activity on an unwilling or uninterested partner. And as with children's games, it ought to stop when it turns nasty.

All things considered, what contributes most to the enhancement of sexual pleasure is evidence of the partner's enjoyment. If your lover trembles and melts in your arms, you must be doing something right, and that makes you feel sexually worthy. It is such mutual enjoyment that catapults lovers to the heights of sexual ecstasy and makes sexual interaction among the most cherished of human experiences, whose memory persists despite the erosion of time. In the twelfth century Heloise wrote in a letter to her husband, Abelard, after years of enforced separation: "Truly the joys of love, which we experienced together, were so dear to my soul that I can never lose delight in them, nor can they vanish from the mirror of my remembrance. Wheresoe'er I turn they arise before me and old desires awake."[44]

SUGGESTED READING

Comfort, A. *The Joy of Sex* (1972) and *More Joy* (1974). The most sophisticated, sensible, and attractively illustrated of modern sex manuals.

Chang, J. *The Tao of Love and Sex* (1977). The erotic wisdom of ancient China made accessible to modern readers.

Sinha, I. translation of the *Kama Sutra* (1980). The classic Indian love manual, with extracts from the *Koko Shastra, Ananga Ranga,* other Hindu erotic works and beautiful color plates of erotic art.

Melville, R. *Erotic Art of the West* (1973); Dawson, P. *Erotic Art of the East* (1968); Dawson, P. *Primitive Erotic Art* (1973). Profuse illustrations of erotic art from a rich variety of cultures with informative texts.

[43]Goldberg (1983).

[44]Quoted in Van de Velde (1965, p. 231).

Chapter 12

Sex and Interpersonal Relationships

Lovers never get tired of each other, because they are always talking about themselves.

François de La Rochefoucauld

Sexual relations are a form of social interaction. They are a vehicle for the sharing of pleasure, for the expression of affection and love. Yet the common vernacular terms for sexual intercourse ("fuck," "screw") also mean to cheat, trick, take advantage of, and to treat unfairly. Their derivatives ("fucked-up," "screwed-up") stand for inferior, unpleasant, difficult, confusing, blundering, wasteful, disorganized, and neurotic. The same words that express the desire to make love also express dismay, annoyance, or anger. Our language thus betrays much ambivalence and reflects the fact that sexual encounters involve far more than the gratification of our erotic and reproductive needs.

Evolutionary Background of Sexual Relationships

As Frank Beach has stated, "To interpret the sexual behavior of men and women in any society it is necessary to first recognize the nature of any fundamental mammalian pattern and then to discover the ways in which some of its parts have been suppressed as a result of social pressures

brought to bear upon the individual."[1] The reason we discuss the evolutionary perspective first is because the influence of evolutionary factors, when present, antedate the effects of culture and not necessarily because they are more important.

The evolutionary background of our sexual relationships can be ascertained by studying the behavior of living animals and the study of early human fossils. The sexual behavior of *nonhuman primates* is particularly interesting in this regard as long as one avoids ascribing human motives and meanings to animal behavior.

Among lower mammals, sex and reproduction are fairly standardized behaviors largely regulated by hormones centered around the estrous cycle of the female (chapter 4). This pattern largely persists among nonhuman primates, consisting of *monkeys* and *apes* (which include baboons, chimpanzees, and gorillas). Female monkeys and apes have monthy estrous cycles during which they are *in heat* for a period of about 10 days that coincides with ovulation. During this time, the females de-

---

[1]Quoted in Diamond (1965, p. 167).

303

velop swellings and color changes in the genital region (*sex skin*). These visual cues, along with olfactory stimuli from pheromones (chapter 4) elicit sexual arousal in the male.

Sexual interactions are by and large restricted to these periods. For example, when a female baboon enters estrus, she may sexually solicit the younger males, but as she nears her ovulatory peak it is the dominant male baboon in the troop who takes over, and for several days the couple lives in a *consort relationship* with frequent copulations. The association usually ends as estrus subsides, though some consort relationships extend over longer periods. When another female comes into heat, the dominant male now establishes a new consort relationship with her.

Female chimpanzees exhibit similar behavior. During 35-day monthly cycles, they are in estrus for about 6 days, during which they eagerly and indiscriminately copulate with all males in the troup, except their own sons. When signs of estrus subside, their sexual interest wanes although they may continue to be mounted. Gibbons are exceptional among the primates in their *monogamous* associations. They live in family groups marked by closeness between male and female and a lack of clear dominance patterns.[2]

Emergence of the Hominid Sexual Pattern

Unlike other primate females, women have no estrus; they are potentially sexually responsive during any period including menstruation, pregnancy, and nursing. They can reach orgasm readily, whereas the orgasmic ability of other nonhuman female primates is not firmly established. Women ovulate "silently" with no genital swelling. Instead, they have prominent breasts and buttocks, smooth and hairless skin, forward rotated vagina (which facilitates face-to-face intercourse),

a high-pitched voice, and other secondary sexual characteristics to highlight their erotic attraction.

When, why, and how did these transformations come about? Our earliest ancestors (the *protohominids*) lived in the dense forests of East Africa, feeding on fruits and nuts, some 15 to 20 million years ago. They were organized in loosely knit groups whose members more or less fended for themselves, except for mothers who looked after their offspring. About 10 million years ago, woodlands began to replace forests, and early *hominids* were forced to live in more open surroundings. Now vunerable to predators and in need of new sources of foods, they gradually turned to hunting. In this setting emerged the erect posture, larger brain, and the use of tools that compensated for the relative lack of physical prowess of these puny early humans.[3]

The basic attachment among these early humans was between mother and infant. In their tree-dwelling days, mothers could look after themselves and their offspring with relative ease. As they came into estrus they copulated more or less indiscriminately until they got pregnant; beyond that they had no particular need of the male. With the shift to ground dwelling, life became more difficult for mothers, who were now more dependent on males to protect them and share the meat from their hunt while they themselves continued to provide whatever food could be gathered. These associations were facilitated by more selective and stable patterns of male and female relationships known as *pair-bonding,* with attachments that lasted over longer periods of time and enhanced the chances of survival of the infants within their care.

The mechanism of estrus had worked well to attract the male for insemination. But estrus lasted for too short a period to allow the female to get much additional service out of a male. Through natural selection, the *continuous receptivity* of the human female and her orgasmic capacity

---

[2] Accounts of primate sexual behavior can be found in DeVore (1965); Rowell (1972); Hinde (1974); Lancaster (1979); and Symons (1979).

[3]Fisher (1983).

evolved to cement the pair bonds between male and female.[4] The male did not need to wander around anymore looking for females in heat. All he needed to do was stay close to the females at hand and copulate as circumstances permitted.

Evidence from other sources lends credence to the notion that sexuality (especially increased female sexuality) promoted pair-bonding and paternal investment in offspring.[5] Much of human affectional behavior can in turn be viewed as arising from and sustained by these sexual rewards whereby our ancestral "naked ape" developed the capacity for falling in love, for becoming sexually "imprinted" on a single partner, and for evolving pair-bonds.[6] By further extension, such human ties resulted in the formation of *primate societies*. As the male's attachment to the female extended to her offspring, and he came to recognize her child as his, the pair-bond was further cemented. The male thus became domesticated in the service of the family.

This model of sex as personal and social cement is challenged by some primatologists. For example, Symons questions the very assumption that shifts in the hominid pattern of nonreproductive sexual relations evolved to cement pair-bonding or group cohesiveness.[7] The claim that the loss of estrus or constant sexual attraction formed the basis of persistent pairings and social groups is contradicted by the observation that among some primates (such as the Japanese macaque monkey), stable social associations continue in the absence of persistent sexual attractivity or receptivity: some of the shortest and most clearly marked estrous periods occur in animals that also have highly developed social organizations. The seasonality of sexual relations further contradicts the notion of sexual attraction as the basis for persistent social groupings of primates.[8] Therefore,

Figure 12.1 Chimpanzees touching lips. (Baron Hugo van Lawick, © National Geographic Society.)

rather than sex, it is perhaps the division of labor and other social and psychological rewards of associating with members of one's species that have bonded individuals and cemented group ties.

Implications of the Evolutionary Perspective

The sexual behavior of primates may look beguilingly human (figure 12.1). Yet there is no direct way to simply draw inferences from such observations to how humans behave in the contemporary world, let alone to how they behaved during prehistory. Nonetheless, insights gained from evolutionary study may help us trace the continuity of behavioral patterns from the dim past through the last 20,000 years of cultural history.

Sex was neither designed primarily for reproduction (asexual reproduction is more efficient) nor for giving and receiving pleasure (pleasure at best is an inducement to undertake the burdens of reproduction). The main function of sex is the creation of *genetic diversity* in offspring, which allows for greater adaptability and hence higher chances of survival.[9] Reproductive success calls for different strategies for males and females. The investment of a female in her eggs (and resultant off-

[4]Morris (1967).
[5]Hrdy (1981).
[6]Morris (1967, p. 64).
[7]Symons (1979).

[8]Lancaster and Lee (1965).
[9]Wilson (1978).

305

spring) is far greater than that of the male in his sperm, which are produced in the billions. A woman can at most generate several scores of living infants; a man can theoretically father hundreds of children.[10]

This differential investment in offspring by male and female, the sexual dimorphism that favors the male in physical strength, and his greater aggressiveness, are presumably the genetically determined bases of the differences in how male and female approach sexual relationships. As Wilson sums it up, "It pays males to be aggressive, hasty, fickle and indiscriminating. In theory, it is more profitable for females to be coy, to hold back until they can identify the males with the best genes. In species that rear young, it is also important for the females to select males who are more likely to stay with them after insemination."[11]

A variety of calmly reasoned as well as impassioned objections have been raised to these sociobiological assertions. Some question the validity of the observations; others object to the inferences drawn from the data. Many social scientists tend to reject its strong sense of biological determinism. Those concerned with the oppression of women bristle at what they see as tacit vindication for the mistreatment of women by men. Others worry that even if that is not the intent of sociobiologists, there is the potential for the misapplication of their theories to sexist ends.

Sociobiologists disavow the motives ascribed to them and point out that recognizing the residues of our primate heritage does not necessarily mean endorsing them as adaptive, equitable, or desirable in the modern world. But simply because we find certain conclusions socially unpalatable, does not justify their out-of-hand rejection.[12]

Mary Jane Sherfey has offered another evolutionary perspective on female sexuality. On rather shaky grounds, she claims that, *"the human female is sexually insatiable in the presence of the highest degree of sexual satiation."* The rise of modern civilization was contingent on the suppression of this female hypersexuality because it would interfere with maternal responsibilities; furthermore, with the rise of agriculture, property rights, and kinship laws, large families of known parentage could not evolve until the inordinate sexual demands of women were curbed.[13]

The overall conclusions one can draw from the evolutionary perspective are as follows. First, animals do not sexually interact in a random and pointless manner. They behave in the way they do because their behavior has proved adaptive—that is why it has survived.

Second, there is no single way in which all living organisms interact. The polygamy of baboons is no better or worse than the monogamy of gibbons; they are different adaptations to different conditions. Neither pattern in itself says anything about how humans should relate to each other.

Third, there are distinct differences in the strategies of males and females in sexual interaction. Their interests may converge or diverge, calling for cooperation or competition, depending on the circumstances.

Fourth, we exercise far greater control over our environment, ourselves, and each other, than the animals we have descended from. Hence, whatever our evolutionary proclivities, we can largely shape the world that we live in through our culture. But to do this efficiently, and to recognize the biological limits within which we must operate, we need to understand

---

[10]The *Guiness Book of World Records* credits Mrs. Vassiliyev, a Russian peasant, with 69 children born through 27 pregnancies (16 pairs of twins, 7 sets of triplets, and 4 sets of quadruplets) during the 18th century. The paternity record is said to be held by Moulay Ismail, "the Bloodthirsty," former emperor of Morocco (1672–1727) who reputedly fathered 548 sons and 340 daughters (McWhirter, 1982).

[11]Wilson (1978, p. 129).

[12]Hrdy (1981).

[13]Sherfey (1973, p. 112). For a critique of Sherfey's views, see Heiman (1968); Symons (1979); Hrdy (1981).

where we come from, which is what the study of evolution is all about.

Patterns of Human Sexual Interaction

Whatever their evolutionary roots, human patterns of sexual interaction are very much more complex than the behavior of nonhuman primates. The experiences of sexual passion, love, jealousy, and unfulfilled desire have long preoccupied artists and writers. Yet relatively little work has been done by behavioral scientists and sex researchers in this area.

Purposes of Sexual Interaction

There are three main purposes for establishing and sustaining sexual partnership: *procreational*—to have babies; *relational*—to share erotic satisfactions within an affectionate and trusting relationship; and *recreational*—to have fun.[14]

Procreational and Relational Sex For most people, the necessary association of the procreative and relational aspects of sex is quite obvious. The responsibilities of parenthood are most readily discharged in a collaborative relationship, and an elementary sense of responsibility would inhibit people from bringing children into the world unless they can be adequately cared for. There are, of course, single parents who accomplish the tasks of parenting better than some couples. But generally, single parenthood is not a matter of preference but an outcome of divorce, abandonment, or death. Similarly, most women and men still view marriage as the optimal relationship within which to have children. While it is true that "illegitimacy" has lost much of its social stigma, and there are nonmarital relationships that may be just as stable and committed as any marriage, society still makes it much easier for married people to function as parents (chapter 16).

The teachings of the Catholic church require complete linkage between the pro-

creative and relational ("unitive") aspects of sex. These are seen as inseparable in every sexual act, and as legitimately fulfilled only within marriage (chapter 20). Other churches allow for the separation of the procreative from the unitive aspects of sex (as do the majority of Catholics in practice), hence condoning contraception.

People with more secular orientations allow for greater freedom of sexual associations outside of marriage. In more radical departures from the conventional concepts of heterosexual relationships, angry voices have been raised decrying the interlocking of these various aspects of sexual associations. The traditional childbearing role of women as mothers and their male-serving roles as wives, are seen as obstacles to female freedom, equality, and self-fulfillment. Even love is felt to be an instrument of subjugation. In this view, all heterosexual relations are ruled by *sexual politics* in which women are inevitably the losers; the way out is total rejection of such relationships. Women who wish to be mothers can remain single and be artifically inseminated (or adopt a child). The relational and recreational aspects of sex in turn can be fulfilled with other women or by "self-love."[15]

However antithetical these viewpoints may sound, the basic concerns are fundamentally the same; namely, how to ensure fairness and satisfaction in human relationships. One of the basic tenets of Christian theology is the *love of thy neighbor*, which means the people one deals with. In secular contexts, we speak of *contracts* and *commitments* that relate to the same concern. These are issues we discuss at greater length when we deal with sexual morality (chapter 20).

Relational and Recreational Sex The argument in favor of the relational component is that most people (women in particular) find sex more satisfying in the setting of a stable and meaningful relationship. This may be marriage, romantic love,

---

[14]Comfort (1972).

[15]For further discussion of this perspective see Millett (1970); Firestone (1970).

more sedate forms of affection, friendship, or at a minimum, some measure of respect and liking. Sex in such safe and loving relationships can be deeply satisfying. These couples know each others' likes and dislikes, which eliminates fumbling and false starts; they can be counted upon to be trustworthy and forgiving. Sex, affection, and friendship mutually enrich and reinforce each other.

Where sex is preferred without the encumbrances of a close relationship it is usually the "cost" of the relationship (in terms of loss of personal freedom and other considerations) that is at issue rather than a quest for impersonality as such. *Casual sex* (or the famous *one night stand*), which often involves relative strangers, is unsatisfying because it tends to be shallow and dehumanizing; it trivializes the experience. As a result, there is a loss of self-esteem; the next morning one feels cheapened, compromised, used, "like a whore."

Those who defend recreational sex do not deny that there may be unhappy reactions to it but ascribe its liabilities to socially inculcated negative attitudes, the double standard, and irresponsible behavior. Such outcomes represent the failure of recreational sex and should not be used to condemn all forms of it. After all, a great deal of misery results from relational sex as well.

Advocates of recreational sex would concede that the inability to have stable sexual attachments or the exclusive reliance on casual sex betrays a social or psychological deficiency. But what is the harm, they would ask, for two adults who are social equals and mean well, to engage in sex purely for the fun of it without expecting (or precluding) that personal commitments will precede or follow the sexual encounter. Furthermore, why should a man or a woman not have the option of simultaneously engaging in the whole range of these relationships: to have a spouse or a stable mate; good friends to go to bed with, and purely sexual encounters as the opportunity presents itself.

But why would a person want casual sexual outlets when he or she already has satisfying stable relationships? The common explanation is the need for variety in sex, as in other areas; just as people who can afford steak may still yearn for an occasional hamburger, so it may be with sexual experiences.

There is experimental evidence among animals that the introduction of a new sexual partner revives sexual interest (the *Coolidge effect*). This phenomenon is more clearly observed among males than females. Whether a similar mechanism operates among humans has not been established. Nonetheless, the yearning for multiple sexual partners, successive or simultaneous, is a common theme in the literature of many cultures.

If this is more of a male tendency, as seems to be the case, it would fit the sociobiological explanation discussed earlier with regard to reproductive strategy; namely, it is to a man's reproductive advantage to copulate with many women.[16] Other explanations are psychologically based. Sex serves to express dominance. We speak of "sexual conquest," "scoring," and use other metaphors that clearly imply the subjugation of the sexual object. More benignly, new partners may be sought as a way of revalidating one's sexual worth and competence after being sexually desired by a spouse becomes par for the course. A good deal of sexual provocation and flirtation that are not meant to lead to sexual activity fulfills these self-validating functions in more attenuated form. But the line is thin between being seen as a sexually exciting and vibrant person who stays within the bounds (defined by relevant groups) and being a *tease* who is resented because he or she goes too far in one sense but not far enough in another.

The search for transient sexual partners may also reflect deeper neurotic conflicts. This is why *promiscuity,* or the indiscriminate choice of sexual partners, has prompted both moral condemnation as well as psychological judgments (chapter 14).

If the forces of sexual variety are so potent, and given the sexually permissive cli-

[16]Symons (1979, p. 212).

mate of our society, why has the recreational model of sex not taken the country by storm? Why are its exponents mainly among the young (who are experimenting), those in transition between more stable relationships (such as the recently divorced), those who occasionally stumble into it with varying degrees of misgiving, and those incapable of forming lasting ties?

One answer is that we are still hopelessly bound up in traditional conventions and concerns. The day may yet come when unwanted pregnancy and sexually transmitted diseases will cease to be a threat; the double standard of sexual morality will have been demolished, allowing women as free a rein with their sexuality as men, and love and marriage will be devoid of possessiveness.

The counterargument is that this vision of *sexual liberation* is illusory. What stops people from playing at sex the way they would play at some sport is not external constraints. There is something deeper that prompts them to regard sex as not just another game but as something deadly serious. Even when given a free choice, people recognize that one cannot have it both ways in sexual relations; that the relational and recreational aspects at some level become mutually exclusive; what you gain in breadth of experience you lose in depth of commitment. Freedom becomes, in the words of a popular song, "just another word for nothin' left to lose."

Sexual Negotiation

As with any valued commodity, sex cannot be had for the asking; it is not free, no matter what the context of the relationship. Every sexual interaction entails *negotiation*, the terms of which vary widely. Between lovers the terms may be generous and easy; between rapist and victim, they are forced and grimly limited. Sexual negotiations may be subtle and indirect, such as when potential lovers are testing each other, or they can be open and direct when one is bargaining with a prostitute.

Typically, sexual negotiations are concealed within the broader fabric of inter- actions, and are carried out silently. The wife who is angry at her husband may not reject him out loud, but she will go to bed early, pretend to be asleep, or somehow or another convey her unavailability ("I have a headache"). The husband in a similar situation may develop a sudden interest in a late TV program or seek some other phony diversion. When all is well, a glance, a smile, and the couple may be embracing in no time without having exchanged a word.

Stylistic differences in strategy and tactics notwithstanding, sexual negotiations can be subsumed under three categories: cooperation, seduction, and coercion.

Cooperation presupposes two interested partners who feel free to engage in sex. The basic issue is not whether or not to have sex but when, where, and how often. Cooperation does not mean that each side is equally eager (that rarely happens in human affairs) but that they are heading in the same direction.

However willing the partners may be, no sexual interaction occurs unless someone takes the *initiative*. The way in which this may be done is culturally defined. The signaling of sexual interest is often *nonverbal*, relying on facial expressions, gestures, and other cues. When the sexual invitation is more explicit, it takes the form of *propositioning* (making a pass).

When there is an element of enticement, we call it **seduction** ("leading away"). In a benign sense, seduction involves the playful elicitation of sexual interest in a partner. More typically, it involves strategies aimed at exploiting the other person's psychological weaknesses. There may be flattery ("I have never wanted anyone so desperately"), or shaming ("Don't act like a freaked-out child"). Money, gifts, and other favors may be offered to break down resistance. Deception through false declarations of love and promises of marriage have been standard tactics of unscrupulous men.

The element of imposing one's sexual desires on another becomes more patent in various forms of **coercion.** Psychological coercion constitutes *pressuring*; the threat or use of physical force is *rape* (chapter 14).

Figure 12.2   William Hogarth (1697–1764). *Before* (left) and *After* (opposite). (Courtesy of the Metropolitan Museum of Art, Harris Brisbane Dick Fund, 1932.)

Traditionally in our culture as well as many others, there have been gender-related expectations in the respective negotiating postures of men and women: generally the women have been the "sellers" and the men the "buyers" of sexual favors. Similarly, as shown in figure 12.2, men seem more eager to initiate sexual encounters while women more likely want to prolong it.

The fact that men typically initiate sexual overtures, act the role of the seducer, and are the rapists, does not mean that males are more interested in sex than females. The difference is not in terms of sexual interest but in the level of aggres-

sion. Gender differences are further explicable by relational requirements. Given the proper relational context, women can and do take the sexual initiative just as actively as men. If their interests are served by it, women can be as effective seducers as men.

Differences in styles of sexual initiative are already fairly established by the time of adolescence and are manifested in interactions during dating. Among adolescent subjects in one study boys relied on the following strategies: direct approach ("Do you want to?"); declaration of affection ("I love you"); get what you can ("hump and dump"); contrived behavior

("I shouldn't ask you to do this"); persistence; seeking girls who are known to engage in sex ("easy lays").[17]

Adolescent girls may use some of the same direct methods of suggestion, propositioning, or physically initiating sexual activity (one teenage boy reports "Then to my amazement, she sat astride me, grasped my overheated penis as if it were a door-handle, and in a rather business-like fashion lowered herself onto it"). More typically, girls will initiate sexual activity by leading on the male partner in more subtle ways (touching his thigh, letting her legs

[17]Martinson (1976).

or breasts show, playing with his hair, looking at him invitingly, and so on).

Erotic Love

Love carries many meanings ranging from the casual ("I love ice cream") to the personal ("I love you"), and the spiritual ("I love God").[18] Love is a feeling or an emo-

[18]From the Indo-European root *leubh* comes the Old English *lufe* and our term *love* (as well as the German *liebe*). Also from the same source are the Latin *libere* (to be dear) and *libido* (pleasure, desire). The Latin word for love, *amor*, gives rise to the Spanish *amor* and French *amour*, as well as related words in English like *amorous*. The Greek for one form of love, *eros*, is the root for *erotic*.

tional state with regard to another person; it arises from the recognition of attractive qualities in that person or from natural relationships (such as between parent and child) or sympathy (as in friendship). Its manifestations are concern and caring for the well-being of the loved person, delight in his or her presence, and desire for approval and reciprocal affection.

Numerous synonyms for love deal with its qualitatively different aspects: *affection* is a feeling of warm regard; *fondness* is a stronger liking for a person; *devotion* implies a more settled and selfless feeling of dedication. By contrast, *infatuation* is an extravagant attraction or attachment (often sexual) that tends to be transient and reflective of faulty judgment, and *lust* is intense or unrestrained sexual craving. To "want" someone or to be "turned on" by someone are more current ways of expressing sexual interest (as are to "have the hots" and other colloquial expressions).

Further distinctions are based on the degree of intensity of the affectionate relationship. *Liking* someone falls short of loving (although loving often entails a great deal of liking); it may or may not have erotic overtones. More intense attachments that qualify as love have been variously classified.[19] For example, C. S. Lewis distinguishes four kinds. *Affection* (Greek: *storge*) is what typically binds together parents and children, siblings, and other relatives. There is a "built-in" or unmerited character to these bonds; many family members would have no interest in each other if they were not related. *Friendship* (Greek: *philia*) is a tranquil form of personal attachments based on "appreciative love" between kindred spirits who share personal, intellectual, aesthetic, and other values. *Sexual love* (Greek: *eros*) is typified by the experience of "being in love" as well as its more sedate erotic alternatives. *Charity* (equated by some with Greek *agape*) is the most selfless of loves, exemplified in St. Paul's compelling description:

Love is patient, love is kind and envies no one. Love is never boastful, nor conceited, nor rude; never selfish, nor quick to take offense. Love keeps no score of wrongs; does not gloat over other men's sins, but delights in the truth. There is nothing love cannot face; there is no limit to its faith, its life, and its endurance (Corinthians I:13).

Closely related, often inseparable from love, are feelings of **dependency** (literally "to hang from"). Survival of the young would not be possible without the intense attachment of parents or other caretaking adults. This reliance on those who love us persists through life; the phrase that most often follows on the heels of "I love you" is "I need you." Affectional bonds provide the best insurance for the satisfaction of our physical and emotional needs in both our individual as well as our communal relationships. The ultimate test is our willingness to suffer and die for the persons we love.

To isolate the factor of dependency, Lewis differentiates between need-love and gift-love.[20] **Need-love** is part of being human: we are born helpless and remain in need of each other physicially, emotionally, and intellectually. But need-love crosses over into selfishness when it makes excessive and self-serving demands that disregard the needs of the other and fail to reciprocate the favors received; by contrast, **gift-love** is selfless. Abraham Maslow makes a similar distinction between **deficiency-love**, which is selfish, and **being-love**, which is not.[21]

In the sexual sphere, the **love-object** may also be the **sex-object.** For some, the two are inseparable, for others they are quite discrete and sometimes antithetical. "If you love me, you will sleep with me," is countered by, "If you love me you will not insist on sex." Typically, men make the first statement, women the second. Though at-

---

[19]Some attempts at classifying love focus on the object of love. Fromm (1956) distinguishes between *brotherly, motherly, erotic, self-love,* and *love of God.* Harlow et al. (1971) distinguishes five affectional systems: maternal, infant, paternal, peer, and heterosexual (see chapter 9).

[20]Lewis (1960).

[21]Maslow (1968).

titudes may be changing in this regard, the old cultural pattern still lingers whereby men primarily pursue sex while women pursue love.[22]

The notion of dependency may be objectionable to some because it smacks of immaturity, weakness, and helplessness. Hence the more popular concept now is that of **supportiveness:** The proof of affection is to be supportive, which means being helpful, not critical, sustaining those we love in what they want to do, assisting them to become who they want to be.

We take it for granted that we will take care of those whom we care for, but we also recognize the vulnerability we incur in loving someone and resent it if we are taken advantage of. The threat of isolation and loneliness is countered by the threat of subjugation to another person's will and whim. The challenge is to love in such a way that, in the words of John Donne, "Our affections kill us not, nor die."

Falling in Love

The experience of **being in love** is distinguished from other states of loving. Unlike parental love, which is assumed to be naturally "given," and other forms of affection into which we "grow," one "falls" in love in a rather precipitous and involuntary manner. Metaphors involve images of being smitten and struck: by a thunderbolt, ("coup de foudre") say the French, by the arrows of eros said the Greeks (figure 12.3). The element of suddenness ("love at first sight") is compounded by a sense of inevitability and helplessness. Being in love is also distinguished by its exclusivity: we can love many but truly be in love with only one person at a time.

The theme of being in love has long preoccupied writers and artists. Physicians, too, have occasionally dealt with it as a form of transient illness (*lovesickness*) or madness (*erotomania*).[23] Behavioral sci-

Figure 12.3   Adolphe Bouguereau (1825–1905). *Young girl defending herself against Eros.* (The J. Paul Getty Museum, Malibu, California.)

entists have so far not done much with it; the subject seems to wilt under psychological scrutiny.[24]

The trigger for falling in love is often a simple act like a gesture, a glance, or some other expression that suggests interest, or hints at the possibility of reciprocal interest. The eyes communicate this message most eloquently: "Love's tongue is in the eyes" wrote the Jacobian dramatist Phineas Fletcher.[25] People, of course, stare at each other all the time without falling in love; for love to follow there must be a constellation of factors that impart a certain "wholeness" to the experience; everything must be right for the "chemistry" to work.

[22]Safilios-Rothschild (1977). Also see Shaver and Freedman (1976).

[23]An 8th-century Arabian physician called insanity caused by love *Ishik,* after a creeper that twines round a tree causing its death. Marriage was recommended as the best treatment (Balfour 1876).

[24]Tennov (1979) has coined the term *Limerence* for the condition of being in love which she has studied in a sample of several hundred persons. The discussion here is mainly based on that source.

[25]Quoted in Tennov (1979), p. 282.

Box 12.1   Experiences of Being in Love[1]

"Once I fall, really fall, everything about her becomes wonderful, even things that would otherwise mean nothing at all are suddenly capable of evoking curiously positive reactions. I love her clothes, her walk, her handwriting (its illegibility would seem charming, or if it were clear and readable, that would be equally admirable), her car, her cat, her mother. Anything that she liked, I liked; anything that belonged to her acquired a certain magic. *Her* handbag, *her* notebook, *her* pencil. I abhor the sight of toothmarks on a pencil; they disgust me. But not her toothmarks. Hers were sacred; her wonderful mouth had been there."

*

"Yes I knew he gambled, I knew he sometimes drank too much, and I knew he didn't read a book from one year to the next. I knew and I didn't know. I knew it but I didn't incorporate it into the overall image. I dwelt on his wavy hair, the way he looked at me, the thought of his driving to work in the morning, his charm (that I believed must surely affect everyone he met), the flowers he sent, the considerations he had shown to my sister's children at the picnic last summer, the feeling I had when we were in close physical contact, the way he mixed a martini, his laugh, the hair on the back of his hand. Okay! I know it's crazy, that my list of positives sounds so silly, but those are the things I think of, remember, and, yes, want back again!"

[1]From Dorothy Tennov, *Love and Limerence*, Stein and Day (1979, pp. 31–32).

When fully in force, being in love has certain basic components. There is constant thinking about the loved person ("can't get him/her off my mind"), to the point of obsessive preoccupation that pushes all other concerns to the background. An acute longing for reciprocation of one's feelings generates a high susceptibility to detect even the slightest evidence for it in the actions of the beloved, or to imagine it where none exists. The person in love shows an extraordinary ability for dwelling on what is admirable, and denying what is not, in the loved person. His or her faults are recognized at a factual level but then dismissed as unimportant and not allowed to mar one's view. There is thus a simultaneous process of *idealization* (whereby a crooked nose appears straight) and one of *crystallization* (which makes the crooked nose look cute).[26] (see figure 12.4 and box 12.1.)

[26]The term "crystallization" is derived from Stendhal's metaphor whereby a branch cast into a salt mine is transformed by salt crystals into an object of shimmering beauty," while remaining basically a tree branch (Stendhal, 1983).

Figure 12.4   Idealized image of love. François Gerard (1770–1837). *Cupid and Psyche*. (The Louvre.)

Box 12.2   Jealousy

Jealousy (Greek: *zeal*) entails a variety of sentiments; in connection with love, it refers to intolerance of disloyalty or infidelity, and fears of being supplanted in the affection of a loved person. Like love, it has numerous manifestations beyond the erotic, but jealousy tends to be particularly intense in relationships with a sexual component such as marriage or being in love.

There are many gradations of jealousy. At one end, there is the perfectly understandable self-interest in preserving one's stake in a valuable relationship. Like any other partnership, sexual associations are mutually "owned" and either party is entitled to safeguard it by being vigilant that others do not intrude into it; the more formal and stable the association, the greater is the justification for being protective.

This sense of ownership becomes more possessive when it extends to the person as such. Marital partners have tended to behave this way because of the multiple social ties that bind them. Jealousy in these cases may have little to do with love. A husband may not care for his wife sexually but would not want anyone else to touch her because she belongs to him; a wife's possessiveness in similar circumstances is further enhanced by the threat of losing her source of economic support, should the husband be lured away. At the extreme, there is *pathologic jealousy*, where intense suspiciousness without good reason makes the person highly intolerant of any friendly interaction or attention involving his or her partner.

The subject of jealousy, like love, has been widely explored in literature and art (Shakespeare's *Othello* being one of the classics) as well as by some behavioral scientists[1] and clinicians.[2] However painful, unworthy, and humiliating an emotion jealousy may be, it must be recognized and managed. Otherwise it will erode our relationships and cause untold misery (the church has long recognized its destructive potential by including it among the "mortal sins").

What constitutes legitimate grounds for feeling jealous is largely defined by a culture: Some societies do not tolerate even a friendly smile to be directed at someone else's spouse; others allow great latitude depending on the circumstances. Individual idiosyncrasies further complicate where the lines are drawn. So each couple must be reasonably clear as to what is acceptable within the relationship and what is not.

To provoke jealousy in order to invite greater affection and interest is an ancient device, but a double-edged sword. It is a manipulative tactic that may obtain short-term gain at the expense of longer-term trust. There is no room for bluffing between true lovers.

Freedom from all feelings of jealousy is hard to separate from a state of not caring. But jealousy is not a barometer of love. Particularly in its more possessive and violent forms, it is but another form of sexual coercion (chapter 14). To "kill for love" is a contradiction in terms.

[1]See for instance Stephens (1963); Mazur (1973).
[2]Pao (1982).

The fear of rejection induces uncertainty, awkwardness, confusion, and shyness. Yet frustration and adversity may also intensify the limerent person's longings, while vivid fantasies of reciprocal love provide fleeting relief from the pangs of unrequited passion. "Heartache," palpitations or tremors, flushing or pallor, and weakness, accompany moments of uncertainty and doubt; when love is reciprocated, the person feels exhilarated ("walking on air"). Those who are in love are also vulnerable to pangs of *jealousy* that arise from their intense possessiveness of the loved object and the terror of losing it (box 12.2).

The average length of an episode of being in love is two years, but it may range from a few weeks to a lifetime. Sooner or later, the experience comes to an end. The end may be quite sudden: an explosive breakup; an occasional suicide in the depths of despair, or murder in a fit of blinding jealousy. More often, the affair winds down less dramatically as one "lovers' quarrel" after another chips away at the relationship. As the capacity for self-deception is eroded, shortcomings become more visible and then magnified. The breakup may be amicable but more often it is painful and followed by lingering unhappiness.

Love experiences are not always wrecked by failure but succumb to success. Couples who establish compatible relationships, gradually shift to sedate forms of affection, while occasionally stirring the embers of their giddier days. Others settle down for more ambivalent compromises or end up totally disillusioned, finding it hard to comprehend how they could have been so crazy about someone they now can barely stand.

Sexual attraction is an essential ingredient of being in love but the association is not constant. The erotic element may, at one extreme, dominate the experience, coexist with it, or may be seen as an inimical influence to be shunned. Over 90% of Tennov's subjects rejected the statement that, "the best thing about love is sex." The dominant yearning of someone in love is not sex but a "return of feelings"; the wish to be loved in return. In this context the sexual union may epitomize such reciprocity ("giving yourself" to your lover).

The Effect of Love on Sex

Whether the quality of the sexual experience as such is enhanced or hindered by being in love depends on psychological and social attitudes. Much has been made of the Victorian conflict between love and sex (chapter 18): a woman who thinks sex is dirty or vulgar will find her love sullied and soiled by it; a man who thinks likewise may enjoy sex with a woman he does not love but not with one he does.

Such attitudes have been largely abandoned, and love and sex are now seen not only compatible but enhancing of each other. So at least, in principle, sex should be more enjoyable with someone one is in love with. Yet in practice we seem to fall short of this ideal. Among Tennov's subjects 73% of the women but only 51% of the males agreed that "I enjoy sex best when I am in love with my partner;" 14% of the sample found sex disappointing when "very much in love."

A number of factors may account for this. One is the persistence of negative social attitudes whereby sex may represent conquest, seduction, debasement—feelings incompatible with true love. Or more simply, the intense emotional state of being in love, with its attendant hypersensitivity and anxiety to please the other, to always appear in the best possible light, to always perform superlatively, may interfere with the physiological and psychological processes mediating the more down-to-earth sexual responses.

The Effect of Sex on Love

Many factors influence the decision to "go all the way," but whatever the motivations and constraints, there is a clear sense that engaging in coitus entails crossing a threshold that significantly, if not unalterably, changes the nature of the relationship.

Common sense suggests that sex being highly pleasurable, its sharing with another would lead to affectionate relationships, or intensify the affectional bonds of existing relationships. This positive reinforcement model seems to work only in a limited sense. A satisfying sexual experience certainly seeks repetition and engenders a sense of appreciation and affection toward the partner providing it. But purely at the level of physical satisfaction, sex does not seem to generate love anymore than fondness for a particular dish leads one to fall in love with the cook.

Men and women in love go to great lengths to initiate and sustain sexual relationships. Yet just as clearly, sex may be as much a source of agony as it is of ecstasy. This is poignantly expressed in Franz

Kafka's statement, "Coitus is the punishment for the happiness of being together."[27]

Sexual relationships are like delicate flowers that wilt with neglect, but they can also, like hardy weeds, invade and crowd out other worthwhile feelings. One reason for this is that sexuality is often as much intertwined with aggression and hostility as it is with affection, a subject we turn to in chapter 14.

Institutional Contexts of Sexual Relationships

Most sexual relationships are *dyadic*, involving two individuals. Yet they are by no means private affairs carried out in isolation. Like other human interactions, they occur in a social context. The more important of these are *institutional.* We usually think of institutions as organizations such as schools or churches; but sociologists define institutions more broadly as a "set of social processes and activities, including the norms or values they express or embody, focused on some major societal goal (e.g., law, family, war)."[28]

Marriage is the institution that has the most direct bearing on sexual relationships; even sexual behaviors that fall outside it are defined in relation to it, such as premarital or extramarital sex. Therefore, although we are concerned with sexuality rather than with marriage or the family as such, we need to dwell on them at some length to make social sense out of sexual behavior.

Marriage and the Family

Sex, marriage, and the family are intimately bound together even though each serves its own independent purposes as well. There is nonmarital sex as well as sexless marriages; children are born to unmarried couples, while some marriages are childless. Yet for most people, the family provides the main setting for sexual interaction and parenthood. For this reason, all

societies have had to devise means for men and women to form families, facilitating some types of behavior and inhibiting others to that end.

Courtship, marriage, and the family form an integrated unit that Reiss calls the *family system* through which new generations succeed existing ones.[29] *Courtship* provides the means of finding future mates. Its forms are highly varied but it follows two basic patterns: either parents select the mates for their children or the selection is done by the young people themselves.[30] *Marriage* (from "to take a bride") and the *family* (from "dwelling place") have many forms and do not lend themselves to simple definitions. The census bureau defines the family as two or more people related by blood, marriage, or adoption, living together. This adequately describes what prevails in our society, but it does not do justice to the many cross-cultural variations of this key institution (box 12.3).[31]

The state of being married has been widely accepted as the normative condition for adult men and women. In the 1950s, when the United States had entered the most family-oriented period of the century, 96% of people in the childbearing years married.[32] However, traditional family patterns have been recently challenged. The very concept of what the family should be is now in a state of flux, at least for certain segments of the population.

The first significant shift has been toward *later marriage.* In the early days of American society, people married as soon as they could afford to, or could be helped by parents to establish new households. With some fluctuations, this pattern largely persisted into the 1950s, when couples married at the youngest ages in American history.

---

[27]Quoted in Kanfer (1983).
[28]Goode (1977, p. 527).

[29]Reiss (1980).
[30]Derived from the traditions of courtly love in medieval Europe (chapter 18), "courtship" is now used to designate all forms of mate selection, including similar behaviors among animals (box 12.1).
[31]For more details on the family, see Goode (1982); Reiss (1980).
[32]Blumstein and Schwartz (1983).

Box 12.3   Marriage and Family Patterns

In an extensive review of 250 cultures, conducted several decades ago, George P. Murdock found that *polygyny* (one man having several wives) accounted for 75% of marriages; *monogamy* (single spouse) for 24%; and *polyandry* (one woman having several husbands) for 1% of marital patterns.[1] Half of these societies were *patrilineal* (descent traced through the male line); 14% *matrilineal* (descent traced through the female line); and 3% followed the *double descent* form (patrilineal for some purposes, matrilineal for others). Patrilineal societies were usually *patrilocal* (married couple lives with or near the husband's relatives); matrilineal societies, *matrilocal* (married couple lives with or near the wife's relatives).[2]

Though *polygamy* (having multiple spouses) in the form of polygyny appears to be the most common form of marriage, the great majority of the world's population lives in monogamous relationships. Many of Murdock's polygynous societies were small African tribes, some numbering but a few hundred people. And in all polygynous societies, having more than one wife has been an ideal usually attained by only a minority of men in privileged positions. Furthermore, with the advent of modernization, the practice of polygyny has been rapidly dwindling even in societies that do still permit it.

Murdock found the **nuclear family** (consisting of father, mother, and their dependent children) to be universal and the operative unit in all family systems. Thus, even when the members of an **extended family** (consisting of members of several generations) live together, grandparents and their married children would function as separate nuclear family units.

Murdock ascribed four universal functions to the nuclear family: *sexual relations; reproduction; socialization of offspring;* and *eco-nomic cooperation.* This conception of the nuclear family generally holds true for Western societies, but its universality has been challenged on the basis of cultures that do not fit this model. For example, among the *Nayars* of India, fathers have little to do with the socialization of their children or with providing economic support to their wives. *Trobriand islanders,* when studied by Malinowski in the 1920s, also lacked a nuclear family structure. Trobrianders were, like the Nayars, matrilineal, and married couples resided with the husband's mother. The mother's brother was the primary socializing agent for his nephews and nieces—the uncle thus essentially acted as a father. There are other examples from American Indian tribes, where the status of the nuclear family is dubious.

The nuclear family has been central and highly stable in American society throughout much of our history. Typically it is *neolocal* (the married couple lives in a new location apart from either side of the family) and recognizes descent from both the husband's and wife's lineage, although children take their father's family name (*patronymy*). American families tend to be *endogamous* (marry within the group) with respect to factors like race and religion, and *exogamous* (marry outside the group) with respect to other factors (such as the desire to marry into a higher class). Generally there is a strong tendency towards *homogamy* whereby people choose mates with whom they share many social characteristics such as race, ethnicity, social class, religious affiliation and the like.

[1]Murdock (1949).

[2]*Patriarchy* and *matriarchy* are used to refer to social dominance by males and females respectively. In that sense, there are no matriarchal societies now and probably none ever existed (Reiss, 1980).

The next decade saw major shifts. In 1960 28% of women between the ages of 20 and 24 had not yet married. By 1979 this figure had gone up to 49%; by 1981 52% of women in this age group were single.[33]

Another major change has been in the increase in *single-parent* households due to high divorce rates. In 1981 one quarter of American children aged below 18 were living with only one parent; 90% with the mother. This means over 10 million children have single parents and over 1 million a year are involved in divorces. Demographers project that half of first marriages now taking place will break up; the divorce rate for second marriages is now higher than for first marriages.

One way of restructuring intimate relationships that has gained in popularity is *cohabitation.* When dating was the governing pattern, couples who got serious about each other would *go steady.* With or without sex, this relationship implied a commitment that could lead to engagement and marriage. Today, *living together* or cohabitation fulfills similar functions. It may be a temporary arrangement, like going steady; a period of trial or preparation for marriage, like engagement; or a substitute for marriage itself. The key differences are that cohabitation implies sex at all levels of relational intensity, and it rejects the traditional focus on social and legal commitments.

An estimated 4% of unmarried heterosexual adults in the United States (two million couples) are now living together.[34] The practice is especially prevalent among college students, a quarter of whom have reportedly tried it.[35] Some are choosing this alternative because of disenchantment with the more traditional form of marriage. It allows a woman the advantages of a stable and affectionate relationship without compromising her professional aspirations, sense of independence, and identity. Men like it because it provides many of the advantages of a marital relationship without its legal and economic ties and obligations. It gives both sexes an extended chance to test the relationship: If it works, one can enter marriage with more confidence; if it does not, one parts company without the necessity of going through a divorce.

Yet living together is no panacea. As a permanent alternative to marriage, it makes the tasks of parenthood far more complicated in a society whose customs and laws are built around marriage. Cohabitation now works mainly as a *childless lifestyle:* most of these couples have no children living with them. Even though some have custody of children from previous marriages, most couples are reluctant to bring children into the new relationship without getting married.

Cohabitation may lead to procrastination in making definite commitments and perpetrates a state of uncertainty for the future (often more disadvantageous to the women).[36] And while it lasts, cohabitation does not spare the couple from facing many of the same joys and sorrows of married people.

Nonetheless, there is now wider recognition of cohabitation as an acceptable alternative to marriage. Through *palimony* settlements (which compensate the financially disadvantaged member of a cohabiting couple) the courts have tacitly recognized cohabitation as a form of legitimate partnership even if it falls short of marriage. A similar recognition has been given in the past to *common-law marriages*, a union entered into by common agreement but without benefit of a formal marriage. The law and much of society remains unwilling as yet to give similar recognition to homosexual couples even though substantial numbers of overtly gay men and women now live together and some would like to get formally married (chapter 13).

Attempts to constitute **group marriage**

---

[33]For the sources of these statistics and their possible causes, see Blumstein and Schwartz (1983).

[34]Spanier (1983).

[35]Bower and Christopherson (1977).

[36]The expectation to marry is fairly close among cohabitating men (38%) and women (35%). However, a higher proportion of women is eager to marry; more so, if they have never been married before (68%) than if they have (61%) (Blumstein and Schwartz, 1983).

have been tried sporadically since ancient times, and *communes* have been part of American culture for centuries. A good example was the Oneida Community in upstate New York founded in 1841 by the radical clergyman, John Humphrey Noyes, and his wife.[37] Such experiments had a short-lived revival in the late 1960s in the form of *counterculture* communes that ranged from the highly idealistic to the patently bizarre (chapter 16).

The basis for such associations is not purely sexual, although sex is an important incentive. The broader purposes may be economic, ideological, or religious; sex may or may not always be part of what is shared in such communal living. Members of group marriages have access to each other sexually but they do not necessarily engage in group sex (discussed farther on).

The term *multilateral marriage* is now also used to describe group marriage defined as a voluntary family association of three or more individuals based on affectionate bonds and sexual intimacy. Typically these arrangements involve two couples who may have been married to other people earlier and may have children; members of the group "marriage" are however not married to each other in the legal sense of the term.

To accommodate these changed and shifting marital patterns, Reiss proposes a new definition of the family as a "small kinship-structured unit" whose key function is the "nuturant socialization of the newborn."[38] In this context, "kinship" does not imply a biological connection ("consanguinal kinship") or marital tie ("affinal kinship") between the couple, nor does it specify that they belong to opposite sexes.

Marital Sex

There is a clear understanding that people who get married will make themselves sexually available to each other. A marriage that is not sexually consummated is not legally valid, and the failure or the un-willingness to be reasonably accommodating as a sexual partner has been usually accepted as grounds for dissolution of the marriage even in circumstances where divorce and annulment are granted reluctantly such as within the Catholic church.

Sexual Activity in Marriage   Coitus is the predominant but not the only sexual outlet for married people; marital coitus accounted for 85% of the total sexual outlets of married males in the Kinsey sample. The remaining 15% of orgasms were obtained from extramarital coitus, homosexual contacts, masturbation, nocturnal emissions, and animal contacts.[39] Individuals, of course, varied in this regard: some had only marital coitus, others had one or more additional outlets. For women younger than the middle thirties, the proportion of orgasms resulting from conjugal sex was similar to those of married men. For older women, other outlets then became more prominent; by their late forties only slightly more than 70% of female orgasms resulted from marital coitus.

The prominence of sex in marriage varies widely as reflected in the frequency of coitus. The median frequency of marital coitus in the younger couples Kinsey surveyed was about three times a week and steadily declined with age. Though this pattern of decline with age continues to persist (see figure 15.2), recent studies indicate that married couples now engage in intercourse more frequently, spend more time during it, rely on more coital variation, and obtain greater satisfaction from it.[40]

Sexual Satisfaction in Marriage   The impact of sex on the marital relationship is mixed; for some couples sex is the icing on the marital cake, for others the sore point of an otherwise compatible relationship. People who enjoy a harmonious relation-

---

[37]Talese (1980) gives a brief account of these communities.
[38]Reiss (1980).

[39]Kinsey et al. (1948); (1953).
[40]Hunt (1974). Also see Westoff (1974); Levin and Levin (1975); Trussell and Westoff (1980); Blumstein and Schwartz (1983). For factors affecting coital frequency, see Burgoyne (1974).

ship are more likely to share sexual pleasures while a happy sexual life endears people to each other. In other words, good sex makes a good marital relationship better, while poor sex makes a bad relationship worse.

In the Hunt survey there was a close correlation between people rating their marriage "very close" and marital sex as "very pleasurable," as well as between the level of satisfaction with marital sex and the frequency of intercourse. Among women in the *Redbook* survey who were having marital sex 16 or more times a month, over 90% described their sex life as "good" or "very good"; when marital coitus was nonexistent, 83% said their sex life was "poor" or "very poor." Yet an active sex life is no guarantee of sexual happiness: some 9% of those in the sexually active group were unhappy with their sex life; paradoxically enough, a similar percentage of women in the sexually inactive group expressed satisfaction with the sexual aspect of their marital life. There was a similar association between orgasm and sexual satisfaction: 81% of the women who were regularly orgasmic considered the sexual aspect of their marriage to be good as against 52% of those who were only occasionally orgasmic and 29% who never were orgasmic (or were uncertain about it).[41] Sexual satisfaction is difficult to assess since so much of it depends on personal and cultural expectations. Yet sex therapists report that there is a good deal of overt or smoldering sexual discontent in many marriages (chapter 15).

Extramarital Sex

Sex outside marriage may be *pre*marital, *extra*marital, or *post*marital. We discussed premarital sex in chapter 9. The sexual behavior patterns of the divorced are fairly similar to those who have never married. The widowed tend to be older, and the particular problems of their lives are dealt with in chapter 15. We therefore mainly focus here on extramarital sex.

Patterns of Extramarital Sex   *Extramarital sex* is the descriptive and nonjudgmental term used by behavioral scientists. *Adultery* (Latin: to pollute) is the legal and formal designation for sex by a married person with someone other than the spouse. *Infidelity* ("being unfaithful") implies a breach of faith that may or may not characterize such behavior, hence the alternative term *comarital sex*. (Colloquial terms include "cheating," "double-timing," "having an affair," and so on.)

People seek sexual partners outside their marriage for a host of reasons. Some are compensatory: lengthy separations, illness, lack of interest on the part of the spouse, or any other reason that makes sex no longer available within marriage.

Under a second set of conditions, marital sex is available but unsatisfactory. This may result from the effects of aging, the monotony of habit, mismatches in intensity of desire, and the need for variety. They may be motivated by anger or in retaliation to a spouse's infidelity or as a means of settling other scores. Extramarital sex can also fulfill a variety of other purposes: it may be done as a favor to a valued friend, to assert one's independence, as a source of extra income; as a way to advance one's career (and sometimes that of one's spouse).

There may be certain life phases that predispose people to turn to extramarital outlets. For instance, there may be an increased tendency for extramarital sex during the transition into middle age ("midlife crisis") when the person tries to fulfill unresolved yearnings of youth "before it is too late," or hopes to magically avert the onslaught of age by clinging or regressing to youthful activities epitomized by sexual adventurism.

Finally, there are those who neither lack sex nor profess to be dissatisfied with their marital partners, yet they see monogamy as unduly restrictive for the full development of their human potential both individually and as a couple. So they establish an *open marriage* that by mutual consent does not exclude other intimate relations.[42]

---

[41]Levin and Levin (1975).

[42]This may or may not include sexual license. See O'Neill and O'Neill (1972); Libby and Whitehurst (1977); Knapp and Whitehurst (1978).

In these arrangements a *primary relationship* is distinguished from *secondary relationships:* one's basic loyalty is to the former; the latter are permissible and desirable to the extent that they do not interfere with the former.

Though the desire for sexual variety or better sex has been a classic justification offered for engaging in extramarital sex, its physical satisfactions often seem to fall short. In the Hunt sample, 53% of women regularly reached orgasm with their husbands, but only 39% did so in extramarital coitus; only 7% of women never reached orgasm with their husbands against 35% in extramarital affairs; two-thirds of married males rated coitus with their wives as "very pleasurable," but fewer than half rated the extramarital experience as highly.

Adultery as a breach of marital trust is a common ground for divorce in most cultures, but such violations are not defined uniformly. There are some forms of extramarital relations that have been socially approved, usually for men but not for women. The *concubine* in ancient times had a formal status (chapter 17). During more recent periods, the *mistress* fulfilled a somewhat similar role in an exclusive sexual relationship with a married man who supported her. Far less common is the *kept man* retained by an older woman (or a homosexual lover). Such arrangements may be indistinguishable from prostitution in some cases, but they may also entail affectionate associations in which the monetary considerations are secondary to the relationship.

Extramarial sex can also occur in the context of **spousal exchange.** Some cultures have practiced *wife-lending* either as a temporary accommodation for a visitor or as a longer-term exchange of spouses.[43] A modern counterpart in our culture, though by no means a socially approved one, is *wife-swapping* or *swinging*, a form of consensual extramarital sex in which husband and wife participate together (box 12.4). A related variant is *group sex*, which involves three or more individuals having sex together. When such sexual partnerships become more stable, they are referred to as *ménage à trois* (French: household of three).

Prevalence of Extramarital Sex   In the Kinsey sample, about half of married men and a quarter of women admitted that by age 40 they had had at least one extramarital affair. Like premarital coitus, this outlet was generally irregular, occurring only once every 1 to 3 months, and accounted for fewer than 10% of total sexual outlets.

More current findings show no evidence of major change with respect to the acceptability of extramarital sex: 80 to 90% of couples in the Hunt survey found the prospect of infidelity by the spouse objectionable and most of these couples refrained from extramarital sex because it was "sinful," "wrong," and "dishonest" (as well as out of fear of disease and pregnancy). Nonetheless, half of married men and 20% of married women in the Hunt sample have had affairs by the time they reached 45 years (with female rates approaching male rates in younger groups). The *Redbook* survey found that 38% of wives 35 to 39 years old had extramarital experiences. The *Cosmopolitan* survey figures were even higher: half of wives 18 to 34 years old and over two-thirds of those 35 or older reported such experience. It would be necessary, however, to have more reliable evidence than magazine surveys to confirm this trend since such surveys reflect little more than the views of their readership. For example, with respect to female extramarital affairs, a 1983 *Playboy* survey of its readership reports that by age 50, nearly 65% of wives had had affairs; a similar poll by the *Ladies Home Journal* reports only 21%. As a public pollster aptly put it, "If *Reader's Digest* surveyed its readers, it would probably find that *nobody* had any extramarital affairs.[44]

Effects of Extramarital Sex on Marriage
The effect of extramarital sex on marriage

---

[43]For example, the Chukchee of Siberia, with the consent of their wives, made such reciprocal arrangements to allow themselves the comforts of home during their extended travels (Ford and Beach, 1951).

[44]Quoted in *Time,* January 31, 1983, p. 80.

Box 12.4   Swinging

Swinging typically involves two married couples exchanging partners to engage in sex. In some cases, a married couple swings by adding one other person of either sex or getting together with more people to engage in group sex. Typically, it is the husband who draws his wife into swinging with people they do not otherwise know or plan to become intimate with. A basic rule of swinging is not to become emotionally involved, and people usually conceal their true identities.

Swingers are predominantly middle-class, white, married couples who otherwise lead fairly conventional lives. It is estimated that about 2% of married couples have tried swinging on at least one occasion.[1] Many never repeat the experience; only a minority of these couples continues to swing occasionally. The *Redbook* survey similarly found that about 4% had tried swinging, but half of these had done so only once.[2] Given the nature of these self-selected samples, these figures are probably higher than what prevails in the general population. The likelihood of couples experimenting with swinging may have further declined since the 1970s.

Swingers locate prospective couples at swingers clubs, through personal referral, by advertising in underground papers and swinger magazines. After an exchange of pictures and preliminaries, the sexual encounter occurs at a motel or the home of one of the couples. The couples may switch partners and go to separate rooms (*closed swinging*) or engage in sexual encounters in the same room (*open swinging*). On these occasions or during group sex, homosexual contact among males is rare, but quite common among females, who are often pressed to it by the men.[3]

There have been attempts to institutionalize such activity. In the 1970s the *Sandstone Retreat* near Los Angeles operated for several years as a private club.[4] *Plato's Retreat* in New York City is a more commercialized attempt to cater to such tastes. Such swingers clubs also exist in other major cities.

The primary justification offered for swinging is that it provides sexual variety for both partners in an open and safe form, free of deception. It is said to expand the sexual horizon of a marriage without threatening the primary relationship ("the family that swings together clings together").[5] It provides for women what has been traditionally available to men under the double standard.

Yet swinging seems riddled with problems for many of its participants. The requirement that it entail no emotional involvement, so as to make it safe for the primary relationship, makes these sexual encounters shallow and impersonal. Women, who are supposed to be the equal beneficiaries of swinging, are generally not interested in it but participate as a favor to their husbands and friends. If women become enthusiastic about it, their husbands seem to resent it.

Despite the sexual freedom or abandon expected, these encounters can be quite anxiety-provoking. There is the problem of invidious comparisons: women are likely to worry about the attractiveness of their bodies; men about their potency. Not to be wanted by anyone in a sexual free-for-all can be devastating. And there are the practical problems of catching a sexually transmitted disease, fear of discovery, blackmail, and so on.[6]

[1]Hunt (1974).
[2]Tavris and Sadd (1977).
[3]For further discussion of swinging, see Denfield and Gordon (1970); Bartell (1971); and Gilmartin (1977).
[4]Talese (1980) presents a revealing picture of the place in its heyday.
[5]Denfield and Gordon (1970).
[6]See Talese (1980) and Bartell (1971) for examples of some of the negative consequences of swinging.

is difficult to assess since such behavior involves varied motivations, forms, and ethical judgments. For example, a couple that allows each other sexual freedom represents a situation different from a couple whose dealings are veiled in deception or based on one spouse coercing the other into grudging tolerance. The effects of quietly supplementing marital sex with due regard to the sensibilities of the spouse are not the same as aggressively flaunting extramarital affairs to offend and humiliate. A person who gets involved after due thought and understanding of what he or she is doing differs from the uncontrollably driven individual who gets embroiled in one senseless liason after another.[45]

One of the classical justifications for extramarital sex is that they sustain marriages that would otherwise be intolerable or tide couples over difficult times. If one of the spouses feels shortchanged in the marriage, the opportunity to go outside may rectify the balance in that person's mind with respect to the sort of "deal" he or she is getting in the marriage.

These considerations notwithstanding, extramarital sex constitutes a problematic and guilt-ridden activity for many of its participants. Its practice takes time and energy away from the marital relationship. Its concealment causes anxiety and its discovery results in turmoil that frequently endangers the marriage.

It could be argued that these liabilities need not exist: If there is no deception, there need be no guilt; husbands need not feel cuckolded, nor wives deceived, if extramarital relations cease to reflect on the spouse's desirability as a lover. Perhaps if society changed its rules, there would be no penalties to pay.

But what if we are not dealing here with arbitrary rules? What if there are irreconcilable conflicts between primary affectionate bonds and their dissipation in secondary sexual relationships? How would a change in societal values accommodate a compromise between these competing tendencies? "I believe that both the discipline of fidelity and a measured use of the untruth offer more alternatives for the future" says Francine du Plessix Gray, "than the brutal sincerity of open marriage."[46]

Prostitution

Simply stated, *prostitution* (Latin: "to expose publicly") is engaging in sex for money. Yet in fact, prostitution is a complex institution with ancient roots ("the world's oldest profession") that has taken many forms in various cultures. We discuss the historical aspects of prostitution later on (chapters 17, 18) and consider its social and legal aspects separately (chapter 19). Our concern at this point is with its common forms and interpersonal aspects.

Sex with prostitutes represents, at least in principle, *nonrelational sex* in its starkest form. The prostitute provides sex with no questions asked; the client pays for it with no further obligations. Typically, the encounter takes place once only. Yet in practice, those who go to prostitutes often have other, more personal needs, and such interactions are not always devoid of a relational component. Occasionally, longer-term relationships are established with prostitutes, including marriage.

Types of Prostitutes A prostitute (*hooker*, *whore*) is usually a woman who sells her sexual services to men. Male prostitutes (*hustlers*) similarly cater to homosexual men (chapter 13). Very few women buy the sexual favors of men or of other women. The *gigolo* (French: dance-hall pick-up) is a young man who lives with and provides older women with social companionship and sometimes with sex in exchange for financial support.

Female prostitutes operate in several modes, but the services they provide and the nature of the relationship with the customer (called the "John") are basically the same. The **streetwalker** solicits men in

---

[45]For discussions of the distinction between "healthy" and "unhealthy" reasons for extramarital relations, see A. Ellis (1969); Neubeck (1969).

[46]Plessix Gray (1977).

Figure 12.5 A streetwalker in New York City soliciting a potential customer. (Photo by Burt Ginn, Magnum Photos, Inc. Reprinted by permission.)

public places such as streets and bars (figure 12.5); sex takes place in a hotel room, house, car, or some other secluded spot. The *housegirl* works in a *brothel* ("whorehouse," "house of pleasure") along with other prostitutes. The *call girl* is more exclusive and meets her clients by special appointment, usually in hotel rooms.

Depending on the level of public tolerance, prostitutes also use a variety of guises and fronts. For instance, *massage parlors, dating services, sex clubs,* and so on offer various sex services (legitimate massage establishments now advertise their services as providing "nonerotic massage"). Massage parlors may provide a variety of sexual services, the simplest of which is masturbation (called a "local"). Women who provide such services are called *hand*

whores.[47] Prostitutes may also provide a wide range of specialized sexual activities that cater to the needs of paraphiliacs such as sadomasochists (chapter 14).

The personal and professional relationship of prostitutes with their "employers" vary. Streetwalkers usually work for a *pimp,* who runs a "stable" of women. The pimp offers these women protection, procures customers, provides a measure of sexual and social companionship, and a sense of belonging. In exchange, he takes most of their earnings. The relationship may be voluntary but typically it is highly coercive, with the pimp using psychological enticement, intimidation, drugs, and physical violence to keep and control the woman.

Prostitutes who work in brothels have a more autonomous existence, sharing about half their earnings with the brothel owner or manager (*madame*). However, at least in the past, women have been virtual prisoners in these houses. The call girl relies on referral agencies for customers—and comes closest to being self-employed.[48]

The definition of prostitution is sometimes expanded to encompass other sexual favors which entail financial gain. Expensive gifts that men give their girlfriends, or even traditional marital relationships are thought by some to represent forms of prostitution.

"Saving sex for my lover/husband was my gift to him in exchange for economic security—called meaningful relationship or marriage . . . With that romanticized image of sex, in a society that does not have economic equality between the sexes, I was forced to bargain with my cunt for any life of financial security. Marriage under these

[47]See Talese (1980) for descriptions of the women who work in these places, and Simpson and Schill (1977) of the customers they serve.

[48]There is an extensive literature on prostitution. See, for instance, Esselstyn (1968); Sanford (1975); Whittaker (1974). Bullough (1977) provides an extensive bibliography.

circumstances is a form of prostitution."[49]

Psychosexual Aspects of Sex with Prostitutes

The phrase *turning a trick* succinctly summarizes the bare-bones minimum of sex with prostitutes: man pays the fee, ejaculates, goes on his way. Similarly, the reasons men seek prostitutes are often reduced to a few basic types exemplified by the sailor on shore leave, the traveling businessman, men who are physically unattractive, shy, sexually incompetent, dysfunctional, or "kinky"—these are the common customers that are thought to keep the business of prostitution going. Closer scrutiny reveals a more complex picture whereby prostitutes fulfill as many emotional needs as they do sexual ones, hence the necessity to seek the psychological explanations of sex with prostitutes, not just the sexual (box 12.5).

Prevalence of Sex with Prostitutes

Some 70% of males in the Kinsey sample reported having had at least one contact with a prostitute (the rates being inversely proportional with level of education). The Hunt survey showed a marked decrease in the proportion of American males who were sexually initiated by prostitutes, had sex with them, and the average frequency with which they did so. Thus the estimated use of prostitutes by single males in the 1970s appears to be no more than half as widespread as it was in the 1940s. This change is ascribed to the greater sexual freedom in sexual relationships that has come about in recent years.

Nonetheless, prostitution remains very much alive. Although reliable estimates of the number of active prostitutes are difficult to obtain, numbers range from a low of just under 100,000 based on arrest records to as high as 250,000.[50]

Many more women engage in prostitution on a part-time or occasional basis. Assuming that a full-time prostitute averages 20 clients a week, the number of men using prostitutes during a given week would range from two to five million.

Why Women Become Prostitutes

Much has been written about the psychological and social roots of prostitution and the exploitation of women for such purposes ("white slavery"). The use of teenage girls, and even children, is particularly appalling, an issue we take up in chapter 19.

The majority of prostitutes come from lower-class backgrounds and broken homes. Poverty and the prospect of making a better living (or just staying alive) for women with little education and no special skills are clearly major motivating factors. The hope of romance, glamor, sexual excitement, enticement, and the coercion by pimps are additional incentives. At the more personal level, adolescent rebellion, family conflicts, the inability to establish mature emotional ties, hostility toward men, and masochistic needs may be among the factors that decide why some women but not others under comparable social circumstances gravitate to prostitution. In all of this, the wide diversity among prostitutes must be kept in mind. The sophisticated call girl financing her education, the streetwise hooker making a living, and the pathetic teenage drug addict supporting her habit, are all prostitutes but live very different lives and present very different challenges to society.

Celibacy

If prostitution represents one end of a spectrum where sex is stripped from its usual relational considerations, celibacy would constitute the other end where sex is excluded completely from intimate personal associations. *Celibate* means "unmarried," but its modern usage refers more broadly to anyone who voluntarily refrains from sexual relations, or more strictly defined, from all forms of sexual experience (*chaste*). To *abstain* from sex means avoiding it in more general terms and not necessarily because of moral considerations.

There are two broad categories of celibacy: **religious celibacy** ("sacerdotal celi-

---

[49]Betty Dodson in Talese (1980, pp. 577–578).
[50]Sheehy (1973).

Box 12.5   Patterns of Prostitute-Client Interaction

In the client-prostitute relationship, it is primarily the needs of the man that define the pattern of interaction. Yet the prostitute may also select customers with an eye to her own psychological predilections. Interactions between prostitute and client thus entail a set of *roles* that are necessary to *enact a scene*. These have been characterized by Martha Stein as paired associations that fulfill special purposes.

The *opportunist-hooker* combination is to provide the man *sexual release*. Such release may underlie other pairings as well but for many men it is the primary objective. The sexual opportunist is attracted to the convenience, availability, variety, and lack of commitment entailed in the "quickies" provided by hookers. The most common activity is fellatio followed by coitus ("half and half").

Adventurer-playmate combinations seek *sexual expansion*. Some men in this category are looking for a sexual education. Others want to experience socially forbidden practices or give vent to their paraphiliac fantasies. Specially popular are troilism with a pair of prostitutes, watching lesbian interactions, anal stimulation, fetishistic and sadomasochistic practices. The experienced prostitute can marshal a wide selection of erotic aids, including porno movies, pictures, scents, lotions, erotic costumes, drugs, dildos, vibrators, whips, restraints and so on. Most importantly, the seasoned prostitute has fascinating tales to tell of her real or imaginary sexual experiences.

In the *lover-romantic partner* association, the men (often middle-aged) are keen on professing love and expect reciprocal expressions of affection. *Friend-confidente* roles depend on companionship. Unlike lovers who want to project a flattering image of themselves, friends are more interested in ventilating their problems and getting sympathy and reassurance. The *slave-dominatrix* relationship involves make-believe relinquishing of control on the part of the man, and the setting up of various bondage and masochistic scenarios. The *guardian-daughter figure* roles call for a young and naive (or at least naive-acting) prostitute on whom the guardian bestows his attentions like a wise and benign uncle. By contrast, the *juvenile-mother figure* pairing calls for a older, more imposing, full-figured woman who can provide comfort and concern as well as sexual pleasure. In the right hands, these emotionally immature men may be able to get useful help and instruction.

Sociosexual entertainment is the aim of the *fraternizer-party girl* linkage. Stag parties, unlisted attractions at conventions, and similar settings for sexual adventure, rely on prostitutes. In some ways, these are extensions of youthful practices like adolescent hell-raising and fraternity bashes, which allow men in groups to live out sexual fantasies. Group sex, involving 3 or 30 partners, is the main staple.

Finally, prostitutes cater to *status-enhancement* through *promotor-business woman* partnerships. It is not unusual for businesses to add sex to the food and drink provided to potential buyers. To maximize the benefits of this approach, some men establish working relationships with select prostitutes whose services are paid for by the company. These women then act as the business partners when dealing with representatives of other companies whose behavior they try to influence through sexual favors.

bacy"), exemplified by the vows of chastity of Catholic orders of priests, monks, and nuns; and *secular celibacy*, where men and women refrain from sex either temporarily or permanently for a variety of personal reasons. It is with the latter category that we are concerned here.

Most celebate people refrain from sex because they do not have acceptable partners. Countless widows lead sexually inactive lives because of the scarcity of eligible men. Widowers are more often held back by illness and problems of sexual potency. With young adults, abstinence is more often elective. Some men and women would like to engage in sex but many prefer to wait until they are married or fall in love, or until the right person and the right circumstance come together. It is not unusual for some people to feel not yet ready to engage in a sexual relationship without quite knowing why. Others decide that sexual relationships are too distracting, risky, and troublesome. They may have more absorbing interests, or wish to lead quiet lives. Casual sex may seem distasteful; more intense attachments too burdensome. They believe they can get more out of their relationships with their friends by keeping sex out of it.

Celibacy can also reflect more negative reasons for rejecting sex. Those who feel sexually unattractive may want to avoid the humiliation of not being wanted. Others who have been hurt in love affairs want to avoid repetition of the experience. It is not unusual for rape victims to shun sex until they feel whole again. Those who suffer from sexual dysfunction may decide to give up the struggle. The inability to initiate or sustain sexual relations may be but a facet of more basic failures in the capacity for intimacy.

Modern perceptions of sex as a good thing have tended to cast all forms of celibacy in a negative light. "The Victorian nice man or woman was guilty if he or she did experience sex," says Rollo May, "now we feel guilty if we don't."[51] But there is a growing perception that perhaps we have inflated the importance of sex in human relations, that its overemphasis is detrimental to broader human ties and eventually to the fuller enjoyment of sex itself. So a new form of celibacy is said to be emerging whereby people do not basically reject sex but nonetheless decide to forego its experience at least for a while.[52] They do this in order to regain perspective, to deepen their pesonal relationships without being distracted by sex, to attend to other neglected aspects of their lives, or even to become rejuvenated sexually.

The rapid changes in the status and self-perception of women have also generated some hesitancy. The "old rules" of sexual interaction have become suspect, yet there are as yet no clear alternatives. This sense of transition leads to an attitude of wait-and-see. So once again, the universals in human sexual relations must come to terms with the particulars of today's cultural realities to reshape old patterns and to create new ones to provide us with the social matrix within which to shape our sexual lives.

SUGGESTED READINGS

Symons, D. *The Evolution of Human Sexuality* (1979); and Hrdy, S. B. *The Woman That Never Evolved* (1981). Fascinating accounts of the evolutionary background of sexual interactional patterns.

Tennov, D. *Love and Limerance* (1979). The experience of falling in love studied from a psychological perspective. Pope, K., et al. *On Love and Loving* (1980). Contributions by a group of behavioral scientists.

Masters, W. H., and Johnson, V. E. *The Pleasure Bond* (1976); Offit, A. K. *The Sexual Self* (1977). Human sexual relationships seen from the perspective of sex therapists.

Reiss, I. L. *Family Systems in America* (1980); Goode, W. J. *The Family* (1982). Comprehensive accounts of the nature and functions of the family.

[51]May (1969).
[52]Brown (1980).

Chapter 13

Homosexuality

Not all things are black nor all things white . . . The living world is a continuum in each and every one of its aspects. The sooner we learn this concerning human sexual behavior the sooner we shall reach a sound understanding of the realities of sex.

Alfred C. Kinsey

Despite an enormous amount of writing and speculation, there are still a great many uncertainties over the subject of homosexuality.[1] There have been dramatic changes in social attitudes toward homosexuals and in the self-perceptions of gay men and women. Yet homosexuality still evokes a good deal of ambivalence and condemnation. In a 1983 public poll, 65% of Americans thought gays should have equal rights in terms of employment opportunities but close to 60% disapproved of homosexuality as an alternative life-style.[2]

When we refer to homosexuals, we are talking about millions of men and women of all shapes and forms, who may be rich or poor, educated or ignorant, powerful or powerless, smart or dumb. Homosexuals can be found among all races, classes, ethnic groups, and religions, living in every city and town, across the face of the nation. They are people we all meet all the time whether we realize it or not.

Homosexuality as a Behavioral Entity

We defined homosexuality earlier as a sexual attraction and activity between members of the same sex (chapter 8). But neither this nor the other definitions are entirely satisfactory for reasons that will be explained shortly.

Defining Homosexuality

The word *homosexual* is made up of the Greek prefix for "the same" (not to be confused with Latin *homo:* man) and the Latin for "sex." Although sexual encounters between same-sex persons have been known throughout history, the concept of *homosexuality* as an entity is more recent. When sexual acts between men, such as anal intercourse, were condemned in the past, there was no implication that people who engaged in them were fundamentally different from heterosexuals. As the study of personality development and sexual behavior became established in the 19th century, homosexuality came to be seen as an entity and the search began for its causes and treatment. Our legal system actually never quite adopted this viewpoint; it continued to forbid same-sexed behaviors, but homosexuality, as such, in terms of having a particular sexual orientation, was never proscribed by law (chapter 19).

[1]The literature on homosexuality is very extensive. Weinberg and Bell (1972) provide an annotated bibliography; Brewer and Wright (1979) provide extensive references. Major studies, like Bell et al. (1981), have more select bibliographies on various aspects of homosexuality.

[2]*Newsweek,* August 8, 1983, p. 33.

Sexual acts between homosexuals were designated in the past with terms like *sodomy* and *buggery*, which primarily referred to anal intercourse. In the 19th century, *sexual inversion* was commonly used in medicine and sexology until it was superceded by *homosexuality* in the 1930s.[3]

Though descriptively neutral, this term too became imbued with derogatory associations in common usage. In reaction, the word *gay* emerged as the preferred designation by homosexuals, while heterosexuals came to be referred to as *straight*. Although this usage of "gay" goes back to the 13th century, it is only recently that the term has become popularized. It has now replaced most other colloquial terms, though some continue to be used perjoratively.[4] However, it has not yet been accepted into the clinical and scholarly literature; hence, *male homosexual* and *female homosexual* (or *lesbian*)[5] continue to be the terms used in formal discussions. *Homophile* is the more formal term preferred by some gay organizations.

Unlike other sexual behaviors like masturbation or coitus, there is no *homosexual act* as such; fellatio, anal intercourse, or anything else that homosexuals may do can also be done by heterosexuals; the reverse is also true, except for penile-vaginal intercourse. So it is not the nature of a sexual act but the same-sexed character of the partners that defines the behavior as homosexual.

What is then a *homosexual person?* If we define him or her as someone who engages in homosexual acts, how many such acts are necessary to cross the line—1, 10, 100? Suppose a man or a woman is strongly attracted, or sexually oriented, toward members of the same sex but does not act on these wishes and never engages in homosexual behavior: do we nonetheless call that person homosexual? Finally, suppose someone is both attracted to and engages in sexual behavior with same-sexed persons but does not think of himself or herself as being homosexual, does not have a *homosexual identity:* is that person homosexual? In short, what is it that marks a person as homosexual—behavior, orientation, identity, or all three?

It is quite common to think of homosexuality as a single entity—a cluster of behavioral and personality characteristics that defines a person as belonging to a sexual category. But studies of gay men and women show tremendous diversity in how homosexuals look, think, feel, and behave sexually. There is no one description that characterizes their sexual lives. This is why the plural term *homosexualities* is sometimes used for this cluster of sexual behaviors.[6]

The Homosexual-Heterosexual Polarity

When discussing gender (chapter 9), we noted the tendency to view it in *unidimensional* and *bipolar* terms. Sexual orientation has been placed into a similar model whereby homosexuality and heterosexuality are seen as the polar opposites of a single personality trait: a person is either one or the other; the less of one, the more of the other.

But it so happens that only a small minority of homosexuals are so exclusively oriented to members of the same sex that they have no heterosexual leanings whatsoever; the great majority have varying degrees of heterosexual interest and experience as well. Hence, most of the people with homosexual inclinations are in effect *bisexual* or *ambisexual* (Latin *ambo:* both).[7] There is clearly an overlap of homosexual

---

[3]The term "homosexuality" was first used by an obscure Hungarian doctor, called Benkert, in 1869. It became popularized in German by Magnus Hirschfeld and in the English-speaking world by Havelock Ellis (Ellis, 1942, vol. 1, pp. 2–4).

[4]Common colloquial expressions include queer, fag, faggot, fruit, nellie, homo, cocksucker, pansy, queen (for males), and dyke, lez, lessie, femme (for females). (Haeberle, 1978). For a longer list see Rodgers (1972).

[5]"Lesbian" is derived from the Greek island of Lesbos, home of the female poet Sappho (see chapter 17).

[6]Bell and Weinberg (1978).
Colloquial names: AC/DC; switch hitter.

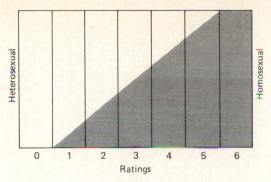

Figure 13.1 Heterosexual-homosexual rating scale. (From Kinsey et al., 1953. Courtesy of the Institute for Sex Research.)

and heterosexual interests in significant numbers of people. Even where this is not translated into behavior, the prospect is present in their fantasies. Just as homosexual thoughts occur to many heterosexuals, 24% of homosexual males and 35% of lesbians have heterosexual fantasies, and 34% of male homosexuals and 54% of lesbians have heterosexual dreams.[8]

To represent these gradations, Kinsey devised a seven-point **heterosexual-homosexual scale** (from 0 to 6) to reflect a person's level of homosexual behavior (see figure 13.1). Those categorized as group 6 are *exclusively homosexual;* those categorized as group 0 are *exclusively heterosexual.* Those with equal interest in homosexual and heterosexual activity are at the midpoint of the continuum (category 3). Those in categories 1 and 5 have predominantly heterosexual or homosexual orientations, with only incidental interest in the other direction. Categories 2 and 4 include those in whom a clear preference for one sex coexists with a lesser but still active interest in the other.

This rating scale does not indicate the amount of sexual activity, only the *ratio* between the two sexual orientations. A person with 100 heterosexual and 10 homosexual experiences would be placed close to the heterosexual end of the scale, whereas a person with 10 homosexual and

no heterosexual experiences would be at the other end of the continuum. Furthermore, people are not evenly distributed across the scale. Table 13.1 summarizes the percentage distribution of the Kinsey sample in these categories. The figures are given in ranges because different subgroups of the sample (defined by age, marital status, education, and so on) have varying rates. The first category (0) shows people who have never had any homosexual experiences. The next five (categories 1 to 5) show the percentage of people who have had increasing degrees of homosexual contact. Category 6 is for those who are exclusively homosexual. In most categories the proportion of female homosexuals is only about a third as great as that of the male group.

Kinsey's motivation in devising this scale was to get away from labeling people homosexual or heterosexual and to focus instead on their behavior. But Kinsey's continuum has been criticized for promising more than it delivers. What is the real difference between saying that a woman engages only in heterosexual behavior and calling her heterosexual; or placing a man in category 6 and calling him homosexual? It is thus argued that Kinsey's scheme adds little to our understanding of sexual orientation.[9]

The Problem of Identification

There is much confusion about whether you can tell from looks and mannerisms if a person is homosexual. Are homosexual men effeminate and lesbians mannish? Do they dress and act in distinctive ways? From what we have already said about variability within the homosexual population, it should be clear that there are no simple answers to these questions. But certain aspects of this issue need further clarification.

Sexual Orientation and Gender Identity
Homosexuality, unlike transsexualism (chapter 9), is not a gender disorder. But there appears to be a strong link between

[8]Bell and Weinberg (1978, p. 289).

[9]Robinson (1976).

Table 13.1   Heterosexual-Homosexual Ratings (ages 20–35)

Category	In females (percent)	In males (percent)
X No experience with either sex	14–19	3–4
O Entirely heterosexual experience		
Single	61–72	53–78
Married	89–90	90–92
Previously married	75–80	
1-6 At least some homosexual experience	11–20	18–42
2-6 More than incidental homosexual experience	6–14	13–38
3–6 As much or more homosexual experience than heterosexual experience	4–11	9–32
4–6 Mostly homosexual experience	2–6	5–22
6 Exclusively homosexual experience	1–3	3–16

Data from Kinsey et al. (1953).

gender nonconformity and the development of homosexuality, as we shall discuss in connection with the development of homosexuality.

About half of homosexual men and three quarters of heterosexual men are typically "masculine" in their identity, interests, and appearance; half of homosexual men (and a quarter of heterosexual men) do not conform to this image. Likewise, a fifth of lesbians but a third of heterosexual women are typically "feminine."[10] Therefore, gender nonconformity is suggestive but nonconclusive in defining sexual orientation; hence many gay men and women are indistinguishable from their heterosexual counterparts on this count (figure 13.2).

The tilt of gender nonconformity towards homosexual orientations is further accentuated by the gay counterparts of gender stereotyping. This is where we get the exaggerated manifestations of the gender attributes of masculinity and femininity in the well-known homosexual types: the limp-wristed, lisping, bejewelled, and perfumed male *queen* with the mincing gait; the mannish, swaggering, tough-talking female *dyke* (or *butch*); the passive and meek

femme (or *nelly*); and so on. These characterizations are real but applicable to only part of the gay population. Homosexuals also reveal themselves through distinctive forms of dress, ornamentation, speech, and so on. But these are matters of style and fashion as well as signals of gay communication. It is therefore no wonder that one is struck by the similarities between people encountered in gay neighborhoods.

Figure 13.2   Gay friends. (Gerard Koskovich, Gay and Lesbian Alliance at Stanford.)

[10]Bell et al. (1981).

Latent and Pseudohomosexuality   At a deeper level, psychoanalysts have put forth the concept of *latent homosexuality*. Unlike the ordinary homosexual who is conscious of his or her sexual orientation, the latent homosexual is only dimly conscious of such deep-seated impulses. This concept has not found general acceptance.[11]

Another concept is that of *pseudohomosexuality* to characterize homosexual thoughts, images, dreams and fantasies experienced by heterosexuals without concomitant sexual arousal.[12] This label may be useful to reassure persons who are frightened by such thoughts. But one cannot totally exclude their having a homosexual inclination that they are trying to deny.[13]

Elective and Obligatory Homosexuality
Those who prefer homosexual relationships even though they have ample opportunity for heterosexual outlets are said to be *obligatory* homosexuals in the sense that their choice is set (but not in the sense that they are obliged to act on these impulses). By contrast, other homosexual behaviors may be *elective* (or *facultative*) by being more a matter of choice and circumstance. For example homosexual acts may be *compensatory* or substitutes for heterosexual outlets that are unavailable. Prisons, single-sex boarding schools, and other situations may prompt some to seek homosexual contacts even when they would prefer an opposite-sex partner. Under some of these conditions, homosexual acts may be manifestations of social dominance. Just as in certain cultures a man of high status may be expected to have a mistress, his counterpart in prison acquires a "girl"

(usually an effeminate younger inmate). Others may engage in homosexual activities out of curiosity, adventurousness, or social defiance, or as an act of ideological solidarity.

Whether or not elective forms of homosexual behavior constitute "true homosexuality" is open to debate. Such behaviors may lend some credence to the notion of latent homosexuality; since only some people and not others elect such outlets, whatever the sexual privations and circumstances, it could be argued that they were predisposed or had hidden needs in this direction in the first place. The counterargument is that the basis of choice in sexual orientation is not always necessarily sexual; other personality factors (having to do with dominance, for instance) may play the determinative role over who will do what under similar cirumstances.

The Problem of Bias

The condemnation of homosexuality through most of history has led to misperceptions and deliberate falsification of the realities of their lives. Homosexuality has also had its apologists who have skewed the facts in the opposite direction.

Effects of Homosexual Labeling   That the label of homosexuality has been socially stigmatizing is well known. Because society has for so long regarded such persons as *deviants,* there has been a tendency for homosexuals to think of themselves as different, marginal, or abnormal. But beyond that, the very fact of defining someone on the basis of sexual orientation tends to overemphasize the sexual component in that person's character and distorts our perception of who that person really is. If you get to know someone quite well and then find out that he or she is gay, your impressions of that person are likely to be quite different than if you were to know that the person is gay at the onset.

Another outcome of the excessive focus on the sexual element in homosexuals is the perception that they are *hypersexual,* or more driven by uncontrollable impulses

[11]For current views of the concept of latent homosexuality, see Salzman (1980).

[12]Ovesey (1969).

[13]Homosexual impulses and fantasies can be highly disturbing to those who find such behavior objectionable. The force of homosexual impulses trying to break into consciousness is at times so strong as to cause acute anxiety ("homosexual panic") although the person may not quite recognize the cause of the distress.

than heterosexuals. But homosexuals are not demonstrably any more sexually active than their heterosexual counterparts. On the average, homosexual men have sex about 2 or 3 times per week: lesbians about 1 or 2 times a week. There is, of course, considerable variation within the homosexual population (as there is among heterosexuals): about 20% of gay men and women have sex once a week; 13% of men and 21% of women once or fewer times a month; 17% of men and 13% of women more than 4 times a week.[14]

Homophobia   People are entitled to their sexual preferences and moral judgments. But hidden behind such ostensibly rational views there may be a pervasive dread of homosexuality that cannot be justified on realistic grounds. Such irrational fear and rejection of homosexuality is referred to as *homophobia*.

Long before this concept became popular, Freud spelled out the psychodynamics of how such attitudes may develop and their association with certain forms of paranoia.[15] Freud ascribed the *neurotic* fear of homosexuality to repressed homosexual wishes. Because homosexual affection is not tolerated in our culture, specially for males, a man cannot admit to himself that he *loves* another man, in a sexual sense. Therefore, he unconsciously changes it to the conviction that he *hates* the man. To further exonerate himself, he may then accuse the man of hating and persecuting him.

Current proponents of the concept of homophobia also find *religious prejudice* to be responsible for homophobic attitudes because the Judeo-Christian tradition has condemned homosexuality over the centuries. Homosexuals may be resented because they represent a *threat to established values*. Homosexuality frightens people because it would compete with and undermine the reproductive functions of marriage and the family. Finally, heterosexuals are said to *envy* homosexuals because they lead such free and easy sexual lives without the burdens of marriage and parenthood.[16] These arguments are important as a means to help people see their bias. But the usefulness of the concept of homophobia is vitiated if it is used to attack everyone who has anything negative to say about one or another aspect of homosexuality.

Homosexual Behavior

Genital exploration and mounting can be observed between same-sexed animals in many mammalian species.[17] But such interactions are often more of a social than a sexual nature. For instance, mounting serves to define dominance among nonhuman primates. In Japanese macaques, male-male mounting is also part of amicable play and a way of reaffirming social bonds when the social group is under stress.

More direct forms of sexual interaction between same-sexed animals have been observed among certain monkeys in captivity that appear to be of a clearly erotic nature (figure 13.3). These interactions usually involve animals with friendly relationships and occur in amicable contacts. But it is not clear whether such behavior represents a response to captive conditions or is part of the normal sexual repertory of these animals.[18]

Same-sexed interactions among females involving mounting have also been observed, but in these cases there are no signs of sexual excitement, bodily contact is minimal, and the act is terminated quickly.[19]

---

[14]Bell and Weinberg (1978). Other studies report somewhat higher or lower rates. See for instance, Westwood (1960); Weinberg and Williams (1974); Saghir and Robins (1973). The impression of hypersexuality may also come from the fact that some male homosexuals lead highly prolific sexual lives with hundreds of sexual partners.

[15]The psychoanalytic classic on this topic is Freud's analysis of the memoirs of Daniel Schreber, a judge in Dresden.

[16]Weinberg (1972).
[17]Ford and Beach (1951); Denniston (1980).
[18]Lancaster (1979).
[19]Ford and Beach (1951).

Mounting with unilateral manual genital stimulation between two males.

The female-female homosexual mounting position.

Mounting with mutual oral genital stimulation between two males.

A supine position observed during both heterosexual coitus and female-female homosexual interactions.

Mounting with unilateral oral genital stimulation between two males.

Mutual presentations with manual genital stimulation between two males.

Figure 13.3 The variety of heterosexual and homosexual interactions observed in captive Macaca arctoides. (Chevalier-Skolnikoff, 1974.)

Prevalence of Homosexual Behavior

We do not know with any accuracy how prevalent homosexual behavior is in the general population in our society. When the Kinsey statistics first appeared, they seemed astonishingly high: among males 37% have had at least one homosexual experience leading to orgasm following puberty; 10% had significant levels of such experience over at least 3 years; 4 to 6% led primarily or exclusively homosexual lives.[20] The corresponding female figures

[20]Kinsey et al. (1948).

Figure 13.4   Erotic scenes depicted on a red-figure cup by the Nikosthenes. Early 5th century B.C. (Museum of Fine Arts, Boston, Perkins Collection.)

were half to one-third as high, with 13% of women having had at least one such experience.[21]

One explanation for the probable inflation of the Kinsey figures is that since members of the Kinsey team interviewed all comers and were sympathetic to people irrespective of their sexual preferences, homosexuals may have flocked to them, thus skewing the sample. Be that as it may, a central problem of the Kinsey and similar surveys is the way cases are counted. Do we tally a single episode of mutual masturbation by two young men the same way as a persistent pattern of oral or anal sex throughout adulthood? Kinsey did: half of all homosexual experiences reported to him was accounted for by single sexual acts in youth.

Even by the most conservative estimates, several million adults are assumed to be more or less exclusively homosexual.

If persons with bisexual orientations were to be included, these numbers would be very much larger. The extent to which homosexual orgasm contributes to the total outlet of each sex at various ages and in connection with factors like marriage and social class was discussed in chapter 8.

Patterns Homosexual Interactions

Homosexuals engage in all of the sexual activities that heterosexuals do, except for vaginal intercourse. They *kiss* and caress as well as engage in *manual stimulation, oral sex,* and *anal intercourse* (figure 13.4). Homosexuals may also engage in *simulated coitus* involving the apposition of the genitalia or their friction against other body parts (figure 13.5). Less frequently, dildos and vibrators may be used vaginally or anally (more often by gay men than lesbians). Some men may also engage in more dangerous practices like inserting the hand into the rectum ("fist fucking").

[21]Kinsey et al. (1953).

Figure 13.5   G. Courbet. *Sleep, 1867.* (Petit Palais, Paris. Photo Bulloz.)

At an individual level, homosexuals do not engage in all conceivable activities any more than heterosexuals do. For instance, some gays object to anal or oral sex on aesthetic, health, or other grounds. The differences in prevalence between heterosexual and homosexual couples are not that striking. For instance, 5% of heterosexuals engage in fellatio as against 17% of gay men, every time they engage in sex; similarly, 6% of heterosexuals use cunnilingus as against 12% for lesbians, every time they engage in sex.[22] Manual stimulation of the genitals and breasts is the most common form of lesbian sex; cunnilingus the preferred technique for reaching orgasm. For men, fellatio is the most commonly preferred technique of arousal and orgasm.[23]

For same-sex couples the boundaries between the foreplay and the main sexual activity are hazy. Except for homosexual men, who in typical male fashion rush to oral or anal intromission and orgasm, sexual encounters between gay men and especially gay women, tend to be more protracted, with alternating sequences of sexual stimulation, "teasing" withdrawal, and further stimulation.[24]

Same-sex lovers are generally more effective sexual performers, since they have the advantage of knowing from experience with their own body what is likely to be pleasurable. Nonetheless, there are sex manuals that specifically address means of enhancing sexual encounters among gay men and women.[25]

Active and Passive Roles   In traditional heterosexual relationships, the man takes the active role: it is he who stimulates and

[22]Blumstein and Schwartz (1983, p. 236).
[23]Bell and Weinberg (1978).

[24]Masters and Johnson (1979).
[25]See the *Joy of Lesbian Sex* by Sisley and Harris (1977); the *Joy of Gay Sex* by Silverstein (1978). For a more general work on anal pleasure, see Morin (1981).

Box 13.1   Homosexuality in Cross-Cultural Perspective

In their cross-cultural survey, Ford and Beach found 76 societies for which data were available on homosexuality (mostly of males). In one form or another, homosexuality was permissible in 64% of them. In the other 36% that did not condone homosexual activity (and some punished it severely) there was nonetheless some clandestine homosexual activity that went on.[1] Homosexual behavior is never the dominant form of sexual activity in any culture, and even groups which permit homosexual activity do not sanction exclusive homosexuality.

In societies where access to women is rigidly controlled, male homosexuality is tacitly accepted as a substitute to coitus. There is no dishonor attached to such activity as long as one takes the "active" role, such as being the inserter in anal intercourse. Effeminate males in these cultures are under some pressure to submit.[2]

A number of societies that have institutionalized male transvestism also permitted homosexual activity by these individuals (but the two should not be equated in all cases). These include Indian societies in North and South America and island societies in Polynesia. In these contexts, some men put on female dress and take on female functions, including sexual activity with their "husbands"; Devereux's Mohave informants reported both oral and anal intercourse to occur.[3] In other situations, such as in Tahiti, men who undertake gender role reversal remain unattached but interact sexually with other men on occasion.

Homosexuality can also be ritualized and restricted for certain phases of life. Some societies in the highlands of New Guinea incorporated homosexual acts in their initiation ceremonies. Typically, this involved the transfer of semen, anally or orally, from an adult to a teenage boy to ensure proper growth and masculinity. Semen, the masculine essence, is presumed to have magical powers as a source of virility and strength.

Gilbert Herdt has done an extensive study of these practices among the Sambia, a warrior group living in New Guinea. These people believe that the ingestion of semen by boys is essential for their development into men and undertake to accomplish this

then penetrates the female partner. Based on this model, and on the assumption that homosexuals imitate heterosexuals, it has been assumed that gay men and women take on similar *active* and *passive* roles. In actual practice, many homosexuals do not sort themselves out along such lines. Preferences are more personal and shared. Gay men and women typically take turns, or simultaneously stimulate each other in whatever activity they are engaged.

This does not mean, however, that certain gay men and women do not deliberately cultivate active-passive, masculine-feminine roles; some homosexuals may even convince themselves that they are engaging in homosexual behavior only in some roles but not in others. Especially in cases of power-based homosexual encounters, such as in prison, the dominant male who penetrates other men may be quite convinced that he is not engaging in a homosexual act and he will resist the prospect of submitting to anal penetration ("get fucked"). Similar attitudes prevailed in the ancient world (chapter 17).

Social Perspectives on Homosexuality

Differing social attitudes toward homosexuality can be discerned in many cultures (box 13.1). In our society, such considerations can be traced to the origins of Western culture in the beliefs and practices of Judaism and the classical world (chapter 17).

Box 13.1   Homosexuality in Cross-Cultural Perspective *(Continued)*

through secret rituals. Those whose semen is ingested during fellatio are young bachelor men. Yet such extensive socialization with homosexual practices does not produce a society of homosexuals. Once the Sambia male marries, homosexual practices are replaced with marital cohabitation and fatherhood.[4]

Little is known about lesbian sexuality in preliterate societies, in part because anthropologists have been mostly male and have had limited access to information about the private lives of women. It is also possible that female homosexuality is, in fact, less common. Lesbian practices were reported in 17 of the 76 societies analyzed by Ford and Beach, but details on these activities were limited.

Cross-cultural comparisons of homosexuality usually dwell on their diversity. A recent comparative study of homosexual communities in the United States, Guatemala, Brazil, and the Philippines offers six tentative conclusions about the commonalities of such experience, which are as follows: (1) homosexual persons appear in all societies; (2) the percentage of homosexuals in all so-

cieties seems to be about the same and remains stable over time; (3) social norms do not impede or facilitate the emergence of homosexual orientation; (4) homosexual subcultures appear in all societies, given sufficient aggregates of people; (5) homosexuals in different societies tend to resemble each other with respect to certain behavioral interests and occupational choices; and (6) all societies produce similar continua from overtly masculine to overtly feminine homosexuals.[5]

Additional insights into homosexuality are available from historical studies of ancient cultures, especially that of classical Greece, where certain forms of homosexual relations were institutionalized (discussed in chapter 17).

[1]Ford and Beach (1951).
[2]See Carrier (1980) for sources.
[3]Devereux (1937).
[4]Herdt (1981).
[5]Whitman (1983).

Social Judgments of Homosexuality

After long neglect, the history of the treatment of homosexuals in our culture is now receiving due attention by serious scholars.[26] The story is by no means a tale of relentless condemnation. Yet on balance, homosexual behavior has been viewed negatively by religious institutions as being *immoral;* by the law, as a *crime;* by medicine, as an *illness;* and by public opinion, as a *social deviance.* Opposed to these perceptions is the current view that homosexuality is a *life style.*

We discuss the historical perspective on homosexuality in chapters 17 and 18; deal with its legal and moral aspects in chapters 19 and 20; and review the recent development of the gay liberation movement in chapter 16. In the rest of this chapter we discuss homosexuality as an illness and as a way of life.

As we discussed in chapter 1, by the 19th century homosexuality was firmly established as a diagnostic entity within psychiatry. Compared to earlier views of homosexuals as sinners and criminals, the perception that they were *sick* was seen as a humane alternative since it called for help and understanding rather than condemnation and persecution. But increasing numbers of homosexuals gradually came to view this approach as less than a bless-

[26]See, for instance, Boswell (1980); Dover (1978). Earlier attempts at historical overviews were made by Karlen (1971) and Bullough (1977).

ing, since to be called sick was in some ways just as stigmatizing as other judgments.

Homosexuality continued to be listed as a disease entity in the *American Psychiatric Diagnostic Manual* until 1974. When this diagnosis was deleted, homosexuality ceased to be considered officially as a psychiatric disorder.[27] This change was viewed by its opponents as more of a political than medical action; a subsequent survey of a sample of the APA membership showed that two-thirds of psychiatrists still considered homosexuality a disorder.[28] But if the exclusion of homosexuality from the category of illness was a political act, it could be argued that its inclusion in the first place was also a reflection of social, not medical, judgments; that physicians merely reflected the prejudices of society against homosexuals.

The current official medical stance strikes a compromise. *Ego-dystonic homosexuality* is defined as a psychosexual disorder, but homosexuality as such is explicitly excluded as a diagnostic category.[29] "Ego-dystonic" means that the behavior is unacceptable to the person and his or her desire to change the homosexual orientation is a matter of "consistent concern." So according to this approach, if a homosexual is significantly unhappy with his or her sexual orientation (on intrinsic grounds, not merely because society is making life difficult) then the person has a psychosexual disorder; if not, then there is no psychiatric illness involved. This position is quite similar to that taken by Freud almost 50 years ago (see box 13.2).[30]

There is an extensive clinical literature attesting to the presence of psychological conflict and distress among homosexuals seeking therapy.[31] Homosexuals who seek help may not be representative, of course, of the broader gay population. Yet studies undertaken with those who are not in therapy also report considerable conflict among them: the rate of suicide attempts among homosexual males is 20%; among heterosexual men 5%;[32] But even if homosexuals are demonstrably more unhappy, is it their homosexuality as such or the negative societal reaction to it that makes life harder for them?

Other studies show the majority of homosexuals to be well-adjusted people with no evidence of psychiatric illness.[33] Behavioral scientists are not even able to differentiate homosexual from heterosexual men in detailed life histories and psychological tests.[34] Sociologists suggest that when homosexuals behave differently, it is because they are responding to the role of deviants to which society assigns them and to which they respond by living up to societal stereotypes of themselves.[35]

These contradictory findings can be reduced to three alternatives: first, homosexuals are psychologically no worse off than heterosexuals despite social pressures; second, homosexuals suffer because of the social judgments that condemn them to a state of deviancy; third, homosexuals have psychological problems quite apart from the problems imposed by society.

One hardly needs to be a behavioral scientist to appreciate the enormous burden that a homosexual orientation has entailed and still entails for so many. There are certainly those to whom homosexuality poses no social threat; but to most others, especially for people in public life, the threat of disclosure is very serious. It is hard to imagine how such circumstances would not affect, to a lesser or greater extent, peoples' lives, personalities, and behaviors. And as long as these circumstances exist,

[27]American Psychiatric Association (1980).

[28]Lief (1977). For an account of how the American Psychiatric Association dealt with this issue see Bayer (1981).

[29]American Psychiatric Association (1980, p. 281).

[30]For further discussion of these diagnostic shifts with regard to homosexuality, see Spitzer (1981).

[31]See, for instance, Bieber (1962); Ovesey (1969); Hatterer (1970); Stoller (1975); Socarides (1975).

[32]Bell and Weinberg (1978).

[33]Saghir and Robbins (1973). Also see Chang and Block (1960); Freedman (1967); Wilson and Greene (1971); Hooker (1975).

[34]Hooker (1975).

[35]Gagnon and Simon (1967); Weinberg and Williams (1974).

Box 13.2   Freud's Letter to Mother of a Homosexual[1]

April 9, 1935

Dear Mrs. X,

I gather from your letter that your son is a homosexual. I am most impressed by the fact that you do not mention the term yourself in your information about him. May I question you, why do you avoid it? Homosexuality is assuredly no advantage, but it is nothing to be ashamed of, no vice, no degradation, it cannot be classified as an illness; we consider it to be a variation of sexual function produced by a certain arrest of sexual development. Many highly respectable individuals of ancient and modern times have been homosexuals, several of the greatest among them (Plato, Michelangelo, Leonardo da Vinci, etc.). It is a great injustice to persecute homosexuality as a crime, and cruelty too. If you do not believe me, read the books of Havelock Ellis.

By asking me if I can help, you mean, I suppose, if I can abolish homosexuality and make normal heterosexuality take its place. The answer is, in a general way, we cannot promise to achieve it. In a certain number of cases we succeed in developing the blighted germs of heterosexual tendencies which are present in every homosexual, in the majority of cases it is no more possible. It is a question of the quality and the age of the individual. The result of the treatment cannot be predicted.

What analysis can do for your son runs in a different line. If he is unhappy, neurotic, torn by conflicts, inhibited in his social life, analysis may bring him harmony, peace of mind, full efficiency, whether he remains a homosexual or gets changed. If you make up your mind he should have analysis with me (I don't expect you will!!) he has to come over to Vienna. I have no intention of leaving here. However, don't neglect to give me your answer.

Sincerely yours with kind wishes.

Freud

P.S. I did not find it difficult to read your handwriting. Hope you will not find my writing and my English a harder task.

[1]From Freud, S. "Letter to an American Mother."

attempts to discern the true burden of being a homosexual (if any) will remain conjectural.

The Causes of Homosexuality   Those who accept homosexuality as an illness seek to find its *causes*. Those who do not, object to this approach since the question of "cause" implies that there is something wrong; why not seek the "causes" of heterosexuality, they ask. Polemics aside, the search for possible biological and psychosocial origins of sexual orientation as such is certainly a legitimate quest, and we come back to it when we deal with the development of homosexual orientation.

Treatment of Homosexuality   Those who dismiss the notion of homosexuality being an illness deride *treatment* as a form of brainwashing or "conversion" to the majority sexual value system.[36] For those who seek help, the results of treatment have not been particularly effective, whether based on psychotherapy, behavior modification, or other methods. There are several possible reasons for this.

Many homosexuals have sought therapy with no real intention of changing their sexual orientations; they may acquiesce to therapy because they are pressured into it by their families or driven to it by anxiety

[36]Asked if they would use a "magic heterosexual pill" to effortlessly change to heterosexuality, only 5% of lesbian and 14% of male homosexuals said they would use such a pill (Bell and Weinberg, 1978, pp. 124 and 129).

and depression. So they play along or seek relief; but very few really want to change. Even those who are sincerely motivated have not usually succeeded in changing and instead have chosen to use therapy to adjust better to their sexual circumstances. But it is not true that homosexuals cannot ever become heterosexual. There have been reports of such shifts through various therapeutic methods.

The bias in the past was to push homosexuals into treatment irrespective of whether they needed or wanted it. There is now the contrary pressure not to seek treatment, because that would mean admitting that there is something wrong with being homosexual. Homosexuals who are in distress, but do not wish to change their sexual orientation, seek gay therapists so that their sexual orientation does not become the focus of their treatment; there are also heterosexual therapists who are quite capable of not imposing their own sexual values on their patients.[37]

Homosexuality as a Life Style

Over the past decade or so, professional opinion has shifted to favor the view that homosexuality is a *way of life* or an *alternative life style*, a perspective shared by large numbers of gay men and women participating in the homosexual subculture.

The Homosexual Subculture

A *subculture* is a subdivision of a national culture based on some common feature, such as ethnicity, that provides a group with a separate sense of identity, and a set of normative values. When applied to an age group, like the "youth subculture," the concept tends to become diffuse; this is even more so the case with respect to a more diverse group like homosexuals.

Nonetheless, the concept of a *gay subculture* is useful to delineate the common world and life style of at least a subseg-

ment of the gay population. It must be clear however, that simply being homosexual does not amount to membership in its subculture. To be a part of it, one must participate in it in some active, visible way.

Homosexual subcultures have existed throughout history in many cultures.[38] The estimated percentages of homosexuals at the turn of the century are not that different from current figures; before World War I, Berlin was said to have between 1,000 and 2,000 male prostitutes and about 40 homosexual hang-outs.[39] Nonetheless, Victorian society could be highly punitive when it chose to, as in the case of the famous trial of Oscar Wilde (chapter 18).

Until the emergence of the *gay liberation movement* (chapter 16), homosexuals in America were largely connected through *underground networks* and furtive associations. When the existence of such a group came to light, the result often led to scandal.[40] Most major American cities now have sizable gay communities with distinctive life styles.[41] Nonetheless the majority of gays continue to live within the general population.

The primary reason for the existence of the gay subculture is the opportunity it provides for sexual interaction. But around this nucleus are clustered multiple other activities and functions such as networks of friendship, mutual support, political solidarity, recreation, commerce, and all else that constitutes the fabric of urban life.

An important aspect of a gay community that sets it apart from other subcultural enclaves is that those who live there do so by their own choice; it is not like a ghetto where people are stuck. For this reason, gay communities have a wide representation of occupational levels. But they are almost entirely male enclaves.

[37]For an overview of the relationship of homosexuality and mental illness, see Marmor (1980).

[38]Boswell (1980).

[39]Karlen (1971).

[40]Boise, Idaho, was shaken in 1955 with the discovery of a homosexual underworld that involved some of the city's most prominent citizens. For the story, see Gerassi (1967).

[41]For a city-by-city tour of the scenes and the lives of gay men, see White (1981).

Lesbians do not congregate in similar communities. Unlike males, they attract little attention living as couples within ordinary residential areas (and not all women who live together are lesbians). They also tend to lead more private lives and are less dependent on public places of gay life, as we describe shortly. The majority of gay men similarly lives within the general population.

The Nature of Homosexual Relationships

There is no such thing as a single *homosexual life style* or a standard way in which homosexuals interact with one another. Like heterosexuals, homosexuals live and relate in many ways that are shaped by a host of economic and social factors as well as their sexual orientation. There are distinctive features to some homosexual styles, but what is most flamboyant must not be equated with what is typical. In addition to the more general influences exerted by factors like social class, two factors, more than any other, determine the relationships and life styles of homosexuals: one is whether their homosexuality is *overt* or *covert*; the other is being *male* or *female*.

Covert Homosexuals   Covert or *closet* homosexuals are found in all sectors of our society, where they usually pass for heterosexuals in their business and social relationships. They may be married, with or without children, and in most other respects remain indistinguishable from the rest of the population. They may lead *double lives*, simultaneously engaging in heterosexual and homosexual relations; they may restrict their homosexual behavior to periods when they are away from home; or may suppress their homosexual desires altogether.

We know relatively little about the nature of the relationships of covert homosexuals; these persons are hard to identify and resist revealing the details of their lives, for understandable reasons. Because we do not know how many covert homosexuals there are, we cannot tell what proportion of the population is gay. Since they keep their sexual lives largely hidden, the true range of homosexual behavior becomes dominated by the activities of overt homosexuals.[42]

Coming Out   When a covert homosexual admits his sexual orientation to outsiders, he or she is said to have *come out of the closet*. Some cross this line early in life, others late; some do it voluntarily, others are coerced. Whatever the circumstances, it is never easy.

Those who come out when still young must confront the frequent (but not inevitable) surprise, dismay, and anger of parents. There is often a terrible sense of failure on the part of parents who search the past for what they did wrong. Even though their gay sons or daughters may seem perfectly content, parents worry about their leading lonely and sterile lives and regret foregoing the pleasure of seeing their children married, becoming grandparents, and so on. On the other hand there are also parents who face this prospect with equanimity and throw their love and support behind their offspring's "choice."

Those who come out later in life might have the difficult task of confronting their own spouses and children as well as business associates and friends. The outcomes vary from the breakup of families to a readjustment of the relationships; what happens in the work place often depends on the type of job the person holds.

Coming out does not mean total revelation. A 1982 marketing survey of urban gay males showed that of those who considered themselves "publicly out," only one in five had informed business associates and an equal proportion still hid the truth from their families.[43]

[42]Covert homosexuals are resented by their fellow homosexuals and given derogatory nams like "canned fruit," "crushed fruit" (because they are crushed by society's mores) and so on (Rodgers, 1972).

[43]*Newsweek*, August 8, 1983, p. 34.

Overt Homosexuals Overt homosexuals constitute a far more accessible group. They have given up all pretense and openly rely on the homosexual community for gratification of their sexual needs. Because of their social visibility, they set the tone and convey the public images of what it is like to be homosexual.

In Bell and Weinberg's study 71% of overt homosexuals could be placed in five general patterns of personal relationships; the remaining 29% were too diverse to be categorized.[44]

About 28% of lesbians and 10% of male homosexuals lived as *close couples,* similar to married heterosexuals. One third of these lesbian couples had been living together for 4 or more years, as had 38% of these men. The close-couple homosexuals rarely cruised for new sexual contacts in gay bars and other public places. They had the fewest sexual, social, or psychological problems. They were happy with their partners and enjoyed spending their leisure time at home in warm and caring relationships.

About 17% of lesbians and 18% of male homosexuals lived as *open couples.* These people were less happy with their partners, less deeply committed to them, and more likely to look for social and sexual gratification outside their relationship. They cruised more often, and on the whole they were less happy, less self-accepting, and more lonely than the close-couple homosexuals. Lesbians found the open-couple pattern less acceptable than male homosexuals did.

Homosexuals who were well adjusted, had few if any sexual problems, did not regularly live with a partner, and followed a "swinging singles" lifestyle were designated as *functionals.* Some 15% of the homosexual males and 10% of the lesbians were in this group. These people had very high levels of sexual activity. They were much more interested in having multiple partners than in finding someone with whom to establish a close relationship. Much of their lives centered around sexual

activities. They were highly involved in the gay community and cruised a great deal. They had the least amount of regrets about being homosexual and were energetic, friendly, and self-reliant; their overall social and psychological adjustment was second only to the close-couple homosexuals.

Homosexuals who had many partners, were not living as couples, and had psychological and sexual problems were designated as *dysfunctionals.* About 12% of male homosexuals and 5% of lesbians were in this group. These people were prone to worry, found little gratification in life, and showed a poorer overall adjustment than the previous three types. They were most likely to regret being homosexual and worried about their sexual inadequacies, their inability to form an affectionate relationship, and their lack of fulfillment. They came closest to fitting the stereotype of the "tormented homosexual."

Some homosexuals were found to have a low level of sexual activity, few partners, and no close relationships. These people, designated as *asexuals*, were characterized by their general lack of involvement with others. Some 16% of male homosexuals and 11% of lesbians followed this solitary life style. They tended to be older than the people in the other categories. Although they described themselves as lonely, their general psychological adjustment was about the same as the homosexual population taken as a whole.

The typical, or perhaps stereotypical, image of the overt male homosexual that comes across in his public behavior shows a far greater preoccupation with sex than do lesbians. In Bell and Weinberg's sample, 57% of lesbians had had fewer than 10 partners in their lives, whereas 57% of the male homosexuals had had more than 250 partners; 45% of lesbians were currently living as couples, compared with 28% of the male homosexuals; 74% of the men said that more than half of their partners were strangers in contrast to 6% for lesbians. Seventy percent of males reported having sex once a week or more, against only 54% of females; only 3% of lesbians

[44]Bell and Weinberg (1978).

Box 13.3   Two Poems by C.P. Cavafy

DECEMBER, 1903

And if I can't speak about my love—
if I don't talk about your hair, your
    lips, your eyes,
still your face that I keep within my
    heart,
the sound of your voice that I keep
    within my mind,
the days of September rising in my
    dreams,
give shape and color to my words, my
    sentences,
whatever theme I touch, whatever
    thought I utter.

ON THE STAIRS

As I was going down those ill-famed
    stairs
you were coming in the door, and for
    a second

I saw your unfamiliar face and you
    saw mine.
Then I hid so you wouldn't see me
    again,
you hurried past me, hiding your
    face,
and slipped inside the ill-famed house
where you couldn't have found
    pleasure any more than I did.

And yet the love you were looking
    for, I had to give you;
the love I was looking for—so your
    tired, knowing eyes implied—
you had to give me.
Our bodies sensed and sought each
    other;
our blood and skin understood.

But we both hid ourselves, flustered.

From *C.P. Kavafy Collected Poems* Translated by E. Keeley and P. Sherrard. Princeton, N.J.: Princeton University Press, 1975, pp. 178–179.

cruised once a week or more for partners, as against 42% of males.[45]

In their search for sexual partners, gay men tend to focus more on physical attractiveness and youth than do gay women.[46] The former are more genitally focused in their sexual interactions than the latter.

Homosexual males (like their heterosexual counterparts) are more interested than women in sexual variety, including the paraphilias (discussed in chapter 14). The interaction of male wih male also greatly enhances the potential for violence. Rapes and sadomasochistic practices are more common in male than female homosexual encounters. These differences are not specific to homosexuals; they also hold true for heterosexuals. In sexual interactions, gender overrides orientation—lesbians behave more like heterosexual women than they do like homosexual men, and the same is true for homosexual men, who are more like heterosexual men than lesbians.

Studies of homosexuality dwell on issues such as sexual behavior, interpersonal relationships, and the gay subculture, but say relatively little about homosexual love.[47] The joys and sorrows

[45]Bell and Weinberg (1978), p. 308.
[46]The emphasis on youth for sexual contact makes life harder on aging homosexuals. However, though more likely to be living alone and with fewer sexual involvements, older gays do not seem to be psychologically worse off (Weinberg and Williams, 1974).
[47]Silverstein (1981) deals with this issue with sensitivity.

of such love, as evident in homoerotic poetry, are basically indistinguishable from its heterosexual counterpart (box 13.3).[48]

Public Places of Gay Life

The sexual associations of heterosexuals are integrated within the larger network of their social relationships. They meet potential sexual partners during their everyday life, at the places they study, work, and socialize. Every culture, no matter how sexually restrictive, has some means of bringing together its unattached men and women.

Homosexuals have been deprived of such socially sanctioned means of finding sexual partners. When a heterosexual man makes a pass at a woman, he may be at worst rebuffed; when a homosexual man makes a pass at a man, he may be insulted, disgraced, or even beaten up. Unless, of course, the other man is also gay. But how does one tell? To resolve this problem, gays rely on two measures: various means of recognizing each other; meeting in special places.

Making Contact   Homosexuals have mannerisms, gestures, dress codes, and so on to signal their sexual orientation. To be useful, these cues must be distinct enough to be picked up readily yet not so obvious that they will compromise the user.

Visual signals are the most versatile means of nonverbal communication. Homosexuals use them most effectively (as do heterosexuals) in their courtship. Unlike set messages conveyed through appearance, or overt gestures, which can be picked up by outsiders, searching looks and meaningful glances can be more effectively focused on the quarry, and instantly disowned if necessary. The meaning of these eye signals becomes more explicit as they become ritualized; slowly scanning the other person's body, or glancing backwards, make their meanings quite clear.

Another classical maneuver is *lingering* near the other person while ostensibly preoccupied with some other interest, like window shopping (it is not an accident that there are ordinances against loitering). Once the ice is broken, verbal exchanges settle matters further.

These methods of communication are utilized mainly during **cruising,** the process of searching for sexual partners. Cruising is most effective when combined with going to places where gays congregate. These include *gay bars, public baths, rest rooms,* and designated (though shifting) parts of streets and parks known as meeting places. The common element to all these sites is that they are most often used by males looking for casual sex.[49]

Gay Bars   In cities that tolerate them (and most major cities now do) the gay bar is the main public place where gay men meet. Like the heterosexual "singles bar," it functions as a *sexual marketplace* for picking up casual sex partners, and occasionally a more permanent relationship. Gay bars however are also important for homosexuals as a place to socialize, exchange news, and partake in a feeling of solidarity; unlike heterosexuals, they do not have many other intitutionalized settings for these purposes. Some gay bars cater to special groups, such as leather-jacketed motorcyclist types, or serve as hangouts for male prostitutes; other bars feature drag shows or male go-go dancers.

The clientele of gay bars is virtually all male. Very few women frequent these bars, be they lesbian or straight. Heterosexual women (called "fruit flies") go to gay bars because gay men provide good company without "hassling" them sexually. Gay men, in turn, welcome these women because they do not compete with them for partners and add a certain measure of novelty, interest, and gentility to the group. Lesbians have their own bars (where men are not welcome) but they infrequently go to such bars looking for a pickup; they usually go with a friend to socialize, ex-

---

[48]For an overview of the theme of homosexuality in literature, see Purdy (1975); Kiell (1976).

[49]Hoffman (1968).

change gossip, have a few drinks, and dance.

The second striking feature of clients of gay bars is their apparent youthfulness and fastidious grooming. These men are either young or take pains to look young. Typically, they are dressed like college undergraduates with sport shirts, jeans, and loafers. But unlike a typical gathering of college students, these men are meticulously groomed, and collectively project a certain aura of muted effeminacy, showing little of the boisterous ways and rowdiness of younger heterosexual men. Taken individually, a few of the patrons are readily identifiable as gay; but most of the rest would arouse no suspicion in another setting.

There is very little by way of sexual activity in most gay bars. Instead, there is a great deal of milling about, superficial conversation, and cruising with the eyes. Outsiders are specially perplexed at the shadowy figures who stand about silently watching; one observer has likened it to guests at a reception waiting for the arrival of an important figure. The important figure in this case is the perfect young lover, the object of everyone's fantasy, who rarely shows up.

Others conduct their business more briskly: scan the group, identify a likely partner, reach agreement, leave together. The agreement usually is to have sex, with no strings attached. The particular type of sex is not discussed; most gay men are versatile enough to find some mutually desirable sexual activity.

Gay Baths   Steam baths that cater to gay men (there are none for lesbians) are the most streamlined settings for fast, multiple, anonymous sexual encounters. These steam baths consist of a public area ("orgy room") where men may engage in oral sex or pair up before heading for private cubicles. Or a man can simply wait in one of these rooms with the door ajar; the position he takes on the bed signals the sexual activity desired; lying on the stomach will mean, for instance, that he wants to be penetrated anally.

Public Sites for Sexual Encounters   Designated parts of streets and parks are usually only suitable to establish contacts, but not private enough to engage in sex. They are riskier than bars and baths but they cost nothing; besides not every city has a bath or bar, but public pick-up sites have been long in existence.

Public toilets have been among the most commonly used locations to make contact or have sex on the premises.[50] Under these circumstances, the sexual interaction must be particularly fast and furtive for fear of apprehension. This very element of danger, as well as perhaps the excretory functions associated with toilets, appears to impart to the sexual activity an added sense of excitement.

[50]Authorities have tried to suppress such encounters by removing toilet doors, employing police decoys, or even setting up observation posts behind one-way mirrors.

Figure 13.6   Gay hustler. Berlin, 1930's. (Drawing by Christian Schad. *Bilder Lexicon*.)

Box 13.4   The Hustler[1]

There is a terrific, terrible excitement in getting paid by another man for sex. A great psychological release, a feeling that this is where real sexual power lies—not only to be desired by one's own sex but to be paid for being desired, and if one chooses that strict role, not to reciprocate in those encounters, a feeling of emotional detachment as freedom—these are some of the lures; lures implicitly acknowledged as desirable by the very special place the male hustler occupies in the gay world, entirely different from that of the female prostitute in the straight. Even when he is disdained by those who would never pay for sex, he is still an object of admiration to most, at times an object of jealousy. To "look like a hustler" in gay jargon is to look very, very good.

*

Outside of a busy coffeeshop where hustlers gather in clusters throughout the night, an older man in a bright-new car parked and waited during a recent buyers' night. Youngmen solicited anxiously in turns, stepping into the car, being rejected grandly by him, stepping out, replaced by another eager or desperate youngman. Smiling meanly, the older man—one of that breed of corrupted, corrupt, corrupting old men—turned down one after the other, finally driving off contemptuously alone, leaving behind raised middle fingers and a squad of deliberately rejected hustlers—some skinny, desolate little teenagers among the more experienced, cocky, older others; skinny boys, yes, sadly, progressively younger, lining the hustling streets; prostitutes before their boyhood has been played out, some still exhibiting the vestiges of innocence, some already corrupted, corrupt, corrupting—an increasing breed of the young, with no options but the streets—which is when it is all mean and ugly, when it is not a matter of choice; wanted for no other reason than their youth, their boyhood . . . And yet, later that very night, I met a man as old as the contemptuous other one—but, this one, sweet, sweet, eager to be "liked," just liked, desperate for whatever warmth he might squeeze out, if only in his imagination, in a paid encounter, eager to "pay more"—to elicit it—simply for being allowed to suck a cock . . . Hustling is all too often involved in mutual exploitation and slaughter, of the young and the old, the beautiful and the unattractive.

[1]From, Rechy J., *The Sexual Outlaw*, Grove Press, 1977, pp. 153–154.

The Homosexual Prostitute

Homosexual prostitutes are predominantly male and are called *hustlers* (figure 13.6). Lesbians rarely sell or buy sex. Generally quite young, gay prostitutes fall into four categories: full-time street and bar hustlers; full-time call boys or kept boys; part-time hustlers; and delinquents who use prostitution as an extension of other illegal acts like robbery.[51]

Street hustlers often deny being gay and pretend to be in the business for the money only. To reinforce their claim of being straight, they may have girls tagging along. They may also consent only to certain forms of sex such as being the insertor in anal intercourse. Hustlers work in known locations and are picked up by men who are often in cars; or they may make contact in bars and other public locations (box 13.4).[52]

[51]Allen (1980). Also see Drew and Drake (1969).

[52]Rodgers (1972).

Male models or *call boys* more readily admit being gay. Like the street hustlers, they bank on their youth and good looks. The appearance of masculinity and having large genitals (being "well hung") are important selling points.

Call boys may work through matchmaking agencies that transmit phone messages or provide more elaborate services such as rooms or entire male brothels with specialized sexual services.[53]

The Development of Sexual Orientation

One of the key components in sexual development is the shaping of the person's *sexual orientation.* Although most people have not one but a set of sexual outlets or a variety of ways of attaining sexual gratification, this term has come to signify mainly the choice between heterosexual and homosexual alternatives.

There is no generally acceptable answer to the question why some people develop heterosexual, others homosexual or bisexual identities and sexual preferences. Some sex researchers since Krafft-Ebing have regarded homosexuality as having a "constitutional" origin or some other biological cause. Yet, most behavioral scientists have attributed homosexual preference to social factors, especially to childhood experiences with parents. The debate is by no means settled.

Evidence of Biological Determinants

Those who favor a biological explanation for homosexual development range from those who suspect the presence of predisposing factors to others who are certain that prenatal "programming" decides the matter regardless of social circumstances.[54]

Genetic Factors   A hereditary basis for homosexuality has long been hypothesized but it was not until the mid-twentieth century that attempts were made for empirical confirmation. These studies have focused on issues like sex-ratios among siblings of homosexuals, their position in the birth order and studies of twins.

Kallmann reported a higher degree of *concordance* among identical (*monozygotic*) twins, than among fraternal (*dizygotic*) twins, or unrelated men.[55] On the face of it, this would point to a genetic linkage, except that Kallmann's identical twins were not reared apart; hence shared experiences rather than genes could as well explain the outcome. The same explanation may hold true for the fact that 25% of the brothers of homosexual men are reported to be also homosexual.[56]

Hormonal Factors   Given the crucial role of sex hormones in sexual differentiation, the possibility of their similarly influencing sexual orientation has attracted much attention.[57] According to some studies, homosexual men tend to have lower testosterone levels than heterosexual men; lesbians have higher levels of testosterone than heterosexual women; women whose mothers were injected with androgenic substances during pregnancy show a higher bisexual or homosexual potential; exclusively homosexual men, when injected with estrogen, respond metabolically in a female pattern. Other investigators have noted differences in physical structures, blood chemistry, and sleep patterns between homosexual and heterosexual men; likewise, lesbian and heterosexual women have been found to differ with respect to physical structures.[58]

Inconsistencies in methodology and outcomes make these studies inconclusive. For instance, instead of being higher, testosterone levels have also been found to be lower among homosexuals, while other

[53]Rechy (1963); (1977) provide insightful portrayals of the life of hustlers.

[54]For an annotated bibliography of studies of biological factors in homosexuality, see Weinberg and Bell (1972), references 1–67.

[55]Kallman (1952).
[56]Pillard et al (1982).
[57]Money (1980).
[58]Bell et al (1981).

studies have failed to find any significant difference between the testosterone levels of homosexual and heterosexual men.[59] Some of these differences may also be the results of homosexual life styles rather than the causes for it.

Brain Differences   In the most extreme view, a few investigators claim to have identified hypothalamic brain centers which govern "male" and "female" sexual response patterns. Based on animal studies, Dörner argues that just as androgenization determines male and female mating behavior in lower animals, a similar process operates among humans: the effect of androgens in prenatal life leads to the development of a sexual orientation toward females; the deficiency of prenatal androgens, or insensitivity of tissues to its effect, leads to a sexual orientation toward males, irrespective of the genetic sex of the individual.[60]

This prenatal theory of sexual orientation has been criticized on theoretical and empirical grounds. A major obstacle to such research at the experimental level is that the animal world does not have clear parallels to human sexual orientation, although same-sex sexual interactions occur in many species.

At the clinical level, the available evidence also fails to support the contention that hormonal exposure defines sexual orientation. For example, women with *congenital adrenal hyperplasia* who are exposed to excess androgen in prenatal and postnatal life develop heterosexual, not homosexual, orientations in the majority of cases. Similarly, *hypoandrogenized* males who are reared as males usually develop a heterosexual orientation. Sexual orientation thus seems to follow mainly the sex of rearing. However, because the available evidence in this area is limited, the present view among investigators is that the presence of biological factors in the genesis of sexual orientation cannot be dismissed.[61] This perspective has gained further support from the failure to demonstrate persuasively the existence of psychosocial etiological variables in the development of homosexuality, as we discuss next.

Evidence for Psychosocial Determinants

There is little dispute over the pervasive influence of psychosocial factors in the development of sexual orientation. The questions pertain to whether or not such factors are sufficient in themselves to explain why some people become heterosexual and others homosexual.

Psychoanalytic Views   Freud hypothesized that everyone is born with a *bisexual potential* and during psychosexual development children can develop either a heterosexual or a homosexual orientation depending on the way the oedipal conflict and related issues are resolved. Homosexuality can thus be thought of as a *fixation* at a stage of development or a *regression* to such a stage. Though Freud clearly considered heterosexual genitality the normative development (chapter 9), he was quite lenient in his view of homosexuality (box 13.2).

Based on these premises, various explanations have been elaborated by psychoanalysts as to why and how an individual becomes homosexual. For example, during the oedipal phase, *castration anxiety* is said to turn some boys away from the mother as a sexual object and subsequently from all women as well. Or, faced with a harsh and rejecting mother; the boy turns to the father (hence to men) for love and erotic satisfaction. The mechanisms proposed for the development of lesbianism are somewhat more convoluted but basically follow the same principles.[62] The fear and hatred of men, in particular, is commonly cited as to why some women turn to other women for sex and love.

---

[59]Tourney (1980).
[60]Dörner (1976).
[61]Ehrhardt and Meyer-Bahlburg (1981).

[62]Deutsch (1944). For further psychoanalytic perspectives on homosexual development, see Salzman (1968); Marmor (1965); Marmor (1980).

In an attempt to provide more systematic support for these claims, Bieber and a group of psychoanalysts studied 106 male homosexual patients and concluded that the most significant factors in the family constellation of these men are a detached and hostile father and a seductive and overly attached mother who dominates and depreciates her husband. Such parents inhibit the expression of masculine behavior in their boys. The domineering mother discourages her son's heterosexual impulses except when they are directed at herself, and she is jealous of any interest that her son shows in any other female. The boy whose father is aloof or openly hostile lacks a masculine figure with whom he can readily identify and whose behavior he can emulate. Hence, in later life the boy retains a fear of heterosexual relationships and a frustrated need for the masculine (paternal) love that he failed to receive as a child.[63]

Social Learning Approach   Though many behavioral scientists disagree with psychoanalytic constructs on methodological and conceptual grounds, they agree by and large that homosexual orientations emerge because of learning through social experience (chapter 9).

Sociological approaches have focused on *peer relationships* (presumed to be poor among homosexuals); the role of *fortuitous labeling* (people are called "homosexual" or treated as sexually different, hence they begin to act that way); the consequences of *atypical sexual experiences* (such as lack of opportunity to interact with members of the opposite sex); and the effect of *homosexual seduction* in childhood.[64]

An extensive investigation of homosexual development by Bell, Weinberg, and Hammersmith (involving 979 homosexual and 477 heterosexual men and women living in the San Francisco Bay Area in 1969–1970) has failed to substantiate the effects of any of the psychosocial causes referred to above or has found their influence to be far weaker than has been supposed.[65] Some of the explanations offered earlier (including Bieber's model) appear applicable to one homosexual subgroup or another, but no set of familial or other social factors have proved to have general explanatory value for the development of homosexuality as a whole.

These investigators followed a *path analytic* model whereby they traced the flow of possible causal influences in parental and familial associations through childhood and adolescent peer relationships and sexual experiences, to the final outcome of adult homosexuality or heterosexuality. Through this model they could see how one factor (such as a boy's lack of identification with the father) could lead to a dislike of traditional male activities and proceed along a chain of subsequent developments; or how that chain could be interrupted at any given stage. The main conclusions these investigators reached were as follows:

1. By the time boys and girls reach adolescence, their sexual preference is likely to be already determined, even though they may not yet have become sexually very active.
2. Homosexuality was indicated or reinforced by sexual feelings that typically occurred 3 years or so before the first "advanced" homosexual activity, and it was these feelings, more than homosexual activities, that appear to have been crucial in the development of adult homosexuality.
3. The homosexual men and women in the study were not particularly lacking in heterosexual experiences during their childhood and adolescent years. They are distinguished from their heterosexual counterparts, however, in finding such experiences ungratifying.
4. Among both the men and the women, there is a powerful link between gender nonconformity and the development of homosexuality.

---

[63]Bieber et al. (1962).
[64]Bell et al. (1981). For an extensive list of sources *see* Weinberg and Bell (1972).

[65]Bell et al. (1981).

5. Respondents' identification with their opposite-sex parent while they were growing up appears to have had no significant impact on whether they turned out to be homosexual or heterosexual.
6. For both male and female respondents, identification with the parent of the same sex appears to have had a relatively weak connection to the development of their sexual orientation.
7. For both the men and the women, poor relationships with fathers seem more important than whatever relationships they may have had with their mothers.
8. Insofar as differences can be identified between male and female psychosexual development, gender nonconformity appears somewhat more salient for males than for females, and family relationships are more salient for females than for males.

Bell and his associates did not set out to look for biological roots of homosexuality, but they were led to the conclusion that their findings were "not inconsistent with what one would expect to find, if indeed, there were a biological basis for sexual preference."[66] But that is far from saying that a biological basis for sexual preference has been established.

The Predictive Value of Childhood Experience   The research of Bell and his associates has two key findings that strongly link childhood characteristics to the development of a homosexual orientation. First, is the association of *gender nonconformity* and sexual orientation; second is the presence of *homosexual feelings* long before the occurrence of homosexual activity; such feelings are typically present about two years before any genital contact occurs with same-sex peers. These homosexual feelings are not so much sexual urges but romantic attachments and experiences of falling in love with another person of the same sex.[67]

These conclusions confirm earlier findings linking gender-nonconformity and homosexual orientation. For example, two-thirds of male homosexuals studied by Saghir and Robins recalled a "girl-like" syndrome during boyhood (including a preference for girls' toys and games, female playmates, and avoidance of rough-and-tumble play). Likewise, two-thirds of female homosexuals, but only 20% of heterosexuals, recalled "tomboyish" behavior in their preteens. Such behavior persisted into adolescence among more than half of the homosexual versus none of the heterosexual group.[68]

Among the subjects studied by Whitman, 47% of male homosexuals had been interested in doll play and 47% in cross dressing (none of the heterosexual men had been interested in either); 42% of homosexuals and 1.5% of heterosexuals preferred female playmates; 29% of homosexuals and 1.5% of heterosexuals were regarded as "sissies"; 80% of homosexuals preferred childhood "sex play" with other boys; about the same proportion of heterosexuals preferred "sex play" with girls.[69]

The association between gender behavior in childhood and sexual orientation can be explained in several ways. One is with respect to developmental continuity. A boy who acts more like a girl will grow up to choose a sexual partner as a woman would, namely a man; it has been frequently observed that boys whose best friends were girls in childhood are more likely to have male lovers in adulthood, and vice versa. This process may depend on the boy's identifying with girls and being socialized by them. Or boys who get attached to a female peer group may be socially stigmatized by other boys (and rejected by their fathers) hence, become "starved" for male affection, which they then remedy in adulthood by becoming homosexual.[70]

The association of gender nonconformity with a homosexual orientation in pre-

---

[66]Bell et al. (1981, p. 216).
[67]Bell et al. (1981).

[68]Saghir and Robins (1973).
[69]Whitman (1977).
[70]Green (1980).

adolescence may indicate that if there is a biological basis for one, it probably accounts for the other as well. This could also mean the variety of familial factors that have been suggested as explanations for homosexuality, may themselves be the result of the prehomosexual son or daughter being, in fact, "different." So it is not the "distant" father who "makes" a homosexual out of the son but rather the gender-atypical prehomosexual who "makes" the father distant.[71]

Most importantly, since gender identity and affectional ties seem to be the basic issues that define sexual orientation, it is these processes that we need to understand if we are to unravel the mystery of sexual object choice. Sex, as such, once

again must take the back seat in explaining how this crucial aspect of our sexual life is defined.

[71]Bell et al. (1981).

SUGGESTED READINGS

Marmor, J. (ed) *Homosexual Behavior* (1980). Multidisciplinary contributions representing the current mainstream of professional viewpoints on homosexuality.

Bell, A. P., et al. *Sexual Preference—Its Development in Men and Women* (1981). Exploration of the roots of the development of homosexual orientation.

Rechy, J. *City of Night* (1963) and *The Sexual Outlaw* (1977). Literary and autobiographical accounts of the world of the hustler.

Silverstein, C. *The Joy of Gay Sex* (1978); Sisley, E. L., and Harris, B. *The Joy of Lesbian Sex* (1977). Techniques for the enhancement of gay sexual interactions.

Rodgers, B. *Gay Talk* (1972). Fascinating insights into the gay subculture through its language.

Chapter 14

Paraphilias and Sexual Aggression

Homo sum; humani nihil a me alienum puto. (I am human; there is nothing human that is alien to me.)

Terence

The classification of mental and behavioral disorders subsumes disturbances related to sexuality under the heading of *psychosexual disorders.*[1] It includes three major categories: *gender identity disorders; paraphilias;* and *psychosexual dysfunctions.* In the first instance, the problem involves gender disorders (chapter 9); in the second, abnormal sexual behaviors, in the third, disturbances of sexual performance and satisfaction (chapter 15).

Sexual aggression is not part of this scheme. Coercive sexual behaviors such as rape are not diagnostic entities (just as theft and murder are not). This does not mean that such behaviors are healthy or desirable; but that they are symptomatic of other underlying problems and do not constitute discrete forms of psychological disturbance or mental illness in themselves. We include coercive sex in this chapter on the assumption that it has close similarities to the paraphilias where the underlying mo-

tivation is also hostility. The central theme of this chapter is therefore the association of sex with **aggression** and **hatred**—emotions that are just as critical to our understanding of sexuality as are love and affection.

Sex, Dominance, and Aggression

Aggression (Latin: to step forward) in one sense, denotes vigor and initiative; in another, it represents attack and hostility. Both meanings of aggression are applicable to sexuality, and the boundary between the two is often not clear. *Dominance,* or mastery, of some individuals over others generates social hierarchies. Dominance is achieved through aggression, or more often, the threat of aggression; like aggression, it is also closely linked to sexual interaction.

Sex and Dominance

In most animal species and human societies males are dominant over females. Within each sex, individuals rank them-

[1] American Psychiatric Association (1980).

354

selves with regard to one another as well in *dominance hierarchies*. An individual's position in such hierarchies is fairly stable but not fixed; it must be maintained through competition, threats, and continuous self-assertion.

Animals settle dominance disputes mainly through *threat gestures*. Different species have their characteristic threat gestures as well as special physical features that enhance such displays. Since primates have no antlers, claws, or other elaborate fighting tools, they convey threat by frontal displays and facial expressions. For instance, the *stare* (which often precedes attack) is intimidating while the shifting of the eyes indicates fear and submissiveness. It is hypothesized that the erect penis fulfills a similar secondary function. Monkeys and apes guarding territory may display erections, presumably to look more fierce. When primates use mounting for nonsexual purposes, it is usually to convey dominance (chapter 11). In fact, it is through such behavior that copulatory patterns are learned in childhood (see figures 9.12 and 9.13 in chapter 9).

The possible relevance of all this to humans is not certain though quite interesting. Facial hair indicates maturity and the flowing white beards of patriarchs command respect. The square chin (Dick Tracy style), broad shoulders, piercing stare, bushy eyebrows, are all part of the stereotypical image of the man who is dominant and virile, even though in reality none of these features have any bearing on either.

The male preoccupation with penis size may be a residue of its earlier threat potential. The human penis is larger than those of other primates and may have been selected during evolution because of its role in male-to-male competition. The *codpiece*, a pouch at the crotch of tight-fitting breeches, worn by men in the 15th and 16th centuries, made the genitals look prominent. Gourds have been used as *penis sheaths* and as artificial extensions of the penis by men in New Guinea as well as in early African and American cultures.[2] The

penis as a symbol of dominance also explains the peculiar headdress of certain Ethiopian village chiefs (figure 14.1). In many cultures, these practices are evident in covert form in various manifestations of **phallic symbolism.**

The association of status and sex appeal is patently clear. At least part of the power struggle between men is, at some level, in the service of enhancing their sexual potential.[3] An important reason why older men are still able to attract younger women is that they have power and status. In short, dominance status has a profound influence on sexual relationships. It influences who initiates sex, what form it takes, even the positions of coital partners, and many other related factors.[4]

In the broader social context of sexual interactions men and women compete for the most desirable partners and, within a given relationship, they try to obtain maximally the sexual satisfactions they want. These selfish aims are tempered by considerations of affection and the willingness to share with the sexual partner all that one has to offer.

Sexual associations in turn influence status. The more attractive a man's wife, mistress, or date, the more it reflects his worth. The same is true for women: the more attractive and socially desirable her man, the greater her status.

With the political and social emancipation of women, profound changes are taking place in redefining the association of sex and dominance. Many women are not willing anymore to be ornaments and testimonials to the social status of men. Yet the use of social dominance to obtain sexual aims and the use of sex to enhance one's social standing continue to be very much part of the pattern of our interpersonal relationships.

---

[2]Heider (1973); Wickler (1972).

[3]Henry Kissinger is quoted as saying, "Without an office, you have no power, and I love power because it attracts women." (*Washington Post,* January 7, 1977.)

[4]Blumstein and Schwartz (1983).

Figure 14.1 Phallic brow ornament made of metal worn by a southern Ethiopian as an insignia of rank. (From Wickler, W., *The Sexual Code*, 1972, p. 54.)

Sex and Hostility

We referred in chapter 8 to the use of colloquial terms for coitus such as "fuck" and "screw" to convey anger and hostility. There are just as eloquent nonverbal gestures which use phallic imagery to express sexual insults.[5] Though typically used by men, gestures like "giving the finger" (figure 14.2) may be used by women as well to convey the same meaning ("Fuck you!").

There are countless ways in which men and women hurt each other by using sex as the vehicle of their hostility. Usually, such interactions involve not physical violence but psychological devices like sexual indifference, deprecation, ridicule, and other means of injuring the partner's self-esteem. Such factors often underlie various forms of sexual dysfunction and are further discussed in that connection (chapter 15).

The relationship of sex and hostility operates both ways: Sexual frustration, dis-satisfaction, and jealousy provoke anger, whereas hostility based on other grounds is expressed through sexual unresponsiveness, sabotage, and violence. The closer we feel to someone, the more vulnerable we are to that person. Lack of reciprocation in love or sexual interest by the object of our affections is deeply frustrating. Real or imagined fickleness or infidelity generates intense jealousy and anger: 84% of all female murderers kill men and 87% of all female victims are murdered by men. One out of three murder victims are spouses, mistresses, lovers, or sexual rivals of the murderer. Seven percent of homicides follow a "lover's quarrel."[6]

There are other forms of physical violence (such as hitting, shoving, arm twisting, hair pulling) that seem to occur quite often in intimate relationships. The *battered-wife* has become the focus of much attention in recent years but such violence is by no means restricted to the marital situation or to males battering females (though women remain by far the more frequent victims). There are reports of *premarital abuse*, which on some college campuses involve over 60% of single respondents who report having been subjected to abusive or aggressive behavior or to have inflicted such abuse during courtship.[7] Such violence is more common in relationships that have a high level of involvement than those that are more casual. Ironically, such quarrels among lovers also convey a certain sense of caring; it is not unusual to precipitate an argument just to get a response out of the other person. When one deeply cares for someone, any show of emotion, even anger and abuse, may be preferable to indifference.

The seemingly pervasive presence of hostility in eroticism prompted Freud to argue that most men are driven to debase their partner (he called it "the most prevalent form of degradation in erotic life"). Robert Stoller, elaborating on this theme, proposes that *hostility* is the basis of all *sex-*

[5]Morris (1977) has numerous illustrations of such gestures.

[6]See Tennov (1979, p. 289) for sources.
[7]Athanasiou et al. (1970).

Figure 14.2   Phallic gesture: the middle-finger jerk.

ual excitement. The keys to our sexual inner life are our fantasies. These are the coded scripts by which our mind silently seeks to work out lingering problems from childhood and the requirements according to which we choose our sexual partners and behaviors. Since there is much hurt and frustration from our childhood, their replay and attempted resolution require a victim. The desire to hurt the sexual partner is in retaliation of having been hurt oneself and an attempt to obtain mastery over the experience by reliving it. During this process, says Stoller, "Triumph, rage, revenge, fear, anxiety, risk are all condensed with a complex buzz called 'sexual excitement.'" It should be clear that what is at issue here is the person's fantasy life; these urges need not, and often are not, carried out, and people can overcome their hostility in affectionate and caring relationships.

Paraphilias

In chapter 8, we defined paraphilias as sexual behaviors characterized by variant choices of *sexual object* as manifested in pedophilia, zoophilia, fetishism, necrophilia; and variant choices in the *means* of sexual gratification such as in voyeurism, exhibitionism, and sexual sadomasochism.

The most important feature of paraphilias is the reliance on unusual and bizarre imagery or acts for sexual excitement. Once arousal is achieved, the consummatory response is not unusual; the paraphiliac attains orgasm through masturbation or coitus like everyone else.[8] Paraphiliac fantasies and acts tend to be insistently and involuntarily *repetitive.* They usually involve either the use of nonhuman objects, the imposition or suffering of pain and humiliation, or sexual activities with nonconsenting partners.[9]

The *sexual imagery* of the paraphiliac, however odd, is not uncommon. For instance, many ordinary people have fantasies involving sadomasochistic activities or sex with animals (a theme elaborated in the art and literature of many cultures). Similarly, voyeuristic, exhibitionistic, and fetishistic activities in mild and socially acceptable form are part of our culture; the portrayal of seminudity in advertising, the erotic attraction of female lingerie or high-heeled shoes, and so on, are among the many examples. Most men and women who indulge in such fantasies and practices are not paraphiliacs because in their case, their activities do not constitute ends in themselves; they enhance rather than compete with their primary sexual aims. Thus the man who is turned on by frilly black panties is after the woman who wears them, not the panties themselves, and the woman who exhibits herself in them wants to arouse the man, not to shock him.

Robert Stoller divides the paraphilias (which he calls *sexual aberrations*) into two categories: *variants* and *perversions.* Both are erotic techniques that constitute discrete sexual acts (rather than being mere complements to other acts, like coitus) and as atypical entities depart from the culture's

---

[8]Beach (1977).
[9]American Psychiatric Association (1980).

avowed definitions of normality. Variants are aberrations that do not entail the staging of fantasies of harming others but may be engaged in out of curiosity or as a compensatory substitute; a perversion is a fantasy experienced as such or acted out, constituting "a habitual, preferred aberration necessary for one's full satisfaction, primarily motivated by hostility." It is an *erotic form of hatred*. Therefore, it is not the act as such but the hostile intent that marks it as a perversion.[10]

The prevalence and variety of paraphilias are greater among men than women. Especially if one looks at paraphiliacs who have run afoul of the law and become **sexual offenders**, then the overwhelming majority of cases will be seen to be male (chapter 19).

One explanation is that men have a stronger sexual drive, which spills over into paraphiliac activities. Or that men are socially less constrained than women hence more likely to act on their paraphiliac fantasies. If one accepts the premise that hostility is the basis of the paraphilias then the fact that males are generally more aggressive than females, offers a plausible linkage.

It is hard to speak of a "relationship" in paraphiliac interactions. For a man pursuing such aims, the woman is merely an erotic vehicle, not a sexual partner: she is like a mannequin, modeling the boots and the underwear, the arm that wields the whip, the buttocks that receive the blows. The woman's own needs, preferences are irrelevant; she is there to enact a role.

Given this dehumanizing context and the odd expectations that are entailed, it is hard for paraphiliacs to find ordinary women to interact with; hence, they must rely on prostitutes to stage their fantasies of "kinky sex," or impose them on unwilling victims. As a result, the paraphilias are disapproved socially with varying degrees of intensity (chapters 8 and 19).

Pedophilia

The *pedophiliac* or *pedophile* (Greek: lover of boys) is attracted to and engages in sex-ual activity with children of either sex as a preferred or exclusive form of erotic gratification. To differentiate it from sexual activities of children with each other, the diagnosis of pedophilia requires that the offender be at least 10 years older than the child.[11] In chapter 9 we discussed these behaviors from the perspective of the child; here we shall approach it from that of the adult.

In most cultures, prepubescent children are excluded as sexual partners. But after sexual maturity is attained, when a person is given adult status is quite arbitrary. The laws of some states define child molesting and statutory rape to include sex with persons as old as 16, 18, or even 21 (chapter 19). To place such judgments on a more reasonable basis, sex researchers define a *child* as younger than 12 years old; a *minor* as between 13 and 15 years old; and an *adult* as more than 16 years old.[12]

Though there may be good reasons why an adult should not engage in sex with those who are legally underage, only sexual encounters with children or minors can be reasonably construed as manifestations of a paraphilia. As with other paraphilias, much of our knowledge regarding pedophilia comes from studies of convicted offenders.[13] Most of these pedophiliacs are male and are more commonly interested in prepubescent girls (usually 8 to 10 years old) than in prepubescent boys (who are somewhat older).

Heterosexual Pedophiles   Though often pictured as a stranger who lurks about the school playground seducing or abducting unsuspecting little girls, in about 85% of cases the pedophile is a relative, family friend, neighbor, or acquaintance. He makes advances to the child either in her

---

[10]Stoller (1975).

[11]American Psychiatric Association (1980).

[12]Gebhard et al. (1965). Some of the famous lovers in history may strike us as shockingly young: Romeo and Juliet were teenagers; Dante fell in love with Beatrice when she was 9; Petrarch with Laura when she was 12.

[13]A study of 1,500 convicted sex offenders and two control groups undertaken by the staff of Institute for Sex Research in the 1960s, remains a major source in this area (Gebhard et al, 1965).

home, where he is visiting or living (as an uncle, stepfather, grandfather, or boarder), or in his own home, where the child is used to visiting him or is enticed by promises of various treats. Seventy-nine percent of such contacts occur in homes, 13% in public places, and 8% in cars.

The sexual encounter with the child is often quite brief, though there may be a series of such episodes; it is unusual for a prolonged or intimate relationship to develop. Sexual contact most often consists of the man taking the child on his lap and fondling her genitals. Intercourse is rarely attempted because, among other factors, these men are often impotent. (Coitus is attempted in about 6% but intromission is achieved in only about 2% of such cases.) Physical harm to the molested child occurs in about 2% of the instances, although threats of force or some degree of physical restraint is present in about one-third of cases where the person is convicted.

Heterosexual offenders against children tend to be older than any other group of sex offenders, with the exception of those involved in incest. The average age at conviction is 35, and 25% are over 45. Many of them seem immature and relate poorly to adult women, about 5% are senile, and 15 to 20% mentally retarded.[14] Three quarters of these men have been married at some time in their lives. They profess to be conservative and moralistic. Some require alcohol before they can commit the offense, others claim alcohol as an excuse. They usually do not have criminal records, other than previous convictions for molesting children.

Homosexual Pedophiles   The homosexual pedophile ("chicken hawk") sexually abuses young and teenage boys. The activities include fondling and masturbation of the boy, mutual masturbation, fellatio, and sometimes anal intercourse. These men are scorned by the majority of homosexuals who confine their sexual activities to other adults.

Homosexual pedophiles do not usually molest strangers, unless they are boy prostitutes. The children with whom they become involved are most often relatives or the sons of acquaintances. In addition, contacts are sometimes made through youth organizations, which they infiltrate as counselors.

Homosexual offenders against children generally show serious deficiencies in socialization and interpersonal relationships. They often say that they prefer the company of boys because they feel uneasy around adults. Their average age at the time of the offense is about 30 years. Only 16% are married at the time, and their sexual experiences have usually been predominantly, but not exclusively, homosexual.

The child molester, especially when he uses force, arouses tremendous aversion and anger in the community. Even in jail, he is so resented by other inmates that he must be secluded for his own safety. Some pedophiles are quite dangerous and torture and kill their victims. Partly based on this threat, the law treats pedophilia as a very serious offense (chapter 19).

Incestuous Pedophiles   From a social perspective, *incest* is a particularly disturbing form of sexual behavior. There is nothing peculiar about the physical aspects of coitus between close relatives, and incestuous relations among adults do not, as such, constitute a paraphilia, however socially disapproved or undesirable such relationships may be (chapter 16).

Incest is considered a paraphilia when it involves children and adolescents. We have already dealt with this issue from the child's perspective (chapter 9); now we consider it from that of the adult perpetrator of incest.

Prevalence figures on incest vary widely from a few cases per million a year based on men who are convicted of the offense, to 33% among one sample of female psychiatric patients.[15] Sex between cousins is

[14]From a psychiatric perspective, isolated acts of sex with children and such behavior accompanying serious mental incompetence would not warrant the diagnosis of pedophilia.

[15]Rosenfeld (1979a). The recollections of psychiatric patients must be interpreted with particular caution since fears, wishes, or fantasies of childhood sexual experience may be confused with the reality of such occurrences.

more common than among siblings; same-generational or sibling incest is far more common than cross-generational sex between parents and children; father-daughter incest is by far more common than mother-son incest.[16] In a study of court cases of incest 90% involved fathers and daughters, stepfathers and stepdaughters, grandfathers and granddaughters; and 5% involved fathers and sons.[17] Only about one in 100 cases of reported incest involve mothers and sons.[18] Mothers who are participants in these acts that come to public attention tend to be much more disturbed psychologically (often psychotic) than is the case with father-daughter incest. Incest is presumed to be more common among the socio-economic lower classes living in overcrowded and chaotic social conditions and in isolated communities, but it also occurs (though it is less often reported) in middle-class or upper-class families.[19]

Much of our information on incest comes from convicted offenders, by no means a representative group. The only data on offenders of incest laws available from the Gebhard study are on father-daughter relationships. In these cases, the ages of the daughters at the time of the offense were correlated with certain characteristics of the fathers. Men who became sexually involved with prepubertal daughters tended to be ineffectual persons who drank excessively. The sexual contacts involved either fondling the external genitals or oral-genital contacts. Offenders with sexually mature daughters tended to be rigidly moralistic persons who were poorly educated and had a median age of 46 at the time of the offense. Sexual interaction with older daughters culminated in coitus in the majority of instances.[20]

Incest usually reflects the presence of serious *family pathology*.[21] Typically, the marriage is an unhappy match between a dominating and impulsive man and a weak and passive woman, neither of whom is able to function adequately either as spouse or as parent. The involved daughter is fearful and submissive to the father's wishes to avoid further family dissension.[22] A conspiracy of silence on the part of all concerned allows the incestuous arrangement to persist (box 14.1 presents examples of such interactions). Charges are pressed usually by the wife when some new development disrupts the compromise, though some women report it as soon as they become aware of it. Or the daughter may rebel against the incestuous relationship as she gets older.[23]

Dealing with incestuous families is difficult. Particularly where younger children are involved, the problem must be handled with discretion and sensitivity (chapter 9). Sometimes there is no choice but to remove the father from the family, but generally a therapeutic approach to the family as a whole may work better than heavy-handed punitive measures directed at the offender alone. However indignant one may be at a father, his incarceration may result in so much guilt and hardship for the child and the rest of the family, that it may be better for him to continue living at home, provided sufficient safeguards are taken to protect the child and help resolve the issues that make the family dysfunctional.[24]

Zoophilia

Sexual contact with animals, *zoophilia* or *bestiality*, is defined as a paraphilia when the act or fantasy of engaging in sexual activity with animals is a repeatedly preferred or exclusive method of obtaining sexual arousal and gratification. Otherwise, fantasies of sex with animals are quite common. Recourse to animals, when other outlets are not available or by way of occasional experimentation, would not also as such be considered a paraphilia.

[16]Finkelhor (1979). For a report of a rare case of mother-son incest, see Shengold (1980).

[17]Maisch (1973).

[18]Weinberg (1955).

[19]Hunt (1974); Butler (1978).

[20]Gebhard (1965).

[21]Rosenfeld (1979b).

[22]Selby et al (1980).

[23]Finkelhor (1979).

[24]For further discussion and less common forms of incest, see Meiselman (1978).

Box 14.1   Father-Daughter Incest[1]

Marcia's father began genital manipulation and oral-genital relations with her when she was five years old and continued these activities on a regular basis for nearly two years. She thinks that she must have known there was something wrong with the relationship because she was afraid that her mother would find out about it. The father, who was alcoholic and amoral to the point of bringing prostitutes home with him from time to time, once insisted that she go to bed with him while her mother was in the same room. When he asked her to get under the covers with him, her mother laughed and thought it was "very cute" to see them in bed together. While her mother stood by, greatly amused, her father began to fondle her, and she became so disturbed by the situation that she ran from the room in tears. Her mother never did realize that incest was occurring. However, an older sister who had also had an incest affair with the father recognized the clues to incest and ordered the father to stop it, threatening him with disclosure to legal authorities.

*

Daughter (raising voice): You don't see how we could have done it?

Mother: No, un-unh. No.

Daughter (angrily): We went to the dump! We went out into the sticks! Right out there in the cow pasture! OK, you went away! Everybody was away from the house! We've had it in your bed! We've had it in my bed! We've had it in the bathroom on the floor!

(Mother utters a loud moan.)

Daughter: We've had it down in the basement! In my bedroom down there, and also in the furnace room!

Mother (shakily): I just can't believe it, I just can't, just can't . . .

Daughter: Mom. . .

Mother (with a trembling voice): Just can't see how anything like this could possibly happen and how you could treat me this way.

Daughter: Because . . .

Mother (shouting and weeping): After all I've done for you! I've tried to be a mother to you, I've tried to be a respectable mother, and you accuse your father of something so horrible, that's . . .

Daughter (shouting): Mother, it's true! You've got to believe it . . .

Mother: She's my daughter, and I love her, but I cannot (to therapist) believe this!

*

Pamela was molested by her stepfather over a period of several months, and although she disliked being touched by him she did not tell her mother immediately because she "didn't want to upset Mom." When she did tell, the mother insisted that her husband leave the home and moved swiftly to get a legal injunction to keep him from harassing them. Since he was unemployed and had been living on her earnings, she was in a position of considerable power in the home. When seen in therapy, the daughter seemed to idolize her mother and spoke of her as if she were a movie star, and the mother was relieved that the marriage had ended.

*

Una was ten years old when she was raped by her father while he was acutely psychotic. Her mother arranged for his hospitalization and provided comfort and emotional support to her daughter, carefully explaining to her that her father was not himself when the rape occurred. She credited her mother with helping her to understand and forgive her father, thus diminishing the traumatic effect of the incident.

[1]From Meiselman, K.C., *Incest*, Jossey-Bass, 1978, pp. 169.

The animals most often used are those found on farms and in homes as pets. Cows and sheep are the animals used by men, while dogs are made to service women. Other animals, like donkeys and horses, have been used by both sexes, in the case of women usually in the form of sexual exhibitions. The more common activities are masturbating the animal, inducing it to perform mouth-genital stimulation, and where feasible, engaging in coitus.

Animal contacts were by far the least prevalent of the six components of total sexual outlet studied by Kinsey. Even though 8% of adult males and 3% of females reported such contacts with animals, their activities accounted for a fraction of 1% of the total sexual outlets. In some groups, however, such behavior was relatively common: Among boys reared on farms, as many as 17% had had at least one orgasm through animal contact after puberty.

Humans have a long history of intimate associations with animals in many aspects of life, including the sexual (figure 14.3). This theme has been an important source of fantasy and an inspiration for many works of art. Classical mythology abounds with tales of sexual contacts between the gods disguised as beasts and apparently unsuspecting goddesses and mortals (chapter 17). Zeus, for instance, ap-proached Europa as a bull, Leda as a swan (figure 14.4), and Persephone as a serpent.

Historical references to zoophilia are also plentiful. Herodotus mentions the goats of the Egyptian temple at Mendes, which were specially trained for copulating with human beings. More often, we find that references to animals as sexual objects are framed as prohibitions. The Hittite code, the Old Testament, and the Talmud specifically prohibit such contacts for males; and the Talmud extends the prohibition to females. Penalties for such sexual contacts tended to be severe, often mandating death of both human and animal participants (Leviticus 20). Such sanctions remained in effect in Europe throughout the Middle Ages and, indeed, until fairly recently. In 1944, an American soldier was convicted at a general court-martial of sodomy with a cow and sentenced to a dishonorable discharge and 3 years at hard labor.

Fetishism

In fetishism the preferred or exclusive source of sexual arousal is an *inanimate object* or *body part*. Anthropologists call *fetish* those objects that were believed by preliterate cultures to have magical powers. In a more general sense, a fetish is any article that is valued beyond its intrinsic worth because of superstitious or other meanings ascribed to it. The term **erotic fetishism** was coined by Alfred Binet, the pioneer of intelligence tests.

The boundary between ordinary erotogenic objects or parts of the body and fetishes is quite nebulous. Many heteorsexual males, for instance, are aroused by the sight of female underwear and such underwear is often designed for that purpose. Most people are also partial to various parts of the anatomy of the opposite sex besides the genitals. Heterosexual men are typically fascinated by female breasts and legs; many women are partial to slim male buttocks and broad shoulders. Blond female hair and dark male eyes seem to have special erotic attraction in our society. Common and individual preferences with respect to such features are a normal

Figure 14.3 Petroglyph in a cave in northern Italy. (From Anati, *Camonica Valley*, 1961, p. 128.)

Figure 14.4 *Leda and the Swan* (after Michelangelo). (Courtesy of the National Gallery, London.)

part of cultural patterns of sexual preference. It is therefore not very meaningful to say that American men have a "breast fetish." One can, however, consider a man to be a breast fetishist if his interest in breasts supersedes his attraction to the woman who has the breasts.

Fetishes are often symbolic substitutes for the genital organs but the genitals themselves can be thought of as fetishes if they get in the way of perceiving the sexual partner in his or her totality. The preoccupation with the penis seen among some gay men may thus represent a form of fetishism. Hoffman associates this with the dehumanized sexual interactions that take place in public places, like toilets, where the focus of interest is the penis and little else.[25] Psychoanalysts may explain such phallic preoccupation as a defence against *castration anxiety.*

Physical defects, such as a missing limb or cross-eyes, may become the object of fetishistic fixation. To be sexually attracted to an amputee or someone with a physical deformity is not as such a sign of fetishism, but to be attracted to such persons because of their defects may be fetishistically motivated.

In principle, any article or body part can become endowed with fetishistic meaning, but some objects are more commonly chosen than others. The attraction may arise from the shape, texture, color, or smell of the article, or from a combination of these features. Most fetishes fall into two categories: *soft fetishes* are "feminine" objects that are soft, furry, lacy, pastel-colored (like panties, stockings, and garters); *hard fetishes* are "masculine" objects that are

[25]Hoffman (1968).

Figure 14.5 *The Damned* (1969). Male actor imitating Marlene Dietrich's classic pose in *The Blue Angel,* displaying garter, black stockings, and shoes. (Courtesy of the Museum of Modern Art Stills Archive.)

Figure 14.6 Jack Lemmon and Tony Curtis in *Some Like it Hot* (1959). (Courtesy of the Museum of Modern Art Stills Archive.)

smooth, harsh, or black (like spike-heeled shoes, gloves, and other garments and objects made of leather or rubber) (figure 14.5). The latter category is more often associated with sadomasochistic fantasies and practices.

Fetishes are sexual symbols with both shared and private meanings. For instance, high-heeled women's shoes are regarded as sexy in our culture; but the shoe fetishist will in addition be attracted to particular types of shoes and may collect them in staggering numbers, like a substitute harem.[26] Others may do the same with gloves or some other item. The fetishist uses these objects for sexual arousal either in conjunction with some other sexual activity or to masturbate with.

Transvestism

Transvestism (Latin: *trans,* across; *vestia,* dress) or *cross-dressing* is the recurrent and persistent practice, usually on the part of males, to dress in the clothing of the op-

[26]Stekel (1964).

posite sex for purposes of sexual excitement. In a society with *unisex* clothing, this condition would presumably not exist. But in our society, and most other cultures, male and female garb is distinctive, although the degree of overlap varies. Women have more latitude in what they wear than men do (pants and jackets are, for instance, worn by many women, but American men do not wear skirts). It is also not uncommon for males to dress up as females in playful situations, dramatic skits, and comedies (figure 14.6).

There are other situations where wearing female clothing by men would also not qualify as transvestism. As discussed earlier, transsexualism is one instance. Homosexual men who dress up as women to attract other men, or who masquerade in "drag" for reasons other than sexual arousal are not transvestites; nor are female impersonators who act this way to earn a living. To qualify as transvestism, there must be a compulsion to cross-dress and intense frustration when the practice is interfered with.

Most transvestites, in the more strict sense of the term, are heterosexual males. Their cross-dressing may be *partial*, in which case they are only attracted to certain items of female clothing. These men are hard to separate from cloth fetishists. They may wear, masturbate with, or incorporate the article of female clothing (panties, stockings, bras, and so on) in coital interactions or paraphilias such as sadomasochism. The *complete* cross-dresser puts on female garments either as an occasional activity or all the time. Some of these men carry on this practice on their own, others participate actively in the transvestite subculture.[27]

Cross-dressing typically begins in childhood or early adolescence. The majority of children who play at it do not become transvestites, or obtain erotic gratification from it. Those who do, usually begin to cross-dress in public after they reach adulthood.[28] The wives of transvestites may learn to live with their husbands' cross-dressing habits or even assist them in their practice. Others are distressed by the situation and seek help or break the relationship.[29] Female transvestism is very rare. For whatever other reasons women may choose to wear men's clothing, apparently few do so for the purpose of sexual arousal.[30]

Necrophilia

The most bizarre choice of sexual object is the body of a dead person. The "love" or sexual use of corpses known as *necrophilia* is a very rare entity usually involving psychotic men. In a highly unusual case, a 23-year-old woman who worked as an apprentice embalmer, was convicted in 1982 of sexually molesting the corpse of a 33-year old man. She admitted to have had sexual contact with up to 40 other bodies. A letter that she left with the corpse after her last episode read, "I've written this with what's left of my broken heart. If you read this, don't hate me. I was once like you. I laughed, I loved, but something went wrong. I give it all to know. But please remember me as I was, not as I am now."[31]

There are many references in art and literature to love and death, but these are usually not intended to be linked in an erotic sense. The work of some authors, nonetheless, has an aura of eroticizing death, as in Edgar Allan Poe's lines:

> I could not love except where death
> was mingling his with beauty's
> breath.

Voyeurism

It is said that Leofric, lord of Coventry in the eleventh century, agreed to remit an oppressive tax if his wife, Lady Godiva, would ride through the town naked on a white horse. Lady Godiva, a benefactress of monasteries and of the poor, consented. Out of respect and gratitude everyone in town stayed behind shuttered windows during her ride—everyone except Tom the tailor, who peeped and went blind: hence the name *Peeping Tom*.

The sight of women who are scantily dressed, inadvertently exposed, or in the process of undressing is erotically arousing to many men. Although women have traditionally been assumed not to have similar interests, many more now report that they too like to look at nude or partially clad male bodies. Hence, along with typically male pastimes like "girl watching," "girlie" magazines, burlesque shows, and topless dancing, there are now women's magazines which feature nude male photographs and male strip shows for women. Nonetheless, the nature of such female behavior is a bit uncertain. The circulation of men's magazines like *Playboy* is very much larger than those of its coun-

[27]In Southern California, there is an organization called *Chic* which consists exclusively of heterosexual cross-dressers. Its members are mostly professional men who gather for dinner and socializing while elaborately dressed and made-up as women, many of them accompanied by their wives.

[28]Croughan et al. (1981); Wise (1982).

[29]Wise et al. (1981).

[30]Stoller (1982).

[31]*San Francisco Chronicle*, April 25, 1982, p. 2.

Figure 14.7 Chinese bathing scene (18th century?). Artist unknown.

Figure 14.8 Thomas Rowlandson (1756–1827). Untitled.

terparts like *Playgirl*. Furthermore, it is said that the number of women who read *Playboy* are far fewer than the number of gay men who read it.

Such behavior is loosely called *voyeurism* (French: to look) but it does not constitute a paraphilia. To qualify as such, the voyeuristic behavior must be repetitive and compulsive; the person being observed must be unaware of it, and the person engaging in the behavior should have no further interest or contact with the person being watched.[32]

By these more stringent criteria, virtually all voyeurs are men who secretly watch women undressing, bathing, in the nude, or engaged in sex (figures 14.7 and 14.8). They typically masturbate during or following the act; as a rule they do not physically molest their victim. Many a man who happens to come upon such a scene will not avert his gaze; but the voyeur seeks such occasions. He will not only look at an open window from the sidewalk but will cross the lawn and hide in the bushes to do so. Part of the excitement of the act comes from the element of danger and the satisfaction derived from the fact that the

woman would be frightened and angry should she realize her privacy is being violated.[33] This is why the typical voyeur is not interested in ogling his own wife, female friend, or even strangers in a nudist colony.

Voyeurs tend to be younger men (the average age at first conviction is 23.8 years).[34] Two-thirds of them are unmarried, one-fourth are married, and the rest are either divorced, widowed, or separated. Very few show evidence of serious mental disorders, and alcohol or drugs are usually not involved in their deviant behavior: only 16% of the Gebhard sample was drunk at the time of the offense; none were under the influence of drugs. In intelligence, occupational choice, and family

[32]American Psychiatric Association (1980, p. 273).

[33]"It's as bad as being raped . . . At least when you are raped you know who your attacker is" was said by a woman whose male coworkers had drilled a hole in the wall of the women's bathroom in a West Virginia coal mine (*San Francisco Chronicle*, July 2, 1981, p. 7).

[34]Gebhard et al. (1965). Also see Stoller (1977).

background, "peepers" tend to be a heterogeneous group whose single most common characteristic is a history of highly deficient heterosexual relationships. Most voyeurs do not have serious criminal records, but many have histories of misdemeanors.

Exhibitionism

To expose or *exhibit* (Latin: to hold out) one's body or genitals for purposes of erotic stimulation is a common component in sexual interaction. The term *exhibitionistic* is also applied to a variety of behaviors whereby the person dresses or acts in ways that are sexually provocative. A great deal of exhibitionistic behavior is commercialized. This occurs most flagrantly through pornographic shows, books, and movies, but there is also a great deal of exposure, specially of the female body, in advertising.

Exhibitionism (also called "indecent exposure") as a paraphilia has a much more restricted meaning. It represents in many ways the mirror image of voyeurism. Instead of peeping, the exhibitionist ("flasher," "flagwaver") obtains sexual gratification from exposing his genitals to women or children who are involuntary observers, usually complete strangers (figure 14.9).

In a typical sequence the exhibitionist suddenly confronts a woman with his genitals exposed. He usually has an erection and may openly masturbate. He may wear a coat and expose himself periodically while riding on a subway or bus, or he may stand in a park and pretend that he is urinating. The exhibitionist does not usually molest or seek further sexual contact with the victim. His gratification comes from observing the victim's surprise, fear, or disgust; women who keep calm foil his intent. There is a compulsive quality to the exhibitionist's behavior that may be triggered by states of excitement, fear, restlessness, and sexual arousal. Once in this mood, he is driven to find relief.

The average age of the exhibitionist at conviction is 30 years. About 30% of ar-

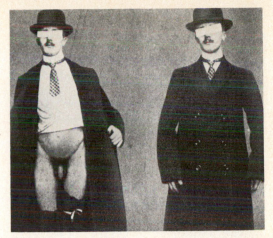

Figure 14.9 Exhibitionist. (Photograph from *Bilder Lexicon.*)

rested exhibitionists are married, another 30% are separated, divorced, or widowed, and 40% have never been married. Despite previous arrests and convictions, exhibitionists tend to repeat their behavior. One-third of the offenders in this category in the Gebhard study had four to six previous convictions, and 10% had been convicted seven or more times.[35] Exhibitionists do not usually show signs of severe mental disorder, and alcohol and drugs are involved only in rare instances.[36]

Obscene Calls The shock value of sex is also used in many other ways. Off-color humor or *dirty jokes*, sexual innuendo and the like that are meant to embarrass can represent subtle forms of exhibitionism. More flagrant is the practice of *obscene telephone calls*. The caller is usually a sexually inadequate man who tries to have an erotically charged interchange with a woman through this apparently safe method (progressively less safe, as detection mechanisms are improved). His pleasure is derived from eliciting embarrassment as his intent becomes clear to the women.

There are several types of obscene call-

[35]Gebhard et al (1965). For a review of the literature see Gaylord (1981).

[36]For discussion of the etiology of exhibitionism see Blair and Langon (1981).

ers. Some will talk about their own sexual experiences, others will seek sexual information by pretending to be conducting surveys for underwear manufacturers, or carrying out a sex study (an amazing number of women fall for this ploy). Telephone companies recommend that the recipient of an obscene call remain calm and either hang up immediately or, if possible, alert the operator by means of another line, to trace the call and notify the police.

The telephone has also been put to the opposite purpose of titillating callers. Like the Dial-It service for time and the weather, one can now listen to women delivering rauchy recorded "messages." One such number in New York City attracts 500,000 callers who earn the producers $10,000 and the telephone company $25,000 a day.[37]

[37]*Time,* May 9, 1983, p. 39.

Figure 14.10 Flagellation scene, c. 1770. Etching from Caylus, *Therese Philosophe.*

Sadomasochism (S-M)

The infliction of mild pain through scratching, pinching, slapping, and so on forms part of the sexual repertory of sexually adventurous couples. Teeth marks and purplish bruises ("hickies") about the neck, caused by passionate kissing, are classical signs of ardent lovemaking.

Defined as paraphilia, *sexual sadism* entails the infliction of physical of psychological suffering on another person; *sexual masochism* is the enjoyment of such suffering for erotic excitement. These themes have been widely explored in art (figures 14.10 and 14.11) and literature. The terms *sadism* and *masochism* are derived from the names of two historical figures who wrote on these respective themes. The first is the nobleman Donatien Alphonse François, marquis de Sade (1740–1814); the second, the Austrian author, Leopold von Sacher-Masoch (1836-1895). The works, as well as the biographies, of both men are widely known. A sample of de Sade's work (box 14.2) makes clear the basis of his reputation (discussed further in chapter 18).[38]

[38]For a summary of their lives, see Ellis, H. (1942), vol. 1, part 2, pp. 116–119.

Figure 14.11 Mauron, *The Cully Flaug'd.* Eighteenth-century illustration for *Fanny Hill.*

Box 14.2   The Literature of Sadomasochism[1]

Upon the spot La Rose opens a closet and draws out a cross made of gnarled, thorny, spiny wood. 'Tis thereon the infamous debauchee wishes to place me, but by means of what episode will he improve his cruel punishment? Before attaching me, Cardoville inserts into my behind a silver-colored ball the size of an egg; he lubricates it and drives it home: it disappears. Immediately it is in my body and I feel it enlarge and begin to burn; without heeding my complaints, I am lashed securely to this thorn-studded frame; Cardoville penetrates as he fastens himself to me: He presses my back, my flanks, my buttocks on the protuberances upon which they are suspended. Julien fits himself into Cardoville; obliged to bear the weight of these two bodies, and having nothing to support myself but these accursed knots and knurs which gouge into my flesh, you may easily conceive what I suffered; the more I thrust up against those who press down upon me, the more I am driven upon the irregularities which stab and lacerate me. Meanwhile the terrible globe has worked its way deep into my bowels and is cramping them, burning them, tearing them; I scream again and again: no words exist which can de-

scribe what I am undergoing; all the same and all the while, my murderer frolics joyfully, his mouth glued to mine, he seems to inhale my pain in order that it may magnify his pleasures: his intoxication is not to be rendered; but, as in his friend's instance, he feels his force about to desert him, and like Saint-Florent wants to taste everything before they are gone entirely. I am turned over again, am made to eject the ardent sphere, and it is set to producing in the vagina itself, the same conflagration it ignited in the place whence it has just been flushed; the ball enters, sears, scorches the matrix of its depths; I am not spared, they fasten me belly-down upon the perfidious cross, and far more delicate parts of me are exposed to molestation by the thorny excrescences awaiting them. Cardoville penetrates into the forbidden passage; he perforates it while another enjoys him in similar wise: and at last delirum holds my persecutor in its grasp, his appalling shrieks announce the crime's completion; I am inundated, then untied.

[1]From *Justine,* by the marquis de Sade. Grove Press, 1965, p. 732. Reprinted by permission of Grove Press, Inc. Translated from the French by Austryn Wainhouse & Richard Seaver. Copyright © 1966 by Austryn Wainhouse and Richard Seaver.

Sadomasochism demonstrates better than any other paraphilia the intrusion of *dominance* and *aggression* into sexual interactions. The role of dominance is most clear in the practices of *bondage and domination* (B-D) where a nonconsenting victim or a willing collaborator is bound, gagged, and immobilized, hence at the mercy of the *master* or *mistress* ("dominatrix"). It is not uncommon for these activities to utilize fetishistic trappings like black leather garments and high-heeled shoes, along with the classic chains, shackles, harnesses, and other paraphernalia.

The use of aggression is manifest in the verbal humiliations dished out as well as the whippings administered to the victim.

These may range from perfunctory blows delivered mainly for their symbolic effect, to more severe flagellation that raises welts and draws blood. Rarely, more extensive sadomasochistic practices are carried to the point of causing mutilation and death.

S-M activities primarily involve males. Women may be excited by them but mostly do it to please their male partners, for money, or as involuntary victims. The practice may be more common and more violent among gay men probably because the aggressivity of a male interacting with a male compounds violence.

Prevalence figures on these practices are hard to get, and given the wide variety of behaviors, one would not know what to count. A great deal of violence pervades our culture, some of which is clearly sexualized. Rock groups, fashion photographers, and other followers of S-M chic are part of the mainstream of society. Couples who buy S-M paraphernalia at sex shops and stage bondage shows in their bedrooms are playing at S-M, not practicing it.

Those who are more earnest (most of whom are male) use prostitutes or go to special S-M establishments to experience bondage, humiliation, and some pain; but they are not interested in getting really hurt. A substantial proportion of these men are said to be affluent professionals and executives who seek safe ways of fulfilling their masochistic needs. The Chateau in Los Angeles reportedly offers over 1,000 affluent clients the personalized services of 13 women (6 dominants, 3 submissives, 4 switch-hitters). These mistresses consider themselves "sex therapists" of sorts who provide men mainly with the "mental trips" of sadomasochistic excitement in a safe and "loving" setting.[39]

Many more men obtain their enjoyment through reading specialized magazines featuring photographs and cartoons of women dressed in black leather with spiked-heeled shoes trying to look menacing as they gag, bind, whip, and variously "torture" their victims. There is no evidence that the readers of such material are particularly prone to carry out their violent fantasies on unwilling others.

Finally, there are those who really mean business. Serious injury may result from their practices by miscalculation, getting carried away, or because of pure viciousness. Whereas the participants of the former practices are voluntary, the true sadists in this category will inflict pain on involuntary victims who fall in their hands. The pathology underlying such behavior goes deeper than the paraphilias of sexual sadism.

Many explanations have been offered for why men and women court pain and punishment in their sexual interactions. A common answer is that sex makes them

feel guilty. Therefore, the suffering, either before or after sexual pleasure, expiates for this sense of guilt and makes the experience more tolerable.

Atypical Paraphilias

A number of paraphilias that cannot be readily classified with other categories are subsumed under this heading. Several of them are closely linked to fetishism whereby instead of body parts, the discharges of the body acquire erotic significance. They include erotic fascination with feces (*coprophilia*), enemas (*klismaphilia*), urine (*urophilia*), and filth (*mysophilia*).[40] These stimuli may be used as excitatory aids in connection with some other sexual activity or as erotic ends in and of themselves.

These paraphilias often serve a sadomasochistic function. For example, during rapes (especially gang rapes) the assailants may urinate and defecate on the victim as an additional means of degradation. Ejaculation itself may become a means of *soiling*, which is why semen may be spurted at the victim's face, hair, and body.

Alternatively, some paraphiliacs obtain enjoyment out of being urinated on ("golden showers"). Others sniff dirty socks, lick unwashed feet, drink urine and taste feces.[41]

There are certain behaviors where the sexual intent is more concealed. In *pyromania*, the person setting fire may actually feel aroused and masturbate. In *kleptomania*, where the person steals for the excitement of the act, the stolen objects may have sexual significance in a symbolic sense. In principle, virtually any act may become eroticized to constitute a paraphilia.

Promiscuity

In discussing recreational sex (chapter 12), we raised the possibility that the indiscriminate choice of sexual partners may reflect neurotic conflicts and hence represent a form of psychosexual disturbance.

The term **promiscuity** (Latin: being

---

[39]*Time*, May 4, 1981, p. 73.

[40]Money (1980, p. 82) has a longer list

[41]Krafft-Ebing (1978) provides numerous case histories.

mixed) is currently in disfavor because of its vagueness and judgmental connotations. Nor is it recognized formally as a paraphilia. Yet, properly defined, it has some features that may link it to the behaviors discussed in this chapter.

To determine the meaning of such behaviors one must ask why a person seeks a succession of sexual partners or multiple partners simultaneously. Lorna and Philip Sarrel differentiate between "sleeping around," as a phase that increasing numbers of young people are going through, from promiscuity that entails *indiscriminate* and *compulsive* seeking of transient sexual encounters that have no relational or affectionate context.[42] The former, which serves purposes of sexual experimentation and discovery, may or may not involve concomitant close relationships with the sexual partner (which makes some women worry about becoming "promiscuous"). In contrast, the truly promiscuous, though they may externally behave no differently, have a driven, uncontrolled, and destructive element to their sexual encounters (they are also more likely to mismanage contraception). Sex for them is serving some neurotic need of which they are usually not quite aware.

There is a parallel to the "sleeping around" of young people in the behavior of the recently separated and divorced. In the aftermath of such traumatic break-ups, it is not uncommon for these women and men to have a succession of superficial affairs. Here again, we are not dealing with a paraphilia but an attempt at restitution. New sexual partners hold out for these people a means of reaffirming their desirability to counter rejections they have suffered; to vent their anger and frustration; to distract themselves from their grief; to make up for lost time, and so on.[43] Casual sex may be useful under these circumstances as a temporary respite but is not an adequate substitute to "working through" the residual problems of the disrupted relationship and reestablishing a more satisfactory one.

Group sex offers particularly good prospects for fulfilling paraphiliac needs. Why else would the image of a mound of humanity with intertwined limbs and sweating torsos so excite the erotic imagination? If sex with 1 person is good, is sex with 12 a dozen times better? Do we stuff our mouths with meat, fish, and fowl all at once to get the cumulative benefit of their tastes?

The hidden nature of such activity is often quite thinly veiled. People are reduced to bodies and genitals (fetishism); one sees (voyeurism) and shows (exhibitionism) the forbidden; there is real and vicarious homosexual gratification; it is an oedipal dream come true; the fulfillment of the polymorphously perverse yearnings of childhood. As the ultimate erotic fantasy, this is the stuff of pornography at its purest (chapter 19).

Promiscuous paraphiliacs make great literary characters. The character of Don Juan is widely represented in literature, art, and music. His female counterpart is the demonic woman who seduces, vanquishes, and destroys men.[44] Emile Zola's *Nana* is a classic example. As a highly coveted courtesan, she is seemingly driven by greed. Yet the more her lovers lavish money and gifts on her, the more she scorns them. Don Juan and Nana are the penis and vagina personified; not as reproductive emblems or as instruments of pleasure, but as weapons of destruction and hatred. It is the element of hatred along with compulsion that links these behaviors with the paraphilias. And it is to the more naked expressions of sexual aggression and hatred that we now turn.

Coercive sex

The use of coercion in sex represents one of the most serious problems in human sexual interactions. Its most extreme manifestation is *rape*, but there are many other forms of sexual coercion as well. Any activity that forces sex on another person, in any shape or form, constitutes *sexual coer-*

---

[42]Sarrel and Sarrel (1979).
[43]Scarf (1980).

[44]For example, Byron's *Don Juan* (1824); Mozart's *Don Giovanni* (1787); Richard Strauss' *Don Juan* (1888); and Shaw's *Man and Superman* (1903).

cion. To further broaden the definition, one can say that to interfere with another person's freedom to act or not to act sexually (except to prevent harm to oneself or to others as in rape) may be another form of sexual coercion.

It is difficult to lay down hard and fast rules as to when forceful persuasion crosses over into coercion. Even between the most loving couples, the sexual interests of the partners do not always coincide. So when people ask, argue, cajole, plead, and demand sex, the line between negotiation, persuasion, and coercion may become hard to draw.

Sexual Harassment

While rape has been long recognized, the phenomenon of sexual harassment has attracted public attention only recently. There are two reasons for this. First, there has been a general male tendency to view women as passive sexual objects; fair game to be had, if one can get away with it. Second, respectable women have been expected to be defensive about sex, saying "no" even when they mean "yes." This has meant that if couples are to get anywhere sexually, men should push, women should resist, but eventually yield. As a result, *sexual pressuring* has earned a certain social legitimacy, even when women have clearly felt annoyed, exploited, and humiliated by it.

Many women still like to be courted subtly and tastefully even under circumstances that hold little prospect for the establishment of an erotic relationship. Some may encourage such attention by acting in a coquettish and flirtatious manner either with seductive intent or by way of "innocent fun."

Men often have trouble reading these signals correctly. Some men interpret every gesture of friendship as a form of sexual interest. They do not understand that even if a women is quite candidly interested in sex, she may still be uninterested in them personally. That is a hard pill to swallow, so they doggedly persevere, do not know when to stop, and are insensitive to the fact that women do not want uninvited attention, let alone being pawed, clawed, and slobbered over.[45] Nor are women above pestering men with their attentions, even when these are clearly unreciprocated and unwelcome.

Irritating as these interactions may be, the term *sexual harassment* is best reserved for situations when unwelcome sexual attention is applied to someone who is in a more *vulnerable* position. The boss making unwanted passes at an employee; the professor making advances to a student while holding the prospect of special favors or the threat of retribution; these are the everyday examples of sexual harassment that have now become a source of wide social concern. The reason that men are usually the offenders is because men are more often in positions of power and this is the sort of predatory behavior which is typically male. But this situation is now changing, and men working for women may also be vulnerable to sexual harassment.

Dealing with sexual harassment is a delicate issue. Blanket rules forbidding sexual interactions in the workplace or within institutions are likely to be resented and unenforceable, though they may serve a useful function by sensitizing people to these problems.

One can nevertheless safely expound the general principle that sexual interactions work best in symmetrical relationships: for instance, in the professor-student case, the latter is subject to exploitation, the former to blackmail.

The more practical way to deal with sexual harassment is not to dwell on the right of people to pair off but to focus on the harassment component itself. That seems best dealt with first by the person being harassed giving a discrete and clear message (such as through a private letter) that the sexual attentions of the other person are not welcome and must stop; if that does not work, then the person may

---

[45]Men who press against women and rub their genitals against their bodies in crowded places are known as *frotteurs* (French: to rub).

have to take recourse to more formal institutional means to be free of the harassment.[46]

Rape

When sexual coercion relies on the use or threat of physical force, it is referred to as *sexual assault*, the most flagrant example of which is *rape* (Latin: to seize). Rape is a topic of great psychological and social importance whose significance has been further elucidated recently through our heightened awareness of the exploitation of women. It is a complex behavior that can be best understood in its various relevant contexts. We are primarily concerned here with its descriptive aspects; in chapter 19 we discuss rape as a legal offense and consider its broader social implications.

The Nature of Rape   In formal terms, rape is a *legal* entity rather than a paraphilia. The law generally defines rape as an act of sexual intercourse with a woman other than one's wife, against her will, and through the use of threat or force or by taking advantage of circumstances that render her uncapable of giving consent (chapter 19). When the victim is below the legally defined *age of consent*, the act constitutes *statutory rape*, irrespecive of the presence or absence of coercion. Although the great majority of rapes are perpetrated by men on women, and which is what we shall be mainly concerned with here, there are also cases when men are the rape victims (box 14.3).

In discussions of rape, the tendency is now to define it in broader terms. For example, Susan Brownmiller characterizes rape as "A sexual invasion of the body by force, an incursion into the private, personal inner space without consent—in short, an internal assault from one of several avenues by one of several methods—constitutes a deliberate violation of emotional, physical, and rational integrity and is a hostile, degrading act of violence that deserves the name of rape."[47]

An expanded definition of rape would subsume acts other than coitus; such as insertions of fingers or foreign objects into bodily orifices. Even more broadly, any act that sexually debases women or robs them of their sexual autonomy is sometimes referred to as "rape," such as voyeurism, pornography ("visual rape"), and "compulsory pregnancy" (meaning lack of choice for abortion). The political purpose of extending the meaning of rape may be understandable, but by applying the term to situations that do not even entail physical contact with the victim, the term loses its specificity and much of its force.

Another critical issue has to do with the sexual nature of rape. It was assumed in the past that rape was perpetrated by lusty men who had no other sexual outlets. Hence, prostitution was justified on the grounds that it safeguards the virtue of decent women by accommodating male lust.[48] In this perspective, rape is like theft; if a man cannot get sex lawfully, he will take it by force.

This is challenged by the view that rape is not a function of the availability of sexual outlets; that it is not a sexual act at all, but a form of violence.[49] This makes the experience hardly any less horrifying for the victim. But it does, in a sense, "clear" sex from blame. Since rape is bad and sex is good, then rape cannot be sex. There are two problems with this view. First, to claim that coitus can be a nonsexual act undermines the concept of sexual behavior as an objectively definable entity. Second, it perpetrates a Polyannaish image of everything about sex being wonderful, when in reality, sex can also be awful, and rape is

---

[46]For an overview of sexual harassment, particularly from a legal perspective see MacKinnon (1979).

[47]Brownmiller (1975, p. 422).

[48]After the Australian government closed down brothels in Queensland, there was an increase of almost 150% in rapes and attempted rapes (Barber, 1969). Increases in rape in this country appear unrelated to the extent of prostitution (Geis, 1977).

[49]Brownmiller (1975) calls it "a conscious process of intimidation by which *all men* keep *all women* in a state of fear."

Box 14.3   Men as Rape Victims

Men are almost always raped by other men, and most such cases of *homosexual rape* occur in prisons.[1] Younger inmates who are physically weak, timid, and unable to defend themselves become common targets, but they are by no means the only victims. To avoid repeated assaults, some young inmates attach themselves to one of the more dominant men by becoming his "girl." While sexual deprivation in prison is a self-evident fact, power and dominance are also major motivations in such attacks, especially when they are of a particularly violent nature. The victim is often overpowered and gang raped by a group of inmates. The presence of guards does not entirely deter such assaults which may even occur when inmates are being transported in police vans.[2] The fear of retaliation often inhibits victims from "snitching" to the authorities.

Outside of prison, adult male rapes are very rare, accounting for very few of the cases reported to the police or seen in rape crisis centers. Victims are often hitchhikers who are overpowered by force or the threat of weapons. These rapists are usually strangers but there have also been cases when an employer, even a brother or father has been the assailant. The most common acts in male rape are sodomizing the victim and forcing him to perform fellatio. The reactions of the male victims are quite similar to those of females subjected to rape.

It has been long assumed that women are incapable of raping men and that their involvement in rape is usually in the role of an accomplice who restrains, or subdues another woman who is then raped by the man. In gang rapes, female partners may actively participate in the sexual degradation of the victim. Legally, a woman may be charged with rape if she aids and abets in the commitment of the crime. Lesbian sexual assaults are rare and confined to prison settings.

Recent evidence shows that women can and do occasionally rape men. These usually involve one or several armed women, who force a man to perform cunnilingus and to have sexual intercourse. These victims can also be abused by being urinated on and injured (in one case, castrated). There are only a few cases of heterosexual male rape on record, because such cases are rare and also men who undergo such an experience may be even more reluctant to report it than are women rape victims. As a consequence of being raped, men develop a "post-assault syndrome" characterized by depression, sexual aversion, and dysfunction; reactions that are quite similar to the experience of women following rape.[4]

Women are very much less likely than men to be the assailants in rape cases for much the same reasons that make women commit far fewer violent crimes than men. In addition, the typical way for a women to use sex to express anger has been to withhold sex rather than force it on a man. Whether or not changing conceptions of sex roles and the greater willingness of women to take the sexual initiative will also be reflected in more of them acting like male assailants, remains to be seen. But it seems highly unlikely that being raped by women will ever present a comparable threat to the safety and sexual integrity of men, as it does for women.

[1]Weiss and Friar (1974).
[2]Davis (1968).
[3]Cited in Groth (1979, p. 119).
[4]Groth (1979); Sarrel and Masters (1982).

a good example of it. Hence, it is more realistic to view the hostile and sexual elements in rape as inseparable, even though one or the other may predominate in a given case.

Prevalence of Rape   Despite greater public awareness of rape, we only have an incomplete picture of its prevalence. The sexual assaults that gain the most notoriety through the media are by no means representative of the great majority of rapes that occur. Rape is among the most underreported crimes: Estimates of its true incidence range from 5 to 20 times the reported figures.[50] A national survey administered by the Census Bureau showed that half of all rapes in 1979 were unreported.[51]

Even when rapes are reported, only half of the offenders are apprehended; of those, only three out of four are prosecuted, and of these only half are found guilty. Since the national rape statistics put out by the FBI and most studies of rapists are based on convicted rapists, it is easy to see why the conclusions that emanate from such data cannot be extended to rapists at large.

Rape is also the most rapidly increasing violent crime in the United States. Reported cases have increased from 17,190 in 1960 to 63,020 in 1977, and to 82,088 in 1980 (amounting to 73 rapes per 100,000 women).[52] Estimates based on general population surveys (albeit nonrepresentative samples) show even higher figures. Nine percent of the males responding to a 1977 *Redbook* questionnaire reported that they had used force, or the threat of force, to make a woman engage in sex.[53] One out of every four of the women in the *Cosmopolitan* survey had had at least one experience of rape or sexual molestation.[54] These figures are appallingly higher than the prevalance of rape in most other countries. Yet there are some cultures, such as the Gusii of Africa, where rape is even more a part of the sexual life of a society than in the United States.[55]

The picture is further complicated by the inclusion of **marital rape.** Legally and traditionally, husbands have been exempt from the charge of rape, since wives have been expected to submit to their sexual wishes. The particular problems that this form of rape represents will be dealt with further in chapter 19.

The common locations for reported rapes vary somewhat from one to another source. In a 13-city survey, about one-fifth occurred in the victim's residence; an additional 14% nearby (such as yard, driveway, sidewalk). About half of all rapes took place in an isolated area (street, park, playground, parking lot) where the victim was either accosted or taken to. All in all, some 65% of rapes occurred outdoors.[56]

Sociological Profile of the Heterosexual Rapist   Sociological studies based on crime statistics and independent surveys provide broad sketches of the men who rape while attempts at more intimate profiles have come from psychiatric and psychological studies of individual rapists. Nonetheless there is as yet no coherent picture that emerges from any of these approaches.

Police records and sociological samples of apprehended rapists place the majority of these men within a *subculture of violence.* Most of these men are poor, with minimal education, simple vocational skills, of low social status, angry at the dominant culture, and prone to violence to obtain what they cannot otherwise get.

One of the best known sociological studies of rape was conducted by Menachem Amir during 1958–1960, in Philadelphia.[57] Based on police records, his salient findings were as follows: The majority of rapists and their victims were 15 to 19 years old, and mostly unmarried;[58] 90% of offenders belong to the "lower part of the occupational scale" (from skilled workers down to the unemployed); blacks were overrepresented both among offenders and

---

[50]Amir (1971).

[51]U.S. Department of Justice, National Crime Survey (1981).

[52]Federal Bureau of Investigation (1981). The major violent crimes are murder, assault, and robbery. While the rates of all these crimes have been on the rise, that of rape has been going up three times higher than those of other crimes combined (Hindelang and Davis, 1977).

[53]In addition, 8% of the males reported using less direct forms of force, such as threatening a woman with loss of job or monetary assistance (Tavris, (1978).

[54]Wolfe (1980, p. 238).

[55]LeVine (1959).

[56]Hindelang and Davis (1977).

[57]Amir (1971).

[58]This is true for violent crimes in general. In California nearly 90% of those arrested for homicide in 1981 were men, half of them younger than 25 years.

victims;[59] in 82% of cases, offender and victim lived within the same area; 57% of offenders had a previous police record (as did 20% of the victims); alcohol use was present in one third of cases (and in 63% of these cases, it was present in both offender and victim);[60] 53% of rapes occurred on weekends (highest number occurred on Fridays between 8:00 P.M. and midnight); 71% of the rapes were planned and often took place indoors, including the residence of one of the participants.

The extent to which rape is committed by strangers rather than by persons known to the victim differs in various studies. A national task force has reported 53% of assailants to be strangers, 30% to be only slightly acquainted with the victim, and only 10% to have had a close present or past relationship with her.[61] A study conducted in Memphis found strangers to account for 73% of rapes with the qualification that reports of rapes committed by strangers are more likely to be taken seriously by the police.[62] In the *Cosmopolitan* survey, 33% respondents who had been "sexually assaulted" (which is a broader category than rape) had been victimized by strangers, as against 46% by friends and acquaintances, and 7% by husbands.[63]

There is a common perception that rape is typically a crime committed by a solitary male. Yet in Amir's study, 43% of the victims had been subjected to *multiple rape* where two ("pair rape") or more men ("group rape" or "gang bang") had raped a woman; altogether 71% of these offenders had been involved in multiple rapes (55% group rape; 16% pair rape). Men who participated in group rape tended to be younger and to have arrest records. These rapes were usually planned (even if the victim was not always preselected) and perpetrated on a woman who lived in the neighborhood.

Gang rapes tend to be particularly vicious because the victim is totally defenseless and her assailants try to outdo each other in violating and degrading her. It is not unusual for female members of gangs to participate in these acts by holding down the victim or even more actively participating in her torment.

Psychological Profile of the Rapist   Psychological studies have shown rapists to have had behavioral problems in childhood and to have grown up in chaotic family environments marked by deprivation, neglect, extreme sibling jealousy, and sexual abuse.[64] They come across as immature characters who act younger than their age. They seem lacking in self-esteem, uncertain of their sense of masculinity, and intent on reassuring themselves on both counts.

Psychoanalysts see rapists as sociopathic characters with poorly developed superegos or as persons lacking control of their aggressive impulses. Some show evidence of displacement of hostility from the mother, or some other significant female figure, to the rape victim. Other rapists are thought to suffer from deep-seated sexual anxieties that they try to dispel by demonstrating phallic aggressivity.

There have been attempts made to devise a *typology of rapists.*[65] One common type is the *amoral*, predatory, and aggressive young man who seeks whatever sexual gratification he can find, regardless of the consequences to others. These men violate women in the same way that they help themselves to other people's property, and use whatever force is necessary to achieve their aim. They tend to be the most opportunistic; if they encounter a

---

[59]Amir's finding that about 80% of rapists and victims were black reflects at least in part, the large proportion of blacks in Philadelphia; national statistics show about half of arrested rapists to be black. Historically, black men themselves have been the most frequent victims of false accusations of rape in racist white communities.

[60]Rada (1975) reports 50% of rapists to have been drinking at the time, 43% of them heavily (10 beers or more). Some 12% were on drugs with or without alcohol.

[61]Mulvihill et al. (1969).

[62]Quoted in Brownmiller (1975, p. 393).

[63]Wolfe (1980).

[64]Delin (1978).

[65]Burgess and Holmstrom (1974).

woman in a vulnerable situation (such as during a burglary) they often take advantage of it.

The second type of rapist is *sexually inadequate*. His primary aim is to obtain sexual gratification, which he seems unable to get through nonviolent means. Shy and inept in his dealings with women, he compensates for his inadequacies through paraphilias and fantasies of being a great lover. These rapists are most likely to select their victims carefully and strike at an opportune moment. Alcohol is often used to boost their self-confidence. If the victim fights back vigorously, they are likely to give up and flee. If they do proceed with the rape they are likely to prove impotent, fail to ejaculate, or ejaculate prematurely.[66]

Unlike the first two, the *hostile* type of rapist has a primary interest in hurting his victim. His hostility may be aimed at a particular woman, or women in general, whom he sees as sexually frustrating, untrustworthy, and hostile. Though the roots of his rejection of women and all things feminine may not be apparent to him, his pursuit of exaggerately "masculine" activities and behavior is already evident in adolescence. These men are more likely to rape older women. Their acts of rape are typically triggered by an upsetting experience with a woman who is important to the rapist, but the victim is usually a stranger on whom the rapist seems to displace his rage.

The fourth type is the *sadistic* rapist whose sexual satisfaction comes primarily from tormenting his victim. Such a man relishes the struggles and suffering of the woman and may be further aroused by her anger and resistance. This group is psychologically the most disturbed, with long histories of cruel behavior toward children and animals. Beyond inflicting the trauma of rape, they are the most likely to seriously injure their victim (one in 500 rapes ends in death) (figure 14.12). Some will also mutilate the corpse and have further sexual contact with it. Every decade or two a Jack

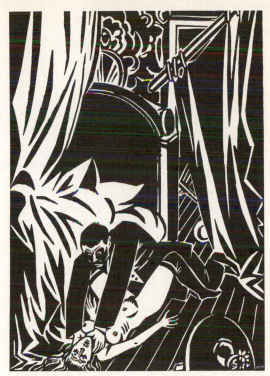

Figure 14.12 Franz Masareel, *Sex Murder*. (Book illustration, Ginzberg *L'Enfer*, p. 185.)

the Ripper type murderer terrorizes entire cities.[67] Though these criminals attain the most public notoriety, they represent the smallest proportion of rapists.[68]

Such categorizations do not mean that rapists come in pure types. Issues of power, anger, and sexuality operate in every rape. Above all, the rapist has a need to feel dominant over his victim by placing her in an inferior, degrading position. He dehumanizes the woman, keeps her anonymous, and sees her as an object rather than a person.[69]

---

[66]Groth and Burgess (1977) found 34% of rapists to have suffered such dysfunction during the assault.

[67]Jack the Ripper was an unidentified man who murdered and mutilated prostitutes in London, in 1888. A modern counterpart is Albert DeSalvo, widely assumed to have been the Boston Strangler, who during the early 1960s strangled, stabbed and sexually mutilated 11 women, ranging in age from their 20s to their 70s (Frank 1966).

[68]Groth et al. (1977); Groth (1979).

[69]Brownmiller (1975).

The Heterosexual Rape Victim   Rape is a highly traumatic experience usually with immediate and long-term consequences (box 14.4). The rape victim is typically a young unmarried woman, who lives in an urban, lower-class neighborhood. The rape victim generally shares the demographic characteristics of her assailant.

These statistical probabilities notwithstanding, the prospect of rape is not restricted to any particular age group or type of women. Rape victims have ranged in age from infants to women in their eighties, including members of every ethnic group, social class, marital status, and vocational category. Women living in middle-class neighborhoods and on college campuses have recently become more frequent targets of rape. The inclusion of marital rape would make the victim population even more heterogeneous.

Beyond such descriptive features, there are no typologies of victims comparable to those of rapists that we discussed earlier. It was traditionally assumed that most women who get sexually assaulted somehow bring it on themselves through incaution ("she should have known better") or provocation ("she was asking for it"). Furthermore, it has been thought that rapists succeed only when a woman deep down wants to be raped ("no woman can be raped against her will"). Such assumptions add insult to the grievous injury experienced by rape victims.

The circumstances under which high-risk victims live do not allow them much choice in being exposed to sexual assault. And once an assault takes place, most victims usually have limited choices. This is not to say that women in these situations should feel helpless and meekly yield to their assailant (box 14.5). On the contrary, one reason why one out of four cases of rape are not completed is due to the woman's reactions during the assault. Nonetheless, when a woman has to contend with a group of men, a rapist who is armed, or bent on hurting her, her willingness to undergo the degradation of rape, rather than get seriously injured or killed, is by no means reflective of a desire to be a willing partici-

pant in the act. Even in the least harrowing of these experiences, women may be understandably paralysed by fear and numbed by the enormity of what is happening to them.[70]

The treatment of the rape victim and of the rapist is difficult each in its own way. After years of neglect, we are now beginning to develop more humane and effective ways of helping women cope with the experience of rape (box 14.4). How to deal with rapists is more problematic; we return to this issue in chapter 19 (see in particular box 19.6).

Rape victims have been dealt with for so long with so much suspicion and callousness by the public, police, courts, and even physicians, that there is now a strong reaction to any attempts at examining their behavior for fear that such inquiry may once again shift the blame onto the victim. Nonetheless, *victim precipitation* is a concept used in criminology, which focuses on contributory behavior on the part of the victim.[71] It seeks answers to the question of whether the crime might have been avoided if the victim had behaved differently. It does not blame the victim for the crime nor does it justify the crime irrespective of the victim's behavior.

The concept of victim precipitation has a self-evident validity when applied to other crimes. For example, if the victim of an assault was the first to use physical force, or used abusive language against the assailant, his or her behavior can be said to have been provocative. But what should constitute provocation in case of rape? If a woman has prominent breasts that incite the rapist's lust, is she "precipitating" the assault? If a woman dresses and behaves in a sexually provocative manner because she enjoys doing so or wants to attract the

---

[70]The possibility of erotic response on the part of the victim raises special problems. Among the 50 victims investigated by Becker (cited in Geis, 1977) and 90 cases seen by Burgess and Holmstrom (1976), none of the women had experienced orgasm during rape. Should a woman feel aroused she is likely to experience additional guilt over the experience.

[71]Wolfgang (1958).

Box 14.4   The Aftermath of Rape: The Victim's Experience

The possible aftereffects of rape include *emotional trauma, physical injury, pregnancy,* and *sexually transmitted diseases.* Each of these concerns must be dealt with promptly, effectively, and compassionately.

A woman who is raped should contact the police and receive medical attention immediately before doing anything else, such as washing, changing clothes, or clearing up the rape scene. Contacting a Rape Crisis Center can be very helpful; such centers have personnel who provide emotional support, transportation, clothes, information, and numerous other services. Close friends and relatives can also be a source of much comfort.

In a study of the experience of 92 rape victims all of the women underwent an *acute disorganization phase.*[1] Immediately after the rape, most women displayed either feelings of fear, anxiety, and anger, or a controlled external appearance that obscured their true emotions.

Many women reported various problems in the first weeks after the rape, including physical pain from injury during the rape, tension and tiredness, upset stomach, and gynecological problems. The most common emotion mentioned during this time was fear of death and physical violence.

Generally about 2 to 3 weeks after the rape the women began the second or *reorganization phase.* At this time half the victims moved to a new location. Others took a trip. Many changed their telephone numbers. Some of the women reported nightmares and fears of such situations as being outdoors, indoors, alone, or in crowds.

Masters and Johnson have identified several sexual problems that occur in women after they have been raped.[2] The exact frequency of these problems is unknown, since Masters and Johnson made their observations on a limited sample of women who came to them for help with sexual problems.

The most common complication noted after rape was the development of an *aversion* for sexual activity and a diminishing interest in sex. This was more common when the rape had become public knowledge. This aversion to sex most commonly developed in women who were made to participate in a number of sexual acts with one or more men over an extended period of time. An aversion to sex was also more likely to develop if a woman was forced to perform a sexual act that she had never done before, such as oral or anal intercourse.

Second, *vaginismus* sometimes developed, particularly in those women who had been raped a number of times by one or more males. This condition was more common among rape victims who experienced severe physical injury in the genital area.

Third, some rape victims lost their ability to have *orgasms.* The women most likely to lose orgasmic function were the ones who were most committed to the double standard, being totally dependent on men and viewing their sexual role as one of service.

Fourth, some women experienced severe *relational problems.* The most common one involved a married woman's relationship with her husband. Some husbands were unable to deal with the rape, especially if the rape was public knowledge; they felt that everyone now knew that their wife had been "dirtied." Some husbands wanted to seek revenge on the rapist. Such male reactions added greatly to the wife's distress and increased the likelihood that she would develop sexual problems. In some cases the husbands of rape victims developed sexual dysfunctions themselves. However, most husbands were very understanding and helpful to their wives.

[1]Burgess and Holmstrom (1974).
[2]Masters and Johnson (1976).

Box 14.5 Preventing Rape

Rapes, like accidents, are not entirely avoidable without seriously restricting one's freedom of movement. Yet taking precautions may reduce the risk of their occurrence.

Since many rapes occur in the victim's home, there should be locks on doors and secure windows. Doors to the outside should be equipped with a dead-bolt lock and a chain lock and always kept locked when a woman is alone. To prevent strangers from getting into the house doors should not be opened to unknown people. Requesting identification from service people may deter a potential rapist.

When away from home, it is best to avoid deserted, enclosed stairwells. Car keys should be out and ready before arriving at the car to minimize the time needed to open the door. The back seat should be checked to make sure no one is there. It is important to keep the car doors locked while driving. Picking up strangers is very dangerous, as is hitchhiking. At her job, a woman should refuse to work alone in a deserted building. When outside it is wise to avoid walking alone at night especially in the dark, deserted streets or open areas. If a woman has to walk outside at night, she should stay close to the curb on well-lighted streets, walk quickly, and carry something in the hand that could be used to disable an attacker temporarily. A handbag, umbrella, books, keys, and other suitable objects can be used effectively for this purpose. A freon horn that creates a loud noise can be specially useful. Taking a course or reading books on self-defense may prove valuable, provided one becomes reasonably proficient. A woman can also bite, scratch,

and kick an attacker without being an expert in martial arts.

What should a woman do in an actual rape encounter? Unfortunately there is no single right answer that applies to all rape encounters. Every woman must judge for herself what is the best course of action is in each potential rape situation. In some cases it is possible to talk a rapist out of committing a rape. Otherwise, a woman should scream and run if possible; shouting "Fire" is more effective than "Rape" in attracting help.

An assertive tone of voice and defiant posture will dissuade some rapists, since some potential rapists do not want to become involved with a woman who resists. If the rapist attacks, fighting back can increase the woman's chances of escaping in some circumstances, but resisting may also cause the rapist to become violent and hurt the victim more. Urinating or defecating into one's pants may discourage some rapists, infuriate others. The use of weapons like guns and knives is very risky since these can be seized and turned against the victim. Their use is also illegal, and a woman may be liable for harming her assailant.

As in any crisis situation, what comes naturally to the person tends to work best. The worst thing is to become paralysed with fear. There is always something that a woman can try and it may work. The objective in these situations is to gain the initiative and choose the best strategy available. The point is to get away, not to win a battle.[1]

[1]For an elaboration of these suggestions on rape prevention, see Horos (1974); Conroy (1975); Burgess and Holmstrom (1979), and materials made availble in rape prevention programs.

attention of men she finds congenial, does that mean that any man should feel free to force his attention on her? If a college student hitches a ride, or a housewife lets a delivery man into her house, are they inviting sexual assault? One is likely to get varied answers to these questions. These dilemmas and ambiguities that may cloud certain acts of rape have been skillfully explored by writers and artists (figure 14.13).

Figure 14.13 Kurosawa's *Rashomom* (1950) presents three conflicting versions of the rape of a woman as seen from the perspectives of the rapist, the husband, and an uninvolved witness. (Courtesy of Janus Films and the Museum of Modern Art Stills Archives.)

Amir considered one in five rapes to have been victim-precipitated on the basis of a variety of criteria, such as the victim going to the offender's house to drink and getting into other "risky situations marred by sexuality." When more incriminating criteria are used (such as the victim making sexual advances or agreeing to sexual relations but then changing her mind), victim precipitation emerges as a factor in 4% of cases (as against 22% in homicides and 14% in assaults).[72]

The notion of victim precipitation appears to be widely held among adolescents as well. In a survey in Los Angeles, teenage boys considered forcing a girl to have sex to be acceptable behavior under certain circumstances, and 42% of the girls agreed.

The circumstances which justified the use of force involved sexual encouragement by the girl ("to get him sexually excited," " to lead him on," "to say yes") followed by her refusal to have sex. This was especially the case if the girl was known to have slept with other boys or showed up a party where she knew there would be drinking and drugs. In other situations, girls who meant to be stylish by going braless or wearing tight jeans, were misunderstood by boys who interpreted such behavior as a sexual invitation. (There was general agreement that spending money on a girl did not give a boy a right to her body.) Race, age, sexual experience, socioeconomic status, made little or no difference in the attitudes of these youth.[73]

[72]Mulvihill et al. (1969).

[73]*Los Angeles Times*, September 30, 1982.

That coercion is already part of the sexual interactions of substantial numbers of adolescents, in the mainstream of American society, betrays a serious failing in the socialization of the nation's youth. This is not only a problem which concerns women but a matter to be dealt with in a larger societal context, an issue we return to in chapter 16.

SUGGESTED READINGS

Stoller, R. J. *Perversions: The Erotic Form of Hatred* (1975). Insightful account of the manifestation and psychodynamics of the paraphilias.

Meiselman, K. D. *Incest* (1978). An overview of incest and report of a group of cases.

Groth, A. N. *Men Who Rape* (1979); Brownmiller, S. *Against our Will* (1975). The first, a psychological analysis of the motivations and behavior patterns of rapists; the second, a thought-provoking social and historical overview of rape.

Chapter 15

Sexual Dysfunction and Therapy

If your life at night is good, you think you have
Everything; but, if in that quarter things go wrong,
You will consider your best and truest interests
Most hateful.

Euripides, *Medea*

The wish for sexual enhancement is understandable when there is evidence of sexual dysfunction such as impotence or the failure to reach orgasm. But many who have no such symptoms still feel that their sexual life is somehow lacking; others, though basically content, aspire to ascend to greater heights of erotic ecstacy. So we have the sexually hungry, the sated but bored, and the devotees of gourmet sex, all wishing to improve their sexual lives.

Paradoxically, there are countless other men and women whose sexual life would seem to be sadly lacking, but who act as if they are quite satisfied with it. Some say these people do not know any better and should be awakened to that fact. Others say they should be left alone; we have created enough unhappiness, as is, by setting up unrealistic sexual expectations.

Sexuality and Aging

Aging is not a disease, but disabilities that result from it may lead to outcomes, such as impotence, that are similar to those caused by illness. This is why we discuss more fully the effects of aging on sex in this chapter without implying that they entail illness or pathology.

Our sexual images of the elderly are laden with stereotypes. Elderly males are seen either as sexually depleted or as lecherous "dirty old men." The older woman is stereotypically the sexless "little old lady" or the lustful granny of *Playboy* cartoons. Elderly couples have been long thought of as dozing serenely side by side, their sexual lives fading memories; more recently, they have been rehabilitated into tireless lovers copulating the night away. Sex was said to die off with age; now we are told it gets better with age.

There are many reasons for this confusion. One source is the lag in recognizing that older people are now significantly different than before: they enjoy better health, are more active, more independent, and live longer. But an excess in optimism comes from generalizing from self-selected samples and the tendency of well-meaning sexual enthusiasts to paint everything sexual in rosy hues.

Even more basic is our confusion over what it means to grow old. The process of aging is real enough, but there is no discrete phase of life called "old age" that could be chronologically delimited and consistently characterized. It is better therefore to think of this phase of life as *late adulthood*—continuous with, rather than qualitatively different from, earlier phases of adulthood—and to consider *middle age* (chapter 9) as the time when the effects and concerns over aging begin to become salient.

Aging entails distinctive, albeit not fully understood, physiological changes in the sexual response cycle (chapter 3). To restate briefly, as men approach 50, they require longer and more direct physical stimulation to attain erection; the erect penis is not as firm and points more downward than it did earlier; arousal leads less often to orgasm; amount and force of ejaculation lessen; the recovery phase is longer. Menopausal women undergo thinning of the vaginal lining and less lubrication during arousal (chapter 7). In both sexes, the physiological responses to sexual stimulation remain basically the same but are significantly attenuated in intensity.

These physiological changes coupled with concomitant shifts in sexual desire typically result in a reduction of sexual behavior and a *decline* in sexual capacity, rather than sexual *dysfunction*. This is no different from changes in other physical capacities, such as the ability to run, as one gets older.

Sex in Midlife

After decades of preoccupation with adolescence as a period of developmental turmoil, emphasis has shifted lately to the *midlife crisis* as a time of stress and turmoil in adulthood.

The *midlife transition* (the years between 40 and 50) appears to have special significance for sexuality. In the past, attention was mostly focused on the menopause. It was widely assumed that with the changes in reproductive function experienced at this time, women lost sexual interest. We now know that this is not the case. At least on physiological grounds, there is no reason for any decline in sexual interest, and the alterations of genital function (such as decreased vaginal lubricatory response) can

be readily dealt with by simple means like artificial lubricants (chapter 7).

For increasing numbers of women the menopause may actually signify a time of enhanced interest in sex. There is total freedom from concerns over pregnancy and contraception. Women become more self-assured and inclined to take greater charge of their lives. There may even be a physiological basis for the increased sexual drive, the result of the decline in levels of estrogen and progesterone and hence in their dampening effect on the erotogenic action of androgens, which continue to be secreted at their premenopausal levels.

Nevertheless, the menopause continues to be a source of concern to many women because of psychological and social considerations. As a result of inevitable physical changes, a woman may feel less attractive, since there is such a strong association between youthfulness and sexual attractiveness in most people's minds (more so with respect to women than for men).

The midlife transition may present special problems for men as well. The fear of declining potency is an important source of anxiety and sometimes the cause of sexual dysfunction. This is a time of taking stock with regard to vocational, relational, and sexual concerns. Some middle-aged men slowly and quietly give up sex, others exhibit an adolescent-like urge to sleep around or fall head over heels in love one last time. Less often, it is the middle-aged woman who initiates extra-marital relations on a temporary basis, seeks a new stable relationship, or simply loses interest in sex for the rest of her life. It is no wonder that marital relationships come under special stress at these times. Most people emerge from the turmoil of midlife with some restructuring and reorientation of their sexual life: some go on to live sexually fulfilling lives; others become casualties, with their sexual relationships disrupted or gone stale.

Sex in Late Adulthood

Sexual activity, like vigorous physical effort, is often assumed to be a prerogative

Figure 15.1 Chinese coital scene. Nineteenth century (?). (Photo by Dellenback. The Kinsey Institute, Indiana University.)

of youth. Having performed their reproductive functions and obtained whatever satisfaction they could out of it, older individuals have been expected to have no further interest in sex or not be able to act on it because of loss of sexual attractiveness or capability. Whenever older people, especially women, have behaved differently, society has tended to view their persistent sexuality as unnatural and undignified.

Such attitudes are by no means common to all cultures, many of which unabashedly accept the sexuality of the elderly (figure 15.1).[1] American society is also becoming more accepting and encouraging of sexual experiences among the elderly. Of

particular importance has been the realization that older individuals actually have more active sex lives than is generally realized.

In a sample of 101 men free of physical ailments likely to interfere with potency, 65% of those between 56 and 69 years, and 34% of those over 70, were still sexually active.[2] Other investigations report that more than 75% of the men aged 61 to 71 were engaging in coitus at least once a month; also 37% of those aged 61 to 65 and 28% of those aged 66 to 71 had coitus at least once a week. Among the women, 39% of those between 61 and 65 and 27% of those between 66 and 71 were still engaging in intercourse. Among those aged 66 to 71, only 10% of the men and 5% of the women claimed to have no sexual desire. When men gave up coitus, 40% did so because of impotence, 17% because of illness, and 14% because of loss of interest. When women stopped engaging in sex, the main reason given related to their husbands: 36% of them had died; 12% were no longer living with their wives; 22% had become impotent; and 20% were ill.[3]

In a survey of 4,246 men and women aged 50 and older (the largest such sample studied to date), the prevalence and variety of sexual activity is even more impressive. But this is a study involving a self-selected sample of respondents to a questionnaire sponsored by *Consumer Reports* (response rate 0.2%). As with other magazine surveys (chapter 8), its findings cannot be generalized; in all likelihood, the respondents are sexually more open and active than the population at large.

The *Consumer Reports* sample shows that though there is a decade-by-decade decline in sexual activity, among those aged 70 and older, 81% of the married and 50% of the unmarried women remain sexually active, as do 81% and 75% of the corresponding groups of men; 50% of these sexually active women and 58% of the men have sex at least once a week; 61% of the

---

[1]A cross-cultural survey shows sexual activity to be an important part of the lives of older men in 70%, and of older women in 85% of these cultures (Winn and Newton, 1982).

[2]Finkle (1959).
[3]Pfeiffer et al. (1968), (1969), (1972); Pfeiffer and Davis (1972).

women and 75% of the men report "high enjoyment of sex." Masturbation is reported by 43% of the men and 33% of the women aged 70 or older. About half of the subjects have oral sex; some have experimented with anal sex and vibrators.[4]

Vicissitudes in social attitudes notwithstanding, sexual activity does undergo significant change in the later years of adulthood. There is a distinct decline in the frequency of orgasm with aging. However, longitudinal data suggest that levels of sexual activity for given individuals remain more stable over time than previously assumed.[5] Furthermore, this decline does not begin with what we commonly think of as old age; in the Kinsey sample the decline was already in progress by the end of the third decade for males and somewhat later for females (see figures 8.4 and 8.5). More current evidence confirms these findings. In a study of 628 men between the ages of 20 and 90, coital activity was found to decrease after the 30s: weekly frequencies declined from 2.2 at ages 30–34 to 0.7 at ages 60–64; to 0.4 at 65–74, and to 0.3 at 75–79.[6] *Consumer Reports* respondents also show a decline with age beyond the fifth decade.

It is important to bear in mind that while such statistical trends are valid for large groups, they need not apply to everyone within them. Thus getting older does not necessarily mean that there is an inexorable decline in sexual activity for each and everyone. And even when there is a fall in coital frequency, this need not necessarily reflect a failure of all other forms of sexual satisfaction. It is important to differentiate the effects of aging, as such, from the effects of illness; the chances of getting sick become proportionately higher as one grows older, but aging as such is not a disease. Thus, sexual dysfunction should not be necessarily attributed to the aging process simply because it happens to coincide with it. Furthermore, important as biological factors may be, psychosocial considerations may also affect sexual behavior and function in equally profound ways.

The potential causes for the decline of sexual interest and activity in older years are many: physiological changes; medical problems (heart disease; stroke, diabetes, arthritis, anemia, prostatic conditions, gynecological conditions); drugs used to treat these conditions; psychological problems (fear of impotence, expectation of sexual decline, feelings of sexual unattractiveness, interpersonal conflicts, depression, grief, guilt, shame); and practical considerations (constraints in sexual interaction among the elderly, especially those in nursing homes and other institutions).

Demographic factors confront women with particularly serious problems because of their relative longevity and hence their larger numbers. In the 40- to 44-year age group, there are 213 single, separated and divorced women for every 100 men. In addition to the 7% of women who never marry, there are 644 widows to 100 widowers. Age is also a serious impediment to their remarriage: women divorced in their 20s have a 76% chance of remarriage; for those in their 50s or older the chances are less than 12%.[7] There are simply not enough eligible older men, and the prospects of finding younger sexual partners are even worse because of cultural and other constraints. As a result of these factors, the sexual options available to older women can be severely limited.

Another impediment to sex in the older years comes from concerns that sexual activity is itself debilitating. There are indeed certain illnesses, including acute heart ailments, during which sexual activity, like other kinds of activity, must be restricted; but by and large sex helps maintain older individuals in better health, rather than being a detriment to good health.[8]

A critical requirement in maintaining good sexual function in the older years is the sustenance of sexual activity through-

[4]Brecher (1984).
[5]George (1981).
[6]Martin (1977).

[7]Blumstein and Schwartz (1983, p. 32).
[8]Butler and Lewis (1976).

out adulthood. The aging woman who is sexually inactive experiences greater shrinkage of vaginal tissues; the aging male has greater difficulty in resuming sexual activity after a long period of abstinence. Thus to retain sexual function in the older years it is important that one remain active sexually.[9] This, in turn, is facilitated by keeping fit through exercise, proper nutrition, rest, attention to personal appearance, and keeping up with intellectual interests and community life.

The "use it or lose it" dictum reflects current professional thinking in the field of geriatric sexuality.[10] But it must be qualified on two counts. First, using it does not mean you may not lose at least some of it; it means you have a better chance of not losing more of it. Second, using it may not be the cause of not losing it; instead, both may be a function of a more basic underlying variable. In a study of 60- to 79-year-old married men, the level of sexual activity—whether high or low—seemed to be fairly consistent through life. These patterns were not correlated with factors like marital adjustment, sexual attractiveness of wives, sexual attitudes or the demographic features of the marital history. Instead, frequency of sex, erotic responsiveness to visual stimuli, and "time comfortable without sex" were closely interrelated and thus possibly linked to the strength of a person's basic sexual motivation. Furthermore, those with potency problems seemed to be quite free of performance anxiety, feelings of sexual deprivation, and loss of self-esteem.[11]

The recognition that each phase of life has its particular joys and sorrows facilitates sexual satisfaction in the older years. The sexuality of youth is urgent and impulsive, aimed at genital pleasure. That of young adulthood is generative, bound by intimacy. As one grows older, sex can serve an increasingly broader range of emotional needs, such as enduring affection and sharing the pleasures of being touched, caressed, and embraced. Awareness of the approaching sunset of life, in particular, can impart a special poignancy to each act of love.

While it is all to the good that there is now a strong impetus to accept and encourage sex among the elderly, there is an attendant danger that a new set of sexual expectations will develop that may become burdensome. Just as it is unjustifiable to inhibit people from expressing their sexual urge in the older years, it is equally pointless to push people into such activity if they are not so inclined. After all, the fact that sex is a good thing does not mean that one is not entitled to have had enough of it. Or if one's spouse is no longer available through illness or death, one may well wish to give up sex rather than turn to others. In short, it should be possible to spend the latter years of one's life in a happy and satisfying manner with or without sex, whichever way one may choose, as long as that choice remains available.[12]

Prevalence and Types of Sexual Dysfunction

No human function works flawlessly all of the time, and sex is no exception. Yet it is difficult to define a given level of sexual activity as inadequate or problematic since there is so much variation in ordinary sexual function and so much of sexual satisfaction is a matter of subjective judgment. Nonetheless it is possible to characterize more significant failures of sexual function and satisfaction as *sexual dysfunction* ("dys" meaning faulty or difficult). Such dysfunction is differentiated from diseases of sex organs (such as STDs); reproductive failure (sterility); and aberrations of sexual behavior (paraphilias).

Sexual dysfunction may be *organic* in nature. Yet in the majority of cases where there are no demonstrable physical causes it is considered to be of *psychogenic* origin (caused by psychological or emotional factors) or a manifestation of *psychosomatic ill-*

[9]Kolodny et al. (1979); Comfort (1976).
[10]Masters and Johnson (1982).
[11]Martin (1981).

[12]The literature on the sexuality of the elderly is reviewed in Ludeman (1981).

ness (conditions caused by the interplay of psychological and somatic variables that appear as physical symptoms).

This *illness model* for sexual dysfunction is questioned by some on practical and conceptual grounds.[13] There is no disagreement that in cases where there is a demonstrable physical cause, such as a painful infection, we are indeed dealing with a disease process; the same consideration would be extended to psychiatric illness, such as severe depression interfering with sexual function.

Yet there are many other instances where such causative factors are absent and the sexuality disability is due to ignorance, culturally inculcated negative attitudes, faulty sexual learning, or some traumatic experience in the past. It is argued that in these cases the sexual problem may be just as well thought of as a *learning disability* and dealt with as such, rather than as symptomatic of some deep-seated psychological conflict.

Definitions of Sexual Dysfunction

Until recently, most male sexual disorders were lumped together as *impotence* ("lacking in power") and female disorders as *frigidity* ("coldness"). "Impotence" remains in current use but in a more restricted sense; "frigidity" has been discarded because of its lack of precision and the pejorative connotation that such women have a cold and rejecting character, whereas there is no necessary correlation between personal warmth and orgasmic responsiveness. With the advent of new methods of sex therapy the classification of sexual dysfunction has become more sophisticated, though it is still evolving and not entirely satisfactory.

In current medical classification, disorders of sexual function are called *psychosexual dysfunctions.*[14] The major entities subsumed under psychosexual

dysfunctions are *inhibited sexual desire,* manifested by a persistent and pervasive lack of sexual interest; *inhibited sexual excitement,* manifested by problems with erections in males and vaginal lubricatory responses in the female; and *inhibited orgasm,* which consists of recurrent and persistent failure or undue delay in reaching orgasm despite the attainment of a normal excitement phase in the female and erection in the male.

Three other conditions are more gender specific. *Premature ejaculation* refers to the lack of voluntary control in the male over the onset of ejaculation, usually manifested by the inability to sustain coitus for a desired period. In the female, "premature orgasm" does not constitute a dysfunctional entity: if the woman is multiorgasmic, she can have additional orgams; if she is content with one orgasm and reaches her climax before her partner does, she can continue having coitus until he reaches orgasm. *Vaginismus* is an exclusively female problem consisting of involuntary spasm of the musculature of the vaginal opening that interferes with coitus. *Dyspareunia* is painful coitus, which is also a more common female than male condition.

In order to qualify for one of these diagnositc entities, the sexual problem must fulfill two conditions. First, it must be *functional,* or constitute an independent entity rather than be the symptom of an identified organic disease (like diabetes) or discrete psychiatric illness (like depression). Furthermore, the sexual problem must be *recurrent or persistent:* occasional failure does not count. Normal functioning entails considerable fluctuation in sexual desire, and it is expected that anyone's performance will occasionally falter. No one is expected to be able to have sex at will with anyone, anywhere, anytime.

This classification reflects the thinking of most psychiatrists and sex therapists, but its specific terms are by no means the only ones used to designate these entities. For instance, inhibition of sexual excitement in the male is referred to by some as *erectile dysfunction.* Other terms for inhibited or-

---

[13]Masters and Johnson (1970). A more extreme critic is Szasz (1980).

[14]American Psychiatric Association (1980).

gasm are *retarded ejaculation* or *ejaculatory incompetence* in men and *orgasmic dysfunction* or *anorgasmia* in women.[15]

The more useful classifications of illnesses are those that are based on causation (such as gonorrhea or syphilis). This cannot yet be done with sexual dsyfunction, but most of these dysfunctional conditions can be linked to failures in the fundamental physiological events underlying sexual function: vasocongestion and myotonia (chapter 3). Thus, vasocongestive failure would account for problems of erection in the male and difficulties with sexual arousal in the female; disturbances in myotonia would subsume the various difficulties related to orgasm.[16]

Since competence in sexual performance is not an all-or-none phenomenon, there is no absolute scale on which to evaluate or quantify it; hence there are no accurate figures about the prevalence of sexual dysfunction. The consensus among therapists is that there is a great deal of it. Masters and Johnson have estimated that half of all marriages are to various degrees sexually dysfunctional. Whatever the true figures, there is probably no other human disturbance that entails more silent suffering. Since no one dies of sexual distress and fertility may not be disturbed, the only loss seems to be erotic gratification; yet the toll in unhappy marriages and unfulfilled lives is incalculable.

Statistics on sexual dysfunction usually come from studies by sex therapists and reflect the situation among those who seek help. But even more interesting is the prevalence of disordered sexual functions among couples at large who are not in sex therapy. The results of one study are presented in box 15.1.

Failures in sexual performance as such are reasonably discrete entities. More difficult to ascertain are problems that arise from lack of sexual interest or *inhibition of sexual desire*. In the study shown in box 15.1, we find that a third of the couples have intercourse less than two to three times a month. Is that too little, too much, or just right? Who is to decide what is the optimal frequency for a given couple? The answers are relative to several dimensions: the present compared with the past ("I am not as interested in sex as I used to be"); the level of interest of one person as against another ("my wife/husband is not as interested in sex as I am"); comparisons with some external norm ("we are not having as much sex as our friends"). In practice, help is usually sought because of discrepancies in sexual desire between marital partners, or couples in some other stable relationships; the unattached are more likely to seek more compatible partners. In the past, the sexual interest of wives usually lagged behind those of their husbands; currently apathy is reportedly a primary complaint of men in therapy as well.[17]

In its mildest form, sexual apathy is simple indifference. A person forgoes sexual intercourse for extended periods because he or she does not feel like having it, despite propitious circumstances and willing partners. Sometimes a person is so preoccupied with other events and activities that sex is understandably set aside for a while. But it is also possible that these other activities are at least unconsciously aimed at avoiding sex. In treating marital difficulties, a therapist must determine when a person is making unreasonable demands on a harrassed marital partner and when a sexually apathetic partner is deliberately becoming exhausted at bedtime. Given the relativity of judgments in matters of sexual desire, the issue of socially defined levels of expectation is of particular importance. If general expectations of what constitutes a normal or desirable pattern of sexual activity are unrealistically high, people who would otherwise be quite content can be made to feel inadequate.

---

[15]The term *preorgasmic* is more of an expression of a hope than the description of a condition.

[16]Kaplan (1974). Also see Masters and Johnson (1970); LoPiccolo and LoPiccolo (1978); and Leiblum and Pervin (1980).

[17]Botwin (1979).

Box 15.1   Prevalence of Sexual Dysfunction in "Normal" Couples

The following tables are from a study of 100 couples who are predominantly white, well educated, and middle class. Eighty percent claimed to have happy or satisfying marriages. The information was gathered through a 15-page self-report questionnaire, only part of which dealt with sex. None of these couples were at the time in marital therapy or counseling, although 12 percent had had such help in the past.

Frequency of Self-Reported Sexual Problems

Problem	Women (%)	Men (%)	P*
Sexual dysfunctions—women:			
Difficulty getting excited	48		
Difficulty maintaining excitement	33		
Reaching orgasm too quickly	11		
Difficulty in reaching orgasm	46		
Inability to have an orgasm	15		
Sexual dysfunctions—men:			
Difficulty getting an erection		7	
Difficulty maintaining an erection		9	
Ejaculating too quickly		36	
Difficulty in ejaculating		4	
Inability to ejaculate		0	
Comparison of sexual difficulties reported by men and women:			
Partner chooses inconvenient time	31	16	0.05
Inability to relax	47	12	0.001
Attraction to person(s) other than mate	14	21	NS
Disinterest	35	16	0.01
Attraction to person(s) of same sex	1	0	NS

Disorders of sexual desire in the form of *inhibition* is now the main form of dysfunction recognized in this area.[18] There was a time, not so long ago, when the opposite of apathy, *hypersexuality*, was considered pathological. Designated as *satyriasis* for men and *nymphomania* for women, these vaguely defined entities were often reflective of societal anxieties over unbridled sexuality rather than based on any solid medical foundation. Frenetic sexual activity, like apathy, can certainly result from psychological problems (and sometimes brain dysfunction), but it is not the frequency or intensity of sexual activity but its underlying causation and its social manifestations that define it as problematic.

Erectile Dysfunction (Impotence)

The inability to attain and maintain penile erection is the most incapacitating of all

[18]Kaplan (1979).

Box 15.1   Prevalence of Sexual Dysfunction in "Normal" Couples *(Continued)*

Problem	Women (%)	Men (%)	P*
Different sexual practices or habits	10	12	NS
"Turned off"	28	10	0.01
Too little foreplay before intercourse	38	21	0.05
Too little "tenderness" after intercourse	25	17	NS

*Significance of t-test comparing men and women. NS: not significant.

Ratings of Sexual Satisfaction

Question	Women (%)	Men (%)
How satisfying are your sexual relations?		
Very satisfying	40	42
Moderately satisfying	46	43
Not very satisfying	12	13
Not satisfying at all	2	2
How satisfactory have your sexual relations with your spouse been in comparison to other aspects of your marital life?		
Better than the rest	19	24
About the same	63	60
Worse than the rest	18	16
Do you have any of the following specific complaints about your marriage?		
Sexual dissatisfaction:		
Yes	21	33
No	79	67

From E. Frank, C. Anderson, and D. Rubinstein, *New England Journal of Medicine,* 299 (1978), 111–115. By permission.

male coital dysfunctions. *Primary impotence* exists when a man has never been able to have coitus; *secondary impotence* refers to failure in a previously functional person and is ten times more common than primary impotence.[19] Others use the primary versus secondary distinction to separate cases where there is no demonstrable physical cause from others where there is (estimated to be 10 to 15% of cases).[20]

[19]Masters and Johnson (1970).
[20]Kolodny et al. (1979, p. 375).

Occasional inability to have an erection is exceedingly common, especially as a man gets older. Such transient failure is not pathological. Nor is there any absolute scale against which to evaluate the strength of potency. Obviously a man must keep his erection long enough to enter the vagina if he is to have coitus. Beyond such minimal criteria, judgments are relative: a penis that is powerless in one instance may perform adequately in another. For clinical purposes a man is considered impotent if his attempts at coitus fail in one out of four instances.[21]

Erectile impotence affected about one of every hundred males under 35 years of age in the Kinsey sample, but the inadequacy was seriously incapacitating in only some of them. At age 70 about one in four males was impotent. The progressive decline of potency with age is fairly general, though some men retain their potency well into old age. Among the subjects studied by Masters and Johnson, 31% had problems of potency, of which 13% were primary and 87% of the secondary type.[22]

For many men, it is difficult to imagine a more humiliating problem than impotence. The damage is far more serious than loss of sexual pleasure. Male notions of masculinity and personal worth are so closely linked to potency that serious dysfunction is likely to shatter self-esteem. In most cultures impotent men have been objects of derision. The classic Arabian love manual, *The Perfumed Garden*, has this to say about impotent men:

> When such a man has a bout with a woman, he does not do his business with vigor and in a manner to give her enjoyment . . . He gets upon her before she has begun to long for pleasure, and then he introduces with infinite trouble a member soft and nerveless. Scarcely has he commenced when he is already done for; he makes one or two movements, and then sinks upon the woman's breast to spend his sperm; and that is the most he can do. This done he withdraws his affair, and makes all haste to get down again from her . . . Qualities like these are no recommendation with women.[23]

More rational and compassionate views of impotence would result if loss of sexual potency were not made to reflect on any other aspect of the person. Furthermore, a less genital focus would foster a broader perception of sex. There are many physically handicapped persons, for instance, who are incapable of erection yet able to engage in other pleasurable forms of sexual relations without thinking of themselves as being less of a man (box 15.2).

Ejaculatory Dysfunction

The most prevalent form of ejaculatory disturbances is *premature ejaculation.* Armed with the finding that three out of four men reach orgasm within two minutes of intromission, and that most male animals do so even sooner, Kinsey made light of premature ejaculation as a form of sexual dysfunction. Yet there is no question that a significant number of men (and their partners) complain of the inability to delay ejaculation until sufficient mutual enjoyment has been obtained; it is of small comfort to them to learn that nonhuman primates ejaculate even sooner.

Although premature ejaculation is a less frequent reason than impotence for seeking help, Masters and Johnson believe it to be the most common male sexual dysfunction in the general population: an estimated 15 to 20% of men have significant difficulty controlling rapid ejaculation (although less than 20% of this group consider it a serious enough problem to seek help for it).[24]

For clinical purposes, premature ejaculation has been defined as the inability to control ejaculation long enough to satisfy a normally functional female partner in at least 50% of coital encounters.[25] It has also been suggested that this issue is not a problem if a couple agrees that the quality of their sexual relations is not influenced by when the male ejaculates.[26]

The opposite of premature ejaculation is *retarded ejaculation,* which consists of difficulty in ejaculating intravaginally. This is the least frequent of all male sexual dys-

---

[21]Masters and Johnson (1970, p. 157). For overviews of male dysfunction, see Friedman (1968); Zilbergeld (1978); Kolodny et al. (1979).

[22]Masters and Johnson (1970, p. 358).

[23]Nefzawi (1964 ed., p. 110).

[24]Masters and Johnson (1970); Kolodny et al. (1979).

[25]Masters and Johnson (1970).

[26]LoPiccolo (1977).

Box 15.2   Sex and the Handicapped

It is all too easy to think that the physically handicapped are neither capable of nor interested in sex, or that their disabilities are so overwhelming that the lack of sex in their lives should be the least of their problems. But that is not the case. A good deal can be done to sexually rehabilitate the handicapped and to educate the public.[1]

Serious physical handicaps usually develop because of chronic illness or through acute injury to the spinal cord. Severe arthritis over the years may seriously limit mobility and the ability to engage in sex without discomfort. Confined to beds and wheelchairs, these usually elderly persons do not always lose their sexual interest although society and circumstances make its fulfillment very difficult.

Even more compelling is the plight of thousands of younger men and women who have sustained spinal cord injuries leading to paralysis of the legs (*paraplegia*) or of the legs and arms (*quadriplegia*). The most common causes for such injuries are accidents and war injuries; hence these conditions are more common among men than women.

Paraplegics usually lack all sensation below the level of the spinal lesion and lose normal bladder and bowel control. Sexual functions are similarly seriously disrupted but by no means uniformly absent. Two-thirds of men with cord lesions may retain some erectile ability in response to physical stimuli (without necessarily being able to feel it). Ejaculatory disturbance is more severe; usually less than 5% can ejaculate and they are usually infertile.[2] Paraplegic women likewise lack genital responsiveness and coital orgasmic ability, but they are more likely to

retain fertility. They may also attain orgasm through stimulation of the nipples, lips, and other erogenous areas. Both males and females can experience erotic dreams with orgasm ("phantom orgasm").

Quite understandably, some seriously handicapped persons may give up sex. Yet others manage to adapt to their circumstances and show extraordinary resilience in maintaining an active and joyful interest in it. They learn to use whatever parts of their bodies allow them to give and receive sexual pleasure. Sexuality for them takes on a broader meaning than the mere coupling of genitals. As one man put it, "I can't always be genital, but that gives me more permission to be gentle."[3]

The following guidelines intended for the physically handicapped may hold a larger lesson for all of us.

A stiff penis does not make a solid relationship, nor does a wet vagina. Absence of sensation does not mean absence of feeling. Inability to move does not mean inability to learn. The presence of deformities does not mean the absence of desire. Inability to perform does not mean inability to enjoy. Loss of genitals does not mean loss of sexuality.[4]

[1]For various approaches to dealing with the consequences of disability, see Comfort (1978); Webster (1983).
[2]Bors and Comarr (1960).
[3]Cole (1982).
[4]Anderson and Cole (1975).

functions and was recognized as an entity only recently. Many males occasionally experience a temporary inability to ejaculate during coitus that is overcome by more vigorous thrusting or some other form of heightening arousal. Those with retarded ejaculation cannot accomplish this except with much more difficulty. Those who are totally unable to reach orgasm intravaginally are said to suffer from *ejaculatory in-*

competence. The majority of these men can have orgasm through masturbation, but in some 15% orgasm does not occur extra-vaginally as well. In these cases arousal subsides slowly without the climactic release of orgasm.

Given the emphasis on prolonging coitus, one would think this condition would be taken as a blessing. Yet beyond a certain point coitus ceases to be enjoyable for either partner. The man experiences a sense of failure and frustration over his lack of control; the woman may feel responsible for not being sufficiently exciting, or irritated at having to go on beyond the point of enjoyment (similar to how men feel when they have to go on seemingly endlessly before the woman reaches orgasm).

Female Orgasmic Dysfunction

Serious attention to female sexual dysfunction is a relatively recent phenomenon.[27] Problems of sexual arousal in women during coitus are harder to separate from those of orgasm. Since premature orgasm does not usually constitute a problem, *coital anorgasmia* is by far the most common form of female sexual dysfunction, accounting for some 90% of all cases.[28]

The prevalence of primary anorgasmia (when a woman never reaches orgasm during coitus) is about 10%; a similar percentage of women experience orgasm only sporadically. This means that 20%, or at least one out of five women, have difficulty obtaining orgasmic response during coitus. There are some indications, however, that the problem of female coital inadequacy may be significantly receding. In the Hunt survey, 53% of women who had been married for 15 years reported that they always or nearly always reached orgasm in coitus: the corresponding figure in the Kinsey survey two decades earlier was only 45%. Likewise, the proportion of wives who

never reached orgasm had gone down from 28% in the Kinsey sample to 15% in the Hunt sample. The 1983 *Family Circle* poll shows some 85% of wives basically satisfied with their sex lives; but that does not necessarily mean that all of these wives are orgasmic.[29]

Anorgasmic women have been less disturbed by their failure to reach coital orgasm than men have been by impotency. Not only are these women able to engage in sex and conceive without orgasm but at least some say that they enjoy coitus even if they do not reach orgasm. Whether there is a physiological explanation for this or women are simply willing to settle for less is not clear.

Dyspareunia and Vaginismus

Discomfort in coitus is a fairly common complaint among women, but in only a small percentage is *dyspareunia* or painful coitus severe enough to interfere with sexual function. Pain may be experienced during vaginal penetration, during intercourse, or following it. Pain and fear of pain may be indistinguishable in the first instance. When pain occurs only during intercourse, it is less likely to be confused with fear of pain. Pain after coitus usually takes the form of a dull pelvic ache.

When pain is felt or anticipated, a certain amount of muscular tension is inevitable. As the muscles surrounding the introitus become spastic, they cause *vaginismus,* which makes vaginal penetration difficult if not impossible. Vaginismus occurs more often in response to psychological than physical causes and affects 2 to 3% of adult women.[30] Generally these women have no difficulty with sexual arousal and can attain orgasm through noncoital means.

Causes of Sexual Dysfunction

Sexual function is vulnerable to disruption by biological and psychological causes,

---

[27]For an overview of female dysfunction, see Musaph and Abraham (1977); Kolodny et al. (1979).

[28]For more detailed sources on anorgasmia, see Kinsey et al. (1953); Masters and Johnson (1970); Fisher (1973); Kaplan (1974); Hunt (1975); Levin and Levin (1975).

[29]*Family Circle,* January 1983 (in *Time* January 31, 1983, p. 80).

[30]Kolodny et al. (1979); Spark et al. (1980); Spark (1983).

as well as cultural attitudes. In principle, all three variables may interact in a case; in practice, we identify the disorder according to the dominant cause. The causes of sexual dysfunction can thus be subsumed under several categories: *organic, psychogenic,* and *cultural.*

In most cases specific correlation between cause and effect is not possible, especially where causes are psychogenic. Thus the same psychological problem may cause sexual apathy in one woman and anorgasmia in another; impotence in one man and premature ejaculation in another. Where organic causes are present, more specific linkages may be possible. For example, a drug that disrupts parasympathetic function will cause impotence; one that interferes with sympathetic function will interfere with orgasm.

Organic Causes

The majority of cases of sexual dysfunction are said to result from psychogenic, not organic, causes. However, in a significant number of cases (perhaps as high as one third) there may be a physical basis underlying the sexual disorder, more so for some conditions than others.[31] For example, impotence is more often due to an organic cause than premature ejaculation; deep pelvic pain during coitus is much more likely to have a physical basis than sexual apathy.[32]

The sexual organs are well designed to fit (figure 15.2). Despite variations in the penis and vagina, disparities in size and shape are almost never a cause of coital difficulty. As an elastic structure, the relaxed and lubricated normal vagina expands to whatever extent necessary to accommodate the erect penis.

Sexual functions can suffer as a result of *acute* or *chronic, local* or *systemic* causes. Debilitating illness such as cancer, degenerative diseases, severe infections, or systemic disorders involving various bodily systems may affect sexual function indirectly. Local disturbances may interfere

more directly. The common element in a lot of these conditions is that they cause pain, which has a dampening effect on sexual interest, arousal, and enjoyment.[33]

Sexual functions are particularly vulnerable to disturbances of the *circulatory, nervous,* and *endocrine* systems. Diseases specifically affecting male genitals include conditions that cause pain during intercourse (inflammation, Peyronie's disease); interfere with intromission (congenital anomalies); or affect testicular function (tumors). Among women there are even more frequent causes for pain during coitus (inflammations, clitoral adhesions, deep pelvic infections, various forms of vaginitis).

Surgical conditions that can be at fault include procedures that damage the male genitals and their nerve supply (some types of prostatectomy, lumbar sympathectomy). Castration will result in loss of libido and impotence (chapter 4). Likewise, female surgical procedures that damage the sexual organs (obstetrical trauma, poorly repaired episiotomies) often cause pain during coitus. Conditions that interfere with the blood supply to the genitals (thrombosis) also cause difficulty, among males in particular.

Endocrine disorders of significance include hypopituitarism, hypogonadism, hypothyroidism, diabetes, and certain liver diseases like hepatitis and cirrhosis. Testosterone deficiency is now reported to be a more prevalent cause of sexual dysfunction than previously suspected. In one group of 105 consecutive cases of impotence, 35% had abnormalities of the hypothalamic-pituitary-gonadal axis. It is thus important that organic factors first be ruled out in all cases of sexual dysfunction.

Finally, there are neurological disorders that may seriously influence sexual functions in both sexes. These include diseases of the frontal and temporal lobes of the brain, such as tumors, vascular damage, trauma, and disturbances in the spinal cord, such as congenital anomalies, degenera-

[31]Munjack and Oziel (1980).
[32]Kolodny et al. (1979).

[33]Two thirds of patients with chronic pain report a deterioration of their sexual relations (Maruta and McHardy, 1983).

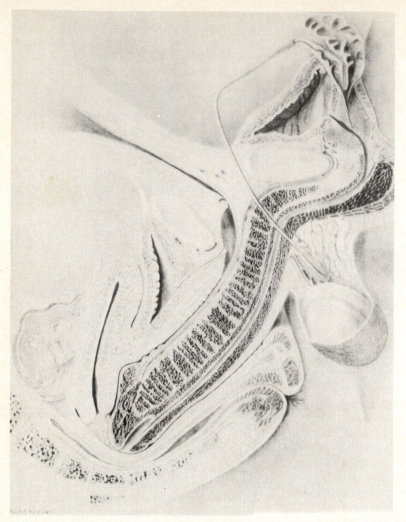

Figure 15.2 Male and female sex organs during coitus. (From Dickinson, R.L., *Atlas of Human Sex Anatomy*, 2nd ed. Williams & Wilkins, 1949, figure 142.)

tive conditions, and injury. Epilepsy usually does not cause sexual problems.[34]

When there are no obvious ailments, the differentiation of organic from psychogenic sexual dyfunction may not be easy. If the dysfunction occurs in one but not another setting, then the cause is more likely to be psychological. With regard to impotence, a helpful test is to monitor erections during sleep, since virtually all healthy males have erections in their sleep (closely associated with REM periods) (chapter 10).

The diagnostic procedure is carried out in a sleep laboratory where the pattern of sleep can be followed with an electroencephalographic recording and simultaneous monitoring of eye movements. Penile erections occurring during sleep are detected by strain gauges placed at the base of the penis that accurately record the magnitude, duration, and pattern of erection and whether it coincides with REM

[34]For more detailed discussion of organic causes, see Kaplan (1974); Kolodny et al. (1979).

sleep. Complete absence or seriously deficient nocturnal penile tumescence suggests an organic cause. If *nocturnal penile tumescence* (NPT) is normal, psychogenic factors are much more likely to be the underlying problem of the impotency.[35]

Alcohol   *Alcohol* is a common cause of sexual dysfunction: 40% of chronic alcoholics have problems with potency and 15% with orgasm.[36] Chronic alcohol usage interferes with hormone production, liver function, and nutrition, and causes nerve damage, all of which are detrimental to sexual function. These organic effects are compounded by the psychological and social problems caused by alcoholism.

Although alcohol is widely believed to enhance sexual activity, in fact it has a physiologically dampening effect on sexual arousal and performance (box 11.7). Even well below levels of intoxication, alcohol has been shown to inhibit erections and female arousal.[37]

Although alcohol intoxication exerts a progressively greater physiological depressant effect in retarding orgasm, women who do reach orgasm despite it report a heightened sense of pleasure; men report decreased arousal and less pleasurable orgasm under its influence. Alcohol can also be the trigger for psychogenic impotence; a man may fail to have an erection after having too much to drink; that event may in turn lead to anxiety and subsequent failures even when sober.[38]

Drugs   Drugs are another important source of difficulty. This is especially the case with substances like *sedatives* (such as barbiturates) and *narcotics* (such as heroin).

Marijuana, like alcohol, has a widespread reputation as an enhancer of sexual experience (box 11.7), but marijuana use has also been linked with erectile problems, lowered testosterone production, and disturbance in sperm production.[39] Men who smoke four or more marijuana cigarettes per week have significant decreases in testosterone production. The decrease in testosterone is related to the amount smoked; that is, heavier usage produces even lower levels of hormone production.

Another class of drugs that may impair libido and sexual response are the *antiandrogens*, which include estrogen, adrenal steroids such as cortisone, and ACTH (often used to treat allergies and inflammatory reactions). There are also experimental drugs like *cyproterone acetate* that have been employed in the suppression of certain compulsive sexual behaviors (box 20.5).

Other drugs act peripherally by blocking the nervous impulses to the genitalia. Drugs that block the effects of the parasympathetic system interfere with arousal and erection, those that block the effect of the sympathetic system interfere with orgasm (chapter 3). It is on this basis that *antihypertensive drugs* frequently impair sexual arousal in both sexes and cause impotence. Commonly used *tranquilizers* like Valium and Librium can also cause transient sexual dysfunction. Once the use of these drugs is discontinued, their effects subside. The main thing to remember is that whenever a person is using a drug—any drug—and develops sexual problems, the drug must be evaluated as the first possible cause of the difficulty.

Psychogenic Causes

Psychogenic factors are more difficult to identify and classify than organic causes, especially if they pertain to deep-rooted intrapsychic conflicts. In the past, treatment of sexual dysfunction mainly focused on fundamental personality problems. In more recent approaches to sex therapy (the "new sex therapy") the emphasis is on the identification of more immediate psychological factors in sexual dysfunction.

[35]Fisher et al. (1975); Karacan (1978); Bohlen (1981); Marshall et al. (1981).

[36]Kolodny et al. (1979).

[37]Wilson and Lawson (1978); Malatesta (1979).

[38]For an annotated bibliography on alcohol and sexuality, see O'Farrell et al. (1983).

[39]For a review of studies in this area, see Kolodny et al. (1979, pp. 336–340).

Sexual anxiety based on the fear of failure is possibly the most common immediate cause of psychogenic impotence. Other sources of sexual problems are the *demand for performance* and the excessive *need to please* the partner. Such attitudes elicit resentment and anger, which interfere with sexual enjoyment and function. Another important cause of difficulty is *spectatoring*, whereby a person anxiously and obsessively watches over his or her reactions during sex. Because satisfactory sex requires that one be able to lose oneself in the erotic interaction, such spectatoring can distract the person and prevent the appropriate sexual responses from building up to orgasm.

The *failure to communicate* forces a couple to guess what is desired and what is ineffective or objectionable in the sexual interaction. Communication that is clear and appropriate to the occasion is necessary to provide information as well as reassurance. Even when failure to communicate is not the primary cause of the dysfunction, it helps to perpetuate other existing pathological sexual interactional patterns.[40]

The deeper causes of sexual malfunction in both sexes are rooted in internal conflicts primarily related to past experiences. When these conflicts dominate a person's sex life to the extent that inadequate performance is the rule, then the causes can be considered to be primarily *intrapsychic*. On the other hand, when the sexual problem seems to be part of a larger pattern of conflict between two people, it is more convenient to label it as *interpersonal*. This distinction, though arbitrary, has practical merit in treatment. Intrapsychic causes must be dealt with as such. As veterans of multiple divorces discover, even though marital partners change, the conflicts may remain the same. Interpersonal conflicts also ultimately result from intrapsychic problems, but the latter may be more circumscribed and require no special attention.

Treatment may then be focused on the relationship rather than on the individuals themselves.

Learning theorists have proposed a variety of models to explain the genesis of sexual malfunction. Central to many is the mechanism of *conditioning*, in which the unpleasant feelings associated with an experience determine one's future reaction to a similar situation (chapter 8). Sometimes the antecedents of the experience are easy to trace. For instance if a man suffers a heart attack during coitus, thereafter the very thought of sex makes him anxious and unable to perform. Similarly, after a rape experience, a woman may find coitus difficult even with a loving partner.[41]

More often the causes are a complex series of long-forgotten learning experiences. The transmission of certain sexual attitudes and values to children—like sex being dirty or dangerous—is one example (chapter 9). A person may not remember specific or implied parental admonitions and punishments, but their damaging effects on his sexual performance persist.

Psychoanalysts offer explanations of sexual malfunctioning based on infantile conflicts that remain unconscious. For instance, conflicts arising from unresolved *oedipal wishes* may be major causes of difficulties in both sexes. *Castration anxiety* is a common explanation for failures of potency among males. A man who retains repressed and unresolved incestuous wishes from the oedipal period may unconsciously reexperience these forbidden and threatening desires when he attempts to engage in coitus. Under these circumstances he cannot perform, and by losing his potency he obviates the necessity for engaging in the symbolically incestuous act and its dire consequences. When a man is impotent with his wife but not with a prostitute, he may be unconsciously equating his wife with his mother. Men who distinguish between "respectable" women (to be loved and respected) and "degraded" women (to be enjoyed sexually) are said to have a "madonna-prostitute complex."

[40]Kaplan (1974). See chapter 7 for further discussion of immediate causes. Fay (1977) considers sexual problems related to poor communication.

[41]Wolpe and Lazarus (1966).

The female counterparts to these conflicts involve the father. As certain types of men may be unconsciously identified with the father, coitus with that type of man, and by extension any man, elicits guilt and dysfunction. By failing to enjoy the experience, the woman feels less guilty about her incestuous wishes.

Another psychological factor is the threat of *loss of control*. As orgasm implies self-abandonment, it raises the possibility that aggressive impulses will also be triggered. The fears that the erect penis will tear apart the vagina or that the penis will be trapped and choked by the vagina are not uncommon. Such concerns may be experienced consciously but more often are unconscious: the man simply fails to have an erection, or the woman fails to reach orgasm, neither quite realizing why.

Interpersonal conflicts are extensions of intrapsychic problems, but sometimes the problems arise only in a particular type of relationship. All that we have discussed with regard to sex and interpersonal relationships (chapter 12) is pertinent here. Intense disappointment, muted hostility, or overt anger may obviously poison sexual interactions. Subtle insults are just as detrimental. Women, for instance, are quite sensitive about being "used." If a man seems to be interested predominantly in a woman's body and neglects her person, she will feel that she is being reduced to the level of an inanimate object. Some women associate coitus with being exploited, subjugated, and degraded, and rebel against it by failing to respond.

The attitudes most detrimental to the male's enjoyment are those that threaten his masculinity. Lack of response on the part of the woman, nagging criticism, and open or covert derision rarely fail to have an impact on male enjoyment. A man who feels overburdened by a dependent wife may also react negatively if her sexual requirements also appear to be endless.

Other dyadic causes include *contractual disappointments*. When people establish friendships or get married, their sexual expectations are rarely overtly communicated or negotiated. There is thus much room for misunderstanding and anger over the partner's not living up to what was expected. Or, as people change, new needs and new preferences alter the nature of the original relationship to which the couple must adjust.

In power struggles, many forms of *sexual sabotage* can be seen to operate: One partner will pressure the other into sex at inopportune moments; sex is withheld as punishment; people make themselves undesirable or even repulsive through neglect of their bodies, cultivating the wrong physical image, or behaving in ways that are irritating and embarrassing to the other. The intended and sometimes unintended victim of all this is the sexual life of the couple.[42]

Cultural Causes   Societies influence sexual function through the shaping of sexual values during development and in defining the institutional contexts of sexual relationships (chapter 12). It would be foolish to deny the impact of prevalent sexual values on individuals, but sweeping indictments of a culture's negative effects are difficult to substantiate. For instance, a great deal has been made of the damaging effects of Victorian attitudes on female sexuality or the new assertiveness of women in sapping men's virility (the "new impotence"), often with little concrete evidence.[43]

In our progressively secular society it is claimed that *shame* is replacing guilt as the primary social inhibitor.[44] Yet others think that as a society we have become *shameless*.[45] Some argue that the premium we put on competence and success, combined with an overemphasis on sex, creates a formidable hurdle to enjoyment. Orgasm becomes a challenge, rather than the natural climax of coitus. Inability to achieve it or a certain form of it (multiple, vaginal, mutual, and so on) becomes not only a signal

---

[42]For further discussion of interpersonal conflicts affecting sex, see Kaplan (1974, chapter 9).

[43]See Ginsburg, et al. (1972); Gilder (1973).

[44]Ellis and Abarbanel, eds. (1967, p. 451).

[45]Tyrmand (1970).

of sexual incompetence but also a reflection of personal inadequacy. Ironically, as we become freer about sex, new problems are beginning to arise related to excessive demands for performance and wholly unrealistic expectations of what sex can be and do in one's life.

Sex Therapy

Sexual dysfunction may be mild and transient, requiring no therapy, or it may present formidable challenges to treatment. Its remedies range from fairly simple short-term programs to highly specialized, intensive treatments, involving psychological methods, drugs, and surgery.

Until about a decade or so ago, the treatment of sexual dysfunction was in the hands of generalists. Medical practitioners, psychiatrists, psychologists, marriage counselors, and others who treated physical and psychological problems and marital conflicts also dealt with sexual dysfunction. Following the work of Masters and Johnson new methods of treating sexual dysfunctions have developed that appear more efficient. The application of these newer methods is now referred to as *sex therapy*.[46] Its focus is on the elimination of sexual symptoms rather than on personality change or the restructuring of relationships outside the sexual realm.[47]

Treatment Strategies

Sex therapy so far has no particular theoretical base of its own. It combines medical and psychotherapeutic approaches with techniques of behavior modification. After organic factors are ruled out, the key tasks are to have the couple accept sex as a natural function that requires no heroic effort, but only a relaxed, accepting attitude. The focus is on the couple, not the individual.[48] No one is at fault, no one is sick; there are merely inhibitions to overcome. Each must learn to give, as well as to receive, sexual pleasure, to get actively involved rather than to be a spectator or a passive participant.

Many of the techniques of sex therapy are basically the same as the means of sexual enhancement we discussed earlier (chapter 11). The main difference is that with sex therapy, a couple is put through the paces in more graduated fashion and under guidance, rather than being left to the couple's own resources to manage and improvise. By the same token, ordinary couples with no dysfunction can find the techniques of sex therapy useful for enhancing their sexual experiences.

The Masters and Johnson sex therapy format has a number of general procedures which are supplemented by more specific strategies for particular conditions. Treatment progresses along two complementary tracks. The first involves verbal interactions between the couple and the male and female co-therapists: detailed sexual histories are taken from each patient; roundtable sessions explore past experiences, conflicts, feelings, attitudes on both sexual and nonsexual matters that have a bearing on the dysfunction; successes and failures in the ongoing treatment are analysed, and so on. Concurrently, the couple go through a sequence of sexual assignments in private that eventually culminate in mutually satisfactory sexual intercourse, if the treatment is successful.

Of particular importance in the early phase of therapy are *sensate focus* exercises whereby the ground is prepared for

---

[46]This method was originally described in Masters and Johnson (1970). For a summary in non-technical language, see Belliveau and Richter (1970).

[47]Helen Kaplan (1974) coined the term the "new sex therapy" to designate these newer approaches. Her book that carries that title has a comprehensive exposition of the causes, forms, and treatments of sexual dysfunction. The literature on sex therapy has been expanding rapidly. See Kolodny et al. (1980); Barbach (1980); Leiblum and Pervin (1980); Kaplan (1979); LoPiccolo and LoPiccolo (1978); Caird and Wincze (1977).

[48]Earlier in their work, Masters or Johnson provided *partner surrogates* to some of their women patients. Legal and other considerations make this practice problematic.

Figure 15.3   The training position for the treatment of female sexual dysfunction.

the subsequent tackling of the more specific problems. These exercises are aimed to overcome the common immediate causes of sexual dysfunction: anxiety, spectatoring, demand for performance, noncommunication. They start with activities focused on touch awareness and *pleasuring* rather than sexual arousal. The couple is positioned so that the helping partner has easy access to the body of the person with the primary problem. Thus, when the symptom is female anorgasmia, the man is seated behind the woman (figure 15.3); in impotence, the position is reversed.

Treatment of Erectile Dysfunction   In these cases, the woman takes the initiative to touch, caress, or gently fondle the man's body, initially keeping away from the genitals. The man guides his partner by his verbal or nonverbal responses as well as by leading her hand (*handriding*). In subsequent stages, the couple alternates in

pleasuring one another, with more direct communication to guide each other's actions.

After the couple becomes adept at general physical enjoyment, they move on to more explicitly erotic techniques, with the woman now stimulating the man's genitals, along with the rest of his body (figure 15.4). Yet the interaction continues in a relaxed, nondemanding manner; the man is not expected to have an erection at any particular moment and neither person is allowed to rush to coitus. When erections do occur in the natural course of events (which is often quite soon in treatment) they are allowed to come and go to instill in the man the conviction that he does not "will" an erection but lets it happen; and should it subside, there need be no sense of failure. Intercourse is forbidden at this stage; it is to be attempted only when erections become frequent and stable during sensate focus exercises.

Figure 15.4   The training position for the treatment of male sexual dysfunction.

The transition from sensate focus exercise to coitus is gradual. When the man has gained enough confidence, the woman positions herself astride him and inserts his penis into her vagina, thus relieving him of taking responsibility and risking fumbling and failure (see figure 11.7). Intercourse then proceeds with the primary aim of providing the man with adequate satisfaction: the man is encouraged to be "selfish" during these periods; the sharing of pleasure follows later. In sum the basic aim in the treatment of erectile dysfunction is to reduce anxiety so that physiological reflexes can take their natural course.

Treatment of Anorgasmia   The procedure for dealing with female anorgasmia follows the same basic procedure as above. This time the man takes the initiative with early sensate focus exercises, but when it comes to coitus, it is once again the woman who assumes the above posture. This allows her to be in charge, thus counteracting her fears of being hurt, used, or put down. It also enables her to gauge the depth of intromission and the force and frequency of coital thrusts, all to suit her needs and help achieve orgasm.

Other strategies focus on more particular issues. For instance, women are more likely than men to have concerns over their physical attractiveness, to be more inhibited in allowing themselves sexual pleasure, or reticent to expose their erotic needs. These and similar issues are dealt with both in discussion and practice. Women who have never experienced orgasm must be made to have that experience first, however it is attained. She may thus be encouraged to stimulate her own body and genitals or be brought to orgasm by the partner manually or orally (some therapists also advocate the use of vibrators). Once the inhibiting barrier is broken, the woman will know what it feels like to have an orgasm and can be gradually helped to transfer the experience to the coital encounter. Other "bridging" techniques consist of clitoral stimulation by the man or woman concurrent with coitus (figures 11.8 and 11.9). The basic aim in treating failures of female sexual arousal and orgasm is to help the woman to "let go" and remove the inhibitions that block orgasm.

Treatment of Ejaculatory Incompetence
The first goal in the treatment of male ejaculatory incompetence is similarly aimed at

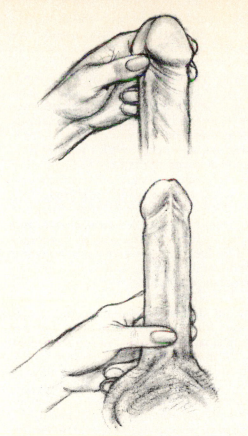

Figure 15.5   The "squeeze technique," glans version (top); basilar version (below).

bringing him to erection. When the man feels that ejaculation is imminent, she uses the *squeeze technique* to avert ejaculation. This consists of squeezing the *glans* between thumb and forefinger or putting pressure at the *base* of the penis (figure 15.5). She must be gentle yet firm in applying pressure at the front and back of the glans for 3 to 4 seconds, being careful not to hurt. Why this maneuver works is not known. It results in partial loss of erection, which is then brought back with further stimulation, and the cycle repeated.[49]

Coitus takes place in the woman-above position. During it, the squeeze cycle is repeated a number of times; as he senses the approach of ejaculation she takes out the penis, squeezes the glans, and reinserts it. This is followed by slow thrusting, and the cycle is repeated until the man wishes to ejaculate.

To avoid coital disengagement, pressure on the base of the penis may be substituted for the squeeze technique or a *stop-and-go* alternative substituted: as excitement mounts, thrusting stops; after a few moments of rest, when the sensation of ejaculatory imminence is passed, thrusting is resumed. The basic aim in treating premature ejaculation is to train the man to anticipate orgasm and modulate his level of arousal accordingly.

Treatment of Vaginismus   The treatment of vaginismus utilizes the process of desensitization. The nature of the problem is explained to the patient, who is made aware of the muscular tension at the vaginal opening. Then in a relaxed and safe setting (with the male partner present) a small plastic dilator is gently inserted in her vagina, as she watches in a mirror. The woman then uses these dilators, moving from the smallest to the largest, which is about the size of an erect penis, leaving each in place for short periods of time. When the couple moves on to coitus, she takes charge, inserting the penis herself, controlling the thrusts, and so on. The basic

crossing the orgasmic barrier. The man is helped to reach orgasm first through solitary masturbation; then he does it in the partner's presence, or she masturbates him (having him ejaculate at her genitalia). Coitus is attempted last with her bringing him to pitch excitement and inserting the penis as he is about to ejaculate. This is the only instance when the man takes the above-woman position (figure 11.6) to maximize his arousal, while she continues to stimulate the base of his penis manually.

Treatment of Premature Ejaculation   With premature ejaculators, there is no problem with arousal; treatment is focused on helping the man gain ejaculatory control. This is done by the woman assuming the stimulating position (figure 15.4) and gently

[49]The basic procedure in this approach was reported by Semans (1956).

aim in treating vaginismus is the elimination of the vaginal spasm.

These treatment strategies sound quite straightforward, yet in order to work they must operate in a broader therapeutic context. Their success calls for skillful therapists, motivated patients, and resourceful and compassionate sexual partners. These are not easy tasks for the squeamish.

Psychotherapy and Behavior Therapy

Psychotherapeutic methods of treating sexual dysfunction long antedate sex therapy. Their practitioners constitute a diverse group of specialists with various theoretical orientations; they may be psychoanalysts, psychiatrists, clinical psychologists, psychiatric social workers, and marriage and family therapists. The common element in their approach is the reliance on verbal interchange as the primary means of understanding and resolving personality problems and interpersonal conflicts.

Psychotherapy   Psychotherapists frequently deal with sexual issues, but most of them do not think of themselves as "sex therapists"; the methods they use to treat sexual problems are the same methods they use to treat any other kind of problem. *Psychoanalysis,* the most intensive form of psychotherapy, is highly focused on sexuality, given the centrality of sex in psychoanalytic concepts of psychopathology (chapter 8). So whether a patient seeks help because of a sexual problem or any other problem, much of the psychoanalytic work involves analyzing erotic fantasies, dreams, events, and past and present sexual relationships. Improvements in the patient's condition are contingent on the acquisition of insight into the origins of the difficulties. A scaled-down version of psychoanalysis is *insight-oriented therapy,* which pursues more limited therapeutic goals.[50]

Psychoanalysis has been the dominant influence in past psychotherapeutic approaches, but it is by no means the only one (and not all psychoanalytic approaches are Freudian; there are also Jungian, Adlerian, and other schools of psychoanalysis). A quite different psychotherapeutic approach, for instance, is the ***client-centered therapy*** developed by the psychologist Carl Rogers, which aims at clarification of the client's current feelings rather than fathoming unconscious conflicts and links to the past. ***Supportive therapy*** similarly focuses on current and conscious matters and aims at alleviating anxiety and distress by allowing the patient to express painful feelings, receive psychological support and comfort, and help negotiate interpersonal conflicts.

Behavior Therapy   Behavior therapy is another form of psychotherapy, but for historical reasons it is more often referred to as *behavior modification.* Its focus is the problem behavior itself, which it aims to change by techniques derived from learning theory (chapter 8) without concerning itself with underlying psychodynamics.[51]

Behavior therapy, like psychotherapy, subsumes a variety of techniques. *Desensitization* is based on the principle of *reciprocal inhibition,* which makes use of the fact that contrary emotional states are mutually exclusive. For instance, one cannot be simultaneously anxious and calm; if anxiety interferes with a certain function, its effects can be counteracted or inhibited through relaxation.

Anxiety-producing situations cause distress at various gradations. For example, a woman may find it very difficult to have intercourse with her husband in the nude, although coitus when partially clothed may be more tolerable. The behavior therapist and the patient draw up a detailed list that hierarchically expresses the degrees of displeasure anticipated in these sexual situa-

---

[50]For a succinct discussion of the principles of insight-oriented psychotherapy, see DeWald (1971).

[51]For general texts on behavior therapy, see Wolpe and Lazarus (1966); Wolpe (1958); Lazarus (1977); and Bancroft (1977). For a text that exclusively deals with behavioral approaches to sexual dysfunction, see Jehu (1979).

tions, all the way to behavior that the patient can contemplate without discomfort, for instance, being kissed on the cheek. The patient is then trained in the techniques of deep-muscle *relaxation* (which is incompatible with anxiety), and desensitization proceeds as follows: when the woman is fully relaxed, she is asked to fantasize the least fearful of the anxiety-provoking situations on the list; if she gets anxious, she switches back to the relaxation routine, otherwise she tackles the next troublesome scene. This process is repeated until the patient is able to confront in fantasy the anxiety-provoking situation in full (coitus in the nude). The next step is to transfer this mastery of the anxiety-producing stimulus to real-life situations.

Group Approaches   The therapeutic modalities we have described so far deal with individual patients or couples. In *group therapy* approaches, a small number of individuals (usually about six to eight) work with one or two therapists. The members of a group may be chosen on the basis of having similar or different problems; they may be unrelated or couples. In addition to the economy in the therapist's time, this approach allows members of the group to share and learn from their problems and lend each other emotional support. On the other hand, this format does not allow for as much individual attention, and especially with respect to a topic like sexual dysfunction, the group may have an inhibiting effect on candor.

Lonnie Barbach has reported good results working with groups of five to seven women with orgasmic problems. These women meet with two female therapists for discussions and also carry out daily assignments aimed at "getting in touch" with their own body and sexual feelings. Orgasm is initially attained through masturbation and eventually through coitus (the sexual partners are not directly involved in the treatment program). After 5 weeks, 93% of these women reach orgasm through masturbation and 3 months later, half are orgasmic during coitus.[52]

[52]Barbach (1975).

Treatment with Drugs

Drugs are among the oldest forms of medical treatment. Their application to sexual dysfunction is but another facet of the search for aphrodisiacs discussed earlier (box 11.7). Prior to the advance of modern pharmacology, the drugs available to physicians to treat sexual disorders were neither effective nor safe. A well-known example is *Spanish fly,* a powder made from dried cantharis beetles, hence also called *cantharides.* When ingested, it causes urethral irritation and penile vasocongestion that may result in erection, along with dangerous side effects.

Yohimbine (from the bark of the African Yohimbe tree) is a nervous stimulant that may cause erections, but its effects are not reliable. Yohimbine is usually combined with *nux vomica* (which contains minute amounts of *strychnine,* a central nervous stimulant and lethal poison) and testosterone in a drug called *Afrodex.*

The great chemical hope for the treatment of sexual dysfunction, especially impotence, has resided in sex hormones; a hope that has been largely unfulfilled. Given their indispensable role in the organization and activation of sexual functions, it was quite reasonable to expect that hormones would enhance sexual drive and potency. To this end, testicular extracts began to be used in the 19th century.[53] Furthermore, since sexual decline and aging seemed to be closely associated, such treatments were extended to purposes of *rejuvenation,* or regaining youthfulness. Though there has never been any convincing proof of their efficacy, extracts of testicular, placental, and embryonal tissue from animals continue to be used in various parts of the world in futile attempts to fight the effects of aging.

The popularity of **testosterone treatment** has waxed and waned over the years. There is little question that it can be effective in

[53]The renowned French physiologist, Charles Brown-Sequard, administered an animal testicular extract to himself in 1889 and became convinced of its invigorating effect; yet now we know that the extract he used was devoid of hormones (Gilman et al. 1980).

restoring libido and potency in males who have an androgen deficiency. There is also some evidence that testosterone may enhance sexual function in cases of psychogenic impotence by giving the body a temporary physiological boost.[54] Testosterone is therefore a legitimate drug for the treatment of select cases. But there is no convincing evidence that it has any effect on counteracting the effects of aging on sexual decline. Furthermore, the use of testosterone by older persons poses certain risks because of its exacerbating effect on prostatic hypertrophy and cancer (chapter 7).

Estrogens have no erotogenic effect; on the contrary, they tend to dampen libido in both sexes. Yet the use of estrogen in menopausal women can have a salutory effect on sexual function by counteracting the changes in the vaginal lining and the drying out of the lubricatory response that results from estrogen deprivation (chapter 7). The effect of androgens on women's sex drive is ambiguous (chapter 4), and their virilizing side effects so patently unacceptable that testosterone is not used to treat female sexual dysfunction.

Various other drugs have occasionally appeared on the scene with incidentally aphrodisiacal effects without any of them becoming established as a reliable and safe therapeutic agent for sexual dysfunction. Examples are *L-dopa*, which is widely used in treatment of Parkinson's disease (a neurological disorder); *PCPA* (p-chlorophenylalanine), an experimental psychiatric drug; and *vitamin E*, used in some cases of infertility.[55]

Tranquilizers like Valium and Librium may occasionally help by reducing anxiety but are not specific agents for treating sexual dysfunction. In short, compared with antibiotics, which have had a revolutionary effect in controlling sexually transmitted diseases, nothing remotely approaching their specificity and effectiveness has yet appeared with regard to the chemical treatment of sexual dysfunction.

Physical Treatment Methods

Kegel Exercises   It is not uncommon for women, following childbirth, to develop *stress incontinence,* involuntary loss of spurts of urine on coughing or straining (which raises the intra-abdominal pressure on the bladder). The cause of this is loss of tone of the pubococcygeal muscle. To correct this problem, a set of exercises were devised consisting of squeezing and relaxing the vaginal orifice, which gradually strengthened the muscles in the region of the vaginal and urethral orifices. In the 1950s the gynecologist Arnold Kegel discovered that these exercises simultaneously helped improve his patients' sexual responsiveness and orgasmic ability. Hence, these *Kegel exercises* have been a part of various sex therapy regimens.[56]

The great merit of Kegel exercises is their simplicity. The woman first learns to control the pubococcygeal muscle by squeezing around a finger inserted in the vagina, or by interrupting and releasing the stream of urine. She is then instructed to tighten and relax these muscles in a prescribed manner (such as tighten for 3 seconds, relax for 3 seconds, in sets of 10, three times a day).[57] The use of a *perineometer* (see box 3.2) allows for more accurate assessments of vaginal muscle tone. It also provides an aid to Kegel exercises by allowing the woman to monitor the effect of contracting her pubococcygeal muscle.

Despite much clinical evidence attesting to the value of strengthening the perineal musculature, empirical data on the correlation of vaginal muscle tone with orgasmic capability have been lacking. A study with normal subjects did not find pubococcygeal strength to be positively correlated to frequency or self-reported intensity of orgasm.[58]

Masturbation Training   *Masturbation,* with or without *vibrators,* is recommended by

[54]Kaplan (1974).

[55]Gessa and Tagliamonte (1974); Hyppa et al. (1975).

[56]Kegel (1952).

[57]For more specific instructions, see Kline-Graber and Graber (1978).

[58]Chambless (1982).

some therapists, specially in cases that do not yield to more standard approaches. Its main purpose is to allow a woman with primary anorgasmia to experience what it feels like to have an orgasm, which then serves as the goal. As more powerful sources of stimulation, vibrators may bring about orgasm when other means fail.

Various regimens have been devised for such *masturbation training* with the purpose of eventually enabling anorgasmic women to attain coital orgasm.[59] These attempts have so far met with mediocre success. It can be argued that as long as a woman reaches orgasm, by whatever means, she need not be considered dysfunctional. In this view, the attainment of masturbatory orgasm alone would be viewed as a successful outcome.

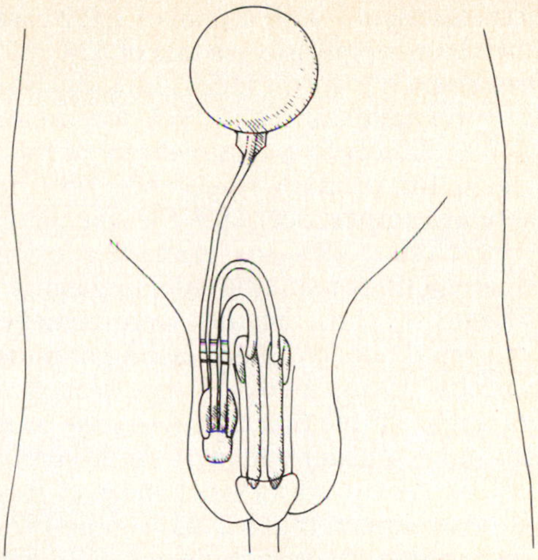

Figure 15.6 An inflatable penile prosthesis. (Courtesy of American Medical Systems.)

Penile Prostheses The use of *splints* to support a limp penis has been known in many cultures. These have been in the form of flat retainers or hollow dildos to house the flaccid penis.

Technological advances in surgery now make it possible to implant penile **prostheses** (Greek: addition). These devices do not cure impotence; they are mechanical devices that make intercourse possible. Currently, their use is more often for those cases of impotence that are due to irreversible organic causes such as lesions of the spinal cord. They are also a last resort for cases that do not respond to sex therapy and otherwise would have no hope of coital function.

One type of penile prosthesis consists of a fixed plastic rod implanted in the penis, providing it with enough rigidity to allow coitus. These devices are relatively simple to install surgically but have the obvious disadvantage of providing the man with a permanently erect penis.

A more sophisticated prosthesis is the inflatable variety (figure 15.6). This device consists of a *reservoir*, filled with a fluid, that is implanted in the abdomen. It is connected by tubes to a small *pump* lodged in the scrotal sac, which in turn has tubes

leading to the inflatable *cylinders* implanted in the penis. To attain erection, the man pumps the fluids into the cylinders; to return the penis to a flaccid state, a valve in the pump is released, sending the fluid in the penile cylinders back to the reservoir. The disadvantages of the inflatable device include the greater technical difficulty entailed in its installation, but the complications that occasionally arise are about equal in the two types of devices.

These procedures are not free from complications (such as infection or leakage of pump fluid) but success rates are now quite high.[60] A mechanically inflated penis may not be the same as a naturally erect one, but given the alternatives, these surgical methods certainly restore a level of function and mutual satisfaction that would be otherwise unavailable to these men.

Treatment Outcomes in Sex Therapy

The results originally reported by Masters and Johnson were quite spectacular.

---

[59]Anderson (1981); LoPiccolo and Lobitz (1978).

[60]In a series of 128 cases reported by Kessler (1981), 125 patients had normally functioning prostheses and were enjoying satisfactory intercourse. Complications occurred in about 16%, but could be remedied in many cases. Also see Kessler (1980).

Their experience with a total of 1,872 cases showed an overall success rate of 82% (85% for males; 78% for females) ranging from 99% for vaginismus and 96% for premature ejaculation to a relatively more modest 67% for primary impotence, 78% for secondary impotence, 76% for ejaculatory incompetence, 79% for primary anorgasmia, and 71% for situational anorgasmia.[61]

Moreover, these results were achieved over periods of treatment lasting a matter of weeks, using methods that looked so beguilingly simple that almost anyone could seemingly master them. And almost everyone tried to do so in a surge of therapeutic optimism that led during the 1970s to the emergence of a whole new field with numerous practitioners and two professional certifying organizations. This very profusion of people with varying credentials in the field requires that one choose a sex therapist with some care (box 15.4).

Not everyone became an overnight convert to the new sex therapy. Those steeped in the tradition of treating causes, not symptoms, reacted with predictable resistance. Others saw it as one more example of ideology parading as science and intruding into human behaviors that require no formal intervention.[62] And even those who readily acknowledged the benefits of the new approaches remained wary of its mechanical panacea aspects.[63] Yet success is hard to argue with.

A decade after the publication of *Human Sexual Inadequacy*, the book with which Masters and Johnson inaugurated the new field of sex therapy, a number of questions have been raised about the effectiveness of their methods.[64] The main thrust of this criticism is that Masters and Johnson do not say enough, and what they say is not said clearly enough to allow other investigators to evaluate and replicate their work. In other hands, the same techniques have produced less spectacular yet quite respectable results. The report of one such serious effort is presented in box 15.4.

Whatever its merits, it is clear that sex therapy is not going to be the answer to all of human sexual ills; sexuality is too vast and too complex an area to be so tamed. Sex therapy has nonetheless clear merits and will continue to be a viable and useful field. But it is important that we remain aware of its definite shortcomings. It has as yet no solid conceptual base, little intellectual substance, and a young clinical tradition. It is likely to be most effective when it is integrated into a broader program of treating psychological problems and interpersonal conflict.[65] It may be some time before the treatment of sexual dysfunction occupies its rightful place among respected medical and behavioral specialties; but that day will surely come, since sexual dysfunction deserves to be attended to as much as any other form of human suffering.

Prevention of Sexual Dysfunction

Better an ounce of prevention than a pound of cure is a truism as applicable to sexual dysfunction as to any other problem. Unfortunately, we do not yet have a clear enough understanding of the causes of sexual dysfunction to speak with confidence about their prevention. Nonetheless, several guidelines may be useful.[66]

Sex is a physiological function whose full enjoyment requires a healthy and vigorous body commensurate with one's age. Prevention, amelioration, and cure of disease and proper sexual hygiene (chapter 7) are thus prerequisites for avoiding sexual dysfunction.

Sex is also a psychological function that is highly vulnerable to anxiety, depression, and interpersonal conflict. Perhaps there is no way to entirely eliminate these painful emotions from human life, but at least their intrusion into the sexual realm

---

[61]Masters et al. (1982).

[62]Szasz (1980).

[63]Offit (1977); (1981).

[64]Zilbergeld and Evans (1980). For further discussion of evaluating treatment outcomes, see John (1979); Hogan (1977); Wright et al. (1977).

[65]Arentewicz and Schmidt (1983).

[66]For more extensive discussion of the prevention of sexual dysfunction, see Qualls et al. (1978).

Box 15.3   Getting Help for Sexual Dysfunction

We usually seek help when pain or distress become more than we can bear, or when we get alarmed that the physical or psychological condition might get worse if not attended to. Sexual dysfunction does not hurt in the same way; hence it is harder to decide when to seek help. This is especially true if the problem is not sexually incapacitating, such as with total impotence or anorgasmia.

If a person experiences a fairly rapid and consistent change in sexual function, it is important to get a medical check-up. Since symptoms of sexual dysfunction may be the early manifestations of an underlying illness, there may be more at stake than disruption of sexual function.

On the other hand, the fact that a sexual problem has existed for many years does not mean that one should simply accept it and learn to live with it. If there is uncertainty as to whether the dysfunction warrants therapy, one can always seek a consultation to help clarify the matter in one's own mind. Seeing a physician or sex therapist does not necessarily mean committing oneself beforehand to undergoing treatment.

Sex therapy, like other forms of specialized care, is, in principle, best sought not directly but through a provider of primary care such as a general practioner or the staff member of a student health center. Unfortunately, with respect to sexual problems, one may feel awkward going to such sources or may discover that these individuals are not particularly knowledgeable or sympathetic to one's sexual concerns.

However, going directly to a sex therapist has its own problems. Unlike other health care professions, the practice of sex therapy is still unregulated. Anybody entitled to see patients or clients can set up shop as a sex therapist; as a result there are many quacks in the business as well as well-meaning persons who simply lack the requisite skills and training to do sex therapy.

One is generally better off seeking help from individuals and clinics affiliated with institutions like hospitals, medical schools, and universities. Even if such institutions themselves do not provide such care, they may refer the person to reputable therapists. Similar advice may be obtained from medical and psychological societies and the directories of professional organizations in this field, such as the American Association of Sex Educators, Counselors and Therapists (AASECT).

Since membership requirements in such groups are not very stringent, one must continue to exercise good judgment before making a final choice. One should only deal with therapists who have a graduate degree from a reputable institution, show clear evidence of having received serious training in the treatment of sexual dysfunction, and are willing to discuss openly their qualifications, methods, time schedules, and fees. If there is any hint that sexual intimacies with the therapist are expected, go elsewhere.

The choice of a competent therapist is important but the success of sex therapy largely depends on the individual or couple. Going through the motions of seeking help to placate a spouse, or hoping that things are somehow going to change without serious effort on one's part, tends to become futile exercises. Joyful enthusiasm may be too much to expect from those contemplating therapy, but a certain measure of motivation and determination is essential for its success.

Box 15.4   Outcomes of Sex Therapy

In a systemic study of sex therapy outcomes, using modifications of the Masters and Johnson techniques, Arentewicz and Schmidt in Germany have reported the following outcomes. The ratings by therapists are shown first, followed by the ratings of patients and their partners.

OD = orgasmic dysfunction;
VA = vaginismus;
ED = ejaculatory dysfunction;
PE = premature ejaculation.

Overall Outcome of Therapy; Ratings by Therapists

	OD	VA	ED	PE	Total
Sexual functioning					
(1) drop-out, separation of couple	6%	4%	0%	7%	4%
(2) drop-out, unimproved	13%	7%	11%	3%	10%
(3) drop-out, slightly improved	5%	0%	2%	3%	3%
(4) therapy completed, unimproved	2%	0%	4%	0%	2%
(5) therapy completed, slightly improved	4%	11%	4%	3%	5%
(6) therapy completed, improved	33%	0%	14%	20%	23%
(7) therapy completed, distinctly improved	17%	11%	16%	23%	17%
(8) therapy completed, cured	19%	67%	49%	40%	35%
Sexual satisfaction					
(0) inapplicable, drop-out	24%	11%	13%	13%	17%
(1) less satisfying than before therapy	0%	0%	2%	0%	0%
(2) just as unsatisfying	5%	0%	6%	3%	4%
(3) somewhat more satisfying than before therapy	28%	43%	28%	24%	29%
(4) satisfying	43%	47%	50%	59%	49%

Overall Outcome of Therapy Reported by Patients/Partners

	OD Patient/ Partner		VA Patient/ Partner		ED Patient/ Partner		PE Patient/ Partner		Total Patient/ Partner	
Our sexual problems are now _____ than before therapy:										
(1) worse	2%	0%	0%	0%	0%	0%	0%	0%	0%	0%
(2) unchanged	3%	3%	0%	6%	2%	5%	4%	0%	3%	4%
(3) slightly better	29%	38%	11%	0%	13%	18%	36%	33%	23%	27%
(4) much better	64%	52%	61%	59%	73%	67%	55%	62%	65%	58%
(5) fully removed	2%	6%	28%	35%	13%	10%	5%	5%	8%	11%

Based on tables 4.11 and 4.20 in Arentewicz, G. and Schmidt, G. *The Treatment of Sexual Disorders* (1983).

could be minimized. The more one can keep sex free from entanglements in other concerns, the better off one will be sexually. Couples who treat sex as a "demilitarized" zone are at a distinct advantage: they can retreat to their sexual relationship as a haven from other areas of conflict so that at least one important facet in their lives remains satisfying and which in turn can ameliorate other aspects of their relationship.

Sex is highly influenced by culture. Shame and guilt serve legitimate social purposes, but they can also reflect unwarrented antisexual prejudice that can cripple sexual development and mar erotic enjoyment. At the moment, our culture does not have a clear and common understanding of how to raise childen who will be sexually healthy and fulfilled adults; it is important that it does develop such understanding.

One of the most important requirements for preventing sexual dysfunction is the active maintenance of sexual activity itself. To keep at it, to work at its enhancement, to deal with its shortcomings, are what it takes to remain sexually alive. Equally important is to place sexual fulfillment in a realistic perspective for each stage of life. The best way to get sexual satisfaction is not to chase it constantly like a rainbow. Sex is one of the greatest joys in life yet sex alone cannot make life fulfilling. On the other hand, even people with inhibited and faltering sexual lives need to remember that it is not sex that gives up on people but people who give up on sex.

SUGGESTED READINGS

Butler, R. N., and Lewis, M. I. *Sex after Sixty* (1976). A concise, clear, and sensible presentation of sexuality in late adulthood.

Masters, W. M., and Johnson, V. E. *Human Sexual Inadequacy* (1979). The book that established the modern field of sex therapy. For a briefer and more readable account of this work, see Belliveau, F., and Richter, L. *Understanding Human Sexual Inadequacy* (1970).

Kaplan, H. S. *The New Sex Therapy* (1974). The most comprehensive and clearly presented work on sex therapy.

Leiblum, S., and Perrin, L., eds. *Principles and Practice of Sex Therapy* (1980). Contribution to various aspects of sexual dysfunctions and their treatment.

Part III

Culture

Chapter 16
Sex and Society

Society is like the air, necessary to breathe, but insufficient to live on.

George Santayana

The task facing us in this section is more complex than the previous ones. Though much remains to be learned about the biology of sex, much of the information that pertains to it is straightforward. The behavioral aspects of sexuality are more convoluted; yet by comparison, its cultural complexities are awesome.

To clarify the relationship between sex and society, we shall approach it first from three perspectives: sociological, anthropological, and historical. This sets the stage for a review of the salient sexual issues in contemporary society, which is followed in subsequent chapters by more extensive discussions of eroticism in Western culture, sex and the law, and sexual morality.

Social Perspectives on Sexuality

We use words like "social" and "cultural" quite freely, yet precise and comprehensive definitions of such terms are difficult to agree on.[1] Simply stated, *society* (Latin: fellowship) is the network of human relationships and institutions that bind together a group of people characterized by a common culture. *Culture* (Latin: cultivate) in turn refers to the characteristic behavior patterns of members of a society

including their language, customs, as well as their intellectual, artistic, and material accomplishments. These terms are often used interchangeably and sometimes combined in the broader term, *sociocultural.* Even more global is *civilization*, which subsumes the social and cultural characteristics of a particular nation, region, epoch, or heritage, usually of an advanced state of development.

Sexuality and its psychosocial derivatives, like gender and sex roles, permeate virtually all aspects of our society and culture: the family, religious institutions, the legal system, networks of personal relationships, and countless other human associations are touched by the sexual element. Gender, marital status, social class, ethnic background, and other variables, in turn, shape and regulate our sexual behavior. Similarly, art, literature, and popular culture (such as movies, television and advertising) are suffused with sexual themes that both reflect and influence our sexual attitudes.

In a pluralistic society like the United States, the enormous variability of behavior patterns would seem to overwhelm any attempt at discerning common characteristics in our sexual behavior. When viewed in an even broader cross-cultural perspective, the range of diversity becomes quite bewildering. Yet it is possible to see underneath all this diversity not only shared

[1]In a review of definitions of culture, Kroeber and Kluckhohn (1952) discuss over 160 alternatives.

elements across cultures but linkages to our evolutionary primate ancestors.

Social Regulation of Sexuality

All societies take an active interest in and regulate the sexual behavior of their members.[2] At the most fundamental level, societal interest in sex is connected with reproduction. Even in instances where societal judgments about a particular sexual behavior may appear unrelated to reproduction, further inquiry usually reveals their linkage. It is only very recently that we have been able, through contraception, to separate sex from reproduction on a voluntary, reliable, and large-scale basis. It is, therefore, difficult to say how societies would have handled sexuality if it had no reproductive consequences. The chances are that it would still be necessary to regulate sex if it continued to arouse such strong passions and to create the potential for conflict.

The only way that a society can preserve itself over any length of time is by ensuring that its ranks are replenished by successive generations of its own people. To this end, society must ensure that its members mate, look after their offspring, and that each new generation is properly integrated into its social structure through *socialization,* "the process by which individuals acquire the knowledge, skills, and dispositions that enable them to participate as members of society"[3] (figure 16-1).

Sexual socialization is an important component of this process (chapter 9). Through it we learn the sexual values and behavioral patterns for our culture. This entails acquiring the knowledge and skills for expressing sexual interest and in eliciting similar interest in those we find sexually attractive. It is through sexual socialization that we learn how to make a pass, propose marriage, or establish other sexual association. Equally important, we acquire the sexual inhibitions and taboos of our society that check the expression of

Figure 16.1 Growing up to be like mother. (Abigail Heyman; Magnum. Reprinted by permission.)

sexual impulses considered socially unacceptable. In short, sexual socialization imparts cultural substance and form to our sexual drive during psychosexual development (chapter 9) and throughout adult life.

Social Norms and Sanctions The rules by which interactions among individuals are defined and patterned are known as *social norms.* In the sexual realm, many of these behavioral expectations are understood *informally.* We learn from family members and peers what forms of sexual behavior are socially acceptable or unacceptable. Or the social norms may be more *formally* expressed in legal statutes and moral codes that specifically state what one ought or, more often ought not, to do.

[2]Ford and Beach (1951).
[3]Brim (1966); Goslin (1969).

Social norms are enforced by *sanctions,* which are socially prescribed rewards (*positive sanctions*) or punishments (*negative sanctions*). Sanctions, like norms, may be informally or formally enforced. The anticipated reactions of people whose opinions matter to us, or in a more general sense, *public opinion,* is a powerful informal force that keeps many of us in line. Penalties imposed by the *law* for prohibited sexual behaviors are examples of formal negative sanctions.

Behaviors like sex are socially regulated through clusters or configurations of norms called *institutions.* Marriage is an excellent example. Through such institutionalization societies attempt to provide ready-made solutions for succeeding generations whereby fundamental individual needs are satisfied in ways that society considers optimally desirable for its broader purposes as well.[4]

All societies institutionalize sex in ways that are common in some respects and different in others. For instance, procreative sex is considered a legitimate and desirable aim in virtually all societies. However, societies differ widely with respect to their norms and sanctions concerning other forms of sexual behavior. The more important distinctions between societies in this regard are not whether or not they have more or less strict rules but how rigorously they enforce their rules.[5]

Sex is among the most highly institutionalized forms of behavior. Not only does society take an interest in how we behave sexually in public, but its norms try to pervade the private recesses of our lives. In addition to determining who can have sex with whom and under what circumstances, society has passed judgment on solitary activities like masturbation or the private sexual interactions between married couples. In addition to moral injunctions, legal attempts to regulate sexual behavior were such, until fairly recently, that

if all the sex laws on the books were enforced, 95% of the male population would end up in jail.[6]

Sources of Social Control   Society controls the sexual behavior of its members in a number of ways. Social institutions provide sets of assumptions and norms that define what is right or wrong. To the extent that members of society internalize such *values,* they constitute the key sources of self-control. Most people refrain from engaging in sexual acts that are socially unacceptable not because they are afraid of going to jail, or ending up in hell, but because they would feel ashamed, guilty, and suffer loss of self-esteem.

Sexual values expressed in moral tradition, the law, or public opinion perform two functions: *guidance* and *social control.* Guidance over the propriety of sexual behavior is part of the broader propensity of people to provide each other with directives based on personal experience or speculation based on the parental model of socializing children. Such guidance can be quite helpful and benign. As with travel guidebooks, if the information provided is sensible, you get the benefit of the experience and expertise of others who have been there before. But by exposing yourself to such views you also run the risk of prejudging matters in the light of someone else's preferences and biases.

Where more direct forms of control are involved, you are told what to do and what not to do, and there are various mechanisms to help ensure that you respond accordingly. One psychological mechanism that we develop in childhood is the feeling of *guilt.* To a variable degree, we all develop a certain capacity to feel guilty when we violate what we consider to be legitimate moral precepts. Guilt is a painful emotion, since it calls into question our sense of integrity and self-worth. It ranges in intensity from a mild prick of the conscience to such overwhelming anguish that it drives some to suicide.

[4]Further discussion of these concepts can be found in introductory texts in sociology. See Goode (1982); Dolby et al. (1973); Smelser (1973).

[5]Ford and Beach (1951).

[6]Kinsey et al. (1948, p. 392).

A related feeling is the sense of *shame.* Shame is experienced as embarrassment, or loss of face in the eyes of significant others or with respect to broader public opinion. Shame over sexual matters takes one of two forms. One has to do with bodily manifestations such as nudity. In this respect, genital exposure or activity engenders the same sort of shame associated with other private functions like urination, menstruation, and defecation. It is possible that aversion to eliminatory functions has contaminated sexuality by proximity; equally plausibly, shame over sex may have been extended to neighboring functions. In another sense, shame is more like guilt in that it is elicited by sexual behaviors considered reprehensible and unworthy.

Feelings of guilt and shame are usually experienced following sexual activities considered wrong. But they also operate in an anticipatory manner whereby they squelch the desire for a prohibited act in our consciousness. This nipping in the bud of forbidden desire before it arises is the most effective means of inhibiting socially unacceptable sexual activity. By contrast sometimes the very forbiddingness of an act makes it exciting; or one may engage in such behavior as a form of rebellion (such as a teenager getting pregnant in order to embarrass parents).

In addition to the effects of *internal* devices like guilt and shame, institutions control behaviors through more *external* means as well. Since we do not want to upset the people who are important to us and on whom we are dependent, individuals in key institutional roles, such as parents and educators, can effectively uphold social norms through informal controls or more formal sanctioning systems.[7]

Especially for adolescents and young adults, the *peer group* is a very important source of social control. Even though most of us eventually do accept the sexual norms prevalent in adult society, adolescence is typically a time during which a variable degree of experimentation and deviation

from adult norms commonly takes place. The adolescent subculture acts as a bridge between the world of childhood and adult society. Children absorb the sexual values of their parents on faith, but for adolescents such uncritical conformity becomes less tenable. Though the presumption of sexual innocence in childhood can no longer be maintained by adults, teenagers are rarely given consistent and workable sexual guidelines, and they become acutely aware of the inconsistencies between adult sexual values and behaviors.

In their transition to adulthood, adolescents must manage two sets of social norms: those of their own *youth subculture,* and those of the *adult world.* Since these two sets of behavioral expectations are to a variable extent contradictory, there is a potential for turmoil. A young man or woman who uncritically embraces all of the normative expectations of the peer culture will be in danger of becoming labeled a *delinquent.* This is more likely to happen on sexual than other grounds since traditionally sexual misbehavior, especially for girls, has been one of the main causes of social disgrace. If, on the other hand, the youngster entirely disregards the sexual norms of peer culture, then he or she is likely to suffer a degree of social *isolation.* In either case, the transition to adulthood is made more difficult.

Most adolescents negotiate this transition quite successfully by going along with the expectations of the peer culture in the short run, while bearing in mind the normative requirements of the adult world in the long run. They are assisted in this by their parents and other institutional representatives who simultaneously uphold traditional sexual virtues while concurrently allowing youngsters considerable autonomy. Parents in fact are just as likely to be upset by a lack of sexual interest in their children as they are in the face of sexually irresponsible behavior. Typically, teenage boys who are not interested in girls have worried their parents over the potential of homosexuality; girls who are indifferent to boys have raised concerns over their failing to find husbands.

[7]DeLamater (1981).

Attaining adulthood does not free us from social controls. On the contrary, what may have been tolerated during youth ("sowing one's wild oats," "raising hell," "playing around") becomes unacceptable for responsible and settled adults. Though we become emancipated from the control of parents and teachers, powerful social forces continue to keep us in line. If employers, friends, and neighbors disapprove of our sexual behavior, we may not get promoted, lose our jobs, fail to receive social invitations, and become unwelcome in the neighborhood; such sexual ostracism may even extend to our children, who may get shunned by their friends under pressure from their parents.

American society makes a constitutional separation between church and state, yet the *Judeo-Christian tradition* has been and remains an important source of social influence regarding our norms of sexual behavior. Even individuals who have no formal affiliation with religious institutions are likely to have their concepts of sexual morality influenced by these traditions; whether one adheres to traditional sexual morality or reacts against it, the relevant sexual norms are still defined in that general context (chapter 20).

Nonetheless, for a lot of people, the social norms of sexual behavior are not so much based on formal theological arguments as on a **folk theory** or common sense view of sexual behavior.[8] Such popularly held perspectives include many assumptions about sexual behavior based on traditional beliefs, derivations from the findings and opinions of social scientists (including sex researchers), and more idiosyncratic colorings based on personal beliefs, convictions, and prejudices.

Given these multiple social controls on individual behavior, one may wonder how much *autonomy* we have as individuals in what we do sexually. Does our sexual behavior reflect any voluntary choice or is it merely the outcome of various biological and social forces that shape us in childhood and buffet and steer us through our

adult lives? These are fundamental questions to which there are no clear-cut answers. It is easy enough to see ourselves as puppets manipulated by biological and social strings (and there is much disagreement about which set of strings pulls harder), yet the enormous variety of sexual behavior even within the confines of a given age group or social class would indicate that at one level or another a measure of individual choice must operate as well. Whether or not such individual variance merely reflects indiosyncratic combinations of various determinants or whether we are endowed as human beings with a more independent sense of *free will,* is a fascinating issue into whose philosophical complexities we cannot delve here.

Social Judgments of Sexual Behavior

Society makes a variety of judgments about what constitutes desirable and undesirable sexual behavior. As we discussed earlier (chapter 8), such judgments are generally made on the basis of four major criteria: *statistical* (How common is the behavior?) *medical* (Is the behavior healthy?); *moral* (Is the behavior ethical?); and *legal* (Is the behavior legitimate?).

Sometimes these four criteria are mutually reinforcing in obvious ways. For example, incest is an infrequent form of sexual behavior that is considered psychologically unhealthy, morally inadmissable, and illegal in most societies. Rape would be another instance. It would be difficult, however, to find many other examples that are quite so clear-cut. In principle, we must judge sexual behaviors primarily according to one or another of these criteria at a time, yet there is a strong tendency to seek corroboration from the rest. In fact, one set of judgments is often predicated on another: An act may considered immoral or illegal because it is unhealthy or offends the majority, or vice versa; an activity that we consider immoral would seem even worse if we assumed it to be indulged in by only a few, to be unhealthy and illegal.

[8]Davenport (1977).

The application of such judgments to sexual behavior in heterogeneous and pluralistic societies has resulted in much confusion. The meaning of the statistical norm has been distorted, partly through ignorance and sometimes deliberately: the prevalence of homosexuality, for instance, tends to be underestimated by its opponents and overestimated by its proponents. Medical judgments on sex have often lacked scientific support; the presumed harmfulness of masturbation is a case in point. Morality has been confused with tradition and the preferences of those in power. Statutes and ordinances have frozen into law may dubious health claims and moral conclusions. Not infrequently, the original determination that an act, such as oral sex, is unhealthy or uncommon turns out to be incorrect or no longer applicable, but the moral and legal judgments based on it persist.

Societies set up norms and assert various judgments on sexual behavior because it is widely assumed that the untrammeled expression of all sexual impulses would prove socially disruptive. It is easy to argue that incest would wreak havoc in families, that we should not sexually exploit children, nor coerce others into sexual interactions. But it is harder to understand why society has made so much fuss over the intricacies of lovemaking even between married couples in the privacy of their bedrooms. Some of our sex laws and the social attitudes they reflect would be laughable if as a result of them people had not been harassed, ostracized, jailed, and condemned to death; all for participating in sexual activities that claimed no victims, elicited no complaints from its participants, and would not even have come to light unless authorities and self-appointed guardians of public morality had not relentlessly unearthed them.

Several factors help explain, if not justify, some of these seeming absurdities that have characterized the treatment of sex in our society as well as in many others. First is a process exemplified by the ancient Talmudic precept of "building fences." Its underlying reasoning is that in order to better defend a fundamental moral principle or to ensure that a key commandment is not violated, one must enact subsidiary rules, like building concentric fences, around the core principles to be defended. In this way people can learn the error of their ways at the lesser cost of minor transgressions committed at the "periphery," and be saved from committing the more grave central sins.

To apply this principle to sexual behavior, let us assume that the central social concern is to protect the integrity of the family. Behaviors like incest or extramarital relations can be said to pose direct threats, hence are clearly forbidden. Premarital relations would represent a less direct danger by presumably jeopardizing marital happiness in the future. Prohibitions are then extended further to drinking, dancing, wearing makeup, provocative clothing, and so on which could sexually arouse people, thereby leading them into those sexual behaviors that are threats to the marital unit.

This is a logical principle but one that is easy to abuse. Moral fences can be extended in ever-widening circles to the point where they cease to bear any evident relationship to the core virtues that they are supposed to defend. The preoccupation with externalities and other secondary concerns becomes so ritualized that upholding them becomes the substitute for the original purposes of moral behavior. For instance, in Victorian England, the observance of the Sabbath became such an all-consuming endeavor for moralists that Christianity was seemingly reduced to doing nothing pleasurable on Sundays.

Another difficulty in the setting of sexual norms is that they seem generally geared at protecting the most vulnerable or the least sensible individuals. For example, one may argue that there are many who are able to engage in premarital sex without harming anyone. That may be so, it would be objected, but what about the countless others under similar circumstances who would be seriously hurt? How can we say that engaging in a given behavior is socially acceptable for some but not for others? What choice do we have but to declare as unacceptable all sexual behaviors where the majority or a substantial number of individuals are likely to get harmed?[9] While there is a compelling logic to this position from the point of view of social responsibility, it does not quite help to dispel the feeling on the part of thoughtful and responsible men and women that the sexual norms of our society seem to have little bearing on their own lives and that they are being held to standards of the lowest common denominator of personal responsibility. And since thoughtless and irresponsible people are not likely to pay attention to social norms anyway, why encumber everyone else with rules of behavior designed for them? Maybe because the more sophisticated need the blanket rules more than they think they do.

Sex and Social Stability

The regulation of sexual behavior—whatever form it takes—and the prohibition of its socially nonacceptable forms—whatever the justifications offered—are intended to safeguard *social stability*. But if all societies attempt to control sexual behavior to this end, why is it that some do it so much more stringently than others?

There are probably many causes that explain this, but one is particularly worth highlighting; namely, the correlation between the rigidity of control exercised within a social system and the tendency toward sexual restrictiveness. Not only does this association hold generally, it also seems independent of differences in political ideology or belief systems. For example, the Catholic and Evangelical churches have similar views about sexual morality despite wide doctrinal differences in other respects. More paradoxically, their common moral perspective corresponds quite closely with the prevailing sexual ideas of Communist states that are stridently atheistic: they all endorse marital sex and condemn

all other sexual outlets. It is thus not an accident that the Chinese People's Republic has imposed upon its people the same sexual standards that the Catholic church has tried to enforce since the Middle Ages. (One important exception is fertility control, which is endorsed by the Chinese state and opposed by the Vatican.)

Sexual Laxity and Social Disintegration
The supposition that unbridled sexuality leads to social instability is widely shared even in democratic societies that allow their citizens a great many other freedoms. With respect to sexual behaviors outside of marriage, grudging tolerance is as far as most contemporary societies will go; there is no active endorsement of nonmarital or other forms of sexual behavior by mainstream social institutions. In other words, even at its most tolerant, society will merely permit its members greater latitude in sexual behavior without making common cause with those elements who would like to promote such activity for one reason or another. All institutionalized perspectives toward sex, therefore, tend to be conservative. The assumption is that people do not need encouragement to engage in sex; what they need are various levels of moderation and restraint.

This social tendency to keep sex in check appears to be based on two presuppositions. The first is that sexual permissiveness leads to sexual license that can disrupt the family, erode the social fabric, sap the strength of the nation, and lead to social disintegration. The decline and fall of the Roman Empire is often held as an example (figure 16.2). Though one can just as plausibly argue that unbridled sexual license is a symptom and not a cause of social disintegration, so ingrained is the belief in the licentiousness-decline-fall sequence that sexual permissiveness, particularly among youth, is interpreted at times in downright conspiratorial terms. Some Americans suspect a Communist plot underneath the loosening of sexual standards. These suspicions are fueled by the fact that some sexually adventurous

---

[9]Saint Paul advanced the more stringent expectation that a Christian should do nothing that would scandalize a fellow Christian.

Figure 16.2 Thomas Couture, *Romans of the Decadence* (1847). A 19th-century Frenchman's view of Roman sexual excess. (The Louvre, Paris.)

youth are also political radicals who openly preach revolution. But Soviet youth who imitate the life style and sexual ways of their Western counterparts are similarly denounced by their political elders for succumbing to bourgeois decadence and playing into the hands of the capitalist enemies of their society. Thus, whatever the dominant ideology, those who are disenchanted with it also seem to favor permissive sexual attitudes. Is this because political dissidents wish to violate the established sexual norms of the Establishment as part of their revolt, or do those who feel sexually oppressed become political dissidents?

The second social concern is that uncontrolled sexual behavior would dissipate people's energy that would otherwise be put to more constructive and creative uses: sex is so much fun that were we to have free access to it, none of us would want to work or strain ourselves to do anything useful. These dual presumptions make sexuality appear like a gushing stream: if

unchecked it will sweep everything on its way; if unharnessed it goes to waste.

On both counts, the more controlling and authoritarian a political system, the more sexually intolerant it is likely to be. Nazi Germany, which must have violated every other code of morality, was nonetheless very keen on upholding family virtues and highly intolerant of sexual behaviors outside monogamous relationships.[10] On the other hand, it does not follow that those who endorse sexual permissiveness are necessarily humane and compassionate people; pimps and pornographers are a case in point.

Among the many thinkers who have addressed these issues, the ideas of Freud became particularly well known during the first half of this century (box 16.1). Even though his social theories are now widely challenged by behavioral scientists, Freud's speculations in this area remain of interest.

[10]For the persecution of homosexuals in particular, see Haeberle (1978).

Box 16.1   Civilization and Its Discontents

The relationship of sex and society greatly interested Freud, particularly toward the latter part of his life, when he turned his attention from narrower clinical concerns to more global cultural issues.[1] Freud's central thesis was that there is an irreconcilible antagonism between *instinctual demands* (erotic and aggressive) and the restrictions of *civilization*. Since free gratification of instinctual needs was incompatible with civilized society, those needs had to be repressed and subordinated to the purposes of *work* and *reproduction*.

Freud saw this process of sexual repression operating simultaneously at the individual level as well as the societal level. The roots of sexual repression could be traced to the development of the individual (*ontogenesis*) as well as to the history of the human race (*phylogenesis*). The social pattern was set in our dim prehistoric past; as individuals, each of us must go through the process during our own life cycle.

Freud theorized that the id, the seat of the sexual instinct, operates on the basis of the *pleasure principle*. The harnessing of this libidinal force is the task of the ego and the superego (representing reason and conscience), which are governed by the *reality principle*. Where the pleasure principle seeks immediate gratification, the reality principle requires restraint and delayed satisfaction. The desire for play, pleasure, and freedom on the one hand is balanced by the requirements of work, productivity, and security on the other. Although these internal forces may appear to be at odds, they work for a common purpose in the long run. As we give up the unrestrained and ephemeral pleasures of the moment, more reliable and lasting pleasures accrue to us in due time. In this way the reality principle safeguards rather than nullifies the pleasure principle. Not only does this process yield more satisfying responses to our sexual needs, but through the process of *sublimation* our libidinal energies are channeled into intellectual and artistic creativity that enriches our lives and lends to the generation of culture and civilization.

In order to sustain this process, society incorporates the reality principle in its institutions. Individuals growing up within a social system learn of the reality principle in the form of norms and other requirements of law and order and then, in turn, transmit these to the next generation. Members of society must diligently sustain their culture because the tendency to go back to a "state of nature" is never vanquished.

"The price of progress in civiliation," wrote Freud, "is paid by forfeiting happiness through the heightening of the sense of guilt."[2] Though he thought the repression of our primitive erotic and aggressive behavior was inevitable for the exercise of our higher interests and capacities, Freud was also keenly concerned with the price exacted in achieving these aims. Since libidinal repression, especially when harsh and excessive, resulted in neurotic disorders and sexual dysfunction, Freud expressed the fear that the social restrictions on sexual behavior that had channeled so much energy into civilization now posed the danger of throttling the very instinctual sources of our life and happiness.

[1] Freud's ideas about the relationship of sex and society pervade many of his writings. See, in particular, *Civilization and Its Discontents*, in vol. 21 of his collected works. For an introduction to Freud's sociological views, see Marcuse (1955); Jones (1957, vol. 3).

[2] Quoted in Jones (1957, vol. 3, p. 342).

Much of Freud's theorizing may have been a reflection of the social context in which he lived. And certainly, a great deal has changed since Freud confronted these issues at the turn of the century. Yet the basic dilemmas in balancing individual sexual freedom against socially imposed restraints are still very much with us. There are some who think that we still have a long way to go in liberating sexuality from social constraints; others think we have already gone too far in unleashing sex. Most people are probably uncertain about where they stand.

Sex and Social Status

Sexual behavior and social status are interdependent at a number of levels. Behavior patterns are typically linked to socioeconomic background (chapter 8) although the correlation has now somewhat weakened. However, what is considered proper sexual behavior as well as what one can get away with have always been and remain linked to social class.

Early in the 19th century, before Hawaii came under the influence of missionaries, the sexual life of the aristocracy bore little resemblance to that of the common people. Aristocratic men and women led a life of great sexual freedom. Without benefit of marriage, they freely engaged in heterosexual and homosexual associations. To keep their lineage pure, siblings were mated with each other, as were uncles and aunts with their nieces and nephews. Infanticide eliminated undesirable offspring. Yet the sexual life of commoners, who were virtual slaves, was rigidly regulated through religious proscriptions.

Hawaiian aristocrats enjoyed sexual freedom as a matter of hereditary privilege. In other cases, people (usually men) acquire such rights by rising to positions of power and prominence. Among Solomon Islanders, adultery and rape are severely disapproved and punished. But men of particular renown could engage in both with relative impunity. Their powers did not make their behavior acceptable, but their victims were powerless to do any-

thing about it. These men would actually indulge in transgressions as a way of testing and reasserting their authority. Similar examples can be found in other societies.

Privileged social status may, however, also place special constraints on sexual freedom and choice. Edward VIII of England was forced to abdicate in 1936 before he could marry Mrs. Simpson, who was a divorceé. Adlai Stevenson's aspirations to the presidency of the United States were not helped by his being divorced. It would be unthinkable, at this time, for an avowed homosexual to be elected or appointed to such high office.

A great deal depends, of course, on the particular area within which a person is prominent. The professional reputation of an Einstein or a Picasso is not likely to be jeopardized by his sexual behavior; stars in some entertainment fields may even have their reputation enhanced by being sexually adventurous. On the other hand, public figures such as judges, executives, and elected officials who are thought to sexually misbehave are likely to face a loss of confidence in their integrity and judgment. Those in sensitive positions with access to classified information are furthermore considered to be subject to blackmail if they have a "secret vice." There are many examples of people in high positions whose careers were effectively ended because of sexual scandals.

Cross-Cultural Perspectives on Sexuality

In chapter 1, we discussed the anthropological approach to the study of sexuality and in subsequent chapters we noted examples of various sexual behaviors in different cultures. Our purpose in this section is to reconsider the cross-cultural perspective on sexuality in a more integrative manner by focusing on the universals of sexual experience, and by examining the way in which other societies operate sexually as functional units in contrast with our own.

Sexual Diversity and Unity

Diversity is common to all human experience but variation with respect to sex is particularly striking. People may have one, two, or three meals a day but everyone eats every day and no one eats ten meals a day. Not so with sex: some have daily orgasms, others go without it for years. The greater uniformity of other physiologically driven behaviors like eating is clearly because of their greater biological imperative. Sex is an activity that is not essential for either sustaining life or bodily health at the individual level.

This greater latitude allows cultures to define what is sexual and what is not beyond genital interactions. For example, our culture does not imbue eating with any overt sexual significance. We eat freely in public with strangers and in the home with members of our family. But in some tribal societies the sharing of food is characteristic of the marital relationship, hence, brothers and sisters may not eat together without appearing to indulge in a sexual act. On the other hand nudity does not have the same erotic connotations in some other cultures that it does in our own.

Embedded within the bewildering array of intercultural differences in sexual attitudes and behaviors, there are a number of elements common to all cultures. The most fundamental constant in sexuality is its *universality*. Sex is present in all societies, it does not have to be discovered or invented. Equally valid is the role that every society takes on to shape and *regulate* the sexual behavior of all its members through socialization in childhood and rewards and punishments throughout adult life. All peoples are convinced that sexual activity is necessary, although reasons for this conviction vary. It is also widely believed that a man's sexual needs exceed those of a woman, and this opinion persists even in cultures that grant equal sexual privileges to both sexes. All societies practice marriage in one form or another, and the institution always carries with it special sexual privileges and obligations. Marriage forms vary but by far the most common

form is *monogamy*. By contrast, *polyandry* is extremely rare, and although many societies approve of polygamy, very few men in any of these cultures actually have more than one wife. The reasons are partly social and partly economic; either a second partner is not available and willing, or the man cannot afford one.[11]

In every society, the culture of sex is anchored in two directions: "In one direction, it is moored to the potentialities and limitations of *biological* inheritance. In the other direction, it is tied to the internal logic and consistency of total culture."[12] In order for a society to maintain a coherent structure, its sexual attitudes and regulations must be integrated with all the other rules and regulations governing other forms of behavior; sexual roles are not meaningful in isolation.

Intercultural similarities are manifested in the common sexual beliefs and behavioral patterns of individuals in various societies; similarities are also present in the relationship of sexuality to the structure and function of society as a whole. Those aspects of sexual behavior that are tied closely to biological foundations show relatively less variance. For instance, of all sexual behaviors only coitus can lead to pregnancy (except for artificial forms of insemination that hardly qualify as sexual behavior). Hence, behaviors centered around reproduction may take numerous forms in the choice of partners, patterns of courtship, coital behaviors, and so on, but by necessity the endpoint is the same whatever the culture.

At the other end, the *symbolic* and intellectual aspects of sex are highly variable; it is these aspects that are more closely tied to the "internal logic" of the culture as a whole. For instance, ideals of sexual attractiveness, moral judgments, and definitions of gender are in large measure manifestations of the way a given culture is generally oriented.

Lying behind all the rules, regulations, sanctions, and taboos are the implicitly or

[11]Beach (1977), p. 116–117.
[12]Davenport (1977, p. 161).

Box 16.2   The Incest Taboo

The term "incest" (Latin for "impure" or "soiled") is commonly used for sexual relations between close relations. Where the defining line is drawn varies in cultures. Sex between parents and offspring is a universal taboo. Relations between siblings and with grandparents, uncles, and aunts are also generally prohibited. The taboo is extended to cousins in some but not in all groups. (The Roman Catholic and Greek Orthodox churches prohibit marriage between first cousins, but such marriages are permitted by many Protestant churches, as well as by Judaism and Islam.)

Given the affectionate bonds between members of a family and the intimate circumstances under which they live together, one could imagine that sexual relations would be common between them. Yet incest is quite rare, just as it is infrequent among nonhuman primates in the wild. The incest taboo is generally assumed to have appeared early in human evolution in an attempt to safeguard the integrity of the family unit, since sexual competition within the family would be highly disruptive. Furthermore, mating outside the family has been important in the formation of larger social units held together by kinship ties.[1] The proposition that the incest taboo arose from recognition of the genetic dangers of inbreeding through concentration of disease-carrying recessive genes is less plausible (although there is a somewhat higher risk entailed in these situations).[2] Whatever its precise origins, prohibitions against incest have become universal. As Ford and Beach conclude:

Among all peoples both partners in a mateship are forbidden to form sexual liaison with their offspring. This prohibition characterizes every human culture. This generalization excludes instances in which mothers or fathers are permitted to masturbate or in some other sexual manner stimulate their very young children. A second exception is represented in the very rare cases in which a society expects a few individuals of special social rank to cohabit with their immediate descendants. The Azande of Africa, for example, insist that the highest chiefs enter into sexual partnership with their own daughters. In no society, however, are such matings permitted to the general population.[3]

explicitly shared value systems upon which a society must be founded, and which it must preserve or perish. These value systems and the cultural mechanisms related to them constitute what Davenport defines as a society's "internal logic and consistency." It is the relation of sexuality to this more general aspect of any society that we must seek to understand.

Universal rules are harder to apply to specific sexual activities without qualification except for a few basic behaviors. Sexual fantasy and self-stimulation are part of the sexual repertoire of every group, if not of every individual within it. Heterosexuality is the dominant sexual orientation and coitus the primary sociosexual behavior in all cultures but never the only one. Marriage is the primary setting for coitus but not all societies attempt to confine coitus to married partners; those that do, vary in their effectiveness in preventing sex outside marriage.

All societies have an *incest taboo* (box 16.2) that prohibits sexual associations between persons of various levels of kinship; parent-child (especially mother-son) incest is generally considered to be the most offensive. Some sexual unions not prohibited by incest taboos may still be looked upon

Box 16.2   The Incest Taboo (Continued)

Similar exceptions for intermarriage between siblings were also made among the Incas and the ancient Egyption pharaohs (Cleopatra was an offspring of such mating and herself a partner in them.) But other societies are stricter than our own: 72% of the societies surveyed by Ford and Beach were found to have more extensive incest prohibitions than are common in the West; sometimes they were broad enough to exclude half the population as potential sex mates for a given individual.

We know relatively little about the prevalence of incestuous behavior in our society. Although Freudian theory claims incestuous wishes to be a normative part of childhood experiences (chapter 8), in actual behavioral terms, incest has been assumed to be very rare. Recent evidence shows that it is perhaps more common than it has been thought to be (chapters 9 and 14).

Though we ascribe the paucity of incest to the social prohibition against it, other factors at the psychological level may also work against it. Fleeting thoughts and fantasies aside, most people have no erotic interest in their children, parents, and siblings despite the bonds of love that bind them together. Living closely with others while growing up

seems to exert an inhibiting effect on erotic interest and falling in love. This operates even when there are no social sanctions involved. For example, children reared together in an Israeli kibbutz rarely fall in love or marry each other despite parental preference and social convenience to the contrary.[4] They think of each other more as brothers and sisters.

Even more compelling is the Chinese experience where child brides brought up by the in-laws along with the boys of the family that they are supposed to marry often fail to develop any erotic interest in each other, becoming like brothers and sisters. Similar feelings seem to develop in coed dorms where men and women become friends but are not likely to be involved in romantic or erotic attachments.

[1]For an analysis of the functional significance of the incest taboo, see Murdock (1949). For a sociological review, see Weinberg (1976); for an anthropological perspective, Gregersen (1982).

[2]Kaye (1983).

[3]Ford and Beach (1951).

[4]In a study of 2,769 marriages of second-generation kibbutzim dwellers, not one occurred among those who had been reared together the first 5 years of their lives (Talmon, 1964).

with disdain. Rules of propriety regulate sexual associations based on certain aspects of kinship other than consanguinity (sexual relations between parents and stepchildren or stepsiblings are, for instance, considered unacceptable in our society). Other considerations relate to race, ethnicity, religious affiliation, age, generational differences, and discrepancies in social class; the general tendency with respect to all these factors is toward *homogamy* whereby people generally affiliate with their own kind (chapter 12).

Despite these similarities at the level of principle, cultures vary widely with re-

spect to specifics and the manner in which these principles are applied. Even with respect to the most universal rules of sexual behavior, there are exceptions to be found: for instance, in matrilineal societies, incest between a maternal uncle and his sister's daughter is a more serious offense than mother-son incest.

Repressiveness and Permissiveness

Another important source of dissimilarity between societies is the level of control they exercise over sexual behavior. In

Box 16.3   Sexually Repressive Cultures

Inis Beag (pseudonym) is a small Irish island community that shows a pattern of extreme denial and suppression of sexuality. As described by John Messenger, the people of Inis Beag shroud in mystery all manifestations of sex as well as other natural functions like menstruation and urination.[1] Nudity is abhorred to the point where a man is embarrassed to even bare his feet in public and reluctant to see a nurse when ill to avoid exposing his body. Marital coitus is the only acceptable sexual outlet; a "duty" that women "endure" and men hazard despite its presumed debilitating effects. Sex is always initiated by the husband; foreplay is restricted to touching and fondling of the buttocks; coitus occurs only in the male-superior position; underclothes are not removed; and it is doubtful if women ever reach orgasm.

The only other discernable sexual activity in Inis Beag is male masturbation and occasional petting encounters involving a few of the island girls and occasional visitors. The men who engage in these acts are so ashamed that they even shun the confessional. The village curate who was hospitalized in a mental institution was assumed to have been driven mad by the frustration of living in the same house with his pretty housekeeper. If any but the most conventional form of sex is ever practiced in Inis Beag, its existence is effectively kept secret from the community, which lives in fear of clerical opprobrium and public censure and ridicule.

It may be tempting to ascribe the prudishness of Inis Beag to Irish Catholicism, but quite similar attitudes can also be discerned in cultures that are neither Irish nor Catholic. For example, in the tribal tradition of the Manus of Papua-New Guinea, coitus between husband and wife was regarded as sinful and degrading, hence to be undertaken only in the strictest secrecy. Women in particular were averse to sex, which they endured occasionally to produce offspring. Intercourse outside of marriage was an even worse offense, which brought on supernaturally ordained punishments. So secretive were Manu women about menstruation that Manu men denied that their women experienced monthly cycles.[2]

For the people on Yap Island in the Pacific the repression of sexuality came not from moral scruples but concerns over health. Yappese men believed coitus caused weakness and susceptibility to illness. Women were similarly vulnerable, particularly for the period of several years following the birth of a child.

[1]Messenger (1971).
[2]Davenport (1977, pp. 123–124).

chapter 9, we discussed cross-cultural variations in levels of *repressiveness* and *permissiveness* with respect to childhood and adolescent sexuality. Similar contrasts can be based on criteria other than socialization patterns, although there is likely to be a close correlation between the manner in which a society rears its children and the eventual behaviors expected of its members. To dramatize the range of these differences, boxes 16.3 and 16.4 present examples from both ends of the spectrum of sexual control.

Cultures also vary in the *emotional tone* with which people engage in sex. Thus, the sexual joyfulness and the exuberance of the Mangaians contrasts with the antagonism that imbues sexual intercourse for the Gusii of southwestern Kenya.[13] For these people, even marital coitus is an act in which the man must overcome the woman's resistance and cause her pain and humiliation and women must retaliate by taunting and frustrating the men. This pat-

[13]LeVine (1959).

Box 16.4   Sexually Permissive Cultures

The best-known examples of sexually permissive cultures come from the Polynesian islands of the central Pacific. Observations of the Tahitians' free-wheeling sexuality go back to Captain Cook, the 18th-century British explorer who discovered these islands. Before the impact of the West, Tahitian children and youth of both sexes were encouraged to masturbate and engage in premarital intercourse with few restrictions. Marital and extramarital coitus was freely engaged in and openly discussed. The life of the community, its modes of aesthetic expression, song and dance, were suffused with eroticism.

Some of this sexual tradition of old Polynesia has persisted on the island of Mangaia described by Marshall.[1] As we discussed in connection with childhood adolescent sexual development (chapter 9), Mangaian boys and girls start masturbation between the ages of seven and ten, and coitus at about 14. Virtually everyone has had substantial sexual experience prior to marriage and keeps up an active sex life following it. Sex is as regular a part of the life of these people as dinner.

Sex for Mangaian men and women means coitus, pure and simple. To that end, foreplay (including mouth-genital stimulation) is freely practiced; "dirty talk," music, scents, and nudity further heighten arousal. Women are active sexual partners psychologically as well as physically. Female orgasm, often multiple, is considered a necessary outcome of successful coitus; its absence is feared to be injurious to the woman's health. Pregnancy does not inhibit Mangaians, who continue to engage in sex up to the onset of labor pains. Parenthood strengthens further the sexual ties of a couple.

But the Mangaian experience also demonstrates the constraints and contradictions that exist even in a society known for its unabashed and unrestrained celebration of sex throught life. Much of premarital sexual activity, for instance, goes on covertly for the sake of maintaining appearances, and there is a certain ambivalence coloring the relationship of sexual partners. Unlike our culture where sex is expected to follow love, the Mangaian start with sex and hope that it will lead to affection. This tends to cast their sexual relations in a more physical and mechanistic mold.

[1]Marshall (1971).

tern of interaction is inculcated during youth. When adolescent males are in seclusion following circumcision, girls are brought to the area to dance in the nude while making disparaging remarks about the boys' freshly mutilated genitals (and arousing them to painful erections).

Historical Perspective: The Sexual Revolution

Historical periods are demarcated by major events like wars and periods of massive social change, such as the Industrial Revolution. More arbitrarily, we depend on chronological landmarks; for example, we speak of the 20th century or the decade of the 1960s, as if events were turned on and off by the calendar. Nonetheless such devices are useful signposts if we remember that salient developments that characterize a given period neither start nor end within it.

Societies are constantly changing and evolving, but since the rate of such change is not uniform we get the impression that not much new seems to be happening over some periods, whereas at other times change appears to be bewilderingly rapid.[14]

[14]For theories and methods in the analysis of social change, see Goode (1977).

World War II is a major watershed in American history. No other epoch experienced as many major changes as the period following it.[15] The sexual culture that exists today was shaped during the past 40 years since the war and therefore constitutes the historical context for our current sexual attitudes and behaviors.

It is widely believed that the decade between 1965 and 1975 was a period of such rapid change in sexual attitudes and behaviors that it amounted to a *sexual revolution.* Others see these events as but another ripple in the constantly shifting social tides and hence unworthy of such a dramatic designation. We need more time to be able to see these very recent events in their proper historical perspective. Besides, one can endlessly quibble over what should qualify as a "revolution" and what should not.

There can be little doubt that very significant changes have taken place recently, leading to widespread questioning of traditional sexual attitudes and shifts in sexual behaviors. Many of these shifts became most evident in the mid-1960s and on, but they did not start then. In the mid-1950s, sociologist Pitirim Sorokin wrote a polemical indictment of the "American sex revolution" that had been in progress "the last few decades." He saw millions of men and women yielding to "sexual freedom" and American society becoming "sexualized."[16] Without knowing the date of Sorokin's book, one would assume he was reacting to the changes of the 1960s and 1970s, rather than writing a decade earlier.

The history of sexuality does not exist as a thing apart; it is part of the larger flow of events. The sexual component of a culture is influenced by economic, social, and political changes and, in turn, influences them. For example, the technological breakthroughs that led to the birth control pill and the establishment of the contraceptive industry became possible because there was a popular demand for such a device and the social climate permitted its development in the postwar period. The availability of the Pill, in turn, exerted a profound influence on our sexual attitudes and behaviors. As this example makes clear, the causes of the sexual revolution cannot be neatly separated from its manifestations. Yet for purposes of discussion it would be convenient to make such a distinction.

Causes of the Sexual Revolution

The roots of the sexual revolution must be sought in the political and economic conditions that prevailed at the end of World War II. As we discuss in chapter 18, the previous major upheaval in American mores and sexual behavior also took place following a major war, which was World War I.

The Impact of World War II on Sexuality America emerged from World War II a changed nation. The social upheavals of wars exert their influence on sexuality in several ways. At the practical level, spouses become separated and circumstances make it difficult for singles to get married; this places pressure to have sex outside of marriage. The enormous stress of wartime, and the uncertainty of the future further propel people to seek whatever emotional comforts and sexual satisfaction they can get while they are alive. The armed forces bring together many men and women of diverse class and sexual backgrounds; even more striking is the exposure to other cultural mores when men are sent overseas. New situations, exposure to the duress of war, and the anonymity of being away from home and in uniform—these prompt people to behave sexually in ways that they ordinarily would refrain from. Having crossed these lines, their sexual attitudes are often changed irrevocably.

World War II had a number of other longer-range influences. The large numbers of women who joined the labor force made a dramatic impact on sex roles, as we discuss later in connection with *fem-*

---

[15]For a historical overview of American society since World War II, see Bailey and Kennedy (1983, chapter 48); Degler (1974).

[16]Sorokin (1956).

inism. The postwar role of the United States as a superpower ultimately led to its involvement in the Vietnam War, which led to widespread disenchantment among youth, as we shall elaborate when we deal with the rise of the *counterculture*.

The *civil rights movement* also contributed to the questioning of conventional values and prejudices. Following the 1954 ruling of the Supreme Court that declared segregation in public schools unconstitutional, this movement gathered enormous momentum. Although aimed at removing the residues of racial discrimination, and unrelated to sex as such, the widespread concerns it raised with respect to the rights of minorities, and the tactics of political protest that it spawned, became important in galvanizing the antiwar movement and subsequent efforts by women and homosexuals in the pursuit of their own civil rights and liberties. This does not mean that the civil rights activists caused the subsequent changes in the form of a chain reaction. Rather the fact that various liberation movements were bunched together points to the existence of a period of pervasive and radical cultural upheaval. But this upheaval was selective: not every attempt at rebellion attracted a following; workers did not take over factories nor were churches overwhelmed by atheists.[17]

The Effect of Economic Factors on Sexuality It has been argued that periods of economic hardship lead to the curtailment of sexual desires, while times of plenty encourage sexual expression. In the face of scarcity, all thoughts and energies must turn to the tasks of economic survival and reconstruction. Sex, especially nonreproductive sex, like other pleasures and frivolities, is deemed incompatible with times of austerity.[18]

The early phase of *industrialization* in Western societies presented such an economic challenge, hence the sexual restrictiveness of the Victorian era. In post World War II America, however, very different conditions prevailed. After a shaky start, the economy experienced a dramatic upsurge in the 1950s, boosting millions of Americans into middle-class affluence. America became a society of *consumers* living in a land of plenty. Not only was there no need for self-denial, frugality was now downright detrimental to an economy that relied on consumers to keep it booming. The satisfaction of the desire for goods and services, and the creation of more and newer needs, became the normative social expectation. In this atmosphere, the satisfaction of sexual needs received tacit approval as well.

As society became more permissive and advertisers more aggressive in their use of the female body and sexual innuendo, sex became an ally of *business*. Not only did sex sell in the form of erotic movies, books, and other entertainment, but entire industries became dedicated to its enhancement and others used it to peddle their wares. Manufacturers of cosmetics, clothing, alcohol, cigarettes, cars, and even more unlikely products based their appeals on the promise of making people sexy and sought after. As sex became commercialized, commerce became sexualized. The net effect was to make sex freer, more open, more available, and more acceptable for a broad spectrum of middle-class, respectable, mainstream Americans. "The business of America is business," had declared Calvin Coolidge. Now that sex became part of that business, Americans took it to their bosom as never before.

The Resurgence of Feminism Some of the most important recent influences on our sexual attitudes and behaviors have come from changes in the family, gender identity, the sex roles of women, and, to a lesser extent, those of men.

World War II played a crucial part in this transformation. As men were drafted into the armed forces, more than 5 million women took their places in various occupations, including heavy industry. A great many of these women stayed on their jobs following the war, and many more joined them subsequently; the percentage of

[17]Harris (1981).
[18]Schmidt (1982).

women in the labor force went up from less than 20% early in the century to 60% in the 1970s. This is a matter of singular importance, since contrary to the belief that women's liberation created the working woman, it was working women—especially the working housewife—who created the women's liberation movement.[19] Once the movement was underway, it in turn prompted other women to seek careers outside the home.

The inequities in job opportunities and pay that had handicapped women became more evident and less tolerable with the massive influx of women into the labor force. Working outside the home further burdened many women with guilt for deviating from the traditional roles of wife and mother (what Betty Friedan called the "feminine mystique").

Women's revolt against sex discrimination was spearheaded by the feminist movement. If *feminism* is defined as the theory and advocacy of the political, economic, and social equality of women and men, then its manifestations can be found throughout history. But as a more focused and political effort, the roots of modern feminism go back to the mid-19th and early 20th centuries (box 16.5).

American feminism, popularly known as the *Women's Liberation Movement*, has followed two main paths during the past two decades. In 1961 the appointment of the Commission on the Status of Women by President John Kennedy instigated a broadly based reformist approach to sexual equality. The *National Organization of Women* (NOW), founded by Betty Friedan in 1963, sustained this thrust, working primarily within the established political order. The other wing of the women's movement emerged from activist groups within the New Left and took a more *radical* stance. Whereas mainstream femininsts generally worked toward sexual equality within heterosexual and marital relationships, the more radical elements among women's liberationists more often opted for *separatism*.[20]

A number of writers proved highly influential in bringing the feminist message to large audiences. Simone de Beauvoir's study, *The Second Sex* (published in French in 1949) set the stage for the rise of a new feminist consciousness by examining the effects of patriarchal societies on women's lives; the ways in which men "act" while women are "acted upon" and relegated to be the "second sex."[21]

In the 1960s, Betty Friedan addressed the predicament of women who were in the "housewife trap"; having relinquished opportunities for higher education, careers, and political power in order to live up to the "feminine mystique"—being "truly feminine" and "sexually successful wife and mother."[22] In the next decade Kate Millett focused on sex and aggression in contemporary literature in *Sexual Politics*,[23] and Germaine Greer, in the *Female Eunuch*, dealt with women who are socially "castrated" to become "feminine" instead of "female."[24] Many other books have since reiterated and expounded on these themes.

The feminist revolt was primarily a *sex role revolution*, not a sexual revolution, primarily focusing on issues of gender and sex roles. Although the quest for sexual freedom and opposition to sexual exploitation have always been important themes in feminist ideology, its more basic concerns are social and its efforts are directed at the elimination of *sexism* (a term patterned after "racism"). In the feminist perspective, women are seen as victims of *institutional discrimination*, as in educational and vocational settings, and of *cultural discrimination*, as manifested in sexist language, personality stereotyping, prevalent

---

[19]Harris (1983). Working women must not be confused with "working-class women." The Women's Movement has been, and remains, primarily a middle-class phenomenon.

[20]For the history of the women's movement, see Freeman (1973); Faust (1980).

[21]deBeauvoir (1952).

[22]Friedan (1964). For Friedan's more current views see Friedan (1981).

[23]Millett (1970).

[24]Greer (1971). For Greer's more current views see Greer (1984).

Box 16.5   Origins of the Feminist Movement

The struggle for women's rights gained momentum by the middle of the 19th century under the leadership of women (many of them Quakers) like Lucretia Coffin Mott (1793–1880). Elizabeth Cady Stanton (1815–1902) (who would not have the word "obey" in her marriage ceremony and was bold enough to advocate suffrage for women), and Susan B. Anthony (1820–1906). These women and their collaborators were tenacious and undaunted social reformers whose concerns extended to working for temperance (to ban the use of alcohol) and the abolition of slavery.[1] Other feminists assailed sex-role restrictions. Elizabeth Blackwell became the first American physician and Margaret Fuller edited a journal. Lucy Stonely retained her maiden name after marriage, and Amelia Bloomer rejected the traditional long dress in favor of short skirts with loose fitting trousers ("bloomers").

The first *Women's Rights Convention,* held in 1848 in Seneca Falls, New York, formally launched the modern women's rights movement. It demanded the right to vote and equal rights for all women. Though these efforts were met with public scorn, hostility, and ridicule, the inexorable march for sexual equality was on.

Many outstanding women worked tirelessly for social reform across a wide spectrum of activities at improving living conditions for the poor, prison reform, fighting alcoholism, cruelty to children, prostitution; legal reform, and abolition of slavery. But the central cause was getting the vote, which was seen as the key to political power and social equality in a democratic society. This was accomplished gradually: Wyoming was the first state to grant women the suffrage in 1869, but the United States as a whole did not do so until 1920. (Several countries still deny women the right to vote.) The attainment of this goal led to an abeyance in the push for other causes of the feminist movement until recent times.

One of the major legislative goals of modern feminists is passage of the 27th amendment to the Constitution (known as the *Equal Rights Amendment, ERA*) which states that, "Equality of rights under the law shall not be denied or abridged by the United States or any State on account of sex." Congress has yet to ratify it.

Early feminists were dedicated to the liberation of women but not to "sexual liberation" as we now understand the term. They were staunch supporters of the integrity of the family, and their battles on behalf of *sexual purity* and domesticity promoted a "rigid and antisexual moral code."[2] But the fact that these women attacked the sexual behavior of men at the time does not mean that they were hostile to sex as such or adverse to it in their own lives.[3]

Though atypical of the mainstream feminist movement, there were also women who were openly in favor of sexual independence and freedom. The irrepressible Victoria Woodhull (1838–1927) proclaimed in a lecture: "Yes, I am a free lover! I have an inalienable, constitutional, and natural right to love whom I may, to love as long or as short a period as I can, to change that love everyday if I please!"[4]

[1]Bailey and Kennedy (1983, p. 191).
[2]Epstein (1981).
[3]Freedman (1982).
[4]For a documentary account of women's lives in the 19th century, see Olafson et al. (1981).

social and behavioral theories, and characterizations in the public media.[25] The key terms in the lexicon of feminism are "choice," "autonomy," and "authenticity"; its central aim is "full equality for women, in truly equal partnership with men"—the goal of the National Organization of Women.[26]

A feminist woman wants the right to choose whether she will be a homemaker or a career woman, whether she will be a mother, whether she will be heterosexual, lesbian, or celibate. Autonomy is her freedom to make these choices instead of accepting the dictates of church, state, spouse, or employer. By making such choices, she becomes an authentic person—she can be herself, instead of playing a role, trying to fit a pattern or a stereotype. To Betty Friedan, authenticity means the end of sex role playing: no longer "playing feminine" or "playing masculine" but sharing the breadwinning and child-rearing functions with men.[27]

Traditional marital and heterosexual relationships are generally seen by feminists as oppressive and exploitative of women, and the *double standard* of sexual morality for men and women discriminatory and unfair. But there are no specific sexual norms that would fairly characterize all feminists, who are mainly women but include many men as well. The majority of these women, like the female population at large, is heterosexual, although there is probably a higher proportion of lesbians among feminist women. With respect to attitudes toward men and marriage, feminists range from those who accept marital relationships redefined along companionate rather than traditional lines, to others who totally reject the institution of marriage, and some who would shun all sexual dealings with men.[28] We shall have more to say on feminist perspectives on sexuality farther on.

The Rise of the Counterculture *Youth* was in the vanguard of the sexual revolution and the group most affected by it. This sexual element, however, was part of a broader new ideology and way of life, which, because of its rejection of the values of the established culture, came to be known as the *counterculture*.[29]

The prominence of youth on the social scene in the 1960s was in part due to its sheer numbers. The *baby boom* that followed World War II generated over 11 million people aged 15 to 24 by 1970 (the birthrate leveled off in the 1960s). This generation was brought up by relaxed and permissive childrearing and educational precepts ("the gospel according to Dr. Spock"). Its members were also more highly educated and affluent than ever before: in the 1970s one in four persons aged 18 to 24 was enrolled in an institution of higher learning; adolescents in the 1960s were spending $20 billion a year, much of it on clothes, rock music, and other entertainment. The combination of these factors made American youth *idealistic, outspoken,* and *self-indulgent,* likely recruits for the social causes and political turmoils facing the nation.

Society has always had its critics; challenge to authority has deep historical roots in American culture. In the 1950s, these attitudes were given public expression by *Beat* writers like Alan Ginsberg and Jack Kerouac, who were particularly disenchanted with the materialism and conformity of their times. Like their *bohemian* predecessors earlier in the century, the *beat generation* of intellectuals and their followers ("beatniks") adopted manners and mores, in sexual and other terms, conspicuously different from those expected or approved by the majority of society.

The rise of the counterculture in the 1960s was in part a further elaboration of these earlier themes of *alienation*. But new de-

[25]Kando (1978).
[26]For further readings on the feminist perspective see Cox (1976); Rossi (1973); *Signs* (1980).
[27]Friedan (1981).
[28]See Koedt (1976); Johnston (1973).

[29]Two popular books that capture the flavor of this era are Roszak (1969) and Reich (1970). Sources that chronicle and analyse the rise of the youth culture of the 1960s include Kenniston (1965), (1968), and (1971); Goodwin (1974); Unger (1974).

Figure 16.3 Counterculture gathering. (© Bonnie Freer 1983, Photo Researchers, Inc.)

velopments fanned the flames of disenchantment. The social liberation movements for civil rights and women's rights, along with the antiwar sentiments, challenged the legitimacy and values of the established order on a broader front than before.

The *antiwar movement* led people into acts of civil disobedience and some to violence. Although adults of all ages were part of it, college youth became its most vocal component. This contributed to the development of a *generation gap* whereby substantial segments of the nation's youth rejected the sexual values and norms of those in authority (the "Establishment") and adults in general ("never trust anyone over thirty").

Some of these political dissenters were social activists who wanted to change the world. But there was also a more passive response on the part of others who wanted to withdraw from conventional society ("turn on, tune in, drop out") and evolve *alternative life styles* based on peace, love, harmony with nature, self-knowledge, and rejection of materialistic values. Seemingly overnight, a whole new subculture, the

*counter*culture, emerged with its own dress, language, music, and philosophy of life. And the quickest way to attain its goals seemed to be through drugs and sexual activity.

The converts to this *psychedelic culture* came to be known by a variety of names like *hippies* and *flower children*. They shared a sexual outlook that rejected conventional morality and advocated individual freedom ("do your own thing"). Their sexual ethic went beyond tolerance and championed sex as an antidote to violence ("Make love not war") (figure 16.3).

The momentum and the more florid manifestations of the counterculture had fizzled by the late 1970s. Though the flower generation wilted away, the decade left its unmistakable mark (and casualties), particularly with respect to premarital sexual permissiveness, tolerance of unconventional sexual behaviors, greater acceptance of sex generally, and the liberalization of public sexual expression in society at large.

The Separation of Sex and Reproduction

Sex has long been used for purposes other

than reproduction, but the control of its reproductive consequences were until recently not very reliable. The contraceptive pill became commercially available in the 1960s, ushering in an unprecedented era of widespread usage of a highly reliable and relatively safe contraceptive. Montagu ranks the social significance of the Pill with that of the discovery of fire, learning how to make tools, the development of urbanism, the growth of scientific medicine, and the harnessing of nuclear energy.[30]

Widespread usage of contraceptives had a major influence on sexual attitudes and behavior, and changes in social attitudes were themselves an important impetus in the development of newer contraceptives. Since moral and legal codes of sexual behavior probably originated in large measure because of concerns about illegitimate offspring, the ability of people to decouple sex from reproduction puts social constraints over sexual behavior in a new light. Without important breakthroughs in contraceptive technology, it is doubtful if many of the subsequent changes in sexual behaviors would have taken place. It would have been particularly difficult, for instance, to make allowance for greater premarital permissiveness without at least some assurance that such sexual experimentation would not generally lead to pregnancy. Similarly, extramarital sex would have been far more problematic if the risk of pregnancy remained a substantial risk in such relationships.

Most importantly, effective contraception enabled women, especially unmarried women, to exercise a far greater choice on a key element in their lives. The ability to delay or altogether avoid motherhood without relinquishing sex opened up the possibility of a far greater range of vocational prospects for women and thus was an important factor in the resurgence of the womens' movement with its own attendant impact on sexual attitudes and behavior.

[30]Montagu (1969, p. 11).

The Manifestations of the Sexual Revolution

Changes in the sexual culture of a society must be ascertained with respect to the *attitudes* and *behaviors* of its members. While closely linked, the two are not the same, particularly with respect to sexuality. What people say and what people do are often quite different. Sexual attitudes reflect the more idealized values of a culture with respect to how one *ought to behave;* sexual behaviors reflect how workable and desirable these values are in meeting the everyday sexual needs of ordinary men and women.

Sexual attitudes and behavior usually change together and influence each other. The more acceptable premarital sex becomes, the greater is the likelihood that young people will engage in it; the more prevalent premarital sex becomes, the more socially acceptable it is likely to be, in the long run.

Changes in attitudes are more likely to occur before shifts in behavior, but such changes need not always follow. For example, attitudes toward homosexuality have become far more liberalized recently, but there is no evidence that more people are engaging in homosexual behavior. Liberalization in attitudes, such as making forbidden sexual materials available, may sometimes actually lead to a diminution in their use.

We have better evidence for the shifts in sexual attitudes than in behaviors. For instance, a 1969 poll of 1,600 Americans in various walks of life revealed that 81% considered a policemen who takes money from a prostitute to be more immoral than the prostitute; 71% thought that a doctor who refuses a house call to a seriously ill patient is less moral than a homosexual; 54% regarded a politician who accepts bribes as more reprehensible than an adulterer.[31] More recent polls show that despite changes in moral attitudes toward

[31]*Time*, June 6, 1969, pp. 26–27.

some aspects of sexual behavior, a substantial majority of Americans still profess their belief in the traditional values of family life. For instance, in a nationwide poll of 1,044 registered voters, almost four out of five persons condemned extramarital sex and two out of three disapproved of teenagers having sex relations.[32] This is of course what people say; what they do may be another matter.

In referring to shifts in sexual attitudes and behaviors, we shall often rely on terms, such as *liberation*, that are commonly used to characterize these events. These are *value-laden* terms: Liberation means freedom from undesirable constraints, hence it makes good sense to the proponents of a trend but not to its opponents. The women's liberation movement thus signifies to some not freedom from undesirable contraints but the loss of the traditional prerogatives enjoyed by women; gay liberation may likewise be seen as condoning "enslavement" of people to undesirable sexual practices.

Liberation of Female Sexuality The major changes of the sexual revolution have primarily influenced women. It is in the sexual behavior of women, within and outside of marriage, that the dramatic shifts have occurred. The behavior of men has not changed that much, and what has changed has been in large part in reaction to alterations in female sexuality.

Women are now sexually more candid, interested, tolerant, active, and fulfilled than probably at any earlier period of the nation's history. No institution, however conservative, is opposed anymore to the right of women to sexual satisfaction. The Catholic church and fundamentalist evangelists are, in principle, just as much in favor of female sexual fulfillment as are feminists and sexual liberationists. Perceptions as to the proper institutional and relational context for such fulfillment, of course, vary sharply among these groups.

While we acknowledge these changes, we should not portray these recent developments as a renaissance emerging from the sexual dark ages of the 1950s or the earlier decades of the century. As we see in chapter 18, even in Victorian times there was a great deal more sexual expression than we may have been led to believe. One may also argue that while sexuality has now become more explicit, the importance and power of the erotic element in our lives has not commensurately increased, and may even have decreased.

Since there is little disagreement over the rights of women to seek and obtain sexual satisfaction we focus here on the differences that now exist about how women should best accomplish these aims. Since feminism represents the newer ideology, let us start by examining the feminist perspectives on this issue.

The centrality of sexuality in feminist ideology has tended to vary. Betty Friedan has downplayed it, maintaining that when women have gained social parity with men, sex will take care of itself; Germaine Greer urged liberated women not to marry (though she is not opposed to heterosexual sex as such); and Shulamith Firestone called for "a revolution in every bedroom."[33] Lesbian feminists have tended to place heavier stress on sexual liberation since they are engaged in a two-pronged struggle: one for sexual equality, another for freedom of sexual choice. For some of them, the fundamental drive of feminism is sexual liberation, which is only possible through total sexual independence from men. The contention that a true commitment to feminism is only compatible with lesbianism, has proven at times to be divisive within the broader feminist movement.[34]

There are other competing trends within feminist attitudes to sexuality. Some women are unabashedly in favor of greater sexual activity, experimentation, and satisfaction.

---

[32]Yankelovich et al. (1977).

[33]Firestone (1970).
[34]Hole and Levine (1971).

They reject the double standard whereby women have had to be faithful to their husbands or at least confine their sexual enjoyment to loving relationships, while men have been freer to enjoy sex irrespective of such constraints. Equity would require women to be permitted to behave as men do, if they so choose, and feel free to engage in casual sex or do whatever else that men have traditionally done to satisfy their sexual interests.

Other feminists reject the double standard not only because it favors men but because of the sort of shallow and exploitive sex that it has been associated with. They are not interested in acquiring equal rights to engage in casual sex, or to buy and sell sex. Instead they want men to stop engaging in these activities because they are demeaning to both men and women.

At its inception, the women's movement appeared to convey conflicting sexual messages. On the one hand, there was the image of the liberated woman who was now ready to meet and beat men at their own games. On the other, some of the more visible radical feminists seemed to repudiate all traditional precepts of femininity and sexiness: they refused to alter or to adorn their bodies (unshaven legs, no makeup), or to dress, act, and talk in ways that would be expected to please and entice men. In 1969 a group of feminists publicly burned padded bras, false eyelashes, steno pads, and copies of *Playboy* while crowning a live sheep "Miss America" ("Ain't she sweet/making profit off her meat") to protest the national beauty pageant being held in Atlantic City. A fashion show for brides was disrupted by the Women's International Terrorist Conspiracy from Hell who sang to the tune of the wedding march, "Here come the slaves/ Off to their graves." Thousands of women marched with placards that warned "male chauvinist pigs" that their world was coming to an end and advised women, "Don't Cook Dinner Tonight—Starve a Rat."[35]

Feminist attitudes to sexuality have evolved considerably over the past dec-ade. There is a broader recognition that the new permissiveness can make sex into "an almost chaotically limitless and therefore unmanageable realm in the life of women." The sexual revolution has robbed women of their traditional right of refusal; therefore, a "new chastity" must now restore the right of women to say no.[36]

Simultaneously, there is an acceptance even by some of the most staunch feminists that while many women remain determined to struggle for their rights, they do not necessarily wish to reject men or forego all of the trappings of femininity.[37] Similarly there is a move back to marriage as the basis of a stable relationship.[38]

The sexual issues that most preoccupy feminists at present are reproductive control, pornography, and rape. The feminist victories won over contraceptive control and a woman's right to get an abortion are law but remain vulnerable. Pornography has mushroomed to unprecedented levels and rape has attained appalling proportions. Betty Freidan thinks the current, "second stage" of the women's movement will break this vicious cycle of sexual violence and usher in a new mode of sexual interaction between men and women.

> The second stage of the women's movement is not unisex. It is human sex, for women as for men—active or passive, responsive, responsible, playful or profound, no longer an acting out of eroticized rage, or manipulation of covert power, joyless dues for economic support, or brutal revenge of love denied.[39]

Large numbers of women now aspire to combine the rewards of marriage with those of having an independent career. Just like men, they want sexual freedom over a period of being unattached (a "singles life style") followed by a stable and committed relationship. These women share many of the goals of feminism but would want no

[35]Harris (1981, p. 76).

[36]Decter (1973, p. 80).
[37]Brownmiller (1984).
[38]Quinn (1981).
[39]Friedan (1981).

part of its more radical, "bra-burning" image. In that respect they are more like traditional women, concerned with physical attractiveness, clothes, having a good time, and finding the right man at the right time.

More ideologically opposed to feminism are writers like Phyllis Schlafly and Maribel Morgan, who reflect more conservative views of male-female relationships. They oppose the erosion of traditional gender roles and marital relationships; otherwise they are ardently in favor of active and satisfying conjugal sex.

Morgan's book, *The Total Woman*, advocates that a wife cater to her husband's special quirks, "whether it be in salads, sex, or sports." Morgan's gender ideals may be traditional but her unabashed endorsement of female sexual initiative is not. Wives are instructed to call their husbands at work an hour before quitting time and say, "I want you to know that I just crave your body!" She also advises that, "A frilly new nighty and heels will probably do the trick as a starter"; "Be prepared mentally and physically for intercourse every night this week. Be sure your attitude matches your costume. Be the seducer, rather than the seducee."[40]

Phyllis Schlafly endorses the *positive woman* whose satisfactions and influence come through the traditional wife-mother role. By contrast, the new morality of the sexual revolution, "robs the women of her virtue, her youth, her beauty, and her love—for nothing, just nothing. It has produced a generation of young women searching for their identity, bored with sexual freedom, and despondent from the loneliness of living without commitment."[41]

Sex and Marriage The American family has changed more in the last 30 years than in the previous 250.[42] In the late 1960s and early 1970s, the rate of fertility began to decline. This was a trend that had set in two centuries ago, but with the Pill and access to abortion, it became more marked.

The marriage rate similarly began to decline to levels that existed at the end of the Depression. Between 1960 and 1970, the proportion of woman who remained single until the ages of 20 to 24 increased by one-third. The divorce rate soared to a level where one out of three marriages of women 30 years of age or younger was ending in divorce (excluding desertions). These changes are persisting in the 1980s but their rates are slowing down.

Are marriage and the family dying in the aftermath of the sexual revolution? As we discussed in chapter 12, that does not seem to be the case. But the nature of marriage and the family is undergoing significant changes and a number of alternatives are being experimented with by substantial numbers of individuals.

Monogamy and *serial monogamy* (a succession of marriages) remain the predominant forms. These are associated in as many as half of the cases with extramarital relationships, usually in a covert manner and occasionally in an "open" marital relationship. More experimental alternatives are singlehood or *celibacy; cohabitation; group marriage; family network systems* (several families living together); and *communes* (larger groupings) that may or may not involve sexual sharing of partners.[43]

It has been generally assumed that sex is one of the main reasons why people get married. Now that women are economically and socially more independent and psychologically more self-reliant, they are far more reluctant to barter their bodies for a roof over their heads. Similarly, men can now have sex with respectable women without having to marry them. Thus, the formula that marriage is the price men pay to get sex and sex is the price women pay to get married is no more generally tenable.

Premarital Sexual Permissiveness One of the more dramatic aspects of the sexual revolution involves changes in attitudes toward premarital sex. Figure 16.4 places these changes in historical perspective by presenting graphically the remarkable in-

[40]Morgan (1973).
[41]Schlafly (1977).
[42]Blumstein and Schwartz (1983).

[43]Stayton (1984).

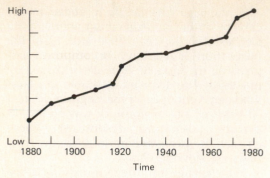

Figure 16.4 Rough estimate of rate of changes in premarital sexual permissiveness—1880–1980. (From Reiss, I. *Family Systems in America*, 1980, p. 169.)

crease in premarital sexual permissiveness over the century between 1880 to 1980. The steady gradient of the curve supports the contention that change has been steady; the sharp upsurge from 1915 to 1925 and from 1965 to 1975 in permarital permissiveness points to the more revolutionary character of these periods.

In an opinion survey of a representative sample of adult Americans conducted in 1963, about 80% stated that premarital intercourse was always wrong: this figure had gone down to 30% in 1975.[44] Attitudes about

premarital sex among college populations are further detailed in table 16.1. The more dramatic shifts in premarital behavior are among women, and the attenuation of the double standard is one of the more striking outcomes of the sexual revolution.

Between 1971 and 1976, the proportion of women aged 15 to 19 who were non-virgins went up from 27% to 35%, an increase of 30% over a period of just 5 years. Reiss estimates that in the 1980s, 75% of all women and 90% of men were entering marriage as nonvirgins.

Though current premarital attitudes among young adults are notably permissive, the dominant norm within the group is *permissiveness with affection*, not promiscuity. One manifestation of this is the increase in the practice of cohabitation. These more liberal patterns of premarital relations are now fairly well accepted among large segments of the population. Many parents have acquiesced, and some may be actually pleased with the greater honesty and responsible freedom with which their offspring seem to be managing their lives.

On the other hand, teenage sexuality continues to raise many concerns. In a nationwide poll conducted in 1977, the majority of adults disapproved of teenagers

[44]For further details, see Reiss (1980), chapter 7.

Table 16.1 Attitudes about Premarital Sex

	Cornell 1940	11 Colleges 1952–1955	18 Colleges 1967	2 Colleges 1971
The percent of women who approved of premarital sex for both men and women.	6	5	21	59
The percent of men who approved of premarital sex for both men and women	15	20	47	70
The percent of women who approved of premarital sex for engaged people only	6	7	23	19
The percent of men who approved of premarital sex for engaged people only.	11	16	22	16

Landis and Landis (1973, p. 145).

having sexual relations.[45] There has been especially serious concern with teenage pregnancy, which involves some one million adolescents each year. Nonetheless, the changes in attitudes and behavior with respect to premarital sex have gone so far that traditionalists who oppose premarital sex on principle have to fight an uphill battle. But it is a battle that continues to be waged.

Reproductive Control   Social regulation of reproductive control primarily concerns contraception, sterlization, and abortion. Over the past two decades, sweeping changes have taken place in American society with respect to the liberalization of individual choice in these matters. There has never been a time in our history when women have been given as complete control of their reproductive options as they are entitled to currently. (The landmark decisions concerning these issues are discussed in chapter 19.)

Most forms of *contraceptive usage* by adults is no longer a serious social concern except for those affected by the continuing opposition of the Catholic church. The controversy over abortion sometimes extends to contraceptive methods like the IUD, which disrupts pregnancy after fertilization has occurred. The provision of contraceptive services to adolescents remains a conflicted issue. Those in favor it point to the urgent necessity to prevent teenage pregnances. Those who oppose the practice perceive in it tacit approval of adolescent sexual activity. Others oppose the provision of contraception to teenagers without parental knowledge, arguing that such practice usurps the parental prerogative to know what their children are doing, especially since contraception carries certain health risks. The counterargument is that mandatory parental notification seriously discourages teenagers from seeking contraceptive help.

A source of even greater contention is *abortion.* Opinions range in this regard from those who would provide pregnant

women with full freedom to abort at will to others who would prohibit abortion on virtually all grounds. *Prolife* and *prochoice* groups are locked in an uncompromising battle over this issue from marching in the streets to political maneuverings in Congress (discussed further in chapter 19).

Gay Liberation   Attitudes towards homosexuality in our society have generally been quite negative, and sometimes exceedingly harsh (figure 16.5). Nontheless there have also been periods of relative tolerance in the history of Western culture (chapter 18).[46]

Homosexual activity can be assumed to have been present throughout human existence, but it was only in the 19th century that the idea of *being homosexual* appears to have taken shape (chapter 13). This meant that a class of individuals could now be singled out based on their sexual preferences, personality characteristics, or even body type. It was this forced segregation that led to the group consciousness of homosexuals and eventually to their liberation movements.

The modern movement for gay rights is another post-World War II phenomenon that has passed through two phases. The first was the homophile movement of the 1950s; the second, the gay liberation movement of the late sixties and the seventies.[47] The ground for the birth of the *homophile movement* was laid down during World War II. The circumstances of war time, discussed earlier, were particularly conducive to shifts in this regard. At its peak, the armed services of the United States held 12 million men. Though there was an attempt to screen out homosexuals during induction exams, the fear of ostracism and patriotism made it easy for the majority of homosexuals to pass the cursory tests. The circumstances of living in close quarters, the stress of combat, and separation from

[45]Yankelovich et al. (1977).

[46]For an historical overview, see Boswell (1980). A collection of primary historical documents can be found in Katz (1976).

[47]For a detailed account of the 1940–1970 period, see D'Emilio (1983).

Figure 16.5 Burning at the stake of homosexual monks in the 16th century in Ghent. (Etching by F. Hogenbergh, 1578. *Bilder Lexicon.*)

families generated close emotional ties between men. Even though in most cases these ties did not include an erotic element, those with homosexual orientations and inclinations had an enhanced opportunity to establish such sexual ties as well. Though the numbers of women in the armed forces were fewer than 150,000, the selection procedures and policies of the Women's Army Corps inadvertently favored the congregation of a large number of lesbians, under circumstances they would not have encountered in civilian life.

The publication in 1948 of the Kinsey report on male sexuality had a major impact in bringing homosexuality into the public consciousness. Not only were the statistics astounding (chapter 13), but Kinsey took a clearly sympathetic stand in favor of homosexuality as an acceptable and normal sexual alternative.

For many years, a "conspiracy of silence" had made the majority of homosexuals invisible; homosexuals themselves laid low, "passing" for heterosexuals, while the public treated the issue mostly as if it did not exist. The minority of homosexuals who came to public attention were those who sought therapy, ran afoul of the law, or lived on the fringes of respectable society: this reinforced the views that homosexuals were sick, criminal, and socially deviant.

Under these changed circumstances of the fifties, homosexuals began to make efforts to get organized. In 1951 the *Mattachine Society* was established in Los Angeles as the first homophile organization in America.[48] The man most responsible for its establishment was Henry Hay, an actor, whose experiences during the Depression years had led him to become a committed member of the American Com-

[48]The name is derived from a medieval French society of masked dancers whose members may have been homosexual.

munist party, Hay's gay collaborators shared his political views, so the Mattachine Society at its conception adopted a Marxist idology and was modeled after the organizations of the Communist party with secret membership, a tight hierarchical structure, and centralized leadership. Its purpose was "to unify isolated homosexuals, educate homosexuals to see themselves as a oppressed minority, and lead them in a struggle for their own emancipation."[49]

The McCarthy era was in full swing in the early 1950s, with the State Department purges of real or presumed Communist sympathizers; homosexuals were especially suspect on the grounds that they were a security risk, being subject to blackmail. As a congressional scrutiny of the Mattachine Society became likely, it was clear that Hay and the other founders could not survive the double jeopardy of their political past and homosexuality. Futhermore, the expanded new membership, consisting now mainly of middle-class men, was oriented toward legal reforms and social acceptability; it also included staunch anticommunists. The leadership of the Mattachine Society, therefore, yielded to a much more cautious and conservative group of men seeking an accommodation with society. This group showed little interest in the issues that more particularly concerned lesbians. Hence, a small group of women, led by Del Martin and Phyllis Lyon, established a separate gay organization for women in 1955 called the *Daughters of Bilitis* (named after an erotic poem by Pierre Louys that purported to be the translation of a poem of Sappho).[50]

The *Gay Liberation* movement was born in a riot that began when police raided the Stonewall Inn in New York's Greenwich Village, on the night of Saturday, June 29, 1969 and the gay customers fought back.

While these events acted as a dramatic catalyst, the shift within the homosexual consciousness had already occurred. After the quiet advocacy of the *homophile movement* to the end of the fifties, a new militancy began to assert itself starting in the sixties, under the leadership of men like Franklin Kameny, which culminated in the gay liberation movement (figure 16.6). The younger men who became the mainstay of the gay liberation movement had not been cowed by living through the McCarthy era. On the contrary, they were part of the broader social upheavals, such as the civil rights movement and the rise of the counterculture, among whose more prominent representatives were gay poets like Ginsberg. Within the feminist movement, in particular, radical lesbians played an important role in shaping the goals and tactics of the women's liberation movement. In short, homosexuals were no more the isolated and beleaguered minority shunned even by other activist groups but allied with and broadly represented within the wider spectrum of social protests of the time.

The gay liberationists differed from their predecessors in the homophile movement (whom they openly disdained for being timid and ineffectual) not only in their more assertive tactics but in their goals as well. Instead of seeking accommodation and compromise, they demanded the full rights of citizenship; rather than dwell on self-help and education, they put their efforts into political action; instead of trying to become like everyone else, they celebrated their differences. As a result, within a decade of the Stonewall riots, the United States had the largest, best-organized, and most visible homosexual minority in the world.[51]

Nonetheless homosexuals continue to be stigmatized and discriminated against in various settings. The majority of Americans continue to disapprove of homosexuality. In a national poll conducted in 1983, only 32% of respondents felt that homosexuality should be considered an accepted alternative life style; only 24% could

[49]D'Emilio (1983), p. 67. The founding of the Mattachine Society is related in an interview with Hay in Katz (1976).

[50]A 1920s lesbian novel, *The Well of Loneliness* by Radclyffe Hall, was a more important influence in shaping the lesbian consciousness.

[51]For an analysis of gay liberation as a reflection of broader shifts in American society, see Harris (1981).

Figure 16.6 Gay parade in New York City. (Arthur Streiber.)

claim homosexual friends or acquaintances; but 65% thought homosexuals should have equal rights in terms of job opportunities.[52] Whether or not homosexuality will be accepted to full parity with heterosexuality as the basis for personal and institutionalized relationships, therefore, still remains to be seen.

Freedom and Abuse of Public Sexual Expression The most important differences between the sexual revolution and earlier periods are not in the sexual behaviors but in the openness with which sexual themes were expressed in public. Whereas sexual discourse was considered improper in the past, there are now whole industries that provide erotic material for public consumption. As a result, our everyday culture is suffused with sexual images, references, and innuendo; whether or not this means our society has become eroticized or merely vulgarized, is open to different interpretations.

In the past, a small group of individuals determined what was fit or unfit for the

public to see and read. Such *censorship* often passed negative judgment on some of the greatest works of contemporary literature as well as a great deal of trash. None of this was generally viewed as being in conflict with First Amendment rights of free speech until the postwar era. During the sexual revolution, the sweeping changes made by the Supreme Court greatly liberalized the public expression of sexual themes. Subsequently, virtually any form of sexual description and depiction became permissible, subject to a few qualifications (chapter 19).

This has had several effects. First, it has allowed serious writers and artists to explore and express sexual themes without fear of reprisals. Sex researchers and educators likewise became more free to study and teach sexuality (chapter 1). Even more broadly, there is now a greater candor and willingness on the part of ordinary people to communicate sexual feelings and thoughts. Sex, as a legitimate subject for public discourse, can now be said to have taken its rightful place in our society.

On another plane, sex has become a major vehicle for entertainment and com-

[52]*Newsweek*, August 8, 1983, p. 33.

merce. This is particularly striking in television and advertising, the two media that reach the largest sectors of the population.[53] This raises two concerns. One is the relentless use of the female body as an attention getter, and the other is the promise of sexual enhancement as a lure to purchase various products. This exploitation of sex for nonsexual ends and false promises that are associated with such use may have far-reaching effects on shaping our sexual and gender attitudes.[54]

Equally important is the influence of popular culture in portraying the interpersonal aspects of sexual and love relationships. Best-selling books and popular movies typically present sexual partners as young and single, physically attractive and healthy. Relationships are short-term; "playing it cool" is the way to go. Contraception, sexually transmitted diseases, and sexual dysfunctions are rarely mentioned, let alone dealt with in any realistic and meaningful way. Sexual passion between ordinary married couples, sexual interest in the elderly, the sexual adaptation of the physically disabled are left out with rare exceptions (such as Jon Voight's portrait of a disabled veteran in the film *Coming Home*).[55] As a result, "sex in the media serves as a fallacious model of human sexual behavior by creating unrealistic and infeasible sexual expectation."[56]

There has always been an intimate association between music and sexuality (chapters 11 and 12). But never before was music so flagrantly eroticized in our culture as was the case during the sexual revolution. *Rock and roll* and other distinctive musical styles of the period were both an outcome of the sexual revolution as well as a significant force in its popularization among youth.[57]

Much of what is portrayed in popular media is geared at attracting viewers and readers by offering them whatever it is that they seem to want. One can fairly say that the mass media reflect rather than generate the sexual attitudes prevalent in society. But, surely, by giving back to the people what they want, we may be reinforcing certain aspects of our culture that leave something to be desired. And by exposing our children to such themes as sexual violence, we could be helping to perpetuate such behavior.

These concerns become more serious with regard to *pornography.* Produced mostly by people of very modest talents who possess even less by way of social responsibility, pornography (hard as it is to define precisely) callously exploits the sexual freedoms that have become available. Its possible consequences are a source of mounting social concern to groups as disparate as feminists and fundamentalist Christians who otherwise disagree on many other issues.

The Conservative Reaction   The majority of Americans seem to welcome greater sexual honesty, openness, acceptance, and equality that have resulted from the sexual revolution. But there is also a pervasive disquiet over one or another of its consequences, and a concern over whether things may have gone too far.

Among a substantial sector this disenchantment is felt very strongly, and people feel a commitment to contain and to reverse the changes that have been brought about. These conservative sentiments have been given their most vocal expressions by the *Moral Majority,* a loose coalition founded by the Reverend Jerry Falwell in 1979. It is hard to know how many people belong to the organization, since estimates have ranged from less than half a million to many times that number.[58] Whatever the

---

[53]In 1983, daily TV viewing for American households averaged 7 hours and 2 minutes—14 minutes more than in 1982 (Rotherberg, 1984).

[54]For the influences of sex on TV, see Himmelweit (1980); Ellis and Sexton (1982). Sexual messages in public environments more generally are discussed in Ladd (1980).

[55]See Abramson and Meeanic (1983) for a sexual analysis of best-selling books and motion pictures over the past 3 decades.

[56]Zilbergeld (1978).

[57]For a history of rock and roll see Miller (1980).

[58]Yankelovich (1981).

hard-core membership, its influence varies widely with the issues at hand.

The Moral Majority reflects the theological beliefs of evangelical Christianity; its sexual values are traditional and its political outlook conservative. It is opposed to almost every change brought about by the sexual revolution and associated events. In short, it is against abortion, premarital sex, easy divorce, pornography, the ERA, and the ideology of "secular humanism" with its emphasis on individual freedom, tolerance within a pluralistic value system, and a questioning attitude toward institutional authority. The Moral Majority is in favor of sexual restraint in youth; marital fidelity; a strong and traditional family with parental authority over children; community censorship of school texts; and influence on television programs to conform to the group's standards and values. To achieve these aims, the Moral Majority is prepared to use the political process and the legal system, as well as to rouse public opinion to assert its views.

Although many people object to various aspects of the Moral Majority ideology and tactics, the alarm it has sounded over the social climate within which children are growing up has resonated with the concerns felt by many parents. Americans are in favor of free speech but most of them do not want their children exposed to obscene magazines in the neighborhood store. Uncertain of their own sexual values, parents are at a loss over how to guide their children sexually. Hence, when a confident voice raises questions over where we are headed as individuals, as parents, as a community, and as a nation, people listen.

The Balance Sheet It is too early to make a reasoned asessment of the net gains and losses that we have incurred through the sexual revolution. *Sexual pessimists* are likely to decry any liberalization of sexuality, turning a blind eye to its benefits. *Sexual optimists* can be counted on to be enthusiastic about everything and anything that is sexual. For them, sex can do no

wrong, and if evidence to the contrary is incontrovertible, they change the definition of what is sexual.

After centuries of struggle, we have finally gained a large measure of sexual freedom and choice; in due time we shall perhaps attain even greater freedom to express our sexual needs without some of the penalties that are still attached to it. But there are also disquieting questions about the price of all this. In overcoming needless guilt, are we losing our capacity for shame? By taking the mystery out of sex, are we robbing it of its deeper meaning and trivializing it?[59] These are sources of concern to a wide spectrum of people, including some who work within the field of sexuality.[60] It is to such questions that we seek answers in subsequent chapters.

SUGGESTED READINGS

Reiss, I. *Family Systems in America,* 3rd ed. (1980). Sociological perspectives on marriage, the family, and the socialization of children approached from a vantage point of particular relevance to sexuality.

Gregersen, E. *Sexual Practices* (1982). Wide-ranging and profusely illustrated overview of cross-cultural aspects of sexual attitudes, customs, and behaviors.

Robinson, P., *The Modernization of Sex* (1976). Addresses the intellectual underpinnings of the sexual revolution by examining the work of the key sex researchers of the era.

Sorokin, P. *The American Sex Revolution* (1956). A highly critical view of the sexual revolution, even before it had fully emerged. Many of the arguments made by the conservative sociologist against sexual liberality are presently repeated by others.

Hurwood, J., ed. *The Whole Sex Catalogue* (1975). A boisterous, uninhibited hodgepodge of sexual materials from the classical to the novel, the curious and trivial, reflecting much that was typical of the sixties. The section on contemporary sexual artists is especially interesting.

D'Emilio, J. *Sexual Politics, Sexual Communities* (1983). A well-researched and sympathetic account of the gay liberation movement in the United States from 1940 to 1970.

[59]Tyrmand (1970).
[60]Schmidt (1982).

Chapter 17

Sexuality in the Ancient World

There is no greater nor keener pleasure than that of bodily love—and none which is more irrational.

Plato

To understand who we are as adults, we need to learn about our childhood. Likewise, to understand the place of sexuality in our culture, we need to learn about its historical origins. Such learning does not come easily since we tend to be pressed by the present, preoccupied with the future, and oblivious to the past.

Sexuality and the Origins of Culture

Sexual representations are among the earliest surviving artifacts of human culture. But the people who produced the statues and cave paintings of the *Paleolithic era* (30,000 to 10,000 B.C) were not erotic artists in the conventional sense of the term. It is conjectured that such art was produced for magical-religious purposes to enhance *fertility*, both within the human group as well as in the herds of wild animals on which these hunters depended.

As with other primitive cultures of more recent times, we assume that sexuality permeated the beliefs and rituals of our earliest forbears and that they used *sexual magic* to promote their welfare and to protect themselves from the forces of evil. Such

magic was believed to work by endowing the created object with a life of its own, which then influenced events in the natural world: by painting figures of bison on a cave wall they hoped to increase the size of the herd; by showing a bison being pierced with a spear, they sought to promote the success of the hunt.

Modern intrepretations of Paleolithic art attribute sexual meaning to a wide variety of images ranging from the obviously sexual (such as representations of the vulva) to the ambiguously abstract. For instance, in the case of the wounded bison, it is conjectured that spear and wound represented phallic and female emblems whose union symbolized the renewal of life through coitus.[1]

Of special interest are Paleolithic **Venus statuettes,** with distinctively exaggerated breasts, buttocks, and midsections (figure 17-1). These are thought to be fertility goddesses, emblems of veneration of female fecundity and nurture. Their male counterparts are ***ithyphallic figures*** (males with erect penises) which are sometimes shown carrying a club or a spear (figure 17-2).

[1] Field (1975).

447

Figure 17.1 "Venus." Lespugne, Haute-Garonne, France. Early Stone Age bonecarving.

Phallic symbols, which may have started as a fertility emblem, gradually had become endowed with added significance with respect to power and authority by the time of the ancient Egyptian civilization. By 3000 B.C. the great civilizations of Egypt and Mesopotamia had emerged, and their written texts express more directly the pervasive importance of sexuality in their religious life and culture (chapter 1).

These early cultures are, in a sense, the source of all subsequent civilizations. But the roots of Western culture are traced more directly to Judaism and the classical cultures of Greece and Rome[2].

Sex and the Judaic Tradition

Judaism is not a monolithic faith with a single established creed or code of sexual behavior but a religious culture spanning

[2] "Western culture" is used here in its commonly used sense of referring to European civilization and its American offshoot. See Chambers et al. (1979, pp.2–3).

some 4,000 years. It is a *way of life* whose precepts, embodied in the law (*Halakhah*), shape the daily routine into a pattern of religious observance and sanctity.[3] The primary sources of the law are the Torah and the Talmud. The **Torah** ("revelation") corresponds to the first five books of the Old Testament and is ascribed to Moses. The **Talmud** ("learning") consists of laws promulgated early in the third century (the *Mishnah*) and rabbinical commentaries on it accumulated over the next several centuries.

During its long history, various trends and moods including "naivete, morbidity, asceticism, pessimism, legalism, practical liberality, pietistic severity, rationality, mysticism" have characterized Judaism.[4] But its overall tone has been one of frank, non-puritanical acceptance of sex, coupled with a self-imposed discipline of sexual restraint. This diversity of Judaic teaching becomes less confusing if particular sexual attitudes and practices are placed in their proper historical context.

Historical Perspective

The early Israelites were a nomadic people. Abraham is regarded as the father of the Israelite nation, and his grandson Jacob organized its people into 12 tribes under his 12 sons. Some of these tribes migrated to Egypt but were eventually led out by Moses (the *Exodus*) sometime before 1200 B.C.[5] Moses received the law and bound his people in a convenant with God (*Yahweh*; mistranslated as "Jehovah").[6]

[3] For definitions of Judaism, see Neusner (1970). Jews call themselves *Israel*, being the descendents of Jacob (who was named Israel as related in Genesis 32:28). *Jew* is from "Judean" (the ancient kingdom of Judah). *Hebrew* (which refers to the Jewish language only) is derived from the Egyptian "hapiru" for driver of caravans (Chambers et al., 1979).

[4] Epstein (1967, p. 24).

[5] Jews substitute BCE ("before the Common Era") for B.C. and CE (Common Era) for A.D.

[6] Circumcision is the sign of the convenant. Jewish male infants are circumcised and named on the eighth day following birth. Female infants are not circumcised; they are named usually on the sabbath following birth.

Moses is thus the first rabbi ("master") or teacher and the prototypical figure of the ideal Jew.[7]

Sex does not seem to have been much of an issue for the nomadic Jews. Sexual intercourse in lawful settings was highly regarded and most forms of nonreproductive sex were frowned upon. But there was no preachment and preoccupation with sexual guilt and sin, and sexual misbehavior was treated no differently from other forms of misconduct. Nor were personal relations lacking in passion and tenderness: Jacob was so smitten with Rachel that he served his prospective father-in-law for seven years "and they seemed to him but a few days, so much did he love her" (Gen. 29:30).

Agriculture and progressive urbanization and commerce, from the 7th century B.C. on, complemented nomadic life. The deportation of Jews to Babylon (by Nebuchadnezar, in 597 B.C.) exposed Jews to a worldly culture. Sexual behavior in the *postexilic* period became more lax; in consequence moral teachings turned more legalistic and ascetic. The innocence and joy of earlier times were now replaced by a pervasive pessimism. People were seen as driven by sexual lust; no one was free of sin; living lost its purpose ("all is vanity": Eccles. 1:2–3). Concurrently, there was a decline in the status of women. They were veiled and confined to the home. Unlike earlier times, when the sexuality of women was accepted, now women were held accountable for leading men astray: "I consider more bitter than death the woman whose heart is snares and nets, and whose hands are fetters," wrote the author of Ecclesiastes (7:26).

The coexistence of corruption and virtue among the Babylonian Jewry is illustrated by the story of Susanna, related in an apocryphal addition to the Book of Daniel. Susanna was the beautiful and pious wife of Joakim, a wealthy Jew in Babylon. Two elderly judges were inflamed with lust for her. When they surprised her at her bath (figure 17.3), she resisted their advances;

[7] Neusner (1970).

Figure 17.2 Bronze Age rock carvings from Sweden.

so they accused her of adultery with a young man and had her condemned to death. But the wise judge Daniel confounded the false witnesses, thus saving Susanna and sending the corrupt elders to their death.

Some of these attitudes persisted into the *Hellenistic period* (4th century B.C. to 2nd century A.D.) and greatly influenced Christian perceptions of sex. To a lesser extent they also lingered within the *rabbinic Judaism* (2nd to 18th centuries) that produced the Talmud.

Judaic Attitudes to Sexuality

Judaic sexual teaching and practice have been generally forthright and accepting of sexuality, although individual rabbis have expressed a wide range of opinions on sexual matters, some more sensible than others. But on the whole, the rabbinic tradition was guided by "the common sense,

Figure 17.3   Rembrandt, *Susanna and the two elders.*

the level-headedness, the close contact with life, and the practical piety of the talmudic sages."[8]

The dominant perspective in Judaism views sex as God's gift. It does not advocate world renunciation and has little taste for asceticism.[9] Orthodox Jews observe the law to its most minute detail, but even at their most strict, they renounce neither the body nor its physical pleasures as evil. Sex becomes unethical when it violates the integrity of human relationships as envis-

aged by the law, not some abstract moral principle.

When a society sets out to regulate the behavior of its members in concrete and precise ways, as traditional Judaism does, rules and regulations to deal with all conceivable contingencies inevitably proliferate. Furthermore, to protect better the core ethical values, additional rules are made as warnings in more peripheral areas on the principal of "building fences" (chapter 16) resulting at times in a sterile and oppressive legalism. Judaism has struggled to free itself from this predicament by formulating broadly applicable ethical principles to guide human conduct. Thus, on the one hand, there are the hundreds of commandments given to Moses, and on the

[8]Epstein (1967, p. 12).

[9]One exception were the *Essenes,* who rejected all pleasures as evil and depreciated women and sexuality. Their views may have been influential in shaping early Christian attitudes.

other, the all-encompassing principle of Leviticus 19:18: "You shall love your neighbor as yourself"; rephrased by Rabbi Hillel in the first century as, "What is hateful to yourself, do not do to your fellowman. That is the whole Torah."[10]

The legal code, even at its most demanding, only sets a minimum standard. Beyond behavioral considerations, one must be concerned about the "purity of mind" where motivation and intent become the area of concern: as a moral Jew, one can never do enough. Some of these expectations were incorporated into Christianity as exemplified by the principle of "sinning in one's heart," whereby one is guilty of impure thoughts even though these are not acted upon.

Sex and Marriage in Judaism

Since institutions facilitate and regulate the satisfaction of sexual needs, a good deal of the law addresses itself to the relationships of people within these contexts. The key institution in Judaism is *marriage*, and it is in this connection that all forms of sexual behavior are evaluated as to their morality and desirability. This centrality of marriage and procreative sex within it are among the key legacies from Judaism that have profoundly influenced sexual morality in Western culture.

As expressed in the story of creation in Genesis, God intended man and woman to complement each other. It is to this end that a man "leaves his father and mother and cleaves to his wife, and they become one flesh" (Gen. 2:24). God's first commandment to Adam and Eve was to be fruitful and multiply (Gen. 1:28). Building on this tradition, Judaism came to view marriage as a religious duty. Husband and wife represented the model of integrated humanity, and one was not fully adult unless married. There is no priestly celibacy in Judaism; all of the prophets and major biblical figures were married, with rare exceptions to allow for special circumstances.

Sex and procreation, the key elements in traditional Jewish marriage, imposed reciprocal obligations. Wives were expected to submit to the sexual desires of their husbands; likewise, each man was obliged to perform his marital duty. For donkey drivers this meant coitus once a week, for camel drivers once in 30 days; for scholars and gentlemen of leisure, every night. Menstruation rendered a woman "unclean" (chapter 20), but pregnancy constituted no obstacle to sex.

Sex is not just a procreative duty in Judaism but was meant to be enjoyed by both sexes: In the World-to-Come, says the Talmud, every man will be called to account for all the legitimate pleasures which he has not enjoyed. There were some who tried to impose restraints of various kinds on sexual enjoyment, or accepted it grudgingly for reproductive purposes only. For example, couples were warned that children conceived while using unorthodox coital postures would be born lame; genital kissing would make the child mute; conversation during coitus would cause deafness, and gazing at each other's nakedness would cause blindness in the offspring. But the more authentic voices of Judaism spoke for the freedom of husband and wife to do as they pleased; to have intercourse at any time and in any manner that they wished. These same sentiments are manifest in the *Song of Songs*, which celebrates the passion of love in lyric poems sung at ancient wedding feasts.[11]

Proscribed Sexual Behavior

Judaism has treated harshly those who transgress laws of proper sexual conduct. Judaic judgments in this regard are based on pragmatic grounds and not on any notions of sex being inherently evil. The core concern is to protect the integrity of marriage; all other considerations emanate from that aim.

---

[10]Neusner (1970, p. 87).

[11]Other interpretations have been offered about the true meaning of this exceptional book. It was included in the Bible because it was considered to be an allegory of the relationship of God with Israel; and in the Christian tradition, that of Christ and the church.

Box 17.1   Jewish Marriage

The term for the Jewish marriage ceremony is *kiddushin* ("sanctification"). It is said that at the "wedding" of Adam and Eve, God acted as the "best man" and plaited Eve's hair to adorn her for her husband.[1] Given that marriage was the preferred state, parents tried to arrange marriages for their children (especially daughters) as soon as possible. For girls, engagement (*betrothal*) became desirable on entering puberty at about age 12; marriage followed a year later.[2] If a girl remained single into adulthood, her father lost the power to contract marriage for her (and to collect the bride-price). She was also liable for her own sexual conduct. Under these circumstances, it is no wonder that daughters were a source of life-long distress for fathers.[3]

Fathers also arranged the marriage of their sons (usually in their late teens) and negotiated the dowry (bride-price). To help a new couple settle down, the law stipulated that "when a man is newly married, he shall not go out with the army or be charged with any business; he shall be free at home one year, to be happy with his wife whom he has taken" (Deut. 24:5).

Monogamy has been the norm in Judaism, but considerable latitude existed for some men of special status. Multiple wives and concubines were maintained by tribal chiefs like Abraham and Jacob, and later on by the nobility (Solomon had several hundred women in his household) to cement political allegiances, to expand progeny, and to provide sexual variety. Polygamy was abolished in Judaism in the 11th century by Rabbi Gershon, followed by the right to keep concubines (*pilegesh*), which was particularly opposed by Maimonides, the 12th-century codifier of the Talmud.[4]

The *good wife* commanded much respect in Judaism. ("A good wife who can find? She is far more precious than jewels," Prov. 31:10.) She is not noted for her beauty (which is "fleeting") nor charm (a "delusion"), but is praised for her God-fearing nature, dedication to her husband and family, and the expert management of her home. "Extoll her for the fruit of all her toil," says Proverbs, "and let her labors bring her honor in the city gate" (Prov. 31:31). For Jews and Christians alike this feminine ideal proved to be a durable model until modern times.

The grounds for divorce were quite liberal for the man. If the wife did not "win his favor" because he found "something shameful in her," he could dismiss her (Deut. 24:5). "Something shameful" was interpreted by the school of Shammai to mean sexual infidelity: the school of Hillel understood it as virtually anything displeasing to the husband (including bad cooking).[5]

Women had some protection against capricious dismissal because the husband was obliged to repay the bride-price, often a substantial sum, on divorce. However, if it could be shown that the wife had been guilty of scandalous behavior, then she could be dismissed without compensation.

[1]Parrinder (1980), p. 192.
[2]Puberty was determined by degree of breast development. In comparison with figs girls were considered "unripe," "ripening," and "ripe."
[3]Cohen (1949).
[4]For the protracted debate over this issue, see Borowitz (1969, pp. 43–49).
[5]Parrinder (1980, p. 194).

Strict laws of consanguinity guarded Jews against *incest.* One was forbidden to commit incest even when threatened by death. The seeming lack of disapproval with which the Old Testament reports Lot's seduction by his two daughters (Gen. 19:32–37) is puzzling (figure 17.4), but this event took place under extraordinary circumstances. Sodom had been destroyed, Lot's wife was dead, he was stranded with his daughters

Figure 17.4 Francois DeTroy, *Lot and his daughters*. (Musio De Arte De Ponce, Puerto Rico)

in the hills, and they had gotten him drunk. Most importantly, the motivation behind the act was to ensure the perpetuation of the family line. Similarly, if the story of creation in Genesis is taken literally, one would have to assume the occurrence of at least brother-sister incest for the propagation of the progeny of Adam and Eve.

Sex Outside Marriage The Judaic concept of *adultery* was based on the violation of the husband's exclusive rights to the sexual and reproductive assets of his wife. A wife committed adultery if she had sex with anyone other than her husband. A husband committed adultery only when he had sex with another man's wife, not when he had sex with anyone other than his wife. This is the original meaning of the Seventh Commandment: "You shall not commit adultery" (Exod. 20:14). The concern over violation of property rights is reiterated in

the Tenth Commandment, which says, "You shall not covet your neighbor's wife, nor his horse, ox, ass, or any other belonging" (Exod. 20:17).

The wife owed faithfulness to her own marriage, the husband to the marriages of his kinsmen ("neighbor").[12] When this faith was violated, the husband was entitled to kill his wife (as well as her lover), or he could cut off her hair, strip her naked ("expose her shame,") and turn her over to a mob to be stoned to death. But she could also be merely humiliated, reduced to servitude, or even forgiven.

Seduction of a virgin required compensation to the father and marrying the maiden if she so desired. *Rape* was an ever-present danger. Ravishing the women of the enemy was taken for granted. But even

[12]Epstein (1967, p. 194).

in peacetime, young women were not always safe. Tamar, the daughter of King David, was violated by one of her own brothers; another brother then killed the offender to avenge her honor (2 Sam.:13). When a group of Benjaminites brutally raped a Levite's concubine, causing her death, the man dismembered the woman's body and sent its pieces to the other tribes, bringing down their wrath on the tribe of Benjamin, which was almost wiped out (Gen. 59:19–21).

The Old Testament does not explicitly forbid premarital intercourse. This is because practically all women got engaged on reaching puberty; so there were not that many unattached young women to have sex with. When sex took place between engaged couples, community reaction tended to be only mildly disapproving since loss of virginity under these circumstances was not stigmatizing.

Since marriage for men came later and not everyone could afford to take a wife, some resorted to prostitutes. Married men who sought sexual variety could do the same without committing adultery. Yet the attitude of Judaism toward **prostitution** was highly ambivalent, not because of some abstract moral principal but because of the nature of prostitution at the time.

Judaism expressly forbade fathers to give their daughters to whoredom. So sex with prostitutes meant consorting with foreign women. Furthermore, there were two kinds of prostitutes. The *zonah* was the ordinary prostitute, the *kadesha* (and her male counterpart, the *kadesh*) was the religious or cult prostitute. These priestesses were "married" to the god in whose temple they served and their duties included engaging in ritual coitus with worshippers and others willing to pay their fee.[13] The real offense in consorting with cult prostitutes was not sex but sacrilege; it meant worshipping a god other than the Lord; the one absolute, unforgivable sin in Judaism.

Nonreproductive Sexual Behaviors   Given the centrality of reproductive sex, one would expect Judaism to frown upon all forms of nonreproductive sexual activity. In general, that has been the case, but Judaism has not been as strict in this regard as, for example, the Catholic church with respect to contraception, sterilization, and abortion.

The Old Testament has no explicit prohibitions of these practices. The sin of Onan (Gen. 38:9) involved coitus interruptus to avoid conception. But the Lord's wrath was aroused at Onan because he was shirking his responsibility to provide heirs to his deceased brother (as required by the *Levirate Law* in Deut. 25:5–6) while enjoying sex with his widow. The Old Testament does not forbid **abortion** and considers the embryo to be part of the mother. Later authorities differed widely about the legitimacy of these practices. **Contraception** was forbidden in the Middles Ages (when Jewish populations were being decimated through persecution). Although some orthodox Jewish authorities still oppose it, Judaism as a whole does not.

The Old Testament repeatedly condemns **bestiality** as a capital offense. The Talmud even prohibits the keeping of pet dogs by widows to avoid suspicion of sexual contact. The case against **masturbation** is far more ambiguous. The standard reference has been to the case of Onan (hence *onanism*); but clearly the act in question had nothing to do with masturbation.

Nocturnal emissions rendered a man "unclean" and required a ritual bath of purification, but the same requirement extended to coitus as well (Lev. 15:16–18). The Israelites were awed by vital bodily fluids like semen, menstural blood, and discharges accompanying chidbirth, since these were associated with the generation of life, and built various rituals and taboos around them, as have other cultures.

---

[13] This practice existed for short periods even within Judaism. For instance, during the reign of Rehoboam, son of Solomon, there were cult prostitutes attached to the Temple of Jerusalem.

There are only two passages in the Old Testament that explicitly condemn *homosexual* acts, both of them in Leviticus: "You shall not lie with a male as with a woman; it is an abomination" (Lev. 18:22); "If a man lies with a male as with a woman, both of them have committed an abomination; they shall be put to death" (Lev. 20:13). Whether these prohibitions were based on ritualistic concerns (as in the case of sex with cult prostitutes) or intended as moral condemnation of the sexual acts as such is a matter of dispute (chapter 20).

Even more problematic is the curious story of the destruction of Sodom (Gen. 18–19). Traditionally it has been assumed that the Lord destroyed Sodom because its people tried to rape the two angels of the Lord whom Lot was sheltering in his home (hence *sodomy*). More recent scholarship has questioned the sexual interpretation of these texts and hence the validity of their representing biblical condemnations of homosexuality.[14]

Judaism has had a profound influence on Christian sexual morality, but this influence has been selective. Christianity owes to it the centrality accorded to marital sex and the condemnation of all other sexual behaviors. The shared belief in the Old Testament has continued to link the two religious traditions in important ways, yet with regard to sexuality, some important differences have emerged. Jews understand the scriptures as a legal system, albeit divine in origin. But as with all laws, there is an inevitable necessity to fit their requirements to changing times. Rabbis have always acted with the authority to teach the law for their own day and age to make possible the continuing faithfulness of the people of Israel.[15] By contrast, the Christian approach to morality has been more abstract, whereby legal precepts from the past have been elevated to absolute and immutable standards that retain their validity independent of changed historical circumstances. This has generally made the

Christian interpretation of the Old Testament with regard to sexual morality far more restrictive than it has been for Jews.

Eros in Greece and Rome

Beginning with its Minoan and Mycenaean origins, Greece went through three periods: the *Archaic* (750–500 B.C.); the *Classical* (500 B.C. to the death of Alexander of Macedon in 323 B.C.); and the *Hellenistic* (the next three centuries, culminating in the establishment of the Roman empire under Augustus in 27 B.C.).

Unlike the nomadic Jews whose character was forged in the desert, the early Greeks were farmers, and hence agricultural concerns shaped their values. Sex for the Jews was primarily a means of reproduction and cementing family bonds. The Greeks also used sex primarily for domestic purposes but allowed far more freedom in its varied expression and celebration.

The 5th-century Athenian attitude to sex was uninhibited, very much like that of contemporary Western societies. That does not mean that people behaved as they pleased; conservative elements coexisted with more liberal and radical factions, as they do today. Yet the Greeks differed from us in their conception of romantic love. Instead of making it the basis of marriage and the justification for sex, the Greeks viewed romantic love and erotic passion as troublesome emotions that, like anger, upset reason.[16] Hence, they were best avoided as the basis of serious decisions like marriage. Sexual pleasure nonetheless was to be had freely both as an end in itself or else sublimated to attain higher philosophical ideals.

[14] See Bailey (1959); Boswell (1980, pp. 92–99).
[15] Borowitz (1969, p. 50).

[16] The irresistibility of love and its demonic force is expressed by the chorus in *Antigone* by Sophocles: "Love, invincible love, you swoop down on our flocks and watch, ever alert, over the fresh faces of our maidens; you float above the waves and across the countryside where the wild beasts crouch. And among the gods themselves, or mortal men, there is no one who can escape you. Whoever touches you is at once thrown into delirium."

Erotic Themes in Greek Literature and Art

The bawdy tales of mythology and Greek art that were rediscovered during the Renaissance became an endless source of inspiration for erotic themes and images in Western art and literature. Even in the Victorian era, works with a classical veneer could pass muster as respectable art. But there was a limit. The figures of satyrs with erect penises which adorned vases (figure 17.5), which were dug up in the 19th century, had their phalli painted out; and in literature whatever did not agree with conventional morality (such as evidence of Greek homosexuality) was often ignored or misinterpreted.

Greek literature abounds in sexual themes and references, including its treatment of the gods (box 17.2). But the fact that the Greek gods were portrayed as fully sexual in *literature* must not be equated with their being so in Greek *religion*. The lurid tales of Zeus pursuing divine and mortal, male and female notwithstanding, the Greeks were quite respectful of their divinities and serious about their religion.

The Greeks treated female nudity in art and daily life with the same acceptance as we do, But they were far more open with the public display of male genitals than we are. This prominence of phallic images and symbols is a striking feature of Greek eroticism. The erect penis was as much a part of the consciousness of the Greeks as the circumcised penis was for the Jews. Giant phalluses made of wood were carried in Dionysiac processions; others hewn out of stone stood on pedestals (figure 17.6). *Herms* (representing the god Hermes) with human head and penis (figure 17.7) were

Figure 17.5 Satyr pursuing a nymph, painting on an Attic red-figure amphora by Oltos, early fifth century B.C. (The Louvre, Paris.)

Figure 17.6 Phallic monument, the Dedication of Karystos, from the Sacred Road on the Island of Delos. Fourth–Third century B.C.

common sights, as were various other phallic monuments guarding the fields and phallic amulets hanging from people's necks (to be discussed in the Roman period).

Yet the use of these phallic symbols was not meant to be obscene, frivolous, or even erotic. The Greeks associated phallic symbols with fertility, rebirth, and immortality. Such sexual emblems were part of their religious life just as other religions rely on other symbols for devotional purposes.

Erotic themes in Greek drama   Just as Homer's *Iliad*, which relates religious myths, is not an attempt at systematic theology but a work of literature, Greek drama, which we think of as literature, has many religious components, with the gods directing events and themselves appearing as characters on stage.

Erotic elements in Greek drama are present both in the tragedies and come-

dies. A major theme in tragedy is incest. The legend of King Oedipus (which provided the name for Freud's *Oedipus complex*) was first related by Homer and retold by the great dramatists, the most renowned being the plays by Sophocles: *King Oedipus* (about 427 B.C.) and *Oedipus at Colonus* (about 408 B.C.) Incest is also the theme in Euripides' *Hippolytus* (428 B.C.), depicting the love of Phaedra for her stepson, Hippolytus, which leads both to their doom.[17]

Greek tragedies were presented in sets of three accompanied by a fourth *satyr play* for comic relief. The revolt of women is a common theme in the comedies of Aristophanes (c.488–380 B.C.) In *Women in Parliament*, they establish a communistic society where property and sex are shared (a parody of Plato's *Republic*); in *Lysistrata* the women seize the Acropolis and declare a sex strike until the men of Greece come to their senses and stop fighting.

The ribaldry in Greek comedy is traceable to its ancient roots in the rituals of *phallic cults*; in the early comedies, actors customarily wore large leather phalluses, as was also the custom in Dionysian festivals. Comedies openly dealt not only with sexual relations between men and women, but also with other forms of sexual behavior such as homosexuality, masturbation, and bestiality, reflecting the Greek prediliction to deal openly with all forms of sexual experience. (Yet the sexually suggestive depictions put on in rock concerts today would probably have shocked the Greeks, given the mixed teenage audience.)

The treatment of love and sex were more subtle in the hands of the lyric poets, prominent among whom was Sappho of Lesbos. We think of her predominantly in connection with Lesbianism, but the Greeks regarded her as one of their greatest poets and called her the "Tenth Muse." Only fragments of her work have survived.

[17] Sophocles used the same story in *Phaedra*, as have others up to modern times. See for instance Eugene O'Neill, *Desire Under the Elms*.

Box 17.2   Erotic Themes in Greek Mythology

In Hesiod's account, *Eros* is the god of love who at the dawn of creation emerges from Darkness and brings all other things to life. Similarly, Hesiod places the birth of *Aphrodite,* the goddess of love, in the misty origins of the world when Earth was embraced by Heaven, and Time (*Kronos*) was born. Kronos turned against his father and castrated him; the amputated genitals swollen with sperm fell into the sea and from their white foam arose the lovely Aphrodite.

In Homeric legends, twelve Olympian gods (a third of them female) conducted their divine business, like a royal family, under *Zeus.*[1] Zeus was married to his sister, *Hera,* but he also pursued other women: he carried off *Europa* by disguising himself as a snow-white bull, seduced *Leda* by taking the form of a swan, and impregnated *Danae* by descending on her in the guise of a shower of gold. Through such unions were born the great heroes like *Heracles* (who, true to his father, deflowered all 50 daughters of King Theseus in a single night). Nor was Zeus partial only to women. One of his most celebrated loves was *Ganymede*, a handsome Trojan youth whom he snatched up while disguised as an eagle and carried off to Olympus to supplant *Hebe* (the goddess of youth) as the cup-bearer to the gods.

Aphrodite, the goddess of love, was married to her lame and ugly half-brother *Hephaestos,* the god of fire and the celestial blacksmith, but became enamored with another half-brother, *Ares,* the splendid god of war. *Helios,* the all-seeing sun-god, leaked the news of the affair to Hephaestos who forged a fine net ensnaring Aphrodite and Ares in bed. Their predicament, however, created more merriment than outrage, when *Hermes* volunteered to change places with Ares. He must have had his chance some other time since from their union was born *Hermaphrodite,* whom the Hellenistic Greeks glorified for representing the complementarity of male and female beauty. Another of Zeus' daughters, *Athena,* was delivered from Zeus' head and became patroness of virgins (and Athens).

The character of *Eros,* the god of love changed over time. From creator of the world in Hesiod, he was later cast as the son of Aphrodite who would smite people with passion at her bidding. Shown as a hand-

---

Eroticism in Greek Art   Love of beauty suffused Greek culture, and its genius could combine the unabashedly sensual with the highest aesthetic and ethical ideals. It makes no sense to speak of *erotic art* in Greece as a thing apart; no meaningful distinction can be made between artists (or writers) who dealt with erotic themes and others who did not. Sex was part of art as it was part of life.

Greek art was inspired by the same mythological themes as literature, but the two were not directly complementary; that is, artists did not produce images as illustrations for texts. Erotic themes in vase paintings usually portray anonymous fig-ures or interactions between satyrs and maenads. Rarely is a deity like Aphrodite or Zeus identified.

The Greeks idealized youth, and it was as often the beauty of the young male as the female that inspired them. Only recently have the accomplishments of classical art in the more frankly sexual realm become accessible to the general public.[18] But more generally, such as in representations of the nude, the profound influence of Greek art has been very much in evidence. The same is true for Greek philosophy, which has greatly influenced

[18] See Boardman and La Rocca (1975); Marcade (1962); and John (1982).

Box 17.2   Erotic Themes in Greek Mythology (*Continued*)

some youth, Eros represented the ideal of male beauty who pursued his own loves, as in the haunting story of *Psyche* (Soul). By the Hellenistic period Eros had taken on his more familiar cherub's form in which he persisted into Roman times as *Amor* and *Cupid*.

Aphrodite's patronage of love and fertility was shared by **Dionysus,** (another issue of Zeus' illicit loves) who discovered the vine and became the god of wine.[2] The association of flowers with Aphrodite and the vine with Dionysus points to the close connections between religion, sex, and agricultural concerns in ancient Greece. The great Dionysian festivals celebrated the passing of seasons and invoked the blessing of deities to ensure good harvests. Sexual activity in these contexts was thus inextricably bound with the broader cycles of life and fecundity in nature.[3]

The companions of Dionysus included *satyrs* or the *sileni* (named after *Silenus*, their "father" who had tutored Dionysus). These creatures represented unbridled, earthy sensuality.[4] They endlessly pursued the forest nymphs (*meneads*), whose songs and dances enchanted them (figure 17.5). Crude and lewd

as the satyrs could be, they were of a benign nature. Not so the *centaurs*, who were brutish (with some exceptions like Chiron, the tutor of Achilles). Where the satyrs seduced, the centaurs raped; the escapades of satyrs were led to sexual pleasure, the frenzy of centaurs to tragedy and death.[5] Centaurs thus represented the more lawless and violent aspects of male sexuality, while satyrs stood for its more hedonistic side.

[1]Of the many contemporary renditions of these myths, see Graves (1959); Holme (1979); Grimal (1981). The last two references include numerous illustrations from the immense body of art inspired by these myths.

[2]*Dionysian* has come to signify that which is ecstatic, sensual, irrational, and suffused with creative energy, as contrasted with *Apollonian*, which stands for the rational, well-ordered, serene, and dignified.

[3]Dionysiac festivals are discussed in Webb (1975); Marcade (1962), and at greater length in Licht (1932).

[4]Greek satyrs were men with horses' ears, tail, and hoofs. In the Hellenistic period, they got mixed up with *Pan* (the protector of flocks and nature) and took on their more familiar image as goat-men.

[5]A famous instance is the battle of the centaurs and Lapiths, ensued when the centaurs tried to rape the bride and guests at a wedding. The scene is depicted on the frieze of the Parthenon and the Temple of Zeus at Olympia.

Western thought but not always with respect to sexual issues (box. 17.3).

Sexual Behavior

Outside of literary and a few historical sources, we have no information about the actual behavior of the ancient Greeks; the ancient world offers no counterparts to our clinical case studies and sex surveys. Therefore, whatever inferences are drawn about Greek sexual behavior must be drawn cautiously.

As in other cultures, most Greek men and women married, had sex, begat chil-

dren, and kept up the cycle of life. But there were two aspects of Greek society that imparted a more special quality to their sexuality; the close association of men with each other, which led to homosexual attachments; and the presence of a slave population that engendered sexual exploitation on a large scale.

Greek Homosexuality   Greece was very much a man's world. Men spent most of their time with other men, especially in time of war (which was a good deal of the time). Their friendships and associations were with other men, including, for some, erotic attachments. Yet Greek homosexuality has

Box 17.3   Plato's Conceptions of Sex and Love

Greek philosophy, intimately linked to both religion and literature, was preoccupied with the same basic issues of life, love, and sex. Philosphical discourse took place wherever philosophically inclined men gathered. Often this was in social settings where men got together to drink, to talk, to be entertained, and sometimes to have sex with each other or female prostitutes. In the most celebrated of these banquets, described by Plato in the *Symposium*, Socrates and half a dozen other friends expound their views on love.[1]

In the *Symposium* (and *Phaedrus*), Plato uses sexual desire as the point of departure in developing his philosophical ideas. Sexual arousal and affection stimulated by physical beauty are the lower-level manifestation of the human impulse to seek understanding of the eternal and immutable *Form* or *Idea* of Beauty (itself an aspect of the Good).

All men desire to be immortal. Commonly this is achieved through bringing forth off-spring. Yet people operating at this level are no different from four-footed beasts reproducing themselves. Others achieve immortality through spiritual heirs—by creating art, philosophy, and similar pursuits. In these instances, the sexual impulse is not gratified directly but expressed through what we would call "sublimated" means.

In his reference to the sexual impulse, Plato does not clearly differentiate between its heterosexual or homosexual fulfillment. The need to reproduce necessitated the physical fulfillment of heterosexual relationships. But the passion generated by homosexual attachments could be redirected to the higher pursuit of Beauty and Goodness.

The central exposition of the meaning of love in the *Symposium* is by Socrates, who ostensibly relates what he has been told in this regard by the priestess Diotima; but since Plato is the actual author, it is hard to know when Plato is reporting the words of Socrates and when he is expressing his own thoughts.

Plato's philosphy opens up a number of issues that go beyond the cheerful acceptance of sex that we have so far ascribed to the Greeks. These issues have been especially important in the subsequent development of Christian conceptions of sexuality.

First, we see a distinct **duality** in Plato's philosophy between the pleasures of the *body* and the higher aspirations of *reason* (which was recast by Christians as the conflict between the *flesh* and the *spirit*). Second, though Plato did not condemn sexual satisfaction as such, he clearly preferred its harnessing to attain immortality through wisdom and beauty, which for Christians became the suppression of sex for spiritual ends.

Plato's pupil, *Aristotle*, in turn condemned sexual intemperance but recognized the legitimacy of sexual affection as such and its potential to lead to virtue. *Epicureanism* sought to avoid the torments of passion in order to preserve serenity. *Stoicism*, in particular, elaborated these concepts in ways that become incorporated in early Christian thought.

[1] *Symposium.* In Edman (1956).

been a source of much confusion. This has been partly the doing of scholars and partly due to the nature of the practice.[19] The

[19] Licht (1969); Dover (1980); and Boswell (1980) point out the denial and distortion of the evidence by some classical scholars; Karlen (1971) refers to the practice of scholars using their research to justify their own disguised homosexuality.

Greeks did not categorize people into heterosexuals and homosexuals (as character types or life styles); they assumed that some men by preference and other by force of circumstance would seek sexual gratification from other males. Such practice was socially approved, tolerated, or disapproved depending on a number of factors.

Figure 17.7 Herm. (J. Paul Getty Museum, Malibu, California.)

When a boy reached puberty, it was customary for him to be attached to an older mentor who helped to guide his development to adulthood. The relationship could be based on respect and affection but without any romantic passion or sex. When the mentor qualified as an *erastes* ("lover") because of his ardent interest in the youth, the youth became the *eromenos* ("beloved"). These same words were used when a man had a "crush" on a woman, but "lover" could also simply mean being an admirer of a youth's beauty, which the Greeks glorified.

The sexual component of the erastes-eromenos relationship was usually left unclear (just as we do not always know if dating couples are having sex). It was assumed that the erastes would seek sexual contact and the eromenos would resist it or yield to it reluctantly and with certain qualifications (figure 17.8). For example, it would not be proper for the eromenos to be used like a prostitute by permitting anal intercourse; he would merely allow ejaculation between his thighs (without himself reaching orgasm). It is questionable, of course, to what extent these niceties were observed in actual practice. At any rate, whatever the nature of the relationship, it ended with the young man attaining adulthood and going on to get married and perhaps becoming an erastes himself. Men also engaged in anal intercourse with women but whether they did this only with

Let us consider first the institution of *pederasty* ("boy love").[20]

[20] The term "boy" is confusing in this regard. The Greek *pais* (or *paidika*) had a number of meanings: child, girl, slave, youth, and the "passive" partner in homosexual relationships (Dover, 1980). In the pederastic context the younger person was an adolescent or youth, never a child.

Figure 17.8 A man courts a boy. Attic black-figure vase, 6th century B.C. (Museum of Fine Arts, Boston.)

prostitutes or with their wives as well, is unclear.

Plato describes episodes and ascribes statements to Socrates that show him as greatly preoccupied with youthful beauty and love. "I can't think of a time when I wasn't in love," says Socrates (when clearly not referring to his wife), and love was the only subject he claimed to know anything about. In the following episode Socrates meets the exceptionally handsome Kharmides. Socrates says:

> Then I just didn't know what to do, and all the confidence that I'd previously felt, in the belief that I'd find it easy to talk to him, was knocked out of me. When Kritias told him that I was the man who knew the cure (sc. for headache), and he looked me in the eye—oh, what a look!—and made as if to ask me, and everyone in the wrestling-school crowded close all round us, that was the moment when I saw inside his cloak, and I was on fire, absolutely beside myself.[21]

If the incident is taken at face value, then the fact that Socrates was sexually aroused by the sight of the youth's body is quite clear. But did such an event take place, or was Plato inventing a story to make a philosophical point? It is such uncertainties that continue to fuel disagreement between classical scholars.

The Greeks sometimes exploited homosexual attachments for military purposes by having pairs of lovers fight side by side. Most famous was the Theban *Sacred Band* whose 150 pairs of lovers fought to death against Philip of Macedon. Moved to tears by their heroism, their conqueror declared, "Woe to them who think evil of such men."[22]

Philip's statement implies that some people did think evil of such men. The Greeks did not indiscriminately approve of all forms of homosexuality and a common insult was to call someone "wide-arsed."[23] In passing such judgment on sexual behaviors, the Greeks were not preoccupied with abstract principles but how an act reflected on a person's manhood and social status. A citizen who prostituted himself was barred from holding public office because a man capable of selling his body was thought likely to compromise the interests of the community as well.[24] Similarly, a man who was penetrated in anal intercourse placed himself in the submissive position of a woman, hence, he could not be trusted on the battlefield (the example of the Theban band to the contrary). What was at issue then was not "homosexuality," but the violation of the Athenian concepts of manhood and dominance that determined social perception of these sexual practices.

We know relatively little about female homosexuality or the attitudes of Greek women to male homosexuality. In Sparta women formed the counterpart of the male eromenos-erastes relationship, but to what extent this entailed a sexual component is unclear. No more certain is the nature of Sappho's relationship with the girls she tutored, to some of whom she was ardently attached. (Sappho lived in the 7th century B.C. but the few biographical details of her life come from Hellenistic times.)

Prostitution   Only a minority of the people in the Greek city-states were citizens. The rest were foreigners and slaves. It was from this sector of the population, rather than the citizenry, that the large number of prostitutes were drawn.[25]

---

[21] Dover (1978, p. 155).

[22] Licht (1932, p. 391). Another celebrated couple was Harmodios and Aristogeiton, who died trying to overthrow the tyrant Hippias and his brother Hipparkhus (who had tried to seduce Harmodios) (Boswell, 1980).

[23] Referring to the Trojan war, Eubolos says: "No one ever set eyes on a single hetaira; they wanked themselves for ten years. It was a poor sort of campaign; for the capture of one city they went home with arses much wider than (the gates of) the city that they took." Quoted in Dover (1980, p. 135).

[24] See chapter 2 in Dover (1980).

[25] The proper role for women of the citizenry was the sheltered life of wife and mother. The Greek word for woman, *gyne*, means bearer of children.

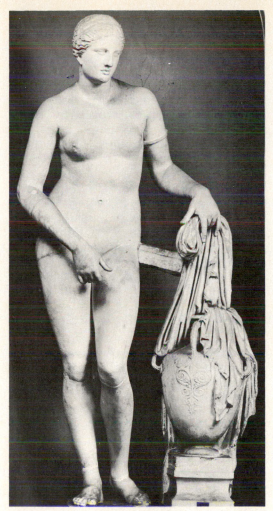

Figure 17.9 Aphrodite, Roman copy after the Aphrodite of Cnidos by Praxiteles, the original dating from ca. 349 B.C. (The Vatican, Rome.)

The common prostitute, called *porne,* sold her sexual favors on her own or in a brothel. Their fees were set by the state, which taxed their revenues. Other prostitutes worked the streets; one enterprising street-walker had the sole of her shoe fitted with large nails that left an imprint that read "Follow me."

More privileged were the *hetairae* ("companions"). Among these women were the sophisticated and ambitious mistresses of some of the most influential per-

sonalities of the time.[26] The inspiration and models of artists, the more famous hetairae had their statues set up in temples and public buildings side by side with those of eminent generals and statesmen. The breathtaking beauty of Phyrne was immortalized in the statue of Aphrodite of Cnidus by Praxiteles (figure 17.9)

Hetairae could also be attached to temples as sacred prostitutes, the *hierodoules,* whose earnings went to the temple treasury. People sometimes donated slaves to temples for this purpose. The Corinthians in particular believed that the prayers of sacred prostitutes to Aphrodite were effective in protecting them in times of danger. It is likely that large numbers of men availed themselves of the services of prostitutes. But extramarital license by men was by no means lightly tolerated by Greek wives. Hence, the motto, "Courtesans for pleasure, concubines for daily needs, and wives to conceive children and be loyal housekeepers" was probably more male wishful thinking than a reflection of the realities of Greek life.[27]

Sexuality in Rome

Rome was the first heir to Greek culture. From its humble beginnings as an agrarian society in the 8th century B.C. it grew to be master of the Mediterranean world well into the 5th century A.D. (The Eastern or Byzantine empire survived for nearly another thousand years after the fall of Rome, but its civilization is secondary to our concerns here.) The Roman empire encompassed a vast conglomerate of diverse societies which lived by their own customs. This obviously makes it impossible to speak of sexuality in the Roman world as if it were a more or less uniform entity; so we shall focus on a few elements of sexual life in Rome proper.

[26] Aspasia was the mistress of Pericles. Alexander the Great had Thais, who following his death married Ptolemy and became queen of Egypt.

[27] Quoted in Marcade (1962, p. 132).

Old Roman gods were credited with the attributes of the Greek deities to whom they most nearly corresponded. Zeus was matched with Jupiter; Hera—Juno; Aphrodite—Venus; Dionysus—Bacchus; Hermes—Mercury; Hephaestros—Vulcan; Ares—Mars; Heracles—Hercules; and so on. Other Roman gods included *Pudor* (modesty), *Fides* (loyalty), and *Priapus* (fertility). Small of body but large of penis, Priapus became popular as guardian of fields and flocks.

Phallic Symbolism   *Phallic worship* was an important part of Roman religious observance since Etruscan times. Phallic emblems were placed in tombs whose walls were decorated with erotic scenes (as in the famous *Tomb of the Bulls* in Tarquinia). Sex was associated with the renewal of life

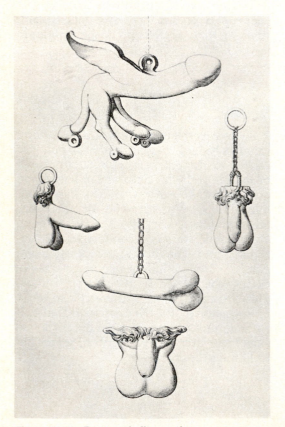

Figure 17.10   Roman phallic amulets.

and was thought to be an antidote to the ravages of death.

Phallic symbols were endowed with magical powers. Widespread in the Roman world were representations of the penis in fantastic forms. Some were hung around the neck as *amulets* (see figure 17.10). Others cast their protective spell from above the doors of homes, shops, and city gates. The reasoning behind these practices was that when the gaze of someone with the *evil eye* fell on a person, the phallic amulet deflected the evil force and spared its wearer.[28]

The Greek tradition of agricultural festivals had its parallel in the Roman *Bacchanalia* and *Liberalia* during which wine, sex, and phallic rituals ensured the fertility of the fields, beasts, and people for another year.[29]

Prostitution   Prostitution was rife in Rome. With their typical administrative flair, the Romans established brothels along major thoroughfares at set intervals so that a man could have his sexual needs serviced while his horse was being attended to. Prostitution lost all its religious associations and became purely a form of commerce, prostitutes now being called *meretrices* ("sellers"; the source of our term "meretricious").

Erotic Art and Literature   In the excavations at Pompeii (which include a high-class brothel) are frescoes portraying various forms of coital activity.[30] Roman statuary art was largely imitations of earlier Hellenistic models (figure 17.11). The Romans completed the shift from the male to the female nude as the ideal human form, a trend already in evidence in the later phase of Greek civilization. Erotic images were also commonplace in Rome as decorations

[28]A legacy of this belief is the word "fascinate," which is derived from *fascinum,* a Latin term for penis that first meant "bewitchment."

[29]Webb (1975).

[30] For extensive illustrations of erotic art from Pompeii, see Grant and Mulas (1975).

Figure 17.11 Roman marble relief of Hercules and nymph. (Museum of Fine Arts, Boston.)

on objects like mirrors and lamps (figure 17.12).

Roman erotic literature shows more originality than art. Ovid's *Art of Love* is a manual of seduction for predatory men and women. The works of Ovid, Catullus, Horace, and other Roman poets (which greatly influenced later writers like Shakespeare) viewed love with a lighthearted and cynical eye and presented women as wayward, mercenary, tormenting, but nevertheless delightful creatures.[31]

With the disintegration of traditional Roman values, life became increasingly more brutal and sex became more sadistic. The combination of boundless power with unbridled license came to be epitomized in the depredations of the more decadent emperors like Nero and Caligula. (Nero committed incest with his mother and eventually killed her by cutting open her abdomen to see the uterus that had contained him.) But these odd characters were by no means representative of Romans generally, and the accounts of this period in the writings of Petronius (*Satyricon*), Martial (*Epigrams*), Juvenal (*Satires*), and Suetonius (*Lives of the Caesars*) are often exaggerated and sensationalized.

[31] Webb (1975).

The Rise of Christianity

A momentous development for Western culture during the Roman era was the rise of Christianity. Jesus was born during the reign of Augustus when the kingdom of Judea was subject to Rome. He began his ministry at about age 30 and was crucified

Figure 17.12 Roman lamp with erotic carvings. (Private collection.)

three years later at the instigation of the Jewish religious authorities in Jerusalum.

The followers of Jesus accepted him as **Christ** (Greek: the annointed) or the **Messiah** (Hebrew: the annointed), the savior promised by God to Israel, and the **Son of God**. So they set about to preach the **gospel** ("good tidings") of Jesus Christ; that is, his birth, ministry of teaching and healing, death, and resurrection.[32] From an insignificant sect in the first century, Christianity became the dominant spiritual, intellectual, and political force in the West over much of the next 20 centuries, and remains today the religion with the largest number of followers in the world.

Jesus left no writings. Information regarding his words and actions comes from his immediate followers (the *apostles*) and their disciples. At first circulated orally, this information was eventually incorporated into the four Gospels of the **New Testament**. There is little in the Gospels pertaining to the deeds and words of Jesus that is explicity of a sexual nature.

Following Jesus, the central figure in Christianity is **Paul,** a Hellenistic Jew (and Roman citizen) from Tarsus in Asia Minor. Though Saint Peter is considered by Catholics to be the first head of the Church, it was Paul (who never met Jesus during his earthly ministry) who played the dominant role in the shaping of early Christianity. His *Epistles* to the early churches, which dealt with key doctrinal and organizational matters, became the basis of much of Christian doctrine. The influence of Paul has been so enormous that he is represented by some as the real founder of Christianity and accused by others of complicating the simple teachings of Jesus.[33] Paul's views of sexuality were especially important in the formation of Christian moral doctrine (chapter 20). To understand these views, it is necessary to place them in the historical and intellectual context that constituted the world of the early Christian church.

[32] Metzger (1973).
[33] Kung (1976, pp. 339–410) discusses Paul's conception of Christianity.

The Early Church

The sexual ideology of early Christianity was influenced by the Judaic attitudes to sex prevalent at the time and by Greek influences. The Bible of the earliest Christians was the Old Testament. The remembered words and deeds of Jesus, which were to become the Gospels, and letters of the apostles to various churches constituted increasingly important sources of guidance; the New Testament in its present form did not appear until the 4th century.

Since Jesus did not repudiate Judaism, and all of the first Christians were Jews, they naturally assumed that in order to be a Christian one had to be or become a Jew, through circumcision and observance of the Judaic law. Paul opposed this view, arguing that the new covenant of Jesus superceded the old, hence non-Jews could become Christians directly, without conversion to Judaism. The fact that Paul prevailed greatly facilitated the extension of Christianity to the Gentiles, and made possible its transformation from a small Jewish sect to a world religion.

Early Christians were nonetheless much affected by the prevailing sexual attitudes in Judaism, which had become more restrictive since early Old Testament times. Philo, the foremost philosopher of the Jewish diaspora in the first century, condemned "the passion of love" and physical attraction as the source of wickedness. Also influential may have been the example of the **Essenes,** a small Jewish sect that adhered to ascetic principles of celibacy and the use of sex for procreative purposes only.

The Influence of Stoicism Hellenistic Jews and Gentiles who were converted to Christianity brought their own intellectual perspectives to their chosen religion, thereby helping shape Christian moral doctrines and sexual attitudes; especially important in this regard were the tenets of **Stoicism.** Founded by Zeno, a Greek, in about 315 B.C., stoicism held *virtue* to be the highest good, best attained by repression of emotion, indifference to pleasure and pain, and patient endurance.[34] Sexual emotions were particularly suspect, whatever the context for their expression. Seneca, a first-century Stoic philosopher, wrote:

> All love of another's wife is shameful; so too, too much love of your own. A wise man ought to love his wife with judgment not affection. Let him control his impulses and not be borne headlong into copulation. Nothing is fouler than to love a wife like an adulteress . . . Let them show themselves to their wives not as lovers, but as husbands.[35]

As Noonan puts it, "Stoicism was in the air the intellectual converts to Christianity breathed. Half consciously, half unconsciously, they accommodated some Christian beliefs to a Stoic sense."[36] Paul himself was influenced by Stoic ideals as were some of the most influential Church figures who followed him. Thus, the notion that there is something shameful and objectionable about sex as such, even in lawful relationships, which came to dominate much of Christian thinking over the centuries, owed its origins not to Judaic roots, not to the teachings of Jesus, but to prevailing currents in Greek and Roman secular thought.

The Apocalyptic Expectation   The Jews believed that at a preordained time, God would bring the world to an end by establishing his kingdom on earth and abolishing injustice, suffering, and death. Jesus shared this *apocalyptic* ("revelation") vision, and certain statements he made were interpreted by his followers to mean that these cataclysmic events would occur during their own lifetime. So the early Christians lived with an acute sense of impermanence, anxiously awaiting the return of the Lord.

This expectation had a major influence on Paul's views. He wrote about the "imminent distress" (1 Cor. 7:26) that the early

Christians faced. Whether what he meant were the persecutions that were to befall Christians or the second coming of Christ that would "end the world," his letters to the churches were written in the anticipation of extraordinary times. Hence the advice he gave on whether or not to get married was formulated in that context rather than in the abstract, for all times (Cor. 1 7:25–28).

More generally, the views of Paul and other early teachers of Christian sexual morality must be understood in the social context within which they were elicited. Paul had to contend with the unbounded sexual license of the Hellenistic society within which Christians lived, as well as having to deal with the special challenges from within particular churches. For example, the challenge from the church in Galatia came from devout and conservative moralists' intent to uphold the letter of the Mosaic law; in the Corinthian church, radical elements claimed that as Christians they were above all laws, hence free to behave as they wished, even if that meant committing incest and having sex with cult prostitutes.

The Patristic Period

The period following the martyrdom of Paul in Rome under Nero (probably in 66 A.D.) to the fall of the Roman Empire in the 5th century, is referred to as the *Patristic Age,* because the development of Christian doctrine was dominated during this time by a select group of men known as the *Fathers of the Church,* several of whom lived in Hellenistic Alexandria, a great center of learning. Included in this group were Ambrose, Origen, Jerome, Clement, Gregory (the Great), and most importantly, Augustine, whose views strongly influenced Catholic doctrine on sexuality over the next thousand years and Protestant attitudes beyond that.

Stoicism continued to exert a powerful influence on Christian sexual morality in the patristic age, leading to the ascetic be-

---

[34] Noonan (1967); Russell (1945).
[35] Quoted in Noonan (1965, p. 67).
[36] Ibid., p. 66.

Figure 17.13 Martin Schongauer, *The Temptation of Saint Anthony*. (National Gallery of Art, Washington, D.C.)

The Challenge of Gnosticism   While suppressing lust, the church had to simultaneously fight those who condemned all sexual experience, including marital intercourse for procreation. The main challenge in this respect was posed by the *Gnostics.* Members of this heretical group claimed superior knowledge (*gnosis*) of spiritual truths and Christian beliefs. Based on a mixture of Iranian myths, Jewish mysticism, Greek philosophy, and Christian doctrines, Gnosticism was a serious threat to the early church (as far back as the time of Paul) because it capitalized on the same principles of moral virtue and chastity endorsed by early Christians.

Paradoxically, the sexual beliefs and practices of the Gnostics placed them at two polar opposites (which they referred to as "left" and "right"). On the extreme right were the Gnostic ascetics who rejected all sex and considered marriage a form of fornication. In support, they cited the celibacy of Jesus and certain statements by Paul such as, "It is well for a man not to touch a woman" (1 Cor. 7:1). At the extreme left, were those who claimed freedom from all moral laws. It is against these *antinomian* elements that Paul's admonitions may have been directed to the effect that Christian liberty not be used as an excuse for sensuality and sexual misbehavior. Yet these practices persisted, and certain factions in the Alexandrian church believed that spouses should be shared. They considered sexual intercourse a form of communion that led to salvation and, according to their detractors, would engage in indiscriminate and perverse sexual practices.

The net effect of the Gnostic doctrines was to negate the concept of marriage as an orderly way of procreation and lawful sexual experience. Hence the Church was pressed to uphold the legitimacy of marriage and procreative sex, at least as a necessity. In simultaneously condemning the asceticsm and sexual license of Gnostics, the Catholic church steered a middle course that it had to defend over and over again from attacks both from the "right" as well as from the "left."

liefs and practices of the Church Fathers. The most extreme forms of self-denial were practiced by the *anchorites* ("those who have fled") who lived alone in the desert. But as exemplified by the experiences of Saint Anthony, these hermits continued to be tormented by sexual visions (figure 17.13). Saint Jerome castrated himself in a fit of sexual self-denial, an act he regretted subsequently.

Many of the teachings of Jerome and Origen were based on Seneca. Abstinence and virginity were glorified. "A virgin," declared Saint Ambrose, "marries God." Saint Jerome considered as adulterers all who were "shameless" with their wives (presumably by enjoying sex). Stoicism also contributed to the Christian doctrine of "natural law," which proved highly influential in condemning nonreproductive sexual behaviors and the use of contraception.

This characterization of the Gnostics has been based mainly on Catholic accounts and judgments of their beliefs, since the Church systematically destroyed the writings of heretics. Recent archeological finds have led to a far more sympathetic evaluation of Gnostic beliefs, revealing their insistence on the primacy of personal experience, pursuit of a solitary path to self-discovery, and distrust of institutional guidance and control. Their rejection or subordination of sex was merely part of a broader attempt to set aside wordly distractions that stood in the way of their internal search for Gnosis.[37]

The Challenge of the Manichees *Manicheanism* was founded by Mani (A.D. 216–217) in Babylon. Largely derived from Gnosticism, Iranian folk religion, and Christianity (Mani referred to himself as an apostle of Jesus Christ), it envisioned a world dominated by the struggle between the forces of *Light* and *Darkness*. All living things, including people, were largely generated by the demons of Darkness, but imprisoned within their bodies was some Light; the Manichean purpose in life was to set this light free.[38]

Since procreation perpetrated the world of Evil, it was itself evil. Hence, sexual desire, which led to procreation, was impure and had to be suppressed, or, if allowed expression, its procreative consequences had to be avoided.[39] As with the Gnostic challenge, the Church was confronted once again with the task of reaffirming the legitimacy of conjugal sex for procreative purpose while simultaneoulsy upholding the virtues of sexual abstinence. The Church therefore accused the Manichees of both denigrating marriage and procreation, as well as sexual perversity. (It was said, for instance, that Manichees spilled semen on grain which then they ate as an eucharist,

Figure 17.14   Simon Marmion, *Saint Augustine*. (The Bettmann Archive.)

the Christian sacrament of the Lord's supper.)[40]

The Manichean challenge was met head on by Saint Augustine (figure 17.14), whose life is sketched in box 17.4. The association between *original sin* and sexuality was made by Clement, but it was through Augustine that it became entrenched in Catholic theology. Yet this was by no means a universal belief within the Church. For example, Ambrose thought Adam's original sin was not sex but pride. Saint Chrysostom also made no connection between original sin

---

[37] Pagels (1979).

[38] Noonan (1967).

[39] To achieve this, the Manichees relied on the infertile period during the menstrual cycle, ironically the only method to be subsequently accepted by the Catholic church.

[40] The belief that eating "released" the Light imprisoned in semen may explain this practice. Whether these rituals were actually performed by all Manichees, by splinter groups like the Cathari or anyone within the group, is open to question.

Box 17.4   Saint Augustine (A.D. 354–430)

Augustine was born in Hellenistic North Africa to a devout and devoted Christian mother and non-Christian father. As a young man, Augustine lived with a woman by whom he fathered a child; but the dominant influence in his life was his mother Monica. At her instigation Augustine became engaged to a young heiress, and dismissed the woman he had loved and lived with, since his professional ambitions as a teacher of rhetoric required that Augustine make a socially advantagious match. Yet Augustine never married and was instead converted to Christianity and a life of celibacy and continence. He went on to become bishop of Hippo, one of the towering figures in Christendom, and its most influential teacher on marriage and sexuality.[1]

Augustine had been a Manichee in his younger years and residues of Manichean sexual attitudes continued to influence his thinking; but more importantly his views developed as reactions to Manichean doctrines. They were further shaped in his rebutting the teachings of Pelagius, who claimed that the original sin of Adam and Eve (which turned sexual intercourse into something shameful) was not transmitted to their descendents. In his book, *Marriage and Concupiscence*, Augustine settled both scores by showing that the Manichees were wrong in declaring marriage sinful and procreation evil, while the Pelagians were wrong in claiming that children born to lawful marriages were not born with the sin of Adam. The Augustinian doctrine that became the Catholic view was that marriage is good, yet "man, born through concupiscence, brings with him original sin."[2]

The concept of *concupiscence* was central to Augustine's concept of sexuality. The term, which literally means "strong desire," referred to the "heat" and "confusion of lust" that accompanies copulation; in short, the libidinal, erotic, pleasurable component in sexual arousal and orgasm. By damning concupiscence, Augustine struck at the very heart of the sexual impulse. He condemned all sexual experience as serving lust and hence shameful. Sexual activities that served no procreative purpose (including coitus with contraception) were especially evil since they were aimed purely at the gratification of lust. Conjugal sex was legitimate in serving procreative ends but tainted with evil to the extent that it was contaminated by concupisc-

and sexuality. Nor did he consider *concupiscence* (box 17.4), as such, to be a sin. Only when immoderate sexual desire failed to stay within the bonds of marriage and led to adultery, did sin occur. It was therefore not sex as such but its abuse that was condemned. If the view of Chrysostom, instead of Augustine, had governed theological developments in the West, a different moral tone and a different way of looking at sex and marriage might have led to different outcomes in Christian conceptions of sexuality through subsequent centuries.[41]

These differences within the Church belie the impression of a single, monolithic Catholic view of sexuality that has persisted through the ages. Furthermore, important as these theological debates may have been, they reveal little about how people actually behaved sexually at the time. We do know that pagan sexual be-

[41] Noonan (1967).

Box 17.4   Saint Augustine (A.D. 354–430) (*Continued*)

ence. The following passage from the *City of God* gives some sense of Augustine's line of reasoning.

> Lust requires for its consummation darkness and secrecy; and this not only when unlawful intercourse is desired, but even such fornication as the earthly city has legalized. Where there is no fear of punishment, these permitted pleasures still shrink from the public eye. Even when provision is made for this lust, secrecy also is provided; and while lust found it easy to remove the prohibitions of law, shamelessness found it impossible to lay aside the veil of retirement. For even shameless men call this shameful; and though they love the pleasure, dare not display it. What! does even conjugal intercourse, sanctioned as it is by law for the propagation of children, legitimate and honourable though it be, does it not seek retirement from every eye? Before the bridegroom fondles his bride, does he not exclude the attendants, and even the paranymphs, and such friends as the closest ties have admitted to the bridal chamber? The greatest master of Roman eloquence says, that all right actions wish to be set in the light, that is, desire to be known. This right action, however, has such a desire to be known, that yet it blushes to be seen. Who does not know what passes between husband and wife that children may be born? Is it not for this purpose that wives are married with such ceremony? And yet, when this well-understood act is gone about for the procreation of children, not even the children themselves, who may already have been born to them, are suffered to be witnesses. This right action seeks the light, in so far it seeks to be known, but yet dreads being seen. And why so, if not because that which is by nature fitting and decent is so done as to be accompanied with a shame-begetting penalty of sin?[3]

[1]For Augustine's biography, see Brown (1967); and for his monumental autobiography, the *Confessions* (Sheed, 1943).
[2]Noonan (1967, p. 169).
[3]Augustine (1934 ed.).

havior in the Hellenistic world was quite unrestrained; the Church could do very little about that. The "barbarian" tribes that infiltrated and toppled the Roman Empire were also only nominally Christian, and mainly followed their own ways. Even the behavior within the fashionable Catholic society in Rome caused Saint Jerome much distress. He deplored the loss of virginity among maidens, the use of contraceptives and abortifacients, and the concealment of unwanted pregnancies, all of which these Christian women justified by saying, "All things are pure to the pure. My conscience is enough for me." Girls paraded about in public with loose-bound hair, flying capes, and loose jointed walk, attracting the eyes of crowds of young men. Saint Jerome admits these young women have their admirers, but he is not one of them.[42]

The rise of Christianity coincided with the eclipse of Rome and may have been

[42] Noonan (1967, p. 131).

partly responsible for it. Numerous reasons have been proposed for what Edward Gibbon called the "decline and fall" of the Roman Empire. Two of the more relevant reasons, for our purposes, are the moral decline and decadence of the late empire and the challenge of Christianity to the more impersonal Roman religion, which offered little comfort to the oppressed masses in this world or in the afterlife. The end of Rome was the start of the making of Western Europe. In the West the Roman Empire might be said to have formally ended in 476, when the last emperor was deposed. It did not, however, come to an abrupt end but was gradually transformed, with the empire of the West passing away while the Byzantine empire in the East (Constantinople) survived for nearly another thousand years.

SUGGESTED READINGS

Epstein, L. M. *Sex Laws and Customs in Judaism* (1967). A comprehensive and interesting presentation of the perspective of Judaism on sexuality.

Larue, G. *Sex and the Bible* (1983). Presents in brief and nontechnical language the discussion of major sexual themes in the Bible.

Boardman, J., and La Rocca, E. *Eros in Greece* (1975). Informative text with splendid illustrations.

Dover, K. J. *Greek Homosexuality* (1978). A scholarly yet readily intelligible account of Greek attitudes and practices with regard to homosexuality.

Chapter 18

Sexuality in Western Culture

Civilized people cannot fully satisfy their sexual instinct without love.
Bertrand Russell

In Western Europe, the civilization that became heir to the classical tradition was largely a peasant society, and its dominant social institution was the Church. It was in the context of this medieval society that Western sexual attitudes and behaviors were forged.

Sexuality in the Middle Ages

The period between the end of the classical period (A.D. 500) and the emergence of modern Europe in the Renaissance (A.D. 1500) is known as the *Middle Ages* of European civilization. The sexual profile of this period is hazy and based mainly on Church sources. Furthermore, since during this millenium major differences in sexual attitudes and behaviors characterized changing social circumstances, it is necessary to distinguish sexual life early in this era from later developments.

The Early Middle Ages

The early Middle Ages were a chaotic and sorry time for Europe. Vast movements of people shaped and reshaped the demographic face of the continent, which was now bereft of any central authority. Whatever survived of the learning and culture of antiquity was preserved in the *monasteries,* Christian enclaves that did not exist for the glorification of sex.[1]

Erotic art and literature had begun to decline during the eclipse of the Roman empire. The Church, which had established itself firmly in the 4th century, effectively suppressed anything that smacked of erotic licentiousness. The fear of obscenity and paganism pervaded all art; when religious themes required the showing of nude bodies (such as with figures of Adam and Eve before the Fall) they were represented without any taint of eroticism. A comparison of the medieval Eve (figure 18.1) with the classical Aphrodite (figure 17.9) compellingly illustrates this change in sexual values.[2]

[1]Christian monasticism had begun with Saint Anthony in the third century in Egypt. As others emulated his ascetic life in the desert, they formed *cenobitic* ("living in common") groups. Monastic life was further formalized in the sixth century by Saint Benedict, whose *Rule* provided the basic guidelines for subsequent religious orders.

[2]Eitner (1975).

Figure 18.1 *Eve*, ca. 1240. Portal sculpture from the Cathedral of Trau, Dalmatia, by Magister Rodovan.

Building on the ascetic values of the Church Fathers and the concept of unquestioned law (*nomos*), the Church had formulated, by the 8th century, a rigidly strict code of sexual morality articulated in *penitential manuals* that detailed lists of sins to be examined during confession and their corresponding penances, consisting of various acts of self-mortification (such as abstaining from sex or eating meat) and of devotion (box 18.1).

Underlying this medieval code was the belief that sexual pleasure was innately sinful, hence, a grudging acceptance of sex as a necessity for reproduction was coupled with prohibitions against all other forms of erotic feeling and expression. Based on these suppositions, sexual restraints could be carried to absurd lengths; some pious couples for instance would have coitus wearing heavy nightshirts (*chemise cagoule*) with a hole in front that allowed penetration while preventing other bodily contact. Intercourse could be performed properly in the man-above position only (the "dog" posture called for seven years of penance).[3] Husband and wife were urged to abstain from sex on Thursdays (in memory of the arrest of the Lord); on Fridays (in commemoration of his death); on Saturdays (in honor of the Blessed Virgin); on Sundays (in honor of the Resurrection); and on Mondays (in honor of the faithful who had died). Continence was necessary for 40 days before Easter, three months before the birth of a child, and two months thereafter.

If marital coitus was subject to such restraints, there is little point in dwelling further on prohibitions regarding other sexual behaviors. But by way of example, even involuntary nocturnal emissions were not exempt from judgment. The offender had to rise at once to say seven penitential psalms followed by thirty in the morning. If the sin was committed while asleep in church, the whole psalter had to be sung in penance.

This bleak erotic landscape of the early medieval world may have reflected the monastic ideal, but it hardly mirrored the reality of sexual behavior of the times. Church law no doubt imparted to the pious

[3]Taylor (1970).

Box 18.1   The Penitential of Theodore, Archbishop of Canterbury, 668–690[1]

II. Of Fornication

1. If anyone commits fornication with a virgin he shall do penance for one year. If with a married woman, he shall do penance for four years, two of these entire, and in the other two during the three forty-day periods and three days a week.

2. He judged that he who often commits fornication with a man or with a beast should do penance for ten years.

3. Another judgment is that he who is joined to beasts shall do penance for fifteen years.

4. He who after his twentieth year defiles himself with a male shall do penance for fifteen years.

5. A male who commits fornication with a male shall do penance for ten years.

6. Sodomites shall do penance for seven years, and the effeminate man as an adulteress.

7. Likewise he who commits this sexual offense once shall do penance for four years. If he has been in the habit of it, as Basil says, fifteen years; but if not, one year less . . . . If he is a boy, two years for the first offense; if he repeats it, four years.

8. If he does this "in femoribus," one year, or the three forty-day periods.

9. If he defiles himself, forty days.

10. He who desires to commit fornication, but is not able, shall do penance for forty or twenty days.

11. As for boys who mutually engage in vice, he judged that they should be whipped.

12. If a woman practices vice with a woman, she shall do penance for three years.

13. If she practices solitary vice, she shall do penance for the same period.

14. The penance of a widow and of a girl is the same. She who has a husband deserves a greater penalty if she commits fornication.

15. "Qui semen in os miserit" shall do penance for seven years: this is the worst of evils. Elsewhere it was his judgment that both participants in this offense shall do penance to the end of life; or twelve years; or as above seven.

16. If one commits fornication with his mother, he shall do penance for fifteen years and never change except on Sundays. But this so impious incest is likewise spoken of by him in another way—that he shall do penance for seven years, with perpetual pilgrimage.

17. He who commits fornication with his sister shall do penance for fifteen years in the way in which it is stated above of his mother. But this penalty he also elsewhere established in a canon as twelve years. Whence it is not unreasonable that the fifteen years that are written apply to the mother.

18. The first canon determined that he who often commits fornication should do penance for ten years; a second canon, seven; but on account of the weakness of man, on deliberation they said he should do penance for three years.

19. If a brother commits fornication with a natural brother, he shall abstain from all kinds of flesh for fifteen years.

20. If a mother initiates acts of fornication with her little son, she shall abstain from flesh for three years and fast one day in the week, that is until vespers.

21. He who amuses himself with libidinous imagination shall do penance until the imagination is overcome.

22. He who loves a woman in his mind shall seek pardon from God; but if he has spoken to her, that is, of love and friendship, but is not received by her, he shall do penance for seven days.

[1]Book 1, Part 2, pp. 184–185, 1938 edition.

feelings of guilt and a sense of gloom suffused sexual themes in moral theology. But at least some segments of the population were untouched. In the eighth century, Boniface despaired of the licentiousness of the English, who "utterly despise matrimony" and "live in lechery and adultery after the manner of neighing horses and braying asses." A century later, Alcuin found the land "absolutely submerged under a flood of fornication, adultery and incest, so that the very semblance of modesty is entirely absent."[4]

Eroticism persisted even within the bosom of the church. Driven underground, erotic art manifested itself in the most unlikely of places, such as in the margins of illuminated devotional books, carvings on choir stalls, and decorations on churches. Ostensibly religious sculptures from the period show provocative female nudes, men with erections, and couples fondling each other, engaging in coitus and having oral sex (in the "sixty-nine" position, no less).[5] Moreover, representations of women with exposed genitals, which were carved on the facades of some medieval Irish churches, suggest a persistence of the belief that genital symbols were protective against evil forces (figure 18.2).[6] Residues of ancient agricultural festivals with their phallic symbols can also be detected in celebrations of spring that took place in May, at which people danced around a decorated wooden *maypole*, a practice that persisted until recent time. Among the Franks in the 10th century, unmarried women danced naked on these occasions in front of the men. Though in time these practices were tamed, they never quite lost their ancient erotic roots.[7]

Much has also been written about the sexual behavior and misbehavior of the medieval clergy. Concurrent with the ideal of chastity expected of monks, parish priests often married until prohibited to do so in

Figure 18.2 *Shelah-na-Gig.* (From Andersen, I. *The Witch on the Wall.* Allen and Unwin, 1977.)

1123 because of the Church's concern over the passing of property by priests to their offspring. But that did not stop some of the clergy from having sexual attachments.

As for life in monasteries, common sense would suggest that they were neither the epitome of sexual purity that the apologists of the Church would have us believe, nor were monasteries the dens of sexual iniquity that the detractors of the Church have painted them to be. The Church would not have survived as long as it has if its words and actions were not reasonably consistent; and monks would not be human if some of them did not yield to temptation.

The image of sexuality that emerges from the early Middle Ages is thus a mixed one. There is the sexually repressive posture of the Church with its ascetic ideals, and then there are the residues of sexual hedonism from the Roman world, the rough ways of the nominally Christian Germanic tribes, and the lapses in clerical discipline. In this

---

[4]Taylor (1970, p. 20).

[5]Webb (1975).

[6]For a discussion of these figures called *Shelah-na-Gig,* see Andersen (1977).

[7]Ellis (1942).

context too the Church had to fight a dual battle. To impose some semblance of order on the chaotic world around it, required repressing sexuality. But the Church also fought fiercely against the tendency toward extremes of asceticism that denigrated all forms of sexual expression. The continuing struggle against the Manichean heresy (reenacted against the *Cathari* and the *Albigenses* in the 12th century) was in part based on opposition to such fanatic asceticism and aimed at preserving the legitimacy of reproductive sex within marriage while simultaneously denying all other expressions of sexuality.

The Late Middle Ages

By the time Europe entered the second millenium of the Christian era expanding populations had settled new lands, commerce had revitalized urban life, government had become more effective, and the Church had begun a significant reform movement, with the finest minds and artistic talents of the period at her service. Centers of monastic learning began to evolve into universities of scholastic distinction. Dante Alighieri wrote the *Divine Comedy*, and magnificent Gothic cathedrals of the 12th and succeeding centuries epitomized the faith and creativity of the period.

As with earlier times, we know little about the sexual life of the common people. But the abundant vernacular literature of the period reveals much about the amorous life of the more privileged classes. Especially pertinent here is the lyric poetry of the *troubadours* which emerged in southern France and entered its age of greatness in the 12th century (box 18.2).

The men and women portrayed in the tales by Chaucer and Boccaccio are witty, worldly, wily, and cynical in their views of sex and marriage. A recurring character is the cuckolded husband, who is sometimes cast as a clod who deserves his fate, at other times the innocent victim of a faithless wife. Whatever the extent to which these figures are representative of the people of the time, they provide interesting glimpses into the sexual relationships of medieval men and women. Similar insights are obtained from medieval art (figure 18.3).

Prostitution never died out in the Middle Ages and became more prevalent with urbanization. Even Saint Thomas Aquinas grudgingly acknowledged that prostitutes served a useful function by helping to preserve the chastity of decent women (he likened it to sewers that keep the streets clean). The Crusaders reintroduced public baths to Europe, some of which also served as brothels.

Hostility toward **homosexuals** had been considerable in the declining years of the Roman empire, and during the early Middle Ages homosexuals do not seem to be much in evidence. With the urban revival in the 11th century, a substantial homosexual community developed, with representatives among prominent and respected figures of European society. Yet in the second half of the 12th century public sentiment once again turned hostile as part of a more generalized intolerance toward minorities in general, including heretics and Jews. These attitudes became incorporated into the theological, moral, and legal documents of the later Middle Ages and continued to influence Western attitudes toward homosexuality for centuries.[8]

The emergence of greater secularization notwithstanding, the Church remained the dominant social institution in the later Middle Ages, legislating sexual morality and behavior. The dominant influence in moral theology in the 13th century was the towering figure of Saint Thomas Aquinas (1225?–1274), the quintessential scholastic theologian, whose aim was a grand synthesis of the faith of Augustine with the reason of Aristotle. Aquinas addressed himself to virtually all forms of sexual behavior, including touching, kissing, fondling, seduction, virginity, marital sex, fornication, adultery, rape, incest, prostitution, homosexuality, and bestiality. Excerpts from his discussion of "nocturnal pollution," though a relatively minor moral

[8]Boswell (1980).

Box 18.2   Erotic Themes in Medieval Literature[1]

In earlier periods heroic epics had been composed for the battle camp where knights and their attendants spent a good deal of their time. What was novel about **troubadour poetry** was its celebration of women and love.[1] As expressed in the tradition of **courtly love,** the troubadours worshiped from afar the idealized and often unattainable noble ladies who managed the family estates during the frequent absences of their men. From the poetic explorations of love by troubadours (such as *The Art of Courtly Love* by Andreas Cappelanus) evolved many of our **romantic** traditions. By separating love from marriage, and by extension, sex from reproduction, the troubadours also helped establish a secular alternative to the exclusively generative purposes of sex as taught by the Church.[2]

Original pre-Christian folktales typically portrayed warriors as violent characters who raped the women they could not seduce. Courtly medieval writers, such as Chretien de Troyes, in France, transformed such warriors into gallant knights, like King Arthur and his court, with Lancelot, Gawain, and the host of the Round Table. Although the ideal is tested in works such as *Sir Gawain and the Green Knight*, knightly behavior continued to be idealized into the 19th century.

In order for a man to become eligible to attend a university during the Middle Ages, he first had to become a member of the clergy, which entailed an obligation to remain celibate (although many clerics left the ecclesiastical state after their university days and led lives as married laymen). Deprived of authorized sexual outlets, some students, particularly at Paris, Orleans, Bologna, and Oxford, led dissolute lives of drinking and whoring. Among this class arose a witty and talented undergraduate literary coterie, called **Goliards** or the sons of Golias (purportedly the Biblical Goliath). They composed in Latin, and occasionally in the vernacular, amusing erotic songs, obscene pieces, and satiric parodies of Scripture and Church legislation.

In the medieval educational system the subject of logic and the technique of debate played very large roles. One result was a spate of clever Latin debate poems, on topics such as "whether a cleric or a knight makes a better lover for a nun," represented particularly well in the outrageous *Love Council of Remiremont*.

In the vernacular tongues the so-called **Latin comedies** abound; these are coarse, funny tales in which the theme of cuckoldry is typical. During the 13th century the "dirty story" (*fabliau*) was raised to an art form in France and the Low Countries. In the next century, Chaucer brilliantly reworked a number of these in *The Canterbury Tales* (ca. 1385). Although this work offers a wide spectrum of medieval secular and religious literary genres, many of its most memorable tales are such *fabliaux*. In using such ribald tales, Chaucer had considerable precedent in Boccaccio's masterpiece, *The Decameron* (ca. 1350); some of its stories are so bold that they are still regularly expurgated in school texts.

[1]This section is based on material provided by George Brown.
[2]Noonan (1967).

---

issue, show the scope of his argumentation (box 18.3).

Medieval eroticism also manifested itself in curious ways through individual and mass *fantasies* and *delusions,* which were interpreted sometimes as divine inspiration and in other cases as the work of the devil. In either instance, women (especially nuns) were more likely to be involved. One must bear in mind that although many of the women who took religious vows were devoted to a life of devotion and service, medieval convents also served as repositories for social re-

Figure 18.3 *Amorous Feast*, engraving by the Master of 1465. (The Louvre, Paris.)

jects, impoverished aristocrats, the morally delinquent, and the mentally disturbed. (Saint Bernardine called them "the scum and vomit of the world.")[9] It is no wonder that in this mixed company, erotic fantasy and mystical piety sometimes got intermingled: Mechthild of Magdeburg felt the "fatherly hand" of the Savior on her bosom and advised "all virgins to follow the most charming of all, the eighteen-year-old Jesus"; Christine Ebner believed herself to be pregnant by Jesus; and Veronica Giuliani took a suckling lamb to her breast in the memory of the lamb of God.[10] Others

[9]Chambers et al. (1974, p. 370).

[10]Taylor (1970, p. 42).

Box 18.3   St. Thomas Aquinas on Nocturnal Pollution[1]

Whether Nocturnal Pollution Is a Mortal Sin?

I answer that, nocturnal pollution may be considered in two ways. First, in itself; and thus it has not the character of a sin. For every sin depends on the judgment of reason, since even the first movement of the sensuality has nothing sinful in it, except in so far as it can be suppressed by reason; wherefore in the absence of reason's judgment, there is no sin in it. Now during sleep reason has not a free judgment. For there is no one who while sleeping does not regard some of the images formed by his imagination as though they were real, as stated above in the First Part (Q. LXXXIV., A. 8, ad 2). Wherefore what a man does while he sleeps and is deprived of reason's judgment, is not imputed to him as a sin, as neither are the actions of a maniac or an imbecile.

Secondly, nocturnal pollution may be considered with reference to its cause. This may be threefold. One is a bodily cause. For when there is excess of seminal humour in the body, or when the humour is disintegrated either through overheating of the body or some other disturbance, the sleeper dreams things that are connected with the discharge of this excessive or disintegrated humour: the same thing happens when nature is cumbered with other superfluities, so that phantasms relating to the discharge of those superfluities are formed in the imagination. Accordingly if this excess of humour be due to a sinful cause (for instance excessive eating or drinking), nocturnal pollution has the character of sin from its cause: whereas if the excess or disintegration of these superfluities be not due to a sinful cause, nocturnal pollution is not sinful, neither in itself nor in its cause.

A second cause of nocturnal pollution is on the part of the soul and the inner man: for instance when it happens to the sleeper on account of some previous thought. For the thought which preceded while he was awake, is sometimes purely speculative, for instance when one thinks about the sins of the flesh for the purpose of discussion; while sometimes it is accompanied by a certain emotion either of concupiscence or of abhorrence. Now nocturnal pollution is more apt to arise from thinking about carnal sins with concupiscence for such pleasures, because this leaves its trace and inclination in the soul, so that the sleeper is more easily led in his imagination to consent to acts productive of pollution. Thus it is evident that nocturnal pollution may be sinful on the part of its cause. On the other hand, it may happen that nocturnal pollution ensues after thoughts about carnal acts, though they were speculative, or accompanied by abhorrence, and then it is not sinful, neither in itself nor in its cause.

The third cause is spiritual and external; for instance when by the work of a devil the sleeper's phantasms are disturbed so as to induce the aforesaid result. Sometimes this is associated with a previous sin, namely the neglect to guard against the wiles of the devil.

On the other hand, this may occur without any fault on man's part, and through the wickedness of the devil alone. Thus we read of a man who was ever wont to suffer from nocturnal pollution on festivals, and that the devil brought this about in order to prevent him from receiving Holy Communion. Hence it is manifest that nocturnal pollution is never a sin, but is sometimes the result of a previous sin.

[1]Excerpts from *The Summa Theologica* Part 2, Question 154, Article 5. 1911–1925 ed.

Figure 18.4 Giorgione (1478?–1511) *The Sleeping Venus* (Courtesy, Staatliche Kunstsammlunger, Dresden.)

felt themselves assailed by devils in the form of *incubi* (as *succubi* similarly tormented males) (figure 10.4). Jeanne Pothière swore that a demon had forced her to copulate 444 times.[11]

This period also had extraordinary manifestations of **sadomasochism.** Asceticism has had a long association with self-inflicted suffering, but in the Middle Ages, *flagellation* as a form of punishment was inflicted on penitents; during the calamitous 14th century, when war, famine, and plague ravaged Europe, hordes of self-flagellants took to the roads in an effort to expiate for their sins. Such practices may provide rich material for psychological studies of the consequences of sexual repression, but it would be an error to take the erotic oddities of an age as representative of the sexual life of its people. Most medieval men and women probably went about their sexual business like others have done before and since, as amply evident in the boisterous scenes painted by Pieter

[11]Cleugh (1963, p. 96).

Brueghel (1520?–1569) and other great artists of later periods.

Sexuality in the Renaissance and the Reformation

Over the 14th and 15th centuries, the Western world went through a major transition. As the civilization of the Middle Ages declined, a period of renewal—the *Renaissance* ("rebirth")—began to reshape European culture, marking the beginning of modern Western civilization. What was "reborn" during this period was a passionate admiration for the classical cultures of Greece and Rome. But the accomplishments of the 14th and 15th centuries went beyond mere imitation of classical antiquity; much of what is distinctive in modern Western culture has its roots in the Renaissance and the Reformation, which followed it in the 16th century.

The men and women of the Renaissance did not rediscover sex, but they did generate a tremendous amount of sexually perti-

Box 18.4   Erotic Art and Literature in the Renaissance[1]

Revival of interest in the sensuous art of the classical world and the growth of sophisticated secular patronage stimulated an extraordinary flowering of erotic art in the 15th and 16th centuries that profoundly influenced the subsequent development of erotic art in the West. Medieval art, as a vehicle of the teachings of the Church, was meant for everyone. The select patronage of Renaissance art, by contrast, preserved it for the more privileged, and its secular patrons were attracted to the very erotic elements absent from medieval art. Nudity and sensuality could now be freely portrayed through the representation of mythological characters and events; with the figure of Venus holding the place of honor (among the more celebrated being the Venuses by Botticelli, Giorgione, and Titian) (figure 18.4).

Michelangelo's sculptures are not explicitly sexual, but a special sensuality endows works like the *Dying Slave*, and the *Pieta* (where a youthful Mary is more of an age to be the mystical "bride" rather than the mother of Christ). There are also a few works of Michelangelo's with overtly erotic content, such as the drawing of a man wearing a penis on his bonnet and other phallic themes (possibly linked to Michelangelo's homosexual interests).[2]

Renaissance artists relied heavily on classical themes such as the amatory escapades of Zeus, the interactions of other Olympian gods, and the erotic antics of satyrs and nymphs (figure 18.5). The erotic portrait as a new form of painting also now makes its appearance with Raphael's *La Fornarina*, portraying his mistress with bared breasts.

The secular interest in the erotic notwithstanding, the church remained the primary patron of the arts, much of which remained of a religious nature. But even here erotic themes found their way through the portrayal of Old Testament episodes such as the temptation of Joseph by Potiphar's wife (Tintoretto, Titian, Veronese), Lot's incest with his daughters (Raphael and Carracci), and Susanna being spied upon at her bath by the elders (Rembrandt and Tintoretto). Raphael also painted erotic frescoes for the bathroom of Cardinal Bibbiena in the Vatican (not open to public view).

The Renaissance preoccupation with the human figure was further pursued in the 16th century by certain followers of Michelangelo and Raphael in a distinctive style known as **mannerism**.[3] Relying on complex metaphorical themes, manneristic painters portrayed the human body in convoluted and contorted forms, often in more explicitly erotic

---

nent culture, more of it in two centuries than did their medieval forebears over 10 (box 18.4). More importantly, the creations of the Renaissance became the prototypical images of most subsequent forms of erotic art and literature, both in its grander tradition and its more vulgar counterparts.

The secularization of Renaissance society and the greater worldliness of the Church diversified the morally tolerable sexual options. This was further expanded by the emphasis on the exercise of the individual will, which was a hallmark of Renaissance mentality and largely respon-

sible for the bewildering mixture of the glorious and the awful within its culture.

The scholastic learning of the Middle Ages had seemed indifferent to the personal and emotional aspects of life. Increasing disaffection with scholasticism found eloquent expression in the Italian poet Francesco Petrarch, who began to articulate, in the late 14th century, a new approach to life through the fullest development of one's physical, intellectual, aesthetic, and spiritual potentials. Such ideas evolved into an intellectual movement based on classical learning and focused on

Box 18.4   Erotic Art and Literature in the Renaissance (*Continued*)

ways than did their Renaissance mentors. A central figure of Italian mannerism was Giulio Romano, one of Raphael's most gifted pupils and an erotic artist of unblushing explicitness. His set of drawings of coital postures became the prototype of "sex-manual" art in the West. Romano's drawings, engraved by Marcantonio Raimondi and embellished with erotic sonnets by Pietro Aretino, were known as the *Sedici Modi* (figure 18.6). Published in 1527, they caused a considerable stir and were much copied (only parts of the original prints have survived). Agostino Carracci painted another famous series of the same type. Developments in mass reproduction (through woodcuts, engravings, and etchings) allowed lesser artists to turn such art into commercial pornography.

In northern Europe a parallel tradition of erotic art developed based on bawdy scenes of everyday life showing the doings of peasants, soldiers, and prostitutes in taverns, markets, and baths. In the hands of artists like Pieter Brueghel, Hieronymus Bosch, Albrecht Durer, and Lucas Cranach, this art, which was aimed at the middle classes, mirrors the erotic aspects of contemporary society even more faithfully than do the works of Italian mannerism.

Two of the greatest writers of the 16th century, François Rabelais and William Shakespeare, were not self-consciously erotic writers but their works are notable for their bawdy humor. Through "Rabelaisian" has come to characterize all that is lustily humorous, the incisive wit of Rabelais (who was both monk and physician) was meant to satirize the lapses of the clergy and the morality of his time. Similarly, though Shakespeare's plays abound with sexual references, sex for him was but one facet of the far more complex emotions of love and hatred with their often tragic consequences. Such pessimism pervades the most famous tragedies of Shakespeare as exemplified by *Romeo and Juliet* (whose plot comes from Boccaccio).

[1]This section based on Eitner (1975) and Webb (1975).
[2]Wilson (1973, p. 17).
[3]"Mannerism" is descriptive of the excessive adherence to the distinctive style or manner characteristic of this art; there is also a pejorative connotation to the term in the same sense that we call people with affectations "manneristic," or mannered.

"the studies of mankind," the *humanities.* This intellectual tradition is the source of the modern approach to the study of human behavior, including sex research, sex education, and sex therapy.

Renaissance Sexuality

The exuberant sexuality of the classical world, long dormant during the Middle Ages, reemerged in Renaissance society. Immense wealth combined with a passion for elegance and luxury led the ruling class to indulge its sexual appetites on a grand scale. The extreme individualism of the young and highly impetuous men who dominated Renaissance society added a characteristically violent aspect to it. It was not unheard of for a band of young men to invade a convent and rape the nuns. Every woman was fair game to the man who could seduce her; he in turn became fair game to her husband's or father's dagger if caught in the act. The sexual behavior of some of the pillars of Renaissance society rivaled the depredations of the most decadent Roman emperors: A prominent Florentine, Sigismondo Malatesta, was

Figure 18.5   Agostino Carracchi (1557–1602), *Satyr and Nymph,* ca. 1590, engraving.

convicted of murder, rape, adultery, incest, sacrilege, perjury, and treason (his son Roberto had to defend himself with a dagger against his father's sexual assaults); Malatesta's nephew, Pandolfo, burned the convent in which the woman he wanted had been sheltered.[12]

Although many churchmen led exemplary lives, some Renaissance cardinals and popes shared in the more outrageous ways of their culture. Numbered among them were great humanists and patrons of art, but also men who were fond of pleasure and luxury. On one occasion, Rodrigo Borgia (Pope Alexander VI), father of the notorious Cesare and Lucrezia, had 50 naked women picking chestnuts off the floor for his dinner guests, who then competed to see who could copulate with the largest number of them.[13]

[12]Johnson (1981).
[13]Taylor (1970).

The Renaissance also saw the revival of another tradition in the rise of the *courtesan* (the common prostitute had never gone out of business). Like her classical predecessors, she was often a woman of intelligence, refined manners, and charm; Veronica Franco, for instance, was a poet and a friend of Tintoretto, prelates, and royalty. Though Renaissance society continued to be ruled primarily by men, the status of women distinctly improved from medieval times, at least for the advantaged classes. Women of the aristocracy were often as cultured, worldly, lascivious, and vicious as the men.

The Reformation and the Protestant Sexual Ethic

The excesses of Renaissance society set in a strong reaction both among the clergy and laity. The depth of offended public piety made possible, for instance, the ascendency in Florence of the Dominican friar Savonarola, who set about containing the hedonism and materialism of the city, prompting its citizens to consign the material trappings of the new paganism to a gigantic bonfire in 1496. The threat of censure here on hung over the heads of artists of even Michelangelo's eminence; in 1559 Pope Paul IV ordered the nudity of some of the figures in the *Last Judgment* to be painted over.

Less dramatic but more lasting were the influence of other reformers who were dismayed at the moral laxity and secularization of the Renaissance Church. As Italian humanists had revived the love for classical learning, Christian humanists, like Thomas More in England and Desiderius Erasmus in Holland, attempted a similar return to the teachings and traditions of the early Christian Church. These early reformers paved the way for Luther's *Protestant revolution,* which irrevocably split the Christian Church in the West.

Martin Luther (1483–1546), who came from a simple working-class German family, was burdened in his youth with a sense of sinfulness. His inability to free himself from that burden as an Augustinian monk

Figure 18.6 *Pair of Lovers.* Late copy after an engraving by Marcantonio Raimondi, based on designs by Giulio Romano.

led to his vision that salvation was only possible through the grace of God by faith alone.

Luther's views on sex and marriage had a profound influence on Western morality. Though less systematic than Aquinas, Luther had an earthier perception of sex and expressed himself with remarkable candor.[14] Luther rejected the Catholic belief that marriage was a sacrament, because he could find no biblical support for it. Instead he considered marriage a divinely ordained duty and privilege, thus harking back to the Judaic tradition of marital relationships. Particularly in his earlier writings, Luther laid much emphasis on the erotic component in marriage. The sexual urge was part of God's creation and there was no point or possibility in resisting it; instead it had to be harnessed and put to good use. (Luther compared marriage to a hospital that cured one of lust and avoided fornication.) Where a marriage failed to fulfill sexual needs, it lost a main purpose for its existence. Hence impotence on the part of the husband and repeated refusal to have sex by the wife constituted grounds for divorce.

Luther initially respected voluntary celibacy while opposing its imposition on the clergy. But eventually he came to regard celibacy and virginity as unnatural, impractical, and false forms of piety. He himself broke his earlier monastic vow of chastity by marrying at age 42 Katharine von Bora ("Katie"), a 29-year-old former nun, and fathered three children.

Protestantism reaffirmed the Catholic emphasis on the reproductive function of

[14]For a summary of Luther's views on sex and marriage, see chapter 6 in Feucht (1961). Among Luther's own work that bears on this area is his lecture on Genesis and *Treatise on Married Life.*

marriage, but it also made much more explicit the function of affection and companionship ("love and honor") between spouses. Even the Catholic church, which vigorously opposed almost everything that Luther stood for, came to adopt (in the Council of Trent in 1563) the reformer's view in this regard, and this has been characteristic of the Western ideal of marriage since then. But Luther cannot be held responsible for our romantic notions of conjugal bliss. He thought husband and wife had to make concessions to the drudgery of daily marital life and the burdens of rearing children; marriage taught couples how to be humble and patient.

Luther and the Protestant reformers, having blessed sex in marriage, categorically disapproved all other forms of sexual expression. Since everyone, including the clergy, could be married, there was no excuse anymore for the existence of other sexual alternatives. That also applied to prostitution; Luther would have none of it (but as a pragmatic politician he cautioned against the precipitous closing down of brothels). As Jeremy Taylor summed up this point of view at a later time, "There is no necessity that men must either debauch matrons or be fornicators; let them marry, for that is the remedy which God hath appointed."[15]

John Calvin (1509–1564) left his own particular imprint on Protestant sexual morality. Where Luther saw marriage as the remedy for sexual desire, *Calvinism* viewed marriage more like a veil that covered sin. To avoid the excessive enjoyment of sex, he urged that a husband approach his wife "with delicacy and propriety" and that she in turn be circumspect so as to avoid touching or looking at his genitals. This mentality extended to disapproving everything that was sensuous, exuberant, or frivolous. The *Puritans* in England further elaborated these attitudes into a distinctively dour form of Protestant sexual morality. Transplanted to New England, these views in turn proved highly influential in shaping the moral ideals of the fledgling American nation.

[15]Quoted in Tannahill (1980, p. 328).

Sexuality in the Age of Reason and the Enlightenment

A number of fundamental shifts in thinking that characterized the 17th and 18th centuries were to have profound repercussions on our views of life. As empirical evidence displaced received beliefs as the basis for reality, skepticism began to undermine the religious faith and intellectual assumptions of the past. In the absence of divine guidance, a *secular morality* had to be found to regulate human conduct. Like orphaned children, people now had to look after each other and come up with some good reasons to restrain their innate selfishness and antisocial behaviors.

Montaigne proposed to make human welfare and self-determination the bases of morality. The purpose of life shifted from the aspiration to eternal life in heaven to attaining happiness on earth. The Declaration of Independence of the United States codified this view by including "the Pursuit of Happiness" among the unalienable rights given by the Creator.

Yet the secular conquest of the human heart and mind proved less than sweeping. Even in the full bloom of the 18th century, people found the methods of science wanting when applied to understanding of people and society. The materialistic implications of Descartes' philosophy seemed to reduce human beings to a set of complicated machines. As Blaise Pascal expressed it, "The heart has its reasons that the mind cannot know."[16]

Manners and Morals

At the turn of the 18th century, the rate of illegitimacy in France was less than 5%; but in England and in New England it may have been as high as 50%. Almost everyone got married (the men at around age 27, women at 25).[17] But we do not have further details of the sexual practices and attitudes of the common people other than

[16]Quoted in Chambers et al. (1979, p. 520).

[17]Such findings come from a new approach in historical research called "family reconstitution," which relies on the analysis of records of parish registers.

what can be surmised from such demographic facts and literary references.

There is no lack of information, however, on the strict manners and loose morals of court life and the more privileged classes. *Manners* rigidly reflected social status, which in turn determined what constituted right or wrong behavior, sexually or otherwise. For instance, many middle-class men wore a simple black or brown coat covering shirt, trousers, and legs. But the elite dressed spectacularly, the men even more so than the women, with elaborate wigs and makeup. Women in richly embroidered robes wore their hair ornately, with ribbons, jewels, and perfumed curls; high-heeled shoes lifted them off the floor, as bodices pushed their bosoms upward. The first fashion magazine appeared in 1672.

Elegance coexisted with crudity. Women did not have an easy task as the arbiters of manners; men spit on the floor and urinated on the stairways of the Louvre. By 1660 guests wiped their fingers on napkins instead of the tablecloth. Social conversation was elegant and witty, touched on all topics including sex; humor could be quite ribald.

In Europe the strictest sexual morals were to be found among the middle classes, they being the most vulnerable as well as the most desirous of improving their social standing. The upper classes felt secure and the poor had little to lose. The law forbade homosexuality on pain of death, but that did not apply to royalty. *Romantic love* was accepted as a relief from marriage, but not as a reason for it. Almost everyone who could afford a mistress had one. Married women felt neglected if no one but their own husbands desired them. The principle of live and let live kept the peace.

Prostitution was rife and accepted, but despised if ill-mannered. Some courtesans reached spectacular heights of prominence and influence. Ninon de Lenclos, the literate and witty daughter of a French nobleman, was almost as famous as Louis XIV. "Love," she said, "is a passion involving no moral obligation." If a man needed religion to conduct himself properly in this world, it was "a sign that he has either a limited mind or a corrupt heart." She kept her lovers in line since they "never quarreled over me; they had confidence in my inconsistency; each awaited his turn."[18] In one family her lovers represented all three generations of father, son, and grandson. Lenclos was kind, generous, and highly trustworthy; men thought nothing of leaving large sums with her for safekeeping. She died in 1705 at the age of 85, in the arms of the Catholic church. Her will called for a simple funeral, but she left a substantial sum to the son of her attorney for the purchase of books. It was a wise investment; the boy became Voltaire.

The 18th century also saw the birth of the condom. It was the age of Casanova, Marquis de Sade, and of the literary character, Don Juan. These personifications of *libertinism* marked a new phase in the ageless game of seduction that now was played out on the stage of the fashionable world. It was a game of calculated viciousness where men selected, seduced, and abandoned a succession of naive and trusting women. In the fevered fantasies of de Sade, sex took on an even more explicitly sadistic character. The literature and art of the period reflect these themes as well as the more general erotic tenor of the times (box 18.5).

Life in America during this period was quite different from Europe.[19] We may think of the Puritan founders of this nation as prudes. Yet the Puritans in the seventeenth century were quite open about sex and viewed it as a natural and joyous part of marriage. One James Mattock was expelled from the First Church of Boston for refusing to sleep with his wife. Puritan maidens with child were married by understanding ministers, as is indicated in the town records of the time. The Puritans were stern and God-fearing people, who could be very harsh with adulterers. Yet their wrath was aroused not because of the sexual element but because such behavior threatened the sanctity and stability of marriage. The lack of a general sexual repressiveness was even apparent to outsiders. A visitor from Maryland who was

[18]Quotations in Durant and Durant (1963, p. 29).
[19]Following section based on Schlesinger (1970).

Box 18.5   Erotic Art and Literature in the Seventeenth and Eighteenth Centuries

By the start of the 17th century the sexual exuberance of the mannerists had given way to the highly ornate *baroque* style.[1] Baroque art flourished under the patronage of the Catholic church and was part of its counter-offensive to the Protestant and secular onslaught. Hence, erotic art in its more licentious manifestations did not thrive during this period.[2] Yet if eroticism is considered outside of the simplistic equation with explictness, then the 17th century could be said to have produced two of the few greatest masters of erotic art in the West: Peter Paul Rubens (1577–1640) and Rembrandt (1606–1669). The powerful eroticism of Rubens is self-evident in his fleshy and sensuous nudes. Rembrandt, the greatest artist of the age, is more subtle in his sexual expression: his nudes in erotic mythology or biblical scenes (see figure 17.3) are convincing representations of real women and more compelling erotically. His etching known as the "Ledikant" (or "The Fourposter Bed") is extraordinary in its frank and tender vision of a couple making love (figure 18.7).

The 18th century witnessed the triumph of more explicit eroticism in art.[3] Sexual tolerance allowed artists of high quality to devote themselves to erotic subjects without fear or embarrassment. Of these artists, Jean Antoine Watteau (1684–1721) was the most gifted. His *fête galante* paintings portrayed contemporary men and women in elaborate costumes celebrating love in a wistful, dreamlike world filled with the trappings of antiquity. Jean Honoré Fragonard (1732–1806) used classical motifs playfully and his erotic subjects are full of movement, wit, and surprise. The work of François Boucher (1703–1770) is the most sexually explicit of the three (figure 18.8) and his use of mythological figures was clearly calculated to exploit their full erotic potential. Countless lesser *galante* artists produced a vast amount of erotica, including etchings to illustrate the flourishing sex literature. By the close of the century there was such a surfeit of licentious pictures that Diderot was moved to write: "I think I have seen enough tits and behinds."

In England the major artist of the first half of the 18th century was William Hogarth (1697–1764). His work frequently touched on sexual matters but in a bitingly critical manner (such as in the *Harlot's Progress*). In a mellower spirit is his pair of paintings *Before* and *After* (figures 12.3). Thomas Rowlandson (1756–1827) was even more explicit in caricaturing the erotic foibles and eccentricities of his contemporaries (figure 14.10).

The great French dramatists of the 17th century—Corneille, Molière, and Racine—explored various aspects of love, often using classical themes. There was hardly a serious

---

in Boston in 1744 reported: "This place abounds with pretty women who appear rather more abroad than they do at New York and dress elegantly. They are for the most part, free and affable as well as pretty. I saw not one prude while I was there."

With the coming of national independence in the eighteenth century, there was a reaction to romantic notions that were associated with Old World feudalism and aristocracy. The republic was pledged to liberty, equality, and rationality. Marriage became more and more a service institution geared to populate the nation and strengthen the labor force. Native as well as visiting observers were quick to comment on this victory of rationality over romantic love. "No author, without a trial," complained Hawthorne, "can conceive of the difficulty of writing a romance about a

Box 18.5   Erotic Art and Literature (*Continued*)

Giacomo Casanova (The Granger Collection).

land's *Memoirs of a Woman of Pleasure* (1749) (commonly known as *Fanny Hill*). Moll Flanders is the prototype of the resourceful prostitute who goes through a turbulent phase ("twelve years a whore, five times a wife . . . , twelve years a thief") eventually settling down to a life of rectitude and prosperity. For Fanny Hill love is paramount, and she is rewarded at the end of her tribulations with a happy marriage. Both works provide moral lessons about the folly of vice while giving explicit examples of its pleasures.

Another 18th-century literary form is erotic autobiography. The prototypical work here is *The History of My Life* (or *The Memoirs*) of Giacomo Giovanni Casanova de Seingalt (1725–1798). Casanova (see accompanying figure) was a Venetian adventurer, who after a lifetime of travel and intrigue retired to the castle of a prominent friend to write and study. His memoirs are regarded as an important historical document.[5]

writer, in fact, who did not touch in one fashion or another upon issues of love and erotic relationships. The 18th century is also well represented in literary forms that deal with sexual themes.[4] Of the literature on prostitution, outstanding examples are Daniel Defoe's *The Fortunes and Misfortunes of the Amorous Moll Flanders* (1722) and John Cle-

[1]Named after the founder of the style, Federigo Barocci (1528–1612), the term has come to designate whatever is ornate or flamboyant in style.

[2]Discussion of art of the period based on Eitner (1975).

[3]"At no other period of Western hisory has erotic art been more warmly cherished and promoted more officially than in the years from about 1720 to about 1780."

[4]This section is based on Purdy (1975a); Eitner (1975).

[5]Casanova (1958).

country where there is no shadow, no antiquity, no mystery, no picturesque and gloomy wrong, nor anything but a commonplace prosperity, in broad and simple daylight, as is happily the case with my dear native land." The French writer, Stendhal, observed that Americans had such a "habit of reason" that abandonment to love became all but impossible. In Europe he wrote, "desire is sharpened by restraint; in America it is blunted by liberty."

The Nineteenth Century

The 19th century, which brings us to the doorsteps of our own era, was characterized by extensive social change and intellectual developments. The *middle class* replaced the aristocracy as the ruling element

Figure 18.7 Rembrandt, *The Fourposter Bed*, 1646.

Figure 18.8 François Boucher, *Reclining Woman*. (Alte Pinakothek, Munich.)

490

in the Western world and the arbiter of its morality.

Victorian manners and morals were already well established when the young Queen Victoria ascended the British throne in 1837, and they outlived her death in 1901.[20] Until recently, Victorian sexual morality was generally seen as rigid, pretentious, oppressive, and hypocritical. Historians have now shown that picture not to be entirely correct. One of the main reasons for the conflicting images from this period is that the life of the middle classes, or the 19th-century bourgeois experience, was highly varied—open and secretive, ordered and chaotic, at the same time. Attitudes toward the show of affection, the supervision of girls, the use of contraceptives, and other significant ingredients of sexuality "differed drastically from decade to decade, from country to country, from stratum to stratum."[21]

Another source of confusion has been the failure to differentiate between sexual *behavior* and sexual *ideology*, or as Carl Degler has phrased it, "What Ought to Be and What Was."[22]

Victorian Sexual Ideology

Sexual repressiveness, and more particularly, the suppression and denial of female sexuality, while by no means characteristic of all of Victorian society, influenced substantial segments within it.[23]

The Repressive Model   The laxity prevalent in 18th-century England had elicited great concern by religious and social reformers, like John Wesley (1703–1791), the founder of the Methodist church and the prototype of the evangelical preacher (what we now call "born again" Christianity). Likewise, the French Revolution, coming on the heels of the loss of the American colonies, posed a profound threat to the ruling classes in England. It was therefore of paramount importance for church and state to find ways of stabilizing society, and moral reform appeared to hold the best prospects. In this climate of urgency the evangelical moralists and the reform societies came into their own.

Public education, which had started in *Sunday schools* for teaching literacy to working children, took on the task of socializing youngsters into the new moral consciousness. Urbanization opened avenues for upward mobility and the willingness to embrace the conservative codes of the middle class. *Public opinion* became a potent force, and among the most prominent arbiters were women like Hannah More, a prolific author of widely read books and morally uplifting tracts.[24]

These measures were largely successful in stabilizing English society but at the cost of making Victorian morality rigid and oppressive, its manners smug and pretentious. The effect of its killjoy mentality was to drive sexual enjoyment underground. Respectable women, by being placed on a pedestal of purity, were asexualized while those relegated to whoredom were dehumanized. Influential physicians fostered the notion that respectable women had no sexual desires or should suppress them if they did (box 18.6). Marital sex became a conjugal duty for procreation that women were supposed to suffer with resignation. By contrast, prostitutes were assumed to be sexually insatiable.

Men had to contend with a similar split; a man could neither enjoy sex freely with his wife whom he was free to love and respect, nor could he love and respect the prostitute with whom he could freely enjoy sex. A burden of guilt or frustration thus became inextricably attached to sex whichever way people turned.

Victorian physicians turned on sex itself because of its presumed health hazards. The venereal diseases, which were wide-

---

[20]Quinlan (1941).
[21]Gay (1984, p. 5).
[22]Degler (1974).
[23]Marcus (1966).

[24]Religious tracts were among the most popular sources of reading for the Victorian public. In 1844 over 15 million copies of such tracts were published in England (Quinlan, 1941, p. 124).

Box 18.6 William Acton on Female Sexual Desire[1]

I should say that the majority of women (happily for society) are not very much troubled with sexual feeling of any kind. What men are habitually, women are only exceptionally. It is too true, I admit, as the Divorce Court shows, that there are some few women who have sexual desires so strong that they surpass those of men, and shock public feeling by their consequences. I admit, of course, the existence of sexual excitement terminating even in nymphomania, a form of insanity that those accustomed to visit lunatic asylums must be fully conversant with; but with these sad exceptions, there can be no doubt that sexual feeling in the female is in the majority of cases in abeyance, and that it requires positive and considerable excitement to be roused at all; and even if roused (which in many instances it never can be) it is very moderate compared with that of the male. Many persons, and particularly young men, form their ideas of women's sensuous feeling from what they notice early in life among loose or, at least, low and immoral women. There is always a certain number of females who, though not ostensibly in the ranks of prostitutes, make a kind of a trade of a pretty face. They are fond of admiration, they like to attract the attention of those immediately above them. Any susceptible boy is easily led to believe, whether he is altogether overcome by the syren or not, that she, and therefore all women, must have at least as strong passions as himself. Such women, however, give a very false idea of the condition of female sexual feeling in general. Association with the loose women of the London streets in casinos and other immoral haunts (who, if they have no sexual feeling, counterfeit it so well that the novice does not suspect but that it is genuine), seems to corroborate such an impression; and as I have stated above, it is from the erroneous notions that so many unmarried men imagine that the marital duties they will have to undertake are beyond their exhausted strength, and from this reason dread and avoid marriage.

Married men—medical men—or married women themselves, would, if appealed to, tell a very different tale, and vindicate female nature from the vile aspersions cast on it by the abandoned conduct and ungoverned lusts of a few of its worst examples.

I am ready to maintain that there are many females who never feel any sexual excitement whatever. Others, again, immediately after each period, do become, to a limited degree, capable of experiencing it; but this capacity is often temporary, and may entirely cease till the next menstrual period. Many of the best mothers, wives, and managers of households, know little of or are careless about sexual indulgences. Love of home, of children, and of domestic duties, are the only passions they feel.

[1]From *The Functions and Disorders of the Reproductive System in Childhood, Youth, Middle Age, and Advanced Life* (1888), p. 208.

spread, were rightfully to be feared, but there was no basis for the imagined sexual and bodily "wasting" presumed to result from the involuntary loss of semen (*spermatorrhea*) and masturbation (box 18.7).

The physicians who espoused these theories were exemplified by Dr. William Acton, whom Steven Marcus characterizes as "a truly representative Victorian: earnest, morally austere yet liberally inclined, sincere, open-minded, possessed by the belief that it was his duty to work toward the alleviation of the endless human misery and suffering which sometimes seem to be the chief constituents of society."[25] But the fact that Acton meant well does not mean that his views were harmless. And as we shall see, his perspective was by no means shared by all other physicians.

Many of the attitudes on sexuality that prevailed in England also characterized American society. While no longer a child of Europe, America remained heir to its culture, with English society serving as the model for its elite and subsequently for the middle classes. However, the influx of immigrants from other cultures and the exigencies of life on its ever-expanding frontier made the United States a far more pluralistic society with its own cultural identity. But that did not protect it entirely from the prudishness of the 19th century. For example, men and women could not visit art galleries together in Philadelphia because of the potential embarrassment that viewing classical statues would cause; refined women used the word "limb" instead of saying "leg." And then there is the story about the school mistress who dressed all four "limbs" of the piano in modest little trousers with frills at the bottom to protect the innocence of the young ladies in her charge. Whether or not such examples reflect the prevailing moral temper of the times or were isolated manifestations is an issue we turn to next.

Control of Male Sexuality   Historians have recently modified the repressive image of Victorian sexuality in two important ways. First, by pointing out that the ideology of the period was by no means inexorably repressive and that people did not always pay heed to the sexual pessimists; second, that the underlying motivation in the control of sexuality was not the suppression of female sexuality or the oppression of women—but the control of male sexuality and the liberation of women.[26]

In the sexual advice literature of the *early* 19th century, coital relations for married couples were placed at four to five times a week, with expectations of mutual orgasm. Female sexual interest was taken for granted; if anything, it was thought to be stronger and more enduring than that of the male.

It was only during the 1840s and 1850s that the more repressive ideology, as exemplified by Acton's work, made its appearance. But such views coexisted with more positive views of sex, which saw it as not detrimental but necessary for health and which recognized and validated the sexual need and capacity of women. Even when these authors referred to the sexual unresponsiveness of women, they put the blame on negative sexual attitudes and the sexual insensitivity and clumsiness of husbands. It was nevertheless the more repressive viewpoint that came to prevail as representative of the age, and its prescriptive statements as to how people should behave were accepted as descriptive of how they did behave.

In the United States, this pessimistic image was reinforced by the highly vocal activities of a number of American social reform organizations that came into being after the Civil War. Known as the *Social Purity* movement, its efforts were aimed at combatting liquor and prostitution, raising the age of consent for girls, and establishing a single standard of sexual behavior for both sexes.

The earnestness and rhetoric of crusaders for social purity made them sound antisexual. Yet their basic purpose was not to *suppress* sex but to *protect* women against the abuse of prostitution and to promote the sexual autonomy of women within marriage. Women played an active and prominent part in these social movements. They recognized that repeated and unwanted pregnancies threatened the health of women and the burdens of a large family precluded their taking on professional and social responsibilities outside of the home.

Given the relative unavailability and unreliability of contraceptives, the only certain assurance against indiscriminate pregnancies was sexual restraint. To this end,

[25]Marcus (1966, p. 2).
[26]Degler (1980) is the primary source for this section.

Box 18.7   Masturbatory Insanity

From the time of Hippocrates physicians have voiced their concern that overindulgence in sex is detrimental to health. Only in the last 250 years, however, has masturbation been singled out as a particularly harmful activity. Before the eighteenth century we can find only occasional reference to masturbation in medical texts. Early in the eighteenth century a book entitled *Onania, or the Heinous Sin of Self-Pollution* appeared in Europe. The probable author was a clergyman turned quack who peddled a remedy along with the book. Although the work became very popular throughout Europe (and is referred to in Voltaire's *Dictionnaire Philosophique*), it appears to have had no immediate impact upon medical opinion.

Then in 1758 appeared *Onania, or a Treatise upon the Disorders Produced by Masturbation* by the distinguished Swiss physician Tissot. It reiterated and amplified the claims of the former work. Tissot's views, coming from an unimpeachable authority, seem to have found ready acceptance. Despite rebuttals and accusations that he was exploiting his medical reputation to further his private moral points of view, the book became a standard reference. By the end of the eighteenth century the *masturbatory hypothesis* of mental disease was well entrenched.[1]

By the early 19th century, these ideas had become accepted in England as well. Sir William Ellis, superintendent of a mental asylum, was writing in 1839 that by far the most frequent cause of "debility of the brain is pernicious habit of masturbation." By midcentury, when this tide reached its high point, "the habit of solitary vice" was found to give rise to hysteria, asthma, epilepsy, melancholia, mania, suicide, dementia, and general paralysis of the insane.[2]

In view of such dire consequences, physicians and parents went to great lengths to prevent children from masturbating. One little girl was made to sleep in sheepskin pants and jacket made into one garment, with her hands tied to a collar about her neck; her feet were tied to the footboard and by a strap about her waist she was fastened to the headboard so she could not slide down in the bed and use her heels; she had been scolded, reasoned with, and whipped, and in spite of it all she managed to keep up the habit.[3]

a variety of arguments and exhortations was marshaled forth by physicians, social reformers, and feminists. Dr. Alice Stockham's objection to a wife's submitting meekly to her husband's sexual demands has a curiously modern ring: "She gives up all *ownership* of herself to her husband, and what is the difference between her life and the life of the public woman? She is sold to one man, and is not half so well paid."[27] Even the "passionlessness" of women may have been useful for them to gain moral parity with men while achieving a measure of control over their reproductive functions.[28] The delicacy and fragility of Victorian women may have been similarly fostered to keep men at bay. The justification of the campaign against prostitution and the condemnation of men who seduced and abandoned women is self-evident in this light.

These efforts were largely successful. The marital fertility rate declined from 7.04 births in 1800 to 3.56 births in 1900. The attempts to legalize prostitution were successfully resisted. Pehaps most significantly, the separation of sex from reproduction began

[27]Stockham (1883, p. 154). Her method for making childbirth easier through diet and exercise ("Tokology") is a good example of the curious mix of sense and nonsense in 19th-century American medicine.

[28]Gordon (1976).

Box 18.7   Masturbatory Insanity (*Continued*)

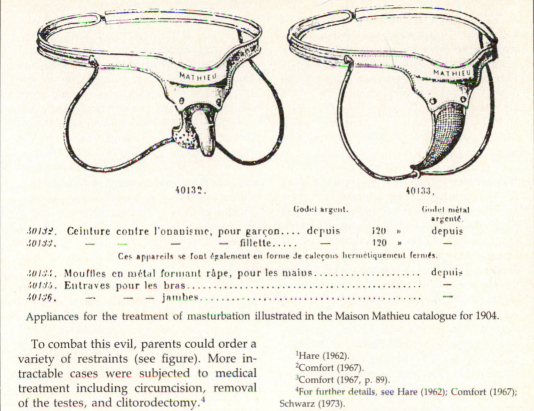

40132.

40133.

Godet argent.   Godet métal
                                     argenté.

40132. Ceinture contre l'onanisme, pour garçon.... depuis   120 »   depuis
40133.   —   —   —   — fillette.....   —   120 »   —
      Ces appareils se font également en forme de caleçons hermétiquement fermés.

40134. Moufles en métal formant râpe, pour les mains.................... depuis
40135. Entraves pour les bras..   —
40136.   —   —   — jambes..   —

Appliances for the treatment of masturbation illustrated in the Maison Mathieu catalogue for 1904.

To combat this evil, parents could order a variety of restraints (see figure). More intractable cases were subjected to medical treatment including circumcision, removal of the testes, and clitorodectomy.[4]

[1]Hare (1962).
[2]Comfort (1967).
[3]Comfort (1967, p. 89).
[4]For further details, see Hare (1962); Comfort (1967); Schwarz (1973).

---

to be established at this time. With better contraceptives becoming more available, abstinence and restraint became less important for the control of fertility. The acceptance of nonreproductive sex was clearly to have enormous repercussions in attitudes toward a variety of sexual behaviors, including sex outside of marriage and homosexuality.[29]

All of these changes were important for the sexual liberalization of subsequent periods. But at the time, the attempts to promote social purity took their repressive toll; by supporting the efforts of men like An-

thony Comstock (chapter 19) the expression of all sexual themes was severely censured; by combatting unrestrained sex, the restraints ended up checking healthy sexual experiences as well. As with William Acton, the fact that Social Purists meant well does not mean that they caused no harm.

Sexual Behavior

Given these cross-currents of sexual ideology, what was the net effect on sexual behavior? It is quite probable that many Victorian men and women were overwhelmed by the guilt and shame engendered by the expectations of abstinence, repression, and self-restraint. But equally

[29]For a review of the ideological implications of these changes, see Freedman (1982).

Box 18.8   Eroticism in Nineteenth Century Art and Literature[1]

The dominant intellectual theme of the first half of the 19th century was *romanticism,* which arose as a reaction against the increasingly arid rationalism of the 18th century. This new movement restored meaning and dignity to subjective experience, but disillusionment with its excessive enthusiasms and idealism in turn led to *realism,* which sought a return to the more plain and possible.

Despite the presumed prudishness of the 19th century a great deal of sexually explicit art was produced during this period, reflecting the contradictions of the age.[1] Among established artists in the romantic tradition, erotic themes are discernible in the works of Theodore Gericault, Eugene Delacroix, Jean Antoine Ingres, and Francisco Goya. Among the realists, erotic elements are particularly notable in the work of Gustave Courbet, who was able to get away with a painting like *Sleeping Women* (the artist called it *Laziness and Sensuality*) in 1866 (figure 13.5), while Eduard Manet's *Olympia* caused a furor not because of nudity but because the fashion-able courtesan, shown reclining on her bed, stared back at the viewer with stark realism. In the same tradition Constantine Guys and his better-known successor Henri de Toulouse-Lautrec documented life in the brothels and places of entertainment patronized by the demimonde. Such forthright treatment of erotic subjects infuriated the respectable classes who much preferred the more teasingly exposed nudes portrayed under classical guise, such as in the work of Cabanel, Bouguereau (figure 18.8), and Alma-Tadema (figure 18.9). In addition, a vast outpouring of frankly pornographic art flooded the market from the Victorian underground purveyed largely by artists of very meager talent with occasional exceptions (figure 18.10).

In the second half of the 19th century, as artists became increasingly alienated from society, they developed a *decadent* art aimed at shocking respectable sensibilities. Aubrey Beardsley and Marquis von Bayros were part of this new realist tradition, as was Felicien Rops, whose work is unparalleled in its brutal

---

plausibly others paid no heed to the prophets of sexual doom. As Alex Comfort has put it, "The astounding resilience of human common sense against the anxiety makers is one of the really cheering aspects of history."[30]

Evidence for a more realistic picture of Victorian sexual behavior comes from historical accounts, autobiographical materials, the art and literature of the period (box 18.8), and surprisingly enough, at least one sex survey conducted by an American woman physician, Dr. Clelia Mosher (box 18.9). One reason that evidence in this area is not more plentiful is the Victorian reluctance to discuss sex publicly. Though one could take that as evidence of sexual re-

pressiveness, the reluctance to engage in sexual discourse cannot be always equated with reluctance to behave sexually.[31]

Marital Sex   The legitimacy of marital coitus was not questioned even by those who voiced concerns over sexual excess. A good deal of evidence attests to the passionate and joyful relationship of many Victorian couples. The Mosher survey further shows that at least for its sample of well-educated, married American women, the ca-

[30]Comfort (1967, p. 113).

[31]According to Foucault (1978), volubility, rather than a conspiracy of silence characterized the 19th-century approach to sex. The issue, however, is not whether the Victorians engaged in sexual discourse at all, but what aspects of it were deemed fit for public consumption.

Box 18.8   Eroticism in Nineteenth Century Art and Literature (*Continued*)

assault on some of the most hallowed images in Western culture. (For example, in a painting by Rops called *Mary Magdalene*, a voluptuous and darkly sinister woman is shown masturbating next to a cross to which is tied a giant phallus.)[2]

In nineteenth-century literature the same excessive sense of propriety coexists with the crudest eroticism of mass pornography. Writers like Charles Dickens brought to life in stark realism the misery of the Victorian social underworld but without explicitly dwelling on its sexual horrors. Romantic poets like Wordsworth, Keats, Shelley, and Byron explored love in lyrical terms, as did Goethe and Balzac in terms of their societal and philosophical ramifications.

Few writers were prepared to challenge frontally the dominant cultural values. "If an author pipe of adultery, fornication, murder, and suicide," wrote one critic, "set him down as a practiser of these crimes."[3] Nevertheless literary challenge and rebellion against the morals and manners of the times were never far beneath the surface, occa-sionally acquiring more concerted expression such as through the *aesthetic movement* with its amoral advocacy of the propagation of beauty, and through *Hellenism* with its tolerant attitudes toward homosexuality. Ironically it was in the 19th century that pornographic writing became an industry (chapter 19).[4]

It was also during this period that translations of the erotic literature of the East were introduced into the English-speaking world. The dominant figure here was Sir Richard Burton, whose numerous accounts of his explorations and his translations of Eastern erotica confronted Victorian society with the freer sexual mores and attitudes of the high cultures of the East. Among Burton's major works are the translations of *The Arabian Nights* and other well-known erotic classics like the *Perfumed Garden* (chapter 11).

[1]Based on Webb (1975); Eitner (1975); Purdy (1975a).
[2]Rops (1975).
[3]Quoted in Curtis (1975, p. 203).
[4]Marcus (1967).

pacity to reach orgasm was not much different from the rates reported during the past decade. These women also practiced birth control methods beyond abstinence and withdrawal, even though the technological inadequacies of the means at their disposal were a source of understandable anxiety.

Even more compelling are the biographical vignettes and amorous correspondence of some couples, which reveal an unabashed celebration of sexual desire and a romantic intensity compared to which today's uninhibited language of love sounds insipid.[32] The period of *engagement* was a time of intense emotional and erotic involvement for Victorian middle-class couples, although it did not often entail coitus. Given the dependence of women on men, the uncertainties of available contraceptions, and sexual convention, premarital sex would have been highly problematic, hence the premium on premarital virginity. Premarital pregnancies were nonetheless fairly prevalent in the United States. Their rates declined from close to 30% of all births in the late 18th century, to less than 10% in the middle of the 19th century, then went up to 20% by the end of the 19th century.[33] The shifts in premarital frequency are ascribed to young people adhering more faithfully to the ethic

[32]"Sweet communions" Mabel Todd called the first nights after her marriage in 1879. "Oh joy! Oh! Bliss unutterable" (Gay, 1984).

[33]Smith and Hindus (1975).

Box 18.9   The Mosher Survey

In 1973 Carl N. Degler discovered in the Stanford University archives a questionnaire survey of the sexual attitudes of 45 American women, 70% of them born before 1870. These handwritten documents, never published at the time, constitute the first known sex survey ever conducted and a unique source on female sexuality in the 19th century.

The survey was the work of Dr. Clelia Duel Mosher, whose own life is an interesting study in Victorian womanhood (see accompanying figure). Mosher was born in Albany, New York, in 1863 into a family of physicians—her father and four of her uncles were doctors. She began her survey when still an undergraduate at the University of Wisconsin in 1892 and did not conclude it until 1920; during this period she had obtained a masters degree from Stanford and an M.D. degree from Johns Hopkins. She spent most of her professional life at Stanford as medical advisor for women and teaching hygiene.[1]

The women surveyed by Mosher were not representative of the general population but probably reflected the views of well-educated, middle-class wives of professional

Dr. Clelia Duel Mosher (1863–1940).

of continence in the first half of the 19th century, then moving away from that ideal later in the century as part of the assertion of their autonomy from family control.

Socially Problematic Behaviors   *Prostitution* was very much a part of Victorian society. Limited employment opportunities for women outside of domestic service (which accounted for more than a quarter of all working women in England) and widespread poverty generated a tremendous supply of potential prostitutes. Typically the *fallen woman* was a working girl, employed at pitiful wages, who supplemented her income by selling her sexual

services. An appalling proportion were teenagers, some barely out of childhood. The Victorian craze for virginity prompted the keepers of bawdy houses to have these pathetic youngsters stitched up over and over so they could be deflowered by wealthy patrons willing to pay premium fees for the pleasure. Doctors attached to these houses provided certificates of virginity (as well as performing the needed surgery). Some Victorian courtesans lived in great luxury, but even they did not quite attain the refinement of their earlier counterparts. The best that women like Cora Pearl could do was dancing nude on a carpet of orchids and bathing in a silver tank

Box 18.9   The Mosher Survey (*Continued*)

men. These responses reveal a remarkable degree of sexual candor and contradict the stereotype of sexually inhibited and unresponsive Victorian femininity.

The majority (35 out of 45) of Mosher's respondents felt the desire for coitus, independent of their husband's interest in sex; only 9 said they never or rarely felt such desire. Mosher must have assumed that women regularly experience orgasm, since she phrased her question not in terms of whether the person experienced orgasm but if she "always" reached orgasm during coitus: 35% said they "always" or "usually" did so; another 40% experienced orgasm "sometimes" or "not always." If the women before 1875 are considered separately, the incidence of orgasm (at least once) is 82%—quite similar to the women in the Kinsey sample born between 1900 and 1920.

The comments these women made on sex are equally revealing: Sexual intercourse "makes more normal people"; "even if there are no children, men love their wives more if they continue this relation, and the highest devotion is based upon it, a beautiful thing, and I am glad nature gave it to us"; "the desire of both husband and wife for this

expression of their union seems to me the first and highest reason for intercourse. The desire for offspring is a secondary, incidental, although entirely worthy motive, but could never make intercourse right unless the mutual desire were also present."[2]

On the other hand, these women also had a number of sexual anxieties. Fear of pregnancy colored their views, since contraceptives were not very reliable. Despite their being well educated, many said they knew nothing about sex before marriage. Of the 45 women, 25 reported having coitus once a week or less; 10 had it once or twice a week, and 9 more frequently. But only 8 out of 37 respondents said they wanted intercourse weekly or more—so over half the women engaged in sex more frequently than they would have liked. Yet in balance, the positive views of these Victorian women are the more impressive findings of the Mosher survey.[3]

[1]For further details of Mosher's life, see Degler's introduction in MaHood and Wenburg (1980).
[2]Quoted in Degler (1980), p. 264.
[3]For the full report of the Mosher survey, see MaHood and Wenburg (1980).

full of champagne in front of dinner guests.[34]

Prostitution was also very much part of the 19th-century American scene. The estimate that there were 20,000 prostitutes in New York City in the 1830s led a social reformer to compute that if each woman received three clients a day, then half of the adult males in the city visited them three times a week. A visiting English journalist in 1867 wrote, "Paris may be subtler, London may be grosser in its vices; but for

largeness of depravity, for domineering insolence of sin, for rowdy callousness to censure, they tell me Atlantic City finds no rivals on earth."[35] As the population of San Francisco swelled from less than a thousand in 1848 to over 25,000 over the next few years, some three thousand prostitutes converged on it from all over the country and from as far away as China and Chile. Other women of pleasure temporarily resided in Washington when Congress was in session.

The ambivalent sexual attitudes of the age also extended to *homosexuality*, which

[34]Tannahill (1980). Nonetheless Pearl could earn 5,000 francs a night when the daily wage of a skilled craftsman was 2 to 4 francs.

[35]Quoted in Tannahill (1980, p. 357).

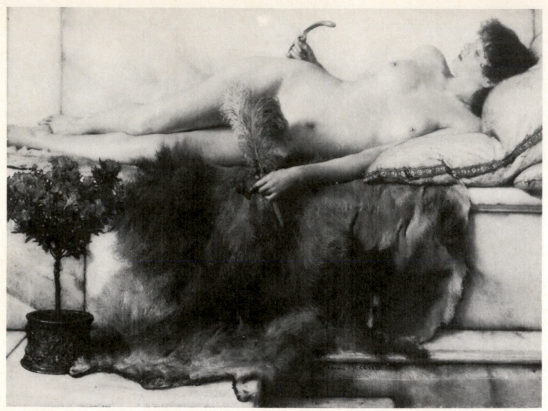

Figure 18.9   Lawrence Alma-Tadema (1836–1912), *In the Tepidarium*. The object in the right hand is an instrument for cleaning the skin after a bath. (Lady Lever Gallery, Port Sunlight, Cheshire.)

was declared criminal, immoral, and indecent. Yet not only male prostitution flourished in London, but a homosexual undercurrent characterized life in elite English all-male educational institutions. Though the privileged classes usually dealt with the matter with tactful disinterest, it was possible to get caught in the grip of public indignation, as Oscar Wilde discovered when he was tried and sent to jail.[36]

The use of the cane for punishing children has been blamed for the "English vice" of flagellation. Sado-masochistic themes are common in Victorian pornography as well as in some of its serious literature.[37] Brothels specialized in offering the joys of being whipped by a woman, whipping her, or simply watching others go through the experience. A small fortune was made by a Mrs. Berkley who invented a contraption that looked like a padded ladder to which the customer would be tied as the "governess" whipped his bare buttocks and another woman fondled his genitals.

Within the United States, prevailing sexual mores were challenged by *utopian communities.* The *Shakers* rejected coitus even for reproductive ends. The *Oneida*

[36]For the circumstances of Oscar Wilde's trial, see Karlen (1971). Shortly before his death, Wilde said, "I never came across anyone in whom the moral sense was dominant who was not heartless, cruel, vindictive, log-stupid and entirely lacking in the smallest sense of humanity. Moral people, as they are termed, are simple beasts. I would sooner have fifty unnatural vices than one unnatural virtue" (p. 225).

[37]See for example, Algernon Swinburne's *The Whippingham Papers* and *A Boy's First Flogging at Birchminster.*

Figure 18.10   by Frederic Bouchot (1800–1842), illustration in *Diabolico Foutromanie*.

community used "coitus reservatus" (chapter 3) to enjoy sex while avoiding conception. The *Mormons* allowed polygamy while limiting sex to reproductive purposes. The *Free Lovers*, whose ideas gained currency toward the end of the century, were more anarchic. They opposed all forms of social control while advocating the rights of women to enjoy sexual pleasure as well as to bear children.

The Modernization of Sex

The turn of the century, which ushered in the modern era, witnessed major trans-

formations in several perspectives. These changes laid the foundations for what came to be accepted as the modern way of thinking about sex or what Paul Robinson has called, the *modernization of sex*.[38]

Sexual modernism represented a reaction to the more repressive elements of Victorianism, or the ultimate ascendance of the more liberal perspectives within Victorianism, which saw sex as a valuable human asset to be managed properly rather

[38]Robinson (1976) is the primary source for this section.

Figure 18.11 Constantin Brancusi, *Princess X* (1916). (Philadelphia Museum of Art.)

than a drain on health or a threat to morality.

The most important exponent of this point of view in the English-speaking world was Henry Havelock Ellis (chapter 1). As the prototype of the **sexual enthusiast,** Ellis saw sex in highly positive terms. Basing his views on cross-species and cross-cultural comparisons, he substituted for the prevailing concepts of sexual normality the notion of a continuum of sexual behaviors that encompassed all the activities that Krafft-Ebing and others viewed as perversions (chapter 14). In this, Ellis anticipated Kinsey and the permissive views of most modern sexologists. But although an unabashed apologist for sexual freedom (and more of a polemicist than scientist), some of Ellis' views, such as on the role of

women, remained rooted in the 19th century.

Early 20th-century society did not become liberalized overnight. But sexual discourse was now in the open and took up subjects that had been shunned previously. The issues addressed went beyond the standard topics pertaining to marital sex and included questions like how to combine the requirements of marriage with the need for sexual variety. Homosexuality and other behaviors considered deviant were now considered to be in the realm of general human experience. Under the impact of Freud's work, the theme of sexuality came out of the closet and began to pervade all aspects of learning and culture that had anything to do with human interactions.

The upheaval of World War I accelerated the changes that were already in ferment. The increased knowledge of contraception and the larger proportion of people living in the greater anonymity of urban life further contributed to the shifts in sexual behavior. The most striking change in this respect was the increase in premarital sex among women between 1916 and 1930: among the women in the Kinsey sample who were born before 1900, less than half as many had had premarital coitus as those born in any subsequent decade (14% as against 36% among women still unmarried by age 25). A similar increase was noted in premarital petting. The most significant change for males was the frequency with which they had sex with prostitutes: reduced to half of what it was in the previous generation.[39]

The *roaring twenties* clearly constituted a sexual revolution among the younger generation. The sense of alarm and exhilaration that these events elicited is captured well in the following passage written in 1929 by two commentators.

> The younger generation is behaving like a crazy man who for one lucid moment has suddenly realized that the physicians in charge are all demented, too. The elders who have for so long been the sacred guardians of civiliza-

[39]Kinsey et al. (1953), p. 298.

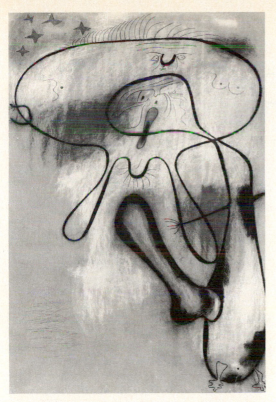

Figure 18.12 Joan Miro, *The Lovers* (1934). (Philadelphia Museum of Art.)

tion have bungled their task so abominably as to have lost irrevocably their influence for sobriety and sanity with the youth of the world. The failure of the church to treat sex and natural impulse with dignity and candor is the largest single fact in that disintegration of personal codes which confronts us in these hectic times: the inevitable swing of the pendulum from concealment to exhibitionism, from repression to expression, from reticence to publicity, from modesty to vulgarity. This revolutionary transition is inevitable and essentially wholesome, for all its crudity and grotesquerie.[40]

Members of the "younger generation" to which this paragraph refers are the grandparents of the population of youth that went through the sexual revolution of the 1960s (chapter 16).

[40]Calverton and Schmalhausen (1929), p. 11.

Developments in art and literature followed the same trends. The main shift was in the incorporation of erotic themes into the cultural mainstream. The autobiographical works of Henry Miller (starting with *Tropic of Cancer* of 1934) were not more explicit than the Victorian *My Secret Life*, but whereas the latter was anonymous and read furtively, Miller's books were taken as serious literature. *Ulysses* (1922) by James Joyce, and other great 20th-century novels, are suffused with eroticism.

Erotic art went through a dramatic transformation. Major artists like Gustav Klimt and Auguste Rodin, whose work straddled the turn of the century, produced highly sensuous works. But shortly thereafter, erotic themes became submerged in the gloomy and contorted visions of artists like Edvard Munch, Egon Schiele, and Jules Pascin.

Modern art tends to be abstract. Though rife with erotic symbolism (figure 18.11) its nonfigurative imagery does not lend itself readily to what most people would find erotically interesting, even when a work has a sexual theme (figure 18.12). For this reason, even though a compulsive eroticism underlies 20th-century art, it has ceased to be an effective form of commercial pornography—a function that has been taken over by film and photography.[41]

The sexual legitimization of erotic themes in art and literature did not come easy, and not until very recently. But after centuries of struggle, eroticism now appears firmly entrenched in Western culture.

SUGGESTED READINGS

Degler, C. *At Odds* (1980). Historical study of women and the family in America from the Revolution to the present. Chapters 11 and 12 are of particular interest for sexuality.

Gay, P. *The Bourgeois Experience* (1984), Vol. 1. Fascinating account of sexuality in the 18th century.

Webb, P., *The Erotic Arts* (1975). Comprehensive, profusely illustrated history of erotic art from ancient times to the present. Highly informative.

Tannahill, R. *Sex in History* (1980). Informative and entertaining overview of sexuality from prehistoric to recent times.

[41]Eitner (1975). For further details, see chapter 6 of Webb (1975).

Chapter 19
Sex and the Law

The welfare of the people is the chief law.
Cicero

In all human interactions, there is a potential for conflict between the individual's freedom of action and its undesirable consequences for others. Hence, all societies have had to enact rules to minimize the likelihood of such conflict while simultaneously trying to protect individual freedom of action. Sexuality has been traditionally considered an area that especially needs regulation because of its reproductive and social consequences. But the propriety of regulating consensual sexual behavior through legal means is a matter of much disagreement (box 19.1).

Laws Pertaining to Marriage and Procreation

Societies are keen on upholding the integrity of the family, even though preferences as to its form and functions may vary (chapter 16). Our concern here is limited to those laws that deal specifically with sexual conduct, although other laws, within the vast body of *family law*, may have an even greater impact on the operation of the marital unit.[1]

Marriage Law

In the United States *monogamy* is the only legal form of marriage and bigamy or polygamy have been criminal offenses since the formation of the union.[2] Given the significance attached to marriage, it is surprising that the law makes it easier to obtain a marriage license than a driver's license.[3] Most states prohibit marriages of minors under the age of 16 or 17, and parental consent is required for marriages of minors under the age of 18 in all states but three.[4] Yet if younger individuals marry by falsifying their ages, the marriage is usually considered valid.

Divorce Law

During the past decade the legal dissolution of marriage through *divorce* has become remarkably easy. In 1970 California was the first state to adopt *no-fault divorce* whereby dissolution of the marriage may

---

[1]A similar consideration applies to issues of sex discrimination, which in recent times has become a prominent legal concern and which also has profound implications to sexual behavior. See Dolgin and Dolgin (1980).

[2]These laws go back to the statute of James I in 1603. The practices of Mormons constituted a challenge to this position, but in 1878 the Supreme Court rejected the claim that antipolygamy legislation violated the religious liberty embodied in the First Amendment. *Reynolds v. United States*, 98 U.S. 145 (1878).

[3]Slovenko (1965).

[4]In Mississippi the minimum age requirement without parental consent is 21; in Nebraska and Wyoming it is 19.

take place if either party asserts "irreconcilable difficulties that have caused irremediable breakdown of the marriage."[5] Thus, either spouse may obtain a divorce regardless of the objections of the other party. It is no longer necessary to show that the other party was guilty of one of the traditional "fault" grounds for divorce such as adultery. Specific acts of conduct are in fact no longer admissible as evidence except where child custody is at issue.[6] In addition, unless agreed otherwise, the community property must be divided evenly.[7] Other states have followed the California lead, and as of 1981 only Illinois and South Dakota still limited divorce to traditional "fault" grounds.

The law views sexual intercourse as a fundamental part of marriage. Hence, if a union is not sexually consummated (that is, intercourse does not take place even once), the law provides grounds for annulment of the marriage. Should either spouse refuse sex subsequent to consummation, the aggrieved party could assert a case of "irreconcilable difference."

Sexual fidelity, particularly on the woman's part, has been one of the key requirements of marriage in the Judeo-Christian tradition. *Adultery* is a criminal offense in a substantial number of states, but the penalties for it are rarely enforced. Currently some 20 states provide misdemeanor penalties punishable by fine. A handful of states have felony penalties with the possibility of lengthy prison sentences; however, as a practical matter, fines are the usual alternative.[8] Although it is now generally recognized that behaviors like adultery should be beyond the reach of the criminal law, legislatures are reluctant to eliminate them.

[5]California Civil Code Section 4506.
[6]California Civil Code 4509.
[7]California Civil Code Section 4800. An interesting side-note to property settlements is the claim for "palimony," which has allowed compensation even where a couple living together were not married to each other.
[8]Massachusetts, for example, provides for a maximum 3-year sentence or a fine of not more than $500 (Massachusetts Ann. Laws, ch. 272, Section 14).

Laws Pertaining to Reproduction

Societal interest in regulating reproductive functions is among the primary reasons for legal preoccupation with the family. This is understandable since no society can trifle with the welfare of its young and hope to exist for long. So in addition to efforts at moral persuasion, laws have been enacted with regard to activities of reproductive consequence, of which we consider three: *contraception*, *sterilization*, and *abortion*.

Contraception   Over most of its history, moral teaching in Western culture has opposed the limitation of offspring; it is only since the 19th century that arguments controlling family size have gained popularity.[9] Contraceptive practice, in the United States, became restricted from 1873, when Congress made it illegal to disseminate contraceptive information, until 1965 when the Supreme Court declared unconstitutional laws restricting the use or discussion of methods of contraception for *married* couples. The Court reached this landmark decision in **Griswold v. Connecticut** on the grounds that the statute invaded the constitutionally protected "right to privacy."[10] Since Connecticut was the only state at the time to prohibit the dissemination of contraceptives to married couples, the importance of the Court's decision was primarily doctrinal; but it proved to have fartherreaching consequences than anticipated, as we point out in connection with abortion.

States were not prohibited from preventing *unmarried* persons from attaining contraceptives until 1972 when the Supreme Court extended the "zone of marital privacy" to them as well.[11] The reasoning was the same: contraception was a matter of individual privacy and it was unconstitutional for the state to invade it.

The dominant figure in the United States waging the battle for contraceptive freedom during the period between the two

[9]For the history of the views of the Catholic church on contraception, see Noonan (1967).
[10]*Griswold v. Connecticut*, 381 U.S. 479 (1965).
[11]*Eisenstadt v. Baird*, 405 U.S. 438 (1972).

Box 19.1   The Nature and Use of Sex Laws

The earliest known legal code goes back to Babylonian King Hammurabi (ca. 1700 B.C.). Emperor Justinian (A.D. 527–565) codified (i.e., collected and systematized) Roman law, which became the basis of *civil law* in 12th-century Europe, supplanting church *canon law*. In the 18th century, the Napoleonic code, based on civil law, gave rise to the laws of continental Europe. England has had its own legal tradition of common law based on decisions of the king's courts.

The laws of the United States, which started with colonial antecedents, come from two main sources: legislatures enact **statute law;** courts build on **common law** based on precedent. Since legislators are elected and judges are appointed by elected governmental executives, the citizenry, in principle, is the keeper of the nations's laws.

Laws pertaining to sexual behavior can be subsumed under several categories: marriage and procreation; consensual behavior between adults; commercial exploitation of sex; and sexual offenses against individuals. The last category of laws is obviously necessary to protect ourselves against behaviors like rape, sexual abuse of children, and activities that constitute public nuisances (such as voyeurism and exhibitionism). But the legitimacy and desirability of the regulation of sexual behaviors subsumed under the three other categories is variously disputed; it is

to them that the following arguments apply.

Those who support the use of the law to regulate sexual behavior avow a strong commitment to the protection of public morals and the integrity of the family. They view sex as potentially highly disruptive to both, and thus they consider the law as an appropriate tool to uphold the majority's moral values. The law should assert itself so that deviant sexual values and practices do not contaminate private and public life and thus weaken the moral fiber of the nation (chapter 16).

Objections to this view are based on several grounds, some procedural, others substantive.[1] Sex laws are often quite vague. Rather than define specifically what is prohibited, they proscribe behaviors which may be described as a "crime against nature," thus begging the question as to what is being prohibited. The difficulty of detecting private behavior between consenting adults may lead to undesirable police practices, while the capricious enforcement of these laws raises questions about arbitrary prosecutions.

The discrepancy between what the law says and what people do in their sexual lives is also very striking: if sex laws were seriously applied and sexual behavior were effectively monitored, the majority of the citizenry would be liable to prosecution. The fact that sex laws are by and large "dead"

---

world wars was Margaret Sanger (see figure 1.9). Her commitment to this cause developed during her work as a nurse among the poor in New York at the turn of the century. Though she was opposed and prosecuted during her early efforts, her spectacular success as an advocate of the birth control movement made her into one of the outstanding public figures of the first half of the 20th century.[12]

[12]For Margaret Sanger's biography and the history of birth control in America, see Kennedy (1970); for her autobiography, see Sanger (1938).

The United States government has been supporting family planning programs since the 1930s, although as late as 1959 President Eisenhower said he could not "imagine anything more emphatically a subject that is not a proper political or governmental activity or function or responsibility"[13] Yet in 1965 he became cochairman of Planned Parenthood (founded by Mrs. Sanger). Two years later over $20 million in federal funds were being spent on con-

[13]Kennedy (1970, p. iix).

Box 19.1   The Nature and Use of Sex Laws (*Continued*)

on the books does not make them wholly innocuous. Widespread violation with impunity of any law, dead or alive, generates disrespect for the law generally; and these laws still retain a certain power for abuse when selectively applied to harass individuals or groups.

The discrepancy between the law and common behavior reflects the law's failure to represent contemporary moral sentiments. This is due, in part, to the law's reliance on precedent, since precedent in sexual matters tends to be conservative and often based on outdated claims of secular harm. It also reflects the fact that few politicians are willing to challenge inactive and antiquated sexual laws because of risk to reputation and careers. For instance, Pennsylvania revised its criminal code in 1972 and in the process deleted its laws against fornication and adultery. The press reported this as a decision "to legalize premarital and extramarital sex"; the laws were reinstated within months. Idaho went through a similar experience at about the same time.

It has also been argued that sex laws reflect a particular religious perspective and therefore are unconstitutional. These "separation of church and state" arguments have so far not found much favor with the Supreme Court. No one denies that the tenets of our sexual morality derive from the Judeo-Christian tradition. But over time, these values have become secularized mirroring the contemporary values of the majority. Quite possibly, they merely reflect the view of persons and groups with the means, the access to the media, and the political connections— the "people who count." These are complex matters that are particularly vexing with regard to sexual behavior, where our knowledge is uncertain, values disparate, and emotion likely to overwhelm reason.

During the two decades a remarkable change has taken place in laws pertaining to the regulation of consensual behaviors, such as homosexual acts in private, as opposed to laws that prohibit sexual offenses against individuals, such as rape. In 1962 the American Law Institute, a private organization composed of prominent judges, lawyers, and law professors, published the *Model Penal Code,* which proposed eliminating criminal penalties for sexual activities between consenting adults (with the exception of prostitution).[2] Based on the recommendations of the Model Code, 22 states have so far removed criminal sanctions on sexual activities between consenting adults.

[1]Packer (1968).
[2]*American Law Institute* (1962).

traceptive programs in the country and $9 million for programs abroad. In 1981, in addition to the much larger sums spent internally, the U.S. provided over $250 million or 16% of the world's family planning funds.[14]

Sterilization   Voluntary sterilization, like other forms of contraception, has no legal restrictions and is now the most popular method of birth control for people over 30

[14]*Time*, April 6, 1981, p. 27.

(see chapter 6). Legal concerns with sterilization persist where the procedure is involuntary or undertaken in circumstances that restrict the person's choice, such as in cases involving the mentally handicapped and sex offenders.

Justification for the sterilization of select cases of mental retardation has been based on preventing genetic transmission of the disability and on grounds that people who are severely handicapped could not be expected to function adequately as parents. In 1927 the Supreme Court upheld a statute under which an institutionalized mentally retarded patient could be sterilized on the order of the superintendent of the state mental hospital.[15] But more recently there was an outcry when two teenage girls, aged 12 and 14, were sterilized through a federally funded family planning program in Montgomery, Alabama, allegedly without either their or their parents' consent.[16] The federal government has now issued regulations calling for standards of *informed consent*, including a 30-day waiting period between the giving of such consent and the date of the procedure. The use of federal funds is also prohibited for sterilizing persons younger than 21, those who are institutionalized, mentally incompetent, and women whose consent is obtained while in labor or undergoing abortion.[17] These regulations are applicable only if federal funds are used.

Currently 10 states authorize compulsory sterilization for the mentally retarded.[18] While there continues to be considerable variance in how this matter is dealt with (for instance, some courts but not others recognize the rights of parents to have their retarded child sterilized) the trend is generally to allow sterilization of a retarded person only after that individual has been accorded due process through a judicial hearing. California outright forbids the sterilization of minors. Only when the retarded person reaches adulthood may the parents (or other guardians of the retarded person) petition a court for sterilization.

Abortion   Abortion is a subject as complex as it is emotionally charged. As Callahan puts it, "Abortion is a nasty problem, a source of social and legal discord, moral uncertainty, medical and psychiatric

confusion, and personal anguish. If many individuals have worked through a position they find satisfactory, the world as a whole, and most societies, have not."[19]

Prior to 1970 various states allowed abortions where pregnancy had resulted from incest, rape, or where it constituted a risk to the physical or mental health of the mother. The breadth and vagueness of the mental health provision in particular made it possible for many women to obtain legal abortions, whatever their real reasons.

The Supreme Court decisions, starting with *Griswold v. Connecticut*, which we cited earlier with regard to contraception, eventually extended the right of privacy "to encompass a woman's decision whether or not to terminate her pregnancy." Thus on January 22, 1973,, the United States Supreme Court, in **Roe v. Wade,** by a 5-to-2 decision, declared unconstitutional all laws that prohibited or restricted abortions during the first trimester of pregnancy.[20] The Court limited state intervention in the second trimester but left the prohibition of third trimester abortions to the separate states. In 1976 the Court further ruled that a woman's decision to terminate her pregnancy could not be made subject to the consent of her spouse or parents. In case of minors she had to be old enough or mature enough to give informed consent.[21]

The legal freedom to have an abortion is not the same as being able to get one. If a woman is poor, she is dependent on inexpensive clinics or financial assistance as well as the availability of physicians and nurses who are willing to perform the operation, since the law does not compel medical personnel to perform abortions if it violates their moral principles. This vulnerability of the indigent became a palpable liability since the states were left free to close their public health facilities to abortions,[22] and free to deny public fund-

[15]*Buck v. Bell,* 274 U.S. 200 (1927).

[16]*Relf v. Weinberger,* 372 F.Supp. 1196 (D.D.C. 1974); and *Relf v. Mathews,* 403 F.Supp. 1235 (D.D.C. 1975).

[17]42 C.F.R. Section 50.201 (1979).

[18]Delaware, Georgia, Maine, Mississippi, North Carolina, Oklahoma, Oregon, South Carolina, Utah, and Virginia.

[19]Callahan (1970, p. 1).

[20]*Roe v. Wade,* 410 U.S. 113 (1973).

[21]*Planned Parenthood v. Danforth,* 428 U.S. 52 (1976).

[22]Poelker v. Doe, 432 U.S. 519 (1977).

ing of abortions.[23] They were also left free to require parental notice (but not consent) for a dependent minor's abortion.[24]

The impact of the law on behavior is dramatically demonstrated by the statistics on abortion. Following the 1973 legislation, the number of legal abortions went up from 744,600 to a million and a half by 1979; teenage abortions almost doubled to 400,000 a year; the proportion of all pregnancies (excluding miscarriages and stillbirths) that end in abortions went up from 19% to 30% during the same period.[25]

The curtailment of federal funding is one way that those who oppose abortion have tried to fight it. Another is to persuade Congress to pass a bill tht would extend the 14th Amendment protection to *persons unborn;* defining "person" in this context to include "unborn offspring at every stage of their biological development" since conception, will then presumably allow states to pass laws defining abortion as murder. As of 1983 efforts to achieve this had failed.[26] A more ambitious alternative is to call a *constitutional convention* to develop an amendment to the Constitution to make abortion illegal. If 18 states, in addition to the 16 that have already done so, make a request to Congress, a constitutional convention must be called to develop such an amendment to the Constitution, an unlikely prospect at the moment. A similar change could come about through a reversal in the majority view of the Supreme Court, a similarly unlikely prospect.

Meanwhile the issue has engaged the passions and committed efforts of various parties to the dispute. Considerations in this regard range from thoughtful discussion to impassioned parades by *prochoice* (figure 19.1) and *prolife* partisans (figure 19.2). Without attempting to recapitulate all of the arguments offered on both sides of this issue, the following considerations are illustrative of the complexities involved.

When does life begin? The question may appear deceptively simple, but it is impossible to answer to everyone's satisfaction. The Supreme Court, in ruling on the issue, admitted its inability to resolve this question, as have numerous experts in biology, medicine, and philosophy. As geneticist Joshua Lederberg has stated, "Modern man knows too much to pretend that life is merely the beating of the heart or the tide of breathing. Nevertheless he would like to ask biology to draw an absolute line that might relieve his confusion. The plea is in vain. There is no single, simple answer to when does life begin?' "[27] Yet Representative Henry Hyde of Illinois claims, "Defining when life begins is the sort of question Congress is designed to answer, competent to answer, must answer."

There is a self-evident validity in the proposition that the fertilized ovum is "alive." But is it any more "alive" than before getting fertilized? The zygote has the potential to grow into an embryo, but so do egg and sperm, which are just one step behind in the generative process. The question becomes even more bewildering when we turn to the issue of the *personhood* of this growing life. Many people have difficulty in seeing a "person" in a cluster of cells; by the same token, it is hard to think of a baby as less of a "person" shortly before birth than after. Where is then the transitional point in this process?[28]

One point of demarcation used is the time when the fetus becomes *viable* outside the uterus. Thus, to interrupt the life of the fetus before it could survive outside of the uterus would be one thing, to do so at a

---

[23]*Maher v. Roe,* 432 U.S. 464 (1977); *Harris v. McRae,* 448 U.S. 197 (1980).

[24]*H.L. v. Matheson,* 450 U.S. 398 (1981).

[25]*Time,* April 6, 1981, p. 22.

[26]Abortion foes in Congress used two tactics. A constitutional amendment, sponsored by Senator Orrin Hatch of Utah, would have given the states and the federal government authority to outlaw abortion; Senator Jesse Helms of North Carolina proposed passing a law that would declare "life" began at conception; this would have made abortion a form of murder and force the Supreme Court to review its past decisions.

[27]Quoted by Leon Rosenberg during Senate hearings (*San Francisco Chronicle,* May 23, 1981, p. 34).

[28]For further discussion of this issue, see Callahan (1970); Noonan (1970).

Figure 19.1   March in support of the right of women to obtain abortions. (Wide World Photos.)

point where it could be kept alive outside the mother would be another. But since our ability to sustain extrauterine life is likely to continue to improve, should the morality or legality of abortion be based on the existing level of medical competence? And what do we do with the more fundamental objection that if we just let the nonviable fetus alone, it would become viable in due time?

Another basic point of contention is the issue of who *owns* the developing organism before birth. The most obvious answer is that the mother owns it; after all, it is an integral part of her own body. But does that mean that the father of the unborn child has no claim on it? Obviously no one should have the right to compel a woman to become pregnant; but having become pregnant by engaging in sex by her own free will, can she treat the matter as if it were entirely her own affair? If that

were to be accepted, could it not be then argued that women should be solely responsible for contraception as well as the consequences of their pregnancy?

Even if one were to concede that the mother alone owns the fetus, it does not necessarily follow that she could do with it as she pleased. We do not allow physicians to remove healthy organs simply because the owner wants to be rid of them; every such procedure requires some additional medical justification. If the fetus is a healthy part of the mother's body, should a similar justification be called for?

Another relevant consideration is that the fetus can be thought of as an *independent entity* in terms that would not apply to any other part of the mother's body. A uterus, for instance, can never be anything but a uterus; but the fetus might eventually become a person with his or her own rights. If society has the right to intrude in cases

Figure 19.2   Demonstration in support of making abortion illegal. (Wide World Photos.)

of child abuse and forbid infanticide, should it not extend such protection to prenatal life as well?

Almost everyone agrees that abortion must be allowed if the mother's life is jeopardized by the pregnancy. But people part company when faced with less compelling indications: health problems that are not life-threatening; psychological considerations; concerns for the impact of the child on the mother's or father's career; the financial burden on the family or society; and some of the other reasons we discussed earlier (chapter 6) that prompt women to seek abortions.

There are also compelling concerns as to how effective antiabortion laws are likely to be, and their likely consequences over time. If abortions become illegal, women will not stop obtaining them. Those who can afford it will go elsewhere and those who cannot, will resort to illegal means. It

is probable that the number of abortions will decline, but it is even more probable that its casualties will increase, and it would be mainly the poor, the young, and the helpless who will pay the penalty.

Laws Pertaining to Adult Consensual Behavior

Why would the law want to regulate the private sexual behavior of a married couple? Why should society care if they engage in anal intercourse? The fundamental legal issue here is whether or not the law has any right intruding into the privacy of sexual relations, irrespective of the marital status and sexual orientation of the partners. The current thrust of the law is that it does not. As a result, criminal statutes governing the relationship between consenting adults have been progressively liberalized during the last few years: some 22 states now entirely exclude from the criminal realm *consensual* adult sexual conduct in *private;* at least another eight states exempt husband and wife from penal sanctions for such behavior. States that retain their criminal statutes with regard to unconventional sexual activity between consenting spouses are unlikely to force a test of their constitutionality in the light of the doctrine of marital privacy established by the Supreme Court.[29] In this light, laws proscribing sexual conduct between married couples are at present of little practical consequence.[30]

Heterosexual Behavior

A more controversial area is sex outside marriage where there is the likelihood of an aggrieved *third* party, or a perceived threat to the institution of marriage. The law distinguishes here between *fornication,* which refers to intercourse between consenting *unmarried* adults, and *adultery,* where at least one of the participants is *married* to someone other than the sexual partner.[31]

Our laws on fornication go back to the Puritans of New England. The tendency over the last several decades has been to decriminalize acts of intercourse between unmarried adults. Thus, in over half of the states, fornication is now no longer a criminal offense; where it is still prohibited by law, it is usually a *misdemeanor* punishable by a fine ($10 in Maryland; $500 in Mississippi). There are further variations in state law whereby some, for instance, punish the offense only when the parties are "living in a state of open and notorious cohabitation."[32] This requirement stems from common law where an illicit sexual relation was considered a secular offense when it became a public nuisance.[33]

An interesting component of fornication laws deals with *seduction.* For example, the California statute reads:

> 268. *Seduction*—Every person who, under promise of marriage, seduces and has sexual intercourse with an unmarried female of previous chaste character, is punishable by imprisonment in the State prison, or by a fine of not more than five thousand dollars ($5,000), or by both such fine and imprisonment.[34]

If pressed, a man could escape prosecution from this charge by marrying the woman (but if she were under age, he could still be prosecuted for statutory rape). Such attempts to protect women now sound quaint but may well have served a useful function when women were far more socially vulnerable in the past.

Homosexual Behaviors

Homosexuality, that is, the admission of having a homosexual preference or orientation, has never been a criminal offense. It is the commission of certain acts considered to be a "crime against nature" that have been unlawful. Thus, a typical statute states: "Every person who shall be convicted of the abominable and detestable crime against nature, either with mankind or with a beast, shall be imprisoned not exceeding twenty (20) years nor less than seven (7) years."[35]

The unspecified criminal behavior is generally understood to be *sodomy* or anal intercourse. Strictly speaking, the legal prohibition is not aimed specifically at

---

[29]Barnett (1973).

[30]In a bizarre case, a man in Indiana who had had anal sex with his wife was convicted in 1965 to the maximum term of 14 years for sodomy, even though his wife, who had accused him, subsequently withdrew the charge. He was released from jail after 3 years, on a technicality. *Cotner v. Henry,* 394 F.2d 873 (7th cir. 1968).

[31]Note the distinction between this definition and the concept of adultery in the Old Testament (Chapter 17).

[32]As in Alaska, Arizona, Illinois, Michigan, New Mexico, and North Carolina.

[33]*Model Penal Code,* Article 213, Comment (Revised 1980).

[34]California Penal Code Section 268.

[35]Rhode Island Section 11-10-1.

homosexuals. But since heterosexual anal intercourse is not likely to be prosecuted, and bestiality is rare and difficult to document, sodomy laws for all practical purposes have primarily dealt with homosexual acts.

Long-standing moral concerns over homosexuality were gradually translated into secular statutes carrying harsh punishments. In the English world, punishment by death for sodomy was decreed by Henry the VIII in 1533, repealed by his daughter Mary in 1553, and revived by his other daughter Elizabeth in 1562. The American colonies perpetuated this tradition. For example, North Carolina retained the death penalty for sodomy until 1869;[36] kept a 60-year maximum prison sentence permissible until 1965, and since then has left imprisonment to the discretion of the courts.[37]

The law has been used capriciously in the past in prosecuting homosexuals. As a result, even though only a few homosexuals were ever prosecuted, the existence of these laws was part of the broader patterns of social harassment to which homosexuals were subjected. Public attitudes toward homosexuality began to shift in the 1960s. An important contribution to this change were the findings and recommendations of *The Wolfenden Report,* issued in 1957 in Great Britain, based on a 10-year study by the Committee on Homosexual Offenses and Prostitution (box 19.2).

The changes in social attitudes toward homosexuals are now also reflected in the law. Twenty-two states have **decriminalized** homosexual acts between consenting adults in private. In those states where homosexual acts are still outlawed, the penalties range from misdemeanor fines to felonies. But in practice, few homosexuals are ever arrested since most homosexual acts are conducted in private and thus protected by the search-and-seizure provision of the United States Constitution. So by and large, adult homosexual consensual

behavior is no longer regulated by the law. But the law remains active with respect to *solicitation* and *loitering* in public places. Most arrests of homosexuals are for such behavior: according to one study, they accounted for 90 to 95% of homosexual arrests in the Los Angeles area.[38] The pertinent California statute reads:

647. Disorderly Conduct Defined Misdemeanor.
Every person who commits any of the following acts shall be guilty of disorderly conduct, a misdemeanor:
(a) Who solicits anyone to engage in or who engages in lewd or dissolute conduct in any public place or in any place open to the public or exposed to public view.
(b) Who loiters in or about any toilet open to the public for the purpose of engaging in or soliciting any lewd or lascivious or any unlawful act.[39]

As is clear from their wording, none of the laws aimed at homosexuals discriminate between men and women but, in effect, lesbians have been practically exempt from legal persecution. For example, between 1930 and 1939, there was a single case of a woman being convicted of a homosexual offense in New York City as against 700 convictions of males on homosexual charges and several thousand cases of males prosecuted for public indecency, solicitation, or other homosexual activity.[40] Kinsey ascribed this discrepancy to a number of factors: homosexual acts are more common among men and more likely to involve behaviors like public solicitation; society is more tolerant toward female homosexuals since it knows less about their activities and is less offended by their stereotypical behavior. Female, unlike male homosexuals, are not perceived as dangerous with regard to seducing or sexually molesting youngsters. Studies such as by Gebhard have failed to substantiate these

[36]North Carolina Chapter 34, Section 6 (1854).
[37]North Carolina Section 14-177. See generally *Model Penal Code,* Article 213, Comment (Revised 1980).

[38]Hoffman (1968, p. 84).
[39]California Penal Code Section 647.
[40]Quoted in Kinsey et al. (1953, p. 485).

Box 19.2   The Wolfenden Report on Homosexuality[1]

53. In considering whether homosexual acts between consenting adults in private should cease to be criminal offenses we have examined the more serious arguments in favor of retaining them as such. We now set out these arguments and our reasons for disagreement with them. In favor of retaining the present law, it has been contended that homosexual behavior between adult males, in private no less than in public, is contrary to the public good on the grounds that—

(i) it menaces the health of society;

(ii) it has damaging effects on family life;

(iii) a man who indulges in these practices with another man may turn his attention to boys.

54. As regards the first of these arguments, it is held that conduct of this kind is a case of the demoralization and decay of civilizations, and that therefore, unless we wish to see our nation degenerate and decay, such conduct must be stopped, by every possible means. We have found no evidence to support this view, and we cannot feel it right to frame the laws which should govern this country in the present age by reference to hypothetical explanations of the history of other peoples in ages distant in time and different in circumstances from our own. In so far as the basis of this argument can be precisely formulated, it is often no more than the expression of revulsion against what is regarded as unnatural, sinful or disgusting. Many people feel this revulsion, for one or more of these reasons. But moral conviction or instinctive feeling, however strong, is not a valid basis for overriding the individual's privacy and for bringing within the ambit of the criminal law private sexual behavior of this kind. It is held also that if such men are employed in certain professions or certain branches of the public service their private habits may render them liable to threats of blackmail or to other pressures which may make them "bad security risks." If this is true, it is true also of some other categories of person: for example, drunkards, gamblers and those who become involved in compromising situations of a heterosexual kind; and while it may be a valid ground for excluding from certain forms of employment men who indulge in homosexual behavior, it does not, in our view, constitute a sufficient reason for making their private sexual behavior an offense in itself.

55. The second contention, that homosexual behavior between males has a damaging effect on family life, may well be true. Indeed, we have had evidence that it often is; cases in which homosexual behavior on the part of the husband has broken up a marriage are by no means rare, and there are also cases in which a man in whom the homosexual component is relatively weak nevertheless derives such satisfaction from homosexual outlets that he does not enter upon a marriage which might have been successfully and happily consummated. We deplore this damage to what we regard as the basic unit of society; . . . We have had no reasons shown to us which would lead us to believe that homosexual behavior between males inflicts any greater damage on family life than adultery, fornication or lesbian behavior. These practices are all reprehensible from the point of view of harm to the family, but it is difficult to see why on this ground male homosexual behavior alone among them should be a criminal offense.

56. We have given anxious consideration to the third argument, that an adult male who has sought as his partner another adult male may turn from such a relationship and seek as his partner a boy or succession of boys. We should certainly not wish to countenance any proposal which might tend to increase offenses against minors.

57. We are authoritatively informed that a man who has homosexual relations with an adult partner seldom turns to boys, and vice versa, though it is apparent from the police reports we have seen and from other

Box 19.2   The Wolfenden Report on Homosexuality[1] (*Continued*)

evidence submitted to us that such cases do happen.

58. In addition, an argument of a more general character in favor of retaining the present law has been put to us by some of our witnesses. It is that to change the law in such a way that homosexual acts between consenting adults in private ceased to be criminal offenses must suggest to the average citizen a degree of toleration by the Legislature of homosexual behavior, and that such a change would "open the floodgates" and result in unbridled license. It is true that a change of this sort would amount to a limited degree of such toleration, but we do not share the fears of our witnesses that the change would have the effect they expect. This expectation seems to us to exaggerate the effect of the law on human behavior. It may well be true that the present law deters from homosexual acts some who would otherwise commit them, and to that extent an increase in homosexual behavior can be expected. But it is no less true that if the amount of homosexual behavior has, in fact, increased in recent years, then the law has failed to act as an effective deterrent. It seems to us that the law itself probably makes little difference to the amount of homosexual behavior which actually occurs; whatever the law may be there will always be strong social forces opposed to homosexual behavior. It is highly improbable that the man to whom homosexual behavior is repugnant would find it any less repugnant because the law permitted it in certain circumstances; so that even if, as has been suggested to us, homosexuals tend to proselytize, there is no valid reason for supposing that any considerable number of conversions would follow the change in the law.

60. We recognize that a proposal to change a law which has operated for many years so as to make legally permissible acts which were formerly unlawful, is open to criticisms which might not be made in relation to a proposal

to omit, from a code of laws being formulated de novo, any provision making these acts illegal. To reverse a long-standing tradition is a serious matter and not to be suggested lightly. But the task entrusted to us, as we conceive it, is to state what we regard as just and equitable law. We therefore do not think it appropriate that consideration of this question should be unduly influenced by a regard for the present law, much of which derives from traditions whose origins are obscure.

61. Further, we feel bound to say this. We have outlined the arguments against a change in the law, and we recognize their weight. We believe, however, that they have been met by the counter-arguments we have already advanced. There remains one additional counter-argument which we believe to be decisive, namely, the importance which society and the law ought to give to individual freedom of choice and action in matters of private morality. Unless a deliberate attempt is to be made by society, acting through the agency of the law, to equate the sphere of crime with that of sin, there must remain a realm of private morality and immorality which is, in brief and crude terms, not the law's business. To say this is not to condone or encourage private immorality. On the contrary, to emphasize the personal and private nature of moral or immoral conduct is to emphasize the personal and private responsibility of the individual for his own actions, and that is a responsibility which a mature agent can properly be expected to carry for himself without the threat of punishment from the law.

62. We accordingly recommend that homosexual behavior between consenting adults in private should no longer be a criminal offense.

[1]*The Wolfenden Report* (1963, pp. 43–48).

fears with regard to male homosexuals as well, who are not more likely to molest youngsters than are heterosexuals.[41]

Despite the liberalization of the law, public suspicion and hostility toward homosexuals persist.[42]

Sexual Offenses against Persons

While it is clear that we need laws with regard to sexual offenses against persons, it is by no means clear what sorts of laws would be best, both from the perspective of their acting as deterrents to such behavior as well as protecting the rights of the offenders.

[41]Gebhard et al. (1965).

[42]The most publicized case so far (partly because of the crusade by singer Anita Bryant) has been the 1977 repeal in Dade County, Florida, of an ordinance to protect homosexuals from discrimination in housing, employment, and public accommodations.

Rape

Common law defines *rape* as unlawful sexual intercourse by a man with a woman other than his wife and without her consent. "By force and against her will" is the common phrase used in statutory definitions, but rape need not, and does not always, involve physical violence. Sexual intercourse under the *threat* of violence or under circumstances when the woman is incapable of giving *consent* (such as when drunk or unconscious) can also constitute rape. If the victim is below the "age of consent," the act is called *statutory rape* and the issue of consent does not enter into the definition of the offense.

Sexual attitudes toward rape have ranged from romanticized views of what the act entails (figure 19.3) to severe condemnation. Rape was punishable by death under Saxon laws and has been generally considered a felony in common law. This English tradition was carried to the colonies where the earliest statutory prohibition of rape

Figure 19.3 Jacopo Tintoretto (1518–1594), *Tarquin and Lucretia*, ca. 1560. (Art Institute of Chicago.)

appeared in 1649 in the laws of Massachusetts: "If any man shall ravish any maid, or single woman, committing carnal copulation with her by force, against her will . . . he shall be punished either with death or some other grievous punishment according to the circumstances."[43]

At present rape is punishable only by imprisonment. For example, in California it carries a sentence of three, six, or eight years, depending on circumstances.[44] One study, based on 1976–77 sentencing data, showed that nationally the average time served for rape was 52 months.[45]

Legal prohibitions notwithstanding, rape continues to be alarmingly prevalent in this country (chapter 14). Yet most rapes go unpunished. For example, of all reported rapes, in 1977 only 51% led to arrests, of which 65% were prosecuted and 47% convicted of rape (and 13% convicted of lesser offenses).[46]

One of the primary problems in rape cases is the determination of the element of *consent*. Since this in turn has been linked to the character and sexual motivations of the woman, rape cases in the courtroom have been tellingly called "man's trial, woman's tribulation" (box 19.3). The wish to be spared this tribulation is one of the reasons so many women let rapes go unreported. Currently, legal practices and attitudes are changing in this regard, but these issues continue to be problematic.[47]

Under traditional rape statutes, courts usually demanded proof of *utmost resistance* by the victim to show that the act was against her will: "A mere tactical surrender in the face of assumed superior physical force is not enough . . . resistance must be to the utmost."[48] In this regard, rape laws have been unique: if a person is being robbed, for example, the law does not require that one put up any resistance. Yet a woman who was sexually assaulted had to demonstrate her lack of consent by resisting in every possible way until overcome by force, exhaustion, fear of death, or great bodily harm. There is an interesting presumption underlying this discrepancy, namely, that while we know no one willingly parts with wallet or purse, the abuse of a woman's body may conceal a hidden wish on her part to give herself to the molester (see discussion of "victim precipitation" in chapter 14).

There have been major recent revisions in rape laws; it is much less likely that a victim will now undergo the sort of humiliating experience that formerly characterized rape trials. The police are more sensitive in dealing with victims, and some states like New York have established special investigation units staffed by women who combine counseling with obtaining evidence. The woman's sexual history may still be admissible in some states, but the majority now have restrictions in this regard. Similarly the tendency is to minimize or eliminate altogether resistance requirements with the recognition that fear can have a paralyzing effect on a woman's behavior and that a woman should not be expected to risk bodily injury in addition to suffering the agony of rape. In Michigan, for instance, the rape statute flatly states that the woman "need not resist."[49] In addition, the majority of states have ruled out testimony about the victim's past sexual relations with third parties.

The severity of punishment for rape may have contributed to the difficulty to get convictions. Therefore it is sometimes argued that if penalties for rape were milder, many more would be convicted and thus the law might have a more effective deter-

[43]Quoted in *Commonwealth v. Chretien* (1981).

[44]California Penal Code Section 264. The sentence, however, can be enhanced if multiple crimes were committed, if a weapon was used, or if great bodily injury was inflicted.

[45]*National Law Journal* (1981, p. 52).

[46]Federal Bureau of Investigation (1981).

[47]For a description of the legal procedure whereby a trial for rape is conducted, see Lasater (1980). The hardships imposed on rape victims by the legal system are discussed in detail by Brownmiller (1975).

[48]*King v. State*, 357 S.W.2d 42, 45 (Tennessee 1962).

[49]Michigan Section 750.520. California removed the resistance requirement in 1980. California Penal Code Section 261(2).

Box 19.3   Excerpts from the Transcript of a Rape Trial[1]

Q. All right. And what, if anything, happened while you were walking in that general area?

A. Three people started to follow us, three men, three fellows . . .

Q. Now what, if anything, did you do at the time that they were walking behind you?

A. They started walking faster and we started walking faster and then we started to run.

Q. Now during the course of this running, what, if anything, happened?

A. Well, Toby stopped and I ran on and then I was grabbed from behind.

Q. How were you grabbed at this time? . . .

A. One hand from behind me around my face and the other hand on my head, my hair, and I was pulled back.

Q. What clothing was ripped off you or taken off you?

A. My—my girdle and—my pants and stockings were ripped to shreds and—my girdle was caught at my ankles so as I was being dragged I couldn't move my feet really . . .

Q. Now, what happened, if anything, on the top of this knoll area on the flat ground?

A. I was raped by all three of them.

Q. You say raped. Specifically, did they have intercourse with you?

A. Yes.

Q. All three of them?

A. Yes.

Q. Was there penetration? In other words, did they actually insert their penis into the vaginal area?

A. Yes.

Q. Were you saying anything or doing anything at this time to them?

A. Yes.

Q. What did you say and what did you do, if anything?

A. I asked them to please leave me alone and I was screaming and I was crying, too.

Q. What, if anything, did any of them do to you during the course of the time that they were having relations with you?

A. One of them hit me (*Crying.*) And they kept saying that if I ever told anybody that they would kill me . . .

Cross-Examination (by defense attorney)

Q. Now, as I recall your testimony . . . you stated that you were dragged from time to time during the course of your being in the company of these three young men. Isn't that correct?

A. That's true, sir.

Q. Were you dragged on your back or on your stomach or on your side?

A. I was standing up, but when my clothes were ripped off me, my girdle was down at my ankles so my feet were together and I wasn't exactly walking. I was, you know, my feet were together because of the girdle.

Q. Had you lost your footing?

A. Had I lost my footing? If I was being pulled, I lost my footing, yes, sir.

Q. So there came a time when you were being dragged along the ground, is that correct?

A. No. I was being held up by the arms like this and my feet were being dragged. I was being moved . . .

Q. Isn't it a fact . . . that you walked the

rent function. Yet the very opposite is also argued by others who would like to see the punishment for rape made more severe, including the reinstatement of the death penalty.

Courts must also be concerned with the rights of the accused. A man could be *falsely accused* of rape, and where there is no sign of struggle or injury, it would be a case of her word against his as to what really happened. The controversial issue of victim-precipitation has contributed to mitigating the rapist's offense by holding the woman at least partly responsible for acting provocatively. In a 1977 case in Madison, Wisconsin, involving the rape of a teenage girl

Box 19.3   Excerpts from the Transcript of a Rape Trial (*Continued*)

entire way from the point where you met the men to the point where you went to the fence, to the opening in the gateway?

A. I don't think you can say that, no.

Q. You were on your feet, weren't you?

A. But I couldn't have walked without someone pulling me with my girdle at my ankles. I would have had to hop.

Q. At what point in your trek across the field did your girdle come off?

A. I can't pinpoint that. I'm sorry.

Q. Isn't it a fact . . . that the point when your girdle came off, you assisted in taking of that girdle off?

A. Assisted? I wouldn't use the word "assisted."

Q. Did you participate in taking that girdle off?

A. No. I might have had my hands down there, yes, but I didn't participate.

Q. Did you in any way pull that girdle down?

A. No I wouldn't say that, sir.

Q. The girdle wasn't ripped, was it?

A. No, it wasn't.

Q. Your dress wasn't ripped, either, was it?

A. No.

Q. Nor were your underpants ripped?

A. No, they were not ripped.

Q. Now, at all times, you had your dress on you, isn't that correct?

A. That is true, sir.

Q. Now, did you assist in taking of your pants off?

A. No, sir.

Q. What effort if any did you make toward taking that girdle off of your body?

A. Effort? I don't understand.

Q. Isn't it a fact . . . that you helped those men take that girdle off your body?

A. That's not true. That's not true.

Q. There were no grass stains on this dress, were there. . . ?

A. I haven't really examined the dress, sir.

Q. At the time . . . you went down the embankment with the young men, isn't it a fact that at one point when the two of you started slipping down this embankment that you said to that young man, "don't hurt yourself," or "be careful," or words to that effect?

A. No. I don't remember ever saying that.

Q. Isn't it a fact . . . at least certainly between the second and the third intercourse you took time out to brush away the excesses of fluid that were around your vaginal area.

A. No, that's not true.

Q. Isn't it a fact . . . that upon the occasion of at least the third intercourse you assisted the third man in the completion of the sexual act to the extent of assisting him in penetrating you?

A. No, that's not true, sir . . .

Q. One more question . . . Is it not a fact that on the occasion of the third intercourse, you said to the man, "come on, come on?" (*Short pause*) Did you hear my question?

A. Yes, I heard your question, sir. (*No answer.*)

[1]*United States v. Thorne* (D.D.C. 1969). Excerpted in Babcock, et al. *Sex Discrimination and the Law* (1975, pp. 164–168). Note: questions by the prosecutor; cross-examination by defense attorney; answers by female plaintiff, a secretary employed by the Department of State.

by another student in the school stairwell, the judge let the offender off with a probation sentence because he considered the victim to have acted provocatively: "Whether women like it or not, they are sex objects. Are we supposed to take an impressionable person, 15 or 16 years of age, and punish that person severely because they react to it normally?"[50] The judge in this case was voted out of office and such considerations are now increasingly viewed as inappropriate.

[50]*New York Times*, May 17, 1977.

Marital rape   Another important recent concern with rape law deals with the issue of *spousal exclusion.* It has been assumed that "marriage constitutes a blanket consent to sexual intimacy which the woman may revoke only by dissolving the marital relationship.[51] As of June, 1980, 37 states did not recognize a woman's right to charge her husband with rape while they were living together (she could charge him with assault if he physically molested her). However, since then, 19 states have enacted laws against marital rape applying to couples living together. Massachusetts was the first to convict a husband for raping his wife.[52] California has a similar provision, but requires that the wife report the violation to an appropriate official within 30 days of its occurrence.[53]

In the Massachusetts case the couple were in the process of getting divorced when the man broke into his wife's home and raped her. The matter is more complex when a woman is raped when living with her husband. The *Rideout* case in Oregon was the first prosecution of this type and focused national attention on the issue.[54] The prosecution of marital rape involving a cohabiting couple raises particularly difficult practical problems of ascertaining consent and providing evidence.

The law presumes women to be incapable of forcing sex on men against their will, hence there are no laws protecting men against heterosexual rape. But some men are in fact raped by women (chapter 14). Similarly rape laws do not extend to homosexual assaults since the law deals with them through sodomy statutes. Currently many states, including California, have adopted statutes that define rape in *sex-neutral* terms, and thus would be equally applicable to homosexual rapes.

---

[51]*Model Penal Code*, Article 213(1), Comment (Revised 1980).

[52]*Commonwealth v. Chretien*, Massachusetts (1981).

[53]California Penal Code Section 262.

[54]*State v. Rideout* (Oregon Cir. Ct. 1978). The husband was charged by his wife of raping her but he was acquitted; shortly after the verdict the couple was reconciled; then they separated again (*New York Times*, March 10, 1979).

Sexual Offenses against the Young

Our society considers unacceptable all sexual interactions between adults and children. The behaviors proscribed by law in this area fall in two categories: sexual interactions with *postpubescent* but legally underage youth; sexual exploitation of *prepubescent* children.

Statutory Rape   The law prohibits sexual intercourse with persons below the age of consent, which in most states is set between the ages of 16 and 18. The age of consent for statutory rape may be as low as 12 years (as in North Carolina), but a man would still be guilty of the lesser offence of "taking indecent liberties with a child" by having sex with a girl younger than 16. Well over half of the states (but not California or New York) have adopted a sexually neutral definition of victim and offender, although the overwhelming majority of adult offenders are male (chapter 9). Rarely is a woman prosecuted for having had sex with a male under the age of consent.[55] In a recent challenge to the California statutory rape law the Supreme Court upheld its constitutionality even though the statute clearly discriminates against men. In a 5-to-4 opinion the Court ruled that such discrimination was justified on the grounds that males, unlike females, were not subject to the consequences of teenage pregnancy.[56]

An important aspect of statutory rape laws is that the male *offender* himself need not be adult. The case just cited involved two teenagers, Sharon, aged 16, and Michael, aged 17, who had intercourse after "an amorous interlude on a park bench." Sharon's family filed the statutory rape charge that got Michael arrested.

The interest of the law in this area reflects societal concerns over the sexual abuse and exploitation of youngsters by adults.

---

[55]In 1978 the New Mexico Supreme Court reversed a court of appeals ruling that had found sexual intercourse permissible between a 23-year-old woman and a 15-year-old boy (*San Francisco Chronicle*, February 26, 1978, p. 3).

[56]*Michael, M. v. Superior Court*, 450 U.S. 464 (1981).

But it also reflects an underlying assumption that sex ought not to be part of pre-adult behavior, however unrealistic that assumption may be. How these age limits are set and why they are changed is itself not always clear. For example, the age of consent in California was raised from 10 to 14 in 1889, to 16 in 1897, and to 18 years in 1913, where it now stands. Meanwhile, the age at puberty has gone down and teenagers have come to enjoy a far greater range of sexual freedom. So the legal assumption that a 16- or 17-year-old woman is incapable of consent seems inconsistent with other freedoms being accorded to them that also have considerable social consequences (such as the license to drive cars).

The dissociation of biological maturity from arbitrarily set limits also raises the possibility of men being deceived by girls who look older and lie about their age. Such considerations could not be used as a defense against a charge of statutory rape in the past, but today a majority of states do not consider as criminal sexual interactions engaged in by men under a reasonable belief that the girl is above the age of consent. Evidence of the girls' promiscuity or her being a prostitute may also be considered relevant in some states.

Child Molestation Sexual offenses involving younger children are severely sanctioned in all states. The California statute in this regard is quite typical:

> 288. *Crime Against Children: Lewd or Lascivious Acts* (a) Any person who shall willfully and lewdly commit any lewd or lascivious act, including any of the acts constituting other crimes provided for in Part I of the Code, upon or with the body, or with any part or member thereof, of a child under the age of fourteen (14) years, with the intent of arousing, appealing to, or gratifying the lust or passions or sexual desires of such persons, or of such child, shall be guilty of a felony and shall be imprisoned in the State prison for a term of three, five, or seven years.[57]

[57]California Penal Code Section 288.

Incest with minors constitutes another form of child molestation (chapters 9 and 14). All states have criminal prohibition against incest, usually dealt with through *consanguinity* laws that restrict marriage and sexual relationships between close relatives (about half the states also forbid marriages between first cousins).[58]

Laws governing incest vary a good deal from one state to another. They also tend to be confusing because incest statutes simultaneously try to accomplish two separate aims: the prevention of consanguineous marriage between relatives and the prevention of the sexual abuse of children. For instance, a fairly typical statute reads:

> *Incest.* Any person who knowingly marries or has sexual relations with an ancestor or descendant, a brother or sister of whole or half blood, or an uncle or aunt, nephew, or niece of whole blood commits incest, which is a Class 5 felony.

> *Aggravated Incest.* (1) Any person who has sexual intercourse with his or her natural child, stepchild, or child by adoption, unless legally married to the stepchild or child by adoption, commits aggravated incest (which is) a Class 4 felony.[59]

The most common penalty for incest in American law is 10 years in prison, though some states (Georgia and Nebraska) have maximum sentences of up to 20 years. The more specific problem of parent-child incest is further dealt with through the same statutory prohibitions involving minors discussed above.

Public Nuisance Offenses

Unwarranted exposure of one's body or furtive attempts to view someone else's exposed body constitute a bothersome intru-

[58]Consanguinity laws go back to Leviticus 18:6–18, 20:11–12. Although incest was not a crime under English common law, it attained criminal status in English statutory law in 1908.

[59]Colorado Rev. Stat. Section 18-60-302.

Box 19.4 The Treatment of Sex Offenders

Even if society were to leave alone all those who engage in consensual sexual behaviors in private as well as those whose sexual quirks pose no significant threat to others, there would still remain a substantial number of individuals whose behavior is socially unacceptable because of its violence, intrusiveness, and exploitation of children. No one seriously doubts that society has an obligation to deal with such violators. The question is how. What are we to do with a sex offender after he is convicted?

Imprisonment has long been the standard way of dealing with lawbreakers on the grounds that it prevents them causing further harm, punishes them and acts as a deterrent to others. But the bizarre and compulsively driven behaviors of sex offenders prompted mental health experts to label them *sick* rather than criminal. The law gradually came to recognize that view as valid and promulgated the **indeterminate sentence.** Under this system, the sexual offender was labeled a *sexual psychopath* and detained for treatment.[1] He would be released back into society when he was deemed no longer a menace.

Hailed at first as an effective and humane approach to criminal punishment, the indeterminate sentence became increasingly unpopular during the 1970s, on several grounds. The first problem was the potential for the capricious use of discretion. Indeterminate sentencing was called "arbitrary, highhanded, and unfair, hypocritically masking retribution and capriciousness behind the mask of scientific objectivity."[2] The problem was compounded by the uncertainty under which a prisoner had to live. Inmates often had no idea when they were to be released, and some were detained for treatment for periods longer than they would have otherwise served on an ordinary sentence. This was certainly not the intent of the law.

The second concern was based on the observation that most of these convicts were not being rehabilitated. Although sexual offenders in some states were placed in special hospitals, more often they were detained with the general prison population. In either case, there was no convincing evidence that they were being cured or reformed in any substantial numbers.

Third and most importantly, psychiatrists and social scientists could not reliably assess the dangerousness of sex offenders, which was a condition of release under the indeterminate sentence. Rather than suffer the wrath of the community by prematurely releasing a violent offender, institutions played safe by keeping many nonviolent offenders locked up.

As a result, disparate groups such as the police, prosecutors, the Prisoners' Union, and the American Civil Liberties Union joined forces to remove the indeterminate sentence provision. In California this occurred in 1977 with the passage of the **Determinate Sentencing Law** (similar measures were put in place in other jurisdictions). Under the new law, even a person found to be a mentally disordered sex offender may not be kept in custody longer than the longest term of imprisonment that could have been imposed for the offense. However, there is one exception: If a person "presents a substantial danger of bodily harm to others," a judicial hearing can provide for additional two-year periods of incarceration.[3]

There are a number of reasons for the difficulty in treating those who commit sex offenses, of which the most common are rape and child molestation. First and foremost reason is the lack of motivation to change. These offenders often deny their actions, even after conviction. They employ rigid defenses against understanding the basis of their behavior and are apt to blame others for their difficulties. Because they are deemed dan-

Box 19.4   The Treatment of Sex Offenders (*Continued*)

gerous, and often are, they must be held in secure settings, such as prisons, which are not primarily geared toward treatment. There is a paucity of well-trained therapists to deal with them, and these therapists must work under difficult circumstances. Funds are scarce, hence the mode of treatment must be rapid and inexpensive. These men relate poorly to others, especially women, yet they need corrective emotional experiences with women that are hard to find. They ask a great deal of therapists but they are more likely than most clients to repel, exhaust, and alienate their therapists.[4]

Insight oriented *psychotherapy* is neither readily available nor particularly effective with sex offenders. Hence most therapeutic approaches rely on various forms of *behavior modification* techniques. More recently, *sex drive reduction* treatment methods have attracted much public attention. Three forms of such treatment are in use: reduction of testosterone by chemical means; reduction of testosterone by castration; and destruction of certain brain centers through surgery.[5]

Brain surgery for this purpose is not yet based on adequate knowledge and therefore frowned upon. *Castration* is rather drastic because of its irreversibility. Nonetheless, a judge in South Carolina presented three young men (who had raped and tortured a 23-year-old woman) with the choice between castration and a 30-year prison sentence. But even if the men had chosen castration, no physician would have been likely to perform it under the circumstances.

The reduction of testosterone levels with *chemicals* ("chemical castration") has many advantages. It is reversible yet more effective than castration for rapidly reducing the sex drive. As we discussed in chapter 4, this may be accomplished by antiandrogens like *cyproterone* or the powerful estrogenic compound *Depo-Provera*.

This approach is now being evaluated, but there is also already a good deal of concern over it from a variety of perspectives. The chances are, no drug therapy will turn out to be a panacea; hence, the treatment of sex offenders is likely to continue to require counseling; but the drug, of course, could make a good deal of difference in enhancing the efficacy of psychological treatments.[6]

The fact that a sex offender who consents to such treatments is likely to draw a light sentence alarms and outrages the victims of their crimes. For example, the rape victim in South Carolina was reportedly terrified at the prospect of having the rapists, now castrated, out on the street; the fact that a man has a reduced sex drive does not preclude his wreaking havoc in vengeance. Finally, we need to consider how such treatments may impinge on the constitutional rights of citizens, even if they are sex offenders. To spend a lifetime in jail or to be castrated is not much of a choice and to present a man with such a choice may constitute cruel and unusual punishment. In view of these considerations, it is as yet unclear if these biomedical treatments offer fresh hope in this area or will prove one more disappointment in dealing with these most difficult problems.[7]

[1]"Sexual psychopath" has neither a precise legal definition nor does it correspond to a psychiatric diagnostic entity. Psychiatrists subsume many of the sex offenses under "paraphilias" (chapter 14).

[2]Report of the Michigan Felony Sentencing Project (1979).

[3]California Penal Code Sections 6316.1 and 6316.2.

[4]Greer (1983). For a review of outcome research in the treatment of paraphilias, see Kilmann et al. (1982).

[5]Freund (1980).

[6]For more details in biomedical methods of treating sex offenders, see Berlin (1983).

[7]See Greer and Stuart (1983) for a collection of readings on the treatment of sex offenders from a variety of perspectives.

sion into people's privacy and are considered offensive; therefore, laws prohibit such conduct. The legal definitions of *voyeurism* and *exhibitionism* are not identical with the broader meaning in which these terms are used in the sexual literature (chapter 14). As legally defined, virtually all exhibitionists and voyeurs are male. The law's concern with female nudity is from the perspective of obscenity, rather than indecent exposure.

Laws are currently highly liberal with regard to *nudity.* People sunbathing in nudist camps or in the privacy of their own backyards are left alone; the concern of the law is with people who offend public sensibilities by making a spectacle of themselves.

The majority of sexual nuisance offenses are committed by male exhibitionists ("flashers"), who constitute almost one-third of all reported sex offenders. Exhibitionists do offend and frighten women and children, but they do not generally molest them physically. Therefore in California the offense is now classified as a misdemeanor, but becomes a felony on repetition of the offense.[60] Some seriously question the appropriateness of any sanctions that involve imprisonment for these offenders. More generally, the problem of how to deal with sex offenders represents a serious challenge to society (box 19.4).

Prostitution

Roman law defined prostitution as the offering of one's body for sale, indiscriminately and without pleasure (*passim et sine dilectu*). As we discussed earlier (chapter 12), the term encompasses a wide range of sexual interactions with varying meanings at the personal and societal levels.[61]

Prostitution is illegal in the United States with the exception of the state of Nevado,

which permits prostitution by local option in counties with populations of less than 250,000.[62] If the success of prostitution laws is to be judged by their efficacy in eliminating it, then clearly they have failed: an estimated 500,000 prostitutes operate full time and an undetermined number of others do so on an occasional basis. Nevertheless, whether or not we would be better off without these laws is an issue that has been debated endlessly and remains an open question.

A particularly disturbing aspect of prostitution is the sexual exploitation of teenagers and children of both sexes. Of the estimated 20,000 runaways aged 16 in New York City, many end up working for some 800 pimps who use persuasion, drugs, coercion, and violence to keep them in their control. In Los Angeles alone there may be as many as 3,000 girls and boys under the age of 14 who are engaged in prostitution. Even small towns have their young prostitutes, some of whom live at home and "turn tricks" for pocket money. Investigators a few years ago identified a traffic pattern of runaway girls from the midwest gravitating to Minneapolis and then being taken to New York City in such large numbers that a section of Eighth Avenue was called the "Minnesota Strip."[63]

Prostitution is a misdemeanor in most states with maximum jail sentences of 90 to 180 days, but the usual penalty is a fine or a few days in jail. This results in a "revolving door" pattern whereby prostitutes who get arrested are back on the streets before the ink has dried on the police blotter and consider the fines as part of the expense of running their business.

Current statutes are gender-neutral and define acts of prostitution in terms broader than coitus. The California statute, for instance, prohibits anyone from soliciting or

[60]California Penal Code Section 314.

[61]There is an extensive literature on the social aspects of prostitution. For a historical overview, see Henriques (1962, 1963, 1968). Bullough et al. (1977) has an extensive bibliography. Also see Bullough (1979).

[62]Nevada Rev. Stat. Section 244.345 (1979). This provision makes prostitution illegal in Nevada's two major cities: Las Vegas and Reno. Although numerous prostitutes operate in these cities, legalization would reinforce their reputation as "sin cities" and possibly scare away tourism and conventions.

[63]*Time*, November 28, 1977, p. 23.

engaging in "any lewd act between persons for money or other consideration."[64] The offense most frequently prosecuted is not prostitution as such but *solicitation,* which is a misdemeanor. A typical ordinance is that of New York State, which makes it illegal to loiter in or near a public place in the following terms:

> Any person who remains or wanders about in a public place and repeatedly beckons to, or repeatedly stops, or repeatedly attempts to stop, or repeatedly attempts to engage passers-by in conversation, or repeatedly stops or attempts to stop motor vehicles, or repeatedly interferes with the free passage of other persons, for the purpose of prostitution, or of patronizing a prostitute shall be guilty of a misdemeanor.[65]

There are more severe penalties in all states for *pimping* (living off the earnings of a prostitute) and *pandering* (recruiting prostitutes). These are usually felonies with maximum sentences ranging from 4 to 10 years. Inducting a child into prostitution is an even more serious offense.

The person least threatened by the law is the customer (or the "John," as prostitutes call him). In some jurisdictions it is not an offense to use the services of a prostitute; even where there are laws against the practice, the customer is much less likely to be arrested. For example, over a period of time, the police in Minneapolis arrested 518 prostitutes but only 141 of their customers.[66] The justification for this is that the customer may be a married and respectable man and prosecuting him would constitute a disproportionately serious penalty, whereas the reputation of the prostitute is not at stake. Apart from the matter of equity, experience has shown that cracking down on customers is a more effective deterrent, yet attempts to do this by the police are generally opposed by those with commercial interests on the grounds that it would be bad for the tourist and convention business.[67] In short, the law and prostitution currently coexist in a truce of sorts marked with elements of hypocrisy and corruption.

Those who support the legal status quo are aware of its shortcomings, yet they regard present laws preferable to such alternatives as *decriminalization,* which would remove criminal sanctions for prostitution as such but would retain penalties for related offenses such as pimping or procurement; *legalization,* which would free the whole area of all legal constraints; and *regulation,* which would impose licensing and other regulatory requirements by the state.

Arguments against Legalization   Some of the arguments for keeping prostitution illegal are based on moral considerations. Given the general assumption that prostitution at best is a necessary evil, a state's formal declaration of its legality would reflect poorly on society's *ethical* standards. Others compare prostitution to slavery (*white slavery*) whereby women are exploited, degraded, and held captive by circumstances beyond their control (figure 19.4); hence, prostitution could not be legitimized in good conscience. If the legal restraints are lifted, it is feared that prostitution will become even more prevalent, with attendant increases in venereal disease and criminal behaviors, like muggings and drug abuse, which often coexist with prostitution.

Other concerns have to do with the potential impact of legalizing prostitution on indigent women who are on welfare. Given the general expectation that people should not be dependent upon the state whenever possible, social agencies may pressure these women into such gainful activity once it is legal. The Welfare Mothers of Nevada has already lodged complaints to this effect.[68]

Any attempt to regulate prostitution will inevitably involve some form of *registration*

[64]California Penal Code Section 647.
[65]New York Penal Law Section 240.37.
[66]*Time,* October 2, 1978, p. 48.

[67]Kaplan (1977).
[68]Kaplan (1977).

Figure 19.4 The reluctant girl. Hishikawa Moronobu (d. 1694), print from Shunga album of twelve sumizuri-E. (Courtesy of the Art Institute of Chicago.)

of prostitutes for certification, health, and tax purposes. Since prostitution is likely to remain a socially despised activity even if legalized, registration procedures may have a seriously *stigmatizing* effect on these women for the rest of their lives. This matter is particularly serious if one realizes that large numbers of women work as prostitutes for limited periods of time and then go on to marry or get into some alternative line of work; in either case, their getting out of prostitution may well be complicated by their being formally branded with this label.

Arguments in Favor of Legalization Those in favor of legalizing prostitution seek a frank and realistic appraisal of where matters stand. *The Wolfenden Committee Report*, which is usually cited with regard to homosexuality, also dealt with prostitution. It made the following assessment:

Prostitution is a social fact deplorable in the eyes of moralists, sociologists, and, we believe, the great majority of ordinary people. But it has persisted in many civilizations throughout many centuries, and the failure of attempts to stamp it out by repressive legislation shows that it cannot be eradicated through the agency of the criminal law. It remains true that without a demand for her services the prostitute could not exist, and that there are enough men who avail themselves of prostitutes to keep the trade alive. It also remains true that there are women who, even when there is no economic need to do so, choose this form of livelihood. For so long as these propositions continue to be true there will be prostitution, and no amount of legislation directed towards its abolition will abolish it.[69]

[69]*Wolfenden Report* (1963).

With regard to the moral argument, the fact that the law does not prohibit a given behavior does not mean that society condones it. That the quality of interpersonal relationships in sex with prostitutes may leave something to be desired, does not mean that it should be the responsibility of the law to improve them. Besides, prostitution does not really compete for the affections of those who are fortunate enough to have warm and loving relationships, but acts as a temporary or permanent substitute for persons lacking such ties. For at least some of their clients, prostitutes may represent the only available sexual outlets.

The money-making potential of prostitution, particularly for those with limited education and skills, gives rise to concern that the market would be flooded with new recruits if the practice were legalized. But this does not seem to have happened in countries where prostitution is legal (such as in Sweden, Holland, and parts of Germany). As long as the moral, social, and psychological constraints against the practice remain as strong as they are now, this is not a likely prospect in this country either.

Other assertions, such as with regard to the health consequences of prostitution, have a dubious basis. For example, prostitution is only a minor vehicle for the spread of venereal disease, being responsible for no more than 5% of cases of STD. The "professionals" in this field are quite aware of the risk of venereal disease and more likely to take effective precautionary measures than the promiscuous amateurs.

Critics of the present situation point out that the problem is not merely with the ineffectiveness of the law but that many of the evils associated with prostitution may themselves be due to the effects of the laws. For instance, having been declared a criminal, there is little left for the prostitute to lose if she then robs her client as well. Since the enterprise needs to be run by someone, criminal elements (including organized crime) inevitably take charge; the prostitute is thus caught between the law that prosecutes her and the pimps who exploit and keep her in bondage. It is thus conceivable that if prostitution were to be cast in a different social light and properly regulated, many of its unsavory characteristics could well disappear. One gets a glimpse of this possibility from the experience with surrogate partners used by some sex therapists (chapter 15). The detractors of this practice may think of these women as a form of prostitute, but the women themselves and their therapeutic sponsors do not, which allows them to maintain a measure of self-esteem and prompts them to deal with their clients in ways that are quite different from interactions with prostitutes.

The debate over prostitution may be cast in abstract terms, but the main practical concern is getting the streetwalkers off the streets. The approach in Britain has been to legalize prostitution but not solicitation; sexual services are made known through discreet cards affixed to doors in select areas. In this country, call girls operate with similar discretion. Another alternative would be to legalize houses of prostitution. This would permit the state to protect the prostitute and the client (and tax the earnings of the house). It would significantly free the prostitute from exploitation and at least some of her dependence on the pimp. In short, prostitution would become one more service industry.

Finally, the argument is made by some in the women's movement that a woman owns her own body and how she uses it is her business including the selling of sexual services.[70] To prevent the exploitation of prostitutes by pimps without shifting control to the state, it is proposed that prostitution be decriminalized, but not regulated; a solution with its own problems.

Pornography

The term *pornography* is derived from the Greek for for "prostitute"; hence pornography literally means *writing about prostitutes*. But in common usage it refers to what is sexually explicit or titillating: a sim-

[70]See Millett (1973).

Box 19.5   Pornography as Pornotopia

Based on his study of the Victorian pornographic novel, Steven Marcus characterizes the sexual fantasies expressed in pornographic novels as utopian and calls it *pornotopia:* It is a vision that regards all human experience as a series of exclusively sexual events or considerations. Pornotopia has no historical or other reality context; it occurs at no particular place or time. The participants have no identity or self outside of their narrowly defined sexual roles; people are reduced to their sexual organs and orifices. The principal natural object in the world of pornotopia is the *female body,* or whatever parts of it are considered erotic. As for man, "he is an enormous erect penis, to which there happens to be attached a human figure."[1] The *penis* in pornography becomes a magical instrument, a source of wonder, an object of worship, which works its effects on women who are helpless, suffering, but eventually ever so grateful for the untold delights it bestows upon them. It is the celebration of the erect penis without conscience or consciousness. The following excerpts illustrate these points.

Cleland, J. *Memoirs of a Woman of Pleasure* (1963) (first published in 1749), pp. 94–95.

I stole my hand upon his thighs, down one of which I could both see and feel a stiff hard body, confined by his breeches, that my fingers could discover no end to. Curious then, and eager to unfold so alarming a mystery, playing, as it were, with his buttons, which were bursting ripe from the active force within, those of his waistband and fore-flap flew open at a touch, when out it started; and now, disengaged from the shirt, I saw, with wonder and surprise, what? not the plaything of a boy, not the weapon of a man, but a maypole of so enormous a standard that, had proportions been observed, it must have belonged to a young giant: yet I could not, without pleasure, behold, and even venture to feel such a length, such a breadth of animated ivory! perfectly well-turned and fashioned, the proud stiffness of which distended its skin, whose smooth polish and velvet softness might vie with that of the most delicate of our sex, and whose exquisite whiteness was not a little set off by a sprout of black curling hair round the root, through the jetty sprigs of which the fair skin showed as in a fine evening you may have remarked the clear light ether through the branchwork of distant trees overtopping the summit of a hill: then the broad and bluish-casted incarnate of the head, and blue serpentines of its veins, altogether composed the most striking assemblage of figure and colours in nature. In short, it stood an object of terror and delight.

ilar sense is conveyed by "lewd," "indecent," "raunchy," "smutty," "bawdy," "ribald," and so on. *Obscene* is derived from the Latin for *repulsive* and can be used to designate anything offensive to accepted standards of decency, though it is usually understood in a sexual sense.

More precise attempts to define pornography have generally turned out to be exercises in futility. Justice Potter Stewart may have spoken for many when he said that he could not define pornography, "but I know it when I see it."[71] But in fact what people "see" as pornographic varies a great deal. For some, obscenity is synonymous with any form of public *sexual expression.* As the chairman of the Georgia State Literature Commission put it, "I don't dis-

[71]*Jacobellis v. Ohio*, 378 U.S. 197 (1964).

sion, which constitute the bulk of the debate in this area, are thus but one manifestation of a more fundamental political concern.

Social perceptions of what is obscene and attempts at its control have varied in Western culture, but prevailing attitudes have generally been restrictive. Heightened public preoccupation with censorship in the 19th century was not merely a reflection of the greater prudishness of the times, but a reaction to technological developments that made possible for sexually explicit materials to come into the hands of the masses (chapter 18).

In the United States, the most significant response to this threat was the passage of Section 1461, Title 18, of the U.S. Code in 1873, which made it a felony to knowingly transmit through the United States mail any "obscene, lewd, lascivious, indecently filthy, or vile article." Popularly known as the **Comstock Act,** this was in part the fruit of the efforts of Anthony Comstock (1844–1915), founder of the New York Society for the Suppression of Vice. As a special agent of the United States Post Office and through the efforts of his organization, Comstock was responsible, over a period of 8 years, for the confiscation of 203,238 pictures and photographs, the destruction of 27,548 pounds of books and 1,376,939 circulars, catalogs, songs, and poems that were considered obscene.[81]

Over the past several decades decisions by the Supreme Court have reversed this trend and placed pornography in a new legal perspective. The first landmark case in recent times involved a New York publisher and book seller, Samuel Roth, who was convicted of mailing a book and other sexually explicit mateials in violation of federal obscenity statutes. In ruling on the case, the Court held obscenity to be outside of the area of the freedom of speech protected by the First Amendment.[82] But the Court also made a distinction between

what is pornographic and what is merely sexually explicit as such, using the criteria of prurience and absence of redeeming social importance. These considerations were further refined in **Miller v. California** in 1973 to set the standards that currently govern decisions in this area. They define obscenity based on:

> (a) whether the average person, applying contemporary community standards would find that the work, taken as a whole, appeals to the prurient interest; (b) whether the work depicts or describes, in a patently offensive way, sexual conduct specifically defined by the applicable state law; and (c) whether the work, taken as a whole, lacks serious literary, artistic, political, or scientific value.[83]

The condition of *prurience* must be met with evident erotic intent (thus a protester carrying a sign that says "fuck the draft" is not legally guilty of obscenity, whatever else he or she may or may not be guilty of).[84] The object of concern is the *average person* and not those who may be overly sensitive or insensitive to these matters; the standards are those of a particular *community*, thus making it possible for something to be obscene in one place but not another.

The second condition requires that the item in question be *patently offensive* to violate the law. State rules in this regard cannot be vague, but must clearly specify what it is they are proscribing. The third and last condition is broader than its historical antecedents that required that the offensive item be shown as being "utterly without redeeming social importance." The burden has now shifted to the character of *the work taken as a whole*. Thus "a quotation from Voltaire in the flyleaf of a book will not constitutionally redeem an otherwise obscene publication."[85] By the same token, the presence of isolated obscene segments

---

[81]Kilpatrick (1960), p. 243.

[82]*Roth v. U.S.*, 354 U.S. 476 (1956). For the oral arguments before the Supreme Court in this and other major obscenity cases, see Friedman (1970).

[83]*Miller v. California*, 413 U.S. 15 (1973).
[84]*Cohen v. California*, 403 U.S. 15 (1971).
[85]*Miller v. California*, supra.

in a work will not necessarily condemn it. Under earlier conditions some of the world's greatest art and literature (even the Bible) could have been declared pornographic because isolated segments could conceivably be called obscene.

Within these constitutional bounds set by the Supreme Court, individual states are free to adopt their own definitions of obscenity. It is also constitutionally permissible for states to restrict a minor's access to sex materials that may not be considered obscene by adult standards. For example, in California, "every person who, with knowledge that a person is a minor, or who fails to exercise reasonable care in ascertaining the true age of a minor, knowingly distributes . . . or exhibits any harmful matter to the minor is guilty of a misdemeanor."[86] In a further attempt to counter "kiddie porn," California has made its distribution a felony punishable by imprisonment for 2, 3, or 4 years, or a fine not exceeding $50,000.[87] Other states have been likewise tightening their laws with respect to child pornography.

We must be clear that the law's concern is only with the *public* aspects of pornography; the possession of obscene matter or its use in private is not a crime.[88] But this right of privacy does not extend beyond the home, and the government has the right to regulate commerce in such materials, including the prohibition of its distribution through the mails.[89]

Pornography places the law in a difficult position as it tries to adjudicate between honest differences in conflicting points of view as well as more self-serving interests. It can count on little help by way of consistent advice from experts. Even if writers and artists could speak on these issues with a united voice (which they do not), their views are likely to be different from those of the "average person" that the courts have to worry about. Nor, as we discussed ear-

lier, is there any consistency in the advice provided by social scientists on the effects of pornography. It is no wonder that in the 33 obscenity cases in which written opinions have been offered, a total of 121 separate opinions have been articulated by the Supreme Court justices.

The Court's record on pornography frustrates many people. There are some who, on principle, object to the law's impinging on any freedom of sexual expression, or if they concede the need for such restraint they have problems with the particular criteria that are used. Others are frustrated at the inability of the law to curtail what seems to them to be so obviously offensive. In 1977, 74% of Americans approved of controlling obscenity in movies, books, and the like; and over half of them had strong feelings on the matter.[90]

Neither the judiciary nor the public are, by and large, concerned anymore with what hangs on museum walls. Rather, their problem is with the vulgar exploitation of every freedom granted by the courts by the purveyors of pornography. Especially distressing is the sort of material that has found its way into television, the most widely shared experience in the country.[91] Such opposition to pornography sometimes results in unusual partnerships between groups with ordinarily disparate perspectives such as politically conservative and feminist groups.

The law is a dynamic and changing institution. Its practices reflect, as well as shape, community beliefs, especially in emotionally charged areas like sexual behaviors that are deemed undesirable or controversial. The law will therefore continue to be a primary focus of political pressure as individuals and groups seek to assert their views and attempt to bend the law to that end.

[86]California Penal Code Section 313.2(a).
[87]California Penal Code Section 311.2.
[88]*Stanley v. Georgia*, 394 U.S. 557 (1969).
[89]*U.S. v. Reidel*, 402 U.S. 351 (1971).

[90]Yankelovich et al. (1977).
[91]The Coalition for Better Television, which claims to represent 400 conservative organizations, is one example of such opposition to what is seen as gratuitous sex, profanity, and violence on TV. *Time*, 1981, pp. 17–20.

SUGGESTED READINGS

MacNamara, D. E. J., and Sagarin, E. *Sex, Crime, and the Law* (1977). An overview of the behavioral legal aspects of socially problematic sexual behaviors and offenses.

Wasserstrom, R. A., ed. *Morality and the Law* (1971). A collection of readings on the question of whether the immorality of a behavior can even, by itself, be sufficient justification for making that kind of behavior illegal.

Callahan, D. *Abortion, Law, Choice, and Morality* (1970). A detailed exposition of the arguments in favor of and against the abortion issue.

Report of the Commission on Obscenity and Pornography (1970). Interesting reading both from the perspective of substance as well as the conflicts it portrays when a diverse group tries to come to grips with the problem of obscenity.

Chapter 20
Sex and Morality

The experience of sin in sexual relations does not wither away even for those who are emancipated from restrictive rules of conduct.

James M. Gustafson

For many people nowadays the topic of morality seems to hold little prospect of interest or excitement. They tend to think of it as something *negative,* exemplified by the biblical "thou shalt not." There is a certain aura of hypocrisy to the term; as if it embodied all those rules of conduct to which people pay lip service without any serious intention of putting them into practice. The notion of morality offends our sense of *autonomy* since it entails someone else telling us what we can and cannot do. And in intellectual discourse "moralistic" is equated with narrow-mindedness and *bigotry.*

These attitudes may be understandable in the light of the abuse of moral principles to suppress sexuality. Nevertheless, such reactions are naive and short-sighted. Moral considerations may not be fun in the short run, but they ensure that we get more satisfaction in the long run. As long as we have to live with each other we have to have rules of conduct, because that is the only way to protect our freedom and independence as individuals and as a society. If some moral laws are regularly broken it may mean that such laws have outlived their usefulness. But where a moral precept continues to be valid, our failure to follow it reflects on our level of responsibility rather than on the legitimacy of the moral code itself.

Moral issues need not be dull. When framed in the language of the day and placed in the context of our immediate lives, they are as fascinating as any other intellectual concern, and they affect our lives more than most. Think of the word "sin." This cornerstone of traditional morality now sounds antiquated to a lot of people. If someone were to ask, "Shall I sin against you?" one is tempted to say, "Sure, sounds like fun." But the question "Shall I exploit you?" does not sound funny. Yet sin and exploitation have much in common.

The words *morality* and *ethics* are derived respectively from the Latin *mores* and Greek *ethos,* terms for habit or *custom.* They thus represent societal judgment as to what is good or bad, right or wrong, virtuous or evil in relation to the actions and character of individual beings. Morality is that aspect of our human nature and experience that "expresses in personal conduct, in society, and in culture ways to order our natural impulses and to guide and govern our actions and relations for the sakes of individual and collective well-being."[1]

The issue of *responsibility* is central to all moral judgments and is in turn based on the concept of *free will.* We can only be held morally responsible for acts over which we have some control, where our

[1]Gustafson (1981, p. 484).

actions are voluntary. Concepts like free will become especially perplexing if we take into account the influence of learning in childhood, unconscious motivational forces, and biological determinants of behavior. These influences would seem to leave people with little autonomy, yet we still need to confront the issue of choice in any discussion of sexual morality.

The most basic choice is the wish to behave morally in the first place. Having decided that, one must choose the basis for moral decisions, be it religious belief, philosphical reflection, or common sense. Finally comes the application of the standard chosen to particular behavioral circumstances.

The Bases of Sexual Morality

Individuals and societies make their moral judgments in a variety of ways but fundamentally these judgments rest either on *religious* or *secular* grounds: the first is based on faith in the will of God; the second, on human principles. Secular and religious morality start with different assumptions, but they have a great deal in common since all ethical considerations ultimately revolve around human welfare. Though the modern world appears to be largely secular, four out of five persons continue to profess some form of religious belief.[2] Nevertheless, secular thought now exerts a powerful moral influence, even when it coexists with religious belief.

Secular Morality

Moral philosophy is the field of knowledge that deals with ethical issues. Typically, the thoughts of individual philosophers have been based on certain assumptions about *human nature* and *common good* which are then expanded through abstract arguments. For example, for Aristotle, the guide to ethical action was the avoidance of extremes and adherence to the "mean." Emmanuel Kant's "categori-

cal imperative" restated the principle of the "Golden Rule."

In modern secular societies, especially those that have broken with past moral traditions, the basis of moral precepts are more diffusely lodged in *political ideology.* For instance, in communist countries, sexual morality is based on Marxist-Leninist principles of the responsibility of individuals to the community.

When sociologists and psychologists deal with issues like attitudes and *values* they address some of the very problems that have preoccupied moral philosophers. Their work is significant for questions of sexual morality especially because empirical knowledge of the causes and effect of behavior clearly has a bearing on ethical concerns. For example, if it could be reliably determined what impact premarital sex has on marital happiness, this information would be of some relevance to moral judgments on sex before marriage. Similarly, the insights of anthropologists and historians provide comparative moral contexts across cultures and time to help us understand better our own moral views on sexual matters today.

The most recent contributions to this field have come from ethologists and sociobiologists who have examined the evolutionary development of *altruism.* There are numerous examples in the animal world where members of a group come to each other's aid, sometimes endangering themselves in the process. Porpoises will circle around females giving birth, thus protecting them from sharks; if one is wounded, others will carry it to the surface where it can breathe. Rhesus monkeys will attack their keeper if a monkey the keeper is holding utters an alarm call.[3] These seemingly altruistic behaviors are assumed to be basically self-serving. By sacrificing themselves for the sake of the group, these animals are indirectly ensuring the survival of their own genes, which are identical to those of their relatives.

When soldiers die for their country and parents go hungry to feed their children,

[2]Christians account for the largest share, accounting for one-third of the world's population in 1980.

[3]For other examples, see Wilson (1975).

to what extent are they behaving like altruistic animals? Whether these novel approaches have any bearing on human morality remains to be seen. So far, notions like the "selfish gene" have proved more confusing than helpful to nonspecialists in these fields.[4] But if such biologically based concepts of altruism prove tenable, they would necessitate a major rethinking of our fundamental precepts about the nature of human free will and morality.

Probably very few people base their moral views and behavior on serious study of any of these areas. Instead, individual morality seems to be guided by a more amorphous public understanding of the limits of acceptable behavior that prevail at any given time. That understanding is shaped by what is filtered down from the more formal attempts to define ethical behavior as well as the interactions of groups with particular moral positions to advance. This public standard is then applied by individuals to specific situations with liberal uses of improvisation based on common sense and assessments of what they can and cannot get away with.

Religious Basis of Morality

There are two fundamental differences between the religious and secular derivation of moral precepts. The secular approach relies on *reason*; the religious approach on *revelation*. The basis for the former is philosphical speculation and empirical observation; for the latter, it is faith. In the secular perspective, human happiness is the final goal of the ethical prescriptions. What this happiness entails may be understood in different ways; the emphasis may be on the individual or on the collective good, yet in all circumstances, "man is the measure." In the Judeo-Christian perspective, God is the measure and all other considerations, including human happiness, are secondary. The religious perspective does not disregard reason nor human welfare and happiness but considers them subservient to devotion to God.

[4]Symons (1979, p. 51).

In the religious approach to morality the basic task is determining *God's will*. To this end, all Christians invoke the Bible as their ultimate authority. But what the Bible says is understood in a variety of ways even by people who in all sincerity may be trying to get its true message.[5] In discerning the Christian basis of sexual morality it is not uncommon to be so bewildered by the diversity of viewpoints as to conclude that there is no Christian perspective on sexual morality, or to presume that there is a clear and consistent body of unchanging knowledge that speaks to these matters, and to label those who may disagree with it as un-Christian. The challenge for the more thoughtful is to distinguish what is "essentially Christian" from the historical artifacts that have accrued to that basic core. That is not an easy task.

In the rest of this chapter we examine three perspectives: the *conservative*, the *liberal*, and the *humanistic*. We shall first state the main premises of each and then present arguments for and against it. Centuries of debate over these issues have yet to exhaust these arguments and counterarguments. Our purpose here is neither to give a comprehensive summary of these nor to adjudicate between them; rather it is to stimulate reflection to help each individual to arrive at his or her own moral conclusions.

The Conservative Moral Perspective

Those who adhere to the conservative view uphold the tenets of **traditional** morality; they "conserve" the ethical outlook that has been tried and found true in the

[5]The 39 books of the Old Testament were written over a span of a thousand years; the 27 books of the New Testament, over a century. The Bible as we know it today dates back to the Council of Trent, which in 1546 decided which of the books used earlier by various churches to include and which to exclude from the final canon. The Old Testament was originally written in Hebrew and Aramaic, the New Testament in Greek. It was not until the 16th century that the first English translation of the Bible was made directly from the original languages.

past. Within Christianity, the conservative view is represented by the **orthodox** sexual values and traditions exemplified by the teachings of the *Catholic church* and *Evangelical Protestants*. Yet sexual conservatism need not be religious or necessarily traditional. For instance, Chinese communism, whose sexual values are highly conservative, is purely secular and often vehemently antitraditional.

The conservative perspective is **absolutist.** Its behavioral codes are predetermined for the individual. Certain sexual behaviors are deemed right, others wrong; individuals do not have a say about the validity of the underlying moral judgment. A Catholic, for instance, has to accept the moral teaching of the church, whatever it is, to remain a Catholic in good standing.

This absolutist approach relies on the promulgation of specific rules to guide conduct. Hence the conservative approach tends to be **legalistic.** We use this term not pejoratively but to indicate that moral precepts in this view are systematically argued and the application of these precepts with regard to sexual behaviors is explicitly spelled out.

The conservative moral view is generally **restrictive** with regard to the range of sexual behaviors allowed. Typically, it accepts heterosexual coitus in marriage as the only morally acceptable form of sex. One could argue that conservative moral principles are not restrictive but, on the contrary, greatly liberating since they allow the fullest expression of sexuality by preventing its dissipation in immoral sexual behaviors. But be that as it may, when looked at with regard to allowing people a range of sexual behaviors, it would still be correct to say that the conservative approach places greater constraints on sexual behavior. To understand better the applications of the conservative moral perspective to sexual behaviors, we mainly focus on the Catholic church, which has addressed these issues over many centuries.

The starting point for Catholic moral theology is the **Bible** as interpreted by the church. The doctrine of *original sin,* based on the fall of Adam and Eve (figure 20.1)

Figure 20.1   Adrien Isenbrandt (active 1510–1551) *Adam and Eve.* (Fine Arts Museum of San Francisco.)

has strongly colored the teachings of the Catholic Church and the sexual experiences of believers (chapter 17). While the Biblical record has remained constant, every age has rearticulated its teachings in the context of its own culture. For example, figure 20.2 illustrates the Old Testament story of David and Bathsheeba showing the figures in medieval garb and settings.

The most authoritative current statement of the Catholic church on sexuality is embodied in the 1968 encyclical letter of Pope Paul VI, *Humanae Vitae* (box 20.1). An *encyclical* is a papal "letter" on a specific subject addressed to the hierarchy and the faithful of the church. Encyclicals are pastoral devices, not formally argued theological documents. Nevertheless they carry enormous weight in the official teaching of the church.

Though primarily addressed to the issue of contraception, *Humanae Vitae* expounds a fundamental moral standard with regard

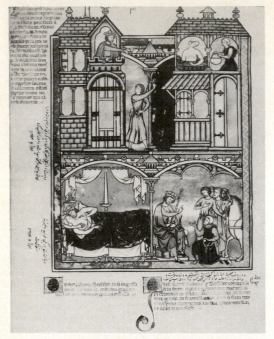

Figure 20.2 *David and Bathsheeba: The Illuminated Book of the Old Testament.* David sees Bathsheeba bathing; takes her to bed; sends her husband Uriah to battle to be killed. (Pierpont Morgan Library, New York.)

to all sexual behaviors. Sexuality is a source of "great joy" but marriage is the only legitimate context for sexual experience that is "honorable and good." Marital sex fulfills two key functions: *procreative* and *unitive.* The procreative aspect involves the conception and rearing of children. The unitive function subsumes all of the relational aspects of the bond between husband and wife based on "true mutual love" and the voluntary sharing of the supreme responsibility of parenthood. There is an "inseparable connection, established by God, which man on his own initiative may not break, between the unitive significance and the procreative significance which are both inherent to the marriage act." The tie between these two aspects of sex is to be maintained not only in general but with regard to each and every act of coitus which

"must retain its natural potential to procreate human life."[6]

In this view of morality, all attempts at contraception and all forms of nonreproductive sexual behaviors are unacceptable since they fail to meet the procreative purpose of sex. Yet within marital relations, not all coital activity that lacks reproductive potential is excluded. It is perfectly lawful, for instance, for men and women who are sterile (from causes not of their making) to engage in coitus. The same would apply to women who are pregnant, or in the "safe" period of their menstrual cycle. So, the moral prohibition is not against every act of coitus that has no reproductive potential but against interfering with the reproductive process through the deliberate use of "artificial" means of birth control, which includes all contraceptive devices other than "natural" means like the rhythm method (chapter 6).

The use of the unitive criterion in turn eliminates all forms of heterosexual intercourse outside of marriage since only the latter is considered as providing the proper context for sexual relationships between a man and woman and the handling of its reproductive consequences. Hence all forms of premarital and extramarital sex are morally unacceptable.

Except for the prohibition against contraceptive usage, this sexual ethic is by no means an exclusively Catholic view. Evangelical and other conservative Protestant churches basically adhere to it, as do orthodox elements in other religions such as Islam and the secular culture of China.

Historically the conservative Christian moral perspective has considered total renunciation of sex to be the highest form of sexual morality. Though not expected of everyone, *virginity* and *celibacy* have been greatly admired virtues in the Catholic church. As a moral ideal, avoidance of sex is marked by two important considerations. First, in order to count as a moral act, abstinence must be voluntary. Second, we must distinguish between those for

[6]*Humanae Vitae* (1972).

Box 20.1   Excerpts from *Humanae Vitae*; Encyclical Letter of Pope Paul VI.[1]

9

In the light of these facts the characteristic features and exigencies of married love are clearly indicated, and it is of the highest importance to evaluate them exactly.

This love is above all fully *human*, a compound of sense and spirit. It is not, then, merely a question of natural instinct or emotional drive. It is also, and above all, an act of the free will, whose dynamism ensures that not only does it endure through the joys and sorrows of daily life, but also that it grows, so that husband and wife become in a way one heart and one soul, and together attain their human fulfilment.

Then it is love which is *total*—that very special form of personal friendship in which husband and wife generously share everything, allowing no unreasonable exceptions or thinking just of their own interests. Whoever really loves his partner loves not only for what he receives, but loves that partner for her own sake, content to be able to enrich the other with the gift of himself.

Again, married love is *faithful* and *exclusive* of all other, and this until death. This is how husband and wife understood it on the day on which, fully aware of what they were doing, they freely vowed themselves to one another in marriage. Though this fidelity of husband and wife sometimes presents difficulties, no one can assert that it is impossible, for it is always honourable and worthy of the highest esteem. The example of so many married persons down through the centuries shows not only that fidelity is co-natural to marriage, but also that it is the source of profound and enduring happiness.

And finally this love is *creative of life*, for it is not exhausted by the loving interchange of husband and wife, but also contrives to go beyond this to bring new life into being. "Marriage and married love are by their character ordained to the procreation and bringing up of children. Children are the outstanding gift of marriage, and contribute in the highest degree to the parents' welfare."

12

This particular doctrine, often expounded by the Magisterium of the Church, is based on the inseparable connection, established by God, which man on his own initiative may not break, between the unitive significance and the procreative significance which are both inherent to the marriage act.

The reason is that the marriage act, because of its fundamental structure, while uniting husband and wife in the closest intimacy, actualizes their capacity to generate new life,—and this as a result of laws written into the actual nature of man and woman. And if each of these essential qualities, the unitive and the procreative, is preserved, the use of marriage fully retains its sense of true mutual love and its ordination to the supreme responsibility of parenthood to which man is called. We believe that our contemporaries are particularly capable of seeing that this teaching is in harmony with human reason.

22

We take this opportunity to address those who are engaged in education and all those whose right and duty it is to provide for the common good of human society. We would call their attention to the need to create an atmosphere favourable to the growth of chastity in such a way that true liberty may prevail over licence and the norms of the moral law be fully safeguarded.

Everything therefore in the modern means of social communication which arouses men's baser passions and encourages low moral standards, likewise every obscenity in the written word and every form of indecency on the stage and screen, should be condemned publicly and unanimously by all those who have at heart the advance of civilization and the safeguarding of the outstanding values of the human spirit. It is quite absurd to defend this kind of depravity in the name of art or of culture or by pleading the liberty which may be allowed in this field by the public authorities.

[1]From Horgan, J., ed. *Humanae Vitae and the Bishops.* (1972, p. 33 ff.) (footnotes omitted).

whom giving up sex is an end in itself (because sex is undesirable) and others for whom its renunciation is for some worthy cause. Throughout its long history, the Catholic church has had to struggle with reconciling its advocacy of sexual renunciation with its endorsement of procreation. Thus even though many major Church figures have maintained asceticism to be the highest form of sexual morality, the complete rejection of sex has never been a central Christian doctrine (chapters 17 and 18).

The Case for the Conservative Moral Perspective

The case for the conservative moral view as expounded by the Catholic church rests on *biblical* grounds, the traditional *teachings* of the church, and its understanding of *natural law.*[7]

[7] See Dennehy (1981); Doherty (1979); Bouyer (1961) for expositions of the orthodox Catholic moral view.

Biblical Grounds   Since Jesus and his disciples were Jews, we may assume that they shared the Judaic sexual traditions of their time. The Old Testament, therefore (but not the Talmud), is part of the primary basis of Christian sexual morality. Even if one sets aside its ritualistic concerns (chapter 17) and the harsh treatment of sexual transgressors (e.g., stoning to death for adultery), the Old Testament can be said to stand for a conservative morality.

The New Testament basis of moral behavior is *love.* Yet it says very little about the views of Jesus on sexual matters. Jesus was chaste and celibate. Sex was not even among the temptations that he confronted in the wilderness. But Jesus did not shun the company of women, including prostitutes, and because of the special circumstances of his life, we cannot assume that he intended his own celibacy to serve as a general model. The fact that Jesus attended the wedding of Cana (where he turned water into wine) is taken as evidence that he approved of marriage; his fondness for children also is clearly attested to in numerous passages.

The few explicit statements Jesus made with respect to sexual behavior dealt with adultery (figure 20.3). When confronted with a woman caught in adultery, Jesus shamed her accusers ("let him who is without sin among you be the first to throw a stone at her") and then told the woman, "Neither do I condemn you; go and do not sin again" (Matt. 8:3–11). But then in the Sermon on the Mount, Jesus seems to be setting an impossibly high standard: "You have heard that it was said, 'You shall not commit adultery.' But I say to you that everyone who looks at a woman lustfully has already committed adultery with her in his heart" (Matt. 5:27–28).

Jesus similarly condemned divorce except on the grounds of "unchastity"; otherwise, every man who divorces his wife "makes her an adulteress; and whoever

Figure 20.3   Gustave Doré (1833–1883), *Jesus and the Woman Taken in Adultery.* (Illustration for *The Bible.*)

marries a divorced woman commits adultery" (Matt. 5:31–32).

These statements convey a highly demanding sense of sexual morality that is further elaborated by Saint Paul, the first major Christian figure to write explicitly on sexual behavior. By his personal example and various statements, Paul advocated chastity and celibacy; his acceptance of marital relations was a concession to human frailty and a safeguard against fornication.

> If any one thinks that he is not behaving properly toward his bethrothed, if his passions are strong, and it has to be, let him do as he wishes: let them marry—it is no sin (1 Cor. 7:36).

> To the unmarried and the widows I say that it is well for them to remain single as I do. But if they cannot exercise self-control, they should marry. For it is better to marry than to be aflame with passion (1 Cor. 7:8—9).

> It is well for a man not to touch a woman. But because of the temptation to immorality, each man should have his own wife and each woman her own husband (1 Cor. 7:1–3).

Paul counseled those who were married to remain together and to "be subject to one another out of reverence for Christ" (Eph. 5:21). "The husband should give to his wife her conjugal rights, and likewise the wife to her husband. For the wife does not rule over her own body, but the husband does; likewise the husband does not rule over his own body, but the wife does" (1 Cor. 7:3–5). Given his qualified endorsement of even marital sex, St. Paul was predictably opposed to all other forms of sexual activity.

Taking the particular texts of the New Testament that form its doctrine on sexuality, Noonan subsumes them under eight themes:

> the superiority of virginity; the institutional goodness of marriage; the sacral character of sexual intercourse; the value of procreation; the significance of desire as well as act; the evil of extramarital intercourse; and the unnaturalness of homosexuality; the connection of Adam's sin and the rebelliousness of the body; the evil of "medicine."[8]

Taken at face value, the New Testament passages dealing with sex provide a strong base of support for conservative morality.

The Teaching of the Church   Christian teachers, Catholic or Protestant, have overwhelmingly espoused a conservative sexual morality. Even the most conservative views now present would be quite liberal compared to those of the Church Fathers (chapters 17 and 18). It is only very recently that Christian voices of any credibility have advocated a more liberal perspective in sexual morality. So Church tradition over 20 centuries is squarely behind sexual conservatism.

Natural Law   Arguments from "natural law" have been an integral part of Catholic moral teaching. The Christian conception of the *law of nature* goes back to Paul ("When the Gentiles who have not the law to do by nature what the law requires, they are a law to themselves," Rom. 2:14). Elaborated by the Stoics, Church Fathers, and systematized by Saint Thomas Aquinas, this concept has been used as a way of integrating theological doctrines and naturalistic observations. Its underlying assumption is that living patterns uncontaminated by human sin and error are *natural*. Thus the sexual processes evident in plant and animal life represent natural models by which to judge human behavior. Similarly the organs of the human body have their self-evident natural functions; eyes are for vision, genitals for reproduction. Based on these models procreation is seen to be the only natural sexual function.

---

[8]Noonan (1967, p. 56). "Medicine" refers to various herbs and drugs associated with sorcery, conceivably including contraceptives and abortifacient potions.

The Church places great importance on the consistency of its views on sexual morality over the ages. But this commitment to conserve fundamental values does not mean that the Church is unwilling to recognize current realities and adapt its teachings to them as long as there is no violation of its hallowed doctrines. In dealing with the issue of birth control, for instance, *Humanae Vitae* takes note of the rapid increase in world population and the difficulty for many to provide for large families.[9]

In sum, the Church would claim that its moral doctrines represent God's will for human sexuality whatever the historical circumstances. Its doctrines are not against sex but against its sinful uses. Furthermore, the Church recognizes human frailty and approaches sexual transgressions with compassion and forgiveness as long as people are willing to recognize the error of their ways.

The defense of the conservative viewpoint among Protestants, such as Evangelical Christians, is based almost exclusively on their understanding of the Bible. But except for their acceptance of contraception, their moral prescriptions would be consistent with the Catholic view: marriage is the only legitimate setting for coitus and within those bounds sex is good and honorable.

The secular arguments in favor of the conservative moral view, such as in communist societies, are based on the general *welfare of the community* and the individuals within it. Since the representatives of the state know best how the interests of the community are to be served, people are expected to live within those bounds.

The Case against the Conservative Moral Perspective

Many reject the conservative moral perspective because they neither accept the authority of the Bible nor of the church. But opposition to traditional sexual ethics

is not always based on wholesale rejection of Christian values. There are many Christians, including some within the Catholic church, who do not accept the conservative stance as representing the authentic Christian message for today's world.

The Biblical Record   There are highly competent theologians, some of them Catholics, who disagree with the interpretations of the moral conservatives about the meaning of various biblical passages dealing with sexuality. Homosexuality provides a good example of such conflicting interpretations of the same biblical texts.

A frequently cited passage in condemnation of homosexuality is the story of the destruction of Sodom in Gen. 19 (figure 20.4).[10] The key phrase on which rests the sexual interpretation of the story is the demand that Lot bring out the two men he is sheltering so that the Sodomites may "know them." If to "know" is interpreted in sexual terms, then there is obviously a homosexual theme to the story. Lot's desperate offer of his two daughters to protect his guests from the mob adds a further sexual tone of the episode. So one may surmise that the "sin of Sodom" that so angered the Lord must have been homosexuality, and the attempt to rape the angels was the final inequity that brought down on the Sodomites the wrath of the Lord.

Other biblical scholars reject the sexual connotation to the story of Sodom (except for the offer of Lot's daughters). Bailey points out that the Hebrew verb "to know" occurs 943 times in the Bible, and only in 10 cases is it used in a sexual sense and in none of them (except for the presumed case of Sodom) with respect to homosexual sex. He interprets "to know" in this case in the literal sense of "becoming familiar with." The people of Sodom were angry at Lot

[9]*Humanae Vitae* (1972), p. 34.

[10]This is where the term *sodomy* comes from. There is no term for *homosexuality* in the Bible, nor in any of the languages in which the original biblical texts were written. (Boswell, 1980, pp. 92–93.)

because he was harboring strangers when he himself was a stranger in town, and thus forced Lot to violate the crucial law of hospitality. So Bailey suggests that the wickedness of Sodom was inhospitality, not homosexuality.[11]

There are two places in the Old Testament that clearly refer to homosexual acts between men. These are Leviticus 18:22 and 20:13, which respectively state: "You shall not lie with a male as with a woman; it is an abomination," and "If a man lies with a male as with a woman, both of them have committed an abomination; they shall be put to death, their blood is upon them."

These unequivocal condemnations are explicit enough, but the question is whether the problem was homosexuality as such or some special aspect of it. In the verses preceeding Leviticus 18:22, there is an equally firm prohibition against approaching "a woman to uncover her nakedness while she is in her menstrual uncleanliness" (Leviticus 18:19). The people of Israel shared with the ancient world a certain awe with regard to sexual discharges, particularly blood and semen, the elements out of which came life. Thus just as the demand for cultic purity was manifested in prohibitions against having sexual intercourse with a woman in her menstrual period, similar considerations may have been applied to contact with semen in the context of homosexual activity between men.[12] The fact that no reference is made to homosexual behavior between women further reinforces this view.

So far as we know, Jesus said nothing about homosexuality (even though he does

Figure 20.4 Gustave Doré, *Destruction of Sodom.* (Illustration for *The Bible Gallery*.)

refer to Sodom). But there are several statements in the writings of Paul that have a clear bearing on this issue: I Cor. 6:9 states, "neither the immoral or idolators, nor adulterers, nor sexual perverts, nor thieves, nor the greedy, nor drunkards, nor revilers, nor robbers will inherit the Kingdom of God"; I Tim. 1:10 similarly cites "Sodomites"; and Rom. 1:26–27 states, "For this reason God gave them up to dishonorable passions. Their women exchanged natural relations for unnatural, and the men likewise gave up natural relations with women and were consumed with passion for one another. Men committing shameless acts with men and receiving in their own persons the due penalty for their error."

For some, these passages clearly indict homosexual acts as immoral. But questions are raised by others as to the precise meaning of key terms in each. For example, the term translated as "sexual per-

---

[11]Bailey (1955). Inhospitality may not sound like such a grave offense to our ears, but within the nomadic tradition of the Middle East it was one of the most crucial norms of behavior. Unless travelers could seek shelter and protection in alien territory, no one would be able to travel very far.

[12]Kosnik et al. (1977, pp. 9–11). Stuhlmueller (1979) rejects this explanation and brings in other biblical evidence to show that condemnations of unaccepted sexual practices were based on perceptions of their intrinsic evils rather than being simply damned because of their associations with cultic purity.

vert" in Cor. 6:9 is a Greek word that means "soft." According to Boswell, it is a common term which occurs elsewhere in the New Testament meaning "sick"; in patristic writings it was used for "cowardly," "refined," "weak-willed," "delicate," "gentle," and "debauched." In more specifically moral contexts it frequently meant "licentious," "loose," or "wanting in self-control." The term may also be broadly translated either as "unrestrained" or "wanton." Boswell concludes that "to assume that either of these concepts necessarily applies to gay people is wholly gratuitous. The word is never used in Greek to designate gay people as a group or even in reference to homosexuals generally and it often occurs in writings contemporary with the Pauline episodes in reference to heterosexual persons or activity."[13]

The Teaching of the Church Traditionally we have sought teachers who have combined experience and wisdom with integrity and dignity. Now we rely more on experts whose primary qualification is empirical knowledge and whose popularity rests on their ability to make that knowledge readily accessible.

Those who reject the moral wisdom of men like Jerome and Augustine may accept its relevance for the past but no longer for the modern world; others would wish that their advice had been disregarded altogether. Here are two men, one could say, who led (what they considered) dissolute lives in their youth and then as a reaction to their own internal conflicts perpetrated an oppressive sexual morality with crippling effects on the lives of others. Or, one could argue, even if they did wonders for their time, are their moral views more relevant to the modern world than pre-Copernican conceptions of the universe to modern astronomy?

Modern teachers of conservative sexual morality are likewise criticized for refusing to adapt moral doctrines to current realities and for failing to sufficiently take into account the opinions and experiences of the Christian men and women whose sexual lives are being regulated. For instance, as far as is known, not a single woman has ever had any significant role in shaping Catholic doctrine on contraception.

Catholic opposition to contraception has been a particularly sore point both within the church and outside it, given the lack of clear biblical authority in support of it (hence the approval of contraception by Evangelical Protestants), and the enormous burden placed by the ban on individuals and society. In allowing for "natural" forms of birth control, the Church appears to some to be compassionate, and to others to be inconsistent. It allows Catholics to avoid pregnancy while remaining faithful to the teaching of the church. But when a couple uses the rhythm method to avoid pregnancy, does that not constitute a deliberate and conscious effort to separate the reproductive from the unitive aspects of sex? What is the difference between shielding yourself from a ball coming at you and ducking it?

Those concerned by the effects of overpopulation are particularly outraged at the Catholic opposition to birth control. Though *Humanae Vitae* acknowledges the problem of overpopulation, it offers no tangible solution to it. Furthermore, the ban on contraception seems to have a thinly veiled secondary purpose of discouraging "immorality," since it is feared that legitimizing birth control would "lead to the way being wide open to marital infidelity and the general lowering of moral standards. Not much experience is needed to be fully aware of human weakness and to understand that men—and especially the young, who are so exposed to temptation—need incentives to keep the moral law, and it is an evil thing to make it easy for them to break the law."[14] But by the same token, should we also not treat venereal disease, because it too can conceivably act as a deterrent to immorality?

---

[13]Boswell (1980, p. 107). For Boswell's views concerning the other references in Paul, see p. 107.

[14]*Humanae Vitae* (1972), p. 43.

Figure 20.5 Gustave Doré, *Devils and Seducers*. (Illustration for Dante's *Divine Comedy*.)

Natural Law Biologists and investigators of animal behavior would not quarrel with the assumption that what prevails in the animal world has relevance to human behavior. But naturalistic observation of animal behavior shows that they engage in all sorts of sexual activities that have no reproductive potential, including masturbation and sexual contacts between members of the same sex. So, if by "natural" we mean what takes place in nature, then these activities must be considered natural for humans as well.

Even within the more limited knowledge of biology of earlier times, the position of the church was not supportable by the available empirical evidence. Likewise, it was deemed acceptable to tamper with nature, such as by damming up rivers, but not when it involved blocking reproduction. It seems, therefore, that the church has used the natural law argument not to seek natural reality but to bolster its theological position by using spurious examples from nature.

The Liberal Moral Perspective

The liberal approach to sexual morality characterizes the perspective of most Prot-estants and Jews, liberal Catholics, and the secular moral beliefs of probably the majority of middle-class Americans. Its overall approach to sex is more flexible, and it rejects the images of gloom and damnation so often associated with traditional views of sex and portrayed in Christian-inspired art (figures 20.5; 20.6; 20.7).

The liberal perspective is more difficult to characterize than the conservative view since it has no common voice, no papal encyclical to speak for it. Sandwiched between the conservative and the humanistic positions, it overlaps with both and hence embodies a wider range of options than either of the other two.

At its conservative end, the liberal view relies on traditional sources of authority like the Bible or church tradition, but it interprets these sources more liberally; that is, behavior continues to be judged by predetermined rules but the rules are less strict. At its liberal end, it relies more on individual judgment to determine the morality of a particular sexual act. Since the exercise of individual choice depends on the situational circumstances or context in which the behavior takes place, this perspective has come to be commonly known as *situation ethics*. As the conservative stand is

Figure 20.6 Gustave Doré, *Paramours and Flatterers Smothered in Filth.* (Illustration for Dante's *Divine Comedy*.)

Figure 20.7 Gustave Doré, *The Lustful.* (Illustration for Dante's *Divine Comedy*.)

traditional, the liberal alternative is said to be *modern;* hence the **new morality** as a synonym for situation ethics. But this does not mean that its adherents denigrate or deny all of the past anymore than conservatives cling to tradition as an anachronism from the past.

In behavioral terms, marital sex re-emerges in the liberal Christian view as the optimal sexual experience, but this time with no strings attached. Procreation continues to be highly valued but there is no obstacle to contraception. Sex remains linked with love, but sexual pleasure as such is also accepted as being a good thing in itself and a couple is free to use whatever reasonable means they choose to attain it.

Masturbation is acceptable for youth, but viewed more ambivalently for adults unless used as an unavoidable substitute for coitus. Premarital sex is viewed with cautious tolerance. Homosexual and extramarital relations are more suspect. In short, nonmarital relationships may be possibly considered moral but the ethical cards are generally stacked against them.

The differences between the conservative and liberal religious perspectives are sharper at the doctrinal level than at their application at the behavioral level: conservative clergy are more flexible in dealing with their parishioners than one would surmise from the teachings of their theological mentors, while liberals behave more conservatively than they talk. The conservative absolutists bend their rules while liberal relativists make up rules, thereby mitigating the less workable aspects of each system.

The Case for the Liberal Perspective

The justification for the liberal perspective starts with the contention that the conservative approach does not work, or that it exacts an inordinate price in human freedom and happiness. The conservative Christian moral perspective is seen as too legalistic. It does not reflect the lucidity and compassion of the original and authentic Christian message. Biblical passages are used to score points rather than to get to the truth; there is too much reliance on theological abstractions and not enough trust in the sincerity and competence of the individual Christian conscience to sort right from wrong. As a result, people either have to behave against their better judgment or violate the precepts of the church: the ban on contraception is a prime example.

To remedy these shortcomings, the liberal perspective focuses on human sexual needs and their most realistic management and satisfaction. Laws are made to serve people instead of people made to serve laws. In defense of this approach, we first examine the views of liberal Catholic theologians, then turn to the perspective of Protestant ethicists.

New Directions in Catholic Morality   Reflective of the ferment within Catholicism for a more liberal approach to sexual morality are the conclusions of a study commissioned by the Catholic Theological Society of America.[15] In a marked departure from the more traditional Catholic approach, the authors of the report took a far less absolutist view: "Morality must never allow itself to be reduced to a simple external conformity to prejudged and pre-specified patterns of behavior. For this reason, we find it woefully inadequate to return to a method of evaluating human sexual behavior based on an abstract absolute predetermination of any sexual expressions as intrinsically evil and always immoral."[16]

[15]The study committee was chaired by Father Anthony Kosnik, a professor of moral theology. For the report, see Kosnik et al. (1977). For a harsher view of traditional Catholic approaches to sexuality see Ginder (1975).

[16]Kosnik et al. (1977), p. 89.

Box 20.2   Sexual Values Conducive to Creative Growth and Personal Integration[1]

(1) *Self-liberating:* Human sexuality flows freely and spontaneously from the depth of a person's being. It is neither fearful nor anxious but rather genuinely expressive of one's authentic self. . . . There is a legitimate self-fulfillment that sexual expression is meant to serve and satisfy. To deny this is unrealistic and a contradiction of universal human experience. Too frequently in theological literature sexual union is seen as a sign and expression of the total gift of self to another. Little attention is given to the element of wholesome self-interest that must be part of authentic human sexual expression. . . . For this reason, it is a serious distortion to speak of the sexual relationship exclusively in terms of expressing a totally altruistic giving of self to another.

(2) *Other-enriching:* Human sexuality gives expression to a generous interest and concern for the well-being of the other. It is sensitive, considerate, thoughtful, compassionate, understanding, and supportive. It forgives and heals and constantly calls forth the best from the other without being demeaning or domineering. This quality calls for more than mere non-manipulation or non-exploitation of others against their will. It insists that wholesome sexuality must con-

tribute positively to the growth process of the others.

(3) *Honest:* Human sexuality expresses openly and candidly and as truthfully as possible the depth of the relationship that exists between people. It avoids pretense, evasion, and deception in every form as a betrayal of the mutual trust that any sexual expression should imply if it is truly creative and integrative.

(4) *Faithful:* Human sexuality is characterized by a consistent pattern of interest and concern that can grow ever deeper and richer. Fidelity facilitates the development of stable relationships, strengthening them against threatening challenges. In marriage, this fidelity is called to a perfection unmatched at any other level and establishing a very special, distinct, and particular relationship. Even this unique relationship, however, should not be understood as totally isolating a spouse from all other relationships thereby opening the way to jealousy, distrust, and crippling possessiveness.

(5) *Socially responsible:* Wholesome human sexuality gives expression not only to individual relationships but in a way also reflects the relationship and responsibility of the individuals to the larger community (family,

---

Yet this "new direction" in Catholic thought stops short of trusting individual judgment with no further prescriptive rules, because that would place morality at the mercy of individual preferences, dispositions, and the mood of the moment.

In an attempt to obviate the pitfalls of a purely *objective* absolutism or a purely *subjective* relativism, the proponents of this view start with a number of presuppositions from which they derive the criteria that should assist the person in making moral judgments. Their basic requirements for any sound system of moral evaluation of sexual behaviors are as follows:

First, we must consider both the "objec-

tive and subjective" aspects of the sexual behavior—what the person thinks and what others think about it. Second, the "complexity and unity" of the person's sexual nature must be acknowledged without establishing hierarchies among its components. The procreative aspects of sex must not be set in competition with its "creative and integrative" components. Third, we must have a constant awareness of the "interpersonal dimension" of behavior, the effects on the other person or persons. In sum, it is the motivation of the person, as well as his or her acts, which provides the basic principles from which to evaluate the morality of sexual behavior.

Box 20.2   Sexual Values Conducive to Creative Growth (*Continued*)

nation, world). Both the historical and empirical data indicate that every society has found it necessary to give direction and impose restrictions on the expression of human sexuality in the interests of the common good. This characteristic, however, goes far beyond what is required for the good order of society. Law cannot be expected to legislate personal morality. What is required here is that people use their sexuality in a way that reveals an awareness of the societal implications of their behavior and in a manner that truly builds the human community. At times, this may mean a willingness to forego personal benefit and growth in order to preserve or promote the greater good of society.

(6) *Life-serving:* Every expression of human sexuality must respect the intimate relationship between the "creative" and "integrative" aspects. And every life style provides means for giving expression to this life-serving quality. For the celibate and the unmarried, human sexuality may find expression in a life of dedicated service to people through church or society. For the married, this life-serving purpose will generally be expressed through the loving procreation and education of children. In most cases, this openness and readiness will lead

marriage partners to become generous cooperators with God in the task of responsibly transmitting life. In some instances, however, this value could mean that a responsible interpretation of God's will leads to a life-serving decision not to beget children. In both instances, the "creative" and "integrative" aspects of human sexuality will be harmonized in the overriding life-serving orientation of the sexual expression. Full sexual expression with an accompanying abortive intent should procreation ensue would be a clear contradiction of this life-serving quality of human sexuality.

(7) *Joyous:* Wholesome sexual expression should be witness to exuberant appreciation of the gift of life and mystery of love. It must never become a mere passive submission to duty or a heartless conformity to expectation. The importance of the erotic element, that is, instinctual desire for pleasure and gratification, deserves to be affirmed and encouraged. Human sexual expression is meant to be enjoyed without feelings of guilt or remorse. It should reflect the passionate celebration of life, which it calls forth.

[1]From Kosnik et al. *Human Sexuality: New Directions in American Catholic Thought.* (1977, pp. 92–95) (footnotes omitted).

In the practical applications of this approach, instead of simply asking or trying to answer the more standard question, "Is this act moral or immoral?" one ought to ask whether the specific sexual behavior in question realizes those values that are "conducive to creative growth and integration of the human person." Among such values, the ones that are deemed particularly significant are presented in box 20.2. These values are meant to allow for the free gratification of one's own sexual needs, physical as well as psychological, while helping provide these same benefits to the partner in sexual relationships that are honest, faithful, socially responsible, "life-

serving," and joyous.[17]

Experiential Bases of Moral Judgment
Protestant moral theologians often apply the essential teachings of Christianity to contemporary realities by combining theological insights with the secular perspectives of the natural and behavioral sciences. A good representative of this approach is James Gustafson, who argues that "there are fundamental bases in human nature and experience which neces-

[17]The behavioral applications of the liberal Catholic approach are discussed under pastoral guidelines in Kosnik et al. (1977, pp. 99–239).

sarily must be taken into account" in formulating or reformulating sexual morality.[18] Traditional Christian ethics of sex and marriage had an experiential foundation; hence whatever modifications that modern circumstances may suggest, ought to be considered in the light of the bases on which the more traditional formulations were made in the first place.

Gustafson proposes three aspects of our experience as sexual beings which must be taken into account in moral judgments. First is the fact that sex is part of our *biological and personal nature*. The biological aspect of sexuality is self-evident but sex is just as importantly a vehicle for expression of psychological needs (chapter 12). A key ethical issue then concerns this relationship between the biological and personal components in human sexuality. Moral codes which fail to take into sufficient account the legitimacy of either are not likely to be successful guides.

The second fundamental issue in sexual ethics pertains to the *reality of sin* in terms of the human propensity to cause harm to others. While the sexual liberation of modern times may have sought to banish the notions of sin and guilt, the reality of human relations continues to attest to their relevance, even though we may now refer to them by other terms like "sexual exploitation." Such terms approximate "one of the most profound traditional meanings of immorality—to use another person for purely selfish ends, to fail to recognize that respect is due to another person as a human agent with capacities to determine his or her own sense of purpose and well-being. It is to make another instrumental to one's own end, and to manipulate others to consent to fulfilling that end. The other becomes an object rather than a person. Sexuality, like other aspects of human relations, is deeply subject to exploitation, and exploitation of others is immoral."[19]

The third issue to be taken into account in any system of sexual ethics is that of *covenant:* the "commitments which persons make to one another, implicit ones that are subject to deception and self-deception and exploitation, and explicit ones such as marriage vows." Such commitments entail the willingness to accept accountability for one another and for the consequences of sexual relations. Rather than simply being externally imposed rituals or contracts, covenants are based on a profound human need, to be related to one another over time in a way that allows the well-being of each and the well-being of both together to be sustained.[20]

Traditional marriage services, as exemplified by the very old service from the Church of England (box 20.3) formalizes the covenant between bride and groom in the presence of the congregation. The steps that are entailed in this process are as follows. Public announcement ("publication of banns") of the impending marriage inquire if anyone has serious objection to the proposed union. During the service, the clergyman's address to the congregation states the basic purposes of marriage: the procreation and nurture of children, the satisfaction of sexual needs; the help, comfort, and companionship to be provided by husband and wife to each other, both in prosperity and adversity. The questions put to the couple and to the bride and groom individually, elicit from them a public acknowledgment of their voluntary commitment and the agreed-upon conditions to love, comfort, honor, and be faithful to each other. The "giving of the bride" signifies the break with the family of origin (a consideration that now applies to the groom as well) and the establishment of an independent family of one's own. The vows undertaken by the couple finalize their marital commitment, which is symbolized by the exchange of rings.

This ceremony is cast within the beliefs and symbols of Christianity and reflects the historical and social contexts within which it was formed. But, as Gustafson argues, the ends it is meant to serve are not arbitrary. Even if changed social circumstances require the revision and modification of

---

[18]Gustafson (1981, p. 490).
[19]Gustafson (1981, p. 490).

[20]Gustafson (1981, p. 491).

various aspects of it, its fundamental purposes remain valid with or without its religious justification.

Such commitments provide a setting of reliable trust and confidence as well as an understanding of the duties and obligations that are essential to our well-being. Even though they also function as a means of checking and controlling sexual behavior, they are meant to facilitate and not hinder human sexual relations. As with the other two bases for sexual ethics, covenants are based upon human experience and biological, social, and personal needs that remain constant even though the particulars of the social conventions and institutions through which we obtain our sexual satisfactions may change.[21]

Situation Ethics   In the late 1960s a widely read Christian exponent of the liberal approach to sexuality was Joseph Fletcher, an Episcopalian clergyman. The title of his first book, *Situation Ethics: The New Morality*, provided the labels, and its message, oversimplified by others, lent a moral voice to what many youths were already practicing at the time.[22]

Fletcher places the main burden of moral choice on the individual. The morality of choices are relative to their contexts; hence, situation ethics is also known as **contextual ethics.** In other words, whether a given sexual behavior is right or wrong cannot be determined in the abstract; it all depends on the circumstances. In principle, situation ethics is liberal only in the sense of allowing freedom of choice. In practice, it also turns out to be liberal in the sense of being sexually permissive, since the freer people are to chose, the more they choose to behave sexually freely. Sexual rules, like behavioral rules in general, are basically prohibitive—they tell us what not to do. So the more rules there are in an ethical system, the more sexually restrictive that system is likely to be. However, the fact that Fletcher sets no predetermined norms

as to whether a given sexual behavior is good or bad, does not mean that he expects people to do as they please. To help steer sexual conduct in the right direction, he provides certain guidelines within which to exercise moral judgment.

Fletcher builds his defense of situation ethics on four general principles. The first is *pragmatism:* what is considered lawful must be expedient, constructive, and workable; not some lofty principle that cannot be reasonably put to practice in everyday life by most people. Second, a moral code should be *relativistic,* meaningful only in a specific context, subject to given circumstances. The third principle is *positivism:* moral decisions are "posited" or affirmed by faith; one starts with faith in God and a commitment to love others; reason then determines the application of this commitment to sexual behavior. The fourth presupposition is *personalism:* people, not abstract values, are the center of concern. Laws are obeyed when they serve a good purpose, broken when they do not.

Buttressed by these presuppositions, the cardinal principle of the situational ethical approach is *love.* This love is defined as *agape,* the selfless Christian love of "neighbor"; not *erotic* love, not even the *philia* of friendship.[23]

Selfless love that seeks nothing in return is intrinsically good. Nothing else, as such, has any value, any ethical quality. If love is the only good, then only love and nothing else can be the norm ruling behavior. Love thus becomes the single regulative principle of Christian ethics, the only principle that should guide our conscience. Thus, the only "rule" to follow in seeking to act morally is to be motivated by love toward the people who are involved in the sexual interaction or who stand to be affected by it.

Love and justice are the same; "justice is love distributed, nothing else." To serve as justice, love must be rational, discriminating, and careful, as well as caring. Many moral dilemmas arise from conflicts be-

---

[21]For other Protestant views see Thielicke (1964); *Towards a Quaker View of Sex* (1964); Feucht (1961).

[22]Fletcher (1966).

[23]Fletcher (1966); (1967).

Box 20.3   The Form of Solemnization of Matrimony[1]

Dearly beloved, we are gathered together here in the sight of God, and in the face of this company, to join together this Man and this Woman in holy Matrimony; which is an honourable estate, instituted of God, signifying unto us the mystical union that is betwixt Christ and his Church: which holy estate Christ adorned and beautified with his presence and first miracle that he wrought in Cana of Galilee, and is commended of Saint Paul to be honourable among all men: and therefore is not by any to be entered into unadvisedly or lightly; but reverently, discreetly, advisedly, soberly, and in the fear of God. Into this holy estate these two persons present come now to be joined. If any man can show just cause, why they may not lawfully be joined together, let him now speak, or else hereafter for ever hold his peace.

*

I require and charge you both, as ye will answer at the dreadful day of judgment when the secrets of all hearts shall be disclosed, that if either of you know any impediment, why ye may not be lawfully joined together in Matrimony, ye do now confess it. For be ye well assured, that if any persons are joined together otherwise than as God's Word doth allow, their marriage is not lawful.

*

N. wilt thou have this Woman to thy wedded wife, to live together after God's ordinance in the holy estate of Matrimony? Wilt thou love her, comfort her, honour, and keep her in sickness and in health; and, forsaking all others, keep thee only unto her, so long as ye both shall live?

I will.

N. wilt thou have this Man to thy wedded husband, to live together after God's ordinance in the holy estate of Matrimony? Wilt thou love him, comfort him, honour, and keep him in sickness and in health; and, forsaking all others, keep thee only unto him, so long as ye both shall live?

I will.

I N. take thee N. to my wedded Wife, to have and to hold from this day forward, for better for worse, for richer for poorer, in sickness and in health, to love and to cherish, till death us do part, according to God's holy ordinance; and thereto I plight thee my troth.

I N. take thee N. to my wedded Husband, to have and to hold from this day forward, for better for worse, for richer for poorer, in sickness and in health, to love and to cherish, till death us do part, according to God's

---

tween two equally desirable aims. When one loves two persons simultaneously, such as in extramarital relations, how does one do justice to both? Morality in such cases involves cost-benefit analyses where the integrity and interest of everyone involved is protected as best as possible. Finally, love as an end justifies its means. This runs counter to the more traditional view that the end does not justify the means. Actions, as such, are neither moral nor immoral; their morality is determined by their motivation and their anticipated consequences. In sexual relations, it is not the act but why one engages in it and whether or not it serves the criterion of its being a loving act that determine its morality.[24]

Secular moralists within the liberal perspective similarly invoke "love" as the basis for moral action, but they use the term in the sense of commitment and loyalty, not as agape. Respect for the rights of

[24]Fletcher (1966).

Box 20.3   The Form of Solemnization of Matrimony[1] (*Continued*)

holy ordinance; and thereto I give thee my troth.

*

With this Ring I thee wed: In the Name of the Father, and of the Son, and of the Holy Ghost. Amen.

Bless, O Lord, this Ring, that he who gives it and she who wears it may abide in thy peace, and continue in thy favour, unto their life's end; through Jesus Christ our Lord. Amen.

*

O Eternal God, Creator and Preserver of all mankind, Giver of all spiritual grace, the author of everlasting life; Send thy blessing upon these thy servants, this man and this woman, whom we bless in thy Name; that they, living faithfully together, may surely perform and keep the vow and covenant betwixt them made, (whereof this Ring given and received is a token and pledge), and may ever remain in perfect love and peace together, and live according to thy laws; through Jesus Christ our Lord. *Amen.*

*

O almighty God, Creator of mankind, who only art the well-spring of life; Bestow upon these thy servants, if it be thy will, the gift and heritage of children; and grant that they may see their children brought up in thy faith and fear, to the honour and glory of thy

Name; through Jesus Christ our Lord. *Amen.*

*

O God, who hast so consecrated the state of Matrimony that in it is represented the spiritual marriage and unity betwixt Christ and his Church; Look mercifully upon these thy servants, that they may love, honour, and cherish each other, and so live together in faithfulness and patience, in wisdom and true godliness, that their home may be a haven of blessing and of peace; through the same Jesus Christ our Lord, who liveth and reigneth with thee and the Holy Spirit ever, one god, world without end. *Amen.*

Those whom God hath joined together let no man put asunder.

*

Forasmuch, as *N.* and *N.* have consented together in holy wedlock, and have witnessed the same before God and this company, and thereto have given and pledged their troth, each to the other, and have declared the same by giving and receiving a Ring, and by joining hands; I pronounce that they are Man and Wife, In the Name of the Father, and of the Son, and of the Holy Ghost. Amen.

[1]Excerpts from the *Book of Common Prayer*, (Oxford University Press, n.d., pp. 300–304). Instructions to the couple and related comments omitted.

---

others, "no one getting hurt," and similar considerations further aid their moral decisions.

Less stringent in its demands and less formal in articulating its principles, the liberal secular code operates more as a do-it-yourself sort of morality where personal interests are tempered by altruism, feelings of decency, and fear of social disapproval. Additional constraints on sexual behavior come from secular ideologies (democratic respect for individual free-

dom, Marxist responsibility to the community, feminist opposition to the exploitation of women); class-based behavioral codes (acting like a "lady" or "gentleman"), loyalty to national, ethnic and family tradition ("It's un-American"; "it is not macho"; "our family does not do that sort of thing") and conventional wisdom ("don't mess around with the wives/husbands of your friends, or the friends of your wife/husband"; "don't look for your honey where you make your money").

The Case against the Liberal Perspective

Being in the middle gives the liberal perspective the advantage of playing the golden mean, but it also makes it vulnerable to attack from both flanks.

The Conservative Critique   Conservatives fault the liberal religious perspective on both conceptual and practical grounds. They point out that secular methods of analysis cannot be indiscriminately applied to the Bible without compromising its religious integrity. What the Bible says about sexual morality must be taken at face value as representing eternal truths that are just as valid today as when they were enunciated. Otherwise, why bother with it? Why not let each generation figure out its own rules? If there is going to be a Christian sexual ethic, it must have a consistent core of moral teaching, rather than a collection of shifting cultural views. Given the repeated attempts over 20 centuries to dilute and distort the fundamentals of the Christian message, the church could hardly be blamed for putting matters in as unambiguous ways as possible to guard the integrity of its moral teaching.

The liberal contention that we must judge *motivation* and not *acts* makes no practical sense: Acts are the only concrete entities we can judge; motivation is at best nebulous. It is what people do, not why they do it, that matters. How else can moral judgments function except by declaring some behaviors good and others not? Moral laws exist for the purpose of avoiding sin and suffering; how can they be faulted for not being person-oriented?

Granted that the church is not the only expert on sexual behavior, who else is? What exactly are secular experts saying on sexual morality that the church is not listening to? Do their views derive from any special knowledge, or are their personal moral preferences put forth dressed up as professional expertise? By accepting these opinions, are those with a liberal moral perspective not being duped in giving up what is morally tried and true?

Christian love is indeed the basis of all moral choice and the exercise of individual conscience is the most reliable guardian we have to save us from error. But such reliance requires that there be a mature Christian conscience to begin with (or a well-developed ethical sense in secular terms). How many people have these attributes? Do teenagers know what *agape* is? Particularly when sexual passion giddies reason, how can we seriously expect people to judge their motivations and circumstances reliably? Without clear and concrete ethical guidance to bring to these situations, will people not improvise as they go along? And is not the effect of such equivocation on the part of moral teachers particularly damaging to the young, whose sexual drives are strong, whose self-control is in need of all the help it can get, and whose experiences are likely to affect them for the rest of their lives?

In sum, conservative moralists charge that the liberal viewpoint of sexual ethics substitutes secular concerns in place of Christian ethics and that in practical terms it reduces morality into a set of personal preferences. With all the talk about *agape* and Christian love notwithstanding, it simply makes it moral for people to behave as they please.

The Humanistic Critique   The main criticism from the humanistic point of view (discussed next) would be that the liberal approach is at heart not much different from its conservative counterpart. Whether one sets up a list of ethical rules or a guiding principle like love, the net effect is an unwarranted intrusion into people's lives to control and inhibit their sexual behavior.

If the conservative Catholic position makes the enjoyment of sex morally impossible, its liberal alternative makes it merely very difficult. Consider, for instance, the moral criteria proposed as the liberal Catholic yardstick. Pious affirmations aside, who can in all honesty measure up to them in actual practice? What other human endeavor is there which sets the moral threshold that high? As a thoughtful (and by no means radical) Catholic couple has put it, "There are few human acts of any kind that can be at a given moment self-liberating, other-enriching, honest, faithful, socially repsonsible, life-serving and joyous."[25] If applied to a business, such expectations would lead it to bankruptcy. So if this is the "new direction" for sexual morality, can it possibly point to anything other than failure and guilt with an added measure of shame, since one is not able to measure up to even the "liberal" standard?

The liberal perspective rejects the making of rules then substitutes for it more subtle and ambiguous forms of rule-making by replacing "You shall not" with "You probably should not." What is needed is not sexual reform but revolution; not just a slackening of rules, but the dumping of the whole baggage of ecclesiastical prerogative that presumes to tell people how to lead their sexual lives against their wishes and against all common sense.

The Humanistic Perspective

There have always been individuals who have chafed under the moral rules imposed by the church and society, but their lonely views have carried little weight in the social arena. Countless others however have simply disobeyed the expectation of conservative and liberal morality without making a public show of it. One can therefore say that the strength of the sexually restrictive codes has been in their articulation, while that of the opposing camp is in its implementation.

During the sexual revolution of the sixties (chapter 16) the voices of sexual dissidents and liberationists became more concerted. Like others who rejected the traditional values in society in other areas, they adopted the word *humanistic* to indicate the primacy of human beings in their alternative life styles and moral orientations.

The humanistic sexual perspective has its own religious and secular components. Radical Protestant and Catholic clergy (many of whom have left the church) have

been among the vanguard of the movement. The excerpt in Box 20.4, which represents the views of a Unitarian minister and sex educators, is fairly representative of this moral viewpoint. Nonetheless, the humanistic approach is primarily secular in thrust and its proponents are concentrated among the younger generation, among groups (like gays) whose sexual preferences run counter to traditional morality, and opponents of traditional gender roles. This perspective also appears to be the dominant moral orientation among professionals working within the field of human sexuality. But within this ethos of tolerance, one would find considerable variation in the level of permissiveness with which professionals in the field approach moral issues in their work and personal lives.

The humanistic perspective is, first and foremost, *libertarian* in outlook.[26] The right of freedom of individual action is extended far beyond the liberal position. One's primary responsibility is to be true to one's own needs and desire ("do your thing") and society has no right to impinge on that privilege except for preventing palpable harm to others.

The humanistic perspective is *pluralistic* in outlook at several levels. It does not assume the dominant sexual values within Western society to be superior to those of other cultures (box 20.5). In our own society, it rejects the imposition of white, middle class morality on other ethnic or socioeconomic subgroups. Similarly, it does not accept heterosexuality as the only normative sexual orientation but considers homosexuality as an equally moral form of sexual relationship. The increased social acceptance of this viewpoint is manifested in the public display of art works celebrating gay relationships (figure 20.8).[27]

The opposition to social control of sexual

[26]*Libertarian* should not be confused with *libertine*, which signifies irresponsible and hedonistic pursuit of sexual pleasure without moral restraints.

[27]That there is persistent resistance to this perspective became evident when the art work in figure 20.8 was vandalized shortly after being placed on the campus of Stanford University.

[25]Mohs and Mohs (1979, p. 160).

Box 20.4   The Humanistic Perspective on Marital Fidelity[1]

We consider traditional monogamy, with its rigid requirement for exclusive devotion and affection, even though hallowed by the theological concept of fidelity, to be a culturally approved mass neurosis. It should be clearly understood that we do not deny anyone the freedom to enter conventional marriage—an absolute convenant between two persons of the opposite sex. Each person has the freedom to make decisions for life affirmation according to his/her deepest convictions and highest priorities, and sometimes what may outwardly appear to be merely a conventional form of marriage may be for the two people involved the epitome of love, joy, and hope. It is possible for a man and a woman to be content and happy only with each other and family members. Indeed, we support all the efforts of various human relations disciplines to strengthen and make more rewarding conventional monogamy. What we do emphatically reject, however, is our society's sanction of this marital model as normative and supreme. We believe all civic and constitutional rights should be extended to personal lifestyles. We prefer a model of monogamy which celebrates co-marital intimacy and does not equate fidelity with sexual exclusiveness. For too long, traditional moralists have been passively allowed to pre-empt other conscientious lifestyles by propagating the unproven assumptions that we cannot love more than one person (of the opposite sex) concurrently; that co-marital or extramarital sex always destroys a marriage; that "good" marriages are totally self-contained and self-restrictive and sufficient; that only emotionally unstable people seek and need intimate relationships outside the husband-wife bond. We repudiate these assumptions and consider them half-truths at best. When these assumptions are dogmatically upheld by society as eternal truths we consider the phenomenon to be a cultural neurosis in the sense that the issue is predetermined, all nonconformists are castigated, and there is no openness to new experience in new contexts.

The semantics of conformity are intimidating. Relational innovators are constantly accosted with negative terms such as promiscuity, adultery, and infidelity. The word "promiscuous," for example, refers to peo-

behavior by rules and regulations gives the humanistic perspective an *antinomian* ("opposing the law") flavor. There is a refusal to morally condemn behaviors that are merely unconventional. Diagnostic terms and distinctions between "normal" and "abnormal" are rejected in favor of terms like "sexual minorities" to characterize people with unusual sexual preferences (chapter 8).

The humanistic perspective prides itself in being *sex-positive*. Sex is valued as a good thing in all its manifestations. Masturbation is not only condoned but encouraged. Premarital sex is taken for granted. Extramarital sex is accepted subject to a number of qualifications. Homosexuality is given full parity with heterosexuality. Non-coercive paraphilias are tolerated. Behaviors like rape are excluded from the realm of sexual behaviors and condemned. Differences in sexual desire are acknowledged and respected, but there is a clear preference for the sexually more active over the less active; virginity is no virtue; continence is no occasion for moral celebration.

The primary purposes of sex in the humanistic perspective are *erotic pleasure* and the enhancement of *personal intimacy* ("the

Box 20.4   The Humanistic Perspective on Marital Fidelity[1] (*Continued*)

ple who lack standards of selection, who are indiscriminate in sexual relations. It should be obvious that it's possible for a person to be sexually intimate with any number of persons chosen according to conscientious standards. The word is commonly used, however, as a judgment against anyone who has more than one socially approved sexual relationship, and especially in a double-standard way against women. This shows a mistaken emphasis on the quantity rather than the quality of interpersonal relationships. It also insists that people cannot have casual sexual experiences. Not all intimate relationships must have the same intensity. Millions of men and women are able to make rapid appraisals of others with whom they can exchange warmth without subsequent emotional strings attached.

Even the term "extramarital" is misleading in the context of open-ended marriage. For it is precisely *within* marriage rather than outside it that open-ended marriage incorporates the freedom for two spouses to enjoy multilateral sexual and friendship relations. "Extramarital" is an all-encompassing term referring to all forms of relationship,

usually sexual, with partners other than the spouse. "Co-marital" is a more appropriate term for open-ended marriages because it at least carries the connotations of togetherness and co-operation within the structure of marriage. Within such marriages the possibility of adultery is totally absent because such exclusion, possessiveness, and jealousy have no place in the relationship. "Adultery" is a theological judgment which can apply only to the restrictive type of covenant. When one partner breaks the vow of "to thee only do I promise to keep myself," a relationship of trust is broken and he or she is unfaithful. But it's also possible to create a model of marriage—a covenant—monogamous in the sense that it's based upon an intended lifetime commitment between two, but which nevertheless is open-ended because it does not exclude the freedom to have any number of intimate relationships with others.

[1]From Mazur, R. *The New Intimacy* (1973, pp. 12–14).

pleasure bond"). Reproduction is valued but no more than contraceptive responsibility. The guiding ethical principle is *permissiveness with affection.* The key terms in its ethical vocabulary are pleasure, acceptance, sharing, communication, negotiation, personal enhancement, nonsexist, nonexploitative, honest, affectionate, and commitment.

Attitudes toward marriage vary. Though by no means a precondition to either erotic enjoyment or sexual intimacy, it is still seen by many as a good way to stabilize an egalitarian relationship with sex an important component of it. Living together accom-

plishes the same for others. Some reject marriage outright, many others would want to extend it to homosexuals as well.

The Case for the Humanistic Perspective

The basic premise of the humanistic perspective rests on the libertarian view of the unalienable human right to be *free.* Adults are endowed with a free will and know what is in their best interest, hence it should be up to them, and no one else, to decide why, when, where, and how to behave sexually.

Figure 20.8   George Segal, *Gay Liberation*, 1980. (On loan to Stanford University from the Mildred Andrews Fund.)

The history of Western culture shows the absurdity of sexual rules both in their relentless application as well as their flagrant violation. Western society has in particular oppressed women and endowed female sexuality with guilt and shame for reasons that are totally at variance with human needs and reason. If the ideal of monogamous sex made good sense in times past, the perpetuation of that ideal, as a sexual norm rather than a personal choice, makes no sense anymore in the modern world.

Although the humanistic perspective is usually equated with the sexual revolution of the sixties, sexual liberationists like Albert Ellis were already proclaiming its aims in the late fifties. The following paragraphs with which Ellis concluded his book *Sex without Guilt* state this perspective quite succinctly:

To conclude: Every human being should certainly refrain from sex participations which needlessly, forcefully, or unfairly harm others. And each of us, if we are to remain rational and non-neurotic, should cease and desist from all sex behavior which is clearly self-defeating.

Over and above this, everyone has a human right to sex-love involvement of his own taste, preference, and inclination. And the more he speaks and fights for that right, the more, in practice, he (or she!) is likely to realize and enjoy it.[28]

The roots of these arguments, which sound so modern, go back to early Christianity. As we discussed earlier, there were some in the early Church who maintained that their salvation by divine grace superceded all allegiances to moral law. Christian orthodoxy condemned them as heretics and effectively stamped them out (chapter 17). But that victory does not mean that the Church was right. Perhaps the Gnostics and others following them rejected formal moral precepts not because they wanted to indulge themselves but because the existence of such rules contradicted their honest belief that the living Lord within them ruled their lives and would help determine in any specific context the ethical course of action. Such antinomian sentiments have continued to surface throughout Christian history and cannot be dismissed as passing aberrations.

In secular thinking, the antinomian ethic has been prominent in existentialist philosophy. As articulated by Jean Paul Sartre, a person making a moral choice has no excuse or justification. Reality is incoherent and thus does not admit of any belief in coherent belief systems or rules of moral behavior. Each existential moment stands in isolation with no connection between one experience and another. In the face of such radical discontinuity, no generalizing moral law or principle can exist: one can only deal with "what is," not with "what should be," and in that context we should understand all and forgive all.

Less esoterically, the humanistic perspective draws support from scientific arguments. Given the interplay of biology and culture in shaping sexual behavior,

Figure 20.9 Indian Goddess. Yakshi figure from Stupa at Sanchi. (Courtesy John La Plante.)

there is not much room for judgments of personal culpability. Does anyone have a choice in "deciding" one's sexual orientation? If not, how can people be held morally responsible for behaviors over which they have no control? Besides, given the variety of moral codes across cultures, all of them set up with the best of intentions, who is to determine by which code to judge someone else's behavior?

The humanistic argument from nature, unlike that of "natural law," is based on empirical evidence. Naturalistic observation shows many sexual activities proscribed for human beings manifested by animals. That makes "unnatural" only those sex acts "that can't be performed," as Kinsey put it.

[28]Ellis (1965), p. 184. The "Playboy philosophy" expounded by Hugh Hefner is also reflective of this perspective.

Box 20.5   Eroticism in Eastern Religions

Comparisons between attitudes towards sex in various religions can be quite illuminating as well as potentially misleading. Religious traditions have much in common in their basic human concerns and cannot be simplistically labelled as "sex-positive" or "sex-negative." One belief system may be permissive with respect to one issue and restrictive with respect to another: *Confucianism* accepts sex as a natural function untainted by any innate sense of guilt, yet traditional Chinese society has been extraordinarily reticent in its public expressions of sexuality. *Islam* shelters its women but does not deny the legitimacy of female sexuality. Japanese *Buddhism* has strict rules governing the interactions of men and women, yet Japan has produced very fine erotic art (*Shunga*).

Particularly instructive is the experience of *Hinduism* with its extremes of asceticism and religiously inspired voluptuousness. Compared to the far more neutral sexual images of Christian iconography, the sights of Indian goddesses (figure 20.9) and the erotic statuary of medieval temples (figure 1.1) are quite startling. "Nowhere," says Parrinder, "have the close relationships of religion and sex been displayed more clearly than in India and, with divine and human models of sexual activity, sacramental views of sex more

abundantly illustrated."[1] The sexual experience has been seen so exalted that the Indian eroticism begins at every point with God.[2]

Shiva is one of the principal and most complex gods of India.[3] Of particular interest are those episodes of his life that deal with his relationship with his wife, *Parvati*. It was Parvati who approached Shiva when he was meditating in the Himalayas, wanting to marry him. *Kama,* the God of Love, tried to kindle Shiva's desire for Parvati, but Shiva churlishly reduced him to ashes with the fire from his eye in the forehead (but then revived him). Shiva then appeared before Parvati as an ascetic and, to test her resolve, described in most unpleasant terms the austere circumstances of his home in the cremation grounds. Parvati was undaunted and Shiva married her.

The passionate sexual encounters with Parvati weakened Shiva's powers, so he retreated to the forest to revive himself. As he danced around naked in full erection, the wives of the forest sages fell in love with him. Their husbands cursed Shiva's penis, which fell to the ground generating a terrible fire. Peace was finally restored when all agreed to worship Shiva's penis symbolized by the *lingam,* hence the explanation for this widespread Indian practice which has persisted to modern times.

An important impetus to the growth of the humanistic perspective has come from feminism (chapter 16). Traditional sexual morality has been especially oppressive of female sexuality and has perpetrated a double standard. It has presumed women to be sexual objects whose purity must be defended in virginity and whose fecundity must be cultivated in marriage. They are also said to have tempted men, inflamed

their lust, and led to their downfall since Eve got Adam into trouble. Venerated or vilified, women have been molded in every conceivable sexual image but given scarce chance to express how they themselves feel and what they want—all this in the name of sexual morality. Nonetheless, feminists do not subscribe to a sexual ethic where anything goes. While tolerant of sexual diversity, they are as harsh as any sexual

Box 20.5   Eroticism in Eastern Religions (*Continued*)

Like Shiva, there are many aspects to Parvati. She is the benevolent Mother Goddess whose genital symbol, the *yoni*, joined with Shiva's lingam, is worshipped as the sources of life (figure 20.10). There is also a terrible aspect to Parvati when she appears as *Durga*, or *Kali*, the goddess of death shown as a naked figure with a garland of skulls, trampling demons, standing on the prostrate body of Shiva in the cremation grounds or copulating with his erect penis.

Vishnu is another great deity whose life manifests many episodes involving love and sex. From his early Vedic origins, Vishnu became particularly prominent after his reincarnation as the popular god *Krishna*. Some of the most popular episodes of Krishna's life were his amorous adventures with the *Gopis* (women who herded cows), among whom he lived for a while. As a divinely handsome youth, Krishna captured the hearts of these lowly women and played erotic pranks on them. Once, while the Gopis were bathing in the nude, Krishna gathered their clothes and climbed a tree: they had to approach him with their hands raised, uncovering their full nudity, in order to get back their garments. When Krishna played his flute, the Gopis flocked to him in bewitched fascination. He led them in a round dance and then made love to each and every one

of them. Yet Krishna made each woman feel that she was dancing and making love with him alone. As legend embroidered variations on this theme, the number of women grew into the hundreds and thousands, as Krishna continued to blissfully satisfy all of them.

Though many of Krishna's amorous adventures, such as his seduction of married women, contradicted moral precepts, these were justified by the primacy of divine love over all other forms of human attachments. For instance, the story of the baring of the bodies by the Gopis was interpreted as symbolic of the nakedness of the soul in the divine presence. In more human terms, Krishna's multiple loves and simultaneous devotion to Radha seemed to provide men with the best of two worlds. Most importantly, the unabashed celebration of sex by the loftiest of divinities vindicated it to Krishna's countless followers.[4]

[1]Parrinder (1980 p. 5).
[2]Rawson (1968 p. 29).
[3]O'Flaherty (1973).
[4]Archer (1957); Dimock and Levertov (1964); and Rawson (1968) provide further details on the worship of Krishna.

conservative in their condemnation of activities that debase and hurt women as epitomized by pornography and rape.

The Case against the Humanistic Perspective

No one quarrels with the ideals of the humanistic view; neither the pope nor anyone else is against the joy to be derived

from sexual intimacy. Everyone is in favor of sharing, honesty, and personal enhancement, and opposed to exploitation in sexual relationships. The problem is not with principle but with practice; not with what people say but what they do; not with the hoped-for benefits but with the actual outcome of humanistic practice.

The lessons of the sexual revolution of a decade ago during which freedom was

Figure 20.10   Lingam-yoni altar. Stone; Eastern India; modern. (From Philip Rawson, *Erotic Art of the East.* New York: G.P. Putnam's Sons, 1968.)

rampant, are still fresh. Love was celebrated in song and dance, proclaimed by gurus and T-shirts. But there was little evidence of it in the lives of forlorn youth, drugged out of their minds, wandering the streets like the walking wounded. There was plenty of sex, of any kind, for free. There were no rules to obey, no shame or guilt to stop anyone. It was a freedom in which the vicious preyed on the lost.

The liberation of female sexuality was to have been one of the benefits of the sexual revolution, and sexual equality that of radical politics. But women in these movements soon found themselves mimeographing flyers during the day and servicing the sexual needs of their comrades during the night. Women continued to be sexually used as they had always been, but now they had lost the right even to say no.

We can thank the sexual revolution for spawning an industry of pornographic films. For 25 cents we can now watch through a tiny window an apathetic man copulating with a bored woman. For a dollar, a nude women in a glass booth will listen through a phone to whatever a man has to say. Singles bars and gay bathhouses are like meat racks for prospective sexual partners with no strings attached. These and other depressing sights clearly show that sexual behavior cannot be left at the mercy of unbridled choice and untutored freedom.

Libertarian principles may sound fine but they do not work. The simultaneous advocacy of freedom of individual action and defense of the rights of others becomes unmanageable in any consistent application of the libertarian principles; we soon run into trouble over defining where one person's untrammeled freedom impinges on another's unalienable rights. It is easy to understand why one would not be permitted to rape another. But what if a couple decides to have sexual intercourse in public and others find this offensive? How do you make "private" the operation of pornographic movies, and sex shops, the selling of sexually explicit magazines in general stores? How will prostitutes find customers if their existence is hidden? Clearly we have to draw a line sooner or later and devise moral codes to define it and laws to defend it.

Envoi

We have come full circle back to the advocacy of conservative morality.

Does that mean that our society has to go back to traditional sexual values? Not necessarily, but it is important that we know what the institutions with the longest experience have to say on the matter so we spare ourselves the trouble of reinventing the moral wheel. We need not repeat the mistakes of the past. Most importantly, we must be aware of the changed circumstances of today's world and rethink our moral precepts in the light of today's realities. We should then have the good sense either to reaffirm the ethical tenets of the past, if we find them tenable, or to have the courage to change them, if we do not.

At the individual level, one is generally better off respecting the sexual values that are prevalent without necessarily swallowing whole all of the moral dictates of society, if they make no sense. Rules are devised for the average: you may not be average. But if you are, and most of us are, do not delude yourself into thinking you are not. A few of those who get far ahead of their times are deemed prophets; most of the rest are called fools.

Moral decisions require reflection, but morality is not an intellectual exercise. It is a commitment to seek what is right and to act on it. It is neither necessary nor feasible for each of us to develop our own unique ethical system; but unless we make our own the moral precepts we choose, they will be of little help to us to avoid doing not only the wrong deed but also, as T. S. Eliot has said:

The last temptation is the greatest treason:
To do the right deed for the wrong reason.[29]

SUGGESTED READINGS

Dennehey, R., ed. *Christian Married Love,* (1981). A series of essays representative of the conservative Catholic perspective. Includes a summary of Karol Wojtyla's (Pope John Paul II) "Love and Responsibility."

Kosnik et al. *Human Sexuality: New Directions in American Catholic Thought* (1977). Liberal reconsideration of traditional Catholic thinking on sexuality.

Fletcher, J. *Situation Ethics* (1966). The original and still best statement of the "new morality."

Mazur, R. *The New Intimacy* (1973). Though not specifically addressed to the issue of sexual morality, this slim volume and *Common-sense Sex* by the same author (1968) convey a good sense of the ethical outlook of the humanistic perspective.

[29]Eliot (1935, part I, p. 44).

References

Abplanalp, J.M., Rose, R.M., Donnelly, A.F., and Livingston-Vaughan, L. "Psychoendocrinology of the menstrual cycle: II. "The relationship between enjoyment of activities, moods, and reproductive hormones." *Psychosomatic Medicine* 41:605, 1979.

Abramson, P., and Mechanic, M. "Sex and the media: Three decades of best-selling books and major motion pictures." *Archives of Sexual Behavior* 12(3), 1983.

Ackman, C.F.D., MacIsaac, S.G., and Schual, R. "Vasectomy: Benefits and risks." *International Journal of Gynecology and Obstetrics* 16:493–496, 1979.

Addiego, F., Belzer, E.G., Comolli, J., Moger, W., Perry, J.D., and Whipple, B. "Female ejaculation: A case study." *The Journal of Sex Research* 17(1):13–21, 1981.

Adelson, J. (ed.) *Handbook of adolescent psychology*. New York: Wiley, 1980

Aitken, R.J. "Contraceptive research and development." *British Medical Bulletin* 35(2): 199–204, 1979.

Allen, D.M. "Young male prostitutes: A psychosocial study." *Archives of Sexual Behavior* 9(5):399–426, 1980.

Altman, L.D. "Measuring the benefits of the pill." *San Francisco Chronicle*, July 22, p. 23, 1982.

Altschuler, G. Chapter 43 in Danforth, D.N. (ed.) *Obstetrics and gynecology*. New York: Harper & Row, 1982.

American Law Institute, Model penal code: Proposed official draft. Philadelphia, 1962.

American Psychiatric Association. *Diagnostic and statistical manual of mental disorders*. (3rd ed.) Washington, D.C.: APA, 1980.

American Social Health Association. "Gays and STDs." Pamphlet, 1982.

Amir, M. *Patterns of forcible rape*. Chicago: University of Chicago Press, 1971.

Anderson, J. *The witch on the wall*: Medieval erotic sculpture in the British Isles. London: Allen & Unwin, 1977.

Anderson, C.L. "What are psychology departments doing about sex education." *Teaching of Psychology* 2(1), 24–27, 1975.

Anderson, G.G., and Speroff, L. "Prostaglandins." *Science* 171:502–504, 1971.

Anderson, J. "Planned and unplanned births in the United States: Planning status of marital births, 1975–1976." *Family Planning Perspectives* 13(2), 1981.

Ansbacher, R. "Artificial insemination with frozen spermatozoa." *Fertility and Sterility* 29:375–379, 1978.

Aquinas, St. Thomas *The summa theologica.* (Fathers of the English Dominican Province, trs.) New York: Benziger Bros., 1911–1925.

Arafat, I., and Cotton, W.L. "Masturbation practices of males and females." *Journal of Sex Research* 10:293–307, 1974.

Arafat, I.S., and Cotton, W. "Masturbatory practices of college males and females." In DeMartino, M.F., (ed.) *Human autoerotic practices*. New York: Human Sciences Press, 1979.

Arentewicz, G., and Schmidt, G. (eds.) *The treatment of sexual disorders*. New York: Basic, 1983.

Arey, L.B. *Developmental anatomy* (7th ed.) Philadelphia: Saunders, 1974.

Arey, L.B. *Developmental anatomy* (revised 7th ed.) Philadelphia: Saunders, 1974.

Aries, P. *Centuries of childhood*. New York: Knopf, 1962.

Arms, S. *Immaculate deception*. Boston: Houghton Mifflin, 1975.

Athanasiou, R., and Sarkin, R. "Premarital sexual behavior and postmarital adjustment." *Archives of Sexual Behavior* 3(3):207–225, 1974.

Athanasiou, R., Shaver, P., and Travis, C. "Sex (A report to *Psychology Today* readers)." *Psychology Today* 4:37–52, 1970.

Atkins, J. *Sex in literature*. Vol. II (1973), Vol. III (1978). London: Calder.

Augustine, St. *The city of God, book XIV.* (J. Healey, tr.) New York: Dutton, 1934.

Augustine, St. *Treatises on marriage and other subjects*. (Wilcox, C., et al., trs.) New York: Fathers of the Church, 1955.

Babcock, B.A., Freedman, A.E., Norton, A.E., and Ross, S.C. *Sex discrimination and the law: Causes and remedies*. Boston: Little, Brown, 1975.

Bailey, D.S. *Homosexuality and the western Christian tradition.* London: Shoestring, 1955.

Bailey, S. *Sexual relations in Christian thought.* New York: Harper & Row, 1959.

Bailey, S. *Sexual ethics: A Christian view.* New York: Macmillan, 1963.

Bailey, T., and Kennedy, D. *The American pageant,* (6th ed.) Lexington, Mass.: Heath, 1979.

Bakwin, H. "Erotic feelings in infants and young children." *Medical Aspects of Human Sexuality* 8(10):200–215, 1974.

Balfour, J.G. "An Arab physician on insanity." *The Journal of Mental Science* 98:241–249, 1876.

Bancroft, J. "The behavioral approach to treatment." Money and Musaph, (eds.) In *Handbook of sexology.* New York: Elsevier, 1977.

Bancroft, J. "Endocrinology of sexual function." *Clinics in Obstetrics and Gynecology* 7(2):253–281, 1980.

Bandura, A. *Principles of behavior modification.* New York: Holt, Rinehart and Winston, 1969.

Barbach, L.G. *For yourself: The fulfillment of female sexuality.* New York: Doubleday, 1975.

Barbach, L.G. *Women discover orgasm.* New York: Free Press, 1980.

Barber, R.N. "Prostitution and the increasing number of convictions for rape in Queensland." *Australian and New Zealand Journal of Criminology* 2:169–174, 1969.

Bardin, C.W., and Catterall, J.F. "Testosterone: A major determinant of extragenital sexual dimorphism." *Science* 211:1285–1294, 1981.

Bardwick, J. *Psychology of women.* New York: Harper & Row, 1971.

Barnett, W. *Sexual freedom and the constitution.* Albuquerque: University of New Mexico Press, 1973.

Bartell, G.D. *Group sex.* New York: Wyden, 1971.

Bates, M. *Gluttons and libertines.* New York: Vintage, 1967.

Baucom, D., and Hoffman, J. "Common mistakes spouses make in communicating." *Medical Aspects of Human Sexuality* 17(11), 1983.

Bauman, K.D., and Wilson, R.R. "Sexual behavior of unmarried university students in 1968 and 1972." *Journal of Sex Research* 10:327–333, 1974.

Bayer, R. *Homosexuality and American psychiatry: The politics of diagnosis.* New York: Basic, 1981.

Bazell, R. "The history of an epidemic." *The New Republic,* August 1983.

Beach, F. "A review of physiological and psychological studies of sexual behavior in mammals." *Physiological Review* 27(2):240–305, 1947.

Beach, F.A. (ed.) *Sex and behavior.* New York: Wiley, 1965.

Beach, F.A. "Hormonal factors controlling the differentiation, development, and display of copulatory behavior in the ramstergig and related species." in Tobach, E., Aronson, L.R., and Shaw, E., (eds.) *The biopsychology of development.* New York: Academic, 1971, pp. 249–295.

Beach, F.A. "Cross-species comparisons and the human heritage." *Archives of Sexual Behavior* 5(5):469–485, 1976.

Beach, F.A. *Human sexuality in four perspectives.* Baltimore: Johns Hopkins University Press, 1977.

Bell, A.P., and Weinberg, M.S. *Homosexualities.* New York: Simon & Schuster, 1978.

Bell, A.P., et al. *Sexual preference: Its development in men and women.* Bloomington: Indiana University Press, 1981.

Bell, R., and Bell, P. "Sexual satisfaction among married women." *Medical Aspects of Human Sexuality* 6(12):136–14, 1972.

Bell, A.P. "Sexual preference: A postscript." *SIECUS Report* 11(2), 1982.

Bell, R.R., and Gordon, M. (eds.) *The social dimension of human sexuality.* Boston: Little, Brown, 1972.

Belliveau, F., and Richter, L. *Understanding human sexual inadequacy.* New York: Bantam, 1970.

Belzer, E.J., Jr. "Orgasmic expulsions of women: A review and heuristic inquiry." *The Journal of Sex Research* 17:1–12, 1981.

Bem, S. "Sex role adaptability: One consequence of psychological androgyny." *Journal of Personality and Social Psychology* 31(4):634–643, 1975.

Bem, S. *Psychosexual functions in women.* New York: Ronald, 1952.

Benson, G.S., McConnell, J.A., and Schmidt, W.A. "Penile polsters: Functional structures or atherosclerotic changes?" *Journal of Urology* 125(6):800–803, 1981.

Bentler, P.M., and Peeler, W.H. "Models of female orgasm." *Archives of Sexual Behavior* 8:405–423, 1979.

Bentler, P.M., and Abrahamson, P.R. "The science of sex research: Some methodological considerations." *Archives of Sexual Behavior* 10(3), 1981.

Berlin, F. "Sex offenders: A biomedical perspective and a status report on biomedical treatment." In Greer, J., and Stuart, I. (eds.) *The sexual aggressor: Current perspectives on treatment.* New York: Van Nostrand Reinhold, 1983.

Bermant, G., and Davidson, J.M. *Biological bases of sexual behavior.* New York: Harper & Row, 1974.

Bernstein, R. "The Y chromosome and primary sexual differentiation." *Journal of the American Medical Association* 245(19):1953–1956, 1981.

Betancourt, J. *Am I normal? An illustrated guide to your changing body.* New York: Avon, 1983.

Betancourt, J. *Dear diary: An illustrated guide to your changing body.* New York: Avon, 1983.

Bieber, I., et al. *Homosexuality: A psychoanalytic study.* New York: Basic, 1962.

Billings, E.L., and Billings, J.J. *Atlas of the ovulation method.* Collegeville, Minn.: The Liturgical Press, 1974.

Bimock, E.C., and Levertov, D. *The praise of Krishna: Songs from the Bengali.* Garden City, N.J.: Doubleday, 1964.

Bingham, H.C. "Sex development in apes." *Comparative Psychological Monographs* 5:1–165.

Bishop, N. "The great Oneida love-in." *American Heritage* 20:14–17, 86–92, 1969.

Blair, C.D., and Langon, R.I. "Exhibitionism: Etiology and treatment." *Psychological Bulletin* 89(3):439–463, 1971.

Block, J. *Lives through time.* Berkeley, CA.: Bancroft, 1971.

Bloom, W., and Fawcett, D.W. *A textbook of histology.* New York: Saunders, 1975.

Blos, P. *On adolescence.* New York: Free Press, 1962.

Blos, P. "Modifications in the traditional psychoanalytic theory of female adolescent development." *Adolescent Psychiatry* 8:8–24, 1980.

Blumstein, P., and Schwartz, P. *American couples.* New York: Morrow, 1983.

Boardman, J., and LaRocca, E. *Eros in Greece.* New York: Erotic Art Book Society, 1975.

Boccaccio, G. *The decameron.* Payne, J. (tr.) New York: Modern Library, n.d.

Bohlen, J.G. "Sleep erection monitoring in the evaluation of male erectile failure." *Urological Clinics of North America* 8(1):119–134, 1981.

Bohlen, J.G., Held, J.P., and Sanderson, M.O. "The male orgasm: Pelvic contractions measured by anal probe." *Archives of Sexual Behavior* 9:403–521, 1980.

Bohlen, J.F., Held, J.P., and Sanderson, M.O. "Response of the circumvaginal musculature during masturbation." In Garber, B. (ed.) *Circumvaginal musculature in sexual function.* New York: Krager, 1982.

Bohlen, J., Held, J., Sanderson, M., and Ahlgren, A. "The female orgasm: Pelvic contractions." *Archives of Sexual Behavior* 2(5), 1982.

Bonaparte, M. *Female sexuality.* London: Imago, 1953.

Bonsall, R.W., and Michael, R.P. "Volatile odoriferous acids in vaginal fluid." In E.S.E. Hafez and Evans, T.N. (eds.) *The human vagina.* New York: Elsevier, 1978, pp. 167–177.

Book of Common Prayer. London: Oxford University Press, no date.

Borneman, E. "Progress in empirical research on children's sexuality." *SIECUS Report* 12(2), 1983.

Boswell, J. *Christianity, social tolerance, and homosexuality.* Chicago: University of Chicago Press, 1980.

Boston Women's Health Book Collective *Our bodies, ourselves* (2nd ed.) New York: Simon & Schuster, 1976.

Borowitz, E.B. *Choosing a sex ethic: A Jewish inquiry.* New York: Schocken, 1969.

Bors, E., and Comarr, A.E. "Neurological disturbances of sexual function with special reference to 529 patients with spinal cord injury." *Urological Survey* 10:191–222.

Botwin, C. "Is there sex after marriage?" *New York Times Magazine,* September 16, 1979, pp. 108–112.

Bouyer, L. *Introduction to spirituality.* Collegeville, Minn.: Liturgical Press, 1961.

Bower, D.W., and Christopherson, V.A. "University student cohabitation: A regional comparison of selected attitudes and behavior." *Journal of Marriage and the Family* 39:447–452, 1977.

Bowlby, J. *Attachment: Attachment and loss,* vol. I. New York: Basic, 1969.

Bowlby, J. *Separation: Attachment and loss*, vol. II. New York: Basic, 1973.

Breasted, M. *Oh! Sex education!* New York: New American Library, 1970.

Brecher, E.M. *The sex researchers*. Boston: Little, Brown, 1969.

Brecher, E.M. *Sex, love and aging*. Boston: Little, Brown, 1984.

Brenhen, M. *Sex and your heart*. New York: Armand, 1968.

Brenton, M. *Sex and your heart*. New York: Award Books, 1968.

Brewer, J.S., and Wright, R.S. *Sex research*. Phoenix, Arizona: Oryx, 1979.

Brim, O.G. "Socialization through the life cycle." In Brim, O.G., and Wheeler, S. *Socialization after childhood*. New York: Wiley, 1966.

Brim, O.G., and Kagan, J. (eds.) *Constancy and change in human development*. Cambridge: Harvard University Press, 1980.

Brobeck, J.R. (ed.) *Best and Taylor's physiological basis of medical practice*. (10th ed.) Baltimore: Williams and Wilkins, 1979.

Brown, G. *The new celibacy*. New York: McGraw-Hill, 1980.

Brown, L. *Sex education in the eighties*. New York: Plenum, 1981.

Brown, L., and Holder, W. "The nature and extent of sexual abuse in contemporary American society." In Greenberg, N.H. (ed.) *Incest: In search of understanding*. Washington, D.C.: National Center on Child Abuse and Neglect, 1979.

Brown, P. *Augustine of Hippo*. Berkeley: University of California Press, 1967.

Brownmiller, S. *Against our will: Men, women and rape*. New York: Simon & Schuster, 1975.

Brownmiller, S. *Femininity*. New York: Simon & Schuster, 1984.

Budoff, P. *No more menstrual cramps and other good news*. New York: Putnam, 1980.

Bullough, V. "Prostitution, psychiatry and history." In Bullough, V. (ed.) *The frontiers of sex research*. Buffalo, N.Y.: Prometheus, 1979.

Bullough, V., et al. *A bibliography of prostitution*. New York: Garland, 1977.

Burgess, A.W., and Holmstrom, L.L. "Rape trauma syndrome." *American Journal of Psychiatry* 131:981–986, 1974.

Burgess, A.W., and Holmstrom, L.L. *Rape: Victims of crisis*. Bowie, Md.: Brady, 1974.

Burgess, A.W., and Holmstrom, L.L. *Rape: Crisis and recovery*. Bowie, Md.: Brady, 1979.

Burgess, A.W., and Holmstrom, L.L. "Coping behavior of the rape victim." *American Journal of Psychiatry* 133(4):413–418, 1976.

Burgoyne, D.S. "Factors affecting coital frequency." *Medical Aspects of Human Sexuality* 8(4):143–156, 1974.

Burnham, J.C. "American historians and the subject of sex." *Societas 2:* 307–316, 1972.

Burton, R.F. (tr.) *The thousand and one nights*. "Printed by the Burton Club for subscribers only." n.d.

Butler, R.N., and Lewis, M.J. *Aging and mental health*. St. Louis: Mosby, 1976.

Butler, S. *Conspiracy of silence*. San Francisco: New Glide Publications, 1978.

Caird, W., and Wincze, J.P. *Sex therapy: A behavioral approach*. Hagerstown, Md.: Harper & Row, 1977.

Calder-Marshall, A. *The sage of sex*. New York: Putnam, 1959.

Calhoun, L., et al. "The influence of pregnancy on sexuality: A review of current evidence." *Journal of Sex Research* 17(2):139–151, 1981.

Callahan, D. *Abortion: Law, choice and moraltiy*. New York: Macmilla, 1970.

Calverton, V.F., and Schmalhausen, S.D. *Sex in civilization*. New York: Citadel, 1929.

Calvin, J. *Institutes of the Christian religion*. McNeill, J.T. (ed.) London: S.C.M. Press, 1960.

Cameron, P. "Note on time spent thinking about sex." *Psychological Reports* 20, 741–742, 1970.

Cameron, P. and Biber, H. "Sexual thought throughout the life-span." *Gerontologist* 13(2):144–147, 1973.

Cameron, P., et al. "Consciousness: Thoughts about world and social problems, death, and sex by the generations." Paper read at Kentucky Psychological Association, Sept. 25, 1970.

Campbell, B. "Neurophysiology of the clitoris." In Lowry, T.P., and Lowry, T.S. (eds.) *The clitoris*. St. Louis: Green, 1976.

Capraro, V., Ridgers, D., and Rodgers, B. "Abnormal vaginal discharge." *Medical Aspects of Human Sexuality* 17(8), 1983.

Carey, J.T. "Changing courtship patterns in the popular song." *American Journal of Sociology* 74:720–731, 1979.

Carrier, J.M. "Homosexual behavior in cross-cultural perspective." In Marmor, J. (ed.) *Homosexual behavior.* New York: Basic, 1980.

Casanova de Seingalt, G.G. *Memoires.* Paris: Gallimard, 1958–1960.

Centers for Disease Control Task Force on Kaposi's Sarcoma and Opportunistic Infections. "Epidemiologic aspects of the current outbreak of Kaposi's sarcoma and opportunistic infections." *New England Journal of Medicine* 306:248–252, 1982.

Chall, L.P. "The reception of the Kinsey report in the periodicals of the United States: 1947–1949." In Himelhock, J., and Fava, S.F. (eds.) New York: Norton, 1955, pp. 364–378.

Chambers, M., et al. *The Western experience.* (2nd ed.) New York: Knopf, 1979.

Chambless, D., et al. "The pubococcygeus and female orgasm: A correlational study with normal subjects." *Archives of Sexual Behavior* 11(6), 1982.

Chang, J. *The Tao of love and sex.* New York: Dutton, 1977.

Chang, J., and Block, J. "A study of identification in male homosexuals." *Journal of Consulting Psychology* 24(4):307–310, 1960.

Chappel, D., Geis, R., and Geis, G. (eds.) *Forcible rape: The crime, the victim, and the offender.* New York: Columbia University Press, 1977.

Cherry, S. *For women of all ages.* New York: Macmillan, 1979.

Chilman, C. *Adolescent sexuality in a changing American society: Social and psychological perspectives for the human services professions.* New York: Wiley, 1982.

Christenson, C.V. *Kinsey: A biography.* Bloomington: Indiana University Press, 1971.

Christensen, H. "Scandanavian and American sex norms: Some comparisons with sociological implications." *Journal of Social Issues* 22:60–75, 1966.

Clarkson, T.B.: and Alexander, N.J. "Long-term vasectomy: Effects on the occurrence and extent of atherosclerosis in rhesus monkeys." *Journal of Clinical Investigation* 65(1):15–25, 1980.

Clayton, R.R., and Voss, H.L. "Shacking up: Cohabitation in the 1970's." *Journal of Marriage and Family* 39:273–283, 1977.

Cleland, J. *Memoirs of a woman of pleasure.* New York: Putnam, 1963.

Clugh, J. *Love locked out.* London: Spring Books, 1963.

Cochran, W.G., Mosteller, F., and Tukey, J.W. *Statistical problems of the Kinsey report of sexual behavior in the human male.* Washington, D.C.: American Statistical Association, 1954.

Cohen, A. *Everyman's Talmud.* New York: Dutton, 1949.

Coleman, S., Piotrow, P.T., and Rinehart, W. "Tobacco: Hazards to health and human reproduction." *Population Reports,* Series L (1): March 1979.

Collis, J.S. *Havelock Ellis: Artist of life.* New York: Sloane, 1959.

Comfort, A. *The Kokashastra.* New York: Stein & Day, 1965.

Comfort, A. *The anxiety makers.* New York: Delta, 1967.

Comfort, A. "Likelihood of human pheromones." *Nature* 230:432–433, 1971.

Comfort, A. *The joy of sex.* New York: Crown, 1972.

Comfort, A. *More joy.* New York: Crown, 1974.

Comfort, A. *A good age.* New York: Simon & Schuster, 1976.

Comfort, A. *Sexual consequences of disability.* Philadelphia: George F. Shickley, 1978.

Comfort, A., and Comfort J. *The facts of love.* New York: Crown, 1979.

Commission on Obscenity and Pornography. *The report of the commission on obscenity and pornography.* Washington, D.C.: U.S. Government Printing Office, 1970.

Conn, J. and Kanner, L. "Spontaneous erections in childhood." *Journal of Pediatrics* 16:337–240, 1940.

Connell, E.B. "Barrier methods of contraception: A reappraisal." *International Journal of Gynecology and Obstetrics* 16:479–481, 1979.

Conroy, M. *The rational woman's guide to self-defense.* New York: Grosset & Dunlap, 1975.

Constantine, L.L., and Constantine, J.M. *Group marriage: A study of contemporary multilateral marriage.* New York: Macmillan, 1973.

Constantinople, A. "Masculinity-femininity: An exception to a famous dictum?" *Psychological Bulletin* 80:389–407, 1973.

Cooper, J.F. *Technique of contraception.* New York: Day-Nichols, 1928.

Cooper, P.T., Cumber, B., and Hartner, R. "Decision-making patterns and post-decision adjustment of child-free husbands and wives." *Alternative Lifestyles* 1(1), 71–94, 1978.

Cox, S. (ed.) *Female psychology: The emerging self.* Palo Alto, CA.: *Science Research Associates,* 1976.

Crain, I.J. "Afterplay." *Medical Aspects of Human Sexuality* 12:72–85, 1978.

Crepault, C., Abraham, G., Porto, R., and Couture, M. "Erotic imagery in women." In *Progress in sexology.* New York: Plenum, 1977.

Croughan, J., Saghir, M., Cohen, R., and Robins, E. "A comparison of treated and untreated male cross-dressers." *Archives of Sexual Behavior* 10(6), 1981.

Dalton, K. *The menstrual cycle.* New York: Pantheon, 1969.

Dalton, K. *Once a month.* New York: Hunter, 1979.

Daly, M., and Wilson, M. *Sex evolution and behavior.* Belmont, CA.: Wadsworth, 1978.

D'Andrade, R.G. "Sex differences and cultural institutions." In Maccoby, E.E. (ed.) *The development of sex differences.* Stanford: Stanford University Press, 1966.

Danforth, D.N. (ed.) *Obstetrics and gynecology.* (4th ed.) Philadelphia: Harper & Row, 1982.

Davenport, W.H. "Sex in cross-cultural perspective." In Beach, F.A. (ed.) *Human sexuality in four perspectives.* Baltimore: Johns Hopkins University Press, 1977, 115–163.

David, H.P. "Abortion: A continuing debate." *Family Planning Perspectives* 10:313–316, 1978.

Davidson, J.M. "Neurohormonal bases of male sexual behavior." In Greek, R.O. (ed.) *Reproductive Physiology II,* 13. Baltimore: University Park Press, 1977.

Davidson, J.M. "The psychology of sexual experience." In Davidson, J.M., and Davidson, R.J. (eds.) *The psychobiology of consciousness.* New York: Plenum Press, 1980.

Davidson, J.M. "The orgasmic connection." *Psychology Today* 15, July 1981.

Davidson, J.M., et al. "Effects of androgen on sexual behavior in hypogonadal men." *Journal of Clinical Endocrinology and Metabolism* 48(6), 1979.

Davis, A.J. "Sexual assault in Philadelphia prisons and sheriff's vans." *Trans-action* 6:8–17, 1968.

Dawson, D.A., Meny, D.J., and Ridley, J.C. "Fertility control in the United States before the contraceptive revolution." *Family Planning Perspectives* 12(2):76–86, 1980.

De Beauvoir, S. *The second sex.* New York: Knopf, 1952.

Debrovner, C.H. "Premenstrual syndrome." *Medical Aspects of Human Sexuality* 215–216, 1983.

Decter, M. *The new chastity and other arguments against women's liberation.* London: Wildwood, 1973.

D'Emilio, J. *Sexual politics, sexual communities.* Chicago: University of Chicago Press, 1983.

Degler, C. "What ought to be and what was: Women's sexuality in the nineteenth century." *American Historical Review* 79(5), 1974.

Degler, C. *At odds: Women and the family in America from the Revolution to the present.* New York: Oxford University Press, 1980, pp. 181–189.

deGroat, W.C., and Booth, A.M. "Physiology of male sexual function." *Annals of Internal Medicine* 92(2):329–331, 1980.

DeLacoste-Utamsing, C., and Halloway, R.L. "Sexual dimorphism in the human corpus callosum." *Science* 216:215–216, 1982.

DeLamater, J. "The social control of sexuality." *Annual Reviews of Sociology* 7:263–290, 1981.

Delaney, J., et al. *The curse.* New York: Dutton, 1976.

Delin, B. *The sex offender.* Boston: Beacon, 1978.

De Martino, M.F. (ed.) *Human autoerotic practices.* New York: Human Sciences Press, 1979.

Dement, W.C. "An essay on dreams." In *New directions in psychology II.* New York: Holt, Rinehart and Winston, 1965, pp. 135–257.

Dement, W.C. *Some must watch while some must sleep.* Stanford, CA.: Stanford Alumni Association, 1972.

Deneld, D., and Gordon, M. "The sociology of mate swapping: Or the family that swings together clings together." *Journal of Sex Research* 6:85–100, 1970.

Dennehy, R. (ed.) *Christian married love.* San Francisco: Ignatius Press, 1981.

Dennerstein, L., et al. "Hormones and sexuality: Effect of estrogen and progestogen." *Obstetrics and Gynecology* 56(3):316–322, 1980.

Denniston, R.H. "Ambisexuality in animals." In Marmor, J. (ed.) *Homosexual behavior.* New York: Basic, 1980.

De Rougemont, D. *Love in the western world.* New York: Pantheon, 1956.

De Sade, D.-A.-F. *Justine,* in *The Marquis de Sade: Three Complete Novels.* New York: Grove, 1965.

Deutsch, H. *The psychology of women.* New York: Grune and Stratton, 1944–1945.

Devereux, G. "Institutionalized homosexuality of the Mohave Indians." In *Human Biology,* 9. Detroit: Wayne State University Press, 1937, pp. 498–527.

DeVore, I. (ed.) *Primate behavior: Field studies of monkeys and apes.* New York: Holt, Rinehart and Winston, 1965.

DeWald, P.A. *Psychotherapy: A dynamic approach.* (2nd ed.) New York: Basic, 1971.

Diamond, M. "A critical evaluation of the ontogeny of human sexual behavior." *Quarterly Review of Biology* 40:147–175, 1965.

Diamond, M. "Human sexual development: Biological foundations for sexual development." In Beach, F. (ed.) *Human sexuality in four perspectives.* Baltimore: Johns Hopkins University Press, 1976.

Diamond, M. "Sexual identity and sex roles." In Bullough, V. (ed.) *The frontiers of sex research.* Buffalo, N.Y.: Prometheus, 1979.

Diamond, M. "Sexual identity: Monozygotic twins reared in discordant sex roles and a BBC follow-up." *Archives of Sexual Behavior* 11(2), 1982.

Diamond, M., et al. "Sexuality, birth control and abortion: A decision-making sequence." *Biosocial Science* 5:347–361, 1973.

Dickinson, R.L. *Atlas of human sex anatomy.* (2nd ed.) Baltimore: Williams & Wilkins, 1949.

Dick-Read, G. *Childbirth without fear.* New York: Harper & Row, 1932.

Diczfalusy, E. "Improved long-acting fertility regulating agents: What are the problems?" *Journal of Steroid Biochemistry* 2:443–448, 1979.

Dienhart, C.M. *Basic human anatomy and physiology.* Philadelphia: Saunders, 1967.

Djerassi, C. *The politics of contraception.* New York: Norton, 1981.

Dodson, B. *Liberating masturbation: A meditation on self-love.* New York: Bodysex Designs, 1974.

Doherty, D. (ed.) *Dimensions of human sexuality.* Garden City, N.Y.: Doubleday, 1979.

Dolgin, J.L., and Dolgin, B.L. "Sex and the law." In Wolman, B.B. and Money, J. (eds.) *Handbook of human sexuality.* Englewood Cliffs, N.J.: Prentice-Hall, 1980.

Donnerstein, E., and Linz, D. "Sexual violence in the media: A warning." *Psychology Today* 18(1):14–15, 1984.

Dornbusch, S.M., et al. "Sex development, age, and dating: A comparison of biological and social influences upon one set of behaviors." *Child Development* 52(1):179–185, 1981.

Dorner, G. "Hormones and sexual differentiation of the brain." In *Sex hormones and behavior.* Ciba Foundation Symposium. Amsterday: Excerpta Medica, 1962.

Dorner, G. *Hormones and brain differentiation.* Amsterdam: Elsevier, 1976.

Doty, R.L., et al. "Changes in the intensity and pleasantness of human vaginal odors during the menstrual cycle." *Science* 190:1316–1318, 1975.

Dover, K.J. *Greek homosexuality.* New York: Vintage, 1978.

Draper, N.E.E. "Birth control." In *Encyclopaedia Britannica* 2:1065–1073. Chicago: Benton, 1976.

Drew, D., and Drake, J. *Boys for sale: A sociological study of boy prostitution.* New York: Brown Book, 1969.

Droegemueller, W., and Bressler, R. "Effectiveness and risks of contraception." *Annual Review of Medicine* 31:329–343, 1980.

Dunn, H.G., et al. "Maternal cigarette smoking during pregnancy and the child's subsequent development: II. Neurological and intellectual maturation to the age of 6½ years." *Canadian Journal of Public Health* 68:43–50, 1977.

Durant, W., and Durant, A. *The age of Louis XIV.* New York: Simon & Schuster, 1963.

Dworkin, A. *Pornography: Men possessing women.* New York: Putnam, 1981.

Easterling, W.E., and Herbert, W.N.P. "The puerperium." In Danforth, D.N. (ed.) *Obstetrics and gynecology.* (4th ed.) Philadelphia: Harper & Row, 1982, pp. 787–799.

Edman, I. (ed.) *The works of Plato.* (The Jowett Translation). New York: Modern Library, 1956.

Ehrhardt, A., and Meyer-Bahlburg, H. "Effects of prenatal sex hormones on gender-related behavior." *Science* 211(4488):1312–1318, 1981.

Ehrhardt, A.A., and Money, J. "Progestin-induced hermaphroditism: I.Q. and psychosexual identity in a study of ten girls." *Journal of Sex Research* 3:83–100, 1967.

Ehrhardt, A.A., Epstein, R. and Money, J. "Fetal androgens and female gender identity in the early treated adrenogenital syndrome." *Johns Hopkins Medical Journal* 122:160–167, 1968.

Ehrlich, P., and Ehrlich, A. *Population, resources, environment.* (2nd ed.) San Francisco: Freeman, 1972.

Eitner, L. "The erotic in art." In Katchadourian, H., and Lunde, D. *Fundamentals of human sexuality.* (2nd ed.) New York: Holt, Rinehart and Winston, 1975.

Elias, J., and Gebhard, P. "Sexuality and sexual learning in childhood." *Phi Delta Kappa* 401–405, March 1969.

Ellis, A. *Studies in the psychology of sex.* New York: Random, 1942.

Ellis, A. *The art and science of love.* New York: Dell, 1965.

Ellis A. "Healthy and disturbed reasons for having extramarital relations." In Neubeck, G. (ed.)*Extramarital relations.* Englewood Cliffs, N.J.: Prentice-Hall, 1969: pp. 153–161.

Ellis, A. Foreword. In De Martino, M.F. (ed.) *Human autoerotic practices.* New York: Human Sciences Press, 1979.

Ellis, A., and Abarbanel, A. (eds.) *The encyclopedia of sexual behavior.* New York: Hawthorn, 1967.

Ellis, G., and Sexton, D. "Sex on TV." *Medical Aspects of Human Sexuality* 16(6), 1982.

Ellis, H. *My life.* Boston: Houghton Mifflin, 1939.

Ellis, H. *Psychology of sex.* New York: Emerson, 1938.

Ellis, H. *Studies in the psychoogy of sex.* New York: Random, 1942.

Engle, G.L. *Psychological development in health and disease.* Philadelphia: Saunders, 1962.

Engelmann, G. *Labor among primitive peoples.* St. Louis: Chambers, 1883.

Epstein, B. *The politics of domesticity: Women, evangelism and temperance in nineteenth century America.* Middletown, Conn.: Wesleyan University Press, 1981.

Epstein, L.M. *Sex laws and customs in Judaism.* (Rev. ed.) New York: Ktav Publishing, 1967.

Erikson, E.H. (ed.) "Identity and the life cycle." *Psychological Issues* 1(1). New York: International Universities, 1959.

Erikson, E.H. *Childhood and society.* New York: Norton, 1963.

Erikson, E.H. (ed.) *Youth: Change and challenge.* New York: Basic, 1963.

Erikson, E.H. *Identity: Youth and crisis.* New York: Norton, 1968.

Erikson, E.H. (ed.) *Adulthood.* New York: Norton, 1978.

Esselstyn, T.C. "Prostitution in the United States." *Annals of the American Academy of Political and Social Science* 376:123–135, 1968.

Evans, J.R., et al. "Teenagers: Fertility control behavior and attitudes before and after abortion, childbearing or negative pregnancy test." *Family Planning Perspectives* 8:192–200, 1976.

Fairbanks, B., Scharfman, B. "The cervical cap: Past and current experience." *Women vs. Health* 5(3):61–80, 1980.

Fawcett, J.T. *Psychology and population: Behavioral research issues in fertility and family planning.* New York: Population Council, 1970.

Fay, A. "Sexual problems related to poor communication." *Medical Aspects of Human Sexuality* 3:48–62, 1977.

Feder, H.H. "Perinatal hormones and their role in the development of sexually dimorphic behaviors." In Adler, N.T. (ed.) *Neuroendocrinology of reproduction, physiology and behavior.* New York: Plenum, 1981.

Fenichel, O. *The psychoanalytic theory of neurosis.* New York: Norton, 1945.

Feucht, O.E. (ed.) *Sex and the church.* St. Louis, Mo.: Concordia, 1961.

Field, J.H. "Sexual themes in ancient and primitive art." In Webb, P. *The erotic arts.* Boston: New York Graphic Society, 1975.

Finch, S. "Sexual disturbances in children." *Medical Aspects of Human Sexuality* 16(5), 1982.

Finkelhor, D. "Sex among siblings: A survey on prevalence, variety, and effects." *Archives of Sexual Behavior* 9(3):171–194, 1980.

Finkelhor, D. *Sexually victimized children.* New York: Free Press, 1979.

Finkle, A.L. "Potency among a sample of men aged between fifty-six and eighty-six." *Journal of the American Medical Association* 170:1391–1393, 1959.

Firestone, S. *The dialectic of sex: The case of a feminist revolution.* New York: Morrow, 1970.

Fisher, C., et al. "Patterns of female sexual arousal during sleep and waking: Vaginal thermo-conductance studies." *Archives of Sexual Behavior* 12(2), 1983.

Fisher, C., Schnari, R., Lear, H., Edwards, A., Davis, D.M., and Wilkin, A.P. "The assessment of nocturnal REM erection in the differential diagnosis of sexual impotence." *Journal of Sex and Marital Therapy* 1:277–289, 1965.

Fisher, H.E. *The sex contract: The evolution of human behavior.* New York: Quill, 1980.

Fisher, S. *The female orgasm.* New York: Basic, 1973.

Fiumara, N.J. "Gonococcal pharyngitis." *Medical Aspects of Human Sexuality* 5:195–209, 1971.

Fletcher, J. *Situation ethics: The new morality.* Philadelphia: Westminster, 1966.

Fletcher, J. *Moral responsibility: Situation ethics at work.* Philadelphia: Westminster, 1967.

Ford, C.S., and Beach, F.A. *Patterns of sexual behavior.* New York: Harper & Row, 1951.

Ford, K. "Contraceptive use in the United States, 1973–1976." *Family Planning Perspectives* 10(5):264–269, 1978

Ford, K., Zelnik, M., and Kanter, J.F. "Sexual behavior and contraceptive use among socioeconomic groups of young women in the United States." *Journal of Biosocial Science* 13(1):31–45, 1981.

Foucault, M. *The history of sexuality* Vol. 1. New York: Pantheon, 1978.

Fox, C.A., and Fox, B. "Blood pressure and respiratory patterns during human coitus." *Journal of Reproduction and Fertility* 19(3):405–415, 1969.

Fox, C.A., and Fox, B. "Uterine suction during orgasm." *British Medical Journal* 1:300, 1967.

Frank, E., Anderson, C., and Rubinstein, D. "Frequency of sexual dysfunction in 'normal' couples." *New England Journal of Medicine* 299:111–115, 1978.

Frank, G. *The Boston strangler.* New York: New American Library, 1966.

Freedman, E.B. "Sexuality in nineteenth-century America: Behavior, ideology, and politics." *Review of American History* 10(4), 1982.

Freedman, M.J. "Homosexuality among women and psychological adjustment." Ph.D dissertation, Case Western Reserve University, 1967.

Freeman, D. *Margaret Mead and Samoa.* Cambridge, Mass.: Harvard University Press, 1983.

Freeman, E.D. "Abortion: Subjective attitudes and feelings." *Family Planning Perspectives* 10:150–155, 1978.

Freeman, J. "The origins of the Women's Liberation Movement." *American Journal of Sociology* 78:792–811, 1973.

Friedan, B. *The feminine mystique.* New York: Dell, 1964.

Friedan, B. *The second stage.* New York: Summit, 1981.

Friedman, D. "The treatment of impotency by Brietal relaxation therapy." *Behavior Research and Therapy* 6:257–262, 1968.

Friedman, L. *Obscenity.* New York: Chelsea House, 1970.

Frisch, R.E. "Critical weight at menarche, initiation of the adolescent growth spurt, and control of puberty." In Grumbach, M.M., et al. (eds.) *Control of the onset of puberty.* New York: Wiley, 1974.

Freud, S. "Letter to an American mother." *American Journal of Psychiatry* 107:787, 1951.

Freud, S. *The standard edition of the complete psychological works of Sigmund Freud* (Strachey, J., ed.) London: Hogarth Press and Institute of Psychoanalysis, 1957–1964.

Freund, K. "Therapeutic sex drive reduction." *Acta Psychiatr. Scand.* 62:5–38, 1980.

Friday, N. *My secret garden.* New York: Pocket Books, 1973.

Friday, N. *Forbidden flowers.* New York: Pocket Books, 1975.

Friday, N. *Men in love.* New York: Delacorte, 1980.

Fromm, E. *The art of loving.* New York: Harper & Row, 1956.

Frosch, J. "The role of unconscious homosexuality in the paranoid constellation." *Psychoanalytic Quarterly* 50(4):587–613, 1981.

Gagnon, J. "Female child victims of sex offenders." *Social Problems* 13:176–192, 1965.

Gagnon, J.H. "Sexuality and sexual learning in the child." *Psychiatry* 28:212–228, 1965.

Gagnon, J.H. "Scripts and coordination of sexual conduct." In Cole, J.K., and Deinstbrier, R. (eds.) *Nebraska symposium on motivation* 21. Lincoln: University of Nebraska Press, 1974.

Gagnon, J. *Human sexualities.* Glenview, Ill.: Scott, Foresman, 1977.

Gagnon, J.H., and Simon, W. (eds.) *The sexual scene.* Chicago: Aldine, 1970.

Gagnon, J. and Simon, W. *Sexual conduct: The social sources of human sexuality.* Chicago: Aldine, 1973.

Galenson, E., and Rolphe, H. "Some suggested revisions concerning early female development." *Journal of the American Psychoanalytic Association* 24(5):29–57, 1976.

Galenson, H. "Early sexual differences and development discussion." In Adelson, E. (ed.) *Sexuality and psychoanalysis.* New York: Brunner-Mazel, 1975.

Gandy, P., and Deisher, R. "Young male prostitutes." *Journal of the American Medical Association*, 212:1661–1666, 1970.

Gawin, F.H. "Pharmacologic enhancement of the erotic: Implications of an expanded definition of aphrodisiacs." *The Journal of Sex Research* 14(2):107–117, 1978.

Gay, G.R., Newmeyer, J.A., Elion, R.A., and Wieder, S. "Drug-sex practice in the Haight-Ashbury or 'the sensuous hippie.'" In Sandler, M., and Gessa, G.L. (eds.) *Sexual behavior: Pharmacology and biochemistry*. New York: Raven, 1975.

Gay, P. *The bourgeois experience: Education of the senses*. New York: Oxford, 1984.

Gaylord, J.J. "Indecent exposure: A review of the literature." *Medical Science Law* 21(4): 233–242, 1981.

Gebhard, P.H. "Human sexual behavior: A summary statement." In Marshall, D.S., and Suggs, R.C. (eds.) *Human sexual behavior*. Englewood Cliffs, N.J.: Prentice-Hall, 1973, pp. 206–217.

Gebhard, P., and Elias, J. "Sexuality and sexual learning in childhood." *Phi Delta Kappa* 50:401–405, 1969.

Gebhard, P.H., Gagnon, J.H., Pomeroy, W.B., and Christenson, C.V. *Sex offenders*. New York: Harper & Row, 1965.

Geer, J.H., Morokoff, P., and Greenwood, P. "Sexual arousal in women: The development of a measuring device for vaginal blood flow." *Archives of Sexual Behavior* 3:559–564, 1974.

Geis, G. "Forcible rape: An introduction." In Chappell, D., Geis, R., and Geis, G. (eds.) *Forcible rape: The crime, the victim, and the offender*. New York: Columbia University Press, 1977.

George, L.K. "Sexuality in middle and late life: The effects of age, cohort and gender." *Archives of General Psychiatry* 38(8):919–923, 1981.

Gerassi, J. *Boys of Boise*. New York: Macmillan, 1967.

Gessa, G.L., and Tagliamonte, A. "Role of brain monoamines in male sexual behavior." *Life Science* 14(3):425–436, 1974.

Giarretto, H. "Humanistic treatment of father-daughter incest." In Helfer, R.E., and Kemfe, C.H. (eds.) *Child abuse and neglect: The family and the community*. Cambridge, Mass.: Ballinger, 1976, Chapter 8.

Gilbaugh, J.H., Jr., and Fuchs, P.C. "The gonococcus and the toilet seat." *New England Journal of Medicine* 301:91–93, 1979.

Gilbert, F.S., and Bailis, K.L. "Sex education in the home: An empirical task analysis." *Journal of Sex Research* 16:148–161, 1980.

Gilder, G.F. *Sexual suicide*. New York: Quadrangle, 1973.

Gillan, P., and Brindley, G.S. "Vaginal and pelvic floor responses to sexual stimulation." *Psychophysiology* 16:471–481, 1979.

Gilman, A.G., et al. *Goodman and Gilman's The pharmacological basis of therapeutics*. New York: Macmillan, 1980.

Gilmartin, B.G. "Swinging: Who gets involved and how?" In Libby, R.W., and Whitehurst, R.N. (eds.) *Marriage and alternatives: Exploring intimate relationships*. Glenview, Ill.: Scott, Foresman, 1977, pp. 161–185.

Ginder, R. *Binding with briars: Sex and sin in the Catholic church*. Englewood Cliffs, N.J.: Prentice-Hall, 1975.

Ginsburg, G.L., et al. "The new impotence." *Archives of General Psychology* 28:218, 1972.

Gleitman, H. *Basic psychology*. New York: Norton, 1983.

Glynn, P. *Skin to skin*. New York: Oxford University Press, 1982.

Gold, A.R., and Adams, D.B. "Measuring the cycles of female sexuality." *Contemporary Obstetrics and Gynecology* 12:147–156, 1978.

Goldberg, M. "Importance of little messages in marriage." *Medical Aspects of Human Sexuality* 17(12), 1983.

Goldfoot, D.A., et al. "Lack of effect of vaginal lavages and aliphatic acids on ejaculatory responses in rhesus monkeys: behavioral and chemical analyses." *Hormones and Behavior* 7:1–27, 1976.

Goldstein, B. *Human sexuality*. New York: McGraw-Hill, 1976.

Goldstein, M. *Pornography and sexual deviance*: University of California Press, 1974.

Gonzalez, E.R. "Contraceptive vaccine research: Still an art news." *Journal of the American Medical Association* 244(13):1414–1415, 1419, 1980.

Goode, W.J. *The family* (2nd ed.) Englewood Cliffs, N.J.: Prentice-Hall, 1982.

Gordon, J.W. and Ruddle, F.H. "Mammalian gonadal determination and gametogenesis." *Science* 211:1265, 1981.

Gordon, L. *Woman's body, woman's right: A social theory of birth control in America.* New York: Grossman, 1976.

Gordon, S., and Dickman, I.R. *Sex education: The parent's role.* New York: Public Affairs Pamphlets No. 549, 1977.

Goslin, D.A. (ed.) *Handbook of socialization theory and research.* Chicago: Rand McNally, 1969.

Gottlieb, M.S., Schroff, R., Schanker, H.M., et al. "Pneumocystis curinii pneumonia and mucosal candidiasis in previously healthy homosexual men: Evidence of a new acquired cellular immunodeficiency. *New England Journal of Medicine* 305:1425–1431, 1981.

Gould, R.L. *Transformations.* New York: Simon & Schuster, 1978.

Goy, R.W., and Goldfoot, D.A. In Schmitt, F.O., and Worden, F.G. (eds.) *The neurosciences.* Cambridge, Mass.: MIT Press, 1974.

Goy, R.W. and McEwen, B.S. *Sexual differentiation in the brain.* Cambridge, Mass.: MIT Press, 1980.

Grafenberg, E. "The role of urethra in female orgasm." *International Journal of Sexology* 3(3), 1950.

Grant, M., and Mulas, U. *Eros in Pompeii.* New York: Morrow, 1975.

Graves, R. *The Greek myths.* New York: Braziller, 1959.

Gray, D.S., and Gorzalka, B.B. "Adrenal steroid interactions in female sexual behavior: A review." *Psychoneuroendocrinology* 5(2):157–175, 1980.

Green, R. *Sexual identity conflict in children and adults.* New York: Basic, 1974.

Green, R. "One-hundred ten feminine and masculine boys: Behavioral contrasts and demographic similarities." *Archives of Sexual Behavior* 5:425–466, 1976.

Green, R. *Sexual identity conflict in children and adults.* New York: Basic, 1980.

Green, R., and Money, J. (eds.) *Transsexualism and sex reassignment.* Baltimore: Johns Hopkins University Press, 1969.

Green, R., Williams, K., and Goodman, M. "Ninety-nine tomboys and non-tomboys: Behavioral contrasts and demographic similarities." *Archives of Sexual Behavior* 11(3), 1982.

Greenberg, J.S. "The masturbatory behavior of college students." *Psychology in the Schools,* 9(4):427–432, 1972.

Greer, G. *The female eunuch.* New York: McGraw-Hill, 1971.

Greer, G. *Sex and destiny.* New York: Harper & Row, 1984.

Greer, J., and Stuart, I. *The sexual aggressor: Current perspectives on treatment.* New York: Van Nostrand Reinhold, 1983.

Greer, J. "The sex offender: Theories and therapies, programs and policies." In Greer, J., and Stuart, I. (eds.) *The sexual aggressor: Current perspectives on treatment.* New York: Van Nostrand Reinhold, 1983.

Gregerson, E. *Sexual practices.* New York: Watts, 1983.

Grimal, P. (ed.) *Larousse world mythology.* New York: Excalibur, 1981.

Grof, S. *Realms of the human unconscious: Observations from LSD research.* New York: Viking, 1975.

Grosskurth, P. *Havelock Ellis.* New York: Knopf, 1980.

Groth, A.N., and Burgess, A.W. "Male rape: Offenders and victims." *American Journal of Psychiatry* 137(7):806–810, 1980.

Groth, A.N., with Birnbaum, H.J. *Men who rape: The psychology of the offender.* New York: Plenum, 1979.

Groth, A.N., Burgess, A.W., and Holmstrom, L. "Rape: Power, anger and sexuality." *American Journal of Psychiatry* 134:1239–1243, 1977.

Groth, A.N., and Burgess, A.W. "Sexual dysfunction during rape." *The New England Journal of Medicine* 297:764–766, 1977.

Grumbach, M.M. "The neuroendocrinology of puberty." *Hospital Practice* 15(3): 51–60, 1980.

Grumbach, M.M., et al. "Hypothalamic-pituitary regulation of puberty: Evidence and concepts derived from clinical research." In Grumbach, M.M., Grave, G.D., and Mayer, F.E. (eds.) *Control of the onset of puberty.* New York: Wiley, 1974, Chapter 6.

Guerrero, R. "Type and time of insemination within the menstrual cycle and the human sex ratio." *Studies in Family Planning* 6(10): 367–371, 1975.

Guillemin, R., and Burgus, R. The harmonies of the hypothalamus." *Scientific American* 227(5): 24–33, 1972.

Gulevich, G., and Zarcone, V. "Nocturnal erection and dreams." *Medical Aspects of Human Sexuality* 3(4): 105–109, 1969.

Gulic, R.H.V. *Sexual life in ancient China.* Leiden: Brill, 1974.

Gustafson, J.M. "Nature, sin, and covenant: Three bases for sexual ethics." *Perspectives in Biology and Medicine* Spring 1981, pp. 483–497.

Haeberle, E.J. *The sex atlas.* New York: Seabury, 1978.

Haeberle, E. "Swastika, pink triangle, and yellow star: The destruction of sexology and the persecution of homosexuals in Nazi Germany." *The Journal of Sex Research* 17(3):270–287, 1981.

Haeberle, E. "The Jewish contribution to the development of sexology." *The Journal of Sex Research* 18(4): 305–323, 1982.

Hafez, E.S.E. *Human reproduction* (2nd ed.). New York: Harper & Row, 1980.

Hagen, I.M., and Beach, R.K. "The diaphragm: Its effective use among college women." *Journal of the American College Health Association* 28(5): 263–266, 1980.

Hale, R.W. "Diagnosis of pregnancy and associated conditions." *Current Obstetric and Gynaecologic Diagnosis and Treatment.* Los Altos, CA: Lange, 1980.

Halverson, H.M. "Genital and sphincter behavior of the male infant." *The Journal of Genetic Psychology* 56:95–136, 1940.

Hampson, J.L. "Determinants of psychosexual orientation." In Beach, F.A. (ed.) *Sex and behavior.* New York: Wiley, 1965, pp. 108–132.

Harbison, R.D., and Mantilla-Plata, B. "Prenatal toxicity, maternal distribution and placental transfer of tetrahydrocannabinol." *Journal of Pharmacology and Experimental Therapeutics* 180: 446–453, 1972.

Hare, E.H. "Masturbatory insanity: The history of an idea." *Journal of Mental Science* 452:2–25, 1962.

Hariton, E.B., and Singer, J.L. "Women's fantasies during sexual intercourse: Normative and theoretical implications." *Journal of Counseling and Clinical Psychology* 42:312–322, 1974.

Harlow, H.F. "The nature of love." *American Psychologist* 13:673, 1958.

Harlow, H.F., McGaugh, J.L., and Thompson, R.F. *Psychology.* San Francisco: Albion, 1971.

Harris, C., et al. "Immunodeficiency in female sexual partners of men with the acquired immunodeficiency syndrome." *The New England Journal of Medicine* 308:1181–1184, 1983.

Harris, M. "Why it's not the same old America." *Psychology Today* 15, 1981.

Hart, B. "Reflective mechanisms in copulatory behavior." In McGill, T.E., Dewsbury, D.A., and Sachs, B.D., (eds.)*Sex and behavior.* New York: Plenum, 1978.

Harvey, S.M. "Trends in contraceptive use at one university: 1974–1978." *Family Planning Perspectives* 12(6):301–304, 1980.

Haseltine, F.P., and Ohno, S. "Mechanisms of gonadal differentiation." *Science* 211:1272, 1981.

Hatcher, R.A. *It's your choice.* New York: Irvington, 1982.

Hatcher, R.A., et al. *Contraceptive technology.* New York: Irvington, 1982.

Hatterer, L.J. *Changing homosexuality in the male.* New York: McGraw-Hill, 1970.

Hayes, R.W. "Female genital mutilation, fertility control, and the patrilineage in modern Sudan: A functional analysis." *American Ethnologist* 2:617–633, 1975.

Haynes, D.M. "Course and conduct of normal pregnancy." In Danforth, D.N. (ed.) *Obstetrics and Gynecology.* (4th ed.) Philadelphia: Harper & Row, 1982.

Heath, R.G. "Pleasure and brain activity in males." *Journal of Nervous and Mental Disease,* 154:3–18, 1972.

Heider, C.B. "The penis gourd of New Guinea." *Annals of the Association of American Geographers* 63(3):312–318, 1973.

Heim, N. "Sexual behavior of castrated sex offenders." *Archives of Sexual Behavior* 10:11–19, 1981.

Heim, N., and Hursch, C.J. "Castration for sex offenders: Treatment or punishment? A review and critique of recent European literature." *Archives of Sexual Behavior* 8:281–305, 1979.

Heiman, J. "A psychophysiological exploration of sexual arousal patterns in females and males." *Psychophysiology* 14:266–274, 1977.

Heiman, J.A. "Female sexual response patterns." *Archives of General Psychiatry* 37:1311–1316, 1980.

Heiman, J.R. "Uses of psychophysiology in the assessment and treatment of sexual dysfunction." In LoPiccolo, J., and LoPiccolo, L. (eds.) *Handbook of sex therapy.* New York: Plenum, 1977.

Heiman, J.R. "Women's sexual arousal: The physiology of erotica." *Psychology Today* 8(11):90–94, 1974.

Heiman, M. "Discussion of Sherfey's paper on female sexuality." *Journal of the American Psychoanalytic Association* 16:406–416, 1968.

Hellerstein, E.O., Hume, L.P., and Offen, K.M. *Victorian women.* Stanford, CA: Stanford University Press, 1981.

Henriques, F. *Prostitution and society: A survey.* Vol 1: *Primitive, classical, oriental.* Vol 2: *Prostitution in Europe and the New World.* Vol. 3: *Modern sexuality.* London: MacGibbon & Kee, 1962–1968.

Henshaw, S.K., et al. "Abortion services in the United States, 1979 and 1980." *Family Planning Perspectives* 14(1), 1982.

Henslin, J.M., and Sagarin, E. (eds.) *The sociology of sex.* New York: Schocken, 1978.

Henson, C., Rubin, H.B., and Henson, D.E. "Women's sexual arousal concurrently assessed by three genital measures." *Archives of Sexual Behavior* 8:459–479, 1979.

Herbst, A.L. "Clear cell adenocarcinoma and the current status of DES-exposed females." *Cancer* 48 Suppl (2):484–488, 1981.

Herdt, G.H. *Guardians of the flute.* New York: McGraw-Hill, 1981.

Herman, J., and Hirschman, L. "Families at risk for father-daughter incest." *American Journal of Psychiatry* 138(7):967–70, 1981.

Hess, E.H. "Imprinting in animals." *Scientific American* 198:81–90, 1958.

Hess, E.H. "Imprinting in a natural laboratory." *Scientific American* 277:24–31, 1972.

Hiernaux, J. "Ethnic differences in growth and development." *Eugenics quarterly* 15:12–21, 1968.

Higham, E. "Sexuality in the infant and neonate: Birth to two years. In Wolman, B.B., and Money, J. (eds.) *Handbook of human sexuality.* Englewood Cliffs, N.J.: Prentice-Hall, 1980.

Hilgard, E.R. *Theories of learning and instruction.* New York: Appleton-Century Crofts, 1956.

Himes, N. *Medical history of contraception.* New York: Gambut, 1970.

Himmelweit, H.T., and Bell, N. "Television as a sphere of influence of the child's learning about sexuality." In Brown, L. (ed.) *Childhood sexual learning: The unwritten curriculum.* Cambridge, Mass.: Ballinger, 1980.

Hinde, R.A. *Animal behavior: A synthesis of ethology and comparative psychology.* New York: McGraw-Hill, 1974.

Hindelang, M.J., and Davis, B.J. "Forcible rape in the United States: A statistical profile." In Chappell, D., et al. (eds.) *Forcible rape.* New York: Columbia University Press, 1977, pp. 87–114.

Hite, S. *The Hite report.* New York: Macmillan, 1976.

Hite, S. *The Hite report on male sexuality.* New York: Ballantine, 1981.

Hoenig, J. "The development of sexology during the second half of the 19th century." In Money, J., and Musaph, J. (eds.) *Handbook of Sexology.* Amsterdam: Excerpta Medica, 1977.

Hoffman, C.H. "Sexually transmitted diseases." *Journal of the American Medical Association* 246(15):1709, 1981.

Hoffman, M. *The gay world: Male homosexuality and the social creation of evil.* New York: Basic, 1968.

Hogan, D. R. "The effectiveness of sex therapy: a review of the literature." In LoPiccolo, J., and LoPiccolo, L. (eds.) *Handbook of sex therapy.* New York: Plenum, 1977.

Hole, J. and Levine, E. *Rebirth of feminism.* New York: Quadrangle, 1971.

Hollister, L. "Popularity of amyl nitrite as sexual stimulant." *Medical Aspects of Human Sexuality* 8(4):112, 1974.

Holt, L.H., and Weber, M. *Woman care.* New York: Random, 1982.

Holme, B. *Bulfinch's mythology.* New York: Viking, 1979.

Hooker, E. "The adjustment of the male overt homosexual." *Journal of Projective Techniques* 21(1):18–31, 1975.

Hopkins, J.R. "Sexual behavior in adolescence." *Journal of Social Issues* 33(2):67–85, 1977.

Hopson, J.S. *Scent signals: The silent language of sex.* New York: Morrow, 1979.

Horos, C. V. *Rape.* New Canaan, Conn.: Tobey, 1974.

Horton, D. "The dialogue of courtship in popular songs." *American Journal of Sociology* 62:569–578, 1957.

Houseknecht, S. "Voluntary childlessness." *Alternative Lifestyles* 1(3):379–402, 1978.

Hrdy, S.B. *The woman that never evolved.* Boston: Harvard University Press, 1981.

Huelsman, B.R. "An anthropological view of clitoral and other female genital mutilations." In Lowry, T.P., and Lowry, T.S. (eds.) *The clitoris.* St. Louis, Mo.: Warren H. Green, 1976.

Hull, C.L. *Principles of behavior.* New York: Appleton-Century-Crofts, 1943.

"Humanae Vitae: Encyclical letter of Pope Paul VI." In Horgan, J. (ed.) *Humanae Vitae and the bishops: The encyclical and the statements of the national hierarchies.* Blackrock, Ireland: Irish University Press, 1972.

Humphreys, L. *Tearoom trade: Impersonal sex in public restrooms.* Chicago: Aldine, 1970.

Humphreys, L., and Miller, B. "Identities in the emerging gay culture." In Marmor, J. (ed.) *Homosexual behavior.* New York: Basic, 1980.

Hunt, M. *Sexual behavior in the 1970's.* Chicago: Playboy Press, 1974.

Hunt, M. "Special, today's man: Redbook's exclusive Gallup survey on the emerging male." *Redbook* 147(6):112ff, 1976.

Hussey, H.H. "Vasectomy—A note of concern: Reprise editorial." *Journal of the American Medical Association* 245(22):2333, 1981.

Hyppa, M.T., Falck, S.C., and Rinne, V.K. "Is L-dopa an aphrodisiac in patients with Parkinson's disease?" In Sandler, M. and Gessa, G.L. (eds.) *Sexual behavior: Pharmacology and biochemistry.* New York: Raven, 1975.

Imperato-McGinley, J., Guerrero, L., Gautier, T., and Peterson, R.E. "Steroid 5α-reductase deficiency in man: An inherited form of male pseudohermaphroditism." *Science* 186:1213–1215, 1974.

Imperato-McGinley, J., Guerrero, L., Gautier, T., and Peterson, R.E. "Steroid 5-reductase deficiency in man: An inherited form of male pseudohermaphroditism." *Science* 186:1213–1216, 1979.

"J." *The sensuous woman.* New York: Lyle Stuart, 1969.

Jackson, E., and Potkay, C. "Precollege influences on sexual experiences of coeds." *Journal of Sex Research* 4:150–161, 1973.

James, B. "The 'silent treatment' in marriage." *Medical Aspects of Human Sexuality* 17(2), 1983.

James, J., and Meyerding, J. "Early sexual experience and prostitution." *American Journal of Psychiatry* 134:1381–1385, 1977.

James, W.H. "The distribution of coitus within the human intermenstruum." *Journal of Biosocial Science* 3:159–171, 1971.

Jehu, D. *Sexual dysfunction.* New York: Wiley, 1979.

Jenson, G.D. "Cross-cultural studies and animal studies of sex." In Sadock, B., Kaplan, H., and Freedman, A. *The sexual experience.* Baltimore: Williams and Wilkins, 1976.

Jessor, S., and Jessor, R. "Transition from virginity to nonvirginity among youth: A social-psychological study over time." *Developmental Psychology* 11:473–484, 1975.

Jewelewicz, R. "Psychogenic amenorrhea." *Medical Aspects of Human Sexuality* 17(7), 1983.

Jick, H., Walker, A.M., and Rothman, K.J. "The epidemic of endometrial cancer: A commentary." *American Journal of Public Health* 70(3):264–267, 1980.

Jick, H., et al. "Vaginal spermicides and congenital disorders." *Journal of the American Medical Association* 245(13):1329–1332, 1981.

Johnson, M. *The Borgias.* New York: Holt, Rinehart and Winston, 1981.

Johnston, J. *Lesbian national: The feminist solution.* New York: Simon & Schuster, 1973.

Jolly, A. *The evolution of primate behavior.* New York: Macmillan, 1972.

Jones, C. *Sex or symbol: Erotic images of Greece and Rome.* Austin: University of Texas Press, 1982.

Jones, E. *On the nightmare.* London: Hogarth, 1949.

Jones, E. *The life and work of Sigmund Freud.* 3 vols. New York: Basic, 1957.

Jones, E. *Papers on psychoanalysis.* Boston: Beacon Press, 1966.

Jones, G.S., and Wentz, A.C. "Adolescence, menstruation and the climacteric." In Danforth, D.N. (ed.) *Obstetrics and gynecology.* (4th ed.) Philadelphia: Harper & Row, 1982, pp. 157–181.

Jost, A. "Problems of fetal endocrinology: The gonadal and hypophyseal hormones." *Recent progress in hormone research* 8:379–418, 1953.

Judson, F. "What practical advice can physicians give patients on avoiding genital herpes?" *Medical Aspects of Human Sexuality* 17(8), 1983.

Jung, C. *Animus and anima.* New York: Springer, 1969.

Jung, C.G. "General aspects of dream analysis." In *Structure and dynamics of the psyche,* Vol. 8. (R.F.C. Hull, tr.) New York: Pantheon, 1960.

Jung, C.G. *Archetypes of the collective unconscious.* Princeton, N.J.: Princeton University Press, 1968.

Kahn, J., and Kline, D. "Toward an understanding of sexual learning and communication: An examination of social learning theory and nonschool learning environments." In Roberts, E. (ed.) *Childhood sexual learning: The unwritten curriculum.* Cambridge, Mass.: Ballinger, 1980.

Kallman, F.J. "A comparative twin study on the genetic aspects of male homosexuality." *Journal of Nervous and Mental Disease* 115:283–298, 1952.

Kando, T.M. *Sexual behavior and family life in transition.* New York: Elsevier/North Holland, 1978.

Kanfer, S. *Time,* July 18, 1983, p. 64.

Kanter, J.F., and Zelnik, M. "Contraception and pregnancy: Experience of young unmarried women in the United States." *Family Planning Perspectives* 4(4):9–18, 1972.

Kanter, J.F., and Zelnick, M. "Sexual experience of young unmarried women in the United States." *Family Planning Perspectives,* 4(4):9–18, 1972.

Kaplan, H.S. *The new sex therapy.* New York: Brunner/Mazel, 1974.

Kaplan, H.S. *Disorders of sexual desire.* New York: Brunner/Mazel, 1979.

Kaplan, J. "The Edward G. Donley memorial lecture: Non-victim crime and the regulation of prostitution." *West Virginia Law Review* 79:593–606, 1977.

Karacan, I., Marans, A., Barnet, A., and Lodge, A. "Ontogeny of penile erection during sleep in infants." *Psychophysiology* 4:363–364, 1968.

Karacan, I. "Advances in the psychophysiological evaluation of male erectile incompetence." In LoPiccolo, J., and LoPiccolo, L. (eds.) *Handbook of sex therapy.* New York: Plenum Press, 1978.

Karacan, I., Salis, P.J., Thernby, J.I., and Williams, R.L. "The ontogeny of nocturnal penile tumescence." *Waking and Sleeping* 1:27–44, 1976.

Karlen, A. *Sexuality and homosexuality.* New York: Norton, 1971.

Katchadourian, H. *The biology of adolescence.* San Francisco: Freeman, 1977.

Katchadourian, H. (ed.) *Human sexuality: A comparative and development perspective.* Berkeley: University of California Press, 1979.

Katchadourian, H. "Sex education in college: The Stanford experience." In Brown, L. (ed.) *Sex education in the eighties.* New York: Plenum, 1981.

Katchadourian, H. "Adulthood." In Wyngaarden, J.B., and Smith, L.H. (eds.) *Cecil's Textbook of Medicine.* Philadelphia: Saunders, 1984.

Katchadourian, H., and Martin, J.A. "Analysis of human sexual behavior." In Katchadourian, H. (ed.) *Human sexuality: A comparative and developmental perspective.* Berkeley: University of California Press, 1979.

Katz, J. *Gay American history.* New York: Crowell, 1976.

Kaufman, A., et al. "Male rape victims: Non-institutionalized assault." *American Journal of Psychiatry* 137(2):221–223, 1980.

Kaufman, A., et al. "Recent developments in family planning in China." *Journal of Family Practice* 12(3):581–582, 1981.

Kegel, A. "Sexual functioning of the pubococcygeus muscle." *Western Journal of Surgery, Obstetrics and Gynecology* 60:521–524, 1952.

Kelly, G.F. "Parents as sex educators." In Brown, L. (ed.) *Sex education in the eighties.* New York: Plenum, 1981.

Keniston, K. *The uncommitted.* New York: Dial, 1965

Keniston, K. *Young radicals.* New York: Dial, 1968.

Keniston, K. *Youth and dissent.* New York: Dial, 1971.

Kennedy, D.M. *Birth control in America: The career of Margaret Sanger.* New Haven: Yale University Press, 1970.

Kepecs, J. "Sex and tickling." *Medical Aspects of Human Sexuality* 3(8):58–65, 1969.

Kessler, R. "Surgical experience with the inflatable penile prosthesis." *The Journal of Urology* 124:611–612, 1980.

Kiell, N. *Varieties of sexual experience.* New York: International Universities, 1976.

Kilmann, P., et al. "Sex education: A review of its effects." *Archives of Sexual Behavior* 10(2), 1981.

Kilmann, P., et al. "The treatment of sexual paraphilias: A review of the outcome research." *The Journal of Sex Research* 18(3): 193–252, 1982.

Kilpatrick, J.J. *The smut peddlers.* Garden City, N.Y.: Doubleday, 1960.

Kinch, R.A.H. "Help for patients with premenstrual tension." *Consultant,* April, pp. 187–191, 1979.

Kinsey, A.C., Pomeroy, W.B., and Martin, C.E. *Sexual behavior in the human male.* Philadelphia: Saunders, 1948.

Kinsey, A.C., Pomeroy, W.B., Martin, C.E., and Gebhard, P.H. *Sexual behavior in the human female.* Philadelphia: Saunders, 1953.

Kirby, D. "The mathtech research on adolescent sexuality education programs." *SIECUS Report.* 12(1), 1983.

Kirkendall, L. "Sex education in the United States: A historical perspective." In Brown, L. (ed.) *Sex education in the eighties.* New York: Plenum, 1981.

Klaus, M.H., and Kennell, J.H. *Maternal-infant bonding.* St. Louis: Mosby, 1976.

Kleeman, J.A. "Genital self-stimulation in infant and toddler girls." In Marcus, I.M., and Francis, J.J. (eds.) *Masturbation: From infancy to senescence.* New York: International Universities, 1975.

Kleitman, N. *Sleep and wakefulness.* Chicago: University of Chicago Press, 1963.

Kline-Graber, G., and Graber, B. "Diagnosis and treatment procedures of pubococcygeal deficiencies in women." In LoPiccolo, J., and LoPiccolo, L. (eds.) *Handbook of sex therapy.* New York: Plenum, 1978.

Knapp, J.J., and Whitehurst, R.W. "Sexually open marriage and relationships: Issues and prospects." In Murstein, B.I. (ed.) *Exploring intimate life styles,* New York: Springer, 1978, pp. 35–51.

Koedt, A. "The myth of the vaginal orgasm." In Cox, S. (ed.) *Female psychology: The emerging self.* Palo Alto, Ca.: Science Research Associates, 1976.

Kohlberg, L. *Stages in the development of moral thought and action.* New York: Holt, Rinehart and Winston, 1969a.

Kohlberg, L. "The cognitive developmental approach to socialization." In Goslin, D.A. (ed.) *Handbook of socialization theory and research.* Chicago: Rand McNally, 1969b.

Kolata, G.B. "Gonorrhea: More of a problem but less of a mystery." *Science* 192:244–247, 1976.

Kolata, G.B. "NIH panel urges fewer cesarean births." *Science* 210:176–177, 1980.

Kolodny, R.C. "Adolescent sexuality." Presented at the Michigan Personnel and Guidance Association Annual Convention. Detroit, November 1980.

Kolodny, R.C., et al. *Textbook of human sexuality for nurses.* Boston: Little, Brown, 1979.

Kolodny, R.C., Masters, W.H., and Johnson, V.E. *Textbook of sexual medicine.* Boston: Little, Brown, 1979.

Korner, A.F. "Neonatal startles, smiles, erections and reflex sucks as related to state, sex, and individuality." *Child Development* 40: 1039–1053, 1969.

Kosnik, A., et al. *Human sexuality: New directions in American Catholic thought: A study commissioned by the Catholic Theological Society of America.* New York: Paulist Press, 1977.

Krafft-Ebing, R.V. *Psycopathia sexualis.* New York: Stein and Day, 1978.

Krane, R.J., and Siroky, M.B. "Neurophysiology of erection." *Urologic Clinics of North America* 8(1), 1981.

Krauthammer, C. "The politics of a plague." *The New Republic,* August 1, 1983.

Kreitler, H., and Kreitler, S. "Children's concepts of sexuality and birth." *Child Development* 37:363–378, 1966.

Kroeber, A.L., and Kluckhohn, C. *Culture.* New York: Vintage, 1952.

Kroop, M.S., and Schneidman, B.S. "How long should foreplay last?" *Medical Aspects of Human Sexuality* 3:32–41, 1977.

Kung, H. *On being Christian.* New York: Doubleday, 1976.

Kutchinsky, B. "The effect of easy availability of pornography on the incidence of sex crimes." *Journal of Social Issues* 29:163–182, 1973.

LaBarbera, J.D., and Dozier, J.E. "Psychologic responses of incestuous daughters: Emerging patterns." *Southern Medical Journal* 74(12): 1478–1480, 1981.

Lacoste-Utamsing, C., and Holloway, R.L. "Sexual dimorphism in the human corpus callosum." *Science* 216, 1982.

Ladas, A.K., Whipple, B., and Perry, J.D. *The G spot*. New York: Holt, Rinehart and Winston, 1982.

Ladd, F. *Human Sexuality; Messages in public environments*. In Roberts, E. *Childhood sexual learning: The unwritten curriculum*. Cambridge, Mass.: Ballinger, 1980.

Lamaze, F. *Painless childbirth*. Chicago: Regnery, 1970.

Lancaster, J.B. "Sex and gender in evolutionary perspective." In Katchadourian, H.A. (ed.) *Human sexuality: A comparative and developmental perspective*. Berkeley: University of California Press, 1979, pp. 51–80.

Lancaster, J.B., and Lee, R.B. "The annual reproductive cycle in monkeys and apes." In DeVore, I. (ed.) *Primate behavior: Field studies of monkeys and apes*. New York: Holt, Rinehart and Winston, 1965.

Langman, J. *Medical embryology* (4th ed.) Baltimore: Williams & Wilkins, 1981.

Lasater, M. "Sexual assault: The legal framework." In Warner, C. (ed.) *Rape and sexual assault*. Germantown, Md.: Aspen Systems, 1980, pp. 231–264.

Laub, D., and Dubin, B. "Gender dysphoria." In Grabb, W.C. and Smith, J.W. (eds.) *Plastic surgery*. (3rd ed.) Boston: Little, Brown, 1979.

Laub, D.R., and Gandy, P. (eds.) "Proceedings of the second interdisciplinary symposium on gender dysphoria syndrome." February 2–4, 1973.

Lauersen, N., and Whitney, S. *It's your body*. New York: Grosset and Dunlap, 1977.

Laws, J.L., and Schwartz, P. *Sexual scripts*. Hinsdale, Ill.: Dryden, 1977.

Lazarus, A. "Overcoming sexual inadequacy." In LoPiccolo, J. and LoPiccolo, L. (eds.) *Handbook of sex therapy*. New York: Plenum, 1977.

LeBoeuf, B.J. "Sex and evolution." In McGill, T., Dewsbury, D., and Sachs, B. (eds.) *Sex and behavior: Status and prospectus*. New York: Plenum, 1978.

Leboyer, F. *Birth without violence*. New York: Knopf, 1975.

Legman, G. *Love and death: A study in censorship*. New York: Hacker Art Books, 1963.

Lehrman, D.S. "Semantic and conceptual issues in the nature-nurture problem." In Aronson, L.R., and Tobach, E. (eds.) *Development and evolution of behavior*. New York: Freeman, 1970.

Leiblum, S.R., and Pervin, L.A. *Principles and practice of sex therapy*. London: Tavistock, 1980.

Lein, A. *The cycling female*. San Francisco: Freeman, 1979.

Lesser, H. "The Hirschfeld Institute for Sexology." In Ellis, A., and Abarban, A. (eds.) *Encyclopedia of sexual behavior*. New York: Hawthorn, 1967.

Lessing, D. *The golden notebook*. London: Michael Joseph, 1962.

Levin, R.J. "Masturbation and nocturnal emissions: Possible mechanisms for minimizing teratozoospermia and hyperspermia in man." *Medical Hypothesis* 1:130, 1975.

Levin, R.J., and Levin, A. "Sexual pleasure: The surprising preferences of 100,000 women." *Redbook*, pp. 51–58, September 1975.

Levin, R.J. "The *Redbook* report on premarital and extramarital sex: The end of the double standard?" *Redbook*, pp. 38–44, 190–192, October, 1975.

Levin, R.J. "The physiology of sexual function in women." *Clinics in Obstetrics and Gynecology* 7(2):213–252, 1980.

Levin, R.J. "The female orgasm: A current appraisal." *Journal of Psychosomatic Research* 25(2):119–133, 1981.

Levine, R. "Gusii sex offenses: A study in social control." *American Anthropologist* 61(6): 965–990, 1959.

Levinson, D.J. *The seasons of a man's life*. New York: Knopf, 1978.

Lewis, C.S. *The four loves*. New York: Harcourt, 1960.

Lewis, D.L. "Sex and the automobile: From rumble seats to rockin' vans." *Michigan Quarterly Review* 518–528, Fall 1980.

Lewis, R.A., and Burr, W.R. "Premarital coitus and commitment among college students." *Archives of Sexual Behavior* 4:73–79, 1975.

Libby, R.W. "Social scripts for sexual relationships." In Gordon, S., and Libby, R.W. (eds.) *Sexuality today and tomorrow*. N. Sciutate, Mass.: Dusbury Press, 1976.

Libby, R.W., and Whitehurst, R.N. *Marriage and alternatives: Exploring intimate relationships*. Glenview, Ill.: Scott, Foresman, 1977.

Licht, H. *Sexual life in ancient Greece*. London: Panther, 1969 (first published in 1932).

Lief, H.I. "Sexual survey #4: Current thinking on homosexuality." *Medical Aspects of Human Sexuality* 11(11):110–111, 1977.

Lief, H.I. "Inhibited sexual desire." *Medical Aspects of Human Sexuality* 11(7):94–95, 1977.

Lisk, R.D. In Martinini, L., and Ganong, W.F. (eds.) *Neuroendocrinology.* New York: Academic, 2:197, 1967.

LoPiccolo, J. "Direct treatment of sexual dysfunction in the couple." In Money, J., and Musaph, H. (eds.) *Handbook of sexology.* New York: Elsevier, 1977.

LoPiccolo, J., and Heiman, J. "The role of cultural values in the prevention and treatment of sexual problems." In Qualls, C.B., Wincze, J.P., and Barlow, D.H. (eds.) *The prevention of sexual disorders.* New York: Plenum, 1978, pp. 43–71.

LoPiccolo, J., and Lobitz, W.C. "The role of masturbation in the treatment of orgasmic dysfunction." In LoPiccolo, J., and LoPiccolo, L. (eds.) *Handbook of sex therapy.* New York: Plenum, 1977.

LoPiccolo, J., and LoPiccolo, L. *Handbook of sex therapy.* New York: Plenum, 1978.

Lowry, T.P. (ed.) *The classic clitoris.* Chicago: Nelson-Hall, 1978.

Lowry, T.P., and Lowry, T.S. *The clitoris.* St. Louis: Warren H. Green, 1976.

Ludeman, K. "The sexuality of the older person: Review of the literature." *Gerontologist* 21(2):203–208, 1981.

Luker, K. *Taking chances.* Berkeley: University of California Press, 1975.

"M." *The sensuous man.* New York: Lyle Stuart, 1971.

Maccoby, E. (ed.) *The development of sex differences.* Stanford, CA: Stanford University Press, 1966.

Maccoby, E., and Jacklin, C.N. *The psychology of sex differences.* Stanford, CA.: Stanford University Press, 1974.

MacKinnon, C.A. *Sexual harassment of working women.* New Haven: Yale University Press, 1979.

MacLean, P.D. "Brain mechanisms of elemental sexual functions." In B.J. Sadock, Kaplan, H.I., and Freedman, A.M. (eds.) *The sexual experience.* Baltimore: Williams and Wilkins, 1976.

MacLeod, S.C. "Endocrine effects of oral contraception." *International Journal of Gynecology and Obstetrics* 16:518–524, 1979.

MacLusky, N.J., and Naftolin, F. "Sexual differentiation of the central nervous system." *Science* 211(4488): 1294–1302, 1981.

MacNamara, D., and Sagarin, E. *Sex, crime, and the law.* New York: Free Press, 1977.

Mahood, J., and Wenburg, K. *The Mosher survey.* New York: Arno, 1980.

Maisch, H. *Incest.* London: Andre Deutsch, 1973.

Makepeace, J.M. "Courtship violence among college students." *Family Relations* 30(1):97–102, 1981.

Malatesta, V.J. "Alcohol effects on the orgasmic ejaculatory response in human males." *The Journal of Sex Research* 15:101–107, 1979.

Malinowski, B. *Sex and repression in savage society.* New York: Humanities Press, 1927.

Malinowski, B. *The sexual life of savages in north western Melanesia.* New York: Harcourt Brace, 1929.

Malla, K. *The anaga ranga.* Burton, R.F., and Arbuthnot, F.F. (trs.) New York: Putnam, 1964 ed.

Marcade, J. *Eros Kalos: Essay on erotic elements in Greek Art.* New York: Nagel, 1962.

Marcus, G.E. "One man's Mead?" *The New York Times Book Review,* March 27, 1983, pp. 3, 22.

Marcus, I.M., and Francis, J.J. (eds.) *Masturbation from infancy to senescence.* New York: International Universities, 1975.

Marcus, S. *The other Victorians.* New York: Basic, 1966.

Marcuse, H. *Eros and civilization.* New York: Vintage, 1955.

Marmor, J. (ed.) *Sexual inversion: The multiple roots of homosexuality.* New York: Basic, 1965.

Marmor, J. (ed.) *Modern psychoanalysis: New directions and perspectives.* New York: Basic, 1968.

Marmor, J. "Epilogue: Homosexuality and the issue of mental illness." In Marmor, J. (ed.) *Homosexual behavior.* New York: Basic, 1980.

Marmor, J. (ed.) *Homosexual behavior.* New York: Basic, 1980.

Marshall, D.S. "Sexual behavior in Mangaia." In Marshall, D.S., and Suggs, R.C. (eds.) *Human sexual behavior.* Englewood Cliffs, N.J.: Prentice Hall, 1971.

Marshall, D.S., and Suggs, R.C. (eds.) *Human sexual behavior.* Englewood Cliffs, N.J.: Prentice-Hall, 1971.

Marshall, P., Surridge, D., and Delva, N. "The

role of nocturnal penile tumescence in differentiating between organic and psychogenic impotence: The first stage of validation." *Archives of Sexual Behavior* 10(1), 1981.

Marshall, W.A., and Tanner, J.M. "Puberty." In Douvis, J.A., and Dobbing, J. (eds.) *Scientific foundations of pediatrics*. London: William Heinemaun Medical Books, 1974.

Martin, C. "Factors affecting sexual functioning in 60-79-year-old married males." *Archives of Sexual Behavior* 10(5), 1981.

Martin, C.E. "Sexual activity in the aging male." In Money, J., and Musaph, H. (eds.) *Handbook of sexology*. New York: Elsevier/North Holland, 1977, pp. 813–824.

Martin, D., and Lyon, P. *Lesbian/woman*. San Francisco: Glibe, 1972.

Martinson, F.M. "Eroticism in infancy and childhood." *Journal of Sex Research* 12:251–262, 1976.

Martinson, F.M. "Childhood Sexuality." in Wolman, B.B., and Money, J. (eds.) *Handbook of human sexuality*. Englewood Cliffs, N.J.: Prentice-Hall, 1980.

Maruta, T., and McHardy, M. "Sexual problems in patients with chronic pain." *Medical Aspects of Human Sexuality* 17(2), 1983.

Maslow, A.H. *Toward a psychology of being*. New York: Van Nostrand, 1968.

Masters, W.H. "Update on sexual physiology." Paper presented at the Masters and Johnson Institute's Postgraduate Workshop on Human Sexual Function and Dysfunction. St. Louis, Mo.: October 20, 1980.

Masters, W.H., and Johnson, V.E. *Human sexual response*. Boston: Little, Brown, 1966.

Masters, W.H., and Johnson, V.E. *Human sexual inadequacy*. Boston: Little, Brown, 1970.

Masters, W.H., and Johnson, V.E. "The aftermath of rape." *Redbook* 147(2):74ff, 1976.

Masters, W.H., and Johnson, V.E. *Homosexuality in perspective*. Boston: Little, Brown, 1979.

Masters, W., and Johnson, V. "Sex and the aging process." *Medical Aspects of Human Sexuality* 16(6), 1982.

Masters, W.H., Johnson, V.E., Kolodny, R.C. *Human sexuality*. Boston: Little, Brown, 1982.

Masters, W., Johnson, V., and Levin, R. *The pleasure bond*. New York: Ballentine, 1976.

May, R. *Love and will*. New York: Norton, 1969.

Mazur, R. *The new intimacy*. Boston: Beacon, 1973.

McCance, A.A., Luff, M.C., and Widdowson,

E.C. "Distribution of coitus during the menstrual cycle." *Journal of Hygiene* 37:571–611, 1952.

McCary, J.L. "Teaching the topic of human sexuality." *Teaching of Psychology* 2(1):16–21, 1975.

McCauley, E. and Ehrhardt, A.A. "Female sexual response: Hormonal and behavioral interactions." *Primary Care* 3:455, 1976.

McClintock, M.K. "Menstrual synchrony and suppression." *Nature* 299:244–245, 1971.

McDougall, W. *An introduction to social psychology*. London: Mephuen, 1908.

McQuarrie, H.G., and Flanagan, A.D. "Accuracy of early pregnancy testing at home." Paper presented at the annual meeting of the Association of Planned Parenthood Physicians, San Diego, October 24–27, 1978.

McWhirter, R. (ed.) *Guinness book of world records*. New York: Bantam, 1982.

Mead, M. *Coming of age in Samoa*. New York: Morrow, 1928.

Mead, M. *Sex and temperament in three primitive societies*. New York: Morrow, 1935.

Mead, M. *Male and female*. New York: Morrow, 1949.

Meiselman, K.D. *Incest*. San Francisco: Jossey-Bass, 1978.

Meissner, W.W., Mack, J.E., and Semrad, E.V. "Classical psychoanalysis." In Freedman, A.M., Kaplan, H.I., and Sadock, B.J. (2nd ed.) *Comprehensive textbook of psychiatry*. Baltimore: Williams and Wilkins, 1975, Vol. 1, pp. 482–566.

Melody, G.F. "A case of coitus per urethra." *Medical Aspects of Human Sexuality* 11, December, 1977.

Menning, B. *Infertility: A guide for childless couples*. Englewood Cliffs, N.J.: Prentice-Hall, 1977.

Messenger, J.C. "Sex and repression in an Irish folk community." In Marshall, D.S., and Suggs, R.C. (eds.) *Human sexual behavior*. Englewood Cliffs, N.J.: Prentice-Hall, 1971, pp. 3–37.

Metzger, B.M. "Introduction to the New Testament." In the *New Oxford Bible revised standard version*. New York: Oxford, 1973.

Meyer, A.W. *The rise of embryology*. Stanford, CA.: Stanford University Press, 1939.

Meyer, M.B., and Tonascia, J.A. "Maternal smoking, pregnancy complications and perinatal mortality." *American Journal of Obstetrics and Gynecology* 128:494–502, 1977.

Michael, R.P., Bonsall, R.W., and Warner, P. "Human vaginal secretions: Volatile fatty acid content." *Science* 186:1217–1219, 1974.

Michael, R.P., Bonsall, R.W., and Zumpe, D. "The evidence for chemical communication in primates." *Vitamins and Hormones* 34:137–186, 1976.

Michael, R.P., and Keverne, E.B. "Pheromones in the communication of sexual status in primates." *Nature,* 218:746–749, 1968.

Miller, P.Y., and Simon, W. "Adolescent sexual behavior: Context and change." *Social Problems* 22:58–76, 1974.

Miller, W.B. "Psychological vulnerability to unwanted pregnancy." *Family Planning Perspectives* 5:(4), 1973.

Miller, W.B. "Sexuality, contraception and pregnancy in a high-school population." *California Medicine* 119:14–21, 1973.

Miller, W. "Sexual and contraceptive behavior in young unmarried women." *Primary Care* 3(3):427–453, 1976.

Miller, W.R., and Lief, H.I. "Masturbatory attitudes, knowledge and experience: Data from the sex knowledge and attitude test (SKAT)." *Archives of Sexual Behavior* 5:447–467, 1976.

Millett, K. *Sexual politics.* Garden City, N.Y.: Doubleday, 1970.

Millett, K. *The prostitution papers.* New York: Ballantine, 1973.

Mischel, W. *Introduction to personality.* (3rd ed.) New York: Holt, Rinehart and Winston, 1981.

Mishell, D.R. "Intrauterine devices: Medicated and nonmedicated." *International Journal of Gynecology and Obstetrics* 16:482–487, 1979.

Mishell, D.R. "Control of human reproduction." In Danforth, D.N. (ed.) *Obstetrics and gynecology* (4th ed.) Philadelphia: Harper & Row, 1982, pp. 252–280.

Mitamura, T. *Chinese eunuchs.* Tokyo: C. Tuttle, 1970.

Mitchell, J. *Psychoanalysis and feminism.* New York: Penguin, 1974.

Moghissi, K.S. "Nutrition in obstetrics and gynecology." In Danforth, D.N. (ed.) *Obstetrics and gynecology* (4th ed.) Philadelphia: Harper & Row, 1982, pp. 203–215.

Moghissi, K.S., and Evans, T.N. "Infertility." In Danforth, D.N. (ed.) *Obstetrics and gynecology* (4th ed.) Philadelphia: Harper & Row, 1982.

Mohl, P.C., et al. "Prepuce restoration seekers: Psychiatric aspects." *Archives of Sexual Behavior* 10(4):383–393, 1981.

Money, J. "Sex hormones and other variables in human eroticism." In Young, W.C. (ed.) *Sex and human secretions.* Baltimore: Williams & Wilkins, 1961, pp. 1383–1400.

Money, J. *Sex errors of the body.* Baltimore: The Johns Hopkins University Press, 1968.

Money, J. "Ablatio penis: Normal male infant sex-reassigned as a girl." *Archives of Sexual Behavior* 4(1):65–77, 1975.

Money, J. "Human hermaphroditism." In Beach, F.A. (ed.) *Human sexuality in four perspectives.* Baltimore: The Johns Hopkins University Press, 1976, pp. 62–86.

Money, J. "Genetic and chromosomal aspects of homosexual etiology." In Marmor, J. (ed.) *Homosexual behavior.* New York: Basic, 1980.

Money, J., and Tucker, P. *Sexual signatures.* Boston: Little, Brown, 1975.

Money, J., and Erhardt, A.A. *Man and woman, boy and girl.* Baltimore: The Johns Hopkins University Press, 1972.

Money, J., Hampson, J.C., and Hampson, J.L. "An examination of some basic sexual concepts: The evidence of human hermaphroditism." *Bulletin of Johns Hopkins Hospital* 97:301–319, 1955.

Money, J. *Sex, man and society.* New York: Tower, 1969.

Money, J., and Musaph, H. (eds.) *Handbook of sexology.* New York: Elsevier, 1977.

Moore, K.L. *The developing human.* (3rd ed.) Philadelphia: Saunders, 1982.

Moos, R., and Lunde, D.T., et al. "Fluctuations in symptoms and moods during the menstrual cycle." *Journal of Psychosomatic Research* 13:37–44, 1969.

Morbidity and Mortality Weekly Report "Genital herpes infection—United States, 1966–1979. 31(11), 1982.

Morgan, M. *The total woman.* New York: Basic, 1973.

Morin, J. *Anal pleasure and health.* Burlingame, CA.: Down There Press, 1981.

Morrell, M., Dixen, J., Carter, C.S., and Davidson, J.M. "The influence of age and cycling status on sexual arousability in women."

Morris, D. *The human zoo.* New York: McGraw-Hill, 1967.

Morris, D. *Manwatching: A field guide to human behavior.* New York: Abrams, 1977.

Morris, J. *Conundrum*. New York: Harcourt Brace Jovanovich, 1974.

Moser, C. "A response to Reiss' *Trouble in paradise*.'" *Journal of Sex Research* 19(2):192–195, 1983.

Mosher, B.A., and Whelan, E.M. "Postmenopausal estrogen therapy: A review." *Obstetric and Gynecology Survey* 9:467–475, 1981.

Mosher, W.D. "Contraceptive utilization: United States, 1976." *Vital Health Statistics* (7):1–58, 1981.

Mulvihill, D.J., et al. *Crimes of violence, a staff report to the National Commission on the Causes and Prevention of Violence*. Washington: U.S. Government Printing Office, 1969, Vol. II, p. 217.

Munjack, D.J., and Oziel, L.J. *Sexual medicine and counseling in office practice*. Boston: Little, Brown, 1980.

Munson, M.L. "Wanted and unwanted births reported by mothers 15–44 years of age, United States, 1973. *Advance Data from Vital Health and Statistics*, 9 (10), August 1977.

Murad, F., and Gilman, A.G. "Estrogens and progestins." In Gilman, A.G., et al. (eds.) *Goodman and Gilman's the pharmacological basis of therapeutics*. New York: Macmillan, 1980.

Murdock, G.P. *Social structure*. New York: Macmillan, 1949.

Murdock, G.P. "The social regulation of sexual behavior." In Hoch, P.H., and Zubin, J. (eds.) *Psychosexual development in health and disease*. New York: Grune and Stratton, 1949, pp. 256–266.

Musaph, H., and Abraham, G. "Frigidity or hypogyneisms." In Money, J., and Musaph, H. (eds.) *Handbook of sexology*. New York: Elsevier, 1977.

Mussen, P.H. *Carmichael's manual of child psychology*. New York: Wiley 1970.

My secret life. New York: Grove, 1966.

Nabokov, V. *Lolita*. New York: Putnam, 1955.

Nadelson, C.C. "The emotional impact of abortion." In Notman, M.T., and Nadelson, C.C. (eds.) *The woman patient*. New York: Plenum, 1978, vol. 1, pp. 173–179.

Nakashima, I., and Zakus, G. "Incestuous families." *Pediatric Annual* 8:29, 1979.

Nefzawi *The perfumed garden*. (R. F. Burton, tr.) New York: Putnam, 1964 ed.

Nelson, N.M., et al. "A randomized clinical trial of the Leboyer approach to childbirth." *New England Journal of Medicine* 302:655–660, 1980.

Netter, F.H. *Reproductive system*. The Ciba Collection of Medical Illustrations, Vol. 2. Summit, N.J.: Ciba, 1965.

Netter, F.H. *Endocrine system*. The Ciba Collection of Medical Illustrations, Vol. 4. Summitt, N.J.: Ciba, 1965.

Neubardt, S., and Schulman, H. *Techniques of abortion*. Boston: Little, Brown, 1972.

Neubeck, G. (ed.) *Extramarital relations*. Englewood Cliffs, N.J.: Prentice-Hall, 1969.

Neugarten, B.L. (ed.) *Middle age and aging*. Chicago: University of Chicago Press, 1968.

Neusner, J. *The way of Torah: An introduction to Judaism*. Belmont, CA.: Dickinson, 1970.

Newman, H.F., and Northup, J.D. "Mechanism of human penile erection: An overview." *Urology* 17(5):399–408, 1981.

Nilsson, L. *Behold man*. Boston: Little, Brown, 1973.

Nilsson, L. *A child is born*. New York: Delacorte, 1977.

Noonan, J.T., Jr. *Contraception: A history of its treatment by the Catholic theologians and canonists*. New York: New American Library, 1967.

Noonan, J.T., Jr. (ed.) *The morality of abortion: Legal and historical perspectives*. Cambridge, Mass.: Harvard University Press, 1970.

Novak, E.R., et al. *Novak's textbook of gynecology*. (8th ed.) Baltimore: Williams & Wilkins, 1970.

Nygren, A. *Agape and Eros*. Watson, P.S. (tr.) Chicago: University of Chicago Press, 1939.

Oakley, G. "Natural selection, selection bias and the prevalence of Down's syndrome." *New England Journal of Medicine* 299(19): 1068–1069, 1978.

O'Farrell, T., Weyand, C., and Logan, D. *Alcohol and sexuality: An annotated bibliography on alcohol use, alcoholism, and human sexual behavior*. Phoenix, Arizona: Oryx, 1983.

Offer, D., and Offer, J. *From teenage to manhood: A psychological study*. New York: Basic, 1975.

Offit, A.K. *The sexual self*. New York: Ballantine, 1977.

Offit, A.K. *Night thoughts*. New York: Congdon and Lattes, 1981.

Olds, J. "Pleasure centers in the brain." *Scientific American* 193:105–116, 1956.

Olin, A.A. (ed.) *Sexually transmitted diseases.* NHS Publication No. (CDC) 81-823. Atlanta: USPHS Centers for Disease Control, 1981.

O'Neill, N., and O'Neill, G. *Open marriage.* New York: Evans, 1972.

Opler, M.K. "Cross-cultural aspects of kissing." *Medical Aspects of Human Sexuality* 3(2):11–21, 1969.

Ortner, S., and Whitehead, H. (eds.) *Sexual meanings: The cultural construction of gender and sexuality.* Cambridge: Cambridge University Press, 1981.

Ory, H.W., Rosenfield, A., and Laudman, L.C. "The pill at 20: An assessment." *Family Planning Perspectives* 12(6):278–283, 1980.

Osborn, C.A., and Pollack, R.H. "The effect of two types of erotic literature on physiological and verbal measures of female sexual arousal." *Journal of Sex Research* 13, No. 4:250–256, 1977.

Osser, S., Liedholm, P., and Oberg, S.J. "Risk of pelvic inflammatory disease among users of intrauterine devices, irrespective of previous pregnancy." *American Journal of Obstetrics and Gynecology* 138(7 pt 2):864–867, 1980.

Ostrow, D.G., Sandholzer, T.A., and Feldman, Y.M. (eds.) *Sexually transmitted diseases in homosexual men.* New York: Plenum, 1983.

Ovesey, L. *Homosexuality and pseudohomosexuality.* New York: Science House, 1969.

Ovesey, L., and Woods, S. "Pseudohomosexuality and homosexuality in men: Psychodynamics as a guide to treatment." In Marmor, J. (ed.) *Homosexual behavior.* New York: Basic, 1980.

Packer, H.L. *The limits of the criminal sanction.* Stanford, CA.: Stanford University Press, 1968.

Pagels, E. *The Gnostic Gospels.* New York: Vintage, 1979.

Pao, P. "Pathologic jealousy." *Medical Aspects of Human Sexuality* 16(5), 1982.

Paige, K.E. "The declining taboo against menstrual sex." *Psychology Today* 12(7):50–51, 1978.

Paige, K.E. "The ritual of circumcision." *Human Nature* 1:40–48, 1978.

Parrinder, G. *Sex in the world's religions.* New York: Oxford University Press, 1980.

Pavlov, I.P. *Conditioned reflexes: An investigation of the physiological activity of the cerebral cortex.* London: Oxford University Press, 1927.

Peckham, M. *Art and pornography.* New York: Harper & Row, 1971.

Perkins, R.P. "Sexual behavior and response in relation to complications in pregnancy." *American Journal of Obstetrics and Gynecology* 134:498–505, 1979.

Perlman, D. "Puzzling ailments that may be AIDS." *San Francisco Chronicle* November 5, 1983.

Perry, J.D., and Whipple, B. "Pelvic muscle strength of female ejaculators: Evidence in support of a new theory of orgasm." *The Journal of Sex Research* 17(1):22–39, 1981.

Persky, H. "Psychosexual effects of hormones." *Medical Aspects of Human Sexuality* 17(9), 1983.

Petersen, A.C., and Taylor, B. "The biological approach to adolescence." In J. Adelson (ed.) *Handbook of adolescent psychology.* New York: Wiley, 1980.

Petersen, I., and Stener, I. "An electromyographical study of the striated urethral sphincter, the striated anal sphincter, and the levator ani muscle during ejaculation." *Electromyography* 10:23–44, 1970.

Peterson, J.R., et al. "Playboy readers sex survey." *Playboy* 30(1) January and March 1983.

Peterson, H. *Havelock Ellis, philosopher of love.* Boston: Houghton Mifflin, 1928.

Pfeiffer, E., and Davis, G.S. "Determinants of sexual behavior in middle and old age." *Journal of the American Geriatrics Society* 20:151–158, 1972.

Pfeiffer, E., Verwoerdt, A., and Davis, G.C. "Sexual behavior in middle life." *American Journal of Psychiatry* 128:1262–1267, 1972.

Pfeiffer, E., et al. "Sexual behavior in aged men and women." *Archives of General Psychiatry* 19:753–758, 1968.

Pfeiffer, E., Verwoerdt, A., and Wang, H.S. "The age range and frequency of sexual intercourse of men and women." *Archives of General Psychiatry* 19:753–758, 1968.

Pfeiffer, E., Verwoerdt, A., and Wang, H.S. "The natural history of sexual behavior in a biologically advantaged group of aged individuals." *Journal of Gerontology* 24:193–198, 1969.

Piaget, J. *The origins of intelligence in children.* New York: International Universities, 1952.

Piaget, J. *The child's conception of the world.* Totowa, N.J.: Littlefield, Adams, 1972.

Pillard, R., Poumadere, J., and Carretta, R. "A family study of sexual orientation." *Archives of Sexual Behavior* 11(6), 1982.

Pincus, G. *The control of fertility*. New York: Academic, 1965.

Plessix Gray, F. du "Manners of deceit and the case for lying." Esquire 88(6):134, 194, 1977.

Pohlman, E.G. *Psychology of birth planning*. Cambridge, Mass.: Shenkman, 1968.

Pomeroy, W.B. "The institute for sex research." In Ellis, A. and Abarbanel, A. (eds.) *Encyclopedia of sexual behavior*. New York: Hawthorn, 1967.

Pomeroy, W.B. *Girls and sex*. New York: Delacorte Press, 1969.

Pomeroy, W.B. *Dr. Kinsey and the Institute for Sex Research*. New York: Harper & Row, 1972.

Pomeroy, W.B. *Your child and sex: A guide for parents*. New York: Delacorte, 1976.

Pope, K.S., et al. *On love and loving*. San Francisco: Jossey-Bass, 1980.

Population Reports. Baltimore, Md.: Population Information Program, Johns Hopkins University, 1982.

Pouillet, T. *L'onanisme chez la femme*. Paris: Vigot Freres, 1897. (Originally published in 1876.)

Price, J., and Valdiserri, E. "Childhood sexual abuse: A recent review of the literature." *JAMWA* 36(7), 1981.

Pritchard, J.A., and MacDonald, P.C. *Williams obstetrics* (16th ed.) New York: Appleton-Century-Crofts, 1980.

Purdy, S. "The erotic in literature." In Katchadourian, H., and Lunde, D. *Fundamentals of human sexuality* (2nd ed.) New York: Holt, Rinehart and Winston, 1975a.

Purdy, S. "The erotic in film." In Katchadourian, H., and Lunde, D. *Fundamentals of human sexuality* (2nd ed.) New York: Holt, Rinehart and Winston, 1975b.

Qualls, C.B., Wincze, J.P., and Barlow, D.H. (eds.) *The prevention of sexual disorders*. New York: Plenum, 1978.

Quinlan, M.J. *Victorian prelude: A history of English manners 1700–1830*. New York: Columbia University Press, 1941.

Quinn, S. "Why I, too, finally had to get married." *San Francisco Chronicle*, February 9, 1981, p. 16.

Rada, R.T. "Alcoholism and forcible rape." *American Journal of Psychiatry* 132(4):444–446, 1975.

Raisman, G., and Field, P. "Sexual dimorphism in the preoptic area of the rat." *Science* 173: 731–733, 1971.

Rawson, P. (ed.) *Erotic art of the East*. New York: Putnam, 1969.

Rawson, P. *Tantra*. London: Thames and Hudson, 1973.

Rechy, J. *City of night*. New York: Grove, 1963.

Rechy, J. *The sexual outlaw*. New York: Grove, 1977.

Reich, W. *The function of the orgasm*. New York: Noonday, 1942.

Reich, W. *The sexual revolution*. New York: Farrar, Straus and Giroux, 1969.

Rein, M.F. "Epidemiology of gonococcal infections." In Roberts, R.B. (ed.) *The gonococcus*. New York: Wiley, 1977, pp. 1–31.

Rein, M.F., and Chapel, T.A. "Trichomoniasis, candidiasis, and the minor venereal diseases." *Clinical Obstetrics and Gynecology* 18:73–78, 1975.

Reiss, I.L. "Changing sociosexual mores." In Money, J. and Musaph, H. (eds.) *Handbook of sexology*. Amsterdam: Excerpta Medica, 1977.

Reiss, I.L. *Premarital sexual standards in America*. Glencoe, Ill.: Free Press, 1960.

Reiss, I.L. "How and why America's sex standards are changing." *Trans-Action* 5:26–32, 1968.

Reiss, I.L. *Family systems in America*. (3rd ed.) New York: Holt, Rinehart and Winston, 1980.

Reiss, I. "Trouble in Paradise: The current status of sex research." *Journal of Sex Research* 18:97–113, 1982.

Reiss, I. "Paradise Regained? A reply to Moser." *Journal of Sex Research* 19(2), 1983.

Remy, J. "Mutilations sexuelles: En France aussi." *L'Express* No. 1447: 58–60, 1979.

Rioux, J. "Sterilization of women: Benefits vs. risks." *International Journal of Gynecology and Obstetrics* 16:488–492, 1979.

Robbins, M.B., and Jenson, G.D. "Multiple orgasms in the male." In Gemme, R., and Wheeler, C.C. (eds.) *Progress of sexology*. New York: Plenum, 1976, pp. 323–338.

Roberts, E.J. *Childhood sexual learning: The unwritten curriculum*. Cambridge, Mass.: Ballinger, 1980.

Roberts, E.J., Kline, D., and Gagnon, J. *Family life and sexual learning*. Vol. 1. Cambridge, Mass.: Population Education, 1978.

Robinson, P. *The modernization of sex*. New York: Harper & Row, 1976.

Rodgers, B.   *Gay talk.* New York: Putnam, 1972.

Root, A.W.   "Endocrinology of puberty: Normal sexual maturation." *Journal of Pediatrics* 83:1–19, 1973.

Rose, L. (ed.)   *The menopause book.* New York: Hawthorn, 1977.

Rosenfeld, A.   "Incest among female psychiatric patients." *American Journal of Psychiatry* 136:791–795, June 1979.

Rosenfeld, A.   "Endogamic incest and the victim perpetrator model." *American Journal of Diseases of Children* 133:406, 1979.

Rosenfeld, A., Nadelson, C., and Krieger, M. "Fantasy and reality in patients' reports of incest." *Journal of Clinical Psychiatry* 40(4), 1979.

Rosenfeld, A., Nadelson, C., Krieger, M., and Backman, J.   "Incest and sexual abuse of children." *The American Journal of Child Psychiatry* 16:(2), 1977.

Rosenfeld, A., Smith, D., Wenegrat, A., Brewster, W., and Haavik, D.   "The primal scene: A study of prevalence." *American Journal of Psychiatry* 137:11, 1980.

Rosenfeld, A., Wenegrat, A., Haavik, D., and Wenegrat, B.   "Parents' fears of their children's developing sexuality." *Medical Aspects of Human Sexuality* 16(10), 1982.

Ross, C., and Piotrow, P.T.   "Birth control without contraceptives." *Population Reports,* Series I (1): June 1974.

Rossi, A.S. (ed.)   *The feminist papers: From Adams to Simone de Beauvoir.* New York: Columbia University Press.

Roszak, T.   *The making of a counter-culture.* New York: Doubleday, 1969.

Roth, P.   *Portnoy's complaint.* New York: Random, 1967.

Rothenberg, F.   *San Francisco Chronicle,* January 25, 1984, p. 1.

Rowan, R.L., and Gillette, P.J.   *The gay health guide.* Boston: Little, Brown, 1978.

Rowell, T.   *The social behavior of monkeys.* Baltimore: Penguin, 1972.

Rowse, A.L.   *Homosexuals in history.* New York: Macmillan, 1977.

Rubin, R.T., Reinisch, J.M., and Haskett, R.F. "Postnatal gonadal steroid effects on human behavior." *Science* 211 (4488):1318–1324, 1981.

Ruble, D.   "Premenstrual symptoms: A reinterpretation." *Science* 197:291–292, 1977.

Russell, B.   *A history of Western philosophy.* New York: Simon & Schuster, 1945.

Russell, M.J., Switz, G.M., and Thompson, K. "Olfactory influences on the human menstrual cycle." Delivered at the American Association for the Advancement of Science, San Francisco, 1977.

Sadersten, P., et al.   "Sexual behavior in developing rats." *Hormones and Behavior* 8:320–341, 1979.

Safilios-Rothschild, C.   *Love, sex and sex roles.* Englewood Cliffs, N.J.: Prentice-Hall, 1977.

Saghir, M., and Robins, E.   "Clinical aspects of female homosexuality." In Marmor, J. (ed.) *Homosexual behavior.* New York: Basic, 1980.

Saghir, M.T. and Robins, E.   *Male and female homosexuality: A comprehensive investigation.* Baltimore: Williams & Wilkins, 1973.

Salzman, L.   "Sexuality in psychoanalytic theory." In Marmor, J. (ed.)   *Modern psychoanalysis.* New York, Basic, 1968, pp. 123–145.

Salzman, L.   "Latent homosexuality." In Marmor, J. (ed.)   *Homosexual behavior.* New York: Basic, 1980.

Samuel, T., and Rose, N.R.   "The lessons of vasectomy: A review." *Journal of Clinical Laboratory Immunology* 3(2):77–83, 1980.

Sanford, J.   *Prostitutes: Portraits of people in the sexploitation business.* London: Secker & Warburg, 1975.

Sanger, M.   *Margaret Sanger: An autobiography.* New York: Norton, 1938.

Sappho of Lesbos   *Lyra Graica.* (J.M. Edmonds, tr.) London: Heinemann, 1922.

Sarrel, P.J., and Coplin, H.R.   "A course in human sexuality for the college student." *American Journal of Public Health* 61:1030–1037, 1971.

Sarrel, P., and Masters, W.   "Sexual molestation of men by women." *Archives of Sexual Behavior* 11(2), 1982.

Sarrel, P.J., and Sarrel, P.M.   *Sexual unfolding: Sexual development and sex therapies in late adolescence.* Boston: Little, Brown, 1979.

Scanzoni, L.D.   *Sex is a parent affair: Guide for teaching your children about sex.* New York: Bantam, 1982.

Scanzoni, L., and Scanzoni, J.   *Men, women and change: A sociology of marriage and the family.* New York: McGraw-Hill, 1976.

Schally, A.V.   "Aspects of hypothalamic regulation of the pituitary gland: Its implications for the control of reproductive processes." *Science* 202:18–28, 1978.

Scarf, M. *Unfinished business*. New York: Doubleday, 1980.

Scharfman, M.A. "Birth and the neonate." In Simons, R.C., and Pardes, H. (eds.) *Understanding human behavior in health and illness*. Baltimore: Williams & Wilkins, 1977.

Schlafly, P. *The power of the positive woman*. New York: Arlington, 1977.

Schlesinger, A., Jr. "An informal history of love U.S.A." *Medical Aspects of Human Sexuality* 4(6):64–82, 1970.

Schmidt, G. "Male-female differences in sexual arousal and behavior during and after exposure to sexually explicit stimuli." *Archives of Sexual Behavior* 4:353–364, 1975.

Schmidt, G. "Sex and society in the eighties." *Archives of Sexual Behavior* 11(2), 1982.

Schmidt, G., and Sigusch, V. "Sex differences in responses to psychosexual stimulation by films and slides." *Journal of Sex Research* 6(4):268–283, 1970.

Schneider, R.A. "The sense of smell and human sexuality." *Medical Aspects of Human Sexuality* 5(5), 1971.

Schofield, M. *The sexual behavior of young people*. Boston: Little, Brown, 1965.

Schofield, M. *The sexual behavior of young adults*. Boston: Little, Brown, 1973.

Schroeter, A.L. "Rectal gonnorrhea." *Epidemic Venereal Disease: Proceedings of the Second International Symposium on Venereal Disease*. St. Louis: American Social Health Association and Pfizer Laboratories Division, Pfizer, Inc. 1972: pp. 30–35.

Schultz, T. "Does marriage give today's women what they really want?" *Ladies Home Journal* June: 89–91, 146–155, 1980.

Schwarz, G.S. "Devices to prevent masturbation." *Medical Aspects of Human Sexuality* 7(5):141–153, 1973.

Sears, R.R. "Development of gender role." In Beach, F.A. (ed.) *Sex and behavior*. New York: Wesley, 1965.

Sears, R.R., Maccoby, E.E., and Levin, H. *Patterns of childrearing*. Evanston, Il.: Row, Peterson, 1957.

Seaver, R., and Wainhouse, A. (eds.) *The Marquis de Sade*. New York: Grove, 1965.

Selby, J.W., Calhoun, L.G., Jones, J.M., and Matthews, L. "Families of incest: A collation of clinical impressions." *International Journal of Social Psychiatry* 26(1):7–16, 1980.

Semans, J.H. "Premature ejaculation: A new approach." *Southern Medical Journal* 49:353–357, 1956.

Sevely, J.L., and Bennett, J.W. "Concerning female ejaculation and the female prostate." *Journal of Sex Research* 14:1–20, 1978.

Sex Education and Information Council of the United States. *Sexuality and man*. New York: Scribners, 1970.

Shah, F., Zelnik, M., and Kantner, J.F. "Unprotected intercourse among teenagers." *Family Planning Perspectives* 7:39–44, 1975.

Shahani, S.K., and Hattikudur, N.S. "Immunological consequences of vasectomy." *Archives of Andrology* 7(2):193–199, 1981.

Shane, J.M., Schff, I., and Wilson, E.A. "The infertile couple: Evaluation and treatment." *Clinical Symposia* 28(5), 1976.

Shanor, K. *Fantasy file*. New York: Dial, 1977.

Shapiro, H.I. *The birth control book*. New York: St. Martin's, 1977.

Shaver, P., and Freedman, J. "Your pursuit of happiness." *Psychology Today* 10(3):26–32, 1976.

Sheed, F.J. (tr.) *The Confessions of St. Augustine*. New York: Sheed and Ward, 1943.

Sheehy, G. *Prostitution: Hustling in our wide-open society*. New York: Delacorte, 1973.

Shengold, L. "Some reflections on a case of mother/adolescent son incest." *International Journal of Psychoanalysis* 61 (pt 4):461–476, 1980.

Sheppard, S. "A survey of college-based courses in human sexuality." *Journal of the American College Health Association* 23:14–18, 1974.

Sherfey, M.J. *The nature and evolution of female sexuality*. New York: Vintage, 1973.

Shettles, L.B. "Predetermining children's sex." *Medical Aspects of Human Sexuality* 6:172ff, 1972.

Shilts, R. "How AIDS is changing gay lifestyles." *San Francisco Chronicle*, May 2, 1983.

Shirai, M., and Ishii, N. "Hemodynamics of erection in man." *Archives of Andrology* 1:27–32, 1981.

Shope, D.F., and Broderick, C.B. "Level of sexual experience and predicted adjustment in marriage." *Journal of Marriage and the Family* 29:424–427, 1967.

Short, R.V. "The development of human reproduction." *Regulation de la fecondité*, INSERM, 83:355–366, 1979.

Signs "Women, sex, and sexuality," 6(1), 1980.

Silber, S. *How to get pregnant.* New York: Scribner's, 1980.

Silber, S.J. *The male.* New York: Scribner's, 1981.

Silverstein, C. *The joy of gay sex.* New York: Simon & Schuster, 1978.

Silverstein, C. *Man to man: Gay couples in America.* New York: Morrow, 1981.

Simons, G.L. *A place for pleasure. The history of the brothel.* London: Harwood-Smart, 1975.

Simons, R.C., and Pardes, H. (ed.) *Understanding human behavior in health and illness.* Baltimore: Williams & Wilkins, 1977.

Simpson, M., and Schill, T. "Patrons of massage parlors: Some facts and figures." *Archives of Sexual Behavior* 6(6):521–525, 1977.

Singer, C. *A short history of anatomy and physiology from the Greeks to Harvey.* New York: Dover, 1957.

Singer, I., and Singer, J. "Types of female orgasm." *Journal of Sex Research* 8(11):255–267, 1972.

Singer, J. *The inner world of daydreaming.* New York: Harper & Row, 1975.

Sisley, E.L., and Harris, B. *The joy of lesbian sex.* New York: Simon & Schuster, 1977.

Skinner, B.F. *The behavior of organisms: An experimental analysis.* New York: Appleton-Century-Crofts, 1938.

Slayton, W.R. "Lifestyle spectrum." *Siecus Report* 12(3):1–5, 1984.

Slovenko, R. *Sexual behavior and the law.* Springfield, Ill.: Thomas, 1965.

Small, C.B., Klein, R.S., Friedland, G.H., Moll, B., Emeson, E.E., and Spigland, I. "Community-acquired opportunistic infections and defective cellular immunity in heterosexual drug abusers and homosexual men." *American Journal of Medicine* 74:433–441, 1983.

Smelser, N.J. *Sociology: An introduction.* New York: Wiley, 1973.

Smelser, N.J., and Erikson, E.H. (eds.) *Themes of work and love in adulthood.* Cambridge, Mass.: Harvard University Press, 1980.

Smith, D.S., and Hindus, M.S. "Premarital pregnancy in America, 1940–1971: An overview and interpretation." *Journal of Interdisciplinary History* 5:537–570, 1975.

Smith, G.G. "The use of cervical caps at the University of California, Berkeley." *Journal of American College Health Association* 29(2):93–94, 1980.

Snyder, S., and Gordon, S. *Parents as sexuality educators: An annotated print and audiovisual bibliography for professionals and parents.* New York: Oryx, 1981.

Socarides, C.W. "Homosexuality and medicine." *Journal of the American Medical Association* 212:1199–1202, 1970.

Socarides, C.W. *Beyond sexual freedom.* New York: Quadrangle, 1975.

Sorensen, R.C. *Adolescent sexuality in contemporary America.* New York: World, 1973.

Sorensen, T., and Hertoft, P. "Male and female transsexualism: The Danish experience with 37 patients." *Archives of Sexual Behavior*, 11(2), 1982.

Spanier, G.B. "Married and unmarried cohabitation in the United States 1980." *Journal of Marriage and the Family* 45(2):277-288, 1983.

Spark, R. "A new approach to impotence." *Medical Aspects of Human Sexuality* 17(6), 1983.

Spark, R., White, R., and Connolly, P. "Impotence is not always psychogenic: Newer insights into hypothalamic-pituitary-gonadal dysfunction." *Journal of the American Medical Association* Vol. 243(8):750–755, 1980.

Sparks, R.A., Purrier, B.G., Watt, P.J., and Elstein, M. "Bacteriological colonisation of uterine cavity: Role of tailed intrauterine contraceptive device." *British Medical Journal of Clinical Research* 282(6271):1189-1891, 1981.

Sparling, F.P. "Sexually transmitted diseases." In Wyngaarden, J.B., and Smith, L.H., Jr. (eds.) *Cecil textbook of medicine.* Philadelphia: Saunders, 1982.

Speert, H. "Historical highlights." In Danforth, D.N. (ed.) *Obstetrics and gynecology* (4th ed.) Philadelphia: Harper & Row, 1982, pp. 2-20.

Spence, J. T., and Helmreich, R.L. *Masculinity and femininity: Their psychological dimensions, correlates and antecedents.* Austin: University of Texas Press, 1978.

Spiro, R. *Children of the kibbutz.* New York: Shocken, 1956.

Spitz, R.A. "Autoeroticism: Some empirical findings and hypotheses on three of its manifestations in the first year of life." *The Psychoanalytic Study of the Child* 3-4:85-119, 1949.

Spitz, R., and Wolf, K.D. "Anaclitic depression." An inquiry into the genesis of psychiatric conditions in early childhood, II." *The Psychoanalytic Study of the Child* II, 1947.

Spitzer, R.L. "The diagnostic status of homosexuality in DSM. III: A reformulation of the issues." *American Journal of Psychiatry* 138(2):210-5, 1981.

Stayton, W. "Lifestyle spectrum 1984." *SIECUS Report* 12(3), 1984.

Stein, M.L. "Prostitution." In Money, J., and Musaph, H. (eds.) *Handbook of sexology*. Amsterdam: Excerpta Medica, 1977.

Steinberg, L. *The life cycle*. New York: Columbia University Press, 1981.

Steinem, G. "Erotica and pornography: A clear and present danger." *Ms*. 7:53–54, 1978.

Steinman, D.L., et al. "A comparison of male and female patterns of sexual arousal." *Archives of Sexual Behavior*. 10(6), 1981.

Stekel, W. *Auto-eroticism*. New York: Liveright, 1930.

Stekel, W. *Sexual aberrations*. (S. Parker, tr.) New York: Grove, 1964.

Stendahl, M.H.B. *On love*. New York: Da Capo, 1983.

Stephens, W.N. "The Family in cross-cultural perspective." New York: Holt, Rinehart and Winston, 1963.

Stern, C. *Principles of human genetics* (3rd ed.) San Francisco: Freeman, 1973.

Stockham, A. *Tokology: A book for every woman*. Chicago: Sanitary Publishing, 1883.

Stoller, R.J. *Sex and gender: On the development of masculinity and femininity*. New York: Science House, 1968.

Stoller, R.J. "Gender identity." In Freedman, A.M., Kaplan, H.I., and Sadock, B.J. (eds.) *Comprehensive textbook of psychiatry/ II*. Baltimore: Williams & Wilkins, 1975, pp. 1400–1408.

Stoller, R.J. "Sexual deviations." In Beach, F. (ed.) *Human sexuality in four perspectives*, Baltimore: Johns Hopkins University Press, 1977, pp. 190–214.

Stoller, R.J. *Sexual excitement*. New York: Pantheon, 1979.

Stoller, R.J. "Transvestism in women." *Archives of Sexual Behavior* 11, (2) 1982.

Student Committee on Human Sexuality. *Sex and the Yale student*. New Haven: Yale University Press, 1970.

Sturup, G.K. "Castration: The total treatment." In Resnick, H.L.P., and Wolfgang, M.E. (eds.) *Sexual behavior: Social and legal aspects*. Boston: Little, Brown, 1979, pp. 361–382.

Suitters, B. *The history of contraceptives*. London: International Planned Parenthood Federation, 1967.

Sullivan, H.S. *The interpersonal theory of psychiatry*. New York: Norton, 1953.

Sullivan, P.R. "What is the role of fantasy in sex?" *Medical Aspects of Human Sexuality* 3(4):79–89, 1969.

Sullivan, W. "Violence in pornography elevates aggression, researchers say." *New York Times* Sept. 30, 1980, pp. C1, C4.

Sulloway, F.J. *Freud, biologist of the mind*. New York: Basic, 1979.

Suomi, S.J., Harlow, H.F., and McKinney, W.T. "Monkey psychiatrist." *American Journal of Psychiatry* 128:41–46, 1972.

Symons, D. *The evolution of human sexuality*. Oxford: Oxford University Press, 1979.

Szasz, T. *Sex by prescription*. New York: Anchor Press, 1980.

Taba, A.H. "Female circumcision." *World Health* Geneva: World Health Organization, 1979.

Talese, G. *Thy neighbor's wife*. New York: Doubleday, 1980.

Talmon, Y. "Mate selection in collective settlements." *American Sociological Review* 29:491–508, 1964.

Tannahill, R. *Sex in history*. New York: Stein and Day, 1980.

Tanner, J.M. *Fetus to man*. Cambridge, Mass.: Harvard University Press, 1978.

Tarabulcy, E. "Sexual function in the normal and in paraplegia." *Paraplegia* 10:202–204, 1972.

Tavris, C. "The sex lives of happy men." *Redbook* 150(5):109ff, 1978.

Tavris, C., and Sadd, D. *The Redbook report on female sexuality*. New York: Delacorte, 1977.

Taylor, G.R. *Sex in history*. New York: Harper & Row, 1970.

Tennov, D. *Psychotherapy: The hazardous cure*. New York: Abelard-Schuman, 1975.

Tennov, D. *Love and limerence*. New York: Stein & Day, 1979.

Terman, L.M., and Miles, C. *Sex and personality: Studies in masculinity and femininity*. New York: McGraw-Hill, 1936.

Terzian, H., and Dale-Ore, G. "Syndrome of Kluver and Bucy reproduced in man by bilateral removal of temporal lobes." *Neurology* 5:373–380, 1955.

Thielicke, H. *The ethics of sex*. (J.W. Doberstein, ed.) New York: Harper & Row, 1964.

Thompson, A. "Extramarital sex: A review of the research literature." *The Journal of Sex Research* 19(1):1–22, 1983.

Thompson, J.F. "The vital statistics of reproduction." In Danforth, D.N. (ed.) *Obstetrics and gynecology.* (4th ed.) Philadelphia: Harper & Row, 1982, pp. 281–291.

Tietze, C. "The pill and mortality from cardiovascular disease: Another look." *Family Planning Perspectives* 11:186–188, 1979.

Tietze, C. *Induced abortion: A world review* (5th ed.) New York: Population Council, 1983.

Tietze, C., and Lewit, S. "Life risks associated with reversible methods of fertility regulation." *International Journal of Gynecology and Obstetrics.* 16:456–459, 1979.

Tinbergen, N. *The study of instinct.* Oxford: Clarendon, 1951.

Tolor, A., and DiGrazia, P.V. "Sexual attitudes and behavior patterns during and following pregnancy." *Archives of Sexual Behavior* 5:539–551, 1976.

Tourney, G. "Hormones and homosexuality." In Marmor, J. (ed.) *Homosexual behavior.* New York: Basic, 1980, pp. 41–50.

Towards a Quaker view of sex; An essay by a group of friends. (rev. ed.) London: Friends Home Service Committee, 1964.

Trussel, J., and Westoff, C.F. "Contraceptive practice and trends in coital frequency." *Family Planning Perspectives* 12(5):246–249, 1980.

Truxal, B. "Nocturnal emissions." *Medical Aspects of Human Sexuality* 17(7), 1983.

Tsung, S.H., et al. "A Review: Adverse Effects of Oral Contraceptives." *Journal of the Indiana State Medical Association* 72(8):578–580, 1979.

Twain, M. "The stomach club: Some remarks on the science of onanism." In *The Mammoth Cod.* Milwaukee, Wis.: Maledicta, 1976.

Tyrmand, L. "Permissiveness and rectitude." *The New Yorker* 46:85–86, 1970.

Udry, J.R., and Morris, N.M. "Distribution of coitus in the menstrual cycle." *Nature* 220:593–596, 1968.

U.S. Department of Justice *National crime survey.* Washington, D.C.: U.S. Government Printing Office, 1981.

Vaillant, G.E. *Adaptation to life.* Boston: Little, Brown, 1977.

Vance, E.B., and Wagner, N.N. "Written descriptions of orgasms: A study of sex differences." *Archives of Sexual Behavior* 5:87–98, 1976.

Van de Velde, T.E. *Ideal marriage.* New York: Random, 1965.

VanGennep, A. *The rites of passage.* Chicago: University of Chicago Press, 1960.

Van Lawick-Goodall, J. "Mother offspring relationship in free-ranging chimpanzees." In D. Morris (ed.) *Primate ethology.* Chicago: Aldine, 1967, pp. 207–346.

Vatsyayana. *The Kama Sutra.* (R.F. Burton and F.F. Arbuthnot, trs.) New York: Putnam, 1963.

Vaughan, V.C., McKay, R.J., and Behrman, R.E. *Nelson textbook of pediatrics.* (11th ed.) Philadelphia: Saunders, 1979.

Veevers, J.E. "The life style of voluntarily childless couples." In L. Larson (ed.) *The Canadian family in comparative perspective.* Toronto: Prentice-Hall, 1974.

Vener, A.M., and Stewart, C.S. "Adolescent sexual behavior in middle America revisited: 1970–1973." *Journal of Marriage and the Family* 36:728–735, 1974.

Verwoerdt, A., Pfeiffer, E., and Wang, H.S. "Sexual behavior in senescence, II: Patterns of change in sexual activity and interest." *Geriatrics* 24:137–154, 1969.

Wahl, P., et al. "Effect of estrogen/progestin potency on lipid/lipoprotein cholesterol." *The New England Journal of Medicine* 308(15): 862–867, 1983.

Walsh, F.M., et al. "Autoerotic asphyxial deaths: A medicolegal analysis of forty-three cases." In Wecht, C.H. (ed.) *Legal medicine annual 1977.* New York: Appleton-Century-Crofts, 1977, pp. 157–182.

The way of a man with a maid. New York: Grove, 1968.

Waxenburg, S.E., Drellich, M.G., and Sutherland, A.M. "The role of hormones in human behavior, I: Changes in female sexuality after adrenalectomy." *Journal of Clinical Endocrinology* 19:193–202, 1959.

Webb, P. *The erotic arts.* Boston: New York Graphic Society, 1976.

Webster, G. "Sexual dysfunction in the paraplegic patient." *Medical Aspects of Human Sexuality* 17(1), 1983.

Weideger, P. *Menstruation and menopause: The physiology and psychology, and the myth and the reality.* New York: Knopf, 1976.

Weinberg, S.K.   *Incest behavior* (rev. ed.) Seacaucus, N.J.: Citadel, 1976.

Weinberg, M.S., and Bell, A.   *Homosexuality: An annotated bibliography.* New York: Harper & Row, 1972.

Weinberg, M.S., and Williams, C.J.   *Male homosexuals: Their problems and adaptations.* New York: Penguin, 1974.

Weis, D.   "Reactions of college women to their first coitus." *Medical Aspects of Human Sexuality* 17(2), 1983.

Weisberg, M.   "A note on female ejaculation." *The Journal of Sex Research* 17:90, 1981.

Weiss, H.D.   "Mechanism of erection." *Medical Aspects of Human Sexuality* 7(2):28–40, 1973.

Wertz, R.W., and Wertz, D.C.   *Lying-in: A history of childbirth in America.* New York: Free Press, 1979.

Westoff, C., and Jones, E.F.   "The secularization of U.S. Catholic birth control practices." *Family Planning Perspectives* 9:203–207, 1977.

Westoff, C.F., and Rindfuss, R.R.   "Sex preselection in the United States: Some implications." *Science,* 184:633–636, 1974.

Westoff, L.A., and Westoff, C.F.   *From now to zero.* Boston: Little, Brown, 1971.

Westwood, G.   *A minority: A report on the life of the male homosexual in Great Britain.* London: Longman, Green, 1960.

White, E.   *States of desire.* New York: Bantam, 1981.

White, S., and Reamy, K.   "Sexuality and pregnancy: A review." *Archives of Sexual Behavior* 11(5), 1982.

Whitman, F.   "Culturally invariable properties of male homosexuality: Tentative conclusions from cross-cultural research." *Archives of Sexual Behavior,* 12(3), 1983.

Whitman, F.   "Childhood indicators of male homosexuality." *Archives of Sexual Behavior* 6:89–96, 1977.

Whittaker, P.   *The American way of sex.* New York: Berkley, 1974.

Whyte, L.L.   *The unconscious before Freud.* New York: Basic, 1960.

Wickett, W.H., Jr.   *Herpes: Cause and control.* New York: Pinnacle, 1982.

Wickler, W.   *The sexual code.* Garden City, N.Y.: Doubleday, 1972.

Wiesner, P.J.   "Gonococcal pharyngeal infection." *Clinical Obstetrics and Gynecology* 18:121–129, 1975.

Wilkins, L., Blizzard, R., and Migeon, C.   *The diagnosis and treatment of endocrine disorders in childhood and adolescence.* Springfield, Ill.: Charles C Thomas, 1965.

Williams, P.L., and Warwick, R.   *Gray's anatomy* (36th ed.) Philadelphia: Saunders, 1980.

Williams, R.H.   *Textbook of endocrinology* (6th ed.) Philadelphia: Saunders, 1981.

Wilmore, J.H.   "Inferiority of female athletes: Myth or reality." *Journal of Sports Medicine* 3(1):1–6, 1975.

Wilmore, J.H.   *Athletic training and physical fitness.* Boston: Allyn and Bacon, 1977.

Wilson, E.L.   *Sociobiology: The new synthesis.* Cambridge, Mass.: The Belknap Press of Harvard University Press, 1975.

Wilson, E.O.   *On human nature.* Cambridge, Mass.: Harvard University Press, 1975.

Wilson, J.D., Griffin, J.E., Leshin, M., and George, F.W.   "Role of gonadal hormones in development of the sexual phenotypes." *Human Genetics* 58(1):78–84, 1981.

Wilson, G.T., and Lawson, D.M.   "Effects of alcohol on sexual arousal in women." *Journal of Abnormal Psychology* 87:609–616, 1978.

Wilson, G.T., and Lawson, D.M.   "Effects of alcohol on sexual arousal in male alcoholics." *Journal of Abnormal Psychology* 87:609–616, 1978.

Wilson, M.L., and Greene, R.L.   "Personality characteristics of female homosexuals." *Psychological Reports* 28:407–412, 1971.

Wilson, W.C.   "The distribution of selected sexual attitudes and behaviors among the adult population of the United States." *Journal of Sex Research* 11:46–64, 1975.

Wilson, E.O.   *On human nature.* New York: Bantam, 1978.

Wilson, S., in Melville, R.   *Erotic art of the West.* New York: Putnam, 1973.

Wincze, J., Hoon, P., Hoon, E.   "Sexual arousal in women: A comparison of cognitive and physiological responses by continuous measurement." *Archives of Sexual Behavior* 6:230–245, 1977.

Wise, T.   "Heterosexual men who cross-dress." *Medical Aspects of Human Sexuality* 16(11), 1982.

Wise, T., Dupkin, C., and Meyer, J.K.   "Partners of distressed transvestites." *American Journal of Psychiatry* 138(9):1221–1224, 1981.

Witkin, H.A., et al.   "Criminality in XYY and XXY men." *Science* 193:547–555, 1976.

Wolfe, L.   "The sexual profile of that *Cosmo-*

politan girl." *Cosmopolitan* pp. 254–265, September 1980.

Wolfe, L.   *The Cosmo report.* Toronto: Bantam, 1981.

The Wolfenden report.   New York: Stein & Day, 1963.

Wolfgang, M.E.   *Patterns in criminal homicide.* Philadelphia: University of Pennsylvania Press, 1958.

Wolpe, J.   *Psychotherapy by reciprocal inhibition.* Stanford: Stanford University Press, 1958.

Wolpe, J.   *The practice of behavior therapy.* New York: Pergamon, 1969.

Wolpe, J., and Lazarus, A.A.   *Behavior therapy techniques.* New York: Pergamon, 1966.

Wong, E.   "Nongonococcal urethritis." *Medical Aspects of Human Sexuality* 17(8), 1983.

Wright, J., Perreault, R., and Mathieu, M.   "The treatment of sexual dysfunction: A review." *Archives of General Psychiatry* 34:881–890, 1977.

Wyngaarden, J.B., and Smith, L.H. (eds.)   *Cecil textbook of medicine.* Philadelphia: Saunders, 1984.

Yalom, I.D., et al.   "Postpartum blues syndrome." *Archives of General Psychiatry* 18:16–27, 1968.

Yankelovich, D.   "Raising children in a changing society." The General Mills American Family Report, 1976–1977.

Yankelovich, D.   "Stepchildren of the moral majority." *Psychology Today* 15, November 1981.

Yankelovich, D., and White, H.   "The new morality." *Time* 110(21):111–116ff, 1977.

Yanovsky, B.   *The dark fields of venus.* New York: Harcourt Brace Jovanovich, 1973.

Yilo, K., and Straus, M.A.   "Interpersonal violence among married and cohabiting couples." *Family Relations* 30(3):339–347, 1981.

Yuzpe, A.   "Postcoital contraception." *International Journal of Gynecology and Obstetrics* 16:497–501, 1979.

Zacharias, L., Rand, W.M., and Wurtman, R.J.   "A prospective study of sexual development and growth in American girls: The statistics of menarche." *Obstetrics, gynecological survey* (suppl.) 31:325–337, 1976.

Zelnik, M.   "Sex education and knowledge of pregnancy risk among United States teenage women." *Family Planning Perspectives* 11:335, 1979.

Zelnik, N., and Kantner, J.F.   "Sexuality, contraception and pregnancy among young unwed females in the United States." In Westoff, C.F. and Parke, R., Jr. (eds.)   *Commission on population growth and the American future: Research Reports, Vol 2: Demographic and social aspects of population growth.* Washington, D.C.: Government Printing Office, 1972.

Zelnik, M., and Kantner, J.F.   "Sexual and contraceptive experience of young unmarried women in the U.S. 1976 and 1971." *Family Planning Perspectives* 9:55–71, 1977.

Zelnik, M., and Kantner, J.F.   "Sexual activity, contraceptive use and pregnancy among metropolitan area teenagers: 1971–79." *Family Planning Perspectives* 12(5):230–231, 233–237, 1980.

Zelnik, M., Kim, Y.J., and Kantner, J.F.   "Probabilities of intercourse and conception among U.S. teenage women, 1971 and 1976. *Family Planning Perspectives* 11:177–183, 1979.

Zichy, M.   *The erotic drawings of Mihaly Zichy.* New York: Grove, 1969.

Zilbergeld, B.   *Male sexuality: A guide to sexual fulfillment.* Boston: Little, Brown, 1978.

Zilbergeld, B., and Evans, M.   "The inadequacy of Masters and Johnson." *Psychology Today* 14, 1980.

Zubin, J., and Money, J., (eds.)   *Contemporary sexual behavior: Critical issues in the 1970s.* Baltimore: Johns Hopkins University Press, 1973.

Name Index

Subject Index

Bold-face numbers indicate page on which term is defined or first discussed.

Literary and Photographic Credits

Figs. 1.4 and 18.7 reproduced by courtesy of the Trustees of the British Museum.

Figs. 2.3, 2.4, 2.11, and 4.8 © 1965 CIBA Pharmaceutical Co., div. of CIBA-GRIGY Corp. Reprinted with permission from *The CIBA Collection of Medical Illustrations,* by Frank H. Netter, M.D., All rights reserved.

Box 4.1 figures courtesy of Charles C Thomas, Publisher, Springfield, Ill.

Fig. 5.5 used with permission of Macmillan Publishing Company from *For Women of All Ages* by Sheldon H. Cherry, M.D. Copyright © 1979 by Sheldon H. Cherry, M.D.

Fig. 6.10 from Shapiro, *The Birth Control Book*, St. Martin's Press, Inc., New York. Copyright © 1977 by Howard I. Shapiro.

Fig. 8.9 from *The Study of Instinct* by N. Tinbergen, published by Oxford University Press, 1951, 1969.

Text pp. 252, 274, and 275 reprinted by permission from Mark Twain's *The Mammoth Cod*, Maledicta Press, 1976, Waukesha, Wisconsin.

Fig. 10.3 Detroit Institute of Arts. Gift of Mr. and Mrs. Bert L. Smokler and Mr. and Mrs. Lawrence A. Fleischman.

Box 10.1 Copyright © 1973 by Nancy Friday. Reprinted by permission of Pocket Books, a division of Simon & Schuster, Inc.

Box 10.2 excerpted from the book *Men in Love: Men's Sexual Fantasies: The Triumph of Love over Rage* by Nancy Friday. Copyright © 1980 by N. Friday. Reprinted by permission of Delacorte Press.

Box 13.3 Edmund Keeley and Philip Sherrard, trs. *C.P. Cavafy: Collected Poems,* ed. George Savidis, tr. Copyright © 1975 by Edmund Keeley and Philip Sherrard. Reprinted by permission of Princeton University Press.

Fig. 13.5 courtesy Musée du Petit Palais, Paris.

Fig. 14.5 The Museum of Modern Art/Film Stills Archive, from *The Damned*, courtesy of Warner Bros.

Fig. 14.6 from The Museum of Modern Art/Film Stills Archive, from *Some Like It Hot*, courtesy of United Artists.

Box 15.4 from *The Treatment of Sexual Disorders* by Gerd Arentewicz and Gunter Schmidt. © 1983 by Basic Books, Inc., Publishers. Reprinted by permission of the publisher.

Fig. 17.4 courtesy Col-Ponce Art Museum (The Luis A. Ferré Foundation).

Box 18.7 and figure excerpted from the book *The Anxiety Makers* by Alex Comfort. Copyright © 1967 by Books and Broadcasts Ltd. Used by permission of Delacorte Press.

Fig. 18.9 courtesy The Walker Art Gallery, Liverpool.

Box 18.9 figure courtesy Stanford University Archives.

Figs. 20.3, 20.4, 20.5, 20.6, 20.7 courtesy Department of Special Collections, Stanford University Libraries.

Fig. 20.8, photo by Chuck Painter, News and Publications Service, Stanford University.

Box 20.2 quotations from *Human Sexuality: New Directions in American Catholic Thought* by Anthony Kosnik et al. © 1977 by The Catholic Theological Society of America.

Box 20.4 from *The New Intimacy* by Ronald Mazur. Copyright © 1973 by Ronald M. Mazur. Reprinted by permission of Beacon Press.